Smart Computing Applications in Crowdfunding

Bo Xing
Institute for Intelligent Systems
University of Johannesburg, Johannesburg
South Africa

and

Tshilidzi Marwala
Vice Chancellor and Principal
University of Johannesburg, Johannesburg
South Africa

CRC Press is an imprint of the
Taylor & Francis Group, an **informa** business
A SCIENCE PUBLISHERS BOOK

CRC Press
Taylor & Francis Group
6000 Broken Sound Parkway NW, Suite 300
Boca Raton, FL 33487-2742

First issued in paperback 2020

ISBN-13: 978-1-138-57771-8 (hbk)
ISBN-13: 978-0-367-78056-2 (pbk)

Visit the Taylor & Francis Web site at
http://www.taylorandfrancis.com

and the CRC Press Web site at
http://www.crcpress.com

Foreword

In 2008, the world economy was brought to its knees by the worst financial crisis since the Great Depression. Many countries around the globe, if not every one of them, were impacted in one way or another, some suffered far more severely than others did. All this created a welcome breeding ground for the concept of crowdfunding: a cooperative, Internet-facilitated solution of apportioning funds and resources, across geographical confines, directly to the projects of interest, circumnavigating and complementing the incumbent financial services institutions. Unsurprisingly, crowdfunding is rapidly becoming part of the world's advancement towards a shared and digital economy. If 2008 was signalling a great increase in crowdfunding related activities, 2013 was the year crowdfunding started receiving global recognition and drawing the attention of various bodies, such as the established financial services industry, economists, politicians, and enterprises. The impression of the scale of the crowdfunding market has since attracted numerous large institutions to jump on the bandwagon.

Simultaneously, the scientific research underbuilding the crowdfunding phenomenon has been gaining momentum. Since 2010, crowdfunding relevant topics have been gradually studied from multiple perspectives across distinct disciplines such as psychology, social science, information and communication technology, economics, computer science, engineering, and entrepreneurship. But the establishment of an elaborated research domain is still far away from us. In 2013, the corresponding academic discussion on crowdfunding was still very rare. Only thereafter, some publications had begun to emerge here and there, and several conferences selected crowdfunding as a side topic, but the scope was nearly negligible from an academic viewpoint.

This is what *Smart Computing Applications in Crowdfunding* is about. In this book, Bo and Tshilidzi offer their readers a new angle from which to view crowdfunding, i.e., via less model-based smart computing algorithms which have their roots in engineering, in computer science, and in informatics. Though less mathematically rigorous to some extent, these intelligent algorithms, often nature-inspired, can cope with numerous real-world hard problems efficiently. Thus, the marriage of smart computing and crowdfunding has the potential to spark novel methods, ways, and means of understanding crowdfunding that worth further dissemination and continuous development.

Overall, this book is a welcome addition to the literature of crowdfunding, artificial intelligence, granular computing, and beyond. I wish all of you lots of joy in reading this exciting work. So, read on.

Ben Shenglin, Ph.D.

December 2017

Dean and Professor, Academy of Internet Finance
Zhejiang University, China
Director of International Monetary Institute
Renmin University of China

Preface

Finance has a great impact on real economy's development. From a historical perspective, financial innovation is often coupled with technological advancement. Among various innovations in the financial sector, crowdfunding is a burgeoning and dynamic industry, in which a diverse variety of business models are incorporated. It is, thus, imperative to conduct a comprehensive exploration of the crowdfunding landscape. Meanwhile, the uprising of smart computing-enabled artificial intelligence is also broadly witnessed, which has inspired (or even forced) numerous sectors (including the financial sector) all over the world to react. Motivated by these two noteworthy phenomena, this book covers the key players and critical issues typically encountered in the crowdfunding domain from the smart computing perspective. Under each player, the application of smart computing technique(s) towards the representative issues is elaborated. It is hoped that this book, *Smart Computing Applications in Crowdfunding*, could be a timely publication that may meet the requirements of a wide spectrum of readerships.

The book consists of eleven chapters which are organized into seven parts. The interrelationship of chapters and sections is illustrated in Fig. P.1 (next page).

Acknowledgments

We would like to thank the University of Johannesburg for contributing to the writing of this book. We dedicate this book to the schools that gave us the foundation to always seek excellence in everything we do: The University of Cambridge and the University of Johannesburg.

Bo Xing, D.Ing.
Tshilidzi Marwala, Ph.D.
Johannesburg, South Africa

January 2018

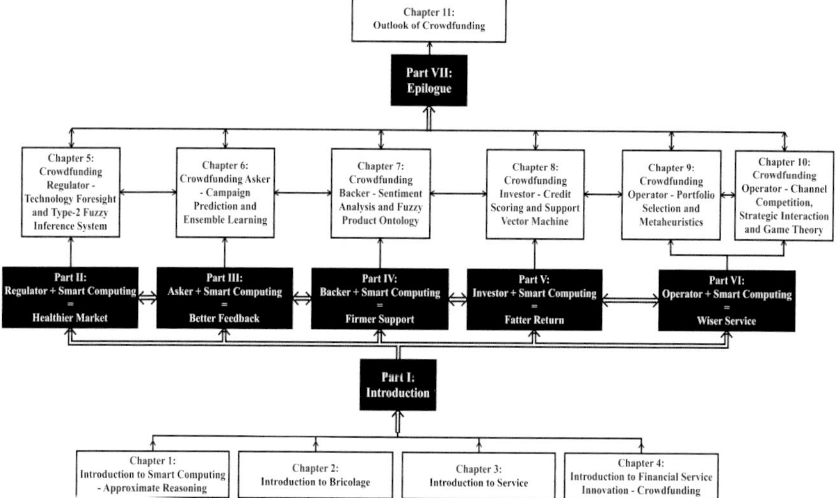
Chapter 11:
Outlook of Crowdfunding
Part VII:
Epilogue
Chapter 5:
Crowdfunding
Regulator -
Technology Foresight
and Type-2 Fuzzy
Inference System
Chapter 6:
Crowdfunding Asker
- Campaign
Prediction and
Ensemble Learning
Chapter 7:
Crowdfunding
Backer - Sentiment
Analysis and Fuzzy
Product Ontology
Chapter 8:
Crowdfunding
Investor - Credit
Scoring and Support
Vector Machine
Chapter 9:
Crowdfunding
Operator - Portfolio
Selection and
Metaheuristics
Chapter 10:
Crowdfunding
Operator - Channel
Competition,
Strategic Interaction
and Game Theory
Part II:
Regulator + Smart Computing
=
Healthier Market
Part III:
Asker + Smart Computing
=
Better Feedback
Part IV:
Backer + Smart Computing
=
Firmer Support
Part V:
Investor + Smart Computing
=
Fatter Return
Part VI:
Operator + Smart Computing
=
Wiser Service
Part I:
Introduction
Chapter 1:
Introduction to Smart Computing
- Approximate Reasoning
Chapter 2:
Introduction to Bricolage
Chapter 3:
Introduction to Service
Chapter 4:
Introduction to Financial Service
Innovation - Crowdfunding

Figure P.1: Interrelationship among different chapters of the book.

Contents

List of Abbreviations

A number of key terms frequently used within this book are defined as below:

1IR	First Industrial Revolution
2IR	Second Industrial Revolution
3IR	Third Industrial Revolution
4IR	Fourth Industrial Revolution
ABM	Adaptive Basis-function Model
AI	Artificial Intelligence
AML	Anti-Money Laundering
ANN	Artificial Neural Network
AoN	All or Nothing
APAC	Asia-Pacific
ATMs	Automated Teller Machines
BMA	Bayes Model Averaging
CA	Correspondence Analysis
CaaS	Computing-as-a-Service
CART	Classification and Regression Tree
CF-IOF	Concept Frequency-Inverse Opinion Frequency
CSA	Canadian Securities Regulatory Authorities
DaaS	Data-as-a-Service
DAO	Decentralized Autonomous Organization
EaaS	Education-as-a-Service
ECOC	Error-Correcting Output Code
ELM	Extreme Learning Machine
EPO	European Patent Office

ESMA	European Securities and Markets Authority
EU	European Union
FINMA	Swiss Financial Market Supervisory Authority
FMCA	The New Zealand's Financial Markets Conduct Act
FSE	Fuzzy Synthetic Evaluation
FSVM	Fuzzy Support Vector Machine
GA	Genetic Algorithm
GT2FIS	General Type-2 Fuzzy Inference System
HC	Hill Climbing
HSI	Hamiltonian of Strategic Interaction
IaaS	Infrastructure-as-a-Service
IAIS	International Association of Insurance Supervisors
ICO	Initial Coin Offering
IFSs	Intuitionistic Fuzzy Sets
IOT	Internet of Things
IPO	Initial Public Offering
IT2FIS	Interval Type-2 Fuzzy Inference System
IVFSs	Interval-Valued Fuzzy Sets
IVIFSs	Interval-Valued Intuitionistic Fuzzy Sets
JOBS Act	Jumpstart Our Business Startups Act
KIA	Keep It All
KL	Kullback-Leibler
KYC	Know Your Customer
LDA	Latent Dirichlet Allocation
LIBSVM	Library for Support Vector Machine
LS-FSVM	Least Squares Fuzzy Support Vector Machine
MCA	Multiple Correspondence Analysis
NLP	Natural Language Processing
NPD	New Product Development
P2P	Peer-to-Peer
PaaS	Platform-as-a-Service
PBoC	People's Bank of China
PC	Personal Computer

PCA	Principal Component Analysis
PFM	Personal Finance Management
PSO	Particle Swarm Optimization
QDT	Quantum Decision Theory
SA	Simulated Annealing
SaaS	Software-as-a-Service
SEC	Securities and Exchange Commission
SMEs	Small and Medium Enterprises
SVM	Support Vector Machine
T1FIS	Type-1 Fuzzy Inference System
T2FIS	Type-2 Fuzzy Inference System
TFM	Technology Foresight Methods
UK	United Kingdom
US	United States
VC	Virtual Currency
WoM	Word-of-Mouth
WoS/K	Web of Science/Knowledge
XaaS	X-as-a-Service

Part I
Introduction

CHAPTER 1

Introduction to Smart Computing Approximate Reasoning

1.1 The Necessity of Computing in Practice: A Brief Reminder

The main aim of this chapter is to introduce smart computing to anyone who intends to apply the corresponding approaches to the interested practical problems. In view of this aim, we begin with simplifying why computations are generally required in practice. Then, we explain the uncertain aspects associated with various practical applications. This will bring us the main theme of this chapter, smart computing.

As quoted by Reed and Dongarra (2015), the English chemist Humphrey Davy once wrote, about two hundred years ago, "*Nothing tends so much to the advancement of knowledge as the application of a new instrument. The native intellectual powers of men in different times are not so much the causes of the different success of their labors, as the peculiar nature of the means and artificial resources in their possession*". Such observation is no less true nowadays that the competitive advantages can be obtained by someone who has the most powerful scientific tools at hand. In 2013, the Nobel Prize in chemistry was shared among three chemists for their remarkable achievements in computational modelling. Actually, computer models simulating real life have turned to be a crucial element for many advancements achieved in numerous domains today, ranging from describing high-energy particle accelerators' advantages (Hamada, 2017), mighty astronomical equipment (e.g., Hubble Space Telescope) (Chen, 2015a), DNA sequencing (Sung, 2017), to agent-based computational economics or artificial economics (LeBaron, 2000; Tesfatsion, 2003; Martinez-Jaramillo, 2007; Marwala, 2013d; Xing et al., 2011; Xing et al., 2012a; Xing et al., 2012b; Marwala, 2013c; Xing et al., 2014; Chen, 2008; Hamill and Gilbert, 2016), and agent-based manufacturing environments (Xing, 2016a; Xing and Gao, 2014gg; Xing and Gao, 2014kk; Xing and Gao, 2014qq; Xing et al., 2014; Xing et al., 2011; Xing and Gao, 2014a). The scientific instruments are becoming increasingly powerful and continuously advancing human knowledge. Here, computing is largely needed inside each one of these instruments in order to facilitate functions such as sensor controlling, data processing, wireless communication, and many more.

In the scientific domain, researchers tend to have many types of tools; however, many of them are often constrained within a specific niche. On the contrary, computing is not just a science augmenter but rather something with inherent computational modelling and data analytics capabilities that can be potentially applied to all patches of science and engineering landscape (Ceruzzi, 2012; Gustafsson, 2011).

1.1.1 *Practical Problems Generalization and Classification*

In order to grasp the necessity of computations in practice, we need to recall what types of practical issues we would like to resolve. Broadly speaking, we can classify the majority of these problems into the following categories:

- **Learning:** We are curious about what is happening around us; in particular, we want to acquire different quantities' numerical values.
- **Predicting:** Upon having these values at hand, we are interested in forecasting the future status of our surroundings along the time axis.
- **Controlling:** By roughly knowing the subsequent states of our environment, we are eager to figure out what changes we can make, if possible, so that the desired future outputs can be obtained.

It should be noted that, more often than not, practical problems involve addressing all three types of tasks. In fact, according to Kreinovich (2008), this categorization can lead us to perceive the differences between science and engineering disciplines.

- **Science Discipline:** The tasks related to learning and predicting world states are usually treated as science.
- **Engineering Discipline:** The tasks of identifying a proper control strategy can be generally regarded as engineering.

A crowdfunding example for explaining such differences can be stated as follows:

1) The problems of measuring fundraising amount at different time intervals and predicting how a particular fundraising campaign will evolve over time belong to the science discipline.
2) The problems of finding the best means to control this development trajectory (e.g., adding new rewards, introducing new investment

bundles, etc.) so that an asker/borrower's funding goal can be better achieved fall into the engineering discipline.

Without exaggeration, computations are required almost everywhere in dealing with both science and engineering problems.

1.1.1.1 Learning the Current State of the World

Suppose some state is characterized by different quantities (denoted by y), the essence of learning is thus to obtain these quantities' numerical values.

- **Measurable Situations:** Sometimes, these values can be easily acquired by directly measuring y. For instance, when we plan to know the current state of a crowdfunding campaign, we can measure its many factors, e.g., fundraising goal, funding raised so far, project start and close date, the number of backers, etc.
- **Unmeasurable Situations:** However, there are numerous quantities that are of our interest but are difficult to measure or are simply unmeasurable, e.g., crowdfunding backers' opinion, evaluation, attitude, emotion, and mood. Under this circumstance, a comprise can be done via the following steps (Kreinovich, 2008):
    1) *Step 1*: Get some relative easier-to-measure quantities (denoted by $x_1,...,x_n$) measured; and then
    2) *Step 2*: Get y estimated according to the measured values (represented by $\tilde{x}_i$) of these auxiliary quantities x_i.
- **How to Compute?** In order to accomplish the above Step 2, i.e., using the approximation of x_i to get an estimate of y, the relationship between y and $x_1,...,x_n$ has to be identified. Here, some algorithm, indicated by $f(x_1,...,x_n)$, is needed to fulfil such task of transforming the values of x_i into an estimate for y. This process can be expressed using Eq. 1.1 (Kreinovich, 2008):

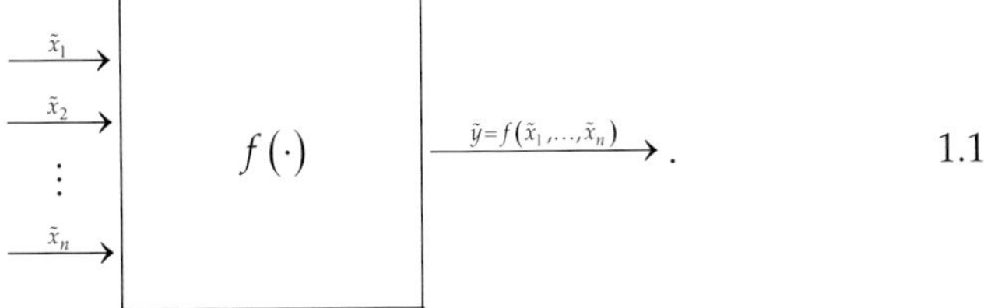

1.1

The complexity of the algorithm $f(x_1,...,x_n)$ varies from situation to situation. In any case, a certain amount of computations are needed to perform learning tasks.

1.1.1.2 Predicting the Future State of the World

Since the current status of the environment is characterized by a set of quantities (denoted by $y_1,...,y_m$), as soon as the values of these quantities are obtained somehow, we can set out to predict their future values. In order to do so, we get to know how the future value z is determined by the current values $y_1,...,y_m$. Likewise, an algorithm, indicated by $g(y_1,...,y_m)$, is further needed to transform the acquired values of y_i into an estimate for z. This is, for example, how a socio-economic system (with no less than 100 million agents involved) is predicted: such simulations and predictions need a lot of computations, therefore, they have to be run on high performance computing equipment.

If we examine the learning and prediction tasks through the lens of computation, some similarities can be identified as follows (Kreinovich, 2008):

- **First**, both problems begin their process by estimating $\tilde{x}_1,...,\tilde{x}_n$ for the quantities of $x_1,...,x_n$; and
- **Second**, a specifically selected algorithm $f(\cdot)$ is always applied to these estimates. The resultant of this operation is an estimate $\tilde{y} = f(\tilde{x}_1,...,\tilde{x}_n)$ for the desired quantity y. In practice, this process shared by both problem classes is often referred to as data processing.

1.1.1.3 Controlling the Desirable State of the World

When the current status is knowable to some extent, and the subsequent future possibilities are partially predictable, we intuitively encounter the third task that is figuring out a means to get guaranteed desirable results. Under this category, two subclasses can be further grouped (Kreinovich, 2008):

- **Constraint Satisfaction:** Practical problems tend to have many constraints, and satisfying all those restrictions (maybe more than one possible design alternative) is often what we want. The goal of this sub-class is, thus, to find any one of these alternatives, and no preferences are imposed.
- **Function Optimization:** For this sub-class there exists an evident preference between different alternative designs (denoted by x). This is why an objective function $F(x)$ is often introduced in order to represent such preference: the larger the value of $F(x)$, the more preferable are design alternatives x. Therefore, optimization means that we would like to find the largest objective function value so that the most preferable design alternative x can be pinpointed.

Both sub-classes inevitably require a great amount of computations (Du and Ko, 2014).

1.1.2 Computational Science and Engineering

Broadly speaking, the main duty of scientists and engineers is to comprehend, develop, or optimize all sorts of 'systems'. Here, the term 'system' represents the object of interest, either a living system (e.g., stem cell), or an artificial technological system (e.g., crowdfunding). Suppose we did not have complex systems like computer processors (Harris and Harris, 2013; Comer, 2017), wind turbines (Hau, 2013), supply chain (Xing et al., 2010b; Xing et al., 2010a; Xing and Gao, 2015c; Xing and Gao, 2015a; Gao et al., 2013a), layout (Xing et al., 2010e), clustering (Xing et al., 2010d), load dispatch (Xing, 2015d), smartphones (Woyke, 2014; Xing and Marwala, 2018f), operating systems (Holcombe and Holcombe, 2012; Xing and Marwala, 2018g), remanufacturing system (Xing and Gao, 2014a; Xing et al., 2010a; Xing and Gao, 2015b), reconfigurable manufacturing system (Xing et al., 2006a; Xing et al., 2009; Xing et al., 2006b), design automation system (Xing and Marwala, 2018e), and robots (Xing and Marwala, 2018n; Xing and Marwala, 2018k; Xing and Marwala, 2018a; Marwala and Hurwitz, 2017; Xing, 2016e) in our life, then engineers/scientists would also not exist.

- **Model:** The underlying reason for the existence of both scientists and engineers is the complexity associated with natural systems and man-made systems. In general, to deal with such complexity, a common practice employed by engineers or scientists is simplification. This means, if something is complicated, it should be made simpler, but not too simple. To put it more formally, a viable simplified system description (i.e., model) is needed in order for engineers and scientists or anyone else to learn about complex systems.
- **Mathematical Model:** In the literature, there are many definitions of a mathematical model available. According to Velten (2009), a more general version can be given as follows: A mathematical model can be represented by a triplet (*System*, *Question*, *Statement*) where a system is denoted by *System*, a question related to the interested system is indicated by *Question*, and *Statement* stands for a set of mathematical statements given by Eq. 1.2 (Velten, 2009):

$$Statements = \{\Sigma_1, \Sigma_2, \ldots, \Sigma_n\}. \qquad 1.2$$

The process of this problem solving scheme can be seen in Fig. 1.1.

1) *Real-world*: Initially, we have a system (*System*) under consideration and a question (*Question*) related to this system.
2) *Mathematical universe*: This hemisphere consists of a set of mathematical statements (*Statement*) together with a possible problem solution (denoted by A^*).

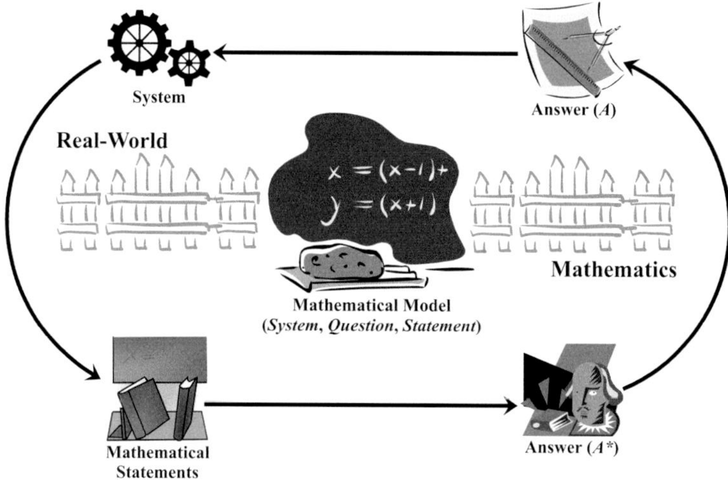

Figure 1.1: Problem solving process.

3) *Bridge*: These two hemispheres are bridged by a mathematical model (*System, Question, Statement*) which can translate the real-world problem into mathematical terms and interpret the obtained solutions in real-world language.

- **Computational Model:** In practice, there is a clear distinction between the following two tasks: (1) Formulating a mathematical model, and (2) Solving the resulting mathematical problem. The former can be done by non-mathematicians, while the latter is often tackled by someone with mathematical expertise (with the aid of advanced software in many cases). When the exact mathematical model of a problem is hard to obtain, an experimental way to find the appropriate solution is very time- and cost-consuming. In this regard, the complex mathematical models' subtleties can be illuminated by computational modelling (Shiflet and Shiflet, 2014).

1.1.3 Key Concepts

In the realm of modern computing, one often talks about computer science, mathematics, logic, and statistics (Paule, 2013). These foundational views can be given a clear technological meaning in the context of smart computing that has an aim of writing algorithms (i.e., mathematical and statistical notions as a base) and using computers for long calculations and verifications. For the rest of this section, we will describe various key concepts for bridging different research endeavours.

1.1.3.1 Sets

Generally speaking, a set refers to a collection of objects (or elements) that share the same properties or satisfy certain equations (Garnier and Taylor, 2002; O'Regan, 2013). In practice, it is a fundamental building block that gives a place to all needed domains in modern societies. For example, in the design of intelligent systems, objects (concrete or abstract) appear naturally (i.e., either definitely in or definitely out of the set) when considering constraints, uncertainties, and design specifications. Furthermore, due to fact that the object of a set need not be real, sets are the most appropriate language to specify several system performances (Veazie, 2017). For instance, we can define a set in asserting an imaginary domain (e.g., the domain of attraction), a conceptual domain (e.g., the domain of credit prediction), or even a specific domain (e.g., the error domain of a proposed algorithm). Accordingly, sets do not only serve as the terms of formulation, but also play a key role in constructing problem solutions (Blanchini and Miani, 2015).

In general, there are four types of sets (Pedrycz, 2013), namely crisp sets (e.g., yes/no, dichotomies), fuzzy sets (e.g., partial memberships), rough sets (e.g., lower and upper boundary), and shadowed sets (e.g., uncertainty regions). A comparative view of these sets is illustrated in Fig. 1.2.

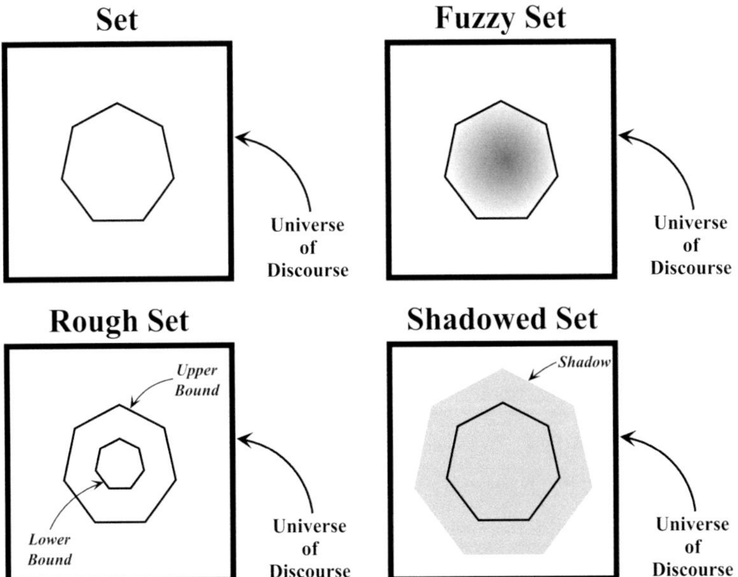

Figure 1.2: A comparative view of sets.

- **Classical (Crisp) Sets:** Classical, two-valued logic (e.g., qualified & unqualified, or true & false) is a basis of traditional mathematics and, in particular, of the crisp set theory. In other words, crisp set theory enables all the objects under consideration to be deterministically classified into two disjoint classes: belonging to a set, or not (Bělohlávek et al., 2017). Its membership function can be defined using Eq. 1.3 (Klir and Yuan, 1995):

$$X_A(x) = \begin{cases} 1 & \text{if } x \in A \\ 0 & \text{if } x \notin A \end{cases}. \quad 1.3$$

  However, although every crisp set is defined by a "sharp" predicate, not every predicate is good enough and thus there is no a perfect classification of a crisp set to which it refers to (Pykacz, 2015; Trillas and Eciolaza, 2015). For example, in many real applications of data mining and image/natural language processing, datasets often contain a large number of features. In these cases, obtaining higher classification accuracy is neither possible nor necessary. Therefore, this practical need leads to the introduction of fuzzy set notions.

- **Fuzzy Sets:** The first publication in fuzzy set theory, written by Zadeh (1965), is over 50 years old. As its name implies, fuzzy set theory refers to a theory of graded concepts, in which an element x of given fuzzy set $\tilde{A}$ has various levels of membership, ranging from 0 (full non-membership) to 1 (full membership) (Zimmermann, 1992). Essentially, this idea intends to imitate the process of human brain solving complex problems by using mathematics (Zadeh, 1965). In general, fuzzy sets can be written in the form of Eq. 1.4 (Zimmermann, 1992):

$$\mu_{\tilde{A}} : x \mapsto \mu_{\tilde{A}}(x) \in [0,1]. \quad 1.4$$

  where $\mu_{\tilde{A}}(x)$ is called membership function or grade of membership.

  It aims to provide a natural way to deal with problems in which imprecise or fuzzy predictions, relations, criteria and phenomena exist. Applications of this theory can be found in areas such as manufacturing (Azadegan et al., 2011; Xing et al., 2010c; Xing and Gao, 2014ii; Xing and Gao, 2014mm), decision policies (Xing, 2016c; Salles et al., 2016; Bosma et al., 2011), and data analysis (Petry and Zhao, 2009; Petkovic, 2014). More information can be found in (Zimmermann, 2001).

  In the literature, fuzzy sets also have three extended versions, namely interval-valued fuzzy sets (IVFSs) (Gozalczany, 1987; Turksen, 1996;

Wang and Li, 1998), intuitionistic fuzzy sets (IFSs) (Atanassov, 1986), and a hybrid of IVFSs and IFSs which is called interval-valued intuitionistic fuzzy sets (IVIFSs) (Atanassov and Gargov, 1989).

1) *Interval-valued fuzzy sets (IVFSs)*: According to Zhang et al. (2009), in many cases, an objective procedure is unfortunately not available in terms of identifying the crisp membership degrees for elements in a fuzzy set. The IVFSs emerged based on these observations. For instance, due to the inherent uncertainties associated with an expert's knowledge, describing the degree of belief (or termed as a membership function's values) in the form of a crisp number (e.g., ranging from 0 to 10) tends to be very hard. More often than not, only a very raw estimation (say between 4 and 6) is obtainable which in turn gives us the values of a membership function falling within an interval of [0.4, 0.6]. More formally, IVFSs belong to type-2 fuzzy sets, in which a fuzzy set A is over a referential set U, i.e., $A: U \rightarrow FS([0, 1])$. More details can be found in (Celik et al., 2015; Mendel, 2017).
2) *Intuitionistic fuzzy sets (IFSs)*: In IFSs, a membership function (denoted by μ) and a non-membership function (indicated by ν) are jointly introduced, in which $\mu + \nu \leq 1$. This formulation relaxes the originally enforced condition $\nu = 1 - \mu$ found in classical fuzzy set theory (Zhang et al., 2009). The basic definition of an IFS in a universe of discourse can be given by Eq. 1.5 (Atanassov, 1986):

$$A = \{\langle x, \mu_A(x), \nu_A(x)\rangle | x \in X\}. \tag{1.5}$$

where $X = \{x_1, x_2, x_3, \cdots, x_n\}$; and $\mu_A(x): X \rightarrow [0, 1]$ and $\nu_A(x): X \rightarrow [0, 1]$ refer to membership degree and non-membership degree, respectively, under the condition of $0 \leq \mu_A(x) + \nu_A(x) \leq 1$. Meanwhile, an intuitionistic index of x to A, termed as the hesitancy degree of the element $\pi_A(x)$, can be defined by Eq. 1.6 (Atanassov, 1986):

$$\pi_A(x) = 1 - \mu_A(x) - \nu_A(x). \tag{1.6}$$

where $\pi_A(x) \in [0, 1]$, $\forall x \in X$. In practice, thanks to non-membership function, IFSs can be used to address the hesitancy that is often caused by information impression (Song et al., 2015). Further discussions in this regard can be found in (Li, 2014).
3) *Interval-valued intuitionistic fuzzy sets (IVIFSs)*: Finally, the IVIFSs act as a further generalization of IFSs in which unit intervals $[\mu_1, \mu_2]$ are employed for membership and non-membership values rather than exact numerical values (Atanassov and Gargov, 1989).

Nowadays, these variations have been widely used in various domains, such as social networking (Chen et al., in press), group decision making (Zhang and Xu, 2015; Chen, 2015b), and technology evaluation (Dereli and Altun, 2013).

- **Rough Sets:** Rough set theory, proposed by Pawlak (1982), serves as a novel mathematical tool for dealing with inconsistency problems (Zhang et al., 2016). A powerful principle underpinning the rough set theory is that hidden patterns in data cannot always be disclosed by precise measurements (Anderson et al., 2000). Accordingly, rough set theory is often regarded as a fundamental concept for artificial intelligence (AI) and cognitive sciences, such as image processing (Sen and Pal, 2009), machine learning (Henry, 2006), data mining (Bae et al., 2010; Fan and Zhong, 2012; Nelwamondo and Marwala, 2007), and knowledge discovery (Ali et al., 2015; Hassan and Tazaki, 2003; Marwala and Lagazio, 2011b).

  In a rough set, the set X is typically approximated via information extracted from B and then constructing the following terms, namely, lower approximation set, upper approximation set, and boundary region by using Eq. 1.7 (Pawlak, 2002):

$$\begin{gathered} \underline{B}X = \{x \mid [x]_B \subseteq X\} \\ \overline{B}X = \{x \mid [x]_B \cap X \neq \varnothing\} \\ BN_B(x) = \overline{B}X - \underline{B}X \end{gathered} \quad 1.7$$

  where the lower and the upper approximations are denoted by $\underline{B}X$ and $\overline{B}X$, respectively, and $BN_B(x)$ represents the B – boundary region of rough set X.

  When a target set involves uncertainty or imprecision, one can use rough set to define such a set approximately via some definable sets (Pawlak and Skowron, 2007b). More specifically, any pair of precise sets can be divided into two parts: (1) The first part is called the lower approximating sub-set including all surely belonged objects, and (2) The second part is called the upper approximating sub-set containing all objects that possibly belong to the set. For those objects that cannot be classified into either upper- or lower sub-set, one can deposit them in the boundary region of a rough set (Pawlak, 2002). More in-depth discussions can be found in (Pawlak and Skowron, 2007a; Polkowski, 2002; Peters and Skowron, 2014; Peters and Skowron, 2016; Peters et al., 2014).

  When compared with fuzzy set theory, we can have the following observations: On the one hand, although extracting rules from data is made possible by rough set theory, a smooth extrapolation across cases is still not allowed. On the other hand, fuzzy set outperforms rough set in terms of smooth extrapolation, but it needs a set of rules to get its process started (Anderson et al., 2000). Based on this, two new models of fuzzy-rough hybridization, i.e., rough fuzzy sets and fuzzy rough sets, are becoming increasingly popular. More theoretical backgrounds can be found in (Hinde and Yang, 2009; Cock et al., 2007; Dubois and Prade, 1990; Morsi and Yakout, 1998; Nanda and Majumdar, 1992; Radzikowska and Kerre, 2002; Nguyen et al., 2014; Lu et al., 2016; Yang and Hinde, 2010) and the corresponding applications can be found in decision making (Anderson et al., 2000), feature selection (Kuncheva, 1992), data mining (Nilavu and Sivakumar, 2015; Srinivasan et al., 2001), to name just a few.

- **Shadowed Sets:** According to Pedrycz (1998), the concept of shadowed sets was proposed to close the gap between fuzzy sets and rough sets. In general, a shadowed set (denoted by S) in a universe of discourse (represented by U) stands for a set-valued mapping $S : U \rightarrow \{0,[0,1],1\}$ and thus has following properties given by Eq. 1.8 (Pedrycz, 1998):

$$\begin{aligned} Core(S) &= \{x \in U : S(x) = 1\} \\ Sh(S) &= \{x \in U : S(x) = [0,1]\} \\ Supp(S) &= cl\{x \in U : S(x) \neq 0\} \end{aligned} \quad . \qquad 1.8$$

where a shadowed set's core is denoted by *Core*(S), in which all objects are fully definable; *Sh*(S) represents the shadow which incorporates uncertainty and imprecision, and a shadowed set's support is indicated by *Supp*(S), in which all elements are incompatible with the rules defined by S.

In general, one can view shadowed sets as an expression of fuzzy sets in a three-way approximation, i.e., {0,1,[0,1]} (Yao et al., 2017), though conceptually, shadowed sets and rough sets are more close to each other, even though their theoretical foundations are indeed very different (Zhou et al., 2011). More specifically, the rough set theory's key concepts (i.e., negative region, lower bound, and boundary region) are associated with shadowed sets' three-logical values (i.e., 0, 1, and [0,1] which correspond to excluded, included, and uncertain characteristics, respectively) (Pedrycz, 2009; Zhou et al., 2011). Further discussions on this matter can be found in (Pedrycz, 1998; Grzegorzewski, 2013; Pedrycz, 2005).

1.1.3.2 Logic

Logic is about reasoning, going from the knowledge assimilation (i.e., premises) to making deductions based on this knowledge (i.e., a conclusion) (Gensler, 2010). In computer science, most theories operate in accordance with a logic system, typically Boolean logic. Typically, a Boolean logic is a set, which consists of at least two distinct special elements 0 and 1, respectively. In its simplest form, any outcome is governed and calculated by following two main laws: (1) The law of contradiction, the possibility that p and $-p$ can co-exist at the same time is zero, hence one side of a contradiction has to be invalid (or false); and (2) The law of the excluded third, nothing can be found between to be and not to be (Moller and Struth, 2013).

Fuzzy logic provides another foundational view for reasoning based on uncertain statements. Generally speaking, this concept was inspired by two notable human abilities: (1) The ability of reasoning and decision-making under the situations like imprecision, information incompleteness, uncertainty, and partiality of truth; (2) The ability to accomplish various perception-based physical/mental tasks with no accurate measurements and computations involved (Pedrycz et al., 2008). In essence, fuzzy logic acts as a novel theory of inference by introducing linguistic IF-THEN rules (Zadeh, 1975a; Zadeh, 1975b; Zadeh, 1975c). The main application domains of fuzzy logic include decision making (Lin and Chen, 2004; Patel and Marwala, 2006), control (Raber, 1994), healthcare (Barro and Marín, 2002; Massad et al., 2008), and image processing (Caponetti and Castellano, 2017).

1.1.3.3 *Probability Theory*

Probability can be loosely defined as "the frequency of occurrence" of an outcome. Typically, the probability is between 0 (i.e., the event cannot occur) and 1 (i.e., the event is guaranteed to occur). In other words, probability theory is about the study of chance. To put probability on firm mathematical ground, we need to first introduce the concept of randomness, which is the central issue in this domain.

- **Randomness:** In general, one can view randomness as a kind of objective uncertainty associated with random variables (Ross, 2014; Schinazi, 2012). Broadly speaking, randomness can be categorized into two classes, namely classical and chaotic randomness (Plotnitsky, 2016; Clegg, 2013).

    1) *Classical randomness*: A classical randomness can be defined as a meaningfully compressed category (i.e., obeying rules) which includes the overall information of a collection of random objects (Clegg, 2013). In other words, the classical randomness depends not on an individual event that carries discernible information but instead on a probability distribution (e.g., cumulative distribution functions) in which we can designate a special pattern or objective

from the whole population (Denker and Woyczyriski, 1998). For example, to measure the randomness in a gambling game, we must not care too much about any particular individual event, but should focus on how often an event comes up (i.e., how different outcomes are distributed). Other examples containing classical randomness also include lottery and stock market.

2) *Chaotic randomness*: On the other hand, when the effect of randomness arises in dynamic systems with only deterministic and well-controlled ingredients, we call this phenomenon chaotic randomness (Denker and Woyczyriski, 1998). Examples in this category include the ball's trajectory on the pinball table, the Brazilian butterfly effect, and the iterations of quadratic maps.

- **Probability Models:** Based on our understanding of randomness, we can now turn to a formal mathematical setting for analysing probability models. In general, any probabilistic model must include the following components: (1) A sample space (denoted by Ω) for an experiment which refers to the set of all possible outcomes of a random process; (2) An event (indicated by B) which represents a subset of the sample space; and (3) A probability function, represented by $\Pr(B)$, which satisfies three properties defined by Eq. 1.9 (Johnson, 2018):

$$\begin{array}{ll} \text{Requirement 1.} & 0 \le \Pr(B) \le 1, \text{ for each event } B \text{ in } \Omega. \\ \text{Requirement 2.} & \Pr(\Omega) = 1 \\ \text{Requirement 3.} & \text{If } A \text{ and } B \text{ are mutually exclusive events in } \Omega, \text{ then} \\ & \Pr(A \cup B) = \Pr(A) + \Pr(B) \end{array} \quad . \qquad 1.9$$

- **Probability Rules:** Bayes' theorem is one of probability theory that defines a relation between certain conditional probabilities (Morin, 2016).

1) *Simple form*: Typically, the 'simple form' of Bayes' theorem is defined by Eq. 1.10 (Morin, 2016):

$$\Pr(A|Z) = \frac{\Pr(Z|A) \cdot \Pr(A)}{\Pr(Z)}. \qquad 1.10$$

2) *Explicit form*: While the 'explicit form' can be given by Eq. 1.11 (Morin, 2016):

$$\Pr(A|Z) = \frac{\Pr(Z|A) \cdot \Pr(A)}{\Pr(Z|A) \cdot \Pr(A) + \Pr(Z|\text{not } A) \cdot \Pr(\text{not } A)}. \qquad 1.11$$

where the first part of denominator, i.e., $\Pr(Z|A)\cdot\Pr(A)$, refers to the combination of the accuracy of the test and the historical data of the problem; while the second part of denominator, i.e., $\Pr(Z|\text{not } A)\cdot\Pr(\text{not } A)$, is still related to the accuracy of the test, denoted by $\Pr(Z|\text{not } A)$, together with new information, represented by $\Pr(\text{not } A)$.

3) *General form*: And the 'general form' of Bayes theorem is given by Eq. 1.12 (Morin, 2016):

$$\Pr(A_k|Z) = \frac{\Pr(Z|A_k)\cdot\Pr(A_k)}{\sum_i \Pr(Z|A_i)\cdot\Pr(A_i)}. \qquad 1.12$$

where A_i stands for a complete and mutually exclusive set of events; $\Pr(A_i)$ denotes the prior probabilities; $\Pr(Z|A_i)$ represents the conditional probability, and $\Pr(A_k|Z)$ indicates the posterior probability.

In practice, Bayes' rule finds application in a wide variety of domains, such as pattern recognition (Marwala, 2007a), militarized interstate dispute (Marwala and Lagazio, 2011a), and finite element model updating (Marwala et al., 2017).

1.1.3.4 Possibility Theory

Through the above discussion, we learn that probability theory provides an acceptable concept of quantitative chance. While in the literature, there is another method called possibility theory which focuses more on the analysis of different types of uncertainty other than chance (Nguyen and Walker, 2006). More specifically, possibility theory is used to deal with problems that have a non-probabilistic character. One example in this regard is to combine possibility theory with linguistic variable concept (found in fuzzy set theory) for the purpose of offering a unified formal framework. Under such framework, a formal management of information with inherent imprecision, vagueness, ambiguity, and uncertainty (Kraft and Colvin, 2017). A crude comparison among possibility theory, probability theory and Boolean algebra was performed by Zimmermann (1992) from different aspects and the results are illustrated in Fig. 1.3.

In practice, applied researchers have learned possibility theory through the lens of measure theory, since it can provide a relatively easy translation between mathematical theory and real-world problems. From a historical perspective, possibility theory is based on two semi-continuous generalized measure theories, i.e., possibility measures and necessity measures (Bělohlávek et al., 2017). The former estimates the feasibility

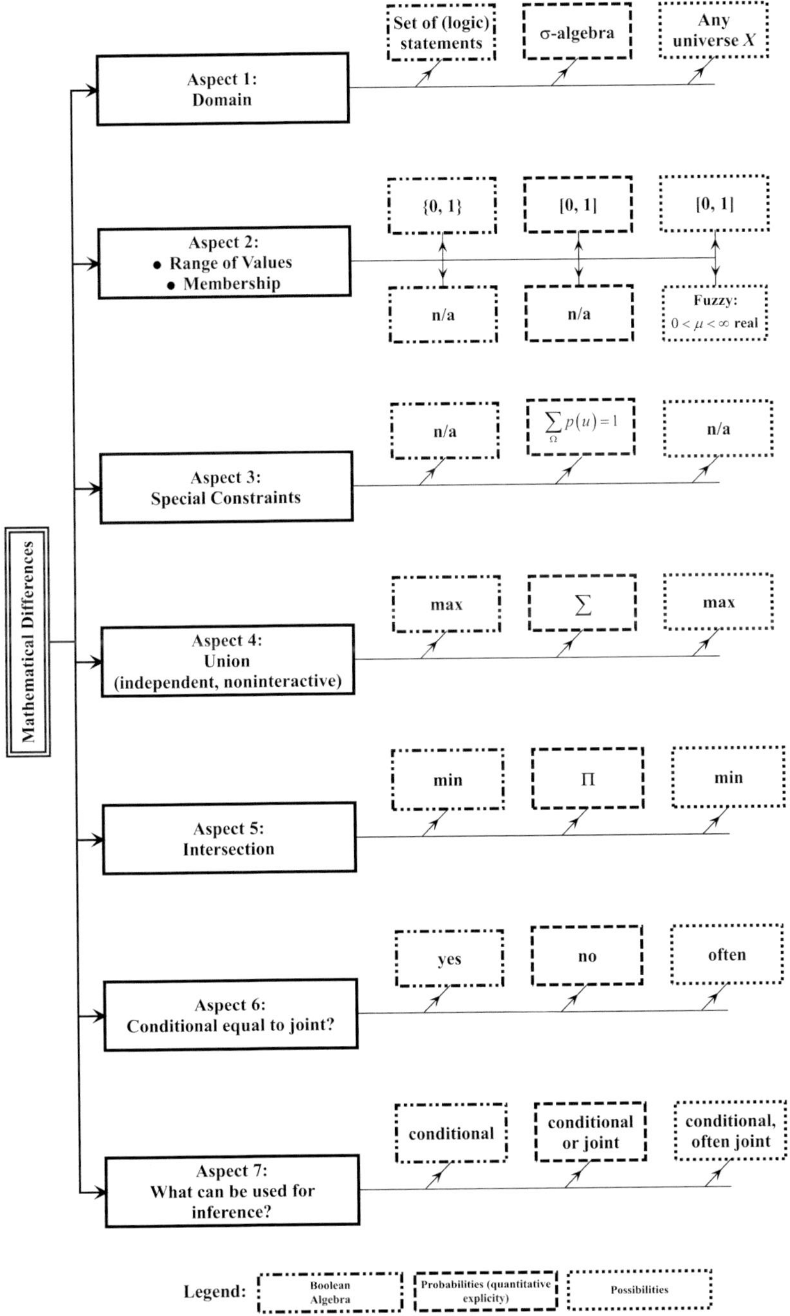

Figure 1.3: Mathematical differences among three areas: Boolean algebra, probabilities, and possibilities.

degrees of alternative options, while the latter indicates priorities (Dubois et al., 1996).

Suppose we have a universe of discourse (denoted by U), then a possibility measure (represented by Pos) is a set function Pos: $P(X) \rightarrow [0,1]$ that meets the properties given by Eq. 1.13 (Zimmermann, 1992; Bělohlávek et al., 2017):

$$\begin{array}{ll} \text{Property 1.} & \text{Pos}(\varnothing)=0,\ \text{Pos}(U)=1 \\ \text{Property 2.} & A \subseteq B \Rightarrow \text{Pos}(A) \leq \text{Pos}(B) \\ \text{Property 3.} & \text{Pos}(\cup_{i\in I} A_i) = \sup_{i\in I} \text{Pos}(A_i) \end{array}. \quad 1.13$$

Given a Pos, the necessity measure (denoted by Nec) can be defined by Eq. 1.14 (Zimmermann, 1992; Bělohlávek et al., 2017):

$$\text{Nec}(A) = 1 - \text{Pos}(\bar{A}). \quad 1.14$$

If $\tilde{A}$ is a fuzzy set in the universe U and π_x denotes a possibility distribution (associated with a variable X that takes value from set U), then the possibility measure can be defined by Eq. 1.15 (Zimmermann, 1992; Bělohlávek et al., 2017):

$$\begin{aligned} \text{Pos}\{X \text{ is } \tilde{A}\} &\hat{=} \pi(\tilde{A}) \\ &\hat{=} \sup_{u\in U} \min\{\mu_{\tilde{A}}(u), \pi_x(u)\} \end{aligned}. \quad 1.15$$

In a similar vein, the necessity measure can also be given by Eq. 1.16 (Bělohlávek et al., 2017):

$$\text{Nec}_F(A) = 1 - \text{Pos}_F(\bar{A}). \quad 1.16$$

where F represents a fuzzy set defined on U.

1.1.3.5 *Interval Analysis*

In general, an interval can be denoted as real numbers with brackets or parentheses, based on whether the end points are included or not. For instance, if $a > b$, then $[b, a]$, $[b, a)$, $(b, a]$, and (b, a) are sets of numbers x that satisfy the conditions given by Eq. 1.17 (Hausdorff, 1962):

$$\begin{array}{l} b \leq x \leq a \\ b \leq x < a \\ b < x \leq a \\ b < x < a \end{array}. \quad 1.17$$

The purpose of interval analysis is to address numerical errors emerging from computation (Moore et al., 2009). In other words, this technique is designed to automatically provide rigorous bounds on all potential errors and uncertainties (Hansen and Walster, 2004). Interested readers should refer to Chakraverty (2014) for more information.

A fuzzy interval typically represents a fuzzy set of real numbers in which the membership function is characterized by unimodal and upper-semi continuous features (Kacprzyk and Pedrycz, 2015). Accordingly, the calculus of fuzzy intervals is an extended version of interval arithmetic built on a possibilistic counterpart of a random variable's computation. For instance, in order to obtain the addition of two fuzzy intervals (denoted by A and B, respectively), one must calculate the membership function of $A \oplus B$ as the possibility degree via the possibility distribution, i.e., $\min(\mu_A(x), \mu_B(y))$, as given by Eq. 1.18 (Kacprzyk and Pedrycz, 2015):

$$\mu_{A \oplus B}(z) = \Pi\left(\left\{(x, y) : x + y = z\right\}\right). \quad 1.18$$

Further discussions about the fuzzy interval can be found in (Dubois et al., 2000; Nguyen et al., 2012).

1.1.3.6 Category Theory

Briefly, a category stands for a labelled directed graph, in which the nodes are termed as objects and the labelled directed edges are called morphisms (Barendregt, 2013). From a conceptual point of view, category theory is a mathematical structure that can be used to formalize high-level concepts (e.g., sets, rings and groups). More in-depth explanations can be found in (Roman, 2017; Awodey, 2006).

1.2 The Unavoidability of Uncertainty in Reality: A Quick Retrospection

The kind of information that we get every day is likely to be as follows: Join us for dinner and the meeting at 7.00 tomorrow. Of course, the uncertainty of such information might lead us into trouble: (1) The first part of the statement is vague with respect to what time the common dinner will be held; and (2) The second part of the statement is ambiguous regarding whether the meeting time is in A.M. or P.M.

In the literature, uncertainty is defined as a measure of the users' understanding of the difference between the information carried by the proposition corresponding to certain phenomena (Galbraith, 1973). Some scholars pointed out that uncertainty is unavoidable and, thus, worth the scrutiny. For example, Faber (2012) examined engineering decision problems that are subject to uncertainty. Kraft and Colvin (2017) treated information retrieval as an uncertain problem and used fuzzy logic to deal with it. Other examples also include learning with uncertainty (Wang and Zhai, 2017), artificial intelligence with uncertainty (Li and Du, 2017), and economic models under uncertainty (Aliyev, 2014).

Traditionally, measures of uncertainty were only related to classical set theory and probability theory (Wierman, 1999; Friedlob and Schleifer, 1999). However, this unique connection is now challenged by many. Among them, several researchers proposed that uncertainty is a multi-dimensional concept (Zadeh, 1965; Pawlak, 1982; Zimmermann, 2001). As a result, a unified perspective on the recent studies regarding uncertainty calls for other theories which can discover different properties of those incomplete and imprecise data, such as fuzziness, roughness, and indefiniteness (Wang and Zhai, 2017; Friedlob and Schleifer, 1999). For the rest of this section, we will briefly discuss some of these properties together with the corresponding mathematical foundations.

1.2.1 *Fuzziness*

Fuzziness is a kind of mathematical way to represent cognitive uncertainty, such as ambiguity and vagueness, in measurements and natural language expressions, imperfectly in experts' thoughts and knowledge, and absence of concepts' boundaries (Colubi and Gonzalez-Rodriguez, 2015; Coppi et al., 2006; Seising, 2008; Zhang, 1998; Klir, 1987). In practice, those imprecise informations are modelled by fuzzy sets based on their associated degrees of membership (Zadeh, 1965; Nguyen and Walker, 2006). Some useful definitions are discussed as follows:

- **Fuzzy Sets:** Suppose we have a non-empty set denoted by $X = \{x_1, x_2, x_3, \cdots, x_n\}$. A fuzzy set (represented by $\tilde{A}$) in X stands for a set of ordered pairs given by Eq. 1.19 (Zimmermann, 1992):

$$\tilde{A} = \left\{ \left(x, \mu_{\tilde{A}}(x) \right) \middle| x \in X \right\}. \qquad 1.19$$

  where the membership degree of x in $\tilde{A}$ is denoted by $\mu_{\tilde{A}}(x)$, and each element x in X maps to a real number belonging to the interval of [0,1].

  1) A support of a fuzzy set $\tilde{A}$, denoted by $S(\tilde{A})$, stands for the crisp set of all $x \in X$ that satisfy $\mu_{\tilde{A}} > 0$.
  2) In classical (crisp) set theory, α-cut means that the degree for elements that belong to a fuzzy set is at least α as given by Eq. 1.20 (Zimmermann, 1992):

$$A_\alpha = \left\{ x \in X \middle| \mu_{\tilde{A}}(x) \geq \alpha \right\}. \qquad 1.20$$

3) A fuzzy set is convex, if for all elements $x_1, x_2 \in X$ and $\lambda \in [0,1]$, the relationship defined by Eq. 1.21 (Zimmermann, 1992) hold:

$$\mu_{\tilde{A}}\left(\lambda x_1 + (1-\lambda)x_2\right) \geq \min\left(\mu_{\tilde{A}}(x_1), \mu_{\tilde{A}}(x_2)\right). \qquad 1.21$$

4) For fuzzy sets, the basic set-theoretic operations are defined by Eq. 1.22 (Zimmermann, 1992):

$$\begin{cases} \text{Fuzzy Union}: & \mu_{\tilde{A}\cup\tilde{B}}(x) = \max\{\mu_{\tilde{A}}(x), \mu_{\tilde{B}}(x)\},\ x \in X \\ \text{Fuzzy Intersection}: & \mu_{\tilde{A}\cap\tilde{B}}(x) = \min\{\mu_{\tilde{A}}(x), \mu_{\tilde{B}}(x)\},\ x \in X. \\ \text{Fuzzy Complement}: & \mu_{\complement\tilde{A}}(x) = 1 - \mu_{\tilde{A}}(x),\ x \in X \end{cases} \qquad 1.22$$

- **Fuzzy Numbers**: A fuzzy number is a fuzzy quantity that stands for a generalization of a real number (Nguyen and Walker, 2006). Typically, there are two types of fuzzy numbers: (1) Triangular fuzzy numbers, and (2) Trapezoidal fuzzy numbers. In practice, triangular fuzzy numbers are the most employed type, in which the fuzzy numbers are characterized by a triangular shape. The general formulation of fuzzy numbers can be given by Eq. 1.23 (Novák et al., 2016):

$$\mu_{\tilde{A}}(x) = \begin{cases} 0 & \text{if } x < a \text{ or } x > c \\ \dfrac{x-a}{b-a} & \text{if } a \leq x \leq b \\ \dfrac{x-c}{b-c} & \text{if } b < x \leq c \\ 1 & \text{if } x = b \end{cases}. \qquad 1.23$$

where $a < b < c$.

1.2.2 *Roughness*

In rough set theory, roughness stands for the uncertainty associated with a target concept that results from its boundary region (Pawlak, 1991). In other words, the uncertainty of rough sets is expressed by means of approximations. Some useful concepts are reviewed as follows:

- **Information System Framework:** The starting point of any rough set is a dataset that is called an information table or information system. Suppose $S = \langle U, A, V, f \rangle$ is introduced to represent an information system, we can have the following (Zhang et al., 2016; Pawlak, 1982):

1) U and A are finite non-empty sets;
2) The former stands for the universe of objects;
3) While the latter indicates the attribute set;
4) $V = \bigcup_{a \in A} V_a$, where V_a denotes the set of values of attribute a; and
5) $f : A \rightarrow V$ denotes a description function.

- **Indiscernible Relation:** Given $\forall B \subseteq A$, there is an associated indiscernibility (or equivalence) relation based on U as defined by Eq. 1.24 (Zhang et al., 2016; Pawlak, 1982):

$$Ind(B) = \left\{(x, y) \in U^2, \forall_{a \in B} \left(a(x) = a(y)\right)\right\}. \quad 1.24$$

Accordingly, the equivalence class of an object ($x \in U$) is denoted by $[x]_{Ind(B)}$ or simply $[x]$, and the pair $(U, [x]_{Ind(B)})$ is termed as the approximation space.

- **Lower- and Upper-Approximation Sets:** For a sub-set $X \subseteq U$, its lower- and upper-approximation sets are given by Eqs. 1.25 and 1.26 (Zhang et al., 2016; Pawlak, 1982):

$$Appr_{lower}(X) = \left\{x \in U \middle| [x] \subseteq X\right\}. \quad 1.25$$

$$Appr_{upper}(X) = \left\{x \in U \middle| [x] \cap X \neq \varnothing\right\}. \quad 1.26$$

where $Appr_{lower}(X)$ stands for an object ($x \in U$) certainly belonging to $X \subseteq U$, while $Appr_{upper}(X)$ denotes an object $x \in U$ possibly belonging to $X \subseteq U$. Furthermore, the set $BND(X) = Appr_{upper}(X) - Appr_{lower}(X)$ represents the boundary region of X.

1) *Rough sets*: If and only if $Appr_{upper}(X) \neq Appr_{lower}(X)$, X can be regarded as a rough set (Zhang et al., 2016; Pawlak, 1982).
2) *Roughness of rough sets*: The roughness of set X is defined by Eq. 1.27 (Zhang et al., 2016; Pawlak, 1982):

$$Roughness(X) = 1 - \frac{\left|Appr_{lower}(X)\right|}{\left|Appr_{upper}(X)\right|} = \frac{\left|Appr_{upper}(X) - Appr_{lower}(X)\right|}{\left|Appr_{upper}(X)\right|}. \quad 1.27$$

1.2.3 *Indefiniteness*

Indefiniteness is a specific concept for the uncertainty of meaning between two or more unclear objects, situations, and/or problems (Wang and Zhai, 2017). Typically, it is associated with a possibility distribution that was proposed by Zadeh (1978). In general, such possibility distribution π can be defined by Eq. 1.28 (Wang and Zhai, 2017; Zimmermann, 1992):

$$\pi(A) = \sup_{x\in A} f(x). \qquad 1.28$$

where $A \subset X$, and X represents a classical set, i.e., $X = \{x_1, x_2, x_3, \cdots, x_n\}$. If and only if $\max_{x\in X} \pi(x) = 1$, the possibility distribution π can be treated as a normalized possibility distribution.

Based on these definitions, the measure of indefiniteness can be given by Eq. 1.29 (Wang and Zhai, 2017):

$$g(\pi) = \sum_{i=1}^{n} \left(\pi_i^* - \pi_{i+1}^*\right) \ln i. \qquad 1.29$$

where $\pi = \{\pi(x) | x \in X\}$ denotes a normalized possibility distribution, π^* represents the possibility distribution's permutation that satisfies $\pi_i^* \geq \pi_{i+1}^*$, for $i = 1,2,\cdots,n$. Since possibility theory focuses more on imprecision and vagueness of linguistic meanings, in order to elicit knowledge from those words or sentences, fuzzy judgements rather than probabilistic values are required.

Suppose we have a fuzzy set ($\tilde{F}$) of universe U and the associated membership function, denoted by $\mu_{\tilde{F}}(u)$, the assignment of the values of variable u to X can be given by Eq. 1.30 (Zimmermann, 1992):

$$X = u : \mu_{\tilde{F}}(u). \qquad 1.30$$

Accordingly, the fuzzy membership function of possibility distribution $\tilde{\pi}$ can be given by Eq. 1.31 (Zimmermann, 1992):

$$\tilde{\pi}_x = \mu_{\tilde{F}}. \qquad 1.31$$

Given $i = 2$, the measure of fuzzy indefiniteness is obtainable by using Eq. 1.32 (Wang and Zhai, 2017):

$$g(\tilde{\pi}) = \begin{cases} \left(\dfrac{\mu_1}{\mu_2}\right) \times \ln 2 & \text{if } 0 \leq \mu_1 < \mu_2 \\ \ln 2 & \text{if } \mu_1 = \mu_2 \\ \left(\dfrac{\mu_2}{\mu_1}\right) \times \ln 2 & \text{others} \end{cases}. \qquad 1.32$$

1.2.4 Relations Underlying the Uncertainties

Since the types of uncertainties vary a lot, different methodologies offer us varied alternatives, listed below.

- **Option 1:** Based on probability theory, randomness measures the uncertainty that the target objects have definite boundaries;
- **Option 2:** Built on fuzzy set theory, fuzziness measures the uncertainty emerging from vagueness;
- **Option 3:** According to rough set concept, roughness measures the uncertainty with respect to imprecise and incomplete information; and
- **Option 4:** With the aid of possibility theory, indefiniteness measures the uncertainty associated with non-specificity of information.

In addition, by combining these fundamental methods with other technologies (Zhang et al., 2016), a large number of new toolkits have become available, such as probabilistic fuzzy sets (Jiang et al., 2017; Kentel and Aral, 2004; Fialho et al., 2016), fuzzy/rough sets with neural networks (Boutalis et al., 2014; Raveendranathan, 2014; Ding et al., 2014), neighbourhood rough sets (Pal et al., 2012), rough/fuzzy sets with clustering algorithms (Zhou et al., 2011, Baraldi and Blonda, 1999a; Baraldi and Blonda, 1999b), fuzzy sets with game theory (Li, 2014; Jiménez-Losada, 2017), rough/fuzzy sets with soft sets (Sun and Ma, 2014; Feng et al., 2010; Das et al., 2017), and fuzzy/rough sets with different smart computing techniques (Xing and Gao, 2014b; Kolokotsa, 2007; Mardani et al., 2015; Mitra and Hayashi, 2000; Kubler et al., 2016; Castillo and Melin, 2015).

1.2.5 *Measure Theory*

In the domain of uncertainty, the introduction of suitable measures, for comparing different information contents carried by distinct uncertainties (e.g., randomness, fuzziness, roughness, and indefiniteness), is ranked as the most attractive topic. Indeed, measure theory stays at the center of addressing uncertainty. In the literature, various types of measures (bearing distinct properties) have been proposed, such as distance, correlation, divergence, entropy and similarity. Amongst them, entropy and similarity are the most frequently used methods for studying the uncertainty/certainty information carried by fuzzy sets. According to Liu (1992), these two measures represent the complementary information towards each other, i.e., certainty (similarity) and uncertainty with respect to the corresponding crisp set.

1.2.5.1 *Entropy Measurement for Uncertainty—Shannon's Entropy for Classical System*

Typically, the measure used for quantifying the uncertainty associated with random variables is called entropy (Wang and Zhai, 2017; Cover and Thomas, 1991; Mitzenmacher and Upfal, 2017). Although this term originated from thermodynamics and was initially used to measure the disorder in a system, it is now widely used in information and communication systems to measure the uncertainty regarding the information content of the system (Zadeh, 1965; De Luca and Termini, 1972), i.e., determining the variation degree of the probability distribution.

Basically, Shannon's entropy emphasizes the measurement of the average uncertainty in bits corresponding to the prediction of a random experiment's outcomes, that is, entropy allows us to learn the distribution function's shape. From the recipient viewpoint, the amount of missed information is known.

- **Discrete Entropy:** In general, entropy relies on a probabilistic description of an event. Suppose we have a discrete random variable (denoted by X) which consists of n instances, represented by $\{X_i$, for $i = 1,2,\ldots,n\}$, the probability mass function of X can thus be given by Eq. 1.33 (Wang and Zhai, 2017):

$$p(X_i) = \Pr(X = X_i). \tag{1.33}$$

Based on this formulation, the entropy in bits of a discrete random variable X can then be defined by Eq. 1.34 (Cover and Thomas, 1991; Li and Du, 2017; Wang and Zhai, 2017):

$$H(X) = -\sum_{i=1}^{n} p(X_i)\log_2 p(X_i). \tag{1.34}$$

If we have two random systems (denoted by X and Y), then the joint entropy of these two systems can be defined by Eq. 1.35 (Wang and Zhai, 2017):

$$H(X,Y) = -\sum_{i=1}^{n}\sum_{i=1}^{n} p(X_i,Y_i)\log_2 p(X_i,Y_i). \tag{1.35}$$

Similarly, we can also compute the conditional entropy $H(X|Y)$ by using Eq. 1.36 (Wang and Zhai, 2017):

$$\begin{aligned} H(X|Y) &= -\sum_{i=1}^{n} p(X_i)H(Y|X = X_i) \\ &= -\sum_{i=1}^{n} p(X_i)\sum_{i=1}^{n} p(Y_i|X_i)\log_2 p(Y_i|X_i) \\ &= -\sum_{i=1}^{n}\sum_{i=1}^{n} p(X_i,Y_i)\log_2 p(Y_i|X_i) \end{aligned} \tag{1.36}$$

where $H(X|Y)$ is, in general, not equal to $H(Y|X)$; while $H(X) - H(X|Y)$ is always equal to $H(Y) - H(Y|X)$.

Since the mutual information can quantify the closeness degree of two random variables, suppose X and Y are discrete random variables, when it comes to measure the relevance of X and Y, we can further define the mutual information by using Eqs. 1.37–1.39 (Wang and Zhai, 2017; Michalowicz et al., 2014):

$$I(X;Y) = H(X) - H(X|Y). \tag{1.37}$$

$$I(X;Y) = H(Y) - H(Y|X). \tag{1.38}$$

$$I(X;Y) = H(X) + H(Y) - H(XY). \tag{1.39}$$

- **Differential Entropy:** The concept of entropy for continuous distribution is called differential entropy. Suppose X represents a continuous random variable with probability density function, denoted by $p_X(x)$, the differential entropy can be defined by Eq. 1.40 (Michalowicz et al., 2014):

$$h_X = -\int_S p_X(x)\log_2(p_X(x))dx. \tag{1.40}$$

where $S = \{x|p_x(x) > 0\}$ denotes the support set of X. While the value of discrete entropy is always non-negative, differential entropy may take any value between ∞ and $-\infty$.

For continuous random variables X and Y, their joint differential entropy, conditional differential entropy, and the mutual information can be further defined by Eqs. 1.41–1.43, respectively (Michalowicz et al., 2014):

$$h_{XY} = \int\int p_{XY}(x,y)\log_2\left(\frac{1}{p_{XY}(x,y)}\right)dxdy. \tag{1.41}$$

$$h_{Y|X} = \int\int p_{XY}(x,y)\log_2\left(\frac{1}{p_Y(y|x)}\right)dxdy. \tag{1.42}$$

$$I(X;Y) = \int\int p_{XY}(x,y)\log_2\left(\frac{p_{XY}(x,y)}{p_X(x)p_Y(y)}\right)dxdy. \tag{1.43}$$

- **Maximum Entropy Estimation Method:** Assume that the density function $p_X(x)$ is unknown, but we know a number of related constraints, such as mean and variance. The maximum entropy estimate of the unknown probability density function is the one that can maximize the entropy, subject to given constraints (Theodoridis and Koutroumbas, 2009).

1.2.5.2 *Entropy Measurement for Uncertainty—Fuzzy Entropy*

A fuzzy set's entropy is to calculate the degree of fuzziness on such fuzzy set (Zadeh, 1968). Among different measures, one measure considered by Zadeh (1968) can be expressed by using Eq. 1.44 (Zimmermann, 1992; Wang and Zhai, 2017):

$$H\left(\tilde{A}\right)=-\sum_{i=1}^{n}\mu_{\tilde{A}}\left(x_i\right)p_i\log_2 p_i \,. \qquad 1.44$$

In addition, another fuzzy entropy considered by De Luca and Termini (1972) can be defined by using Eq. 1.45 (Zimmermann, 1992; Wang and Zhai, 2017; De Luca and Termini, 1972):

$$H\left(\tilde{A}\right)=-K\sum_{i=1}^{n}\left[\mu_{\tilde{A}}\left(x_i\right)\log_2\mu_{\tilde{A}}\left(x_i\right)+\left(1-\mu_{\tilde{A}}\left(x_i\right)\right)\log_2\left(1-\mu_{\tilde{A}}\left(x_i\right)\right)\right]. \qquad 1.45$$

where K represents a positive constant.

Fuzzy entropy is quite different from the classical Shannon's entropy (Jumarie, 1992). The former handles vagueness and ambiguity uncertainties, while the latter deals with randomness uncertainty (i.e., probabilistic). Typically, there are three types of fuzzy entropy, i.e., interval-valued fuzzy entropy, intuitionistic fuzzy entropy, and interval-valued intuitionistic fuzzy entropy (Bustince and Burillo, 1996; Wei et al., 2011; Zhang et al., 2014). Nowadays, those entropy measures have been utilized in dealing with fuzzy systems, such as image processing (Naidu et al., in press), fuzzy decision-making systems (Shi and Yuan, 2015), and fuzzy software testing (Kumar et al., 2012).

1.2.5.3 *Entropy Measurement for Uncertainty—Rough Entropy*

In order to get the a set's knowledge incompleteness quantified, rough entropy was introduced by Beaubouef et al. (1998). In general, it can be formulated according to the uncertainty in granulation and the set's definability (Sen and Pal, 2009). For an information system, denoted by $S = (U, A)$ where $X \subseteq U$, rough entropy of X can be described by using Eq. 1.46 (Liang, 2011):

$$E_A(X) = -\rho_A(X)\left(\sum_{i=1}^{m}\frac{|R_i|}{|U|}\log_2\frac{|R_i|}{|U|}\right). \quad 1.46$$

where $p_A(X)$ represents the rough degree of X. In a similar vein, rough entropy of A can be given by Eq. 1.47 (Liang, 2011).

$$E_r(A) = -\sum_{i=1}^{m}\frac{|R_i|}{|U|}\log_2\frac{1}{|R_i|}. \quad 1.47$$

The relationship between rough entropy and Shannon's entropy can, thus, be defined by Eq. 1.48 (Liang, 2011):

$$E_r(A) + H(A) = \log_2|U|. \quad 1.48$$

1.2.5.4 *Similarity Measurement for Uncertainty—General Similarity Measure*

The entropy measures are introduced in order to address how much uncertainty is associated with non-deterministic phenomena. Yet, what is the data certainty with respect to the deterministic data? In addition, some of the non-deterministic phenomena are expressed in natural language, e.g., pretty large, about 100 km, and quite close. Furthermore, human perception also has inherent uncertainty, which is different from other uncertainties, e.g., the degree of membership value, and group/interval number. In light of this observation, similarity measure was proposed as an alternative (Song et al., 2015; van Eck and Waltman, 2009; Candan and Li, 2001; Khorshidi and Nikfalazar, 2017). Its prominent application domains include pattern recognition (Chen and Chang, 2015; Papacostas et al., 2013; Zeng et al., 2016; Chen et al., 2016; Nguyen, 2016) and decision making (Ye, 2014; Li et al., 2015; Chen, 2015c; Luukka, 2011).

In fact, the study of similarity is an established domain in various research branches of mathematics, such as topology and approximation theory. Typically, a distance function is used in order to identify the similarity between two instances (e.g., patterns, images, and reasoning) quantified.

1.2.5.5 *Similarity Measurement for Uncertainty—Fuzzy Similarity Measure*

Literally, fuzzy similarity measure refers to the calculation of the similarity (or proximity) relationships between fuzzy sets. Zadeh (1971) offered a more formal definition, in which fuzzy similarity measure is considered as the classical equivalence notion's multivalued generalization.

Typically, the methods of fuzzy similarity measures can be broadly categorised into three groups: Set-theoretic, proximity-based and logic-based (Cross and Sudkamp, 2002). Like the fuzzy entropy, fuzzy similarity measure also includes the following three variants:

- Similarity measure for interval-valued fuzzy sets (Chen and Chen, 2009; Ye and Du, in press; Chen, 2015c),
- Similarity measure for intuitionistic fuzzy sets (Li and Cheng, 2002; Liang and Shi, 2003; Farhadinia, 2014; Baccour et al., 2013), and
- Similarity measure for interval-valued intuitionistic fuzzy sets (Xu, 2007).

Interested readers should refer to (Pappis and Karacapilidis, 1993; Xing, 2017b; Wang, 1997; Zwick et al., 1987) for further discussions. In fact, fuzzy similarity measure can be regarded as the dual concept of fuzzy entropy. Accordingly, the relationships between similarity measures and entropy measures have been intensively investigated in the literature (Li et al., 2012; Zhang et al., 2014; Zeng and Li, 2006; Deng et al., 2015; Zeng and Guo, 2008).

1.3 Smart Computing

Recently, the First International Conference on Smart Computing and Informatics was successfully held on 3–4 March 2017 with the aim of offering a unified platform that can incorporate multi-disciplinary and the state-of-the-art research in terms of designing smart computing and information systems. According to (Satapathy et al., 2018), the theme of the conference was to focus on a diverse variety of innovation schemes in system science, artificial intelligence, and sustainable development that can be applied to offer solutions to various problems encountered in society, environment and industries. The scope of smart computing and informatics also consists of the deployment of emerging computational and knowledge transfer methodologies, as well as optimization techniques in distinct disciplines across science, technology, and engineering.

With the engineered system (e.g., financial system) becoming more and more advanced and sophisticated, the associated analysis and synthesis tasks are ever-increasingly demanding. This book borrows the 'smart computing' concept from the literature for the purpose of addressing these issues. Essentially, smart computing comprises different tools developed in other disciplines such as system theory, optimization theory, and computational intelligence. In principle, the process of analysing and synthesizing complex systems via smart computing concept includes: (1) Obtaining each possible action or feasible solution through analysis, (2) Evaluating the obtained outcomes against a certain scale of value or desirability, and (3) Determining the most desired action or optimum solution according to the selected criterion of a system's decision-based goals. This process can be illustrated in Fig. 1.4.

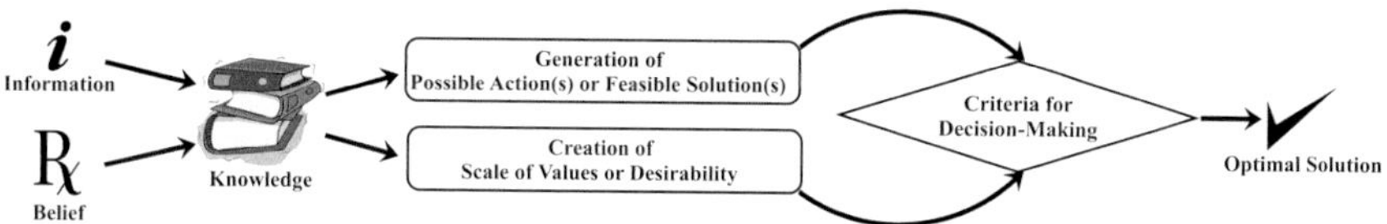

Figure 1.4: Process of smart computing for optimal decision-making.

The development of smart computing has experienced several stages, ranging from the static optimization approaches that are unfortunately incapable of handling or guaranteeing global solutions, through the recent multiple objective optimization methodologies (e.g., particle swarm optimization and ant colony optimization), to the emerging adaptive dynamic stochastic techniques. The evolutionary footprint of smart computing is depicted in Fig. 1.5.

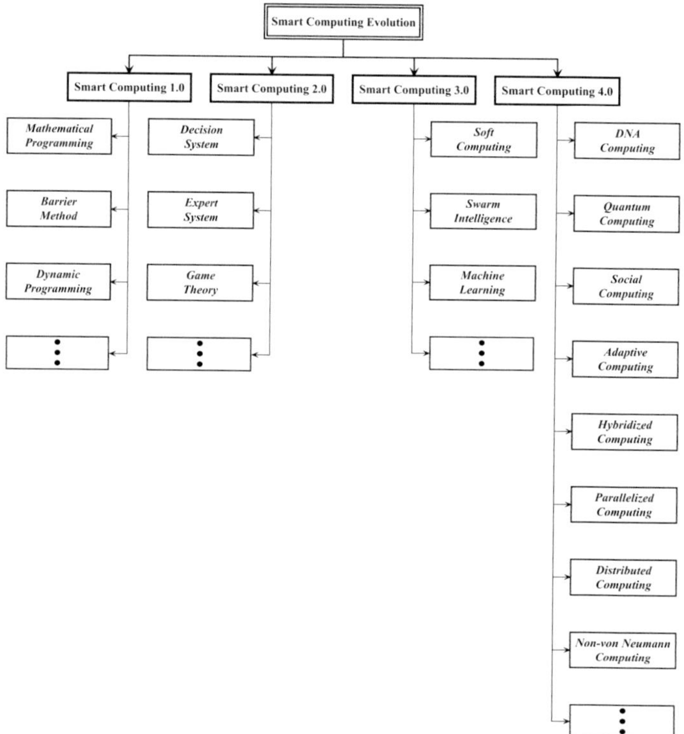

Figure 1.5: Smart computing evolution paradigm.

Regarding the exemplary examples under each category, interested readers should refer to (Suri, 2017; Hageback, 2017; Fister and Fister Jr, 2015; Indiveri, 2015; Polkowski and Artiemjew, 2015; Xing and Gao, 2014b; Chen et al., 2007; McGeoch, 2014; Yu, 2017; Hager and Wellein, 2011; Jeannot and Žilinskas, 2014; Xing, 2017a; Hurwitz et al., 2015; Marinescu, 2013; Xing et al., 2013b; Xing and Gao, 2014m; Xing and Gao, 2014x; Xing and Gao, 2014q; Xing and Marwala, 2018l; Xing, 2015c; Xing, 2014; Xing and Gao, 2014aa; Xing and Gao, 2014bb; Xing and Gao, 2014c; Xing and Gao, 2014cc; Xing and Gao, 2014d; Xing and Gao, 2014dd; Xing and Gao, 2014e; Xing and Gao, 2014f; Xing and Gao, 2014g; Xing and Gao, 2014h; Xing and Gao, 2014i; Xing and Gao, 2014j; Xing and Gao, 2014k; Xing and Gao, 2014l; Xing and Gao, 2014n; Xing and Gao, 2014o; Xing and Gao, 2014p; Xing and Gao, 2014r; Xing and Gao, 2014s; Xing and Gao, 2014t; Xing and Gao, 2014u; Xing and Gao, 2014v; Xing and Gao, 2014w; Xing and Gao, 2014y; Xing and Gao, 2014z; Xing and Gao, 2014ff; Marwala, 2007b; Marwala, 2013a; Marwala and Lagazio, 2011a; Marwala, 2010a; Marwala, 2012a; Marwala, 2014a).

1.4 Data Analytics

Imagine you stroll through any neighbourhood today, when you approach a building, the front door can slide open automatically. When you enter an empty room, a light can flick on by itself. When you jump up and down, a thermostat can trigger the air conditioner that compensates for the gradually warming air around you. When you roam at will, various motion-sensing surveillance cameras can slowly turn to keep you tracked. All these automated electromechanical gadgets work tedious (or even dangerous) jobs that were once performed by human beings.

At the heart of this story is the emergence of data and the corresponding analytics as the means to facilitate our lives. In fact, today's data is being produced so fast that the whole volume tends to be very large. According to one estimation, by the end of 2020 the total amount of data in the world annually will exceed 44 trillion gigabytes (IBM, 2014). Undoubtedly, data brings many benefits to us. However, in the meantime, there are many challenges associated with handling and making the sense of those data (Marwala, 2009; Marwala, 2015).

The basic idea behind data analytics is to collect raw data and convert them into meaningful information that is essential for making decisions. Unfortunately, extracting valuable information from the raw data is not as easy as it may sound. It is difficult to know where to start with the handling of data. To address these challenges, individuals and/or companies must resort to the process of data analytics, which typically includes three phases: Pre-processing, analytics, and post-processing

(Myatt and Johnson, 2014; Verbeke et al., 2018; EMC Education Services, 2015). A detailed overview of this process is depicted in Fig. 1.6.

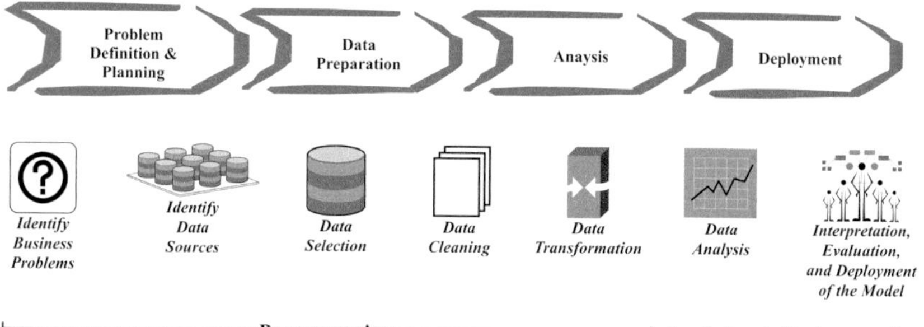

Figure 1.6: General data analytics scheme.

1.4.1 *Define Critical Problem and Generate Working Plan*

In the first step, a clear definition of the critical problem being addressed needs to be obtained, together with the generation of a feasible working plan. To achieve this, one has to take the following factors into account (Myatt and Johnson, 2014): (1) Outline possible deliverables, (2) Identify success causes, (3) Understand useful resources and their constraints, (4) Assemble a suitable team, (5) Come up with a working plan, and (6) Perform an analysis regarding costs and benefits.

1.4.2 *Identify Data Sources*

After defining the problem, we need to identify the data sources in order to support the project. In reality, there are numerous sources with various representations and formats (related to the core problem) that must be figured out. The golden rule here is that the more data there is available, the better the results.

1.4.3 *Select and Explore the Data*

Before a formal data analysis is conducted, a preliminary selection and exploration process towards the collected datasets should be considered. The objective of this step is to help the data analytics teams familiarize themselves with the interested data, i.e., how many cases are covered in the data table, what variables are included and what general hypotheses the data are likely to support.

- **Structured Data Selection and Exploration:** It is commonly agreed that the selection and exploration of structured data can be performed in a relatively controlled manner, e.g., accessing and combining data tables, and summarizing the data by using some descriptive/inferential statistics. The main objective is to improve the reliability of data, thereby ensuring that no substantial different values should be obtained from the repetition of measurement.

1) *First*: The starting point is normally a data table (or called dataset), which consists of the measured or collected data values expressed in the form of numbers or texts. One of the most common ways to deal with de-normalized source data tables is to merge them into a spreadsheet, where the raw data is outlined as rows and columns, denoting observations and variables, respectively. Based on scale, we can classify variables into four classes: Nominal-, ordinal-, interval-, and ratio-scale. Meanwhile, according to the roles they play in the mathematical models, variables can be broadly categorized into two categories, i.e., independent variables and response variables (Myatt and Johnson, 2014).
2) *Second*: Next, in order to get some initial insights with respect to a specific characteristic, we need to summarize the data. Among others, the most commonly reported characteristics for a particular variable include central tendencies, frequency distribution patterns, and something about the real-world estimations deduced from sub-sets of data (e.g., initial hypotheses to test the data) (Myatt and Johnson, 2014). To this end, several descriptive and/or inferential statistical approaches are needed, such as mode, median, mean, variances, standard deviations, confidence intervals, and hypothesis tests. Further discussions on the topic can be found in (Rumsey, 2010; Weiss, 2017; Ott and Longnecker, 2016). In addition, to make data better visualized (e.g., showing trends, outliers, and relationships among data variables), several graphical methods (e.g., bar charts, scatter plots, and box plots) can be employed (Myatt and Johnson, 2014).

- **Unstructured Data Selection and Exploration:** According to IBM's estimation (IBM, 2014), there are about 2.5 quintillion bytes of data being generated every day and among them, 90% are contributed by new technologies (e.g., smart phones and Internet of things) which are characterized by features such as semi-structured, quasi-structured and unstructured (Dobre and Xhafa, 2014; Sivarajah et al., 2017; Wessler, 2013). In general, the structured data format enjoys good predictability, high machine readability and may often be utilized as input to many other applications; while for semi-structured, quasi-structured and/or unstructured type of data, no data tables are ready to be used (Zhai and Massung, 2016; EMC Education Services, 2015), thus the operation involving computationally generated different attributes relevant to the target problem needs to be performed

(Myatt and Johnson, 2014). In this regard, computational intelligence techniques (Xing et al., 2013a; Xing, 2016b; Acharjya et al., 2015; Cacuci et al., 2014; Xing, 2015a), and machine learning methods (Zhou et al., 2017; Liu et al., 2013; Ratner, 2011; Flach, 2012; Witten et al., 2017) are among the most popular tools.

1.4.4 *Clean the Data*

In order to assess the data quality, we have to estimate the potential errors that might occur during the collection and selection processes. To this end, several issues must be considered, such as missing data and outlier/duplicate data (Verbeke et al., 2018).

- **Missing Data:** Missing data is a pervasive problem encountered in many areas, such as engineering (Nelwamondo and Marwala, 2008), economics (Marwala, 2013b), and healthcare (Dhlamini et al., 2006; Faris et al., 2002). The reasons for the emergence of missing values tend to be very different, some examples include sensor faults, uncompleted or inconsistent data, or even fraud. To deal with this problem, three different approaches are commonly utilized, namely neural networks (Marwala, 2000a; Nelwamondo et al., 2007; Abdella and Marwala, 2005), rough sets (Nelwamondo and Marwala, 2007), dynamic programming (Nelwamondo et al., 2009) and decision trees (Ssali and Marwala, 2008). More details can be found in (Marwala, 2009; Marwala, 2014c; Enders, 2010).
- **Outlier/Duplicate Data Detection and Handling:** In general, there are two types of outliers that need to be sought out, namely valid observations (e.g., salary is above $10 million) and invalid observations (e.g., age is over 100 years). To deal with these problems, one can follow the following two steps: Detection and treatment. In addition, data visualization tools can also help in identifying the outlier data (e.g., heavily shifted towards one particular value).

1.4.5 *Transform, Segment, and Load the Data*

After the operations of collection and selection, preliminary exploration, and cleaning are performed, one needs to transform, segment, and load the data at hand into an appropriate form for further processing (e.g., generating consistent scales across variables, converting texts into numbers, binning and combining data, and acquiring additional properties that are often termed as features in raw dataset). In practice, when only a limited amount of samples (i.e., pairs of input and output) is available, in order to learn the whole system's behaviour, we need to define a kind of relationship (typically in the form of mathematical model) that can best represent the connections between the input conditions and the output observations (Strutz, 2011). One of these data transformation techniques is called data fitting. Putting it simply, data fitting refers to the process of establishing a model that can best fit a set of data points, either through interpolated fitting (i.e., connecting the data dots) or via approximated fitting (i.e., accommodating uncertainty degree) (Strutz, 2011).

1.4.6 Model and Analyse the Data

Once various pre-processing steps are completed, a data analytics task team can begin to employ different data analytics tools (e.g., statistics and machine learning) in order to extract useful knowledge. Putting it more formally, data analytics stands for the systematic utilization of technologies and methodologies for deriving insights from data in order to 'see' patterns, processes and drivers for planning, management, operations, measurement, and learning (Aggarwal, 2015). In what follows, we will introduce two statistics-centric analytics methods, namely predictive analytics and descriptive analytics.

- **Predictive Analytics:** Literally, predictive analytics focuses on creating an analytical model for the prediction of a target measure of interest. Generally speaking, the methods involved in this domain, e.g., regression and classification, are characterized by their supervised features.
    1) *Regression*: This technique aims to explain how the outcome of an interested variable is influenced by a set of other variables. Typically, the target variable is continuous in nature (e.g., income and environmental changes). In practice, the means for generating regression models include linear regression, logistic regression, and multiple regression. More details can be found in (Bowerman et al., 2014; Lind et al., 2012; Weiss, 2017; Ross, 2010; Peck and Devore, 2012).
    2) *Classification*: In this domain, being presented with a set of classified examples, a classifier is able to learn how to classify unsorted cases. In other words, classifiers are capable of assigning class labels to new instances. The tools belonging to this category include k-nearest neighbours (Xing and Marwala, 2018i), decision trees (Ssali and Marwala, 2008; Marwala, 2014a), neural networks (Marwala and Hunt, 1999; Marwala, 2000b; Marwala, 2000a; Marwala, 2001; Marwala, 2003), support vector machines (Msiza et al., 2008; Patel and Marwala, 2009), Bayes methods (Marwala et al., 2017) and ensemble learning (e.g., bagging, boosting, and random forest), to name just a few. Further information can be found in (Duda et al., 2000; Ray, 2012; Stoean and Stoean, 2014).

- **Descriptive Analytics:** The aim of descriptive analytics is to discover and quantify relationships hidden in datasets (e.g., people's behavioural patterns). In the literature, it is also called unsupervised learning. Typically, the disclosed patterns can be summarized according to the following operations.
  1) *Association Rules*: These rules are commonly used for mining tractions in databases. Take the association rule {*burger, ketchup*} → {*buns*}, it means that if a customer purchases hamburger meat and ketchup together, the likelihood for him/her to buy buns tends to be high. Interested readers should refer to (EMC Education Services, 2015; Verbeke et al., 2018) for more details.
  2) *Sequence Rules*: These rules intend to, among all possible sequences, identify the maximal sequence(s) that have a certain user-specified minimum support and confidence (Verbeke et al., 2018). The difference between the association rules and sequence rules can be described as follows: (1) The former deals with the intra-transaction patterns, that is, what items show up together at the same time step, and (2) The latter focuses more on inter-transaction patterns, that is, what items show up at distinct time steps.
  3) *Clustering Analysis*: Clustering analysis is one of the popular methods in this category. Unlike classification, clustering is the use of unsupervised methods for grouping similar objects. Putting it more formally, clustering analysis is designed to gain insights about how target objects can be clustered into natural groupings for the purpose of driving specific actions or making recommendations. Typical clustering analysis methods include *k*-means (Kuo et al., 2002; Kuo and Li, 2016), physics-inspired algorithms (Xing and Gao, 2014z; Kundu, 1999; Zheng et al., 2010; Hatamlou et al., 2011; Hatamlou, 2013; Bahrololoum et al., 2012), chemically-based algorithms (Xing and Gao, 2014bb), and bio-inspired algorithms (Xing and Gao, 2014s; Bellaachia and Bari, 2012; Xing and Gao, 2014ll; Xing et al., 2010d; Xing et al., 2010e; Cui et al., 2006; Shafia et al., 2011). More details can be found in (Xing and Gao, 2014b; Xing and Gao, 2014c; Nanda and Panda, 2014).

In summary, data analytics technologies are becoming increasingly powerful with the aid of mathematical and statistical methods. A thorough explanation regarding the use of mathematical tools for data analytics can be found in (Simovici and Djeraba, 2014). In addition, for the models and methods associated with big data analytics, one should refer to (Wessler, 2013).

1.4.7 Interpretation, Evaluation and Deployment

In this step, the analytical models will be trained, tested/evaluated, and deployed. Accordingly, several questions need to be considered, such as (1) Whether the model is valid and accurate on the test data, (2) From the domain experts' viewpoint, does the output of the model make sense? And (3) Do we need to introduce additional data? In practice, the means to deploy data analysis results include generating a report, implementing a standalone or integrated decision-support tool, and measuring business impact.

1.5 Conclusions

Historically, the theory of thermodynamics was inspired by the development of the steam engine, while the fundamental physical principles of electron tubes were only understood properly after millions of them had been utilized during the Second World War. Accordingly, computing is not the very first technology that only received its appropriate theoretical foundations long after being utilized in practice. In this chapter, we briefly outline the necessity of computing and the unavoidability of uncertainty encountered in real-world scenarios which pave the way for the development of smart computing concepts and the associated data analytics. The field of smart computing is broad and is still rapidly expanding as well. At the end of this chapter, some representative applicable areas of smart computing are listed as follows:

- **Finite Element Model Updating:** (Marwala, 2005; Marwala, 2010a; Marwala, 2010c; Marwala, 2010b; Marwala, 2010d; Marwala et al., 2017);
- **Smart Maintenance:** (Xing, 2015a; Xing and Marwala, 2018n; Xing and Marwala, 2018m);
- **Human–Robot Interaction:** (Xing and Marwala, 2018k; Xing and Marwala, 2018e; Xing and Marwala, 2018g; Xing and Marwala, 2018f; Xing and Marwala, 2018d; Xing and Marwala, 2018b; Xing and Marwala, 2018c; Xing and Marwala, 2018j; Xing and Marwala, 2018h; Xing and Marwala, 2018i; Xing and Marwala, 2018a; Xing, 2013; Xing, 2015b; Xing, 2016e; Xing, 2016d; Xing, 2016b);
- **Remanufacturing:** (Xing and Gao, 2014a; Gao et al., 2013b);
- **Condition Monitoring:** (Marwala, 2012a; Marwala, 2012c; Marwala, 2012b); and
- **Militarized Conflict Analysis:** (Marwala and Lagazio, 2011a; Marwala and Lagazio, 2011b; Marwala, 2014b).

References

Abdella M and Marwala T. (2005) The use of genetic algorithms and neural networks to approximate missing data in database. *Proceedings of the IEEE 3rd International Conference on Computational Cybernetics (ICCCC 2005), Mauritius, 13–16 April*, pp. 207–212.

Acharjya DP, Dehuri S and Sanyal S. (2015) *Computational intelligence for big data analysis: Frontier advances and applications*, Springer Cham Heidelberg New York Dordrecht London: Springer International Publishing Switzerland, ISBN 978-3-319-16597-4.

Aggarwal CC. (2015) *Data mining: The Textbook*, Cham Heidelberg New York Dordrecht Lo: Springer International Publishing Switzerland, ISBN 978-3-319-14141-1.

Ali R, Siddiqi MH and Lee S. (2015) Rough set-based approaches for discretization: A compact review. *Artificial Intelligence Review* 44: 235–263.

Aliyev AG. (2014) *Economic-mathematical methods and models under uncertainty*, 3333 Mistwell Crescent, Oakville, ON L6L 0A2, Canada: Apple Academic Press, ISBN-13 978-1-4822-1267-9.

Anderson GT, Zheng J, Wyeth R, et al. (2000) A rough set/fuzzy logic based decision making system for medical applications. *International Journal of General Systems* 29: 879–896.

Atanassov K. (1986) Intuitionistic fuzzy sets. *Fuzzy Sets and Systems* 20: 87–96.

Atanassov K and Gargov G. (1989) Interval-valued intuitionistic fuzzy sets. *Fuzzy Sets and Systems* 31: 343–349.

Awodey S. (2006) *Category theory*, Great Clarendon Street, Oxford OX2 6 DP: Oxford University Press, ISBN 978-0-19-856861-2.

Azadegan A, Porobic L, Ghazinoory S, et al. (2011) Fuzzy logic in manufacturing: A review of literature and a specialized application. *International Journal of Production Economics* 132: 258–270.

Baccour L, Alimi AM and John RI. (2013) Similarity measures for intuitionistic fuzzy sets: State of the art. *International Journal of Fuzzy Systems* 24: 37–49.

Bae C, Yeh W-C, Chung YY, et al. (2010) Feature selection with intelligent dynamic swarm and rough set. *Expert Systems with Applications* 37: 7026–7032.

Bahrololoum A, Nezamabadi-pour H, Bahrololoum H, et al. (2012) A prototype classifier based on gravitational search algorithm. *Applied Soft Computing* 12: 819–825.

Baraldi A and Blonda P. (1999a) A survey of fuzzy clustering algorithms for pattern recognition—part 1. *IEEE Transactions on Systems, Man, and Cybernetics—Part B: Cybernetics* 29: 778–785.

Baraldi A and Blonda P. (1999b) A survey of fuzzy clustering algorithms for pattern recognition—part 2. *IEEE Transactions on Systems, Man, and Cybernetics—Part B: Cybernetics* 29: 786–801.

Barendregt H. (2013) Foundations of mathematics from the perspective of computer verification. In: Paule P (ed) *Mathematics, Computer Science and Logic-a Never Ending Story*. Springer International Publishing Switzerland, ISBN 978-3-319-00965-0, pp. 1–49.

Barro S and Marín R. (2002) *Fuzzy logic in medicine*: Springer-Verlag Berlin Heidelberg, ISBN 978-3-7908-2498-8.

Beaubouef T, Petry FE and Arora G. (1998) Information-theoretic measures of uncertainty for rough sets and rough relational data bases. *Information Sciences* 109: 185–195.

Bellaachia A and Bari A. (2012) Flock by leader: A novel machine learning biologically inspired clustering algorithm. In: Tan Y, Shi Y and Ji Z (eds) *ICSI 2012, Part I, LNCS 7332*, pp. 117–126. Berlin Heidelberg: Springer-Verlag.

Bělohlávek R, Dauben JW and Klir GJ. (2017) *Fuzzy logic and mathematics: A historical perspective*, 198 Madison Avenue, New York, NY 10016, United States of America: Oxford University Press, ISBN 978-0-19020-001-5.

Blanchini F and Miani S. (2015) *Set-theoretic methods in control*, Springer International Publishing Switzerland, ISBN 978-3-319-17932-2.

Bosma R, Kaymak U, Berg Jvd, et al. (2011) Using fuzzy logic modelling to simulate farmers' decision-making on diversification and integration in the Mekong Delta, Vietnam. *Soft Computing* 15: 295–310.

Boutalis Y, Theodoridis D, Kottas T, et al. (2014) *System identification and adaptive control: Theory and applications of the neurofuzzy and fuzzy cognitive network models,* Springer Cham Heidelberg New York Dordrecht London: Springer International Publishing Switzerland, ISBN 978-3-319-06363-8.

Bowerman BL, O'Connell RT and Murphree ES. (2014) *Business statistics in practice,* 1221 Avenue of the Americas, New York, NY, 10020: McGraw-Hill/Irwin, ISBN: 978-0-07-352149-7.

Bustince H and Burillo P. (1996) Entropy on intuitionistic fuzzy sets and on interval-valued fuzzy sets. *Fuzzy Sets and Systems* 78: 305–316.

Cacuci DG, Navon IM and Ionescu-Bujor M. (2014) *Computational methods for data evaluation and assimilation,* 6000 Broken Sound Parkway NW, Suite 300, Boca Raton, FL 33487-2742: CRC Press, Taylor & Francis Group, LLC, ISBN 978-1-58488-736-2.

Candan KS and Li W-S. (2001) On similarity measures for multimedia database applications. *Knowledge and Information Systems* 3: 30–51.

Caponetti L and Castellano G. (2017) *Fuzzy logic for image processing: A gentle introduction using Java*: Springer, ISBN 978-3-319-44128-3.

Castillo O and Melin P. (2015) *Fuzzy logic augmentation of nature-inspired optimization metaheuristics: Theory and applications*: Springer International Publishing Switzerland, ISBN 978-3-319-10959-6.

Celik E, Gul M, Aydin N, et al. (2015) A comprehensive review of multi criteria decision making approaches based on interval type-2 fuzzy sets. *Knowledge-Based Systems* 85: 329–341.

Ceruzzi PE. (2012) *Computing: a concise history,* 55 Hayward Street, Cambridge, MA 02142: The MIT Press, ISBN 978-0-262-51767-6.

Chakraverty S. (2014) *Mathematics of uncertainty modeling in the analysis of engineering and science problems*: IGI Global, ISBN 978-1-4666-4991-0.

Chen G, Church DA, Englert B-G, et al. (2007) *Quantum computing devices: Principles, designs, and analysis,* 6000 Broken Sound Parkway NW, Suite 300, Boca Raton, FL 33487.2742: Chapman & Hall/CRC, Taylor & Francis Group, ISBN 978-1-58488-681-5.

Chen JL. (2015a) *A guide to Hubble Space Telescope objects: Their selection, location, and significance,* Gewerbestrasse 11, 6330 Cham, Switzerland: Springer International Publishing Switzerland, ISBN 978-3-319-18871-3.

Chen S-H. (2008) Computational intelligence in agent-based computational economics. *Studies in Computational Intelligence (SCI).* Berlin Heidelberg: Springer-Verlag, 517–594.

Chen SM and Chen JH. (2009) Fuzzy risk analysis based on similarity measures between interval-valued fuzzy numbers and interval-valued fuzzy number arithmetic operators. *Expert Systems with Applications* 36: 6309–6317.

Chen S-M and Chang C-H. (2015) A novel similarity measure between Atanassov's intuitionistic fuzzy sets based on transformation techniques with applications to pattern recognition. *Information Sciences* 291: 96–114.

Chen S-M, Cheng S-H and Lan T-C. (2016) A novel similarity measure between intuitionistic fuzzy sets based on the centroid points of transformed fuzzy numbers with applications to pattern recognition. *Information Sciences* 343-344: 15–40.

Chen S-M, Randyanto Y and Cheng S-H. (in press) Fuzzy queries processing based on intuitionistic fuzzy social relational networks. *Information Sciences* http://dx.doi.org/10.1016/j.ins.2015.07.054.

Chen T-Y. (2015b) The inclusion-based TOPSIS method with interval-valued intuitionistic fuzzy sets for multiple criteria group decision making. *Applied Soft Computing* 26: 57–73.

Chen T-Y. (2015c) An interval type-2 fuzzy technique for order preference by similarity to ideal solutions using a likelihood-based comparison approach for multiple criteria decision analysis. *Computers & Industrial Engineering* 85: 57–72.

Clegg B. (2013) *Dice world: Science and life in a random universe,* Omnibus Business Centre, 39–41 North Road, London N7 9DP: Icon Books Ltd, ISBN 978-184831-565-5.

Cock MD, Cornelis C and Kerre EE. (2007) Fuzzy rough sets: The forgotten step. *IEEE Transactions on Fuzzy Systems* 15: 121–130.

Colubi A and Gonzalez-Rodriguez G. (2015) Fuzziness in data analysis: Towards accuracy and robustness. *Fuzzy Sets and Systems* 281: 260–271.

Comer D. (2017) *Essentials of computer architecture,* 6000 Broken Sound Parkway NW, Suite 300, Boca Raton, FL 33487-2742: CRC Press, Taylor & Francis Group, LLC, ISBN 978-1-138-62659-1.

Coppi R, Gil MA and Kiers HAL. (2006) The fuzzy approach to statistical analysis. *Computational Statistics & Data Analysis* 51: 1–14.

Cover TM and Thomas JA. (1991) *Elements of information theory,* 605 Third Avenue, New York, NY 10158-0012: John Wiley & Sons, Inc., ISBN 0-471-20061-1.

Cross VV and Sudkamp TA. (2002) *Similarity and compatibility in fuzzy set theory: Assessment and applications*: Springer-Verlag Berlin Heidelberg, ISBN 978-3-7908-2507-7.

Cui X, Gao J and Potok TE. (2006) A flocking based algorithm for document clustering analysis. *Journal of Systems Architecture* 52: 505–515.

Das S, Ghosh S, Kar S, et al. (2017) An algorithmic approach for predicting unknown information in incomplete fuzzy soft set. *Arabian Journal for Science and Engineering* 42: 3563–3571.

De Luca A and Termini S. (1972) A definition of non-probabilistic entropy in the setting of fuzzy sets theory. *Information and Control* 20: 301–312.

Deng G, Jiang Y and Fu J. (2015) Monotonic similarity measures between fuzzy sets and their relationship with entropy and inclusion measure. *Fuzzy Sets and Systems* 316: 348–369.

Denker M and Woyczyriski WA. (1998) *Introductory statistics and random phenomena: Uncertainty, complexity and chaotic behavior in engineering and science*: Springer Science+Business Media New York, ISBN 978-1-4612-7388-2.

Dereli T and Altun K. (2013) Technology evaluation through the use of interval type-2 fuzzy sets and systems. *Computers & Industrial Engineering* 65: 624–633.

Dhlamini SM, Nelwamondo FV and Marwala T. (2006) Condition monitoring of HV bushings in the presence of missing data using evolutionary computing. *Transactions on Power Systems* 1: 280–287.

Ding S, Jia H, Chen J, et al. (2014) Granular neural networks. *Artificial Intelligence* 41: 373–384.

Dobre C and Xhafa F. (2014) Intelligent services for big data science. *Future Generation Computer Systems* 37: 267–281.

Du D-Z and Ko K-I. (2014) *Theory of computational complexity,* 111 River Street, Hoboken, NJ 07030: John Wiley & Sons, Inc, ISBN 978-1-118-30608-6.

Dubois D and Prade H. (1990) Rough fuzzy sets and fuzzy rough sets. *International Journal of General Systems* 17: 191–209.

Dubois D, Fargier H and Prade H. (1996) Possibility theory in constraint satisfaction problems: Handling priority, preference and uncertainty. *Applied Intelligence* 6: 287–309.

Dubois D, Kerre E, Mesiar R, et al. (2000) Fuzzy interval analysis. In: Dubois D and Prade H (eds) *Fundamentals of Fuzzy Sets.* Springer Science+Business Media New York, ISBN 978-1-4613-6994-3, pp. 483–581.

Duda RO, Hart PE and Stork DG. (2000) *Pattern classification,* New York, USA: John Wiley & Sons, Ltd., ISBN 978-0-471-05669-0.

EMC Education Services. (2015) *Data science & big data analytics: Discovering, analyzing, visualizing and presenting Data,* 10475 Crosspoint Boulevard, Indianapolis, IN 46256: John Wiley & Sons, Inc., ISBN 978-1-118-87613-8.

Enders CK. (2010) *Applied missing data analysis,* 72 Spring Street, New York, NY 10012: The Guilford Press, ISBN 978-1-60623-639-0.

Faber MH. (2012) *Statistics and probability theory: In pursuit of engineering decision support,* Springer Dordrecht Heidelberg London New York: Springer Science+Business Media B.V., ISBN 978-94-007-4055-6.

Fan H and Zhong Y. (2012) A rough set approach to feature selection based on wasp swarm optimization. *Journal of Computational Information Systems* 8: 1037–1045.

Farhadinia B. (2014) An efficient similarity measure for intuitionistic fuzzy sets. *Soft Computing* 18: 85–94.

Faris PD, Ghali WA, Brant R, et al. (2002) Multiple imputation versus data enhancement for dealing with missing data in observational health care outcome analyses. *Journal of Clinical Epidemiology* 55: 184–191.

Feng F, Li C, Davvaz B, et al. (2010) Soft sets combined with fuzzy sets and rough sets: A tentative approach. *Soft Computing* 14: 899–911.

Fialho AS, Vieira SM, Kaymak U, et al. (2016) Mortality prediction of septic shock patients using probabilistic fuzzy systems. *Applied Soft Computing* 42: 194–203.

Fister I and Fister Jr I. (2015) *Adaptation and hybridization in computational intelligence,* Springer Cham Heidelberg New York Dordrecht London: Springer International Publishing Switzerland, ISBN 978-3-319-14399-6.

Flach P. (2012) *Machine learning: The art and science of algorithms that make sense of data,* The Edinburgh Building, Cambridge CB2 8RU, UK: Cambridge University Press, ISBN 978-1-107-09639-4.

Friedlob GT and Schleifer LLF. (1999) Fuzzy logic: Application for audit risk and uncertainty. *Managerial Auditing Journal* 14: 127–137.

Galbraith J. (1973) *Designing Complex Organization* Addison-Wesley.

Gao W-J, Xing B and Marwala T. (2013a) Computational intelligence in used products retrieval and reproduction. *International Journal of Swarm Intelligence Research* 4: 78–125.

Gao W-J, Xing B and Marwala T. (2013b) Teaching—learning-based optimization approach for enhancing remanufacturability pre-evaluation system's reliability. *IEEE Symposium Series on Computational Intelligence (IEEE SSCI), 15–19 April, Singapore,* pp. 235–239. IEEE.

Garnier R and Taylor J. (2002) *Discrete mathematics for new technology,* Dirac House, Temple Back, Bristol BS1 6BE, UK: Institute of Physics Publishing (IOP), ISBN 0 7503 0652 1.

Gensler HJ. (2010) *Introduction to logic,* 270 Madison Ave, New York, NY10016: Routledge, ISBN 978-0-415-99650-1.

Gozalczany MB. (1987) A method of inference in approximate reasoning based on interval-valued fuzzy sets. *Fuzzy Sets and Systems* 21: 1–17.

Grzegorzewski P. (2013) Fuzzy number approximation via shadowed sets. *Information Sciences* 225: 35–46.

Gustafsson B. (2011) *Fundamentals of scientific computing,* Berlin Heidelberg: Springer-Verlag, ISBN 978-3-642-19494-8.

Hageback N. (2017) *The virtual mind: Designing the logic to approximate human thinking,* 6000 Broken Sound Parkway NW, Suite 300, Boca Raton, FL 33487-2742: Taylor & Francis Group, LLC, ISBN 978-1-1380-5402-8.

Hager G and Wellein G. (2011) *Introduction to high performance computing for scientists and engineers,* 6000 Broken Sound Parkway NW, Suite 300, Boca Raton, FL 33487-2742: CRC Press, Taylor and Francis Group, LLC, ISBN 978-1-4398-1192-4.

Hamada Y. (2017) *Higgs potential and naturalness after the Higgs discovery,* 152 Beach Road, #22-06/08Gateway East, Singapore 189721, Singapore: Springer Nature Singapore Pte Ltd, ISBN 978-981-10-3417-6.

Hamill L and Gilbert N. (2016) *Agent-based modelling in economics,* The Atrium, Southern Gate, Chichester, West Sussex, PO19 8SQ, United Kingdom: John Wiley & Sons, Ltd, ISBN 978-1-118-45607-1.

Hansen E and Walster GW. (2004) *Global optimization using interval analysis,* 270 Madison Avenue, NewYork, NY 10016, U.S.A.: Marcel Dekker, Inc., ISBN 0-8247-4059-9.

Harris DM and Harris SL. (2013) *Digital design and computer architecture,* 225 Wyman Street, Waltham, MA 02451, USA: Elsevier, ISBN: 978-0-12-394424-5.

Hassan Y and Tazaki E. (2003) Induction of knowledge using evolutionary rought set theory. *Cybernetics and Systems* 34: 617–643.

Hatamlou A, Abdullah S and Nezamabadi-pour H. (2011) Application of gravitational search algorithm on data clustering. *Rough Sets and Knowledge Technology, LNCS 6954,* pp. 337–346. Berlin Heidelberg, Germany: Springer-Verlag.

Hatamlou A. (2013) Black hole: A new heuristic optimization approach for data clustering. *Inf. Sci.* 2322: 175–184.

Hau E. (2013) *Wind turbines: Fundamentals, technologies, application, economics,* Springer Heidelberg New York Dordrecht London: Springer-Verlag, ISBN 978-3-642-27150-2.

Hausdorff F. (1962) *Set theory,* New York: Chelsea Publishing Company.

Henry C. (2006) Reinforcement learning in biologically-inspired collective robotics: A rough set approach. *Department of Electrical and Computer Engineering.* Winnipeg, Manitoba, Canada: University of Manitoba.

Hinde CJ and Yang Y. (2009) A new extension of fuzzy sets using rough sets: R-fuzzy sets. *Information Sciences* 180: 354–365.

Holcombe J and Holcombe C. (2012) *Survey of operating systems,* 1221 Avenue of the Americas, New York, NY, 10020: The McGraw-Hill Companies, Inc., ISBN 978-0-07-351817-6.

Hurwitz J, Kaufman M and Bowles A. (2015) *Cognitive computing and big data analytics,* 10475 Crosspoint Boulevard, Indianapolis, IN 46256: John Wiley & Sons, Inc., ISBN 978-1-118-89662-4.

IBM. (2014) Understanding big data so you can act with confidence. Software Group, Route 100, Somers, NY 10589: IBM Corporation, 1–16.

Indiveri G. (2015) Neuromorphic engineering. In: Kacprzyk J and Pedrycz W (eds) *Springer Handbook of Computational Intelligence.* Dordrecht Heidelberg London New York: Springer-Verlag Berlin Heidelberg, ISBN 978-3-662-43504-5, Part D Neural Networks, Chapter 38, pp. 715–725.

Jeannot E and Žilinskas J. (2014) *High-performance computing on complex environments,* 111 River Street, Hoboken, NJ 07030,: John Wiley & Sons, Inc., ISBN 978-1-118-71205-4.

Jiang H, Kwong CK and Park W-Y. (2017) Probabilistic fuzzy regression approach for preference modeling. *Engineering Applications of Artificial Intelligence* 64: 286–294.

Jiménez-Losada A. (2017) *Models for cooperative games with fuzzy relations among the agents: Fuzzy communications, proximity relation and fuzzy permission,* Gewerbestrasse 11, 6330 Cham, Switzerland: Springer International Publishing AG, ISBN 978-3-319-56471-5.

Johnson RA. (2018) *Miller & Freund's probability and statistics for engineers,* Edinburgh Gate, Harlow, Essex CM20 2JE, England: Pearson Education Limited, ISBN 978-1-292-17601-7.

Jumarie G. (1992) From entropy of fuzzy sets to fuzzy set of entropies: A critical review and new results. *Kybernetes* 21: 33–51.

Kacprzyk J and Pedrycz W. (2015) *Springer handbook of computational intelligence,* Dordrecht Heidelberg London New York: Springer-Verlag Berlin Heidelberg, ISBN 978-3-662-43504-5.

Kentel E and Aral MM. (2004) Probabilistic-fuzzy health risk modeling. *Stochastic Environmental Research and Risk Assessment* 18: 324–338.

Khorshidi HA and Nikfalazar S. (2017) An improved similarity measure for generalized fuzzy numbers and its application to fuzzy risk analysis. *Applied Soft Computing* 52: 478–486.

Klir GJ. (1987) Where do we stand on measures of uncertainty, ambiguity, fuzziness and the like? *Fuzzy Sets and Systems* 24: 141–160.

Klir GJ and Yuan B. (1995) *Fuzzy sets and fuzzy logic: Theory and applications,* Upper Saddle River, NJ 07458: Prentice Hall PTR, ISBN 0-13-101171-5.

Kolokotsa D. (2007) Artificial intelligence in buildings: A review of the application of fuzzy logic. *Advances in Building Energy Research* 1: 29–54.

Kraft DH and Colvin E. (2017) *Fuzzy information retrieval,* www.morganclaypool.com: Morgan and Claypool, ISBN 978-1-62705-952-7.

Kreinovich V. (2008) Interval computation as an important part of granular computing: An introduction. In: Pedrycz W, Skowron A and Kreinovich V (eds) *Handbook of Granular*

Computing. The Atrium, Southern Gate, Chichester, West Sussex PO19 8SQ, England: John Wiley & Sons Ltd, ISBN 978-0-470-03554-2, Chapter 1, pp. 3–31.

Kubler S, Robert J, Derigent W, et al. (2016) A state-of the-art survey & testbed of fuzzy AHP (FAHP) applications. *Expert Systems with Applications* 65: 398–422.

Kumar M, Sharma A and Kumar R. (2012) Fuzzy entropy-based framework for multi-faceted test case classification and selection: An empirical study. *The Institution of Engineering and Technology (IET) Software* 8: 103–112.

Kuncheva LI. (1992) Fuzzy rough sets: Applications to feature selection. *Fuzzy Sets and Systems* 51: 147–153.

Kundu S. (1999) Gravitational clustering: A new approach based on the spatial distribution of the points. *Pattern Recognition* 32: 1149–1160.

Kuo RJ, Ho LM and Hu CM. (2002) Integration of self-organizing feature map and *K*-means algorithm for market segmentation. *Computers & Operations Research* 29: 1475–1493.

Kuo RJ and Li PS. (2016) Taiwanese export trade forecasting using firefly algorithm based K-means algorithm and SVR with wavelet transform. *Computers & Industrial Engineering* 99: 153–161.

LeBaron B. (2000) Agent-based computational finance: Suggested readings and early research. *Journal of Economic Dynamics and Control* 24: 679–702.

Li D, Zeng W and Li J. (2015) New distance and similarity measures on hesitant fuzzy sets and their applications in multiple criteria decision making. *Engineering Applications of Artificial Intelligence* 40: 11–16.

Li D and Du Y. (2017) *Artificial intelligence with uncertainty*, 6000 Broken Sound Parkway NW, Suite 300, Boca Raton, FL 33487-2742: CRC Press, Taylor & Francis Group, LLC, ISBN 978-1-4987-7626-4.

Li DF and Cheng CT. (2002) New similarity measures of intuitionistic fuzzy sets and application to pattern recognitions. *Pattern Recognition Letters* 23: 221–225.

Li D-F. (2014) *Decision and game theory in management with intuitionistic fuzzy sets*: Springer-Verlag Berlin Heidelberg, ISBN 978-3-642-40711-6.

Li JQ, Deng GN, Li HX, et al. (2012) The relationship between similarity measure and entropy of intuitionistic fuzzy sets. *Information Sciences* 205: 314–321.

Liang J. (2011) Uncertainty and feature selection in rough set theory. In: JingTaoYao, Ramanna S, GuoyinWang, et al. (eds) *Rough Sets and Knowledge Technology*. Springer-Verlag Berlin Heidelberg, ISBN 978-3-642-24424-7, Chapter 2, pp. 8–15.

Liang Z and Shi P. (2003) Similarity measures on intuitionistic fuzzy sets. *Pattern Recognition Letters* 24: 2687–2693.

Lin C-T and Chen C-T. (2004) A fuzzy-logic-based approach for new product go/nogo decision at the front end. *IEEE Transaction on Systems, Man, and Cybernetics—Part A: Systems and Humans* 34: 132–142.

Lind DA, Marchal WG and Wathen SA. (2012) *Statistical techniques in business & economics*, 1221Avenue of the Americas, New York, NY, 10020: McGraw-Hill/Irwin, ISBN: 978-0-07-340180-5.

Liu F, Janssens D, Wets G, et al. (2013) Annotating mobile phone location data with activity purposes using machine learning algorithms. *Expert Systems with Applications* 40: 3299–3311.

Liu X. (1992) Entropy, distance measure and similarity measure of fuzzy sets and their relations. *Fuzzy Sets and Systems* 52: 305–318.

Lu J, Li D-Y, Zhai Y-H, et al. (2016) A model for type-2 fuzzy rough sets. *Information Sciences* 328: 359–377.

Luukka P. (2011) Fuzzy similarity in multicriteria decision-making problem applied to supplier evaluation and selection in supply chain management. *Advances in Artificial Intelligence* 2011: 1–9.

Mardani A, Jusoh A and Zavadskas EK. (2015) Fuzzy multiple criteria decision-making techniques and applications—two decades review from 1994 to 2014. *Expert Systems with Applications* 42: 4126–4148.

Marinescu DC. (2013) *Cloud computing: Theory and practice,* 225 Wyman Street, Waltham, 02451, USA: Morgan Kaufmann, Elsevier Inc., ISBN 978-0-12404-627-6.

Martinez-Jaramillo S. (2007) Artificial financial markets: An agent based approach to reproduce stylized facts and to study the red queen effect. *Centre for Computational Finance and Economic Agents.* University of Essex.

Marwala T and Hunt HEM. (1999) Fault identification using finite element modles and neural networks. *Mechanical Systems and Signal Processing* 13: 475–490.

Marwala T. (2000a) Damage identification using a committee of neural networks. *Journal of Engineering Mechanics* 126: 43–50.

Marwala T. (2000b) Fault identification using neural networks and vibration data. *St. John's College.* University of Cambridge.

Marwala T. (2001) Probabilistic fault identification using a committee of neural networks and vibration data. *Journal of Aircraft* 38: 138–146.

Marwala T. (2003) Fault classification using pseudo modal energies and neural networks. *Aeronaut. Astronaut.* 41: 82–89.

Marwala T. (2005) Finite element model updating using particle swarm optimization. *International Journal of Engineering Simulation* 6: 25–30.

Marwala T. (2007a) Bayesian training of neural networks using genetic programming. *Pattern Recognition Letters* 28: 1452–1458.

Marwala T. (2007b) *Computational intelligence for modelling complex systems,* B-2/84, Ground Floor, Sec-16, Rohini, Delhi-110089, India: Research India Publications, ISBN 978-8-1904-3621-2.

Marwala T. (2009) *Computational intelligence for missing data imputation, estimation and management: Knowledge optimization techniques,* New York, USA: IGI Global, ISBN 978-1-60566-336-4.

Marwala T. (2010a) *Finite-element-model updating using computational intelligence techniques: Applications to structural dynamics,* London, UK: Springer-Verlag, ISBN 978-1-84996-322-0.

Marwala T. (2010b) Finite-element-model updating using genetic algorithm. In: Marwala T (ed) *Finite-element-model Updating Using computational Intelligence Techniques: Applications to Structural dynamics.* London, UK: Springer-Verlag, ISBN 978-1-84996-322-0, Chapter 3, pp. 49–66.

Marwala T. (2010c) Finite-element-model updating using particle-swarm optimization. In: Marwala T (ed) *Finite-element-model Updating Using computational Intelligence Techniques: Applications to Structural dynamics.* London, UK: Springer-Verlag, ISBN 978-1-84996-322-0, Chapter 4, pp. 67–84.

Marwala T. (2010d) Finite-element-model updating using simulated annealing. In: Marwala T (ed) *Finite-element-model Updating Using computational Intelligence Techniques: Applications to Structural dynamics.* London, UK: Springer-Verlag, ISBN 978-1-84996-322-0, Chapter 5, pp. 85–102.

Marwala T and Lagazio M. (2011a) *Militarized conflict modeling using computational intelligence,* London, UK: Springer-Verlag, ISBN 978-0-85729-789-1.

Marwala T and Lagazio M. (2011b) Particle swarm optimization and hill-climbing optimized rough sets for modeling interstate conflict. In: Marwala T and Lagazio M (eds) *Militarized Conflict Modeling Using Computational Intelligence.* London, UK: Springer-Verlag, ISBN 978-0-85729-789-1, Chapter 8, pp. 147–164.

Marwala T. (2012a) *Condition monitoring using computational intelligence methods: Applications in mechanical and electrical systems,* London: Springer-Verlag, ISBN 978-1-4471-2379-8.

Marwala T. (2012b) Data processing techniques for condition monitoring. In: Marwala T (ed) *Condition Monitoring Using Computational Intelligence Methods: Applications in Mechanical and Electrical Systems.* London: Springer-Verlag, ISBN 978-1-4471-2379-8, Chapter 2, pp. 27–51.

Marwala T. (2012c) On-line condition monitoring using ensemble learning. In: Marwala T (ed) *Condition Monitoring Using Computational Intelligence Methods: Applications in*

Mechanical and Electrical Systems. London: Springer-Verlag, ISBN 978-1-4471-2379-8, Chapter 11, pp. 211–226.

Marwala T. (2013a) *Economic modeling using artificial intelligence methods,* Springer London Heidelberg New York Dordrecht: Springer-Verlag London, ISBN 978-1-4471-5009-1.

Marwala T. (2013b) Missing data approaches to economic modeling: Optimization approach. In: Marwala T (ed) *Economic Modeling Using Artificial Intelligence Methods.* Springer London Heidelberg New York Dordrecht: Springer-Verlag London, ISBN 978-1-4471-5009-1, Chapter 7, pp. 119–136.

Marwala T. (2013c) Multi-agent approaches to economic modeling: Game theory, ensembles, evolution and the stock market. In: Marwala T (ed) *Economic Modeling Using Artificial Intelligence Methods.* Springer London Heidelberg New York Dordrecht: Springer-Verlag London, ISBN 978-1-4471-5009-1, Chapter 11, pp. 195–213.

Marwala T. (2013d) Real-time approaches to computational economics: Self adaptive economic systems. In: Marwala T (ed) *Economic Modeling Using Artificial Intelligence Methods.* Springer London Heidelberg New York Dordrecht: Springer-Verlag London, ISBN 978-1-4471-5009-1, Chapter 10, pp. 173–193.

Marwala T. (2014a) *Artificial intelligence techniques for rational decision making,* Springer Cham Heidelberg New York Dordrecht London: Springer International Publishing Switzerland, ISBN 978-3-319-11423-1.

Marwala T. (2014b) Causal function for rational decision making: Application to militarized interstate dispute. In: Marwala T (ed) *Artificial Intelligence Techniques for Rational Decision Making.* Springer Cham Heidelberg New York Dordrecht London: Springer International Publishing Switzerland, ISBN 978-3-319-11423-1, Chapter 2, pp. 19–37.

Marwala T. (2014c) Missing data approaches for rational decision making: Application to antenatal data. In: Marwala T (ed) *Artificial Intelligence Techniques for Rational Decision Making.* Springer Cham Heidelberg New York Dordrecht London: Springer International Publishing Switzerland, ISBN 978-3-319-11423-1, Chapter 4, pp. 55–71.

Marwala T. (2015) *Causality, correlation and artificial intelligence for rational decision making,* 5 Toh Tuck Link, Singapore 596224: World Scientific Publishing Co. Pte. Ltd, ISBN 978-9-81463-086-3.

Marwala T, Boulkaibet I and Adhikari S. (2017) *Probabilistic finite element model updating using Bayesian statistics: Applications to aeronautical and mechanical engineering,* The Atrium, Southern Gate, Chichester, West Sussex, PO19 8SQ, United Kingdom: John Wiley & Sons, Ltd, ISBN 978-1-1191-5301-6.

Marwala T and Hurwitz E. (2017) Introduction to man and machines. In: Marwala T and Hurwitz E (eds) *Artificial Intelligence and Economic Theory: Skynet in the Market.* Gewerbestrasse 11, 6330 Cham, Switzerland: Springer International Publishing AG, ISBN 978-3-319-66103-2, Chapter 1, pp. 1–14.

Massad E, Ortega NRS, Barros LCd, et al. (2008) *Fuzzy logic in action: Applications in epidemiology and beyond*: Springer-Verlag Berlin Heidelberg, ISBN 978-3-540-69092-4.

McGeoch CC. (2014) *Adiabatic quantum computation and quantum annealing: Theory and practice*: Morgan & Claypool, ISBN 9781627053358.

Mendel JM. (2017) *Uncertain rule-based fuzzy systems: Introduction and new directions,* Gewerbestrasse 11, 6330 Cham, Switzerland: Springer International Publishing AG, ISBN 978-3-319-51369-0.

Michalowicz JV, Nichols JM and Bucholtz F. (2014) *Handbook of differential entropy,* 6000 Broken Sound Parkway NW, Suite 300, Boca Raton, FL 33487-2742: CRC Press, Taylor & Francis Group, LLC, ISBN 978-1-4665-8317-7.

Mitra S and Hayashi Y. (2000) Neuro-fuzzy rule generation: Survey in soft computing framework. *IEEE Transactions on Neural Networks* 11: 748–768.

Mitzenmacher M and Upfal E. (2017) *Probability and computing: Randomization and probabilistic techniques in algorithms and data analysis,* University Printing House, Cambridge CB2 8BS, United Kingdom: Cambridge University Press, ISBN 978-1-107-15488-9.

Moller F and Struth G. (2013) *Modelling computing systems: Mathematics for computer science*: Springer-Verlag London, ISBN 978-1-84800-321-7.

Moore RE, Kearfott RB and Cloud MJ. (2009) *Introduction to interval analysis*, 3600 Market Street, 6th Floor, Philadelphia, PA,19104-2688 USA: Society for Industrial and Applied Mathematics, ISBN 978-0-898716-69-6.

Morin D. (2016) *Probability: For the enthusiastic beginner*: CreateSpace, ISBN-13 978-1523318674.

Morsi NN and Yakout MM. (1998) Axiomatics for fuzzy rough sets. *Fuzzy Sets and Systems* 100: 327–342.

Msiza IS, Nelwamondo FV and Marwala T. (2008) Water demand prediction using artificial neural networks and support vector regression. *Journal of Computers* 3: 1–8.

Myatt GJ and Johnson WP. (2014) *Making sense of data I: A practical guide to exploratory data analysis and data mining*, 111 River Street, Hoboken, New Jersey: John Wiley & Sons, Inc., ISBN 9781118407417.

Naidu MSR, Kumar PR and Chiranjeevi K. (in press) Shannon and Fuzzy entropy based evolutionary image thresholding for image segmentation. *Alexandria Engineering Journal* http://dx.doi.org/10.1016/j.aej.2017.05.024.

Nanda S and Majumdar S. (1992) Fuzzy rough sets. *Fuzzy Sets and Systems* 45: 157–160.

Nanda SJ and Panda G. (2014) A survey on nature inspired metaheuristic algorithms for partitional clustering. *Swarm and Evolutionary Computation* 16: 1–18.

Nelwamondo FV and Marwala T. (2007) Rough set theory for the treatment of incomplete data. *Proceedings of the IEEE Conference on Fuzzy Systems, London UK*, pp. 338–343.

Nelwamondo FV, Mohamed S and Marwala T. (2007) Missing data: A comparison of neural network and expectation maximization techniques. *Current Science* 93: 1517–1521.

Nelwamondo FV and Marwala T. (2008) Techniques for handling missing data: Applications to online condition monitoring. *International Journal of Innovative Computing Information and Control* 4: 1507–1526.

Nelwamondo FV, Golding D and Marwala T. (2009) A dynamic programming approach to missing data estimation using neural networks. *Information Sciences* doi:10.1016/j.ins.2009.10.008.

Nguyen H. (2016) A novel similarity/dissimilarity measure for intuitionistic fuzzy sets and its application in pattern recognition. *Expert Systems with Applications* 45: 97–107.

Nguyen HT and Walker EA. (2006) *A first course in fuzzy logic*, 6000 Broken Sound Parkway NW, Suite 300, Boca Raton, FL 33487-2742: CRC Press, Taylor & Francis Group, LLC, ISBN 978-1-4200-5710-2.

Nguyen HT, Kreinovich V, Wu B, et al. (2012) *Computing statistics under interval and fuzzy uncertainty: Applications to computer science and engineering*, Berlin Heidelberg: Springer-Verlag, ISBN 978-3-642-24904-4.

Nguyen XT, Nguyen VD and Nguyen DD. (2014) Rough fuzzy relation on two universal sets. *Internatinoal Journal of Intelligent Systems and Applications* 4: 49–55.

Nilavu D and Sivakumar R. (2015) Knowledge representation using Type-2 fuzzy rough ontologies in ontology web language. *Fuzzy Information and Engineering* 7: 73–99.

Novák V, Perfilieva I and Dvořák A. (2016) *Insight into fuzzy modeling*, John Wiley & Sons, Inc., 111 River Street, Hoboken, NJ 07030: John Wiley & Sons, Inc., ISBN 978-1-119-19318-0.

O'Regan G. (2013) *Mathematics in computing: An accessible guide to historical, foundational and application contexts*, London: Springer-Verlag, ISBN 978-1-4471-4533-2.

Ott RL and Longnecker M. (2016) *An introduction to statistical methods and data analysis*, 20 Channel Center Street, Boston, MA 02210, USA: Cengage Learning, ISBN 978-1-305-26947-7.

Pal SK, Meher SK and Dutta S. (2012) Class-dependent rough-fuzzy granular space, dispersion index and classification. *Pattern Recognition* 45: 2690–2707.

Papacostas GA, Hatzimichaillidis AG and Kaburlasos VG. (2013) Distance and similarity measures between intuitionistic fuzzy sets: A comparative analysis from a pattern recognition point view. *Pattern Recognition Letters* 34: 1609–1622.

Pappis CP and Karacapilidis NI. (1993) A comparative assessment of measures of similarity of fuzzy values. *Fuzzy Sets and Systems* 56: 171–174.

Patel PB and Marwala T. (2006) Neural networks, fuzzy inference systems and adaptive-neuro fuzzy inference systems for financial decision making. *Lecture Notes in Computer Science, Vol. 4234*. Berlin Heidelberg: Springer-Verlag, pp. 430–439.

Patel PB and Marwala T. (2009) Genetic algorithms, neural networks, fuzzy inference system, support vector machines for call performance classification. Proceedings of the IEEE international conference on machine learning and applications, Miami, Florida, pp. 415–420.

Paule P. (2013) *Mathematics, computer science and logic—a never ending story*: Springer International Publishing Switzerland, ISBN 978-3-319-00965-0.

Pawlak Z. (1982) Rough sets. *International Journal of Information and Computer Sciences* 11: 341–356.

Pawlak Z. (1991) *Rough sets: Theoretical aspects of reasoning about data*: Springer Science+Business Media Dordrecht, ISBN 978-94-010-5564-2.

Pawlak Z. (2002) Rough set theory and its applications. *Journal of Telecommunications and Information Technology* 3: 7–10.

Pawlak Z and Skowron A. (2007a) Rough sets: Some extensions. *Information Sciences* 177: 28–40.

Pawlak Z and Skowron A. (2007b) Rudiments of rough sets. *Information Sciences* 177: 3–27.

Peck R and Devore JL. (2012) *Statistics: The exploration & analysis of data,* 20 Channel Center Street, Boston, MA 02210, USA: Brooks/Cole, Cengage Learning, ISBN 978-0-8400-5801-0.

Pedrycz W. (1998) Shadowed sets: Representing and processing fuzzy sets. *IEEE Transactions on Systems, Man, and Cybernetics—Part B* 28: 103–109.

Pedrycz W. (2005) Interpretation of clusters in the framework of shadowed sets. *Pattern Recognition Letters* 26: 2439–2449.

Pedrycz W, Skowron A and Kreinovich V. (2008) *Handbook of granular computing,* The Atrium, Southern Gate, Chichester, West Sussex PO19 8SQ, England: John Wiley & Sons Ltd, ISBN 978-0-470-03554-2.

Pedrycz W. (2009) From fuzzy sets to shadowed sets: Interpretation and computing. *International Journal of Intelligent Systems* 24: 48–61.

Pedrycz W. (2013) *Granular computing: Analysis and design of intelligent systems,* 6000 Broken Sound Parkway NW, Suite 300, Boca Raton, FL 33487-2742: CRC Press, Taylor & Francis Group, LLC, ISBN 978-1-4398-8687-8.

Peters JF and Skowron A. (2014) *Transactions on rough sets XVII,* Berlin Heidelberg: Springer-Verlag, ISBN 978-3-642-54755-3.

Peters JF, Skowron A, Li T, et al. (2014) *Transactions on rough sets XVIII,* Berlin Heidelberg: Springer-Verlag, ISBN 978-3-662-44679-9.

Peters JF and Skowron A. (2016) *Transactions on rough sets XX,* Heidelberger Platz 3, 14197 Berlin, Germany: Springer-Verlag GmbH Germany, ISBN 978-3-662-53610-0.

Petkovic D. (2014) Adaptive neuro-fuzzy fusion of sensor data. *Infrared Physics & Technology* 67: 222–228.

Petry FE and Zhao L. (2009) Data mining by attribute generalization with fuzzy hierarchies in fuzzy databases. *Fuzzy Sets and Systems* 160: 2206–2223.

Plotnitsky A. (2016) *The principles of quantum theory, from Planck's quanta to the Higgs boson: The nature of quantum reality and the spirit of copenhagen,* Gewerbestrasse 11, 6330 Cham, Switzerland: Springer International Publishing Switzerland, ISBN 978-3-319-32066-3.

Polkowski L. (2002) *Rough sets: Mathematical foundations,* Berlin Heidelberg: Springer-Verlag, ISBN 978-3-7908-1510-8.

Polkowski L and Artiemjew P. (2015) *Granular computing in decision approximation: An application of rough mereology*: Springer International Publishing Switzerland, ISBN 978-3-319-12879-5.

Pykacz J. (2015) *Quantum physics, fuzzy sets and logic: Steps towards a many-valued interpretation of quantum mechanics,* Cham Heidelberg New York Dordrecht London: Springer International Publishing AG Switzerland, ISBN 978-3-319-19383-0.

Raber R. (1994) Fuzzy in control. *Sensor Review* 14: 26–28.

Radzikowska AM and Kerre EE. (2002) A comparative study of fuzzy rough sets. *Fuzzy Sets and Systems* 126: 137–156.

Ratner B. (2011) *Statistical and machine-learning data mining: Techniques for better predictive modeling and analysis of big data,* 6000 Broken Sound Parkway NW, Suite 300, Boca Raton, FL 33487-2742: CRC Press, Taylor & Francis Group, LLC, ISBN 978-1-4398-6092-2.

Raveendranathan KC. (2014) *Neuro-fuzzy equalizers for mobile cellular channels,* Taylor & Francis Group, 6000 Broken Sound Parkway NW, Suite 300, Boca Raton, FL 33487-2742: Taylor & Francis Group, LLC, ISBN-13 978-1-4665-8155-5.

Ray KS. (2012) *Soft computing approach to pattern classification and object recognitio: A unified concepts,* New York Heidelberg Dordrecht London: Springer Science+Business Media New York, ISBN 978-1-4614-5347-5.

Reed DA and Dongarra J. (2015) Exascale computing and big data. *Communications of the ACM* 58: 56–68.

Roman S. (2017) *An introduction to the language of category theory,* Gewerbestrasse 11, 6330 Cham, Switzerland: Birkhäuser, Springer International Publishing AG, CH, ISBN 978-3-319-41916-9.

Ross SM. (2010) *Introductory statistics,* Elsevier Inc., ISBN 978-0-12-374388-6.

Ross SM. (2014) *Introduction to probability models,* The Boulevard, Langford Lane, Kidlington, Oxford OX5 1GB, UK: Elsevier Inc., ISBN 978-0-12-407948-9.

Rumsey D. (2010) *Statistics essentials for dummies,* 111 River Street, Hoboken, NJ 07030: Wiley Publishing, Inc., ISBN 978-0-470-61839-4.

Salles DCd, Neto ACG and Marujo LG. (2016) Using fuzzy logic to implement decision policies in system dynamics models. *Expert Systems with Applications* 55: 172–183.

Satapathy SC, Bhateja V and Das S. (2018) *Smart computing and informatics,* 152Beach Road, #21-01/04 Gateway East, Singapore 189721, Singapore: Springer Nature Singapore Pte Ltd, ISBN 978-981-10-5543-0.

Schinazi RB. (2012) *Probability with statistical applications,* Springer New York Dordrecht Heidelberg London: Springer Science+Business Media, LLC, ISBN 978-0-8176-8249-1.

Seising R. (2008) On the absence of strict boundaries—vagueness, haziness, and fuzziness in philosophy, science, and medicine. *Applied Soft Computing* 8: 1232–1242.

Sen D and Pal SK. (2009) Generalized rough sets, entropy, and image ambiguity measures. *IEEE Transaction on Systems, Man, and Cybernetics—Part B: Cybernetics* 39: 117–128.

Shafia MA, Moghaddam MR and Tavakolian R. (2011) A hybrid algorithm for data clustering using honey bee algorithm, genetic algorithm and *k*-means method. *Journal of Advanced Computer Science and Technology Research* 1: 110–125.

Shi Y and Yuan X. (2015) Interval entropy of fuzzy sets and the application to fuzzy multiple attribute decision making. *Mathematical Problems in Engineering* http://dx.doi.org/10.1155/2015/451987: 1–21.

Shiflet AB and Shiflet GW. (2014) *Introduction to computational science: Modeling and simulation for the sciences,* 41 William Street, Princeton, New Jersey 08540: Princeton University Press, ISBN 978-0-691-16071-9.

Simovici DA and Djeraba C. (2014) *Mathematical tools for data mining: Set theory, partial orders, combinatorics,* London, UK: Springer-Verlag, ISBN 978-1-4471-6406-7.

Sivarajah U, Kamal MM, Irani Z, et al. (2017) Critical analysis of big data challenges and analytical methods. *Journal of Business Research* 70: 263–286.

Song Y, Wang X, Lei L, et al. (2015) A novel similarity measure on intuitionistic fuzzy sets with its applications. *Applied Intelligence* 42: 252–261.

Srinivasan P, Ruiz ME, Kraft DH, et al. (2001) Vocabulary mining for information retrieval: Rough sets and fuzzy sets. *Information Processing and Management* 37: 15–38.

Ssali G and Marwala T. (2008) Computational intelligence and decision trees for missing data estimation. *Proceedings of the IEEE International Joint Conference on Neural Networks (IJCNN'08), 1–8 June, Hong Kong, China,* pp. 201–207.

Stoean C and Stoean R. (2014) *Support vector machines and evolutionary algorithms for classification: Single or together?*, Springer Cham Heidelberg New York Dordrecht London: Springer International Publishing Switzerland, ISBN 978-3-319-06940-1.

Strutz T. (2011) *Data fitting and uncertainty: A practical introduction to weighted least squares and beyond*, www.viewegteubner.de: Vieweg+Teubner Verlag, ISBN 978-3-8348-1022-9.

Sun B and Ma W. (2014) Soft fuzzy rough sets and its application in decision making. *Artificial Intelligence Review* 41: 67–80.

Sung W-K. (2017) *Algorithms for next-generation sequencing*, 6000 Broken Sound Parkway NW, Suite 300, Boca Raton, FL 33487-2742: CRC Press, Taylor & Francis Group, LLC, ISBN 978-1-4665-6550-0.

Suri M. (2017) *Advances in neuromorphic hardware exploiting emerging nanoscale devices*, 7th Floor, VijayaBuilding, 17 BarakhambaRoad,NewDelhi 110 001, India: Springer (India) Pvt. Ltd., ISBN 978-81-322-3701-3.

Tesfatsion L. (2003) Agent-based computational economics: Modeling economies as complex adaptive systems. *Information Sciences* 149: 263–269.

Theodoridis S and Koutroumbas K. (2009) *Pattern Recognition*: Academic Press (Elsevier), ISBN 978-1-59749-272-0.

Trillas E and Eciolaza L. (2015) *Fuzzy logic: An introductory course for engineering students*: Springer International Publishing Switzerland, ISBN 978-3-319-14202-9.

Turksen IB. (1996) Interval-valued strict preference with Zadeh triples. *Fuzzy Sets and Systems* 78: 183–195.

van Eck NJ and Waltman L. (2009) How to normalize cooccurrence data? An analysis of some well-known similarity measures. *Journal of the American Society for Information Science and Technology* 60: 1635–1651.

Veazie PJ. (2017) *What makes variables random: Probability for the applied researcher*, 6000 Broken Sound Parkway NW, Suite 300, Boca Raton, FL 33487-2742: Taylor & Francis Group, LLC, ISBN-13 978-1-4987-8108-4.

Velten K. (2009) *Mathematical modeling and simulation: Introduction for scientists and engineers*, Weinheim: WILEY-VCH Verlag GmbH & Co. KGaA, ISBN 978-3-527-40758-8.

Verbeke W, Baesens B and Bravo C. (2018) *Profit-driven business analytics: A practitioner's guide to transforming big data into added value*, Hoboken, New Jersey: John Wiley & Sons, Inc., ISBN 9781119286554.

Wang G and Li X. (1998) The applications of interval-valued fuzzy numbers and interval-distribution numbers. *Fuzzy Sets and Systems* 98: 331–335.

Wang WJ. (1997) New similarity measures on fuzzy sets and on elements. *Fuzzy Sets and Systems* 85: 305–309.

Wang X and Zhai J. (2017) *Learning with uncertainty*, 6000 Broken Sound Parkway NW, Suite 300, Boca Raton, FL 33487-2742: CRC Press, Taylor & Francis Group, LLC, ISBN 978-1-4987-2412-8.

Wei CP, Wang P and Zhang YZ. (2011) Entropy, similarity measure of interval valued intuitionistic sets and their applications. *Information Sciences* 181: 4273–4286.

Weiss NA. (2017) *Introductory statistics*, Edinburgh Gate, Harlow, Essex CM20 2JE, England: Pearson Education Limited, ISBN 978-1-29209-972-9.

Wessler M. (2013) *Big data analytics for dummies*, 111 River Street, Hoboken, NJ: John Wiley & Sons, Inc., ISBN 978-1-118-60704-6.

Wierman MJ. (1999) Measuring uncertainty in rough set theory. *International Journal of General Systems* 28: 283–297.

Witten IH, Frank E, Hall MA, et al. (2017) *Data mining: Practical machine learning tools and techniques*, 30 Corporate Drive, Suite 400, Burlington, MA 01803, USA: Morgan Kaufmann, Elsevier Inc., ISBN 978-0-12-804291-5.

Woyke E. (2014) *The smartphone: Anatomy of an industry*, 120 Wall street, 31st floor, new york, ny 10005: the new Press, ISBN 978-1-59558-963-7.

Xing B, Bright G, Tlale NS, et al. (2006a) Reconfigurable manufacturing systems for agile mass customization manufacturing. *Proceedings of the 22nd ISPE International Conference on CAD/CAM, Robotics and Factories of the Future (CARs&FOF 2006), Vellore, India, July 2006.* pp. 473–482.

Xing B, Eganza J, Bright G, et al. (2006b) Reconfigurable manufacturing system for agile manufacturing. *Proceedings of the 12th IFAC Symposium on Information Control Problems in Manufacturing, May 2006, Saint-Etienne, France, pp. on CD.*

Xing B, Nelwamondo FV, Battle K, et al. (2009) Application of artificial intelligence (AI) methods for designing and analysis of reconfigurable cellular manufacturing system (RCMS). *Proceedings of the 2nd International Conference on Adaptive Science & Technology (ICAST), 14–16 December, Accra, Ghana,* pp. 402–409. IEEE.

Xing B, Gao W-J, Nelwamondo FV, et al. (2010a) Ant colony optimization for automated storage and retrieval system. *Proceedings of The Annual IEEE Congress on Evolutionary Computation (IEEE CEC), 18–23 July, CCIB, Barcelona, Spain,* pp. 1133–1139. IEEE.

Xing B, Gao W-J, Nelwamondo FV, et al. (2010b) Artificial intelligence in reverse supply chain management: The state of the art. *Proceedings of the Twenty-First Annual Symposium of the Pattern Recognition Association of South Africa (PRASA), 22–23 November, Stellenbosch, South Africa,* pp. 305–310.

Xing B, Gao W-J, Nelwamondo FV, et al. (2010c) Cellular manufacturing system scheduling under fuzzy constraints: A group technology perspective. *Annual IEEE International Conference on Fuzzy Systems (FUZZ-IEEE), 18–23 July, CCIB, Barcelona, Spain,* pp. 887–894. IEEE.

Xing B, Gao W-J, Nelwamondo FV, et al. (2010d) Part-machine clustering: The comparison between adaptive resonance theory neural network and ant colony system. In: Zeng Z and Wang J (eds) *Advances in Neural Network Research & Applications, LNEE 67,* pp. 747–755. Berlin Heidelberg: Springer-Verlag.

Xing B, Gao W-J, Nelwamondo FV, et al. (2010e) Two-stage inter-cell layout design for cellular manufacturing by using ant colony optimization algorithms. In: Tan Y, Shi Y and Tan KC (eds) *Advances in Swarm Intelligence, Part I, LNCS 6145,* pp. 281–289. Berlin Heidelberg: Springer-Verlag.

Xing B, Gao W-J, Nelwamondo FV, et al. (2011) e-Reverse logistics for remanufacture-to-order: An online auction-based and multi-agent system supported solution. In: Omatu S and Fabri SG (eds) *Fifth International Conference on Advanced Engineering Computing and Applications in Sciences (ADVCOMP), 20–25 November, Lisbon, Portugal,* pp. 78–83. IARIA

Xing B, Gao W-J, Nelwamondo FV, et al. (2012a) The effects of customer perceived disposal hardship on post-consumer product remanufacturing: A multi-agent perspective. In: Tan Y, Shi Y and Ji Z (eds) *ICSI 2012, Part I, LNCS 7332,* pp. 209–216. Berlin Heidelberg: Springer-Verlag.

Xing B, Gao W-J, Nelwamondo FV, et al. (2012b) TAC-RMTO: Trading agent competition in remanufacture-to-order. In: Tan Y, Shi Y and Ji Z (eds) *ICSI 2012, Part I, LNCS 7332,* pp. 519–526. Berlin Heidelberg: Springer-Verlag.

Xing B. (2013) The building and testing of biologically inspired miniature robot for machine condition inspection. Pretoria, South Africa: Department of Mechanical and Aeronautical Engineering, Faculty of Engineering, Built Environment, and Information Technology, University of Pretoria.

Xing B, Gao W-J and Marwala T. (2013a) Intelligent data processing using emerging computational intelligence techniques. *Research Notes in Information Sciences* 12: 10–15.

Xing B, Gao W-J and Marwala T. (2013b) An overview of cuckoo-inspired intelligent algorithms and their applications. *IEEE Symposium Series on Computational Intelligence (IEEE SSCI), 15–19 April, Singapore,* pp. 85–89. IEEE.

Xing B. (2014) Computational intelligence in cross docking. *International Journal of Software Innovation* 4: 78–124.

Xing B, Gao W-J and Marwala T. (2014) Multi-agent framework for distributed leasing based injection mould remanufacturing. In: Memon QA (ed) *Distributed Network Intelligence, Security and Applications.* 6000 Broken Sound Parkway NW, Suite 300, Boca Raton, FL 33487-2742: CRC Press, Taylor & Francis Group, LLC, ISBN 978-1-4665-5958-5, Chapter 11, pp. 267–289.

Xing B and Gao W-J. (2014a) *Computational intelligence in remanufacturing,* 701 E. Chocolate Avenue, Suite 200, Hershey PA 17033: IGI Global, ISBN 978-1-4666-4908-8.

Xing B and Gao W-J. (2014aa) Chemical-reaction optimization algorithm. In: Xing B and Gao W-J (eds) *Innovative Computational Intelligence: A Rough Guide to 134 Clever Algorithms.* Cham Heidelberg New York Dordrecht London: Springer International Publishing Switzerland, ISBN: 978-3-319-03403-4, Chapter 25, pp. 417–428.

Xing B and Gao W-J. (2014b) *Innovative computational intelligence: A rough guide to 134 clever algorithms,* Cham Heidelberg New York Dordrecht London: Springer International Publishing Switzerland, ISBN: 978-3-319-03403-4.

Xing B and Gao W-J. (2014bb) Emerging chemistry-based CI algorithms. In: Xing B and Gao W-J (eds) *Innovative Computational Intelligence: A Rough Guide to 134 Clever Algorithms.* Cham Heidelberg New York Dordrecht London: Springer International Publishing Switzerland, ISBN: 978-3-319-03403-4, Chapter 26, pp. 429–437.

Xing B and Gao W-J. (2014c) Introduction to computational intelligence. In: Xing B and Gao W-J (eds) *Innovative Computational Intelligence: A Rough Guide to 134 Clever Algorithms.* Cham Heidelberg New York Dordrecht London: Springer International Publishing Switzerland, ISBN: 978-3-319-03403-4, Chapter 1, pp. 3–17.

Xing B and Gao W-J. (2014cc) Base optimization algorithm. In: Xing B and Gao W-J (eds) *Innovative Computational Intelligence: A Rough Guide to 134 Clever Algorithms.* Cham Heidelberg New York Dordrecht London: Springer International Publishing Switzerland, ISBN: 978-3-319-03403-4, Chapter 27, pp. 441–444.

Xing B and Gao W-J. (2014d) Bacteria inspired algorithms. In: Xing B and Gao W-J (eds) *Innovative Computational Intelligence: A Rough Guide to 134 Clever Algorithms.* Cham Heidelberg New York Dordrecht London: Springer International Publishing Switzerland, ISBN: 978-3-319-03403-4, Chapter 2, pp. 21–38.

Xing B and Gao W-J. (2014dd) Emerging mathematics-based CI algorithms. In: Xing B and Gao W-J (eds) *Innovative Computational Intelligence: A Rough Guide to 134 Clever Algorithms.* Cham Heidelberg New York Dordrecht London: Springer International Publishing Switzerland, ISBN: 978-3-319-03403-4, Chapter 28, pp. 445–448.

Xing B and Gao W-J. (2014e) Bee inspired algorithms. In: Xing B and Gao W-J (eds) *Innovative Computational Intelligence: A Rough Guide to 134 Clever Algorithms.* Cham Heidelberg New York Dordrecht London: Springer International Publishing Switzerland, ISBN: 978-3-319-03403-4, Chapter 4, pp. 45–80.

Xing B and Gao W-J. (2014f) Bat inspired algorithms. In: Xing B and Gao W-J (eds) *Innovative Computational Intelligence: A Rough Guide to 134 Clever Algorithms.* Cham Heidelberg New York Dordrecht London: Springer International Publishing Switzerland, ISBN: 978-3-319-03403-4, Chapter 3, pp. 39–44.

Xing B and Gao W-J. (2014ff) Overview of computational intelligence. In: Xing B and Gao W-J (eds) *Computational Intelligence in Remanufacturing.* 701 E. Chocolate Avenue, Suite 200, Hershey PA 17033: IGI Global, ISBN 978-1-4666-4908-8, Chapter 2, pp. 18–36.

Xing B and Gao W-J. (2014g) Biogeography-based optimization algorithm. In: Xing B and Gao W-J (eds) *Innovative Computational Intelligence: A Rough Guide to 134 Clever Algorithms.* Cham Heidelberg New York Dordrecht London: Springer International Publishing Switzerland, ISBN: 978-3-319-03403-4, Chapter 5, pp. 81–91.

Xing B and Gao W-J. (2014gg) Used products return pattern analysis using agent-based modelling and simulation. In: Xing B and Gao W-J (eds) *Computational Intelligence in Remanufacturing.* 701 E. Chocolate Avenue, Suite 200, Hershey PA 17033: IGI Global, ISBN 978-1-4666-4908-8, Chapter 3, pp. 38–58.

Xing B and Gao W-J. (2014h) Cat swarm optimization algorithm. In: Xing B and Gao W-J (eds) *Innovative Computational Intelligence: A Rough Guide to 134 Clever Algorithms.* Cham Heidelberg New York Dordrecht London: Springer International Publishing Switzerland, ISBN: 978-3-319-03403-4, Chapter 6, pp. 93–104.

Xing B and Gao W-J. (2014i) Cuckoo inspired algorithms. In: Xing B and Gao W-J (eds) *Innovative Computational Intelligence: A Rough Guide to 134 Clever Algorithms.* Cham

Heidelberg New York Dordrecht London: Springer International Publishing Switzerland, ISBN: 978-3-319-03403-4, Chapter 7, pp. 105–121.

Xing B and Gao W-J. (2014ii) Used product remanufacturability evaluation using fuzzy logic. In: Xing B and Gao W-J (eds) *Computational Intelligence in Remanufacturing*. 701 E. Chocolate Avenue, Suite 200, Hershey PA 17033: IGI Global, ISBN 978-1-4666-4908-8, Chapter 5, pp. 75–94.

Xing B and Gao W-J. (2014j) Luminous insect inspired algorithms. In: Xing B and Gao W-J (eds) *Innovative Computational Intelligence: A Rough Guide to 134 Clever Algorithms*. Cham Heidelberg New York Dordrecht London: Springer International Publishing Switzerland, ISBN: 978-3-319-03403-4, Chapter 8, pp. 123–137.

Xing B and Gao W-J. (2014k) Fish inspired algorithms. In: Xing B and Gao W-J (eds) *Innovative Computational Intelligence: A Rough Guide to 134 Clever Algorithms*. Cham Heidelberg New York Dordrecht London: Springer International Publishing Switzerland, ISBN: 978-3-319-03403-4, Chapter 9, pp. 139–155.

Xing B and Gao W-J. (2014kk) Used product delivery optimization using agent-based modelling and simulation. In: Xing B and Gao W-J (eds) *Computational Intelligence in Remanufacturing*. 701 E. Chocolate Avenue, Suite 200, Hershey PA 17033: IGI Global, ISBN 978-1-4666-4908-8, Chapter 7, pp. 113–133.

Xing B and Gao W-J. (2014l) Frog inspired algorithms. In: Xing B and Gao W-J (eds) *Innovative Computational Intelligence: A Rough Guide to 134 Clever Algorithms*. Cham Heidelberg New York Dordrecht London: Springer International Publishing Switzerland, ISBN: 978-3-319-03403-4, Chapter 10, pp. 157–165.

Xing B and Gao W-J. (2014ll) Post–disassembly part–machine clustering using artificial neural networks and ant colony systems. In: Xing B and Gao W-J (eds) *Computational Intelligence in Remanufacturing*. 701 E. Chocolate Avenue, Suite 200, Hershey PA 17033: IGI Global, ISBN 978-1-4666-4908-8, Chapter 8, pp. 135–150.

Xing B and Gao W-J. (2014m) Fruit fly optimization algorithm. In: Xing B and Gao W-J (eds) *Innovative Computational Intelligence: A Rough Guide to 134 Clever Algorithms*. Cham Heidelberg New York Dordrecht London: Springer International Publishing Switzerland, ISBN: 978-3-319-03403-4, Chapter 11, pp. 167–170.

Xing B and Gao W-J. (2014mm) Reprocessing operations scheduling using fuzzy logic and fuzzy MAX–MIN ant systems. In: Xing B and Gao W-J (eds) *Computational Intelligence in Remanufacturing*. 701 E. Chocolate Avenue, Suite 200, Hershey PA 17033: IGI Global, ISBN 978-1-4666-4908-8, Chapter 9, pp. 151–170.

Xing B and Gao W-J. (2014n) Group search optimization algorithm. In: Xing B and Gao W-J (eds) *Innovative Computational Intelligence: A Rough Guide to 134 Clever Algorithms*. Cham Heidelberg New York Dordrecht London: Springer International Publishing Switzerland, ISBN: 978-3-319-03403-4, Chapter 12, pp. 171–176.

Xing B and Gao W-J. (2014o) Invasive weed optimization algorithm. In: Xing B and Gao W-J (eds) *Innovative Computational Intelligence: A Rough Guide to 134 Clever Algorithms*. Cham Heidelberg New York Dordrecht London: Springer International Publishing Switzerland, ISBN: 978-3-319-03403-4, Chapter 13, pp. 177–181.

Xing B and Gao W-J. (2014p) Music inspired algorithms. In: Xing B and Gao W-J (eds) *Innovative Computational Intelligence: A Rough Guide to 134 Clever Algorithms*. Cham Heidelberg New York Dordrecht London: Springer International Publishing Switzerland, ISBN: 978-3-319-03403-4, Chapter 14, pp. 183–201.

Xing B and Gao W-J. (2014q) Imperialist competitive algorithm. In: Xing B and Gao W-J (eds) *Innovative Computational Intelligence: A Rough Guide to 134 Clever Algorithms*. Cham Heidelberg New York Dordrecht London: Springer International Publishing Switzerland, ISBN: 978-3-319-03403-4, Chapter 15, pp. 203–209.

Xing B and Gao W-J. (2014qq) Complex adaptive logistics system optimization using agent-based modelling and simulation. In: Xing B and Gao W-J (eds) *Computational Intelligence in Remanufacturing*. 701 E. Chocolate Avenue, Suite 200, Hershey PA 17033: IGI Global, ISBN 978-1-4666-4908-8, Chapter 13, pp. 223–236.

Xing B and Gao W-J. (2014r) Teaching–learning-based optimization algorithm. In: Xing B and Gao W-J (eds) *Innovative Computational Intelligence: A Rough Guide to 134 Clever Algorithms.* Cham Heidelberg New York Dordrecht London: Springer International Publishing Switzerland, ISBN: 978-3-319-03403-4, Chapter 16, pp. 211–216.

Xing B and Gao W-J. (2014s) Emerging biology-based CI algorithms. In: Xing B and Gao W-J (eds) *Innovative Computational Intelligence: A Rough Guide to 134 Clever Algorithms.* Cham Heidelberg New York Dordrecht London: Springer International Publishing Switzerland, ISBN: 978-3-319-03403-4, Chapter 17, pp. 217–317.

Xing B and Gao W-J. (2014t) Big bang–big crunch algorithm. In: Xing B and Gao W-J (eds) *Innovative Computational Intelligence: A Rough Guide to 134 Clever Algorithms.* Cham Heidelberg New York Dordrecht London: Springer International Publishing Switzerland, ISBN: 978-3-319-03403-4, Chapter 18, pp. 321–331.

Xing B and Gao W-J. (2014u) Central force optimization algorithm. In: Xing B and Gao W-J (eds) *Innovative Computational Intelligence: A Rough Guide to 134 Clever Algorithms.* Cham Heidelberg New York Dordrecht London: Springer International Publishing Switzerland, ISBN: 978-3-319-03403-4, Chapter 19, pp. 333–337.

Xing B and Gao W-J. (2014v) Charged system search algorithm. In: Xing B and Gao W-J (eds) *Innovative Computational Intelligence: A Rough Guide to 134 Clever Algorithms.* Cham Heidelberg New York Dordrecht London: Springer International Publishing Switzerland, ISBN: 978-3-319-03403-4, Chapter 20, pp. 339–346.

Xing B and Gao W-J. (2014w) Electromagnetism-like mechanism algorithm. In: Xing B and Gao W-J (eds) *Innovative Computational Intelligence: A Rough Guide to 134 Clever Algorithms.* Cham Heidelberg New York Dordrecht London: Springer International Publishing Switzerland, ISBN: 978-3-319-03403-4, Chapter 21, pp. 347–354.

Xing B and Gao W-J. (2014x) Gravitational search algorithm. In: Xing B and Gao W-J (eds) *Innovative Computational Intelligence: A Rough Guide to 134 Clever Algorithms.* Cham Heidelberg New York Dordrecht London: Springer International Publishing Switzerland, ISBN: 978-3-319-03403-4, Chapter 22, pp. 355–364.

Xing B and Gao W-J. (2014y) Intelligent water drops algorithm. In: Xing B and Gao W-J (eds) *Innovative Computational Intelligence: A Rough Guide to 134 Clever Algorithms.* Cham Heidelberg New York Dordrecht London: Springer International Publishing Switzerland, ISBN: 978-3-319-03403-4, Chapter 23, pp. 365–373.

Xing B and Gao W-J. (2014z) Emerging physics-based CI algorithms. In: Xing B and Gao W-J (eds) *Innovative Computational Intelligence: A Rough Guide to 134 Clever Algorithms.* Cham Heidelberg New York Dordrecht London: Springer International Publishing Switzerland, ISBN: 978-3-319-03403-4, Chapter 24, pp. 375–414.

Xing B. (2015a) Graph-based framework for evaluating the feasibility of transition to maintainomics. In: Pedrycz W and Chen S-M (eds) *Information Granularity, Big Data, and Computational Intelligence.* Cham Heidelberg New York Dordrecht London: Springer International Publishing Switzerland, ISBN 978-3-319-08253-0, Chapter 5, pp. 89–119.

Xing B. (2015b) Knowledge management: Intelligent in-pipe inspection robot conceptual design for pipeline infrastructure management. In: Kahraman C and Onar SÇ (eds) *Intelligent Techniques in Engineering Management: Theory and Applications.* Cham Heidelberg New York Dordrecht London: Springer International Publishing Switzerland, ISBN 978-3-319-17905-6, Chapter 6, 129–146.

Xing B. (2015c) Novel nature-derived intelligent algorithms and their applications in antenna optimization. In: Matin MA (ed) *Wideband, Multiband, and Smart Reconfigurable Antennas for Modern Wireless Communications.* 701 E. Chocolate Avenue, Hershey PA 17033: IGI Global, ISBN 978-1-4666-8645-8, Chapter 10, pp. 296–339.

Xing B. (2015d) Optimization in production management: Economic load dispatch of cyber physical power system using artificial bee colony. In: Kahraman C and Onar SÇ (eds) *Intelligent Techniques in Engineering Management: Theory and Applications.* Cham Heidelberg New York Dordrecht London: Springer International Publishing Switzerland, ISBN 978-3-319-17905-6, Chapter 12, 275–293.

Xing B and Gao W-J. (2015a) The applications of swarm intelligence in remanufacturing: A focus on retrieval. In: Khosrow-Pour M (ed) *Encyclopedia of Information Science and Technology*. 3rd ed. New York, USA: Information Science Ref.—IGI Global, ISBN 978-1-4666-5888-2, Chapter 7, pp. 66–74.

Xing B and Gao W-J. (2015b) Offshore remanufacturing. In: Khosrow-Pour M (ed) *Encyclopedia of Information Science and Technology*. 3rd ed. New York, USA: Information Science Ref.—IGI Global, ISBN 978-1-4666-5888-2, Chapter 374, pp. 3795–3804.

Xing B and Gao W-J. (2015c) A SWOT analysis of intelligent product enabled complex adaptive logistics systems. In: Khosrow-Pour M (ed) *Encyclopedia of Information Science and Technology*. 3rd ed. New York, USA: Information Science Ref.—IGI Global, ISBN 978-1-4666-5888-2, Chapter 490, pp. 4970–4979.

Xing B. (2016a) Agent-based machine-to-machine connectivity analysis for the Internet of things environment. In: Mahmood Z (ed) *Connectivity Frameworks for Smart Devices: The Internet of Things from a Distributed Computing Perspective*. Switzerland: Springer International Publishing, ISBN 978-3-319-33122-5, Chapter 3, pp. 43–61.

Xing B. (2016b) An investigation of the use of innovative biology-based computational intelligence in ubiquitous robotics systems: Data mining perspective. In: Ravulakollu KK, Khan MA and Abraham A (eds) *Trends in Ambient Intelligent Systems*. Switzerland: Springer International Publishing Switzerland, ISBN 978-3-319-30184-6, Chapter 6, pp. 139–172.

Xing B. (2016c) Network neutrality debate in the internet of things era: A fuzzy cognitive map extend technology roadmap perspective. In: Mahmood Z (ed) *Connectivity Frameworks for Smart Devices: The IoT Distributed Computing Perspective*. Cham Heidelberg New York Dordrecht London: Springer International Publishing Switzerland, ISBN 978-3-319-33122-5, Chapter 10, pp. 235–257.

Xing B. (2016d) Smart robot control via novel computational intelligence methods for ambient assisted living. In: Ravulakollu KK, Khan MA and Abraham A (eds) *Trends in Ambient Intelligent Systems*. Switzerland: Springer International Publishing Switzerland, ISBN 978-3-319-30184-6, Chapter 2, pp. 29–55.

Xing B. (2016e) The spread of innovatory nature originated metaheuristics in robot swarm control for smart living environments. In: Espinosa HEP (ed) *Nature-Inspired Computing for Control Systems*. Cham Heidelberg New York Dordrecht London: Springer International Publishing Switzerland, ISBN 978-3-319-26228-4, Chapter 3, pp. 39–70.

Xing B. (2017a) Component-based hybrid reference architecture for managing adaptable embedded software development. In: Mahmood Z (ed) *Software Project Management for Distributed Computing: Life-Cycle Methods for Developing Scalable and Reliable Tools*. Gewerbestrasse 11, 6330 Cham, Switzerland: Springer International Publishing AG, ISBN 978-3-319-54324-6, Chapter 6, pp. 119–141.

Xing B. (2017b) Protecting mobile payments security: A case study. In: Meng W, Luo X, Furnell S, et al. (eds) *Protecting Mobile Networks and Devices: Challenges and Solutions*. 6000 Broken Sound Parkway NW, Suite 300, Boca Raton, FL 33487-2742: CRC Press, Taylor & Francis Group, LLC, ISBN 978-1-4987-3583-4, Chapter 11, pp. 261–289.

Xing B and Marwala T. (2018a) Conclusion. In: Xing B and Marwala T (eds) *Smart Maintenance for Human–Robot Interaction: An Intelligent Search Algorithmic Perspective*. Gewerbestrasse 11, 6330 Cham, Switzerland: Springer International Publishing AG, ISBN 978-3-319-67479-7, Chapter 13, pp. 299–305.

Xing B and Marwala T. (2018b) Cyberware capacity–applications layer perspective. In: Xing B and Marwala T (eds) *Smart Maintenance for Human–Robot Interaction: An Intelligent Search Algorithmic Perspective*. Gewerbestrasse 11, 6330 Cham, Switzerland: Springer International Publishing AG, ISBN 978-3-319-67479-7, Chapter 8, pp. 173–191.

Xing B and Marwala T. (2018c) Cyberware capacity–energy autonomy perspective. In: Xing B and Marwala T (eds) *Smart Maintenance for Human–Robot Interaction: An Intelligent Search Algorithmic Perspective*. Gewerbestrasse 11, 6330 Cham, Switzerland: Springer International Publishing AG, ISBN 978-3-319-67479-7, Chapter 9, pp. 193–216.

Xing B and Marwala T. (2018d) Cyberware capacity–platform and middleware layers perspective. In: Xing B and Marwala T (eds) *Smart Maintenance for Human–Robot Interaction: An Intelligent Search Algorithmic Perspective.* Gewerbestrasse 11, 6330 Cham, Switzerland: Springer International Publishing AG, ISBN 978-3-319-67479-7, Chapter 7, pp. 143–171.

Xing B and Marwala T. (2018e) Hardware capacity–beginning of life perspective. In: Xing B and Marwala T (eds) *Smart Maintenance for Human–Robot Interaction: An Intelligent Search Algorithmic Perspective.* Gewerbestrasse 11, 6330 Cham, Switzerland: Springer International Publishing AG, ISBN 978-3-319-67479-7, Chapter 4, pp. 67–91.

Xing B and Marwala T. (2018f) Hardware capacity–end of life perspective. In: Xing B and Marwala T (eds) *Smart Maintenance for Human–Robot Interaction: An Intelligent Search Algorithmic Perspective.* Gewerbestrasse 11, 6330 Cham, Switzerland: Springer International Publishing AG, ISBN 978-3-319-67479-7, Chapter 6, pp. 111–139.

Xing B and Marwala T. (2018g) Hardware capacity–middle of life perspective. In: Xing B and Marwala T (eds) *Smart Maintenance for Human–Robot Interaction: An Intelligent Search Algorithmic Perspective.* Gewerbestrasse 11, 6330 Cham, Switzerland: Springer International Publishing AG, ISBN 978-3-319-67479-7, Chapter 5, pp. 93–110.

Xing B and Marwala T. (2018h) Human capacity–biopsychosocial perspective. In: Xing B and Marwala T (eds) *Smart Maintenance for Human–Robot Interaction: An Intelligent Search Algorithmic Perspective.* Gewerbestrasse 11, 6330 Cham, Switzerland: Springer International Publishing AG, ISBN 978-3-319-67479-7, Chapter 11, pp. 249–270.

Xing B and Marwala T. (2018i) Human capacity–exposome perspective. In: Xing B and Marwala T (eds) *Smart Maintenance for Human–Robot Interaction: An Intelligent Search Algorithmic Perspective.* Gewerbestrasse 11, 6330 Cham, Switzerland: Springer International Publishing AG, ISBN 978-3-319-67479-7, Chapter 12, pp. 271–295.

Xing B and Marwala T. (2018j) Human capacity–physiology perspective. In: Xing B and Marwala T (eds) *Smart Maintenance for Human–Robot Interaction: An Intelligent Search Algorithmic Perspective.* Gewerbestrasse 11, 6330 Cham, Switzerland: Springer International Publishing AG, ISBN 978-3-319-67479-7, Chapter 10, pp. 219–247.

Xing B and Marwala T. (2018k) Introduction to human robot interaction. In: Xing B and Marwala T (eds) *Smart Maintenance for Human–Robot Interaction: An Intelligent Search Algorithmic Perspective.* Gewerbestrasse 11, 6330 Cham, Switzerland: Springer International Publishing AG, ISBN 978-3-319-67479-7, Chapter 1, pp. 3–19.

Xing B and Marwala T. (2018l) Introduction to intelligent search algorithms. In: Xing B and Marwala T (eds) *Smart Maintenance for Human–Robot Interaction: An Intelligent Search Algorithmic Perspective.* Gewerbestrasse 11, 6330 Cham, Switzerland: Springer International Publishing AG, ISBN 978-3-319-67479-7, Chapter 3, pp. 33–64.

Xing B and Marwala T. (2018m) Introduction to smart maintenance. In: Xing B and Marwala T (eds) *Smart Maintenance for Human–Robot Interaction: An Intelligent Search Algorithmic Perspective.* Gewerbestrasse 11, 6330 Cham, Switzerland: Springer International Publishing AG, ISBN 978-3-319-67479-7, Chapter 2, pp. 21–31.

Xing B and Marwala T. (2018n) *Smart maintenance for human–robot interaction: An intelligent search algorithmic perspective,* Gewerbestrasse 11, 6330 Cham, Switzerland: Springer International Publishing AG, ISBN 978-3-319-67479-7.

Xu ZS. (2007) On similarity measures of interval-valued intuitionistic fuzzy sets and their application to pattern recognitions. *Journal of Southeast University* 23: 139–143.

Yang Y and Hinde C. (2010) A new extension of fuzzy sets using rough sets: R-fuzzy sets. *Information Sciences* 180: 354–365.

Yao Y, Wang S and Deng X. (2017) Constructing shadowed sets and three-way approximations of fuzzy sets. *Information Sciences* 412-413: 132–153.

Ye J. (2014) Similarity measures between interval neutrosophic sets and their applications in multicriteria decision making. *Journal of Intelligent & Fuzzy Systems* 26: 165–172.

Ye J and Du S. (in press) Some distances, similarity and entropy measures for interval-valued neutrosophic sets and their relationship. *International Journal of Machine Learning and Cybernetics* DOI 10.1007/s13042-017-0719-z.

Yu S. (2017) *Neuro-inspired computing using resistive synaptic devices*, Gewerbestrasse 11, 6330 Cham, Switzerland: Springer International Publishing AG, ISBN 978-3-319-54312-3.
Zadeh LA. (1965) Fuzzy sets. *Information and Control* 8: 338–353.
Zadeh LA. (1968) Probability measures of fuzzy events. *Journal of Mathematical Analysis and Applications* 23: 421–427.
Zadeh LA. (1971) Similarity relations and fuzzy orderings. *Information Sciences* 3: 177–200.
Zadeh LA. (1975a) The concept of a linguistic variable and its application to approximate reasoning—I. *Information Sciences* 8: 199–249.
Zadeh LA. (1975b) The concept of a linguistic variable and its application to approximate reasoning—II. *Information Sciences* 8: 301–357.
Zadeh LA. (1975c) The concept of a linguistic variable and its application to approximate reasoning—III. *Information Sciences* 9: 43–80.
Zadeh LA. (1978) Fuzzy sets as a basis for a theory of possibility. *Fuzzy Sets and Systems* 1: 3–28.
Zeng W, Li D and Yin Q. (2016) Distance and similarity measures between hesitant fuzzy sets and their application in pattern recognition. *Pattern Recognition Letters* 84: 267–271.
Zeng WY and Li HX. (2006) Inclusion measure, similarity measure and the fuzziness of fuzzy sets and their relations. *International Journal of Intelligent Systems* 21: 639–653.
Zeng WY and Guo P. (2008) Normalize distance, similarity measure, inclusion measure and entropy of interval-valued fuzzy sets and their relationship. *Information Sciences* 178: 1334–1342.
Zhai CX and Massung S. (2016) *Text data management and analysis: A practical introduction to information retrieval and text mining*: Morgan & Claypool Publishers, ISBN 978-1-97000-119-8.
Zhang H, Zhang W and Mei C. (2009) Entropy of interval-valued fuzzy sets based on distance and its relationship with similarity measure. *Knowledge-Based Systems* 22: 449–454.
Zhang Q. (1998) Fuzziness–vagueness–generality–ambiguity. *Journal of Pragmatics* 29: 13–31.
Zhang QH, Xie Q and Wang G. (2016) A survey on rough set theory and its applications. *CAAI Transactions on Intelligence Technology* 1: 323–333.
Zhang QS, Xing HY, Liu FC, et al. (2014) Some new entropy measures for interval-valued intuitionistic fuzzy sets based on distances and their relationships with similarity and inclusion measures. *Information Sciences* 283: 55–69.
Zhang X and Xu Z. (2015) Soft computing based on maximizing consensus and fuzzy TOPSIS approach to interval-valued intuitionistic fuzzy group decision making. *Applied Soft Computing* 26: 42–56.
Zheng M, Liu G-x, Zhou C-g, et al. (2010) Gravitation field algorithm and its application in gene cluster. *Algorithms for Molecular Biology* 5: 1–11.
Zhou J, Pedrycz W and Miao DQ. (2011) Shadowed sets in the characterization of rough-fuzzy clustering. *Pattern Recognition* 44: 1783–1749.
Zhou L, Pan S, Wang J, et al. (2017) Machine learning on big data: Opportunities and challenges. *Neurocomputing* 237: 350–361.
Zimmermann H-J. (1992) *Fuzzy set theory and its applications*, 101 Philip Drive, Assinippi Park, Norwell, Massachusetts 02061 USA: Kluwer Academic Publishers, ISBN 0-7923-9075-X.
Zimmermann H-J. (2001) *Fuzzy set theory and its applications*, Springer Seience+Business Media New York, ISBN 978-94-010-3870-6.
Zwick R, Carlstein E and Budescu DV. (1987) Measures of similarity among fuzzy concepts: A comparative analysis. *International Journal of Approximate Reasoning* 1: 221–242.

CHAPTER 2

Introduction to Bricolage

2.1 Innovation and Economy

The word "innovation" originates from the Latin noun *innovatus*, and its earliest appearance in print can be traced back to the fifteenth century (Shah et al., 2015). Though this word has nowadays been widely utilized (or abused to some extent) in numerous settings with distinct implications, the most modern version of "innovation" was initially interpreted and expounded by the renowned economist Joseph Schumpeter (1883–1950) (Pressman, 2006) in his monograph written in the 1930s (Schumpeter, 1934). Originally trained as a law student, Schumpeter did not do well in his political life and he became penniless after a set of real world speculative investment activities. Schumpeter later returned to academic life where he made a name for himself due to his contributions towards identifying the phases of business cycles and the factors leveraging capitalist economies up and down (Pressman, 2006).

From Schumpeter's viewpoint, capitalism is characterized by the following three simultaneously occurring business cycles (Pressman, 2006):

- **Kitchin Cycles (named after Joseph Kitchin):** These are short-term fluctuations, which normally last three to four years and are caused by business inventory changes. These can be categorized into:
  1) In the first period (one to two years) inventories expand so that businesses can keep pace with the rising sales;
  2) When the growth of sales is slowing down at the second period, businesses spend a year or so cutting back production and clearing up overstock; and
  3) When desired inventory levels are achieved in the final stage, businesses re-expand the relevant inventories in order to meet the uplifted sales.

- **Juglar Cycles (named after Clement Juglar):** The second type of cycle is often associated with the business investment changes regarding physical assets (e.g., new factory and machinery) and these can be categorized as follows:
    1) During the first period (four to five years), expansions are attributed to businesses' desire to modernize their physical assets;
    2) When the initial modernization reaches a suitable level, businesses have little incentive to continue with further investment for this period (another four to five years); and
    3) After the second period, the worn out and obsolete situation of the initial invested physical assets during the first period forces businesses to make another round of investment (typically lasting four to five years).
- **Kondratieff Waves (named after Nikolai Kondratieff):** The third type of cycle concerns the long-run effect, which lasts around forty-five to sixty years. Schumpeter believed that the impetus behind these long-run cycles is invention and innovation. In this regard, the First Industrial Revolution (1IR), characterized by the introduction of steam engines, stimulated a notable Kondratieff wave. The Second Industrial Revolution (2IR) featured the realization of mass production and started a second Kondratieff wave. The Third Industrial Revolution (3IR), represented by the popularization of information and communication technologies, began a third Kondratieff wave. Another unprecedented wave is being promoted by the Fourth Industrial Revolution (4IR) (Xing and Marwala, 2017) which is sparked by the penetration of artificial intelligence (Xing and Gao, 2014aa; Xing and Gao, 2014b; Xing and Gao, 2014bb; Xing and Gao, 2014c; Xing and Gao, 2014cc; Xing and Gao, 2014d; Xing and Gao, 2014dd; Xing and Gao, 2014e; Xing and Gao, 2014f; Xing and Gao, 2014g; Xing and Gao, 2014h; Xing and Gao, 2014i; Xing and Gao, 2014j; Xing and Gao, 2014k; Xing and Gao, 2014l; Xing and Gao, 2014m; Xing and Gao, 2014n; Xing and Gao, 2014o; Xing and Gao, 2014p; Xing and Gao, 2014q; Xing and Gao, 2014r; Xing and Gao, 2014s; Xing and Gao, 2014t; Xing and Gao, 2014u; Xing and Gao, 2014v; Xing and Gao, 2014w; Xing and Gao, 2014x; Xing and Gao, 2014y; Xing and Gao, 2014z; Xing et al., 2013; Xing and Marwala, 2018; Xing and Gao, 2014ff; Marwala, 2015; Marwala, 2014) powered cyber-physical system (Xing, 2015; Xing, 2014a; Xing, 2014b).

Among these different business cycles, gaining an in-depth understanding on the Kondratieff wave in order to achieve a long-term sustainable and flourishing economic development has always been a central topic within economic studies. In this regard, Schumpeter emphasized the crucial role of some non-economic factors (e.g., innovation) in achieving this goal.

2.1.1 Innovation by Models

Faced with global competition, innovation becomes the key driver for modern economic growth. According to Schumpeter (1950), innovation can be understood as a 'creative destruction' process. Later, an extended definition of innovation was proposed by Freeman (1982) where he viewed innovation to be a process of turning an opportunity into new concepts and setting these into extensively used practices in the market or in society. Innovation is, therefore, one of the most important trends for fortifying a defensible position against the forces of competition. However, innovation is not the same as invention. It requires a long, uncertain and expensive process. Moreover, it experiences novel models, i.e., from essentially closed to completely open.

- **Closed Innovation:** For many years, innovation emerged as the 'strategic asset' within a firm. In this regard, the firm sought to discover new technical and operational breakthroughs, update them into products, assemble the new products in its own factories' shop floor and distribute, finance, and service those new products, all within the four walls of the firm (Chesbrough, 2003b). Among others, famous examples include operating systems from Microsoft, Ethernet from Xerox, electric light bulbs from GE, and transistors from Bell Labs (Chesbrough and Appleyard, 2007; Chesbrough, 2003a).

  This model of innovation is usually called 'closed innovation' (see Fig. 2.1) and its philosophy is that "successful innovation requires control" (Chesbrough, 2003a). In our view, it has led to the development of a variety of advanced innovations. However, it also raises some critical issues, as those innovations seem to only create 'individual' technologies and tend to miss deeper and more powerful potential. As a result, they cannot be used to capture more value or offer companies improved competitiveness. Just as Hagel et al. (2013) pointed out that "we cannot grasp the full potential of exponential technologies until we explore the interactions across them".

- **Open Innovation:** Recently, much attention has been paid to 'open innovation' in order to work towards more radical and systemic improvements. The term 'open innovation' (see Fig. 2.2) was first proposed by Chesbrough (2003b), who argued that open innovation is "the use of purposive inflows and outflows of knowledge to accelerate internal innovation and expand the markets for external use of innovation, respectively" (Chesbrough et al., 2006).

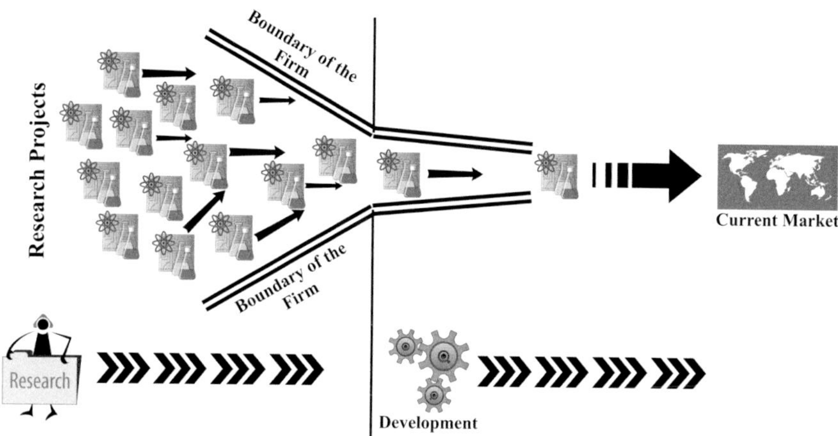

Figure 2.1: Closed innovation.

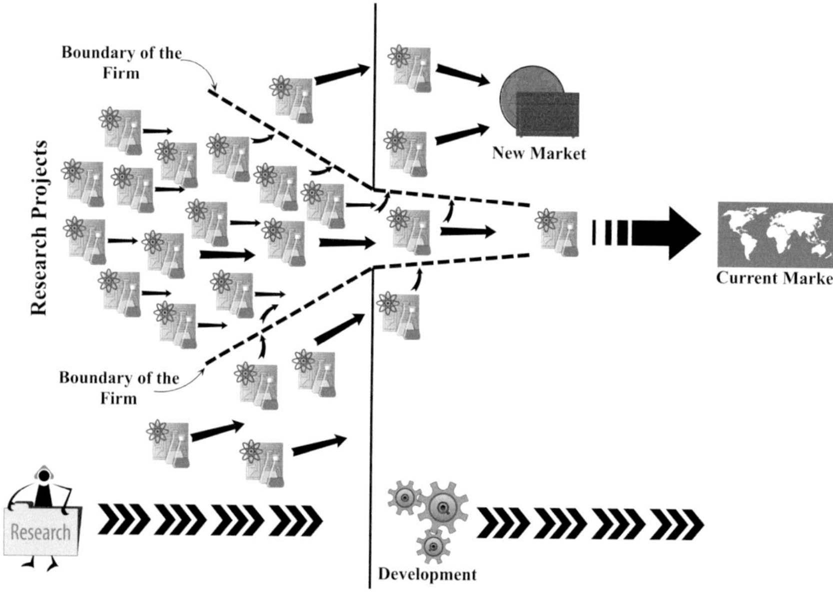

Figure 2.2: Open innovation.

Unlike its counterpart model, the open paradigm emphasizes the emergence of any alternative during the course of research and development. The option set is diverse in terms of when to begin, whether to carry on, how to collaborate, and what to keep or abolish. Some well-known examples include successfully collaborated networks from Hollywood, and SpinBrush from Procter & Gamble

(Chesbrough, 2003a). Moreover, this concept has also been adopted to other growing contemporary areas of research, such as innovation communities (Fleming and Waguespack, 2007), open source software development (Von Krogh et al., 2012; Von Krogh and Spaeth, 2007; West and Gallagher, 2006), and bricolage from entrepreneurship literatures (Senyard et al., 2014). More details should refer to two recently published review articles by Stanko et al. (2017) and Bogers et al. (2017), respectively.

2.1.2 *Innovation by Types*

Furthermore, different types of innovation can be identified into two main forms, namely product-dominant and service-dominant.

- **Product-Dominant Logic:** In the literature, researchers have been focused on the exploration of the impact of innovation on new product development (NPD), i.e., from idea generation to delivery. In general, there are several research streams that are related to open innovation on the NPD process, namely role of resource-poor environments (Cunha et al., 2014), role of idea generation process (Pialot et al., 2012; Spanjol, 2003; Ayag, 2005), role of suppliers in innovation (Das et al., 2006; Henke and Zhang, 2010; Johnsen, 2009), role of purchasing in innovation (Lakemond et al., 2001; Schiele, 2010) and role of supplier and purchasing involvement (Luzzini et al., 2015; Carr and Pearson, 2002). More details should refer to (Noble et al., 2014).
- **Service-Dominant Logic:** Innovation in services-based business models is now a hot topic that is being investigated by developing economies (Chesbrough, 2011). According to Vargo et al. (2010), service-dominant logic highlights the fact that service is essentially the use of one entity's competencies to benefit other entities. Its ultimate aim is to realize service in exchange for service for all economic activities. Several scholars have outlined the benefits associated with this transformation, such as reducing costs (Zott and Amit, 2008), moving toward adjacent spaces (Slywotzky et al., 2003), realizing mutually beneficial outcomes (Maglio and Spohrer, 2013) and catching the advantages of open ecosystems (Chesbrough, 2006). Further information regarding this logic can be found in (Schaefer, 2014; Stevenson, 2013; Witell et al., 2016; Johne and Storey, 1998).

2.2 What is Bricolage?

From economists' point of view, resource scarcity is the core problem in terms of the development of economics (Schiller et al., 2013). Typically, there are four basic factors of production that are used to produce goods and service, namely land, labour, capital, and entrepreneurship (Schiller et al., 2013). However, no matter how those factors are effectively organized, there is not enough to satisfy all our desires. In other words, we cannot have it all, we must decide what we will have and what we must forgo. As a result, we must change our theoretical and epistemological coherences. Among others, from a managerial point of view, one way is through a new conceptual terrain, i.e., the conceptualization of "bricolage" (Kincheloe, 2005; Orlikowski, 2000).

Originally, the term "bricolage" was proposed by the French anthropologist Lévi-Strauss (1962) to describe the contrast between two parallel world views, i.e., engineers and bricoleurs, when both meet technology and product innovation. The former always follow specific procedures to perform their work, while the latter can provide a large number of diverse pathways to achieve innovation. Recently, several researchers pointed out that the concept of "bricolage" can help us frame the imbalances between available resources and our wish list (Baker et al., 2003a; Garud and Karnoe, 2003). According to Baker and Nelson (2005), bricolage refers to "making do with whatever is at hand by reuse and recombination". In this regard, Fuglsang (2010) pointed out that bricolage can be used to characterize the concept of innovation. Senyard et al. (2014) argued that bricolage is a tool that is used to overcome resource-constraints for new firms. An et al. (in press) re-framed bricolage as a concrete activity of opportunity identification and experiential resource learning. Furthermore, the theories of bricolage have been widely used in a range of disciplines and contexts, such as art bricolage, education bricolage, evolutionary bricolage, political bricolage, service bricolage and social bricolage (Domenico et al., 2010; Molecke and Pinkse, 2017). In summary, these studies all converge towards the essence of bricolage characteristics as a mechanism for resource collection, assimilation and re-combination in order to yield something new that has utility (Leadbeater, 2014).

2.2.1 *Bricolage Capabilities*

Rooted theoretically in Lévi-Strauss (1962) work on bricolage, there are several key capabilities underpinning this concept.

2.2.1.1 *Build Something from Nothing*

The work of Lévi-Strauss (1962) suggested that bricolage should function as a unit and should not be constrained by limitations. The concept "build something from nothing" has several specific features which were identified by (Baker and Nelson, 2005), these are that this concept implies that resources are available, and can be recombined for new purposes and improvisation.

2.2.1.2 Ability to Improvise

Another concept that sheds light on the capabilities of bricolage is the ability to improvise (Baker et al., 2003a; Miettinen and Virkkunen, 2005). Typically, the characteristics of improvisation include reacting to the limited resources (i.e., using what is at hand), finding a serendipitous research approach (i.e., what fits), and accepting imperfections and continuing to improvise (i.e., to some degree, critical thinking and problem solving).

2.2.1.3 Networking Ability

In the literature, several researchers used the bricolage concept as a powerful tool to understand the dynamics of organisational networking. For example, Grivins et al. (2017) applied the bricolage concept in order to compare the organizational dynamics of two alternative food networks in Riga and Bristol, respectively. Feyereisen et al. (2017) proposed an innovative solution for the transition of the Belgian fair-trade dairy system. Moreover, Holt and Littlewood (2017) examined several social enterprises (such as NGOs) and micro-entrepreneurs in Kenya that depend on waste materials to produce income, through the abstract perspective of bricolage. Their findings suggested that we cannot speak of bricolage unless the invented new phenomenon can be implemented in an organization with certain networking and economic impacts.

2.2.2 Bricolage and Innovation

For a firm to develop and implement its various strategies, resources are indispensable which typically incorporate two types of assets, namely tangible and intangible (Ray et al., 2004). Traditionally, focus has been largely placed on seeking new means to combine resources optimally, while for innovation, it depends more on managing resources that are often hidden, scattered, or poorly used (Witell et al., 2017). Therefore, service innovation tends to happen in resource-limited situations. Accordingly, the early days of service innovation research mainly focused on the distinctions between product- and service-dominant logic (Snyder et al., 2016; Johne and Storey, 1998), and their possible integration (i.e., product-service system) (Baines et al., 2007; Mont and Tukker, 2006; Hahn and Morner, 2011; Mont, 2002).

2.2.2.1 New Angle of Service Innovation Thinking: Recombinative Thought (Bricolage-Based Viewpoint)

More recently, service research has evolved into a new stage which is rooted in bricolage-based principles in order to create values (Witell et al., 2017). It is often believed that bricoleurs 'innovate' differently to other engineers. In the tone of entrepreneurship, the simplest theoretical relations between bricolage and bricoleur can be expressed in the form of Eq. 2.1 (Frederick et al., 2016):

$$\text{Bricolage} = f\,(\text{bricoleur}). \tag{2.1}$$

where $f(\cdot)$ indicates that bricolage is a function of the bricoleur. An innovator living in a specific cultural context and time period utilizes 'whatsoever he can obtain', either odd or heterogeneous, for the purpose of creative value creation (Witell et al., 2017).

Accordingly, the bricolage-based viewpoint involves recombining existing resources in unique means. Through this new lens, one can view service innovation as (Witell et al., 2017):

- Repackaging diversified resources to form novel resources that are, in certain background, useful for some players (Lusch and Nambisan, 2015); or
- Recombining various practices collaboratively in order to offer innovative solution alternatives for problems that are either new or existing (Vargo et al., 2015).

2.3 Conclusions

Crowdfunding as a new instrument of collaborative financing is not only inherently innovative, but it also serves as the fountainhead of derivative innovations and creativities on many levels, e.g., goods, services, business models. Crowdfunding itself may not qualify as a radical innovation for some reasons, such as the fact that it is incremental in nature and is based on the principle of participating contributor. In this regard, incremental nature implies another evolutionary stage of the conventional financing system with the broadened and deepened utilization of the Internet. Furthermore, participated contributor implies an intuitive transformation from an individual to a crowd and staying close to the center of various radical or disruptive discontinuities from both a macro- and micro-perspective (Leite, 2012). In general, the output of an economy can be enlarged through the following two means: (1) Increasing the number of inputs that enter the productive process (Rosenberg, 2004), and (2) Churning out new forms to harvest more outputs from the unchanged amount of inputs, or in other words, being capable of utilizing inputs more productively (Rosenberg, 1973). As suggested by Leite (2012), crowdfunding appears to be falling within the second category, that is, reorganizing the existing inputs and injecting them into a new processing system for elevating the economy's overall efficiency, competitiveness, and transparency. Based on these understandings, this chapter explored the bricolage literature from the following perspectives and resources:

- **Books/Book Chapters:** (Frederick et al., 2016; Shalley et al., 2015; Sanchez-Burks et al., 2015; Lévi-Strauss, 1962; Leadbeater, 2014);
- **Theses:** (Hutchinson, 2008; Plewe, 2008; Schaefer, 2013; Kuhl, 2014; Senyard, 2015; Stürmer and Onland, 2015; Macfarlane, 2006; Baker, 2012; Alford, 2012; Welter, 2012; Galvanauskaite, 2014; Kupolokun, 2014; Peltone, 2014; Kiss, 2014; Sabdia, 2015; Gabriel, 2015; Bousfiha, 2015; Istanbuli, 2015; Castricum, 2015);
- **Report:** (Phillimore et al., 2016);
- **Reviews:** (Turnbull, 2002; Fisher, 2012; Rogers, 2012; Welter et al., 2016; Visscher et al., in press);
- **Web Resources:** (Silva et al., 2014; Vuletich, 2015); and
- **Articles:**
  1) *Bricolage-Crowdsourcing*: (Schaefer, 2013);
  2) *Bricolage-Business/Entrepreneurship*: (Baker et al., 2003b; Garud and Karnoe, 2003; Baker and Nelson, 2005; Russell and Tyler, 2005; Ferneley and Bell, 2006; Phillips and Tracey, 2007; Banerjee and Campbell, 2009; Senyard et al., 2009; Boxenbaum and Rouleau, 2011; Ilahiane, 2011; Salimath and Jones, 2011; Vanevenhoven et al., 2011; Halme et al., 2012; Desa and Basu, 2013; Rönkkö et al., 2013; Stinchfield et al., 2013; Winkel et al., 2013; Burgers et al., 2014; Benouniche et al., 2014; Feyereisen and Mélard, 2014; Lennefors and Rehn, 2014; Rüling and Duymedjian, 2014; Valliere and Gegenhuber, 2014; Beckett, 2016; Guo et al., 2016; Feyereisen et al., 2017; An et al., in press; Wu et al., 2017; Grivins et al., 2017; Boccardelli and Magnusson, 2006; George et al., 2012; Ernst et al., 2015; Wilf, 2015);
  3) *Bricolage-Design*: (Louridas, 1999; Lewens, 2013);
  4) *Bricolage-Education*: (Hatton, 1989; Andersen, 2008; Warne and McAndrew, 2009; Freeman, 2007; Tam, 2012; Aguinaldo, 2013; Kinn et al., 2013; Campos and Ribeiro, 2016; Hsieh, 2016; Nyika and Murray-Orr, 2017; Bush and Silk, 2010; Wood et al., 2013);
  5) *Bricolage-Financial*: (Engelen et al., 2010; MacKenzie and Pardo-Guerra, 2014);
  6) *Bricolage-Healthcare*: (Hester, 2005; Frit et al., 2014; Renwick, 2014; McMillan, 2015);
  7) *Bricolage-Organization*: (Cunha, 2005; Duymedjian and Rüling, 2010; Perkmann and Spicer, 2014; Bjerregaard and Lauring, 2012; Desa, 2011);

8) *Bricolage-Resources Limitation*: (Cleaver, 2002; Baker, 2007; Mittelman, 2013; Senyard et al., 2014; Desa, 2011; Linna, 2013; Cunha et al., 2014; Stokes, 2014; Bicen and Johnson, 2015);
9) *Bricolage-Service Innovation*: (Fuglsang, 2010; Fuglsang and Sørensen, 2011; Salunke et al., 2013; Witell et al., 2017; Victorino et al., 2005);
10) *Bricolage-Social*: (Domenico et al., 2010; Sundurамurthy et al., 2016; Holt and Littlewood, 2017; Molecke and Pinkse, 2017);
11) *Bricolage-Tourism*: (Baláž and Williams, 2005);
12) *Bricolage-Others*: (Berlo, 1992; Butor and Guynn, 1994; Conville, 1997; Duboule and Wilkins, 1998; Kincheloe, 2001; Kincheloe, 2005; Laurent, 2005; Duncan, 2011; Wibberley, 2012; Manfield and Newey, 2015).

Since a unified holistic framework for analysing crowdfunding is still absent, the authors hope that this preliminary investigation can provide readers with a new angle from which to view crowdfunding.

References

Aguinaldo BE. (2013) Implementing blended learning in an impoverished academic institution using a bricolage approach model. *International Journal of Information and Education Technology* 3: 211–216.

Alford Z. (2012) A peer professional learning group: A professional identity forum. The University of Waikato.

An W, Zhao X, Cao Z, et al. (in press) How bricolage drives corporate entrepreneurship: the roles of opportunity identification and learning orientation. *Journal of Product Innovation Management*: 1–17.

Andersen OJ. (2008) A bottom-up perspective on innovations: Mobilizing knowledge and social capital through innovative processes of bricolage. *Administration & Society* 40: 54–78.

Ayag Z. (2005) An integrated approach to evaluating conceptual design alternatives in a new product development environment. *International Journal of Production Research* 43: 687–713.

Baines TS, Lightfoot HW, Evans S, et al. (2007) State-of-the-art in product-service systems. *Proceedings of the Institution of Mechanical Engineers Part B: J. Engineering Manufacture* 221: 1543–1552.

Baker DJ. (2012) (Re)scripting the self: Subjectivity, creative and critical practice and the pedagogy of writing. *School of Humanities, Arts, Education and Law.* Giffith University.

Baker T, Miner A and Eesley D. (2003a) Improvising firms: Bricolage, account giving and improvisational competency in the founding process. *Research Policy* 32: 255–276.

Baker T, Miner AS and Eesley DT. (2003b) Improvising firms: Bricolage, account giving and improvisational competencies in the founding process. *Research Policy* 32: 255–276.

Baker T and Nelson RE. (2005) Creating something from nothing: Resource construction through entrepreneurial bricolage. *Administrative Science Quarterly* 50: 329–366.

Baker T. (2007) Resources in play: Bricolage in the toy store(y). *Journal of Business Venturing* 22: 694–711.

Baláž V and Williams AM. (2005) International tourism as bricolage: An analysis of central Europe on the brink of European union membership. *International Journal of Tourism Research* 7: 79–93.

Banerjee PM and Campbell BA. (2009) Inventor bricolage and firm technology research and development. *R&D Management* 39: 473–487.

Beckett CR. (2016) Entrepreneurial bricolage: Developing recipes to support innovation. *International Journal of Innovation Management* DOI: 10.1142/S1363919616400107.

Benouniche M, Zwarteveen M and Kuper M. (2014) Bricolage as innovation: Opening the black box of drip irrigation systems. *Irrigation and Drainage* 63: 651–658.

Berlo JC. (1992) Beyond bricolage: Women and aesthetic strategies in Latin American textiles. *RES: Anthropology and Aesthetics* 22: 115–134.

Bicen P and Johnson WHA. (2015) Radical innovation with limited resources in high-turbulent markets: The role of lean innovation capability. *Creativity and Innovation Management* 24: 278–299.

Bjerregaard T and Lauring J. (2012) Entrepreneurship as intitutional change: Strategies of bridging institutional contradictions. *European Management Review* 9: 31–43.

Boccardelli P and Magnusson MG. (2006) Dynamic capabilities in early-phase entrepreneurship. *Knowledge and Process Management* 13: 162–174.

Bogers M, Zobel A-K, Afuah A, et al. (2017) The open innovation research landscape: Established perspectives and emerging themes across different levels of analysis. *Industry and Innovation* 24: 18–40.

Bousfiha M. (2015) New market creation by young entrepreneurial firms. *Department of Technology Management and Economics, Division of Innovation Engineering and Management.* Göteborg, Sverige: Chalmers University of Technology.

Boxenbaum E and Rouleau L. (2011) New knowledge products as bricolage: Metaphors and scripts in organizational theory. *Academy of Management Review* 36: 272–296.

Burgers H, Stuetzer M and Senyard JM. (2014) Antecedents, consequences, and the mediating role of bricolage in corporate entrepreneurship. *Academy of Management Proceedings* 2014: 134–173.

Bush A and Silk M. (2010) Towards an evolving critical consciousness in coaching research: The physical pedagogic bricolage. *International Journal of Sports Science & Coaching* 5: 551–565.

Butor M and Guynn N. (1994) Bricolage: An interview With Michel Butor. *Yale French Studies* 84: 17–26.

Campos LRGd and Ribeiro MRR. (2016) Bricolage in research nursing education: Experience report. *Escola Anna Nery* 20: 1–7.

Carr AS and Pearson JN. (2002) The impact of purchasing and supplier involvement on strategic purchasing and its impact on firm's performance. *International Journal of Operation & Production Management* 22: 1032–1053.

Castricum RP. (2015) Use of the base of the pyramid model to achieve the united nations millennium development goals. *School of Management, College of Businesss.* RMIT University.

Chesbrough H. (2006) *Open business models: How to thrive in the new innovation landscape,* Cambridge MA.: Harvard Business School Press.

Chesbrough H, Vanhaverbeke W and West J. (2006) *Open innovation: Researching a new paradigm*: Oxford University Press, ISBN 9780199290727.

Chesbrough H. (2011) *Open services innovation: Rethinking you business to grow and compete in a new era,* San Francisco: Wiley.

Chesbrough HW. (2003a) The era of open innovation. *MIT Sloan Management Review* 44: 35–41.

Chesbrough HW. (2003b) *Open innovation: The new imperative for creating and profiting from technology,* 60 Harvard Way, Boston, Massachusetts 02163, USA: Harvard Business School Publishing Corporation, ISBN 1-57851-837-7.

Chesbrough HW and Appleyard MM. (2007) Open innovation and strategy. *California Management Review* 50: 57–76.

Cleaver F. (2002) Reinventing institutions: Bricolage and the social embeddedness of natural resource management. *European Journal of Development Research* 14: 11–30.

Conville RL. (1997) Between spearheads: Bricolage and relationships. *Journal of Social and Personal Relationships* 14: 373–386.

Cunha MPe. (2005) Bricolage in organizations. *SSRN Electronic Journal* https://ssrn.com/abstract=882784.

Cunha MPe, Rego A, Oliveira P, et al. (2014) Product innovation in resource-poor environments: Three research streams. *Journal of Product Innovation Management* 31: 202–210.

Das A, Narasimhan R and Talluri S. (2006) Supplier integration—finding an optimal configuration. *Journal of Operations Management* 24: 563–582.

Desa G. (2011) Resource mobilization in international social entrepreneurship: Bricolage as a mechanism of institutional transformation. *Entrepreneurship: Theory & Practice* 36: 727–751.

Desa G and Basu S. (2013) Optimization or bricolage? Overcoming resource constraints in global social entrepreneurship. *Strategic Entrepreneurship Journal* 7: 26–49.

Domenico MD, Haugh H and Tracey P. (2010) Social bricolage: Theorizing social value creation in social enterprises. *Entrepreneurship Theory and Practice* 34: 681–703

Duboule D and Wilkins AS. (1998) The evolution of 'bricolage'. *Trends in Genetics (TIG)* 14: 54–59.

Duncan S. (2011) Personal life, pragmatism and bricolage. *Sociological Research Online* 16: 1–12.

Duymedjian R and Rüling C-C. (2010) Towards a foundation of bricolage in organization and management theory. *Organization Studies* 31: 133–151.

Engelen E, Erturk I, Froud J, et al. (2010) Reconceptualizing financial innovation: Frame conjuncture and bricolage. *Economy and Society* 39: 33–63.

Ernst H, Kahle HN, Dubiel A, et al. (2015) The antecedents and consequences of affordable value innovations from emerging markets. *Journal of Product Innovation Management* 32: 65–79.

Ferneley E and Bell F. (2006) Using bricolage to integrate business and information technology innovation in SMEs. *Technovation* 26: 232–241.

Feyereisen M and Mélard F. (2014) From fair milk to fair enterprise: The consequences of an unexpected bricolage. *Outlook on Agriculture* 43: 207–211.

Feyereisen M, Stassart PM and Mélard F. (2017) Fair trade milk initiative in Belgium: Bricolage as an empowering strategy for change. *Sociologia Ruralis* 57: 297–315.

Fisher G. (2012) Effectuation, causation, and bricolage: A behavioral comparison of emerging theories in entrepreneurship research. *Entrepreneurship Theory and Practice* 36: 1019–1051.

Fleming L and Waguespack DM. (2007) Brokerage, boundary spanning and leadership in open innovation communities. *Organization Science* 18: 165–180.

Frederick H, O'Connor A and Kuratko DF. (2016) *Entrepreneurship: Theory, process, practice,* Level 7, 80 Dorcas Street, South Melbourne, Victoria Australia 3205: Cengage Learning Australia, ISBN 978-0-17-035255-0.

Freeman C. (1982) *The economics of industrial innovation,* London: Frances Pinter.

Freeman R. (2007) Epistemological bricolage: How practitioners make sense of learning. *Administration & Society* 39: 476–496.

Frit P, Barboule N, Yuan Y, et al. (2014) Alternative end-joining pathway(s): Bricolage at DNA breaks. *DNA Repair* 17: 81–97.

Fuglsang L. (2010) Bricolage and invisible innovation in public service innovation. *Journal of Innovation Economics & Management* 3: 67–87.

Fuglsang L and Sørensen F. (2011) The balance between bricolage and innovation: Management dilemmas in sustainable public innovation. *The Service Industries Journal* 31: 581–595.

Gabriel C-A. (2015) How entrepreneurs respond to constraints in developing and emerging countries. New Zealand: University of Otago.

Galvanauskaite I. (2014) Exploring technology's role in social entrepreneurship. *Department of Intercultural Communication and Management.* Copenhagen Business School.

Garud R and Karnoe P. (2003) Bricolage versus breakthrough: Distributed and embedded agency in technology entrepreneurship. *Research Policy* 32: 277–300.

George G, McGahan AM and Prabhu J. (2012) Innovation for inclusive growth: Towards a theoretical framework and a research agenda. *Journal of Management Studies* 49: 661–683.

Grivins M, Keech D, Kunda I, et al. (2017) Bricolage for self-sufficiency: An analysis of alternative food networks. *Sociologia Ruralis* 57: 340–356.

Guo H, Su Z and Ahlstrom D. (2016) Business model innovation: The effects of exploratory orientation, opportunity recognition and entrepreneurial bricolage in an emerging economy. *Asia Pacific Journal of Management* 33: 533–549.

Hagel J, Brown JS, Samoylova T, et al. (2013) From exponential technologies to exponential innovation. Deloitte University Press, 1–28.

Hahn A and Morner M. (2011) Product service bundles: No simple solution. *Journal of Business Strategy* 32: 14–23.

Halme M, Lindeman S and Linna P. (2012) Innovation for inclusive business: Intrapreneurial bricolage in multinational corporations. *Journal of Management Studies* 49: 743–784.

Hatton E. (1989) Levi-Strauss's bricolage and theorizing teachers' work. *Anthropology & Education Quarterly* 20: 74–96.

Henke J and Zhang C. (2010) Increasing supplier-driven innovation. *MIT Sloan Management Review* 51: 41–46.

Hester JS. (2005) Bricolage and bodies of knowledge: Exploring consumer responses to controversy about the third generation oral contraceptive pill. *Body & Society* 11: 77–95.

Holt D and Littlewood D. (2017) Waste livelihoods amongst the poor—through the lens of bricolage. *Business Strategy and the Environment* 26: 253–264.

Hsieh C-C. (2016) A way of policy bricolage or translation: The case of Taiwan's higher education reform of quality assurance. *Policy Futures in Education* 14: 873–888.

Hutchinson SA. (2008) Boundaries, bricolage and student-teacher learning. *Faculty of Education and Language Studies.* The Open University.

Ilahiane H. (2011) Mobile phone use, bricolage, and the transformation of social and economic ties of micro-entrepreneurs in urban Morocco. *International Journal of Business Anthropology* 2: 31.

Istanbuli AD. (2015) The role of Palestinian women entrepreneurs in business development. *Faculty of Economic Sciences and Business Studies.* University of Granada.

Johne A and Storey C. (1998) New service development: A review of the literature and annotated bibliography. *European Journal of Marketing* 32: 184–251.

Johnsen TE. (2009) Supplier involvement in new product development and innovation: Taking stock and looking to the future. *Journal of Purchasing & Supply Management* 15: 187–197.

Kincheloe JL. (2001) Describing the bricolage: Conceptualizing a new rigor in qualitative research. *Qualitative Inquiry* 7: 679–692.

Kincheloe JL. (2005) On to the next level: Continuing the conceptualization of the bricolage. *Qualitative Inquiry* 11: 323–350.

Kinn LG, Holgersen H, Ekeland T-J, et al. (2013) Metasynthesis and bricolage: An artistic exercise of creating a collage of meaning. *Qualitative Health Research* 23: 1285–1292.

Kiss M. (2014) The collective creation and support of touristic landscape. *School of Sociology.* Budapest: Institute of Sociology and Social Policy.

Kuhl DE. (2014) Voices count: Employing a critical narrative research bricolage for insights into Dyscalculia. *The School of Graduate and Postdoctoral Studies.* London, Ontario, Canada: The University of Western Ontario.

Kupolokun O. (2014) For-profit social entrepreneurship: A study of resources, challenges, and competencies in UK. *Goldsmiths College.* University of London.

Lakemond N, Echtelt F and Wynstra F. (2001) A configuration typology for involving purchasing specialists in product development. *The Journal of Supply Chain Management* 37: 11–20.

Laurent P-J. (2005) The process of bricolage between mythic societies and global modernity: Conversion to the assembly of god faith in Burkina Faso. *Social Compass* 52: 309–323.

Leadbeater C. (2014) *The frugal innovator*, 175 Fifth Avenue, New York, NY 10010.: Palgrave Macmillan, ISBN 978-1-137-33536-4.

Leite PdM. (2012) Crowdfunding: critical factors to finance a project successfully. *Faculdade de Economia e Gestão*. Universidade do Porto.

Lennefors TT and Rehn A. (2014) Chance interventions: On bricolage and the state as an entrepreneur in a declining industry. *Culture and Organization* 20: 377–391.

Lévi-Strauss C. (1962) *The savage mind (La Pensée Sauvage)*, 5 Winsley Street, London WI: George Weidenfeld and Nicolson Ltd.

Lewens T. (2013) From bricolage to bioBricks™: synthetic biology and rational design. *Studies in History and Philosophy of Biological and Biomedical Sciences* 44: 641–648.

Linna P. (2013) Bricolage as a means of innovating in a resource-scarce environment: A study of innovator-entrepreneurs at the BOP. *Joumal of Developmental Entrepreneurship* 18: 1–23.

Louridas P. (1999) Design as bricolage: Anthropology meets design thinking. *Design Studies* 20: 517–535.

Lusch RF and Nambisan S. (2015) Service innovation: A service-dominant logic perspective. *MIS Quarterly* 39: 155–175.

Luzzini D, Amann M, Caniato F, et al. (2015) The path of innovation: Purchasing and supplier involvement into new product development. *Industrial Marketing Management*.

Macfarlane K. (2006) An analysis of parental engagement in contemporary Queensland schooling. *Faculty of Education, Centre for Learning Innovation*. Queensland University of Technology.

MacKenzie D and Pardo-Guerra JP. (2014) Insurgent capitalism: Island, bricolage and the re-making of finance. *Economy and Society* 43: 153–183.

Maglio PP and Spohrer J. (2013) A service science perspective on business model innovation. *Industrial Marketing Management* 42: 665–670.

Manfield R and Newey L. (2015) Escaping the collapse trap: Remaining capable without capabilities. *Strategic Change* 24: 373–387.

Marwala T. (2014) *Artificial intelligence techniques for rational decision making*, Springer Cham Heidelberg New York Dordrecht London: Springer International Publishing Switzerland, ISBN 978-3-319-11423-1.

Marwala T. (2015) *Causality, correlation and artificial intelligence for rational decision making*, 5 Toh Tuck Link, Singapore 596224: World Scientific Publishing Co. Pte. Ltd, ISBN 978-9-81463-086-3.

McMillan K. (2015) The critical bricolage: Uniquely advancing organizational and nursing knowledge on the subject of rapid and continuous change in health care. *International Journal of Qualitative Methods* DOI: 10.1177/1609406915611550: 1–8.

Miettinen R and Virkkunen J. (2005) Epistemic objects, artefacts and organizational change. *Organization* 12: 437–456.

Mittelman JH. (2013) Global bricolage: Emerging market powers and polycentric governance. *Third World Quarterly* 34: 23–37.

Molecke G and Pinkse J. (2017) Accountability for social impact: A bricolage perspective on impact measurement in soical enterprises. *Journal of Business Venturing* 32: 550–568.

Mont O and Tukker A. (2006) Product-service systems: Reviewing achievements and refining the research agenda. *Journal of Cleaner Production* 14: 1451–1454.

Mont OK. (2002) Clarifying the concept of product–service system. *Journal of Cleaner Production* 10: 237–245.

Noble CH, Durmusoglu SS and Griffin A. (2014) *Open innovation: New product development essentials from the PDMA*, 111 River Street, Hoboken, NJ 07030: John Wiley & Sons, Inc., ISBN 978-1-118-77077-1.

Nyika L and Murray-Orr A. (2017) Critical race theory-social constructivist bricolage: A health-promoting schools research methodology. *Health Education Journal* 76: 432–441.

Orlikowski WJ. (2000) Using technology and constitution structures: A practice lens for studying technology in organizations. *Organization Science* 11: 404–428.

Peltone J. (2014) Strategic management of entrepreneurial firms during recession. *School of Science, Department of Industrial Engineering and Management.* Aalto University.

Perkmann M and Spicer A. (2014) How emerging organizations take form: The role of imprinting and values in organizational bricolage. *Organization Science* 25: 1785–1806.

Phillimore J, Humphries R, Klaas F, et al. (2016) Bricolage: Potential as a conceptual tool for understanding access to welfare in superdiverse neighbourhoods. *IRIS Working Paper Series, No. 14/2016.* University of Birmingham, 1–20.

Phillips N and Tracey P. (2007) Opportunity recognition, entrepreneurial capabilities and bricolage: Connecting institutional theory and entrepreneurship in strategic organization. *Strategic Organization* 5: 313–320.

Pialot O, Millet D and Tchertchian N. (2012) How to explore scenarios of multiple upgrade cycles for sustainable product innovation: The "Upgrade Cycle Explorer" tool. *Journal of Cleaner Production* 22: 19–31.

Plewe TC. (2008) Besting the tract home: A software-based bricolage approach to affordable custom housing. *Department of Architecture.* Massachusetts Institute of Technology.

Pressman S. (2006) *Fifty major economists,* 2 Park Square, Milton Park, Abingdon, Oxon OX14 4RN: Routledge, Taylor & Francis Group, ISBN 978-0-415-36648-9.

Ray G, Barney JB and Muhanna WA. (2004) Capabilities, business processes and competitive advantage: Choosing the dependent variable in empirical tests of the resource-based view. *Strategic Management Journal* 25: 23–37.

Renwick K. (2014) Bricolage and the health promoting school. *Qualitative Research Journal* 14: 318–332.

Rogers M. (2012) Contextualizing theories and practices of bricolage research. *The Qualitative Report* 17: 1–17.

Rönkkö MJ, Peltonen J and Arenius P. (2013) Selective or parallel? Toward measuring the domains of entrepreneurial bricolage. *Advances in Entrepreneurship, Firm Emergence and Growth* 15: 43–61.

Rosenberg N. (1973) Science, invention and economic growth. *Economic Journal* 84: 90–107.

Rosenberg N. (2004) Innovation and economic growth. OECD, retrieved from https://www.oecd.org/dfe/tourism/34267902.pdf, accessed on 25 December 2017.

Rüling C-C and Duymedjian R. (2014) Digital bricolage: Resources and coordination in the production of digital visual effects. *Technological Forecasting & Social Change* 83: 98–110.

Russell R and Tyler M. (2005) Branding and bricolage: Gender, consumption and transition. *Childhood* 12: 221–237.

Sabdia K. (2015) Effectuation as a construct for new business formation in South Africa. *The Gordon Institute of Business Science.* Pretoria, South Africa: University of Pretoria.

Salimath M and Jones RJ. (2011) Scientific entrepreneurial management: Bricolage, bootstrapping, and the quest for efficiencies. *Journal of Business and Management* 17: 85–104.

Salunke S, Weerawardena J and McColl-Kennedy JR. (2013) Competing through service innovation: The role of bricolage and entrepreneurship in project-oriented firms. *Journal of Business Research* 66: 1085–1097.

Sanchez-Burks J, Karlesky MJ and Lee F. (2015) Psychological bricolage: Integrating social identities to produce creative solutions. In: Shalley CE, Hitt MA and Zhou J (eds) *The Oxford Handbook of Creativity, Innovation and Entrepreneurship.* 198 Madison Avenue, New York, NY 10016: Oxford University Press, ISBN 978-0-19-992767-8, Chapter 6, pp. 93–102.

Schaefer CG. (2014) The impact of service complexity on new service development—a contingency approach. *Durham University Business School.* Durham: Durham University.

Schaefer DJ. (2013) Crowdsourcing as bricolage: A qualitative study of journalists enlisting and using crowdsourcing in social media. *Department of Communication.* University of Colorado at Boulder.

Schiele H. (2010) Early supplier integration: The dual role of purchasing in new product development. *R&D Management* 42: 138–153.

Schiller BR, Hill C and Wall S. (2013) *The macro economy today,* 1221 Avenue of the Americas, New York, NY, 10020: The McGraw-Hill Companies, Inc., ISBN 978-0-07-741647-8.

Schumpeter JA. (1934) *The theory of economic development: An inquiry into profits, capital, credit, interest, and the business cycle,* Cambridge, Massachusetts: Harvard University Press, ISBN 978-0-67487-990-4.

Schumpeter JA. (1950) *Capitalism, socialism and democracy,* New York: Harper.

Senyard J, Baker T and Davidsson P. (2009) Entrepreneurial bricolage: Towards systematic empirical testing. *Frontiers of Entrepreneurship Research* 29: 1–14.

Senyard J, Baker T, Steffens P, et al. (2014) Bricolage as a path to innovativeness for resource-constrained new Firms. *Journal of Product Innovation Management* 31: 211–230.

Senyard J. (2015) Bricolage and early stage firm performance. *Faculty of Business.* Australia: Queensland University of Technology.

Shah R, Gao Z and Mittal H. (2015) *Innovation, entrepreneurship and the economy in the US, China and India: Historical perspectives and future trends,* 32 Jamestown Road, London NW1 7BY, UK: Academic Press, Elsevier Inc., ISBN 978-0-12-801890-3.

Shalley CE, Hitt MA and Zhou J. (2015) *The Oxford handbook of creativity, innovation and entrepreneurship,* 198 Madison Avenue, New York, NY 10016: Oxford University Press, ISBN 978-0-19-992767-8.

Silva LBE, Valente LFO, Motta VB, et al. (2014) *Urban bricolage: Art of temporary and immaterial appropriation of the urban space in Viçosa, MG, retrieved from http://www.nomads.usp.br/virus/virus10/?sec=5#sect51, accessed on 19 August 2017.*

Slywotzky A, Wise R and Weber K. (2003) *How to grow when markets don't,* New York: Warner Business Books.

Snyder H, Witell L, Gustafsson A, et al. (2016) Identifying categories of service innovation: A review and synthesis of the literature. *Journal of Business Research* 69: 2401–2408.

Spanjol J. (2003) Idea generation in new product development—a cognitive framework of the fuzzy front end. *Business Administration.* Urbana, Illinois: University of Illinois at Urbana-Champaign.

Stanko MA, Fisher GJ and Bogers M. (2017) Under the wide umbrella of open innovation. *Journal of Product Innovation Management* 34: 543–558.

Stevenson L. (2013) Philosophical and historical foundations of the concept of innovation: Some implications for contemporary higher education as a service sector. *Australian Centre for Innovation, Faculty of Engineering and Information Technologies.* Sydney: University of Sydney.

Stinchfield BT, Nelson RE and Wood MS. (2013) Learning from Levi-Strauss' legacy: Art, craft, engineering, bricolage and brokerage in entrepreneurship. *Entrepreneurship Theory and Practice* July: 889–921.

Stokes PD. (2014) Crossing disciplines: A constraint-based model of the creative-innovative process. *Journal of Product Innovation Management* 31: 247–258.

Stürmer KB and Onland T. (2015) Towards a dynamic theory causation, effectualtion and bricolage: A study of the process of partnership selection for an open innovation community. *Department of Business Administration, School of Economics and Management.* Sweden: Lund University.

Sunduramurthy C, Zheng C, Musteen M, et al. (2016) Doing more with less, systematically: Bricolage and ingenieuring in successful social ventures. *Journal of World Business* 51: 855–870.

Tam PC. (2012) Children's bricolage under the gaze of teachers in sociodramatic play. *Childhood* 20: 244–259.

Turnbull S. (2002) Bricolage as an alternative approach to human resource development theory building. *Human Resource Development Review* 1: 111–128.

Valliere D and Gegenhuber T. (2014) Entrepreneurial remixing: Bricolage and postmodern resources. *Entrepreneurship and Innovation* 15: 5–15.

Vanevenhoven J, Winkel D, Malewicki D, et al. (2011) Varieties of bricolage and the process of entrepreneurship. *New England Journal of Entrepreneurship* 14: Article 7.

Vargo SL, Lusch RF and Akaka MA. (2010) Advancing service science with service-dominant logic: Clarifications and conceptual development. In: Maglio PP, Kieliszewski CA and Spohrer JC (eds) *Handbook of Service Science.* New York: Springer.

Vargo SL, Wieland H and Akaka MA. (2015) Innovation through institutionalization: A service ecosystems perspective. *Industrial Marketing Management* 44: 63–72.

Victorino L, Verma R, Plaschka G, et al. (2005) Service innovation and customer choices in the hospitality industry. *Managing Service Quality* 15: 555–576.

Visscher K, Heusinkveld S and O'Mahoney J. (in press) Bricolage and identity work. *British Journal of Management*: 1–17.

Von Krogh G and Spaeth S. (2007) The open source software phenomenon: Characteristics that promote research. *The Journal of Strategic Information Systems* 16: 236–253.

Von Krogh G, Haefliger S, Spaeth S, et al. (2012) Carrots and rainbows: Motivation and social practice in open source software development. *MIS Quarterly* 36: 649–676.

Vuletich C. (2015) *Sustainable textile design as bricolage, retrieved form http://mistrafuturefashion.com/wp-content/uploads/2015/12/D3.5h-Bricolage-as-metaphor-FINAL-May-2014.pdf, accessed on 19 August 2017.*

Warne T and McAndrew S. (2009) Constructing a bricolage of nursing research, education and practice. *Nurse Education Today* 29: 855–858.

Welter C, Mauer R and Wuerker RJ. (2016) Bridging behavioral models and theoretical concepts: Effectuation and bricolage in the opportunity creation framework. *Strategic Entrepreneurship Journal* 10: 5–20.

Welter CT. (2012) Distinguishing opportunity types: Why it matters and how to do it. *Graduate Program in Business Administration.* The Ohio State University.

West J and Gallagher S. (2006) Challenges of open innovation: The paradox of firm investment in open-source software. *R&D Management* 36: 319–331.

Wibberley C. (2012) Getting to grips with bricolage: A personal account. *The Qualitative Report* 17: 1–8.

Wilf E. (2015) Routinized business innovation: An undertheorized engine of cultural evolution. *American Anthropologist* 117: 679–692.

Winkel D, Vanevenhoven J, Yu A, et al. (2013) The invisible hand in entrepreneurial process: Bricolage in emerging economies. *International Journal of Entrepreneurship and Innovation Management* 17: 214–223.

Witell L, Snyder H, Gustafsson A, et al. (2016) Defining service innovation: A review and synthesis. *Journal of Business Research* 69: 2863–2872.

Witell L, Gebauer H, Jaakkola E, et al. (2017) A bricolage perspective on service innovation. *Journal of Business Research* 79: 290–298.

Wood NB, Erichsen EA and Anicha CL. (2013) Cultural emergence: Theorizing culture in and from the margins of science education. *Journal of Research in Science Teaching* 50: 122–136.

Wu L, Liu H and Zhang J. (2017) Bricolage effects on new-product development speed and creativity: The moderating role of technological turbulence. *Journal of Business Research* 70: 127–135.

Xing B, Gao W-J and Marwala T. (2013) An overview of cuckoo-inspired intelligent algorithms and their applications. *IEEE Symposium Series on Computational Intelligence (IEEE SSCI), 15–19 April, Singapore*, pp. 85–89. IEEE.

Xing B. (2014a) Novel computational intelligence for optimizing cyber physical pre-evaluation system. In: Khan ZH, Ali ABMS and Riaz Z (eds) *Computational Intelligence for Decision Support in Cyber-Physical Systems.* Singapore Heidelberg New York Dordrecht London: Springer Science+Business Media Singapore, ISBN 978-981-4585-35-4, Chapter 15, pp. 449–464.

Xing B. (2014b) The optimization of computational stock market model based complex adaptive cyber physical logistics system. In: Khan ZH, Ali ABMS and Riaz Z (eds) *Computational Intelligence for Decision Support in Cyber-Physical Systems.* Singapore Heidelberg New York Dordrecht London: Springer Science+Business Media Singapore, ISBN 978-981-4585-35-4, Chapter 12, pp. 357–380.

Xing B and Gao W-J. (2014aa) Chemical-reaction optimization algorithm. In: Xing B and Gao W-J (eds) *Innovative Computational Intelligence: A Rough Guide to 134 Clever Algorithms.* Cham Heidelberg New York Dordrecht London: Springer International Publishing Switzerland, ISBN: 978-3-319-03403-4, Chapter 25, pp. 417–428.

Xing B and Gao W-J. (2014b) *Innovative computational intelligence: A rough guide to 134 clever algorithms,* Cham Heidelberg New York Dordrecht London: Springer International Publishing Switzerland, ISBN: 978-3-319-03403-4.

Xing B and Gao W-J. (2014bb) Emerging chemistry-based CI algorithms. In: Xing B and Gao W-J (eds) *Innovative Computational Intelligence: A Rough Guide to 134 Clever Algorithms.* Cham Heidelberg New York Dordrecht London: Springer International Publishing Switzerland, ISBN: 978-3-319-03403-4, Chapter 26, pp. 429–437.

Xing B and Gao W-J. (2014c) Introduction to computational intelligence. In: Xing B and Gao W-J (eds) *Innovative Computational Intelligence: A Rough Guide to 134 Clever Algorithms.* Cham Heidelberg New York Dordrecht London: Springer International Publishing Switzerland, ISBN: 978-3-319-03403-4, Chapter 1, pp. 3–17.

Xing B and Gao W-J. (2014cc) Base optimization algorithm. In: Xing B and Gao W-J (eds) *Innovative Computational Intelligence: A Rough Guide to 134 Clever Algorithms.* Cham Heidelberg New York Dordrecht London: Springer International Publishing Switzerland, ISBN: 978-3-319-03403-4, Chapter 27, pp. 441–444.

Xing B and Gao W-J. (2014d) Bacteria inspired algorithms. In: Xing B and Gao W-J (eds) *Innovative Computational Intelligence: A Rough Guide to 134 Clever Algorithms.* Cham Heidelberg New York Dordrecht London: Springer International Publishing Switzerland, ISBN: 978-3-319-03403-4, Chapter 2, pp. 21–38.

Xing B and Gao W-J. (2014dd) Emerging mathematics-based CI algorithms. In: Xing B and Gao W-J (eds) *Innovative Computational Intelligence: A Rough Guide to 134 Clever Algorithms.* Cham Heidelberg New York Dordrecht London: Springer International Publishing Switzerland, ISBN: 978-3-319-03403-4, Chapter 28, pp. 445–448.

Xing B and Gao W-J. (2014e) Bee inspired algorithms. In: Xing B and Gao W-J (eds) *Innovative Computational Intelligence: A Rough Guide to 134 Clever Algorithms.* Cham Heidelberg New York Dordrecht London: Springer International Publishing Switzerland, ISBN: 978-3-319-03403-4, Chapter 4, pp. 45–80.

Xing B and Gao W-J. (2014f) Bat-inspired algorithms. In: Xing B and Gao W-J (eds) *Innovative Computational Intelligence: A Rough Guide to 134 Clever Algorithms.* Cham Heidelberg New York Dordrecht London: Springer International Publishing Switzerland, ISBN: 978-3-319-03403-4, Chapter 3, pp. 39–44.

Xing B and Gao W-J. (2014ff) Overview of computational intelligence. In: Xing B and Gao W-J (eds) *Computational Intelligence in Remanufacturing.* 701 E. Chocolate Avenue, Suite 200, Hershey PA 17033: IGI Global, ISBN 978-1-4666-4908-8, Chapter 2, pp. 18–36.

Xing B and Gao W-J. (2014g) Biogeography-based optimization algorithm. In: Xing B and Gao W-J (eds) *Innovative Computational Intelligence: A Rough Guide to 134 Clever Algorithms.* Cham Heidelberg New York Dordrecht London: Springer International Publishing Switzerland, ISBN: 978-3-319-03403-4, Chapter 5, pp. 81–91.

Xing B and Gao W-J. (2014h) Cat swarm optimization algorithm. In: Xing B and Gao W-J (eds) *Innovative Computational Intelligence: A Rough Guide to 134 Clever Algorithms.* Cham Heidelberg New York Dordrecht London: Springer International Publishing Switzerland, ISBN: 978-3-319-03403-4, Chapter 6, pp. 93–104.

Xing B and Gao W-J. (2014i) Cuckoo inspired algorithms. In: Xing B and Gao W-J (eds) *Innovative Computational Intelligence: A Rough Guide to 134 Clever Algorithms.* Cham Heidelberg New York Dordrecht London: Springer International Publishing Switzerland, ISBN: 978-3-319-03403-4, Chapter 7, pp. 105–121.

Xing B and Gao W-J. (2014j) Luminous insect inspired algorithms. In: Xing B and Gao W-J (eds) *Innovative Computational Intelligence: A Rough Guide to 134 Clever Algorithms.* Cham Heidelberg New York Dordrecht London: Springer International Publishing Switzerland, ISBN: 978-3-319-03403-4, Chapter 8, pp. 123–137.

Xing B and Gao W-J. (2014k) Fish inspired algorithms. In: Xing B and Gao W-J (eds) *Innovative Computational Intelligence: A Rough Guide to 134 Clever Algorithms.* Cham Heidelberg New York Dordrecht London: Springer International Publishing Switzerland, ISBN: 978-3-319-03403-4, Chapter 9, pp. 139–155.

Xing B and Gao W-J. (2014l) Frog inspired algorithms. In: Xing B and Gao W-J (eds) *Innovative Computational Intelligence: A Rough Guide to 134 Clever Algorithms.* Cham Heidelberg New York Dordrecht London: Springer International Publishing Switzerland, ISBN: 978-3-319-03403-4, Chapter 10, pp. 157–165.

Xing B and Gao W-J. (2014m) Fruit fly optimization algorithm. In: Xing B and Gao W-J (eds) *Innovative Computational Intelligence: A Rough Guide to 134 Clever Algorithms.* Cham Heidelberg New York Dordrecht London: Springer International Publishing Switzerland, ISBN: 978-3-319-03403-4, Chapter 11, pp. 167–170.

Xing B and Gao W-J. (2014n) Group search optimization algorithm. In: Xing B and Gao W-J (eds) *Innovative Computational Intelligence: A Rough Guide to 134 Clever Algorithms.* Cham Heidelberg New York Dordrecht London: Springer International Publishing Switzerland, ISBN: 978-3-319-03403-4, Chapter 12, pp. 171–176.

Xing B and Gao W-J. (2014o) Invasive weed optimization algorithm. In: Xing B and Gao W-J (eds) *Innovative Computational Intelligence: A Rough Guide to 134 Clever Algorithms.* Cham Heidelberg New York Dordrecht London: Springer International Publishing Switzerland, ISBN: 978-3-319-03403-4, Chapter 13, pp. 177–181.

Xing B and Gao W-J. (2014p) Music inspired algorithms. In: Xing B and Gao W-J (eds) *Innovative Computational Intelligence: A Rough Guide to 134 Clever Algorithms.* Cham Heidelberg New York Dordrecht London: Springer International Publishing Switzerland, ISBN: 978-3-319-03403-4, Chapter 14, pp. 183–201.

Xing B and Gao W-J. (2014q) Imperialist competitive algorithm. In: Xing B and Gao W-J (eds) *Innovative Computational Intelligence: A Rough Guide to 134 Clever Algorithms.* Cham Heidelberg New York Dordrecht London: Springer International Publishing Switzerland, ISBN: 978-3-319-03403-4, Chapter 15, pp. 203–209.

Xing B and Gao W-J. (2014r) Teaching–learning-based optimization algorithm. In: Xing B and Gao W-J (eds) *Innovative Computational Intelligence: A Rough Guide to 134 Clever Algorithms.* Cham Heidelberg New York Dordrecht London: Springer International Publishing Switzerland, ISBN: 978-3-319-03403-4, Chapter 16, pp. 211–216.

Xing B and Gao W-J. (2014s) Emerging biology-based CI algorithms. In: Xing B and Gao W-J (eds) *Innovative Computational Intelligence: A Rough Guide to 134 Clever Algorithms.* Cham Heidelberg New York Dordrecht London: Springer International Publishing Switzerland, ISBN: 978-3-319-03403-4, Chapter 17, pp. 217–317.

Xing B and Gao W-J. (2014t) Big bang–big crunch algorithm. In: Xing B and Gao W-J (eds) *Innovative Computational Intelligence: A Rough Guide to 134 Clever Algorithms.* Cham Heidelberg New York Dordrecht London: Springer International Publishing Switzerland, ISBN: 978-3-319-03403-4, Chapter 18, pp. 321–331.

Xing B and Gao W-J. (2014u) Central force optimization algorithm. In: Xing B and Gao W-J (eds) *Innovative Computational Intelligence: A Rough Guide to 134 Clever Algorithms.* Cham Heidelberg New York Dordrecht London: Springer International Publishing Switzerland, ISBN: 978-3-319-03403-4, Chapter 19, pp. 333–337.

Xing B and Gao W-J. (2014v) Charged system search algorithm. In: Xing B and Gao W-J (eds) *Innovative Computational Intelligence: A Rough Guide to 134 Clever Algorithms.* Cham Heidelberg New York Dordrecht London: Springer International Publishing Switzerland, ISBN: 978-3-319-03403-4, Chapter 20, pp. 339–346.

Xing B and Gao W-J. (2014w) Electromagnetism-like mechanism algorithm. In: Xing B and Gao W-J (eds) *Innovative Computational Intelligence: A Rough Guide to 134 Clever*

Algorithms. Cham Heidelberg New York Dordrecht London: Springer International Publishing Switzerland, ISBN: 978-3-319-03403-4, Chapter 21, pp. 347–354.

Xing B and Gao W-J. (2014x) Gravitational search algorithm. In: Xing B and Gao W-J (eds) *Innovative Computational Intelligence: A Rough Guide to 134 Clever Algorithms.* Cham Heidelberg New York Dordrecht London: Springer International Publishing Switzerland, ISBN: 978-3-319-03403-4, Chapter 22, pp. 355–364.

Xing B and Gao W-J. (2014y) Intelligent water drops algorithm. In: Xing B and Gao W-J (eds) *Innovative Computational Intelligence: A Rough Guide to 134 Clever Algorithms.* Cham Heidelberg New York Dordrecht London: Springer International Publishing Switzerland, ISBN: 978-3-319-03403-4, Chapter 23, pp. 365–373.

Xing B and Gao W-J. (2014z) Emerging physics-based CI algorithms. In: Xing B and Gao W-J (eds) *Innovative Computational Intelligence: A Rough Guide to 134 Clever Algorithms.* Cham Heidelberg New York Dordrecht London: Springer International Publishing Switzerland, ISBN: 978-3-319-03403-4, Chapter 24, pp. 375–414.

Xing B. (2015) Optimization in production management: Economic load dispatch of cyber physical power system using artificial bee colony. In: Kahraman C and Onar SÇ (eds) *Intelligent Techniques in Engineering Management: Theory and Applications.* Cham Heidelberg New York Dordrecht London: Springer International Publishing Switzerland, ISBN 978-3-319-17905-6, Chapter 12, 275–293.

Xing B and Marwala T. (2017) Implications of the fourth industrial age for higher education. *The Thinker: For the Thought Leaders (www.thethinker.co.za).* South Africa: Vusizwe Media, 10–15.

Xing B and Marwala T. (2018) Introduction to intelligent search algorithms. In: Xing B and Marwala T (eds) *Smart Maintenance for Human–Robot Interaction: An Intelligent Search Algorithmic Perspective.* Gewerbestrasse 11, 6330 Cham, Switzerland: Springer International Publishing AG, ISBN 978-3-319-67479-7, Chapter 3, pp. 33–64.

Zott C and Amit R. (2008) The fit between product market strategy and business model: Implications for firm performance. *Strategic Management Journal* 29: 1–26.

CHAPTER 3

Introduction to Service

3.1 Service and Economy

Before we actually talk about service, we have to briefly explore its macro background—the economy. In essence, the economy is about us (Marwala and Hurwitz, 2017a; Marwala and Hurwitz, 2017b; Marwala, 2013). In other words, the economy is an abstracted concept, which refers to the sum of all human production and consumption activities. What a nation's economy produces is the sum of what everyone produces, while what a nation's economy consumes is the sum of what everyone consumes, as illustrated in Eqs. 3.1 and 3.2 (Schiller et al., 2013).

$$X_{\text{an economy produces}} = \sum_{i=1}^{n} x_{\text{people}_i \text{ produces}} . \quad 3.1$$

$$C_{\text{an economy consumes}} = \sum_{i=1}^{n} c_{\text{people}_i \text{ consumes}} . \quad 3.2$$

It would be nice if we would produce everything that we want to consume. However, the harsh reality is that such utopia does not exist because we live in an environment with limited resources. Such scarcity, the shortage of enough resources to meet our insatiable desires, forces every country to contemplate the following three fundamental questions (Schiller et al., 2013):

- **WHAT:** The list of things (goods or services in general terms) that we decide to produce with our restricted resources.
- **HOW:** The means that we choose to produce our selected list of things, i.e., goods or services.
- **WHO:** The beneficiaries that we intend to cover with the produced goods or services.

3.1.1 Factors of Production

Generally speaking, the limited resources that we use to produce things (goods or services) are termed as factors of production. In the real world, we typically have the following four basic factors (Schiller et al., 2013):

- **Land Resources:** Generally speaking, any resources created by nature are involved. Typically, it includes arable lands, minerals, water, oil, and gas reserves.
- **Human Resources:** Human resources have two dimensions, namely quantitative human resources (e.g., the number of human populations) and qualitative human resources (e.g., talents or skills).
- **Capital Resources:** From an economist's point of view, capital refers not just to the manmade equipment (e.g., machineries) but also structures (e.g., buildings, roads, computers, and vehicles) that workers can use to produce things.
- **Entrepreneurship:** Essentially, it is a special type of human resources that drives innovation (Schumpeter, 1950). Broadly speaking, entrepreneurship means the efforts and skills of a person to put all other resources together in a productive venture, such as to produce new or improved products and technologies.

3.1.2 Goods vs. Services

Over multiple decades, macroeconomics has extensively analysed the determinants of the scale of aggregate consumption (i.e., goods or services) (Blanchard, 2017). In a typical market economy, the individual's needs and desires are determining factors regarding which goods are developed and finally produced. As income rises, the consumer moves from "elementary" goods (e.g., water, food, basic clothing and housing) to more "advanced" services, such as culture, healthcare, education or travel. Indeed, this chain reaction is nothing more than the well-known law of consumer's sovereignty, as given by Eq. 3.3 (Wetzstein, 2013):

$$\text{Needs} \rightarrow \text{Desires} \rightarrow \text{Demands} \rightarrow \text{Services}. \tag{3.3}$$

3.1.2.1 Tangible Goods

In economics, the term "tangible goods" normally designates materials that satisfy human wants. Usually, they consist of a foundation (i.e., private goods) and a superstructure (i.e., public goods) (Geuss, 2001).

- **Foundation:** The foundation includes free goods (e.g., air and water), consumable goods (e.g., smartphones and TVs), and commercial goods (e.g., airplanes and tractors).

- **Superstructure:** The superstructures are consumed collectively, these include national defence and roads. Typically, there are two characteristics that are used to define public goods, i.e., non-excludable and non-rivalrous (Kallhoff, 2014).
  1) *Non-excludable*: This feature indicates that nobody can be excluded from the consumption of public goods;
  2) *Non-rivalrous*: This attribute denotes that one individual's consumption of the benefits (brought by public goods) does not deprive another individual's right to use the same public goods.

3.1.2.2 Intangible Services

In contrast to tangible goods, service is often perceived via its intangibility, which leads to diverse meanings across different areas and scenarios. As suggested by Morris and Johnston (1987), for the purpose of deciphering and capturing the intrinsic nature of service in businesses, one have to understand and address its unique variability. In practice, service is often actualized through a set of beneficial activities, such as basic trading services (e.g., shopping malls), food services (e.g., restaurants), recommendatory services (e.g., travel agencies), financial services (e.g., banks), and public services (e.g., transportation).

According to (Qiu, 2014), regardless of the provision and consumption mode, a service is typically characterized by the following five fundamental and key elements:

- **Resources:** Usually, the resources can be categorized as physical resource (e.g., consumption goods), soft resource (e.g., brands), or hybrid resource (e.g., healthcare).
- **Suppliers (or Service Providers):** This element refers to an entity (e.g., an individual, a group, an organization, or governmental agencies) that interacts with clients in order to deliver the desired service.
- **Clients:** Clients are those individuals/organizations that consume, acquire, or utilize a service offered and delivered by a service supplier.
- **Benefits:** The benefits in the service domain generally consists of two classes: (1) value-based benefits (e.g., profit), and (2) need-based benefits (e.g., desire and satisfaction).
- **Time:** Through the lens of time, the service interactions can occur in various time-frames, e.g., an *ad hoc* (e.g., well-controlled interactions), an encounter (e.g., once-off interactions), or a continuum (e.g., clients and service providers repeat interactions continuously).

3.1.2.3 Goods Servitization

In the community of economics research, it is now commonly believed that a good (or a product) should be treated as the conjunction of three feature sets, namely technological features (denoted by [*T*]), service features (denoted by [*S*]), and process features (denoted by [*Z*]) (Gallouj, 2002b).

- **Technological Features** ([*T*])**:** The internal configuration of the technology is often described by [*T*], which indicates the scientific and applied knowledge embedded in myriads of goods in order to offer the desired functionalities (including service function). As mentioned in (Saviotti, 1996), [*T*] is the only feature set that can be directly altered by goods producers. For instance, an in-pipe inspection robot manufacturer can only design and produce the robot based on a set of identified requirements, including purpose, function, behaviour, structure, environment, constraint, working-space, and design descriptions (Xing, 2016b; Xing and Marwala, 2018e), and not directly produce the maintenance of the pipeline, which is rather a service provision (typically provided by a maintenance company) than otherwise (e.g., a manufactured good). Thus, an in-pipe inspection robot's technological features include its motor, wheel, vision system, gear, power supply, control system, and many others (Xing, 2015); while a complex human-robot interactive system's technical features may broadly include its hardware-, cyberware-, and human capacity (Xing and Marwala, 2018m; Xing and Marwala, 2018k; Xing and Marwala, 2018l; Xing and Marwala, 2018e; Xing and Marwala, 2018g; Xing and Marwala, 2018f; Xing and Marwala, 2018d; Xing and Marwala, 2018b; Xing and Marwala, 2018c; Xing and Marwala, 2018j; Xing and Marwala, 2018h; Xing and Marwala, 2018i; Xing and Marwala, 2018a).
- **Service Features** ([*S*])**:** In contrast to [*T*], service features are often perceived from the end users' viewpoint, following a Lancasterian tradition (Lancaster, 1966), as the final or utilization characteristics. In general terms, [*S*] includes various services that are delivered to customers and are facilitated via the good under consideration. For instance, an ambient intelligence scenario's service features may include the facilitation of various interactions, among smart objects, between objects and the surrounding smart environment, and between people and the resident smart objects (Xing and Marwala, 2018i). The notion of service features can be expanded further via hierarchization, which includes primary features, secondary features, and externalities (some unwanted features associated with the good). An interesting

example is a smartphone whose primary function to communicate, the secondary functions can cover entertainment, mobile payments, and so on, while the externalities may include environmental malignance (Xing and Gao, 2014a; Xing and Gao, 2014ee; Xing and Gao, 2014gg; Xing and Gao, 2014hh; Xing and Gao, 2014ii; Xing and Gao, 2014jj; Xing and Gao, 2014kk; Xing and Gao, 2014ll; Xing and Gao, 2014mm; Xing and Gao, 2014nn; Xing and Gao, 2014oo; Xing and Gao, 2014pp; Xing and Gao, 2014qq; Xing and Gao, 2014rr).

- **Process Features** ([Z]): The collection of employed methodologies and deployed organizational modes utilized to manufacture the good is often covered by the process features. More often than not, a clear cut between the product and the corresponding processing techniques is hard to achieve; on the contrary, one tends to find a reasonable fusion in many cases.

As a result, the notion of a product can be expressed by using Eq. 3.4 (Gallouj, 2002b; Saviotti and Metcalfe, 1984).

$$[\text{Product}] = \underbrace{\left\{ \begin{bmatrix} T_1 \\ T_2 \\ \vdots \\ T_j \\ \vdots \\ T_n \end{bmatrix} \oplus \begin{bmatrix} Z_1 \\ Z_2 \\ \vdots \\ Z_j \\ \vdots \\ Z_n \end{bmatrix} \right\}}_{\uparrow \text{ Term 1}} \Leftrightarrow \underbrace{\begin{bmatrix} S_1 \\ S_2 \\ \vdots \\ S_i \\ \vdots \\ S_n \end{bmatrix}}_{\uparrow \text{ Term 2}} . \qquad 3.4$$

where 'Term 1' denotes the fusion of technological and process features, in which $\oplus$ indicates the inseparability; 'Term 2' refers to the service features and the sign $\Leftrightarrow$ stands for a trend of moving from product to service.

Nowadays, servitization has been broadly adopted as a strategy (Lin et al., 2012) of harvesting more profits (Neely, 2008; Malleret, 2006) and competitive advantages via the addition of service offerings (Vandermerwe and Rada, 1988; Baines et al., 2009b) instead of product offerings alone (Schmenner, 2009). The definition of servitization (servitisation or servicisation) varies from different perspectives. Several streams of definitions are provided as follows (Lin et al., 2012):

- **Stream 1–Inception:** Servitization occurs in almost all types of industries throughout the world. Forced by various pressures (e.g., deregulation, technology advancement, globalization, and fierce competition), both goods producers and the incumbent service firms are rushing towards services (Vandermerwe and Rada, 1988). This mostly cited definition drafts the very beginning of servitization's premises.

- **Stream 2–Groundwork Built on Conventional Manufacturing and Operations Management Theories:** Servitization can be regarded as an idea that goes beyond the conventional notion of offering extra services. It takes the bundled offer (the integration of the services and the goods) into consideration (Vandermerwe and Rada, 1988; Robinson et al., 2002), while in (Slack, 2005), the authors believed that, though somewhat undervalued, servitization is a general term that incorporates any strategic attempt at seeking to modify the means of how a product's functionality is delivered to the target markets.
- **Stream 3–Emergent Solution Focusing on Value Creation and Competitive Advantages Obtainment:** In (Åhlström and Nordin, 2006), the authors defined servitization as tactical trials conducted by a manufacturing firm for the purpose of establishing service supply relationships. This is done via introducing additional service offerings in order to augment the original physical product offerings, and thus differentiating the firm itself from the other competitors through the realization of high quality product services. Meanwhile, servitization is defined by Lindberg and Nordin (2008) as a trend that transforms firms from not only producing goods but to supplying services or a goods and services integrated solution. In a similar vein, Neely (2008) regarded servitization as the movement that goods manufacturers take to move beyond their conventional manufacturing confinement and begin to offer service solutions, either via their produced goods or in association with them.
- **Stream 4–Bundled Approach Stressing the Fusion of Goods and Services:** This stream of definitions emphasizes the importance of goods and services combination or integration. Based on this definition, it is not necessary to transform from an absolute product producer to a pure service provider (Pawar et al., 2009; Brax, 2005).
- **Stream 5–Innovative Perspective Emphasizing the Importance of Innovation:** In this stream, innovative service offerings are regarded as the essence in implementing a servitization strategy. Accordingly, defining servitization from the innovation angle (Schmenner, 2009) is a common practice in the literature, e.g., product-service systems (Baines et al., 2007), and product-centric servitization (Baines et al., 2009a).
- **Stream 6–Holistic Value Chain Point of View:** Instead of only targeting individual producers or service providers, this stream of definitions

treats servitization from the entire supply chain perspective (Johnson and Mena, 2008; Lewis and Howard, 2009). Under the umbrella of these definitions, the scope of servitization is extended from purely internal to external as well.

In summation, one can define servitization by incorporating two levels of factors: (1) Strategic level focusing more on innovation, trend, and strategy, (2) Operational level emphasizing the process itself. Generally speaking, as an innovative methodology, servitization is able to generate additional values and realize many competitive advantages. Among all these definitions, one has to remember that different firms may put differentiated levels of attentions on goods and services according to their specific needs.

3.1.3 *Service Economy*

Fundamental alterations in the architecture of "hierarchy of needs" have led to a significant reallocation of all economic activities. Given the rich empirical literature on the service, one can find a new economy transformation is undergoing, i.e., service economy (Gallouj, 2002b; Selya, 1994; Cioban, 2014). In such a situation, where productivity is no longer a main issue, purchasing a service and listening to one of Beethoven's symphonies seem to highly resemble one another (Sundbo, 2002). Take "Ode to Joy" (Jarrett and Day, 2008), Beethoven painted us a landscape with varying rhythms that surprise the listener by veering off unexpectedly or changing in an interesting manner. When the twelfth measure nearly reached its end (through a part of mainly straight quarter notes), Beethoven astonished his listeners by introducing an unexpectedly strong accent on "beat four". Although "beat four" usually serves as a weak pulse (Horvit et al., 2013), the German talent strengthened it by setting out a phrase there, and made it even stronger for the remaining measure by binding the note with "beat one" which is otherwise used for a strong beat. The resultant composition made the listeners feel like they were strolling along and suddenly slipped on a patch of ice, yet this was handled by balancing and carrying on an uninterrupted stride (Jarrett and Day, 2008). Can you imagine the point of requesting for an increased productivity from an orchestra when playing this symphony? An immediate implication from this example lies in the fact that the market in the service economy is no more driven by quantities and prices, but mainly a matter of quality and expectations. As a result, service provisions have always been striving to survive or thrive in a spot of customization (i.e., tailor-made) and standardization (i.e., mass-produced). Though a bit of a contradiction, and commonly under debate, we briefly outline both logics as follows (Sundbo, 2002):

- **Economics of Productivity via Standardization:** Under the classical economic logic, standardization implies the economics of productivity. By following this logic, a producer cares only about prices and quantities. The consumers are capable of assessing the quality and comparing prices of a product. Meanwhile, it is also assumed that customers themselves have the necessary knowledge in terms of classifying a product as service or good. This school of thinking focuses mainly on increasing productivity, which in turn can reduce costs and, thus, lower prices. Under this context, the competitive advantages of a firm increase as the prices fall, which can ultimately lead to more profit via rapid turnover.
  1) The affordable product price can keep customers satisfied;
  2) Quality of a product is high due to its well-refined production process;
  3) Production costs are steerable;
  4) High productivity; and
  5) It is possible to systematize and reproduce innovation, which can lead to a higher return on investment.
- **Economics of Expectations via Customization:** In another economic logic (i.e., service-oriented management and marketing), customization dominates. Since a service product is typically non-storable, it must consume at its production moment, in which consumers tend towards being co-producers. As we can see, the focus of this logic is on customization, which stresses resolving an individual customer's particular problem. In general, the benefits for a service firm to follow customization strategy include (Sundbo, 2002):
  1) Satisfying a customer's wants and, thus, the corresponding contentment degree is lifted;
  2) The service quality can in turn elevate a product's overall perceived quality;
  3) It becomes possible to charge an individualized price (tend to be higher than mass produced counterpart); and
  4) Due to the closeness with the end users, innovation can better address a customer's needs and wants, which can be a great guarantee for a product's success in the marketplace.

From the macroeconomics point of view, the service sector may include communication, transportation, retailing, accommodations, government administration, education, healthcare, community and finance, to name just few. Due to the existence of multiple business forms within service sector, dissimilar resources are normally utilized to deliver services, which in turn generate distinct supplier-customer interactions. Since our perceptions of services are often skewed by these mixtures, the following settings are briefly explained in order to illustrate the diverse meaning of service.

3.1.3.1 Computing-as-a-Service (CaaS)

The fundamental concept of supplying computing resources as a utility is not unfamiliar (Chee and Franklin, 2010; Marinescu, 2013; Srinivasan, 2014) and can be traced back to the mainframe era (Buyya et al., 2009; Pallis, 2010; Zhang et al., 2010; Kleinrock, 2005). As early as 1961, a scene of "*organizing computation somehow in someday as a public utility*" was envisioned by John McCarthy (Pallis, 2010). Later on in 1969, Leonard Kleinrock (the main developer of a miniature version of today's Internet) made a similar assertion (Kleinrock, 2005) about "*...with computer networks becoming technologically matured and sophisticated, we will most likely encounter a new utility—computer utility. It just resembles other commonly used electricity, gas, water, and telephone utilities and will serve individuals and businesses across the nation.*" An abridged evolution roadmap of computing-as-a-service (CaaS) (or more commonly cloud computing) paradigm is depicted in Fig. 3.1.

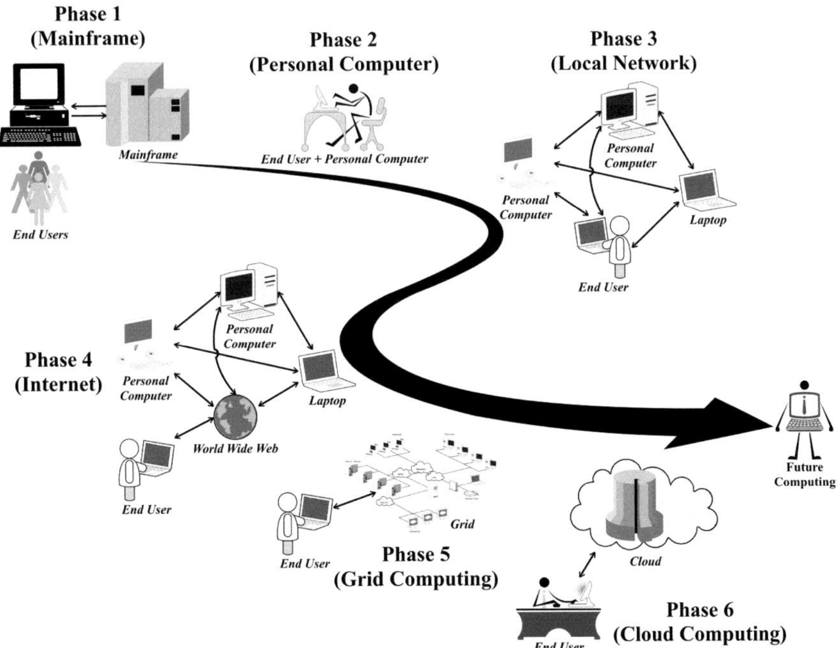

Figure 3.1: Evolution of CaaS.

As illustrated in Fig. 3.1, the evolution of CaaS can be roughly classified into six distinct phases (De, 2016; Voas and Zhang, 2009):

- **Phase I:** The inception of the CaaS concept can be dated back to the 1950s when terminals were introduced in order to connect powerful mainframes and enable sharing among users.
- **Phase II:** During this phase, the Personal Computer (PC) began to penetrate into people's daily life, and the need to share the use of mainframes became pointless.
- **Phase III:** At the third phase, computer networks emerged as a new form of computing paradigm, in which working on one's own PC and sharing resources with other PCs (within the same local network) became a reality.
- **Phase IV:** The fourth phase is dedicated to the Internet era when various local networks could be connected together to form a global network. During this phase, remote applications and resources were accessible via the Internet.
- **Phase V:** The fifth phase witnessed the appearance of the grid computing, which represents collecting computer resources from various locations to achieve a common goal.
- **Phase VI:** As the evolutionary product, cloud computing became available in the sixth phase. In general, a cloud only offers a standard user interface while keeping all other infrastructural resources (e.g., hardware, software, and services) hidden.

The key advantage for customers to adopt the CaaS plan lies in the fact that they do not need to worry about maintaining the complicated onsite computing infrastructure, and can simply enjoy the benefits of pay-per-usage (Holloway, 2017); while on the service provider's side, the advantages include massive scalability, service levels differentiability, economy-of-scale feasibility, dynamic service reconfigurability, and on-demand service deliverability (Foster et al., 2008).

In practice, there are several inter-related technologies; cloud computing, grid computing, utility computing, autonomic computing and network virtualization all serve as the building blocks for realizing the CaaS scheme (De, 2016; Badger et al., 2012; Zhang et al., 2010). Take cloud computing, which works by leveraging virtualization techniques and sharing some merits of autonomic and grid computing; it is often regarded as the next generation of the Internet (De, 2016). The cloud offers a medium through which myriads of things, ranging from varying computing power, complicated computing infrastructure, various business processes, and numerous applications to all sorts of individual interactions, could be provided to consumers as a service whenever requested (Hurwitz et al., 2010; Hurwitz et al., 2012). In general, the service models of cloud computing can be further classified into the following layers (Holloway, 2017; De, 2016):

- **Software-as-a-Service (SaaS):** In SaaS layer, end users are supplied with a set of applications that run on a service provider's infrastructure. These application programmes are, thus, accessible via either a web browser-like thin client interface, or specific program interfaces.
- **Platform-as-a-Service (PaaS):** The PaaS layer offers all types of software development components such as designing, debugging, testing, and deployment tools. Therefore, software designers, debuggers, testers, and developers form the main consumer group of PaaS.
- **Infrastructure-as-a-Service (IaaS):** The IaaS layer is responsible for supplying fundamental computing resources (e.g., processing power, networking capacity, and data warehousing) that end users require in order to complete tasks such as operating systems deployment and applications implement. The IaaS service model enables customers to jump-start a new project via renting computing assets.

These three types of services form a layered structure in a cloud construction as depicted in Fig. 3.2.

As we can see from Fig. 3.2, service layers are positioned on top of two other layers, namely virtual machine management and physical hardware. In principle, when customers move from the IaaS layer through PaaS layer to SaaS layer, the controllability is gradually lost and the manageability is increased.

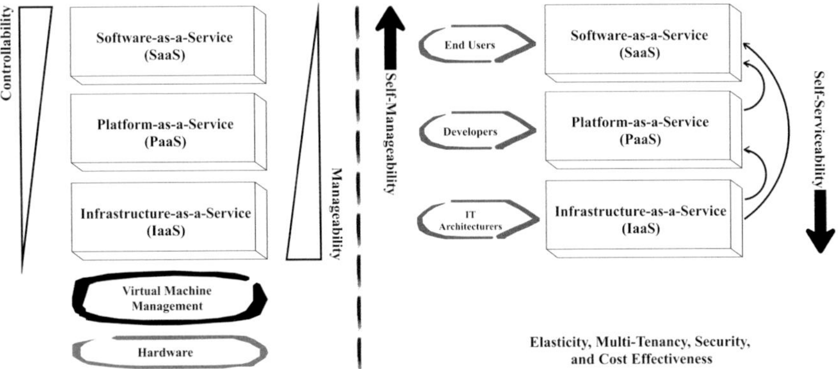

Figure 3.2: Layers of service models and stacks of cloud architecture.

3.1.3.2 X-as-a-Service (XaaS)

In addition to CaaS, we tend to have various X-as-a-service (or XaaS for short) paradigms such as:

- **Data-as-a-Service (DaaS):** In DaaS scenario, a service is often supplied to accredited service users in the data form via a distantly accessible and self-sustained module that can facilitate customers in carrying out their desirable activities. Service purchasers can, thus, acquire the relevant service in a standardized means that is well filed and outlined in a data repository. The tabulated record can help customers determine the availability of a service and its associated functionality (Sarkar, 2015).
- **Education-as-a-Service (EaaS):** Education can be narrowly defined as an instructive process in which a human's cognitive capabilities are enlightened and developed in order to improve competency over a given period (Qiu, 2014). In a typical EaaS scenario, instructors and learners can be regarded as educational service providers and customers, respectively (Xing and Marwala, 2017). A series of teaching and learning activities are carried out between these two parties and are supported by involving administrators and other supporting staff. The institution supplies a range of service products in the form of programmes and courses. The learning results (i.e., service outcomes) could be assessed via different benchmarks.

Overall, agents in a society can benefit from employing a XaaS-like strategy in at least the following five perspectives (Schneider, 2017; FTC, 2016):

- **First**, XaaS gives agents the chance of using capacities that are currently in excess and idled condition. This in turn enables a more productive utilization of the underused or "dead" capital.
- **Second**, by gathering various service providers and purchasers, XaaS intensifies the competition between the sides of demand and supply of a market process and, thus, nurtures better specialization.
- **Third**, by lowering the overheads (e.g., the costs of consumer-supplier matching, the hassles of cutback haggling, and the dilemma of performance monitoring), XaaS reduces transaction charges and expands the scope of businesses.
- **Fourth**, by collecting past consumers' reviews and making them available in front of market novices, XaaS can dramatically mitigate the issue of information asymmetry between customers and suppliers.
- **Fifth**, by opening and maintaining alternative channels, XaaS delivers services to those underserved customers under the existing inefficient and unresponsive incumbent-dominant environment.

3.2 What is Service Science?

According to Qiu (2014), as a meta-science of services, findings from various traditional disciplines must be combined in order to understand the science of service. In this section, we briefly touch on the fundamentals of service science and the associated general laws of service proposed by Qiu (2014).

3.2.1 *Service Science: Newtonian Perspective*

In general, the quality of service is dependent on the efficiency of sociotechnical systems and perceived effectiveness of all service encounters during the course of service lifecycle. From the Newtonian perspective, the 'effort' involved in creating the corresponding service experience can be regarded as the invisible 'force' in physics that can change an object's motion. Based on this understanding, we can summarize the following (Qiu, 2014):

- **The First Law of Service Science:** If the effort (possibly ranging from effortless to effortful) dedicated to a service is unchanged, then the perceived service experience remains constant from an entity's perspective. The goal of this service's first law is to create a generalized reference framework so that other laws can be deduced. In the era of the fourth industrial revolution (Xing and Marwala, 2017), services are often co-produced via the interactions between service suppliers and clients. Accordingly, an entity can be referred to as either living systems, e.g., human physiology (Xing and Marwala, 2018j), biopsychosocial (Xing and Marwala, 2018h), and exposome (Xing and Marwala, 2018i); or non-living systems, e.g., complex adaptive cyber physical logistics system (Xing, 2014; Xing and Gao, 2015), mobile payment system (Xing, 2017), ubiquitous robot system (Xing, 2016a), and pipeline infrastructure maintenance system (Xing, 2015). Therefore, at each instance of service interaction, between service supplier(s) and client(s), experience is often gained from multiple dimensions, which all together will determine how a service is felt.
- **The Second Law of Service Science:** An entity's acceleration (denoted by A) has three properties, namely, (1) directly proportional to the effort change(s) (denoted by E) perceived by an entity; (2) towards the same direction with the E; and (3) inversely proportional to an entity's systemic mass (denoted by M), i.e., $A = E/M$. With the above defined attributes, the service's second law aims at quantitatively determining the differentiation induced by effort changes put on a service. In a typical service dynamics scenario, at any time point t, when a given effort change ($\vec{E}_t'$) is applied to entity one (tiny and uncomplicated

with a systemic mass of $\vec{M}_t^1$), it is assumable to quantify this entity's acceleration as $\vec{A}_t^1$; accordingly, when the same effort change ($\vec{E}_t'$) is applied to entity two (large and complicated with a systemic mass of $\vec{M}_t^2$), its acceleration can be quantified as $\vec{A}_t^2$. The relationship between two accelerations follows $|\vec{A}_t^2|<|\vec{A}_t^1|$, that is, the smaller and less complicated an entity to which the same effort change is imposed, the more considerable the resultant influence is.

- **The Third Law of Service Science:** When entity one applies a certain amount of effort (say E_1) on entity two, the second entity will apply another amount of effort (E_2) in a simultaneous manner. Here, the E_1 and E_2 share the same magnitude but with reversed direction, that is, $E_2 = -E_1$. The essentiality of the third law lies in its mutuality, i.e., interactively and mutually maintaining the equality of efforts to be made by involving entities during service provision. In the context of economics, one can transform an effort into a value of $\vec{V}$. Accordingly, the formulation of $|\vec{V}_1| = |\vec{V}_2|$ holds in service economy environment, that is, a service supplier paid by a client for providing an equivalent value of service.

According to the above discussions, we can perform a series of analogous comparisons between the laws of motion and the laws of service science. The corresponding similarities are illustrated in Fig. 3.3.

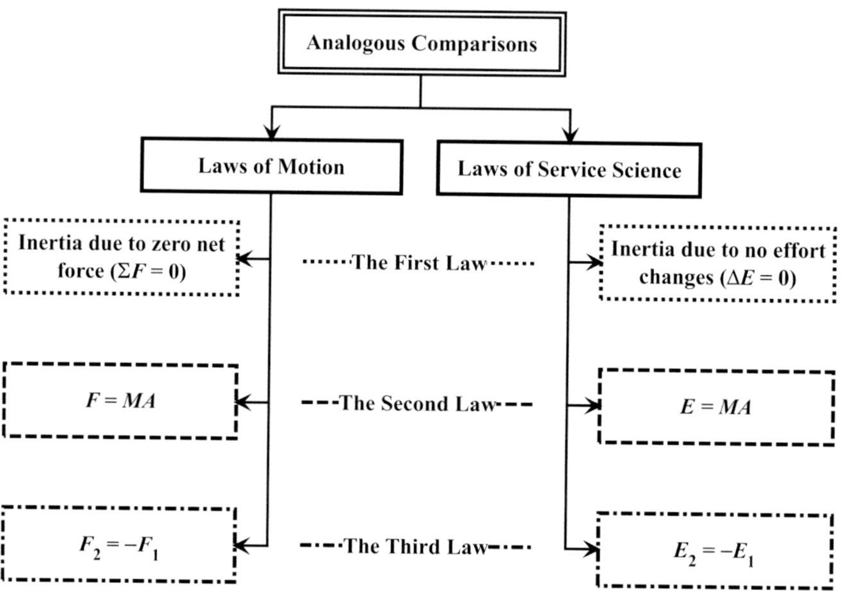

Figure 3.3: Analogous comparisons between the laws of motion and the laws of service science.

3.2.2 Service Science: Systemic Perspective

During the past three centuries, researchers have been actively verifying and demonstrating the effectiveness of Newton's Laws of Motion via numerous observations and experimentations. In contrast to those fathomable, definable and practically measurable physical quantities, the key parameters of a service system (i.e., systemic mass and contributing effort) remain mostly subjective, that is, immeasurable. Built on this understanding, the three laws of service science can be recompiled as follows (Qiu, 2014):

- **The Generalized First Law of Service Science:** If the efforts dedicated to a set of services are unchanged, then the entity's perceived value (denoted by V) after experiencing this set of services remains constant. Mathematically, V is either proportional to E, i.e., $\vec{V} \propto \vec{E}$, or a function of E, that is, $\vec{V} = f(\vec{E})$.
- **The Generalized Second Law of Service Science:** A service system's efficacy A, i.e., the accumulated acceleration, still has three properties, namely (1) directly proportional to the collective effort changes (denoted by E) applied to the system itself; (2) pointing in the same direction with the E; (3) inversely proportional to an entity's systemic resistance (denoted by R), i.e., $A = E/R$. Here, one can view a service system's inertia as its systemic resistance, which can be roughly quantified via the relative complexity and/or inefficacy (caused by multiples reasons, say, the underlying bureaucracy) of a service system. Mathematically, the following formulations hold: $\vec{E} = \vec{R} \times \vec{A}$, i.e., $\vec{V} \propto \vec{R} \times \vec{A}$, or $\vec{V} = f(\vec{R} \times \vec{A})$.
- **The Generalized Third Law of Service Science:** In the simplified service's third law, we have $E_2 = -E_1$. By incorporating the service suppliers' and clients' values within a service system, the following relations hold in the generalized service's third law: $|\vec{V}_1| = |\vec{V}_2|$, $\vec{V}_1 = f(\vec{E}_1)$, and $\vec{V}_2 = f(\vec{E}_2)$.

By performing a series of analogous comparisons between the laws of motion and the laws of service science, we can identify a set of interesting patterns as depicted in Fig. 3.4.

3.3 Service Innovation

Though neoclassical economic theory can somehow explain the underlying reasons regarding the rapid growth of services' demand and supply, it leaves the question of 'how are new services exactly churned out?' largely unanswered (Janssen, 2015). The aim of this section is, thus, to advance

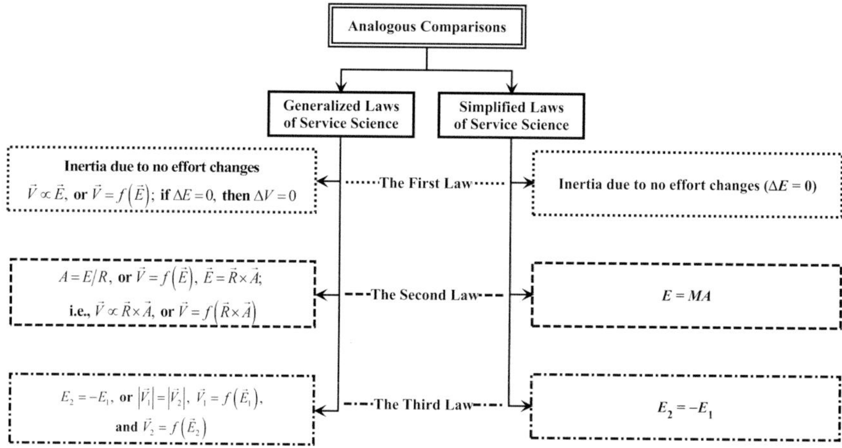

Figure 3.4: Analogous comparisons between the generalized and simplified laws of service science.

our understanding in terms of the nature of service innovation together with its strategic and policy significance.

3.3.1 *Types of Service Innovations*

As suggested by Gallouj and Weinstein (1997), the types of innovations found in the service domain can be clustered into six categories, as illustrated in Fig. 3.5.

Related to the seminal work done by Gallouj and Weinstein (1997), scientific explorations into service innovation can be broadly grouped into three schools of thinking: assimilation (or technology-centered) approach, demarcation (or service-centered) approach, and synthesis (or integration-centered) approach (Shivdas and Sivakumar, 2016; Lancaster, 1966; De Vries, 2006; Coombs and Miles, 2000; Drejer, 2004; Djellal and Gallouj, 2000; Janssen, 2015; Gallouj, 1998).

- **Assimilative Thought (technology-based viewpoint):** This school of thinking views services through the lens of goods manufacturers (De Vries, 2006; Gallouj, 2002a), which is built on an assumption that the "majority of services' economic properties essentially resemble those of their manufacturing counterparts" (Miles, 2007). A noteworthy example goes to (Pavitt, 1984), in which he titled service industries as passive adopters. In (Barras, 1986), the author went further by claiming that a reversed product life cycle policy is typically followed by service industries. Partially due to the unrecognized services' innovativeness, most assimilation researches put services under the existing frameworks and examined them via the same conceptions

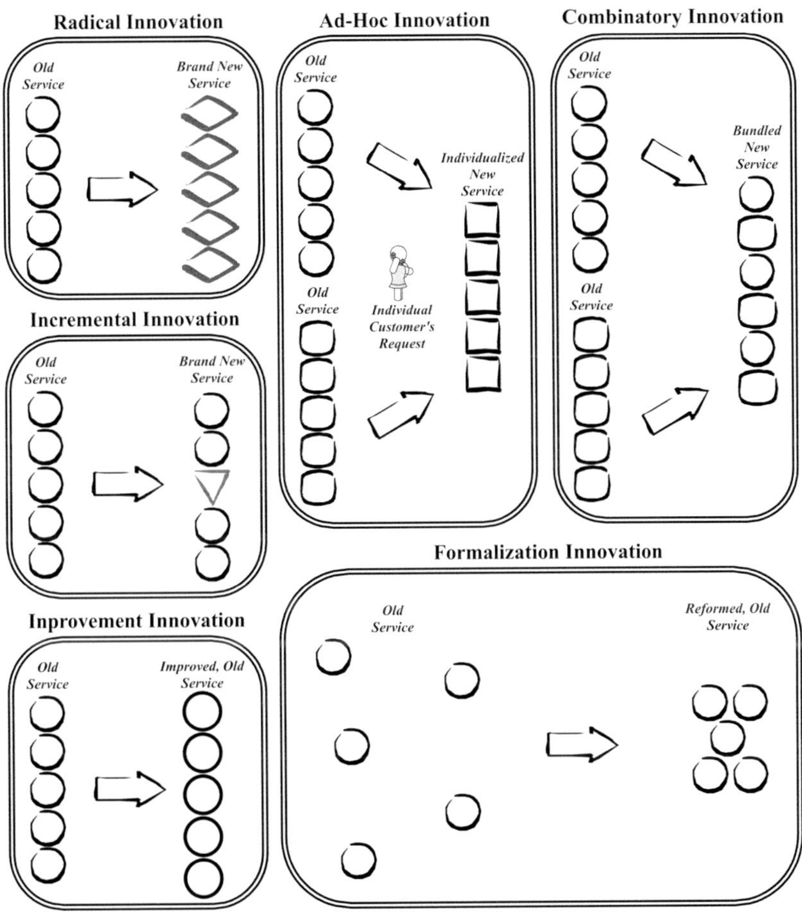

Figure 3.5: Types of service innovations.

and methodologies utilized in manufacturing analysis (Miles, 2005). Accordingly, as concluded in (Janssen, 2015), early works on identifying the role of services in economic development (by focusing on factors such as trading, growth, and productivity) were mostly incapable of embracing service activities' pervasiveness. In summation, this school of thinking performs poorly in discovering the speciality of services and the associated service innovations.

- **Dissimilative Thought (service-based viewpoint):** The demarcative or differentiated school of thinking recognizes the notable differences exhibited by services across different goods manufacturing contexts. Thus, service-based views suggest that new theories are required to understand the nature of service innovations (Coombs and Miles, 2000;

Sundbo, 1997; Gadrey et al., 1995; Van der Aa and Elfring, 2002) since they are characterized by not only their incremental and continuous improvements, but their radical and disruptive influences on causing a leap in customer valuation (De Brentani, 2001; Michel et al., 2008). In particular, special attention is paid to the implications deduced from service peculiarities (e.g., co-produced and intangible). Various insights are obtained through analysing different service industries including banking, tourism, retailing, internet, supply chain, public services, etc. Nevertheless, the high degree of heterogeneity found within the area of services (Janssen, 2015) gradually diverges studies of this category towards a scattered myopic narrow scope which, in turn, makes dissimilative thought suffer from serious localization (Gallouj and Savona, 2009).

- **Synthetic Thought (integration-based viewpoint):** Rather than treating service innovation and goods-dominant innovation logic as simplistically equivalent (assimilative thought) or crudely distinct (dissimilative thought), the synthesis thinking absorbs the merits of former two school of thoughts, and integrates these merits into several overarching models and propositions (Gallouj and Weinstein, 1997). A key motivation for forming servitization logic is that the divide between pure goods manufacturing practices and absolute service offering activities are becoming blurred. On the one hand, manufacturing companies are increasingly shifting towards a services-inclusive business model, while on the other hand, service providers are also gradually resembling the success experiences of manufactured goods (Miles, 2007). This convergence has made service innovation widely accepted as a means of offering precious opportunities to almost every industry within an economy (Mina et al., 2014).

Though each school of thought outlined above takes a different stand, they supplement each other and collectively represent a roadmap that service innovation studies have experienced and evolved (Gallouj and Savona, 2009; Snyder et al., 2016; Carlborg et al., 2014; Dröge, 2010).

3.3.2 *Benefits of Servitization*

According to Spring and Araujo (2009), one can treat servitization as a kind of business model innovation which has the capability of leveraging an accelerated growth (Canton, 1984; Sawhney et al., 2004), an increased profitability (Cohen et al., 2006; Oliva and Kallenberg, 2003; Neely, 2008) and an enhanced economic stability (Lele, 1986; Quinn and Gagnon, 1986). Despite the theoretical divergence, shifting from a tangible goods producer to an intangible service supplier can bring many benefits in practice for manufacturing companies, some of them are introduced as follows (Roos, 2015):

- **Monetary Gains:** Generally speaking, monetary gains include boosting goods sales, mitigating the influence of economic fluctuations on different cash flows (i.e., enhancing economic stability) and adapting to market demand changes (Brax, 2005). More specifically, we can have the following:
  1) Services can generate a new stream of revenue and the associated lower volatility makes it particularly suitable for low-cost economies.
  2) When a company has a large installed product base and/or a relatively long product life cycle, the improvement of service quality is in particular relevant for receiving more profit.
  3) In general, loyal customers are much more profitable than new adopters, especially in the service domain. Accordingly, the economy of loyalty is another important factor.
- **Strategic Gains:** Apart from these financial gains, servitization as a model of innovation can also help companies harvest some strategic opportunities, e.g., additional growth space creation (particularly in matured market environment) (Matthyssens et al., 2006). More specifically, we can have the following:
  1) Since delivering a valuable service is a rather complex system, the imitation of a service model is difficult. This difficulty can not only help the company gain competitive advantage, but also raise the bar of entry.
  2) A well-perceived service can form a closer relationship between manufacturers and users.
  3) By using physical products as a vehicle for delivering services, firms can co-establish their dominant positions in the marketplace.
  4) Servitization strategy can serve a specific user group, such as car rental schemes, which can meet customers' transportation needs without the challenges of owning a car outright.
  5) Pre-sales service can act as a driver for stimulating customers to buy product.
- **Marketing Gains:** Products and services are becoming more and more inseparable. Though product sales tend to be a once-off activity, the service itself can extend the interaction time between the firms and customers.
- **Operations Gains (e.g., Offshoring and/or Outsourcing):** In developed countries, with more and more production operations being offshored or outsourced, the transition from products-driven to service-oriented strategy can help firms receive higher-value gains.

- **Regulation Implementation Gains (e.g., extended producer responsibility):** With the popularity of remanufacturing (Xing and Gao, 2014a), service offerings can help companies put themselves in an advantageous position regarding taking back the end-of-life products.
- **Environmental Contribution Gains:** As a building block of sharing economy, the transformation from ownership to usership via services can be beneficial to the overall environment in multiple senses.

3.4 Conclusions

In this chapter, we have endeavoured to enhance our understanding of service and the associated service innovation. We have observed that the search for new services and the search for goods do not oppose each other diametrically. The difference between the two, never solid initially, begins to blur because of their intermingled co-evolution. Services are beginning to be sold as goods, while goods are starting to look like services. This observation is further proven as we move our discussion into financial services, in Chapter 4.

In conclusion, the literature about service, in particular service innovation, is certainly not coming to an end with the fading of industry boundaries. On the contrary, since knowledge can be commercialized in multiple ways, we can expect that more and more disembodied or 'dematerialized' types of products are on their way. Consumers are increasingly seeking ways to satisfy their needs through either services purchase or equipment rental. Physical artefacts will certainly not disappear but are expected to play a less and less important role in the near future. Eventually, a knowledge-based economy will be characterized by two equally important factors: (1) Advanced and sophisticated technological products, and (2) The performance and qualities of personnel involved in service professions.

References

Åhlström P and Nordin F. (2006) Problems of establishing service supply relationships: Evidence from a high-tech manufacturing company. *Journal of Purchasing and Supply Management* 12: 75–89.

Badger L, Grance T, Patt-Corner R, et al. (2012) Cloud computing synopsis and recommendations. National Institute of Standards and Technology, retrieved from http://nvlpubs.nist.gov/nistpubs/Legacy/SP/nistspecialpublication800-146.pdf, accessed on 07 September 2017.

Baines TS, Lightfoot HW, Evans S, et al. (2007) State-of-the-art in product-service systems. *Proceedings of the Institution of Mechanical Engineers Part B: J. Engineering Manufacture* 221: 1543–1552.

Baines T, Lightfoot H, Peppard J, et al. (2009a) Towards an operations strategy for product-centric servitization. *International Journal of Operations & Production Management* 29: 494–514.

Baines TS, Lightfoot HW, Benedettini O, et al. (2009b) The servitization of manufacturing: A review of literature and reflection on future challenge. *Journal of Manufacturing Technology Management* 20: 547–567.

Barras R. (1986) Towards a theory of innovation in services. *Research Policy* 15: 161–173.

Blanchard O. (2017) *Macroeconomics*, Edinburgh Gate, Harlow, Essex CM20 2JE, England: Pearson Education Limited, ISBN-13 978-1-292-16050-4.

Brax S. (2005) A manufacturer becoming service provider—challenges and a paradox. *Managing Service Quality: An International Journal* 15: 142–155.

Buyya R, Yeo CS, Venugopal S, et al. (2009) Cloud computing and emerging IT platforms: Vision, hype, and reality for delivering computing as the 5th utility. *Future Generation Computer Systems* 25: 599–616.

Canton ID. (1984) Learning to love the service economy. *Harvard Business Review* 62: 89–97.

Carlborg P, Kindström D and Kowalkowski C. (2014) The evolution of service innovation research: A critical review and synthesis. *The Service Industries Journal* 34: 373–398.

Chee BJS and Franklin C. (2010) *Cloud computing: Technologies and strategies of the ubiquitous data center*, Boca Raton, FL: CRC Press, Taylor & Francis Group, ISBN 978-1-4398-0612-8.

Cioban GL. (2014) Towards a "service economy"? *Ecoforum* 3: 111–116.

Cohen MA, Agrawal N and Agrawal V. (2006) Winning in the aftermarket. *Harvard Business Review* 84: 129–138.

Coombs R and Miles I. (2000) Innovation, measurement and services: The new problematique. In: Metcalfe JS and Miles l (eds) *Innovation Systems in the Service Economy: Measurements and Case Study Analysis*. New York: Springer Science+Business Media, ISBN 978-1-4613-6992-9, Chapter 5, pp. 85–103.

De D. (2016) *Mobile cloud computing: Architectures, algorithms and applications*, 6000 Broken Sound Parkway NW, Suite 300, Boca Raton, FL 33487-2742: CRC Press, Taylor & Francis Group, LLC, ISBN 978-1-4822-4284-3.

De Brentani U. (2001) Innovative versus incremental new business services: Different keys for achieving success. *Journal of Product Innovation Management* 18: 169–187.

De Vries EJ. (2006) Innovation in services in networks of organizations and in the distribution of services. *Research Policy* 35: 1037–1051.

Djellal F and Gallouj F. (2000) Innovation surveys for service industries: A review. *Proceedings of the Innovation and Entreprise Creation: Statistics and Indicators, 23–24 November, European Commission, Eurostat, Valbonne, pp. 70–76, retrieved from https://hal.archives-ouvertes.fr/halshs-01113813/document, accessed on 30 August 2017.*

Drejer I. (2004) Identifying innovation in surveys of services: A Schumpeterian perspective. *Research Policy* 33: 551–562.

Dröge H. (2010) Opening up innovation in services: Absorptive capacity in radical and incremental service innovation. *Esade Business School*. Universitat Ramon Llull.

Foster I, Zhao Y, Raicu I, et al. (2008) Cloud computing and grid computing 360-degree compared. *Proceedings of the Grid Computing Environments Workshop (GEC '08), 06 January, Austin, TX, USA*, pp. 1–10.

FTC. (2016) The sharing economy: Issues facing platforms, participants & regulators. USA: Federal Trade Commission, retrieved from https://www.ftc.gov/system/files/documents/reports/sharing-economy-issues-facing-platforms-participants-regulators-federal-trade-commission-staff/p151200_ftc_staff_report_on_the_sharing_economy.pdf, accessed on 5 September 2017.

Gadrey J, Gallouj F and Weinstein O. (1995) New modes of innovation: How services benefit industry. *International Journal of Service Industry Management* 6: 4–16.

Gallouj F and Weinstein O. (1997) Innovation in services. *Research Policy* 26: 537–556.

Gallouj F. (1998) Innovating in reverse: Services and the reverse product cycle. *European Journal of Innovation Management* 1: 123–138.

Gallouj F. (2002a) Innovation in services and the attendant old and new myths. *The Journal of Socio-Economics* 31: 137–154.

Gallouj F. (2002b) *Innovation in the services economy: The new wealth of nations,* 136 West Street, Suite 202 Northampton, Massachusetts 01060 USA: Edward Elgar Publishing, Inc., ISBN 1-84064-670-5.

Gallouj F and Savona M. (2009) Innovation in services: A review of the debate and a research agenda. *Journal of Evolutionary Economics* 19: 149–172.

Geuss R. (2001) *Public goods, private goods,* Princeton, NJ: Princeton University Press.

Holloway M. (2017) *Service level management in cloud computing: Pareto-efficient negotiations, reliable monitoring and robust monitor placement,* Abraham-Lincoln-Str. 46, 65189 Wiesbaden, Germany: Springer Fachmedien Wiesbaden GmbH, ISBN 978-3-658-18 772-9.

Horvit M, Koozin T and Nelson R. (2013) *Music for ear training,* 20 Channel Center Street, Boston, MA 02210, USA: Schirmer, Cengage Learning, ISBN 978-0-8400-2981-2.

Hurwitz J, Bloor R, Kaufman M, et al. (2010) *Cloud computing for Dummies,* 111 River Street, Hoboken, NJ 07030-5774: John Wiley & Sons, Inc., ISBN 978-0-470-48470-8.

Hurwitz J, Kaufman M and Halper F. (2012) *Cloud services for Dummies,* 111 River Street, Hoboken, NJ 07030-5774: John Wiley & Sons, Inc., ISBN 978-1-118-33891-9.

Janssen MJ. (2015) Service innovation in an evolutionary perspective. Eindhoven: Eindhoven University of Technology.

Jarrett S and Day H. (2008) *Music composition for dummies,* 111 River St. Hoboken, NJ, USA: Wiley Publishing, Inc., ISBN 978-0-470-22421-2.

Johnson M and Mena C. (2008) Supply chain management for servitised products: A multi-industry case study. *International Journal of Production Economics* 114: 27–39.

Kallhoff A. (2014) Why societies need public goods. *Critical Review of International Social and Political Philosophy* 17: 635–651.

Kleinrock L. (2005) A vision for the Internet. *ST Journal of Research* 2: 4–5.

Lancaster KJ. (1966) A new approach to consumer theory. *The Journal of Political Economy* 74: 132–157.

Lele MM. (1986) How service needs influence product strategy. *MIT Sloan Management Review* 28: 63–70.

Lewis M and Howard M. (2009) Beyond products and services: Shifting value generation in the automotive supply chain. *International Journal of Automotive Technology and Management* 9: 4–17.

Lin Y, Shi Y and Ma S. (2012) Servitization strategy: Priorities, capabilities and organizational features. In: Xiong G, Liu Z, Liu X-W, et al. (eds) *Service Science, Management, and Engineering: Theory and Applications.* 225 Wyman Street, Waltham, MA 02451, USA: Academic Press, Elsevier Inc., ISBN 978-0-12-397037-4, Chapter 2, pp. 11–35.

Lindberg N and Nordin F. (2008) From products to services and back again: Towards a new service procurement logic. *Industrial Marketing Management* 37: 292–300.

Malleret V. (2006) Value creation through service offers. *European Management Journal* 24: 106–116.

Marinescu DC. (2013) *Cloud computing: Theory and practice,* 225 Wyman Street, Waltham, 02451, USA: Morgan Kaufmann, Elsevier Inc., ISBN 978-0-12404-627-6.

Marwala T. (2013) *Economic modeling using artificial intelligence methods,* Springer London Heidelberg New York Dordrecht: Springer-Verlag London, ISBN 978-1-4471-5009-1.

Marwala T and Hurwitz E. (2017a) *Artificial intelligence and economic theory: Skynet in the market,* Gewerbestrasse 11, 6330 Cham, Switzerland: Springer International Publishing AG, ISBN 978-3-319-66103-2.

Marwala T and Hurwitz E. (2017b) Behavioral economics. In: Marwala T and Hurwitz E (eds) *Artificial Intelligence and Economic Theory: Skynet in the Market.* Gewerbestrasse 11, 6330 Cham, Switzerland: Springer International Publishing AG, ISBN 978-3-319-66103-2, Chapter 5, pp. 51–61.

Matthyssens P, Vandenbempt K and Berghman L. (2006) Value innovation in business markets: Breaking the industry recipe. *Industrial Marketing Management* 35: 751–761.
Michel S, Brown SW and Gallan AS. (2008) An expanded and strategic view of discontinuous innovations: Deploying a service-dominant logic. *Journal of the Academy of Marketing Science* 36: 54–66.
Miles I. (2005) Innovation in services. In: Fagerberg J, Mowery DC and Nelson RR (eds) *The Oxford Handbook of Innovation*. New York: Oxford University Press, ISBN 978-0-19928-680-5, Chapter 16, pp. 433–458.
Miles I. (2007) Research and development (R&D) beyond manufacturing: The strange case of services R&D. *R&D Management* 37: 249–268.
Mina A, Bascavusoglu-Moreau E and Hughes A. (2014) Open service innovation and the firm's search for external knowledge. *Research Policy* 43: 853–866.
Morris B and Johnston R. (1987) Dealing with inherent variability: The difference between manufacturing and service? *International Journal of Operations & Production Management* 7: 13–22.
Neely A. (2008) Exploring the financial consequences of the servitization of manufacturing. *Operations Management Research* 1: 103–118.
Oliva R and Kallenberg R. (2003) Managing the transition from products to services. *International Journal of Service Industry Management* 14: 160–172.
Pallis G. (2010) Cloud computing: The new frontier of Internet computing. *IEEE Internet Computing* 14: 70–73.
Pavitt K. (1984) Sectoral patterns of technical change: Towards a taxonmy and a theory. *Research Policy* 13: 343–373.
Pawar KS, Beltagui A and Riede JCKH. (2009) The PSO triangle: Designing product, service and organisation to create value. *International Journal of Operations & Production Management* 29: 468–493.
Qiu RG. (2014) *Service science: the foundations of service engineering and management*, 111 River Street, Hoboken, NJ 07030: John Wiley & Sons, Inc., ISBN 978-1-118-10823-9.
Quinn JB and Gagnon CE. (1986) Will services follow manufacturing into decline? *Harvard Business Review* 64: 95–103.
Robinson T, Clarke-Hill CM and Clarkson R. (2002) Differentiation through services: A perspective from the commodity chemicals sector. *The Service Industries Journal* 22: 149–166.
Roos G. (2015) Servitization as innovation in manufacturing—a review of the literature. In: Agarwal R, Selen W, Roos G, et al. (eds) *The Handbook of Service Innovation*. London Heidelberg New York Dordrecht: Springer-Verlag London, ISBN 978-1-4471-6589-7, Chapter 19, pp. 403–435.
Sarkar P. (2015) *Data as a service: A framework for providing reusable enterprise data services*, 111 River Street, Hoboken, NJ 07030: John Wiley & Sons, Inc., ISBN 978-1-119-04658-5.
Saviotti PP and Metcalfe JS. (1984) A theoretical approach to the construction of technological output indicators. *Research Policy* 13: 141–151.
Saviotti PP. (1996) *Technological evolution, variety and the economy*, Cheltenham, UK and Brookfield, USA: Edward Elgar Publishing, ISBN 978-1-85278-774-5.
Sawhney M, Balasubramanian S and Krishnan VV. (2004) Creating growth with services. *MIT Sloan Management Review* 45: 34–43.
Schiller BR, Hill C and Wall S. (2013) *The macro economy today*, 1221 Avenue of the Americas, New York, NY, 10020: The McGraw-Hill Companies, Inc., ISBN 978-0-07-741647-8.
Schmenner RW. (2009) Manufacturing, service, and their integration: Some history and theory. *International Journal of Operations & Production Management* 29: 431–443.
Schneider H. (2017) *Uber: Innovation in society*, Gewerbestrasse 11, 6330 Cham, Switzerland: Palgrave Macmillan, ISBN 978-3-319-49513-2.
Schumpeter JA. (1950) *Capitalism, socialism and democracy*, New York: Harper.
Selya RM. (1994) Taiwan as a service economy. *Geoforum* 25: 305–322.

Shivdas PA and Sivakumar S. (2016) Innovation in services: A Lancastrian approach to the field of e-learning. *Education and Information Technologies* 21: 1913–1925.

Slack N. (2005) Operations strategy: Will it ever realize its potential? *GESTÃO & PRODUÇÃO* 12: 323–332.

Snyder H, Witell L, Gustafsson A, et al. (2016) Identifying categories of service innovation: A review and synthesis of the literature. *Journal of Business Research* 69: 2401–2408.

Spring M and Araujo L. (2009) Service, services and products: Rethinking operations strategy. *International Journal of Operations & Production Management* 29: 444–467.

Srinivasan S. (2014) *Cloud computing basics,* Springer New York Heidelberg Dordrecht London: Springer Science+Business Media New York, ISBN 978-1-4614-7698-6.

Sundbo J. (1997) Management of innovation in services. *The Service Industries Journal* 17: 432–455.

Sundbo J. (2002) The service economy: Standardisation or customisation? *The Service Industries Journal* 22: 93–116.

Van der Aa W and Elfring T. (2002) Realizing innovation in services. *Scandinavian Journal of Management* 18: 155–171.

Vandermerwe S and Rada J. (1988) Servitization of business: Adding value by adding services. *European Management Journal* 6: 314–324.

Voas J and Zhang J. (2009) Cloud computing: New wine or just a new bottle? *IT Professional* 11: 15–17.

Wetzstein ME. (2013) *Microeconomic Theory,* 2 Park Square, Milton Park, Abingdon, Oxon OX14 4RN: Routledge, ISBN 978-0-415-60369-0.

Xing B. (2014) The optimization of computational stock market model-based complex adaptive cyber physical logistics system. In: Khan ZH, Ali ABMS and Riaz Z (eds) *Computational Intelligence for Decision Support in Cyber-Physical Systems.* Singapore Heidelberg New York Dordrecht London: Springer Science+Business Media Singapore, ISBN 978-981-4585-35-4, Chapter 12, pp. 357–380.

Xing B and Gao W-J. (2014a) *Computational intelligence in remanufacturing,* 701 E. Chocolate Avenue, Suite 200, Hershey PA 17033: IGI Global, ISBN 978-1-4666-4908-8.

Xing B and Gao W-J. (2014ee) Introduction to remanufacturing and reverse logistics. In: Xing B and Gao W-J (eds) *Computational Intelligence in Remanufacturing.* 701 E. Chocolate Avenue, Suite 200, Hershey PA 17033: IGI Global, ISBN 978-1-4666-4908-8, Chapter 1, pp. 1–17.

Xing B and Gao W-J. (2014gg) Used products return pattern analysis using agent-based modelling and simulation. In: Xing B and Gao W-J (eds) *Computational Intelligence in Remanufacturing.* 701 E. Chocolate Avenue, Suite 200, Hershey PA 17033: IGI Global, ISBN 978-1-4666-4908-8, Chapter 3, pp. 38–58.

Xing B and Gao W-J. (2014hh) Used product collection optimization using genetic algorithms. In: Xing B and Gao W-J (eds) *Computational Intelligence in Remanufacturing.* 701 E. Chocolate Avenue, Suite 200, Hershey PA 17033: IGI Global, ISBN 978-1-4666-4908-8, Chapter 4, pp. 59–74.

Xing B and Gao W-J. (2014ii) Used product remanufacturability evaluation using fuzzy logic. In: Xing B and Gao W-J (eds) *Computational Intelligence in Remanufacturing.* 701 E. Chocolate Avenue, Suite 200, Hershey PA 17033: IGI Global, ISBN 978-1-4666-4908-8, Chapter 5, pp. 75–94.

Xing B and Gao W-J. (2014jj) Used product pre–sorting system optimization using teaching–learning-based optimization. In: Xing B and Gao W-J (eds) *Computational Intelligence in Remanufacturing.* 701 E. Chocolate Avenue, Suite 200, Hershey PA 17033: IGI Global, ISBN 978-1-4666-4908-8, Chapter 6, pp. 95–112.

Xing B and Gao W-J. (2014kk) Used product delivery optimization using agent-based modelling and simulation. In: Xing B and Gao W-J (eds) *Computational Intelligence in Remanufacturing.* 701 E. Chocolate Avenue, Suite 200, Hershey PA 17033: IGI Global, ISBN 978-1-4666-4908-8, Chapter 7, pp. 113–133.

Xing B and Gao W-J. (2014ll) Post–disassembly part–machine clustering using artificial neural networks and ant colony systems. In: Xing B and Gao W-J (eds) *Computational*

Intelligence in Remanufacturing. 701 E. Chocolate Avenue, Suite 200, Hershey PA 17033: IGI Global, ISBN 978-1-4666-4908-8, Chapter 8, pp. 135–150.

Xing B and Gao W-J. (2014mm) Reprocessing operations scheduling using fuzzy logic and fuzzy MAX–MIN ant systems. In: Xing B and Gao W-J (eds) *Computational Intelligence in Remanufacturing*. 701 E. Chocolate Avenue, Suite 200, Hershey PA 17033: IGI Global, ISBN 978-1-4666-4908-8, Chapter 9, pp. 151–170.

Xing B and Gao W-J. (2014nn) Reprocessing cell layout optimization using hybrid ant systems. In: Xing B and Gao W-J (eds) *Computational Intelligence in Remanufacturing*. 701 E. Chocolate Avenue, Suite 200, Hershey PA 17033: IGI Global, ISBN 978-1-4666-4908-8, Chapter 10, pp. 171–185.

Xing B and Gao W-J. (2014oo) Re–machining parameter optimization using firefly algorithms. In: Xing B and Gao W-J (eds) *Computational Intelligence in Remanufacturing*. 701 E. Chocolate Avenue, Suite 200, Hershey PA 17033: IGI Global, ISBN 978-1-4666-4908-8, Chapter 11, pp. 186–202.

Xing B and Gao W-J. (2014pp) Batch order picking optimization using ant system. In: Xing B and Gao W-J (eds) *Computational Intelligence in Remanufacturing*. 701 E. Chocolate Avenue, Suite 200, Hershey PA 17033: IGI Global, ISBN 978-1-4666-4908-8, Chapter 12, pp. 204–222.

Xing B and Gao W-J. (2014qq) Complex adaptive logistics system optimization using agent-based modelling and simulation. In: Xing B and Gao W-J (eds) *Computational Intelligence in Remanufacturing*. 701 E. Chocolate Avenue, Suite 200, Hershey PA 17033: IGI Global, ISBN 978-1-4666-4908-8, Chapter 13, pp. 223–236.

Xing B and Gao W-J. (2014rr) Conclusions and emerging topics. In: Xing B and Gao W-J (eds) *Computational Intelligence in Remanufacturing*. 701 E. Chocolate Avenue, Suite 200, Hershey PA 17033: IGI Global, ISBN 978-1-4666-4908-8, Chapter 14, pp. 238–265.

Xing B. (2015) Knowledge management: Intelligent in-pipe inspection robot conceptual design for pipeline infrastructure management. In: Kahraman C and Onar SÇ (eds) *Intelligent Techniques in Engineering Management: Theory and Applications*. Cham Heidelberg New York Dordrecht London: Springer International Publishing Switzerland, ISBN 978-3-319-17905-6, Chapter 6, 129–146.

Xing B and Gao W-J. (2015) A SWOT analysis of intelligent product enabled complex adaptive logistics systems. In: Khosrow-Pour M (ed) *Encyclopedia of Information Science and Technology* 3rd ed. New York, USA: Information Science Ref. – IGI Global, ISBN 978-1-4666-5888-2, Chapter 490, pp. 4970–4979.

Xing B. (2016a) An investigation of the use of innovative biology-based computational intelligence in ubiquitous robotics systems: Data mining perspective. In: Ravulakollu KK, Khan MA and Abraham A (eds) *Trends in Ambient Intelligent Systems*. Switzerland: Springer International Publishing Switzerland, ISBN 978-3-319-30184-6, Chapter 6, pp. 139–172.

Xing B. (2016b) Ontological framework–assisted embedded system design with security consideration. In: Pathan A-SK (ed) *Securing Cyber-Physical Systems*. 6000 Broken Sound Parkway NW, Suite 300, Boca Raton, FL 33487-2742: CRC Press, Taylor & Francis Group, LLC, ISBN 978-1-4987-0099-3, Chapter 4, pp. 91–118.

Xing B. (2017) Protecting mobile payments security: A case study. In: Meng W, Luo X, Furnell S, et al. (eds) *Protecting Mobile Networks and Devices: Challenges and Solutions*. 6000 Broken Sound Parkway NW, Suite 300, Boca Raton, FL 33487-2742: CRC Press, Taylor & Francis Group, LLC, ISBN 978-1-4987-3583-4, Chapter 11, pp. 261–289.

Xing B and Marwala T. (2017) Implications of the fourth industrial age for higher education. *The Thinker: For the Thought Leaders (www.thethinker.co.za)*. South Africa: Vusizwe Media, 10–15.

Xing B and Marwala T. (2018a) Conclusion. In: Xing B and Marwala T (eds) *Smart Maintenance for Human–Robot Interaction: An Intelligent Search Algorithmic Perspective*. Gewerbestrasse 11, 6330 Cham, Switzerland: Springer International Publishing AG, ISBN 978-3-319-67479-7, Chapter 13, pp. 299–305.

Xing B and Marwala T. (2018b) Cyberware capacity–applications layer perspective. In: Xing B and Marwala T (eds) *Smart Maintenance for Human–Robot Interaction: An Intelligent Search Algorithmic Perspective.* Gewerbestrasse 11, 6330 Cham, Switzerland: Springer International Publishing AG, ISBN 978-3-319-67479-7, Chapter 8, pp. 173–191.

Xing B and Marwala T. (2018c) Cyberware capacity–energy autonomy perspective. In: Xing B and Marwala T (eds) *Smart Maintenance for Human–Robot Interaction: An Intelligent Search Algorithmic Perspective.* Gewerbestrasse 11, 6330 Cham, Switzerland: Springer International Publishing AG, ISBN 978-3-319-67479-7, Chapter 9, pp. 193–216.

Xing B and Marwala T. (2018d) Cyberware capacity–platform and middleware layers perspective. In: Xing B and Marwala T (eds) *Smart Maintenance for Human–Robot Interaction: An Intelligent Search Algorithmic Perspective.* Gewerbestrasse 11, 6330 Cham, Switzerland: Springer International Publishing AG, ISBN 978-3-319-67479-7, Chapter 7, pp. 143–171.

Xing B and Marwala T. (2018e) Hardware capacity–beginning of life perspective. In: Xing B and Marwala T (eds) *Smart Maintenance for Human–Robot Interaction: An Intelligent Search Algorithmic Perspective.* Gewerbestrasse 11, 6330 Cham, Switzerland: Springer International Publishing AG, ISBN 978-3-319-67479-7, Chapter 4, pp. 67–91.

Xing B and Marwala T. (2018f) Hardware capacity–end of life perspective. In: Xing B and Marwala T (eds) *Smart Maintenance for Human–Robot Interaction: An Intelligent Search Algorithmic Perspective.* Gewerbestrasse 11, 6330 Cham, Switzerland: Springer International Publishing AG, ISBN 978-3-319-67479-7, Chapter 6, pp. 111–139.

Xing B and Marwala T. (2018g) Hardware capacity–middle of life perspective. In: Xing B and Marwala T (eds) *Smart Maintenance for Human–Robot Interaction: An Intelligent Search Algorithmic Perspective.* Gewerbestrasse 11, 6330 Cham, Switzerland: Springer International Publishing AG, ISBN 978-3-319-67479-7, Chapter 5, pp. 93–110.

Xing B and Marwala T. (2018h) Human capacity–biopsychosocial perspective. In: Xing B and Marwala T (eds) *Smart Maintenance for Human–Robot Interaction: An Intelligent Search Algorithmic Perspective.* Gewerbestrasse 11, 6330 Cham, Switzerland: Springer International Publishing AG, ISBN 978-3-319-67479-7, Chapter 11, pp. 249–270.

Xing B and Marwala T. (2018i) Human capacity–exposome perspective. In: Xing B and Marwala T (eds) *Smart Maintenance for Human–Robot Interaction: An Intelligent Search Algorithmic Perspective.* Gewerbestrasse 11, 6330 Cham, Switzerland: Springer International Publishing AG, ISBN 978-3-319-67479-7, Chapter 12, pp. 271–295.

Xing B and Marwala T. (2018j) Human capacity–physiology perspective. In: Xing B and Marwala T (eds) *Smart Maintenance for Human–Robot Interaction: An Intelligent Search Algorithmic Perspective.* Gewerbestrasse 11, 6330 Cham, Switzerland: Springer International Publishing AG, ISBN 978-3-319-67479-7, Chapter 10, pp. 219–247.

Xing B and Marwala T. (2018k) Introduction to human robot interaction. In: Xing B and Marwala T (eds) *Smart Maintenance for Human–Robot Interaction: An Intelligent Search Algorithmic Perspective.* Gewerbestrasse 11, 6330 Cham, Switzerland: Springer International Publishing AG, ISBN 978-3-319-67479-7, Chapter 1, pp. 3–19.

Xing B and Marwala T. (2018l) Introduction to smart maintenance. In: Xing B and Marwala T (eds) *Smart Maintenance for Human–Robot Interaction: An Intelligent Search Algorithmic Perspective.* Gewerbestrasse 11, 6330 Cham, Switzerland: Springer International Publishing AG, ISBN 978-3-319-67479-7, Chapter 2, pp. 21–31.

Xing B and Marwala T. (2018m) *Smart maintenance for human–robot interaction: An intelligent search algorithmic perspective,* Gewerbestrasse 11, 6330 Cham, Switzerland: Springer International Publishing AG, ISBN 978-3-319-67479-7.

Zhang Q, Cheng L and Boutaba R. (2010) Cloud computing: State-of-the-art and research challenges. *Internet Services and Applications* 1: 7–18.

CHAPTER 4

Introduction to Financial Service Innovation

Crowdfunding

4.1 Financial Services

The basic principle of financial services is as fundamental financial arrangement associated with several dimensions of economic growth and development, such as education, healthcare, and infrastructure (Bisht and Mishra, 2016; Marwala and Hurwitz, 2017a; Marwala and Hurwitz, 2017b; Marwala, 2013). Usually, they are provided by banks, accounting companies, unit trusts, insurance companies, stock brokerages, and credit card issuers in different finance-related value forms. In recent years, under the umbrella of the Fourth Industrial Revolution (4IR) (Xing and Marwala, 2017), the financial services area has been experiencing a fast-paced growth and evolution period (Scardovi, 2017; Dintrans et al., 2016; Cisco, 2015; Chuen, 2015; Chuen and Deng, 2018). One notable feature is to provide services via new information and communication technologies (ICTs) such as social media (Kamukama and Natamba, 2013), blockchain (Burelli et al., 2015; Cognizant, 2017; Fanning and Centers, 2016), crowdfunding (Gleasure and Feller, 2016), big data (Trelewicz, 2017; Oracle, 2015), and artificial intelligence (AI) (Hamerton-Stove, 2016; Dunis et al., 2016; Marwala, 2014; Marwala, 2015), in order to reshape and transform this domain in new directions.

4.1.1 Goods or Services

The term 'financial services' was coined as a result of Gramm-Leach-Bliley Act which came into practice in the late 1990s (Gramm et al., 1999).

The aim of this act was to offer the American financial institutions more opportunities in terms of merging and enlarging their service offerings. Since then, the debate on whether the financial services are goods or services has been dominated by an attempt to define a clear position of the financial services industries, such as what our business is and what our market is. This dilemma has lasted til the present. However, those distinctions are far from the only concern of the literature. Advertising/promotion plans, marketing techniques, budget analysis, and public communications/relationships are also influenced by the choice (Davis, 2009; Koku, 2014).

4.1.1.1 Financial Services as Goods

According to the neoclassical model, goods can be categorized as three types, i.e., private, public or mixed. Each category has its own characteristics: (1) The features of private goods are characterized by excludability, rivalry, and rejectability, and (2) While true public goods belong to a totally opposite domain, and (3) What about mixed goods? Simply put, they have some but not all of the attributes of private/public goods (Preker et al., 2000).

Applying this definition to the financial sector, when financial services are classed as goods (e.g., credit cards, securities, and pensions), evidences exhibit that the majority of financial services have some degree of those characteristics but with self-definition, i.e., separability, non-perishable, and massiveness (Ehrlich and Fanelli, 2012). In addition, one can also treat financial services as the process of acquiring financial goods (Asmundson, 2011). For instance, in order to cover some of the risks (e.g., automobile accident or a house fire), insurance companies might sell a good (i.e., insurance policy) for the purpose of taking in premiums from customers.

4.1.1.2 Financial Services as Services

Given the fact that modern society's fundamental transactions and exchanges are actually bolstered by finance, financial services are regarded as services that have the potential to spark radical transformations. Compared with goods, services have their own advantages/disadvantages, such as low cost of entry, speed to market and lack of exclusivity (Ehrlich and Fanelli, 2012). Generally speaking, as services, financial services can facilitate six key functionalities, namely payments, market provisions, investment administrations, insurances, deposits and fundraising (World Economic Forum, 2015). From a broader perspective, financial services contain a diverse class of economic activities, such as managing natural disaster risk (Warner et al., 2007), retailing (Ring, 2016) and microfinance

(Moss et al., 2015; Bruton et al., 2014; Jeon and DomenicoMenicucci, 2011; Aggarwal et al., 2015).

In summary, from a macro point of view, financial services are neither goods nor services, but have elements of each (Ehrlich and Fanelli, 2012). In other words, financial services make hybrid offerings, i.e., combining one or more goods and services to offer more benefits to customers than if the goods and services were delivered separately (Schaarschmidt et al., in press).

4.1.2 *Problems and Challenges Faced by Traditional Financial Services*

No matter financial services are classified as goods or services, as a system that involves many different influential factors, they are facing various severely problems and challenges (Sweeting, 2017; García, 2017; Laycock, 2014; Lemieux, 2013; Allen, 2013; Tian, 2017).

- One of the biggest problems within the financial services industries is stability. Stability is regarded as the crucial factor for a smooth functioning of the financial services system (Nicoletti, 2017; Loveland, 2016). To maintain a desirable stability, regulators need to implement a set of clear strategic positioning for distinct financial service offerings (e.g., new business models) or steering towards new directions (e.g., new solvency and capital regulation). In addition, controlling service quality is also an important fact (Ehrlich and Fanelli, 2012). In fact, quality is a competitive weapon that works as a critical differentiator and ensures more durability of the organization. With financial services rapidly becoming available online, the service quality is becoming more and more decisive and challenging (Roy and Balaji, 2015).
- Moreover, cybersecurity, data security and privacy problems also remain demanding (Seo and Park, in press; Rosenberg, 2011; Abend et al., 2014). Many incumbent financial institutions are feeling the pressure, since they are struggling to deliver the higher levels of security services that customers are eagerly looking for (Pegueros, 2012). The associated trust and confidence needs to be rebuilt or reshaped.
- Finally, since financial services are not only about money, but also about information, they inevitably encounter a lot of psychological, social, and technological difficulties, such as consumers' attitudes and behaviours (Chuah and Devlin, 2011), financial knowledge (Königsheim et al., in press), investors' psychology (Richards, 2014) and personalization (Huang and Lin, 2005).

4.2 Digital Transformation in Financial Services

After the financial crisis in 2008, there is a widespread recognition amongst global financial services institutions and companies that new business development opportunities must be churned-out in order to eliminate their operational inefficiencies and increase the corresponding competitiveness. Several authors have argued that one powerful way to solve those issues is to move towards Internetization and digitization (Matt et al., 2015; Zon, 2001; Romãnova and Kudinska, 2016; Gomber et al., 2017; Dapp, 2014; Ben and Chen, 2018; Ben, 2017). Among others, FinTech (the adoption of innovative ICT techniques to enhance the delivery of the financial service) becomes a new norm (McConnell and Nolan, 2017; Ben, 2017).

From policy makers' viewpoint, central to this transformation is the potential that digital technologies' are capable of capturing useful data for governments, bankers, and insurers so as to map the 'risky' populations and to 'nudge' individual behaviour in desired directions (Gabor and Brooks, 2017). While on the investment managers' side, potential benefits are manifold, including high-frequency trading, reducing cost, providing personalized services, and forecasting customers' future behaviours (Nicoletti, 2017).

4.2.1 *What is FinTech?*

FinTech, an abbreviation of the phrase *financial technology*, has disrupted conventional finance realm in recent years and continues to grab headlines. It was implemented mainly after the worldwide financial crisis in 2008 by combining several cutting-edge technologies in order to help drive improvement and innovation in traditional financial services and, perhaps, the entire economy (Puschmann, 2017; Gomber et al., 2017).

Broadly speaking, there are two types of FinTech (Lee, 2015), namely (1) Sustaining FinTech (i.e., existing financial services providers utilize ICT to secure their market role) and (2) Disruptive FinTech (i.e., new entrants offer novel products to challenge incumbents). Since 2010 the investments made in the FinTech sector have been growing dramatically. According to Accenture, a consultancy, in 2015, the total investment hit a new record of $22.3 billion, up by 75% from 2014 (Accenture, 2016). Although North America is the strongest dominant area, the infographic shows that FinTech investment in Asia-Pacific (APAC) has reached $4.3 billion (more than quadrupled in 2015). Among others, China has developed the largest FinTech hub in APAC, accounting for 45% in 2015 (Accenture, 2016).

Indeed, this conclusion that China dominates APAC investments is also confirmed by PwC (2017g) and KPMG (2016a). In addition, KPMG also pointed out that China's FinTech markets mainly focus on four areas, namely (1) Pervasive mobile payments (e.g., Ping++ and ChinaPnR), (2) Online lending (e.g., Lu.com and JD Finance), (3) Internet insurance (e.g., Zhongan Insurance and Ideacome), and (4) Investment (e.g., Zipeiyi and RQuest). More details about FinTech development in China can be found in (HK Financial Services Development Council, 2017; PwC, 2017c; Chen, 2016; Shim and Shin, 2016; Stern et al., 2017; Sheng et al., 2017; PwC, 2016h; Mittal and Lloyd, 2016; Leong et al., 2017; PwC, 2017e).

Meanwhile, readers who are interested in other FinTech markets' overview should refer to Canada (Ernst & Young Global Limited, 2017a; Digital Finance Institute & McCarthy Tétrault LLP, 2016), New Zealand (PwC, 2017a), Russia (Mitina, 2016), Netherlands (PwC, 2016b), Germany (Dorfleitner et al., 2017), Australia (Australian Government, 2016), Switzerland (Diemers, 2017; Ernst & Young Global Limited, 2016), Luxembourg (PwC, 2016c), Ireland (Houlihan and Curneen, 2017; PwC, 2016d), UK (Ernst & Young Global Limited, 2014; Skan et al., 2014; Government Office for Science, 2015), USA (Gach and Gotsch, 2014), Malaysia (PwC, 2016e) and India (PwC, 2017b; PwC, 2017g; KPMG, 2016c).

In addition, the emerging topics that are highly related to FinTech phenomena, such as media perspective on the FinTech (Zavolokina et al., 2016), big data and FinTech (Zhang et al., 2015; Yang et al., in press) and many more, can be found by referring to the following sources:

- **Recently Published Global FinTech Reports:** PwC (PwC, 2017g; PwC, 2017f), KPMG (2017), IMF (He et al., 2017), Bell Pottinger (Bate, 2017), MagnaCarta (Hardie et al., 2017), Deloitte (2017b), The Board of the International Organization of Securities Commissions (IOSCO) (2017), Ernst & Young Global Limited (2017b), GP. Bullhound (2017), Starling Bank (2017), Oliver Wyman (2017), The European Consumer Organisation (BEUC) (2017), Toronto Financial Services Alliance (Yeandle, 2017), and The Economist Intelligence Unit (2017).
- **Books:** (Nicoletti, 2017; Dorfleitner et al., 2017; Sironi, 2016).

4.2.2 The History of FinTech

In fact, financial services sectors have a long history of being supported by technologies (He et al., 2017):

- The first technology-driven innovation appeared in 1950 when credit cards were introduced by Diners Club (Rona-Tas and Guseva, 2014).
- After that, from 1950 through 1970, different innovations were deployed into financial services such as:
  1) Automated teller machines (ATMs) (appeared to replace tellers and branches in 1960);
  2) Quotron system (used to provide stock market quotations in 1960);

3) A global telex network (established in 1966); and
4) In 1973, Carl Reuterskiöld founded the Worldwide Interbank Financial Telecommunication (SWIFT).

- With the development of the Internet, e-trade became available on the online stock exchange site in the 1990s. This remarkable event indicated the beginning of real FinTech. Other notable FinTech success stories include:
  1) In 1994, the first Internet bank emerged;
  2) In 1997, online personal finance management (PFM) software was introduced as a result of a collaboration among Microsoft, Intuit, and CheckFree;
  3) In the meantime, the first mobile payment (i.e., SMS activated Coca Cola vending machine) was made in Helsinki, Finland; and
  4) Finally, around 2000s, many FinTech startups were established to propose solutions for consumers (e.g., PayPal and eBay).
- With the popularity of Web 2.0, a new momentum has been gathered around World Wide Web, which makes it possible to create interactive websites via the cloud. Since then, FinTech entrepreneurs have been working very hard to surpass the traditional financial services.
  1) The world's first peer-to-peer (P2P) lending platform (i.e., Zopa) was put into practice in 2005 in UK;
  2) Bitcoin was introduced in 2008; and
  3) Around 2014, mobile payments became popular in several countries. Indeed, according to one study conducted by CB Insights for Accenture (Skan et al., 2015), global FinTech investment has exploded since 2014 by jumping 201% between 2013 and 2014.

4.2.3 The Future of FinTech

Though the wide spreading of FinTech challenges incumbent financial services to some extent, it provides new ways to reach younger and more tech-savvy customers. According to Belinky et al. (2015), the development of FinTech can be broadly clustered into the following two phases:

- **FinTech 1.0:** During this phase, banking marketplaces are influenced by several minor disruptions, mainly found in areas such as payments, credit, and personal financial advice.
- **FinTech 2.0:** This phase is represented by a 'seamless specialization', which spreads to the value chain's various core elements. During this phase, a wide array of service providers combined/consolidated in order to offer end users cheaper and easier-to-use propositions. The associated changes in customer preferences, technological

advancement and growing investment determine a more radical transformation for the FinTech landscape.

In light of this classification, an analysis from PwC (2017g) suggested that the future of FinTech should focus on the following four broad directions: (1) Financial inclusion, e.g., crowdfunding (Kim and Moor, 2017), (2) Wealth management, e.g., robo-advisors (Sironi, 2016), (3) RegTech (Kavassalis et al., in press; Deloitte, 2015b; Arner et al., 2018) and (4) InsurTech (Yan et al., 2018; Sheng et al., 2016; PwC, 2016f; PwC, 2016g; PwC, 2017d). In another study, Gomber et al. (2017) identified three domains that are currently attracting more and more attention from both academic researchers and practitioners, namely (1) FinTech's ecosystem (to be discussed in Section 4.2.4), (2) Cutting-edge technologies (to be discussed in Section 4.2.5) and (3) FinTech business functions (to be discussed in Section 4.2.6).

4.2.4 *FinTech Ecosystem*

According to Ernst & Young Global Limited (2014), the development of FinTech is typically influenced by three factors: Digital connectivity, economic downturn, and regulatory changes. Therefore, the ecosystem of FinTech (as illustrated in Fig. 4.1) consists of both foundations and superstructures (KPMG, 2016a; Mitina, 2016; Lee, 2016).

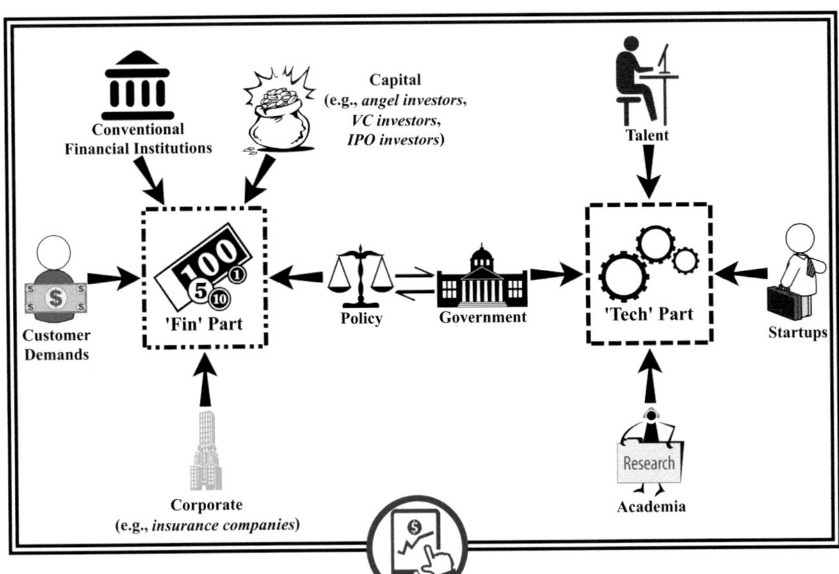

Figure 4.1: FinTech ecosystem.

- **Foundations:** The focus of foundations is on the 'Fin' part, which involves traditional financial institutions (e.g., banks), financial customers (e.g., asset managers), corporates (e.g., insurance companies), and governments. These players, as historical value drivers, are eager to ride the digital wave of transformation.
- **Superstructures:** On the other hand, the superstructures are bolstered by the 'Tech' part, which is comprised of FinTech startups, technology developers (e.g., large existing IT companies) and academia. These actors represent the future value drivers who have the potential to disrupt traditional financial services business models.

A more in-depth analysis regarding the relationship between incumbent financial service providers and the emerging FinTech companies can be found in (Romãnova and Kudinska, 2016; Skan et al., 2015; Kapron and Shaughnessy, 2015; Ghose et al., 2016; Niu, 2016; McKinsey & Company, 2016b; The Economist Intelligence Unit, 2012; PwC, 2016i; CGI, 2016; Drummer et al., 2016).

4.2.5 *Disruptive FinTech Technologies*

Nowadays, AI, big data analytics, Internet of Things (IoT), and blockchain serve as the most representative technologies in terms of enhancing the transformation of traditional financial services (Sheng et al., 2017; Dalager and Jensen, 2016). The corresponding interactive relationship is illustrated in Fig. 4.2.

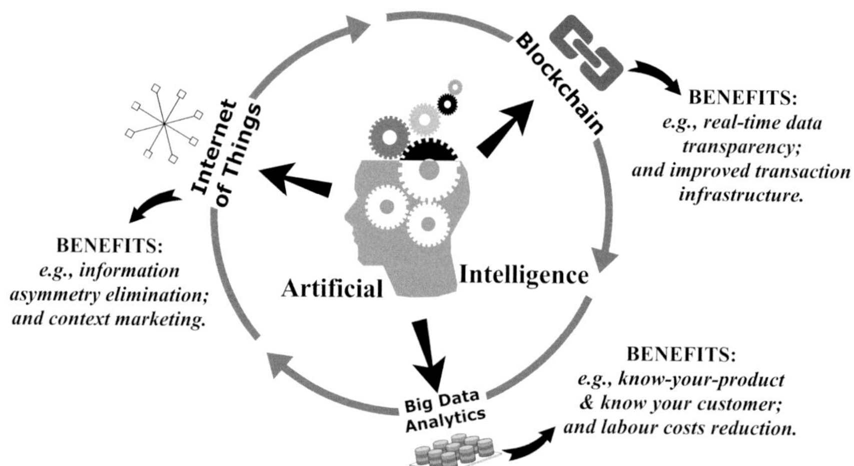

Figure 4.2: Overview of major disruptive FinTech technological drivers.

4.2.5.1 *Artificial Intelligence (AI)*

During the past few years, no technological area has been hotter than AI. According to PitchBook (The Economist, 2017a), a data provider, in the first nine months of 2017, venture-capital investment in AI amounted to $7.6bn, while the full-year figure of 2016 was $5.4bn. From a macro perspective, AI includes massive algorithms (Dalager and Jensen, 2016; Xing and Marwala, 2018l; Xing and Gao, 2014c; Xing and Gao, 2014s; Xing and Gao, 2014z; Xing and Gao, 2014bb; Xing and Gao, 2014dd; Xing and Gao, 2014b; Xing and Gao, 2014q; Xing and Gao, 2014x; Xing and Gao, 2014m; Xing et al., 2013c; Xing and Gao, 2014aa; Xing and Gao, 2014cc; Xing and Gao, 2014d; Xing and Gao, 2014e; Xing and Gao, 2014f; Xing and Gao, 2014g; Xing and Gao, 2014h; Xing and Gao, 2014i; Xing and Gao, 2014j; Xing and Gao, 2014k; Xing and Gao, 2014l; Xing and Gao, 2014n; Xing and Gao, 2014o; Xing and Gao, 2014p; Xing and Gao, 2014r; Xing and Gao, 2014t; Xing and Gao, 2014u; Xing and Gao, 2014v; Xing and Gao, 2014w; Xing and Gao, 2014y) that work together and are applicable in many domains, such as robotics (Xing, 2016b; Xing and Marwala, 2018k; Xing and Marwala, 2018e; Xing and Marwala, 2018f; Xing and Marwala, 2018d; Xing and Marwala, 2018b; Xing and Marwala, 2018c; Xing, 2016d; Xing and Marwala, 2018n; Xing and Marwala, 2018m; Xing and Marwala, 2018g; Xing and Marwala, 2018j; Xing and Marwala, 2018h; Xing and Marwala, 2018i; Xing and Marwala, 2018a; Xing, 2016c), antenna optimization (Xing, 2015a), natural language processing (Yu and Deng, 2015), manufacturing/remanufacturing (Xing and Gao, 2014a; Xing et al., 2010e; Xing et al., 2009; Xing et al., 2010d; Xing and Gao, 2014ee; Xing and Gao, 2014ff; Xing and Gao, 2014gg; Xing and Gao, 2014hh; Xing and Gao, 2014ii; Xing and Gao, 2014jj; Xing and Gao, 2014kk; Xing and Gao, 2014ll; Xing and Gao, 2014mm; Xing and Gao, 2014nn; Xing and Gao, 2014oo; Xing and Gao, 2014pp; Xing and Gao, 2014qq; Xing and Gao, 2014rr; Xing and Gao, 2015b; Gao et al., 2013a; Gao et al., 2013b; Xing et al., 2013a), power system (Xing, 2015b), supply chain (Xing et al., 2010a; Xing et al., 2010b; Xing et al., 2010c; Xing and Gao, 2015a; Xing et al., 2012), cross docking (Xing, 2014), data processing (Xing et al., 2013b) and gesture/facial/speech recognition systems (Xing and Marwala, 2018g).

More recently, the impact of AI in financial services industries has become more and more notable. According to PwC (2017g), a consultancy, there are a growing number of startups that have added AI techniques into the core of their product offerings for fulfilling various tasks such as monitoring trade fraud (Koning, 2016; Šubelj et al., 2011; Quah and Sriganesh, 2008), developing high frequency trading platforms (Hamerton-Stove, 2016; Biais and Foucault, 2014; Chordia et al., 2013;

Menkveld, 2016; O'Hara, 2015), managing financial risks (Dunis et al., 2016) and deploying robo-advisors (Finextra, 2016; Deloitte, 2016; PwC, 2016a). More information in this regard can be found in (Marwala and Hurwitz, 2017a).

4.2.5.2 Internet of Things (IoT)

The term of 'Internet of things' can be defined in a number of different ways, based on the actual deployment areas. Generally speaking, it can be understood as "the network of physical objects or things, typically embedded with different technologies and connectivity, so as to support greater value and service via the Internet" (Xing, 2016a). Adopting IoT brings opportunities as well as challenges across all business domains, such as manufacturing (Löffler and Tschiesner, 2013; Lopez Research, 2014; Mourtzis et al., 2016), supply chain (Xing et al., 2011; Verdouw et al., 2013; Mariani et al., 2015; Xu, 2011) and healthcare (Salahuddin et al., 2017; Bhatt et al., 2017; Santos et al., in press; Kulkarni and Sathe, 2014; Paschou et al., 2013; Santos et al., 2014). In addition, as devices become smarter and smarter, IoT also helps us build new forms of human–robot interactions (Xing and Marwala, 2018n).

Under the umbrella of FinTech, IoT refers to the utilization of mobile and wearable devices/systems for gathering and analysing data from different sources for the purpose of alleviating information asymmetry (Marwala and Hurwitz, 2017c), revealing insights and recommending actions (Capagemini, 2015). In other words, IoT can determine the form of interactions between FinTech companies and the target customer groups. One intuitive example is advising and transacting with customers physically/remotely. Meanwhile, with the aid of IoT, InsurTech firms can offer customers more personalized and flexible products (Berger, 2015; Yan et al., 2018).

4.2.5.3 Blockchain

Blockchain was first introduced in around 2008 by Satoshi Nakamoto and can be briefly defined as a distributed software protocol and decentralized ledger of the cryptocurrency Bitcoin for recording transactions without any intervention (Nakamoto, 2008). As customers are embracing the proposition of 'true source of data', blockchain technology has turned to a real buzzword as it holds out the promise of trust and, thus, has the potential to become a game changer in many levels of financial services (Wang et al., 2016). For example, it is believed that blockchain technology can reduce the costs associated with payments, trade settlement and trade finance, which in turn boosts the efficiency of trusted document storage, e.g., know-your-customer/identity management initiatives (Biella and Zinetti, 2016; World Economic Forum, 2016). Accordingly, based on INNOVALUE (a consultancy) research (Burelli et al., 2015), several large banks have announced their proposals for leveraging blockchain technology. In general, these intentions can be categorized into the following four classes (Burelli et al., 2015):

- **In-house Development (competitive approaches):** practiced by Nasdaq, Citibank, LHVpank, and so on.
- **Investment (acquisition approaches):** employed by Goldman Sachs (invested in CIRCLE); BBVA, NYSE and USAA (invested in Coinbase).
- **Partnership (involvement approaches):** exercised by Western Union, Fidor bank, and Cross River bank (collaborated with Ripple), PayPal (cooperated with Coinbase), bitpay and GoCoin.
- **Accelerators (co-operative approaches):** utilized by UBS (in the case of L39), Barclays (in the case of Blocktrace), Atlas Card and Safello.

In addition, according to a recent study conducted by Deloitte (2015a), blockchain technology startups disrupt the financial services industries in several aspects, such as facilitation (e.g., Provenance and IPFS), smart contracts (e.g., R3's Corda and Ethereum), digital assets (e.g., Factom), exchange platforms (e.g., Epiphyte), payments (e.g., Ripple and Chain), identification (e.g., SlockIt) and other novel dimensions (e.g., Luther Systems for insurance, Enigma for privacy, and Golem for unused computer resources).

4.2.5.4 Big Data Analytics

In order to maintain a competitive edge, financial services sectors have paid a great deal of attention to big data analytics over the past few years (The Economist Intelligence Unit, 2015; Oracle, 2015). Through analytics, search, and visualization, big data technologies can help investors know-your-product, know-your-customer, make better decisions, develop more innovative, business-led data strategies, improve anti-fraud and risk management measures and reduce labour costs (Gutierrez, 2014; Vishnu and Tatseos, 2016; Baesens et al., 2015; Gee, 2015). For example, IBM's InfoSphere (IBM, 2014), a popular big data analytic tool, can be used to help financial services manage all their data and bring a variety of significant insights from them. Meanwhile, according to PwC (2017g), infrastructure typically plays a key role in facilitating in-depth big data analytics. Accordingly, incumbent financial institutions and FinTech startups are intensively making use of protection and scale-resulting functionalities offered by the cloud. Further discussions on how financial services are affected by big data analytics can be found in (Zhang et al., 2015; Yang et al., in press; Gutierrez, 2014; Trelewicz, 2017; Baker, 2015).

4.2.6 Landscape of FinTech Industry

The landscape of Fintech industry is rather broad where we can observe numerous initiatives. A raw classification of these practices is depicted in Fig. 4.3.

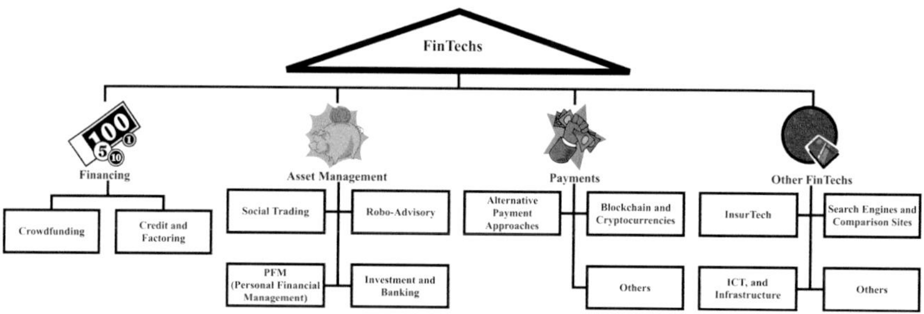

Figure 4.3: Segments of FinTech industry.

Interested readers should refer to several 'FinTech 50' or 'FinTech 100' reports published by KPMG (AWI & FSC & KPMG, 2014; KPMG, 2016a; KPMG, 2016b) and Forbes (Sharf et al., 2015; Sharf et al., 2016), respectively. Amongst this large variety of applications, some are relatively traditional in nature (e.g., use of mobile technology), while others are indeed quite innovative, like AI in banking. For the rest of this section, four main FinTech categories (Gomber et al., 2017; Dorfleitner et al., 2017) are briefly discussed.

4.2.6.1 Financing

Financing can be used to stand for the act of providing funds for SMEs (i.e., small-and medium sized enterprises) and startups (Wang and Yang, 2016; Rossi, 2017). With the introduction of new technologies, this segment can be further divided into two sub-directions, namely crowdfunding and credit and factoring (Dorfleitner et al., 2017). Crowdfunding (or called microfinancing), the main theme of this book and a very popular instrument for startups who wish to raise funds, will be discussed specifically in Section 4.3. Here, we only look at credit and factoring.

As one financing approach, credit and factoring (or called accounts receivable financing), enables an entrepreneur to sell his accounts receivable or invoices to a third party (usually termed as a factor) at a discount, so that the cash needs can be quickly met (Fiordelisi and Molyneux, 2004; Asselbergh, 2002; Soufani, 2000; Klapper, 2006). In other words, the factoring industry offers an intermediate solution for the encountered liquidity problems. In FinTech domain, startups intend to provide innovative factoring solutions, such as managing the account receivables online, providing online instant credit approvals for one's accounts, designing personalized factoring process (i.e., allowing you to choose which customers you would like to factor invoices for) and handling one's collection work (Dorfleitner et al., 2017). Representative practitioners in this category include Advanon in Switzerland, TrustBills and interFin in Germany, BillFront and Milestone in the UK and BlueVine and Kabbage in the USA.

4.2.6.2 Asset Management

In the financial area, the term 'asset management' has two general definitions: (1) One is related to a systematic process of managing clients' assets (both tangible and intangible) in a cost-effective manner (2) The other one stands for advisory services. Under the title of FinTech, asset management has four branches, namely social trading, robo-advice, personal financial management, and investment and banking (Dorfleitner et al., 2017). The first three branches are introduced as follows:

- Social trading represents a network that allows investment decisions to be made based on the information gathered from social networks (Glaser and Risius, in press; Chen et al., 2014). Its main characteristic lies in the fact that it offers an opportunity to copy traders of other users (Amman and Schaub, 2016). More details can be found in (Falkner and Corsthwaite, 2013; Lee and Ma, 2015; Oehler et al., 2016; Wohlgemuth et al., 2016; Pan et al., 2012). Famous startups in this category include eToro, ayondo, and Currensee.
- Robo-advice denotes a user-friendly online application featured by its holistic advices, 24/7 availability, simplified client experience and low costs. Robo-advice is an emerging disruptor and is expected to be successful in the future (Levin et al., 2017; Sommerfeld et al., 2016; McKinsey & Company, 2016c). It aims to automatically grow its client's portfolio (Chia and Shah, 2016). Though young, the market of robo-advisory can be clustered into four segments (as illustrated in Fig. 4.4), depending on the maturity of a robo-advisor's capabilities that have been developed (Deloitte, 2017a).

  According to BBVA (2015a), in 2014, there were about 73,000 customers who experienced such services in the US, with a total volume of $5 billion assets being managed. For instance, a USA-based robo-adviser called Wealthfront (The Economist, 2017b) is able to estimate the cost for a student to study in a given college by taking various factors into account (e.g., the increase in tuition fees and potential financial aid). Wealthfront can then suggest a tax-efficient way of saving for such

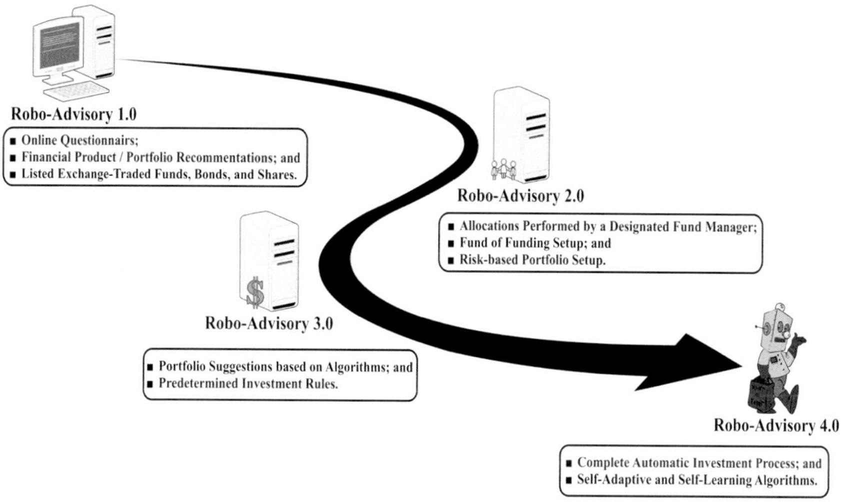

Figure 4.4: Market segments of robo-advisory.

student's parents. Other notable startups also include Future Penny, Stockspot, and Boon.

- Personal financial management (PFM) tools are typically offered by financial advisers via software and apps, with the goal of offering clients private financial services, such as viewing all accounts, transferring or switching between different financial institutions with one mobile application (Anonymous, 2016). According to BBVA (2015b), in 2015, about 82% of clients in the USA claimed that they are currently contracted to such services in one way or another. Notable portals include Mint, Yodlee, Fintonic and MoneyWiz. As the financial hub of the African continent, South Africa also has several PFM FinTech startups such as 22seven, 6Cents and Moneysmart (Anonymous, 2016).

4.2.6.3 Payments

With the ever-increasing penetration rate of mobile devices, alternative payment methods, such as mobile payment and peer-to-peer (P2P) funds transfer, have gradually transformed the way people spend and save money. Indeed, the alternative payments industry is regarded as the largest entity inside the FinTech domain. According to ITWeb (2017), in 2016, $5.5 billion were invested in payments startups, accounting for 22% of global FinTech investment activities. Examples include M-pesa in Kenya (Andiva, 2013), Google Wallet and Apple Pay in USA (Audette, 2014), Alipay and WeChat mobile wallet in China (Galvan, 2017; Xing,

2017; Galvan, 2016; Liu, in press) and Zapper and SnapScan in South Africa (Collett, 2017; Anonymous, 2016). Of course, though most innovative business models are proposed by FinTech startups, incumbent financial institutions do not give up offering traditional banking products in a more effective way. Take South Africa, these products include InstantMoney from Standardbank, MyFinancialLife from Nedbank, Mobile Money from MTN and eWallet from FNB (Anonymous, 2016). More information regarding the global payments landscape can be found in (Chiu, 2017; Shen, 2017; BNY Mellon, 2015; BNY Mellon, 2014; McKinsey & Company, 2016a; McKinsey & Company, 2015; Hayden and Hou, 2015; Life. SREDA VC, 2014; Sage, 2015).

Meanwhile, according to studies conducted by McKinsey (Tapscott et al., 2016) and PwC (2015) separately, another big disruptor within this area is the use of blockchain mechanisms and crypto-currencies (e.g., Bitcoin). Similarly, Gartner Inc., a consultancy, has noticed a sharp rise in the number of enterprises that are devoted to the promotion of the use of blockchain (Furlonger and Valdes, 2017). Theoretically, blockchain technology is able to stimulate a better transparency (e.g., ownership, identity assertions and proofs, encryption keys or device attributes) in financial services.

Overall, in the landscape of the FinTech industry, the payments sector plays a dominating role. As predicted by Juniper research (PwC, 2017g), the global market value of mobile and contactless payments could reach a staggering amount of $95bn by 2018. The emerging topics with respect to payments include the ethics of payments (Angel and McCabe, 2015), security issue (BMO Bank of Montreal et al., 2015; Hu et al., 2005; Arnfield, 2015a; Ortiz-Yepes, 2014; Li et al., in press), regulations (Liu et al., 2015; Chiu, 2017; Andiva, 2013; Arnfield, 2015b) and modelling and analysing mobile payment use cases (Pousttchi, 2008). Interested readers should refer to review articles (Dahlberg et al., 2008; Dahlberg et al., in press) for further discussions.

4.2.6.4 Other FinTechs

The 'other FinTechs' dimension incorporates FinTech practices that are typically beyond the traditional banking function space (i.e., financing, asset management, and payment transactions). Among a variety of exercises, insurance (or called InsurTech) is a remarkable area that is rapidly becoming the next big thing in the FinTech domain. According to the International Association of Insurance Supervisors (IAIS) (2017), the investment in InsurTech has more than tripled, from $800 million (in 2014) to $2.5bn (in 2015). The involved startups are focusing on every aspect of the insurance value chain including marketing, distribution, claim settlement and underwriting and pricing of risks (as illustrated in Fig. 4.5).

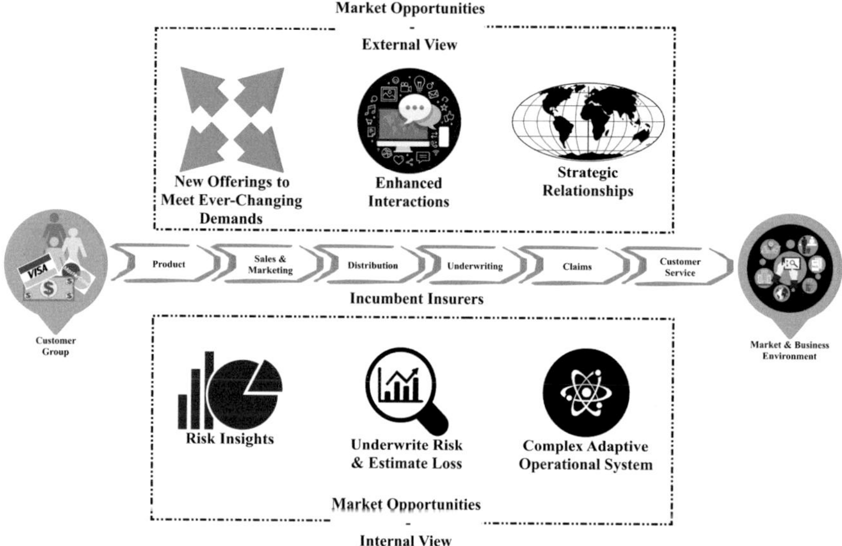

Figure 4.5: The value chain of InsurTech.

Examples of InsurTech include InMyBag, Brolly, Remy and digital Fineprint. More details can be found in several recently published reports (PwC, 2016f; Sheng et al., 2016; PwC, 2017d; PwC, 2016g). In addition, firms dedicated to search engines, comparison sites (and/or the associated techniques) and information technology (together with the corresponding infrastructure) are mainly focused on offering novel technical solutions for financial services providers. Notable cases include Nok Nok Labs, DataFox, Plaid, Zuora and square.

4.3 Crowdfunding

The traditional understanding of fundraising is mainly anchored in banks, angle investors, or other financial services. However, this concept positions artisans (e.g., musicians and artists), scientists and early-stage enterprises at the undesired level of the value chain, which leads them to be less considered. Nowadays, with the popularity of Facebook, LinkedIn and Twitter, digital identity is broadly accepted and this new type of trust in turn fuels the emergence of new financing instruments in digital economy. Among others, crowdfunding appeared as a novel answer to illustrate the financing power of a large number of individuals when tapped for funding through online platforms (Langley, 2016). According to Paulet and Relano (2017), the exponential growth of crowdfunding can be attributed to two factors, namely (1) the shortage of capital, and (2) the development of Web 2.0.

To date, the concept of crowdfunding has attracted a lot of attention globally from multiple dimensions:

- **Big Data and Analytics Companies:** (Nabarro, 2015);
- **Business Sectors:** (McCarthy, 2013; Sheldon and Kupp, 2017);
- **Charitable Enterprises:** (Nesta, 2012; Baeck and Collins, 2013b);
- **Consultancy Services:** (Rechtman and O'Callaghan, 2014; Hartnett and Matan, 2015; Massolution, 2015; AlliedCrowds, 2016; Cuesta et al., 2015; Kasper and Marcoux, 2015);
- **Financial Institutions:** (Baeck et al., 2014; European Banking Authority, 2015);
- **Governments:** (Salazar et al., 2015; European Commission, 2016; Stemler, 2013; Ania and Charlesworth, 2015; Australian Charities and Not-for-Profits Commission, 2017);
- **Law Firms:** (Figliomeni, 2014; Fallone, 2014; Kappel, 2009; Heminway, 2014; CrowdfundingHub, 2016a; Deloitte Legal, 2016);
- **Scientific Research Groups:** (Vachelard et al., 2016; Dahlhausen et al., 2016; Otero, 2015; Savio, 2017);
- **Universities:** (Baskerville and Cordery, 2015; Stagars, 2015; Wieck et al., 2013; Boidus Admin, 2014; Human, 2013; Arnett, 2015; Joly, 2013; Next, 2013); and
- **Venture Capital:** (Green, 2014; WilmerHale, 2016; Kantor, 2014; Beugré and Das, 2013; Segarra, 2013; Hogg, 2014; Gobble, 2012).

Meanwhile, a large number of successors also shared their experiences (Helmer, 2014; Thurston, 2013). For the rest of this section, we will give a brief overview of the whole structure of crowdfunding.

4.3.1 *What is Crowdfunding?*

Crowdfunding—the Internet-based extension of traditional financing by crowds—is an innovative way for people or organizations to finance their projects, causes or startups from the general public over a fixed time period.

- In the broadest sense, crowdfunding can be defined as "*a collective effort made by people who network and pool their money together, usually via the Internet, in order to invest in and support efforts initiated by other people or organizations*" (Ordanini et al., 2011).
- Similarly, Belleflamme et al. (2014) defined crowdfunding as "*an open call, mostly through the Internet, for providing financial resources either in*

the form of a donation or in exchange for the future product or some form of reward".

- From a micro point of view, Mollick (2014) considered crowdfunding as *"the efforts by entrepreneurial individuals and groups—cultural, social, and for-/non-profit—to fund their ventures by drawing on relatively small contributions from a relatively large number of individuals using the Internet, without standard financial intermediaries"*.

As we can see, the above-mentioned three definitions are indeed very similar. Therefore, it is helpful to extract the key elements from the heart of crowdfunding concept as follows (Bouncken et al., 2015):

- **Who:** Each individual (i.e., crowd) who believes in a common effort (e.g., by contributing a relatively small amount of funds) that can catalyse the energy and resources necessary to achieve an ultimately much larger goal.
- **What:** It focuses on financial operations, aims at collecting capitals from a large population of crowd towards the realization of one or more projects.
- **Where:** The Internet serves as the intermediary to facilitate contacts among different players (e.g., investors, stakeholders and communities) in engaging with proponents of the project and monitoring the progress over time.
- **How:** The advancement of Web 2.0 technique plays a key role in leveraging the accessibility of crowdfunding from the crowd perspective.

In light of this analysis, the properties of crowdfunding are different from those mainstream investment banking instruments (e.g., venture capital and angel investing) in several aspects (Attuel-Mendes, 2017; Inter-American Development Bank, 2015; Rubinton, 2011): (1) Crowdfunding can provide new, easy and faster access to investment opportunities in an open environment; (2) Crowdfunding allows for the creation of individual-to-individual links together with the enhanced transparency and accountability; and (3) Crowdfunding engages a highly motivated group to support entrepreneurs in their early stages. These differences are illustrated in Fig. 4.6.

Today, the concept of crowdfunding is being examined in many areas around the globe:

- **Europe:** (Walterus and Williams, 2014; CrowdfundingHub, 2016b; Glasgow Chamber of Commerce et al., 2016; Dushnitsky et al., 2016; European Commission, 2016; Voldere and Zeqo, 2017; Dietrich and Amrein, 2017; Sannajust et al., 2014)

Traditional Funding Scheme

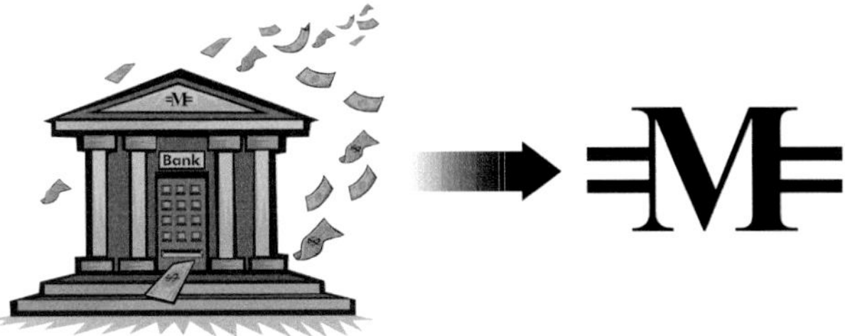

Crowfunding Paradigm

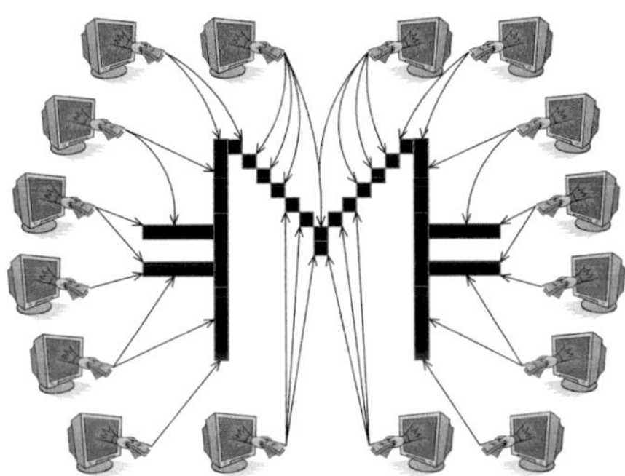

Figure 4.6: Crowdfunding properties.

- **Asia-Pacific:** (Royal and Windsor, 2014; Hu, 2015; Zhang et al., 2014; Beatty et al., 2015; Murray, 2015; Angle, 2017; CGAP, 2017; HK Financial Services Development Council, 2016)
- **North America:** (Freedman and Nutting, 2015a; Buckstein, 2014; Nordicity, 2012; The National Crowdfunding Association (NCFA) of Canada, 2016)
- **Southern America:** (Multiateral Investment Fund, 2015)
- **Africa:** (Boum, 2016; FSDA, 2016; Garvey et al., 2017)
- **Middle-East:** (Mourtada, 2014; Abushaban and Gaza, 2014).

In addition, several cross-cultural comparative studies can also be found in (Cho and Gawon Kim, 2017; Zheng et al., 2014; Gajda and Mason, 2013; Marchand, 2016). Theoretically, anyone (even a company) with an idea of offering a product or a service can seek external financing worldwide through crowdfunding.

Furthermore, the phenomenon of crowdfunding has also attracted a great deal of attention from the academic domain. Several popular research directions can be summarized as follows:

- **Crowdfunding and FinTech:** (Kim and Moor, 2017; Fenwick et al., 2017; CGAP, 2017);
- **The Dynamics of Crowdfunding (e.g., economics, techniques, ethics and cultures):** (Agrawal et al., 2014; Mollick, 2014; Riedl, 2013; Sara, 2013; Snyder, 2016; Kuppuswamy and Bayus, 2013; Valančienė and Jegelevičiūtė, 2013; Valančienė and Jegelevičiūtė, 2014; Pazowski and Czudec, 2014; Josefy et al., 2017; Snyder et al., in press; Voldere and Zeqo, 2017);
- **Discriminations in Crowdfunding (e.g., geography, gender and attractiveness):** (Agrawal et al., 2015; Marom et al., 2014; Lin and Viswanathan, 2016; Barasinska and Schäfer, 2014; Jenq et al., 2015; Ravina, 2012; Jin et al., 2017; Gorbati and Nelson, 2015; Burtch et al., 2015);
- **Motivations of Participation:** (Gerber et al., 2012; Gerber and Hui, 2013; Zhao et al., 2017; Yan et al., in press);
- **Success Factors:** (Yang and Zhang, 2014; Manning and Bejarano, 2017; Fan-Osuala et al., 2018; Burtch et al., 2013; Dushnitsky et al., 2016; Wechsler, 2013; Moutinho and Leite, 2013; Xu et al., 2014; Courtney et al., 2017);
- **Crowdfunding Outcomes:** (Chan and Parhankangas, 2017; Cholakova and Clarysse, 2015; Drover et al., 2017; Mollick and Kuppuswamy, 2014; Deutsch et al., 2017);
- **Sentiment Analysis:** (Parhankangas and Renko, 2017; Wang et al., 2017a; Anglin et al., 2014; Allison et al., 2014; Allison et al., 2017; Mitra and Gilbert, 2014; Larrimore et al., 2011; Davis et al., 2017);
- **Social Capital:** (Mollick, 2016; Colombo et al., 2015; Hui et al., 2014; Kang et al., 2017; Giudici et al., 2013; Lu et al., 2014; Zheng et al., 2014; Marom, 2017; Zvilichovsky et al., 2015; Alfiero et al., 2014; Skirnevskiy et al., 2017; Gleasure and Morgan, in press; Buttice et al., 2017); and
- **Social Entrepreneurship or Social Ventures:** (Meyskens and Bird, 2015; Calic and Mosakowski, 2016; Lehner, 2013; Allison et al., 2014; Ashta et al., 2015; Ashta and Assadi, 2008; Banhatti, 2016; Rana, 2013; Roche-Saunders and Hunt, 2017).

4.3.2 *The History of Crowdfunding*

The birth of the crowdfunding concept is often associated with the time that capitals are constrained. In fact, it is not a brand new idea and shares many commonplaces with crowdsourcing, which means the utilization of collective (or crowd) efforts to add values (idea improvements, instant feedback, and alternative solutions) in achieving an objective (Brabham, 2013). Take the USA's Statue of liberty (one of the world's most recognizable landmarks), it actually was the first big and successful crowdfunding campaign which raised $100,000 from 125,000 citizens for constructing a pedestal at New York harbor in 1884 (Nesta, 2012). Since then, the seed of crowdfunding began to sprout, but mainly in the USA and Europe. Examples include the $12,875 raised by Philadelphia city for planting 15,000 trees and the funds collected by Rotterdam for building a wooden pedestrian bridge (Hughes, 2016; Toren, 2014; Collins, 2013). In 1997, one of the first Internet-based crowdfunding campaigns was launched by Marillion (a British rock group) and the project closed successfully by collecting $60,000 in supporting a trip to US (Hobbs et al., 2016).

Around the year 2008, crowdfunding became a noteworthy financing instrument rather than merely an alternative option. The underlying reasons might be attributed to the worldwide financial crisis, technological innovation and/or the mainstream use of social networks. Another watershed is that, in April 2012, the US President Obama by that time officially signed the JOBS Act (Jumpstart Our Business Startups Act) into law, which opened the door to a new kind of approach under its Title III in terms of raising private investments. According to Zhang et al. (2016), the UK-based crowdfunding market grew to £3.2 billion in 2015, consisting of a great population of participants (about 1.09 million funders and 254,721 fundraisers), and a great pool of projects and businesses involved in funding and fundraising activities. Based on a recent estimate made by the World Bank, the crowdfunding market value could reach $96 billion by 2025 (infoDev and The World Bank, 2013).

4.3.3 Segments of Crowdfunding Platforms

Crowdfunding platforms are actually websites that facilitate the interactions between fundraisers and the crowds for gaining access to additional funding. The platforms typically earn fees as their income for offering all sorts of intermediary services (Belleflamme et al., 2015). The fundamental crowdfunding principle is depicted in Fig. 4.7.

In practice, some platforms allow a variety of projects to be funded through their channels (e.g., Kickstarter and IndieGoGo), while others focus on niche subjects, e.g., RocketHub and Angeldorm for university and college funds, Microryza for scientific research, Foodstart for restaurants, Cruzu and Fundovino for wine making, Appsplit for mobile and tablet apps, Seed&spark for film and TV and Kiva for social and charitable causes (Luzar, 2013). According to Masssolution (2015), a consulting firm, the number of platforms worldwide reached 1,250 in 2015, where the total amount of transacted funds was about $30 bn (doubled since 2014). Generally speaking, based on the aims and forms of return, crowdfunding platforms can be clustered into the following four types: Reward-based, donation-based, equity-based and peer-to-peer-based crowdfunding.

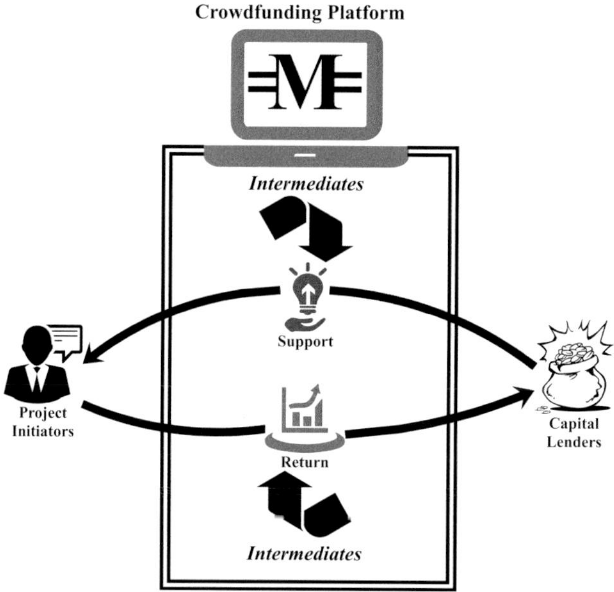

Figure 4.7: The basic crowdfunding principle.

4.3.3.1 *Reward-based Crowdfunding*

As the name implies, a key feature of reward-based crowdfunding is that, instead of offering a stake in the project, the fundraisers offer other forms of rewards to funders (e.g., a personalized copy of final product or pre-ordering of products) (Belleflamme et al., 2014). One of the very first Internet-based reward-based crowdfunding platforms is called ArtistShare, which was launched by Brian Camelio (a musician and computer programmer from Boston) in 2003. Successful campaign cases via this channel include Maria Schneider's jazzy album 'Concert in a Garden', which raised about $130,000, and later won a 2005 Grammy Award for the best large jazz ensemble album (Lewis, 2015). Motivated by its success, more and more rewards-based crowdfunding platforms were launched. Famous examples include Indiegogo (in 2008) and Kickstarter (in 2009). Among various successful stories, Pebble Watch campaign in 2015 (Kickstarter, $20 m) and Ubuntu Edge campaign in 2013 (Indiegogo, $12.8 m) are worth a mention.

Like a recommender system (Gao et al., 2014), the reward-based crowdfunding platforms can provide step-by-step guidance for askers to construct and market their project (e.g., setting the goal, pledging different levels, determining duration time, obtaining feedback, tracking the funding progress and allocating rewards once the campaign has succeeded. Usually, the platform offers two types of fundraising rules, i.e., keep-it-all (KIA) and all-or-nothing (AoN), for operating a project. The KIA rule means that entrepreneurs are free to set a fundraising goal and then keep whatever they have accumulated at the end of the project, regardless of whether or not the initial goal is met; while the AoN rule indicates that the entrepreneurs can only keep the gathered funds if and only if when the desired goal is achieved or exceeded (Cumming et al., 2015). On the one hand, this type of crowdfunding naturally becomes a simple and popular place to meet with consumers (Mollick, 2014); on the other hand, it also enables value discovery and matching (Fan, 2013), market testing and demand measurement (Inter-American Development Bank, 2015; Brown et al., 2017; Moisseyev, 2013; Sheldon and Kupp, 2017), innovation conversation communication (Mollick, 2016; Stanko and Henard, 2016; Stanko and Henard, 2017) and brand message reinforcement of the product (Belleflamme et al., 2014; Cowley, 2016; Bazilian, 2013).

Further discussion regarding this type of crowdfunding can be found in (Kunz et al., 2017; Giudici et al., in press; Steigenberger, 2017; Zheng et al., 2016; Bao and Huang, 2017; Bi et al., 2017; Roma et al., 2017; Frydrych et al., 2014; Thuerridl and Kamleitner, 2016; Buff and Alhadeff, 2013; Wechsler, 2013; Chertkow and Feehan, 2014; Kraus et al., 2016; Lipusch et al., 2016; Gamble et al., 2017; Colistra and Duvall, 2017; Macht and Weatherston, 2015).

4.3.3.2 Donation-based Crowdfunding

The donation-based crowdfunding platforms mainly pursue intangible benefits as a return. Most fundraisers in this area belong to non-profit organizations (e.g., charity or NGO), while funders are often regarded as philanthropists (Flanigan, 2017). In general, the donated projects vary from emergencies (Salazar et al., 2015) to people in need (Berliner and Kenworthy, 2017; Sisler, 2012; Snyder et al., 2016). According to CrowdExpert (2016), the fundraising amount of donation-based crowdfunding has reached $2.85 bn globally in 2015. In this category, GoFundMe is a notable platform launched in 2010. Other platforms also include CrowdCure, Pledgie, JustGiving and Watsi (Meer, 2014; Bekkers and Wiepking, 2011; Zhong and Lin, 2017; Althoff and Leskovec, 2015; Özdemir et al., 2015; Royal and Windsor, 2014; Pitschner and Pitschner-Finn, 2014).

4.3.3.3 Peer-to-Peer-based Crowdfunding

One can treat the P2P based crowdfunding platform as an alternative lending-based microfinance model, in which funders seek loans by offering various interest payments (Marakkath and Attuel-mendes, 2015; Wang et al., 2015). The fundamental work principle of P2P lending is illustrated in Fig. 4.8.

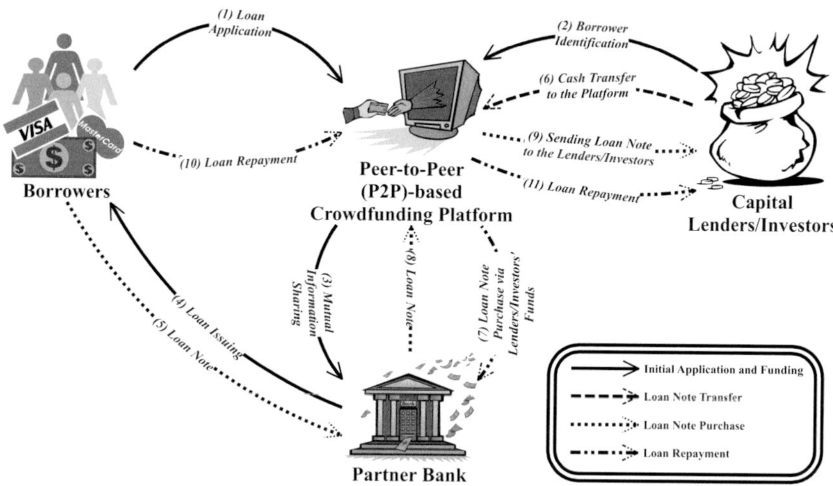

Figure 4.8: The working mechanism of P2P lending.

Generally speaking, there are five kinds of P2P lending (or crowdfunding), namely microfinance lending (e.g., Kiva), social investing (e.g., Microplace and Babyloan), marketplace lending (e.g., Prosper, Lending Club, amd Zopa), peer-to-business lending (e.g., First Circle and Kabbage) and social lending (e.g., Virgin Money). Further discussions in this regard can be found in Chapter 9 of this book.

From an investor's perspective, P2P lending platforms provide a higher rate of return (ranging from 7% to 35%); while from a borrower's viewpoint, they offer loan (or refinance debt) opportunities outside of traditional lending institutions with some merits such as reasonable rates and short time windows (Loureiro and Gonzalez, 2015; Mateescu, 2015). In this regard, Wang et al. (2015) conducted a very good comparative study between P2P lending and conventional bank loans. In addition, further information regarding P2P lending can also be found in (Fong, 2015; Hernando, 2017; Bruton et al., 2014; Bottiglia, 2016; Oxera, 2015; Oxera, 2016; Savarese, 2015).

4.3.3.4 *Equity-based Crowdfunding*

In 2012, Congress amended the USA securities laws in order to enable a new means of financing startups, that is, allowing 'crowds' to enter into the private equity market (Crescenzo, 2016; Dahl, 2012). About three years later, the Securities and Exchange Commission (SEC) of the USA finally adopted the Crowdfunding Regulation, i.e., rules under Title III of the JOBS Act. Since then, firms are allowed to issue and sell securities to all investors (regardless of their income or net worth) via crowdfunding portals (Gelfond and Eren, 2016; Gelfond and Foti, 2012). This practice is regarded as a landmark for equity-based crowdfunding platforms rapidly moving towards a new era. Typically, in comparison with the P2P lending market, equity-based crowdfunding is still small, though it has enjoyed a growth rate of almost 300% since 2014 (Zhang et al., 2016). Take the UK (Dunsby, 2016), there were only 18 equity crowdfunding campaigns launched at Crowdcube in 2015, which had a total value of merely £1 million or so.

In this category, fundraisers commonly offer equity or bond-like shares, while funders usually act like shareholders with the hope of receiving equity or equity-like arrangements (e.g., dividends or a return) on their initial investment (Bradford, 2012; Hornuf and Schwienbacher, in press). Notable platforms include Crowdcube and Seedrs (UK-based), Smart Angels and WiSeed (French-based) and wefunder (USA-based). Compared with the traditional brick-and-mortar style of fundraising practices, equity-based crowdfunding is more time efficient, cost effective (charged between 7.5% and 10%) and has low barriers of entry. More importantly, it is good for startups that are seeking efficient sources of capital with less compromises of their ownership stake or autonomy (Braun and Turke, 2013; Buckstein, 2014; Engstrom, 2015). According to Baeck et al. (2014), among equity-based crowdfunding users, 54% of them are seeking expansion capital, while the remaining 46% are businesses that look for seed or startup capital. Compared with other platforms, the process of equity crowdfunding tends to have the highest complexity. For simplicity, its fundamental working mechanism is illustrated in Fig. 4.9.

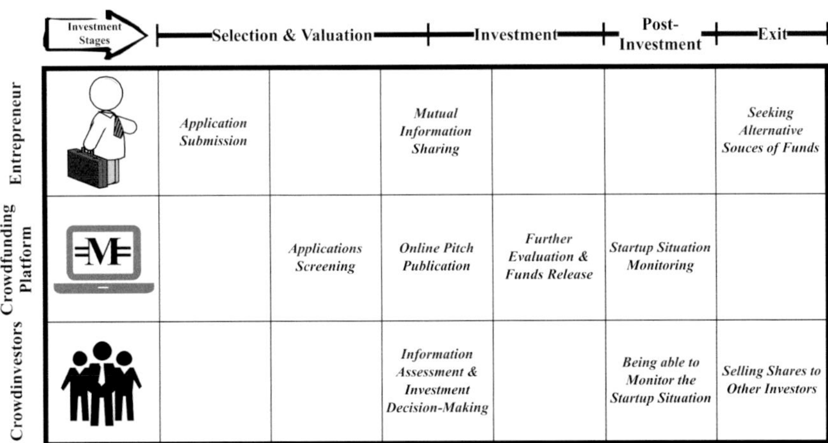

Figure 4.9: The working mechanims of equity crowdfunding platforms.

In the literature, various publications are available for a further exploration on how this type of crowdfunding exactly works.

- **Books:** (Freedman and Nutting, 2015a; Bottiglia and Pichler, 2016);
- **Journal Articles:** (Hornuf and Schwienbacher, 2017; Ahlers et al., 2015; Vismara, 2016b; Vismara, 2016a; Ley and Weaven, 2011);
- **White Papers:** (Lord, 2014);
- **Reports:** (Collins and Pierrakis, 2012; Glover, 2014); and
- **Magazine Articles:** (Tablas, 2013; Saler, 2014; Mandelbaum, 2014; Fields and Gehret, 2012; Freedman and Nutting, 2015b; Quawasmi, 2015; Taylor, 2015; Baker, 2014b; Dahl, 2014; Baker, 2014a; Remus and Mendoza, 2012; Wallace, 2014; Berger, 2014; Sieck, 2012).

All this joint effort contributed by researchers, policy makers, lawyers, entrepreneurs, and even crowdfunders themselves offer a great opportunity in terms of systematizing field knowledge and summarizing the corresponding main characteristics

4.3.3.5 *Mixed Crowdfunding*

This type of crowdfunding platform just operates as its name suggests, that is, mixing tactics learned from other crowdfunding models. Take Crowdbnk and UBREW, these two platforms offer someone a chance to get a reward or start an equity campaign, while Angel.me combines crowdfunding for both rewards and equity. A simplified mapping showing the role of mixed crowdfunding is depicted in Fig. 4.10.

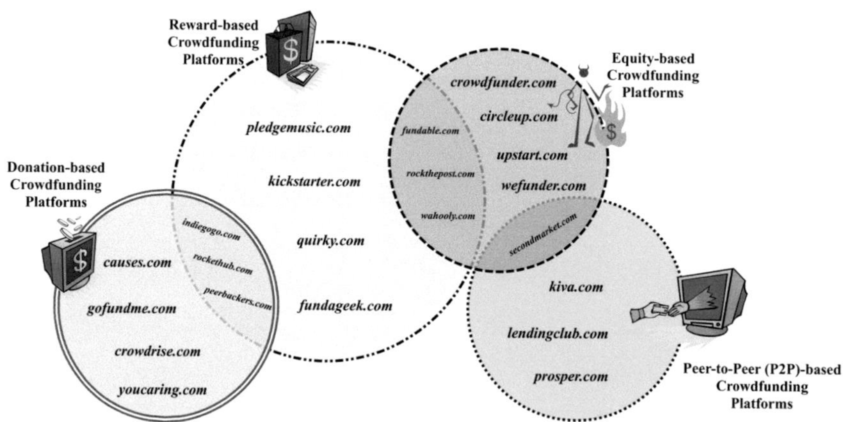

Figure 4.10: The mixed crowdfunding platforms.

4.3.4 *Crowdfunding Ecosystem*

Based on previously discussed classifications, the ecosystem of crowdfunding typically includes the following five key players (Arsenault, 2015; Bouncken et al., 2015; Tomczak and Brem, 2013; Bui, 2017; Agrawal et al., 2014):

- **Regulator:** the policy governor who governs the healthy development of crowdfunding market;
- **Asker:** the person who proposes and runs a crowdfunding project (e.g., an artist, an author, a musician, or an inventor), or someone who wants to get the business promoted (Stemler, 2013).
- **Backer:** the literal meaning of 'backer' denotes individuals or groups that pledge money into crowdfunding campaigns in supporting the interested idea. However, money is not necessarily the sole offering from crowdfunding backers, they can also contribute their opinions and suggestions to the target entrepreneurs. From the angle of early product development (Mollick, 2016; Stanko and Henard, 2017), crowdfunding backers' active role is crucial and they tend to form the earliest possible buyer group (Belleflamme et al., 2014). Last but not least, backers can also stimulate the word of mouth effect for the funded project (Burtch et al., 2014; Bi et al., 2017).
- **Investor:** the term of investor is used specifically to represent those that put money into crowdfunding projects with the expectation of achieving a profit.
- **Operator:** In general, operators stand for the founders of crowdfunding platforms that maintain the functionality of their organizations so that different parties can together to join a crowdfunding campaign.

Accordingly, the main part of this book contains Chaps. 5–10, which are organized as follows:

- **Section II:** Chapter 5 Crowdfunding Regulator—Technology Foresight and Type-2 Fuzzy Inference System;
- **Section III:** Chapter 6 Crowdfunding Asker—Campaign Prediction and Ensemble Learning;
- **Section IV:** Chapter 7 Crowdfunding Backer—Sentiment Analysis and Fuzzy Product Ontology;
- **Section V:** Chapter 8 Crowdfunding Investor—Credit Scoring and Support Vector Machine; and
- **Section VI:** Chapter 9 Crowdfunding Operator—Portfolio Selection and Metaheuristics, and Chapter 10 Crowdfunding Operator—Channel Competition, Strategic Interaction and Game Theory.

4.3.5 Crowdfunding User Manual

Equipped with the fundamental knowledge of the crowdfunding concept, we can now move on to the discussion of how to attract attention and get a campaign successfully funded. Initially, launching a crowdfunding campaign may sound exciting, but in reality, it often becomes a laborious task. Some scholarly insights regarding individual crowdfunding practices can be found in (Belleflamme et al., 2013; Baeck and Collins, 2013a). Take a reward-based crowdfunding campaign, there are generally seven steps (depicted in Fig. 4.11) that one needs to follow (Baeck and Collins, 2013b):

- **Step 1–Prepare an Innovation:** Individuals or entrepreneurs conceive a creative and/or business concept.
- **Step 2–Select a Proper Platform:** According to the ultimate goals, find a proper crowdfunding platform for the project.
- **Step 3–Enrich the Details of a Project:** Tasks may include making a video, writing a narrative, setting a funding goal, and determining the level of rewards.
- **Step 4–Share the Project with the Public:** Promoting the campaign through social media (e.g., Twitter and Facebook) is a popular means.
- **Step 5–Receive Contributions from the Crowd.**
- **Step 6–Develop the Project:** Given the campaign is successful, a project needs to be implemented in accordance with the description. Remember to keep your backers updated with the progress of the project.
- **Step 7–Fulfil the Promise:** e.g., distributing rewards.

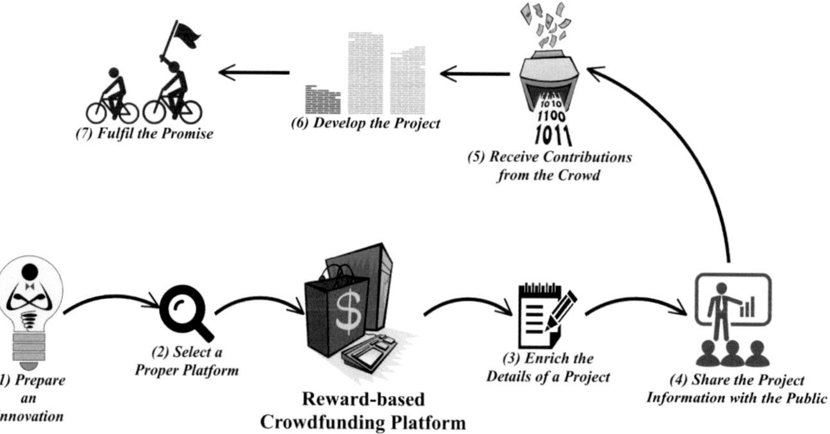

Figure 4.11: The general process of reward-based crowdfunding.

4.4 Conclusions

With its rapid development, the concept of 'crowdfunding' is not only regarded as a sub-category of open innovation, i.e., actively seeking inspirations from outsiders (Stanko and Henard, 2017; Agrawal et al., 2014; Seltzer and Mahmoudi, 2012), but is also becoming one of the mainstream financing instruments for supporting creative ideas in many domains:

- **Art:** (Mollick and Nanda, in press);
- **Civic:** (Stiver et al., 2015; Griffiths, 2017; Glover, 2017; Anonymous, 2012);
- **CleanTech:** (Cumming et al., 2017; Zhu et al., 2017; Zhu et al., 2016; Vasileiadou et al., 2016; Lam and Law, 2016; Hörisch, 2015);
- **Film Making:** (Galuszka and Brzozowska, 2015; Sørensen, 2012; Booth, 2015; Chaudry, 2014; Lewis, 2015);
- **GLAM (i.e., galleries, libraries, archives and museums) Organizations:** (Riley-Huff et al., 2016; Park, 2013);
- **Music:** (Chertkow and Feehan, 2015; Somerford, 2011; Jones, 2013; Lewis, 2015; Fruner et al., 2012; Mixon et al., 2017; McCracken, 2017; Kappel, 2009; Gamble et al., 2017; Martínez-Cañas et al., 2012);
- **PC/Video Games:** (Gilbert, 2017; Cha, in press; Nucciarelli et al., 2017);
- **Publishing:** (Carvajal et al., 2012; Mustafa and Adnan, 2017; Jones, 2014; Reid, 2014; Jian and Usher, 2014);
- **Real Estate:** (Lem, 2014; Levine and Feigin, 2014; IPF, 2016; Marchand, 2016);
- **Scientific Research Projects:** (Li and Pryer, 2014; Cameron et al., 2013; Krittanawong et al., 2018; Mullard, 2015; Dragojlovic and Lynd, 2016; Dragojlovic and Lynd, 2014; Giles, 2012);
- **Startups:** (Polzin et al., in press; Antonenko et al., 2014; Leiber and Tozzi, 2012; Ibrahim and Verliyantina, 2012);
- **Tourism Project:** (Beier and Wagner, 2014; Wang et al., 2017b);
- **Theatre Projects:** (Boeuf et al., 2014; Laughlin, 2013);
- **Wine Production:** (Mariani et al., 2017; Mariani et al., 2014).

References

Abend V, Peretti B, Axelrod CW, et al. (2014) Cyber security for the banking and finance sector. In: Voeller JG (ed) *Cyber Security*. 111 River Street, Hoboken, NJ 07030: John Wiley & Sons, Inc., ISBN 9781118651735, Chapter 9, pp. 97–112.

Abushaban RM and Gaza P. (2014) Crowdfunding as a catapult for innovation in the Middle East: obstacles and possibilities. *Proceedings of IEEE 2014 Global Humanitarian Technology Conference*, pp. 433–440.

Accenture. (2016) Fintech and the evolving landscape: Landing points for the industry. 1–12.

Aggarwal R, Goodell JW and Selleck LJ. (2015) Lending to women in microfinance: Role of social trust. *International Business Review* 24: 55–65.

Agrawal A, Catalini C and Goldfarb A. (2014) Some simple economics of crowdfunding. *Innovation Policy and the Economy* 14: 63–97.

Agrawal A, Catalini C and Goldfarb A. (2015) Crowdfunding: Geography, social networks, and the timing of investment decisions. *Journal of Economics & Management Strategy* 24: 253–274.

Ahlers GKC, Cumming D, Günther C, et al. (2015) Signaling in equity crowdfunding. *Entrepreneurship: Theory and Practice* 39: 955–980.

Alfiero S, Casalegno C, Indelicato A, et al. (2014) Communication as the basis for a sustainable crowdfunding: The Italian case. *International Journal of Humanities and Social Science* 4: 46–55.

Allen S. (2013) *Financial risk management: A practitioner's guide to managing market and credit risk*, 111 River Street, Hoboken, NJ 07030: John Wiley & Sons, Inc., ISBN 978-1-118-17545-3.

AlliedCrowds. (2016) Developing world crowdfunding: Prosperity through crowdfunding. AlliedCrowds, 1–7.

Allison TH, Davis BC, Short JC, et al. (2014) Crowdfunding in a prosocial microlending environment: Examining the role of intrinsic versus extrinsic cues. *Entrepreneurship: Theory and Practice* 39: 53–73.

Allison TH, Davis BC, Webb JW, et al. (2017) Persuasion in crowdfunding: An elaboration likelihood model of crowdfunding performance. *Journal of Business Venturing* 32: 707–725.

Althoff T and Leskovec J. (2015) Donor retention in online crowdfunding communities: A case study of DonorsChoose.org. *Proceedings of the 24th International Conference of World Wide Web, 18–22 May, Florence Italy*, pp. 1–8.

Amman M and Schaub N. (2016) *Social interaction and investing: Evidence form an online social trading network, retrieved from https://www.rsm.nl/fileadmin/home/Department_of_Finance__VG5_/PAM2016/Final_Papers/Nic_Schaub.pdf, accessed on 25 August 2016.*

Andiva B. (2013) Mobile financial services and regulation in Kenya. Kenya Railways Hqs Block D, P. O. Box 36265 00200 Nairobi Kenya: Competition Authority of Kenya.

Angel JJ and McCabe D. (2015) The ethics of payments: Paper, plastic or Bitcoin? *Journal of Business Ethics* 132: 603–611.

Angle P. (2017) Crowd power. *Indian Management* 56: 81–87.

Anglin AH, Allison TH, McKenny AF, et al. (2014) The role of charismatic rhetoric in crowdfunding: An examination with computer-aided text analysis. *Social Entrepreneurship and Research Methods* 9: 19–48.

Ania A and Charlesworth C. (2015) Crowdfunding guide: For nonprofits, charities and social impact projects. 215 Spadina Ave, Suite 400, Toronto, ON, M5T 2C7: HiveWire & Centre for Social Innovation, Document Version: 1701-D.

Anonymous. (2012) Infrastructure meets crowdfunding. *PM Network, 26 August 2012*, p. 15.

Anonymous. (2016) Money (or something) and your cheques for free. *Stuff: South Africa, July~August*. 43–54.

Antonenko PD, Lee BR and Kleinheksel AJ. (2014) Trends in the crowdfunding of educational technology startups. *TechTrends* 58: 36–41.

Arner DW, Barberis J and Buckley RP. (2018) RegTech: Building a better financial system. In: Chuen DLK and Deng R (eds) *Handbook of Blockchain, Digital Finance, and Inclusion: Cryptocurrency, FinTech, InsurTech and Regulation*. 125 London Wall, London EC2Y 5AS, United Kingdom: Elsevier Inc., ISBN 978-0-12-810441-5, Volume 1, Chapter 16, pp. 359–373.

Arnett AA. (2015) Giving grows. *Diverse Issues in Higher Education, July 2, 2015, Vol. 32 Issue 11*, pp. 12.

Arnfield R. (2015a) Mobile payments security 101: how merchants and mobile payment service providers can protect their users against mobile payments fraud. Networld Media Group, 1–39.

Arnfield R. (2015b) Regulation of virtual currencies: A global overview. Networld Media Group, 1–58.

Arsenault A. (2015) Crowdfunding 101: The life cycle of a crowdfunding campaign. Culture Days, 1–43.

Ashta A and Assadi D. (2008) Do social cause and social technology meet? Impact of web 2.0 technologies on peer-to-peer lending transactions. *SSRN Electronic Journal* http://ssrn.com/abstract=1281373.

Ashta A, Assadi D and Marakkath N. (2015) The strategic challenges of a social innovation: The case of Rang De in crowdfunding. *Strategic Change* 24: 1–14.

Asmundson I. (2011) What are financial services. *Finance & Development* March 2011: 46–47.

Asselbergh G. (2002) Financing firms with restricted access to financial markets: The use of trade credit and factoring in Belgium. *The European Journal of Finance* 8: 2–20.

Attuel-Mendes L. (2017) The different ways of collaboration between a retail bank and crowdfunding. *Strategic Change* 26: 213–225.

Audette Y. (2014) Overview of mobile payments for IT cam. KPMG LLP, 1–27.

Australian Charities and Not-for-Profits Commission. (2017) Crowdfunding and charities: Information for charities, donors, and fundraisers about the use of crowdfunding. Australian Government, pp. 1–9.

Australian Government. (2016) Backing Australian FinTech. Commonwealth of Australia, ISBN 978-1-925220-82-7, 1–38.

AWI & FSC & KPMG. (2014) The 50 best FinTech innovators report. AWI, FSC & KPMG Australia, 1–65.

Baeck P and Collins L. (2013a) Working the crowd: A short guide to crowdfunding and how it can work for you. 1 Plough Place, London EC4A 1DE, UK: Nesta, retrieved from http://www.nesta.org.uk/sites/default/files/working_the_crowd.pdf, accessed on 23 June 2014.

Baeck P and Collins L. (2013b) Working the crowd: A short guide to crowdfunding and how it can work for you. 1 Plough Place, London EC4A 1DE, UK: Nesta, 1–19.

Baeck P, Collins L and Zhang B. (2014) Understanding alternative finance: The UK alternative finance industry report 2014. 1 Plough Place, London, EC4A 1DE Nesta, pp. 1–95.

Baesens B, Vlasselaer VV and Verbeke W. (2015) *Fraud analytics using descriptive, predictive and social network techniques: A guide to data science for fraud detection,* 111 River Street, Hoboken, NJ 07030: John Wiley & Sons, Inc., ISBN 978-1-119-13312-4.

Baker G. (2014a) Pulling a crowd. *NZ Business.* 30–33.

Baker H. (2014b) A tasty sum. *Director.* 59–61.

Baker P. (2015) *Dat Divination: Big data strategies,* 20 Channel Center Street, Boston, MA 02210, USA: Cengage Learning, ISBN 978-1-305-11508-8.

Banhatti RD. (2016) Crowdfunding of a social enterprise: The GloW project as a case study. In: Brüntje D and Gajda O (eds) *Crowdfunding in Europe: State of the art in theory and practice.* Springer Cham Heidelberg New York Dordrecht London: Springer International Publishing Switzerland, ISBN 978-3-319-18016-8, Part IV, Chapter 2, pp. 223–239.

Bao Z and Huang T. (2017) External supports in reward-based crowdfunding campaigns: A comparative study focused on cultural and creative projects. *Online Information Review* 41: 626–642.

Barasinska N and Schäfer D. (2014) Is crowdfunding different? Evidence on the relation between gender and funding success from a German peer-to-peer lending platform. *German Economic Review* 15: 436–452.

Baskerville RF and Cordery CJ. (2015) Crowdfunding: A threat or opportunity for university research funding? *Alternative Investment Analyst Review* Spring: 41–52.

Bate C. (2017) The 10 hottest FinTech trends for 2017. Bell Pottinger, 1–13.

Bazilian E. (2013) Brands discover crowdfunding as marketing tool. *Adweek.* 25.

BBVA. (2015a) Financial management: Robo advisors enter the scene. *Fintech Serie By Innovation Edge.* BBVA Innovation Center, 1–25.

BBVA. (2015b) Personal financial management: Easy, quickly and very efficiently are the keys of personal management tools (PFM). *Fintech Serie By Innovation Edge.* BBVA Innovation Center, 1–31.

Beatty A, Lyons R, Tam B, et al. (2015) Crowdfunding to take-off in Australia? *Governance Directions, March 2015, Vol. 67 Issue 2*, pp. 100–102.

Beier M and Wagner K. (2014) Crowdfunding success of tourism projects-evidence from Switzerland. *SSRN Electronic Journal* https://ssrn.com/abstract=2520925.

Bekkers R and Wiepking P. (2011) A literature review of empirical studies of philanthropy: Eight mechanisms that drive charitable giving. *Nonprofit and Voluntary Sector Quarterly* 40: 924–973.

Belinky M, Rennick E and Veitch A. (2015) The Fintech 2.0 paper: Rebooting financial services. Santander InnoVentures, Oliver Wyman & Anthemis Group, 1–19.

Belleflamme P, Lambert T and Schwienbacher A. (2013) Individual crowdfunding practices. *Venture Capital* 15: 313–333.

Belleflamme P, Lambert T and Schwienbacher A. (2014) Crowdfunding: Tapping the right crowd. *Journal of Business Venturing* 29: 585–609.

Belleflamme P, Omrani N and Peitz M. (2015) The economics of crowdfunding platforms. *CORE Discussion Papers.* Voie du Roman Pays, 34, L1.03.01, B-1348 Louvain-la-Neuve, Belgium: Center for Operations Research and econometrics, 1–36.

Ben S. (2017) Making big FinTech footprints. *China Daily: Africa Weekly.* 30.

Ben S and Chen X. (2018) Digital economy offers exciting prospects: From China's experiences, less-developed countries can see how technology changes lives and transforms societies. *China Daily: Africa Weekly.* 11.

Berger R. (2015) Internet of things and insurance. Efma, 1–18.

Berger RC. (2014) Raising capital via crowdfunding: A step closer to reality. *The Central New York Business Journal, 28 March 2014, Vol. 28 Issue 13*, p. 4B.

Berliner LS and Kenworthy NJ. (2017) Producing a worthy illness: Personal crowdfunding amidst financial crisis. *Social Science & Medicine* 187: 233–242.

Beugré CD and Das N. (2013) Limited capital and new venture creation in emerging economies: A model of crowd-capitalism. *SAM Advanced Management Journal* 78: 21–27.

Bhatt C, Dey N and Ashour AS. (2017) Internet of things and big data technologies for next generation healthcare. Springer International Publishing AG, ISBN: 978-3-319-49735-8.

Bi S, Liu Z and Usman K. (2017) The influence of online information on investing decisions of reward-based crowdfunding. *Journal of Business Research* 71: 10–18.

Biais B and Foucault T. (2014) HFT and market quality. *Bankers, Markets & Investors* 128: 5–19.

Biella M and Zinetti V. (2016) Blockchain technology and applications from a financial perspective. UniCredit, 1–33.

Bisht SS and Mishra V. (2016) ICT-driven financial inclusion initiatives for urban poor in a developing economy: Implications for public policy. *Behaviour & Information Technology* 35: 817–832.

BMO Bank of Montreal, CIBC, National Bank of Canada, et al. (2015) Payments security. BMO Bank of Montreal, CIBC, National Bank of Canada, RBC Royal Bank, Scotiabank, & TD Bank Group, 1–52.

BNY Mellon. (2014) Global payments 2020: Transformation and convergence. BNY Mellon, Moorgate & OPUS Advisory Services International Inc., 1–47.

BNY Mellon. (2015) Innovation in payments: The future is FinTech. BNY Mellon, 1–17.

Boeuf B, Darveau J and Legoux R. (2014) Financing creativity: Crowdfunding as a new approach for theatre projects. *International Journal of Arts Management* 16: 33–48.

Boidus Admin. (2014) University of Botswana was crowdfunded. *Boidus Focus, November 2014, Volume 4 Issue 10*, p. 18.

Booth P. (2015) Crowdfunding: A Spimatic application of digital fandom. *New Media & Society* 17: 149–166.

Bottiglia R. (2016) Competitive frontiers in P2P lending crowdfunding. In: Bottiglia R and Pichler F (eds) *Crowdfunding for SMEs: A European Perspective.* Palgrave Macmillan, ISBN 978-1-137-56020-9, Chapter 4, pp. 61–92.

Bottiglia R and Pichler F. (2016) *Crowdfunding for SMEs: A European perspective*: Palgrave Macmillan, ISBN 978-1-137-56020-9.

Boum E. (2016) Crowdfunding in Africa. Afrikstart, 1–70.

Bouncken RB, Komorek M and Kraus S. (2015) Crowdfunding: The current state of research *International Business & Economics Research Journal* 14: 407–415.

Brabham DC. (2013) *Crowdsourcing,* 55 Hayward Street, Cambridge, MA 02142: The MIT Press, ISBN 978-0-262-51847-5.

Bradford SC. (2012) Crowdfunding and the federal securities laws. *Columbia Business Law Review* 1: 1–150.

Braun LM and Turke ZM. (2013) Joining the crowd: Investing profitably in equity crowdfunding. *CSQ: C-Suite Quarterly* Los Angeles & Ventura County, 1–2.

Brown TE, Boon E and Pitt LF. (2017) Seeking funding in order to sell: Crowdfunding as a marketing tool. *Business Horizons* 60: 189–195.

Bruton G, Khavul S, Siegel D, et al. (2014) New financial alternatives in seeding entrepreneurship: Microfinance, crowdfunding, and peer-to-peer innovations. *Entrepreneurship: Theory and Practice* 39: 9–26.

Buckstein J. (2014) Crowdfunding inches toward reality in Canada. *The Bottom Line, Mid-September 2014,* pp. 2–17.

Buff LA and Alhadeff P. (2013) Budgeting for crowdfunding rewards. *MEIEA Journal* 13: 27–44.

Bui HH. (2017) Creating a robust crowdfunding ecosystem in Vietnam. *Bachelor of Business Administration | International Business.* Turku University of Applied Science.

Burelli F, John M, Cenci E, et al. (2015) Blockchain and financial services: Industry snapshot and possible future developments. 3 More London Riverside London, SE1 2RE, UK INNOVALUE Management Advisors Ltd.

Burtch G, Ghose A and Wattal S. (2013) An empirical examination of the antecedents and consequences of investment patterns in crowd-funded markets. *Information Systems Research* 24: 499–519.

Burtch G, Ghose A and Wattal S. (2014) An empirical examination of peer referrals in online crowdfunding. *Proceedings of the 35th International Conference on Information Systems, Auckland, New Zealand,* pp. 1–19.

Burtch G, Ghose A and Wattal S. (2015) The hidden cost of accommodating crowdfunder privacy preferences: A randomized field experiment. *Management Science* 61: 949–962.

Buttice V, Colombo MG and Wright M. (2017) Serial crowdfunding, social capital and project success. *Entrepreneurship Theory and Practice* March: 183–207.

Calic G and Mosakowski E. (2016) Kicking off social entrepreneurship: How a sustainability orientation influences crowdfunding success. *Journal of Management Studies* 53: 738–767.

Cameron P, Corne DW, Mason CE, et al. (2013) Crowdfunding genomics and bioinformatics. *Genome Biology* 14: 1–5.

Capagemini. (2015) The impact of the Internet of thigns on financial services. Capgemini, 1–7.

Carvajal M, García-Avilés JA and González JL. (2012) Crowdfunding and non-profit media: The emergence of new models for public interest journalism. *Journalism Practice* 6: 638–647.

CGAP. (2017) Crowdfunding in China: the financial inclusion dimension. CGAP, pp. 1–4.

CGI. (2016) FinTech disruption in financial services: A consumer perspective. CGI Group Inc., 1–8.

Cha J. (in press) Crowdfunding for video games: Factors that influence the success of and capital pledged for campaigns. *International Journal of Media Management* http://dx.doi.org/10.1080/14241277.2017.1331236.

Chan CSR and Parhankangas A. (2017) Crowdfunding innovative ideas: How incremental and radical innovativeness influence funding outcomes. *Entrepreneurship Theory and Practice* March: 237–263.

Chaudry S. (2014) The impact of the JOBS Act on independent film finance. *DePaul Business & Commercial Law Journal* 12: 215–234.

Chen H, De P, Hu Y, et al. (2014) Wisdom of crowds: The value of stock opinions transmitted through social media. *Review of Financial Studies* 27: 1367–1403.

Chen L. (2016) From Fintech to Finlife: The case of Fintech development in China. *China Economic Journal* 9: 225–239.

Chertkow R and Feehan J. (2014) Setting your crowdfunding rewards. *Electronic Musician*. NewBay Media, LLC, 84–85.

Chertkow R and Feehan J. (2015) Making money from "free": Your music has value, even if you give it away. *Electronic Musician, September 2015, Vol. 31 Issue 9*, pp. 60–64.

Chia WK and Shah Y. (2016) Robo-advisory: Key to the untapped mass affluent market. Synpulse, 1–4.

Chiu IH-Y. (2017) A new era in fintech payment innovations? A perspective from the institutions and regulation of payment systems. *Law, Innovation and Technology* 9: 190–234.

Cho M and Gawon Kim MA. (2017) A cross-cultural comparative analysis of crowdfunding projects in the United States and South Korea. *Computers in Human Behavior* 72: 312–320.

Cholakova M and Clarysse B. (2015) Does the possibility to make equity investments in crowdfunding projects crowd out reward-based investments? *Entrepreneurship Theory and Practice* 39: 145–172.

Chordia T, Goyal A, Lehmann BN, et al. (2013) High-frequency trading. *Journal of Financial Markets* 16: 637–645.

Chuah S-H and Devlin J. (2011) Behavioural economics and financial services marketing: A review. *International Journal of Bank Marketing* 29: 456–469.

Chuen DLK. (2015) *Handbook of digital currency: Bitcoin, innovation, financial instruments and big data*, 125 London Wall, London, EC2Y 5AS, UK: Academic Press, Elsevier Inc., ISBN 978-0-12-802117-0.

Chuen DLK and Deng R. (2018) *Handbook of blockchain, digital finance, and inclusion: Cryptocurrency, FinTech, InsurTech and regulation*, 125 London Wall, London EC2Y 5AS, United Kingdom: Elsevier Inc., ISBN 978-0-12-810441-5, Volume 1.

Cisco. (2015) Seizing opportunity in the new age of financial services. CISCO, 1–7.

Cognizant. (2017) Financial services: Building blockchain one block at a time. 500 Frank W. Burr Blvd,Teaneck, NJ 07666 USA: Cognizant, 1–28.

Colistra R and Duvall K. (2017) Show me the money: Importance of crowdfunding factors on backers' decisions to financially support Kickstarter campaigns. *Social Media + Society* October-December: 1–12.

Collett D. (2017) *Tremendous innovation in sophisticated fintech payments in South Africa, retrieved from http://ventureburn.com/2017/08/tremendous-innovation-sophisticated-fintech-payments-south-africa/, accessed on 11 October 2017.*

Collins K. (2013) *These are our favourite crowdfunding success stories of 2013, retrieved from http://www.wired.co.uk/article/crowdfunding-success-stories, accessed on 30 August 2017.*

Collins L and Pierrakis Y. (2012) The venture crowd: Crowdfunding equity invesment into business. 1 Plough Place, London EC4A 1DE, UK: Nesta, ISBN 978-1-84875-138-5, 1–36.

Colombo MG, Franzoni C and Rossi-Lamastra C. (2015) Internal social capital and the attraction of early contributions in crowdfunding. *Entrepreneurship Theory and Practice* 39: 75–100.

Courtney C, Dutta S and Li Y. (2017) Resolving information asymmetry: Signaling, endorsement, and crowdfunding success. *Entrepreneurship Theory and Practice* March: 265–290.

Cowley S. (2016) *Global brands, taking cue from tinkerers, explore crowdfunding, retrieved from http://www.nytimes.com/2016/01/07/business/global-brands-taking-cue-from-tinkerers-explore-crowdfunding.html?_r=0, accessed on 31 March 2016.*

Crescenzo VD. (2016) The role of equity crowdfunding in financing SMEs: Evidence from a sample of European platforms. In: Bottiglia R and Pichler F (eds) *Crowdfunding for SMEs: A European Perspective*. Palgrave Macmillan, ISBN 978-1-137-56020-9, Chapter 7, pp. 159–183.

CrowdExpert. (2016) *Crowdfunding industry statistics 2015–2016, retrieved from http://crowdexpert.com/crowdfunding-industry-statistics/, accessed on 28 August 2017.*

CrowdfundingHub. (2016a) Crowdfunding crossing borders: An overview of libility risks associated with cross border crowdfunding investments. Keizersgracht 264, 1016 EV Amsterdam, The Netherlands: CrowdfundingHub & FG Lawyers, 1–240.

CrowdfundingHub. (2016b) Current state of crowdfunding in Europe. Keizersgracht 264, 1016 EV Amsterdam, The Netherlands: CrowdfundingHub, 1–67.

Cuesta C, Lis SFd, Roibas I, et al. (2015) Crowdfunding in 360°: Alternative financing for the digital era. *Digital Economy Watch*. BBVA Research, 1–25.

Cumming DJ, Leboeuf G and Schwienbacher A. (2015) Crowdfunding models: Keep-it-all vs. all-or-nothing. *SSRN, https://ssrn.com/abstract=2447567*.

Cumming DJ, Leboeuf G and Schwienbacher A. (2017) Crowdfunding cleantech. *Energy Economics* 65: 292–303.

Dahl D. (2012) More cash for start-ups: New bills could expand crowdfunding options. *Inc., December 2011 / January 2012, Vol. 33 Issue 10*, p. 28.

Dahl D. (2014) Crowdfunding 2.0: Want funding from the crowd? Get ready to bare your soul. *Inc. Magazine, February 2014*, pp. 72.

Dahlberg T, Guo J and Ondrus J. (in press) A critical review of mobile payment research. *Electronic Commerce Research and Applications* http://dx.doi.org/10.1016/j.elerap.2015.07.006.

Dahlberg T, Mallat N, Ondrus J, et al. (2008) Past, present and future of mobile payments research: A literature review. *Electronic Commerce Research and Applications* 7: 165–181.

Dahlhausen K, Krebs BL, Watters JV, et al. (2016) Crowdfunding campaigns help researchers launch projects and generate outreach. *Journal of Microbiology & Biology Education* 17: 32–37.

Dalager B and Jensen MB. (2016) Three technologies that are changing the financial services game and how the workforce must adapt to take advantage of these innovations. Accenture, 1–16.

Dapp TF. (2014) FinTech: The digital (r)evolution in the financial sector. Deutsche Banke AG, ISSN 1612-314X, 1–38.

Davis B, Hmieleski KM, Webb JW, et al. (2017) Funders' positive affective reactions to entrepreneurs' crowdfunding pitches: the influence of perceived product creativity and entrepreneurial passion. *Journal of Business Venturing* 32: 90–106.

Davis GF. (2009) *Managed by the markets: How finance reshaped America*, Great Clarendon Street, Oxford OX2 6DP: Oxford University Press, ISBN 978-0-19-921661-1.

Deloitte. (2015a) Blockchain disrupting the financial services industry? Deloitte, 1–8.

Deloitte. (2015b) RegTech is the new FinTech: How agile regulatory technology is helping firms better understand and manage their risks. Deloitte & Touche, 1–10.

Deloitte. (2016) Robo-advisory in wealth management: Same name, different game—a look at the german robo-advisor landscape. Deloitte, 1–9.

Deloitte. (2017a) The Finnish perspective on robo-advisor: How much influence on wealth management will it have? Deloitte & Touche Oy, Group of Companies, 1–27.

Deloitte. (2017b) A tale of 44 cities: Connecting global FinTech, interim hub review. Deloitte LLP, 1–116.

Deloitte Legal. (2016) Best practice for crowdfunding. Law Firm Deloitte Legal Oü, Roosikrantsi 2, 10119 Tallinn, Estonia: FinanceEstonia & Law Firm Deloitte Legal Estonia, pp. 1–4.

Deutsch J, Epstein GS and Nir A. (2017) Mind the gap: Crowdfunding and the role of seed money. *Managerial and Decision Economics* 38: 53–75.

Diemers D. (2017) Global FinTech report: Switzerland highlights. Partner Financial Services PwC Strategy & Switzerland, 1–2.

Dietrich A and Amrein S. (2017) Crowdfunding monitoring Switzerland 2017. Grafenauweg 10, Postfach 7344, 6302 Zug: Institut für Finanzdienstleistungen Zug (IFZ), ISBN 978-3-906877-11-2.

Digital Finance Institute & McCarthy Tétrault LLP. (2016) FinTech in Canada: towards leading the global financial technology transition. Digital Finance Institute & McCarthy Tétrault LLP, 1–60.

Dintrans P, Bahl M and Anand A. (2016) The work ahead: Seizing the digital advantages in banking and financial services. #5/535, Old Mahabalipuram Road, Okkiyam Pettai, Thoraipakkam, Chennai, 600 096 India: Cognizant, 1–24.

Dorfleitner G, Hornuf L, Schmitt M, et al. (2017) *FinTech in Germany*: Springer International Publishing AG, ISBN 978-3-319-54665-0.

Dragojlovic N and Lynd LD. (2014) Crowdfunding drug development: The state of play in oncology and rare diseases. *Drug Discovery Today* 19: 1775–1780.

Dragojlovic N and Lynd LD. (2016) What will the crowd fund? Preferences of prospective donors for drug development fundraising campaigns. *Drug Discovery Today* 21: 1863–1868.

Drover W, Wood M and Zacharakis A. (2017) Attributes of angel and crowdfunded ventures as determinants of VC screening decisions. *Entrepreneurship Theory and Practice* May: 323–347.

Drummer D, Jerenz A, Siebelt P, et al. (2016) FinTech: Challenges and opportunities. McKinsey & Company, 1–6.

Dunis CL, Middleton PW, Iatos KT, et al. (2016) *Artificial intelligence in financial markets: Cutting-edge applications for risk management, portfolio optimization and economics*: Palgrave Macmillan, ISBN 978-1-137-48879-4.

Dunsby M. (2016) *The 18 biggest UK equity crowdfunding rounds of the last year, retrieved from http://startups.co.uk/crowdfunding, accessed on 30 August 2017.*

Dushnitsky G, Guerini M, Piva E, et al. (2016) Crowdfunding in Europe: Determinants of platform creation across countries. *California Management Review* 58: 44–71.

Ehrlich E and Fanelli D. (2012) *The financial services marketing handbook: Tactics and techniques that produce results,* 111 River Street, Hoboken, NJ 07030: John Wiley & Sons, Inc., ISBN 978-1-118-06571-6.

Engstrom E. (2015) Financing the new innovation economy: Making investment crowdfunding work better for startups and investors. Engine Advocacy, 1–37.

Ernst & Young Global Limited. (2014) Landscaping UK Fintech. Ernst & Young LLP, 1–20.

Ernst & Young Global Limited. (2016) Swiss FinTech report 2016: The role of Switzerland as a FinTech hub. Ernst & Young Global Limited, 1–23.

Ernst & Young Global Limited. (2017a) Accelerating development in Montréal's FinTech ecosystem: Findings and recommendations. Ernst & Young LLP, 1–60.

Ernst & Young Global Limited. (2017b) EY FinTech adoption index: The rapid emergence of FinTech. EYGM Limited, EYG no. 03893-174Gbl, 1–44.

European Banking Authority. (2015) Opinion of the European banking authority on lending-based crowdfunding. European Banking Authority (EBA)/Op/2015/03, pp. 1–40.

European Commission. (2016) Commission staff working document: Crowdfunding in the EU capital markets union. Brussels, Belgium: European Commission.

Falkner R and Corsthwaite D. (2013) *Mirror trading and social media marketing: Regulatory and business risks, retrieved from http://de.slideshare.net/blackswanpartners/black-swan-partners-reedreed-smithsocialtradingmarketingseminarbooklet, accessed on 29 August 2016.*

Fallone EA. (2014) Crowdfunding and sport: How soon until the fans own the franchise? *Marquette Sports Law Review* 25: 7–37.

Fan-Osuala O, Zantedeschi D and Jank W. (2018) Using past contribution patterns to forecast fundraising outcomes in crowdfunding. *International Journal of Forecasting* 34: 30–44.

Fan J. (2013) Study of business model of crowdfunding. *Enterprise Economy* 8: 72–75.

Fanning K and Centers DP. (2016) Blockchain and its coming impact on financial services. *The Journal of Corporate Accounting & Finance* July/August: 53–57.

Fenwick M, McCahery JA and Vermeulen EPM. (2017) Fintech and the financing of entrepreneurs: From crowdfunding to marketplace lending. *SSRN Electronic Journal* http://ssrn.com/abstract_id=2967891.

Fields J and Gehret M. (2012) Crowd control: Preparing for the new world of equity-based crowdfunding. *Utah Business*. 32.

Figliomeni M. (2014) Grassroots capitalism or: How I learned to stop worrying about financial risk in the exempt market and love equity crowdfunding. *Dalhousie Journal of Legal Studies* 23: 105–129.

Finextra. (2016) The future of advisory: Exploring the impact of robo on wealth management. Finextra Research Ltd., 1 Gresham Street, London, EC2V 7BX, United Kingdom: Finextra, 1–33.

Fiordelisi F and Molyneux P. (2004) Efficiency in the factoring industry. *Applied Economics* 36: 947–959.

Flanigan ST. (2017) Crowdfunding and diaspora philanthropy: An integration of the literature and major concepts. *Voluntas* 28: 492–509.

Fong A. (2015) Regulation of peer-to-peer lending in Hong Kong: State of play. *Law and Financial Markets Review* 9: 251–259.

Freedman DM and Nutting MR. (2015a) *Equity crowdfunding for investors: A guide to risks, returns, regulations, funding portals, due diligence, and deal terms*, 111 River Street, Hoboken, NJ 07030: John Wiley & Sons, Inc., ISBN 978-1-1188-5356-6.

Freedman DM and Nutting MR. (2015b) The growth of equity crowdfunding. *Value Examiner*. 6–10.

Fruner S, Oldendorf A, Pentschev M, et al. (2012) Crowdfunding meets music. Hafenstraße 33, 68159 Mannheim, Germany: University of Popular Music and Music Business, 1–176.

Frydrych D, Bock AJ, Kinder T, et al. (2014) Exploring entrepreneurial legitimacy in reward-based crowdfunding. *Venture Capital* 16: 247–269.

FSDA. (2016) East africa crowdfunding landscape study. FSD Africa, 1–32.

Furlonger D and Valdes R. (2017) Practical blockchain: A Gartner trend insight report. Gartner, Inc., 1–17.

Gabor D and Brooks S. (2017) The digital revolution in financial inclusion: International development in the fintech era. *New Political Economy* 22: 423–436.

Gach R and Gotsch M. (2014) The rise of FinTech: New York's opportunity for tech leadership. Accenture, 1–12.

Gajda O and Mason N. (2013) Crowdfunding for impact in Europe and the USA. Toniic llc & Ecn, 1–24.

Galuszka P and Brzozowska B. (2015) Crowdfunding: Towards a redefinition of the artist's role – the case of MegaTotal. *International Journal of Cultural Studies*: 1–15.

Galvan BM. (2016) No cash, no card, no problem. *Beijing Review, March*. 36–37.

Galvan BM. (2017) Dawn of cashless societies. *Beijing Review*. 12–15.

Gamble JR, Brennan M and McAdam R. (2017) A rewarding experience? Exploring how crowdfunding is affecting music industry business models. *Journal of Business Research* 70: 25–36.

Gao W-J, Xing B and Marwala T. (2013a) Computational intelligence in used products retrieval and reproduction. *International Journal of Swarm Intelligence Research* 4: 78–125.

Gao W-J, Xing B and Marwala T. (2013b) Teaching—learning-based optimization approach for enhancing remanufacturability pre-evaluation system's reliability. *IEEE Symposium Series on Computational Intelligence (IEEE SSCI), 15–19 April, Singapore*, pp. 235–239. IEEE.

Gao W-J, Xing B and Marwala T. (2014) Used products return service based on ambient recommender systems to promote sustainable choices. In: Memon QA (ed) *Distributed Network Intelligence, Security and Applications*. 6000 Broken Sound Parkway NW, Suite

300, Boca Raton, FL 33487-2742: CRC Press, Taylor & Francis Group, LLC, ISBN 978-1-4665-5958-5, Chapter 15, pp. 359–378.

García FJP. (2017) *Financial risk management: Identification, measurement and management*: Palgrave Macmillan, ISBN 978-3-319-41365-5.

Garvey K, Ziegler T, Zhang B, et al. (2017) Crowdfunding in East Africa: regulation and policy for market development. Cambridge Centre for Alternative Finance & FSDA, 1–75.

Gee S. (2015) *Fraud and fraud detection: A data analytics approach,* 111 River Street, Hoboken, NJ 07030: John Wiley & Sons, Inc., ISBN 978-1-118-77965-1.

Gelfond S and Eren B. (2016) SEC adopts "crowdfunding" rules for start-up businesses: An easy way to bet on the next Google? *Journal of Investment Compliance* 17: 117–121.

Gelfond SH and Foti AD. (2012) US$500 and a click: Investing the "crowdfunding" way. *Journal of Investment Compliance* 13: 9–13.

Gerber E and Hui J. (2013) Crowdfunding: Motivations and deterrents for participation. *ACM Transactions on Computer-Human Interaction* 20: 1–32.

Gerber EM, Hui JS and Kuo PY. (2012) Crowdfunding: Why people are motivated to post and fund projects on crowdfunding platforms. *Proceedings of the International Workshop on Design, Influence, and Social Techniques, 11–15 February, 2012, Seattle, Washington, USA,* pp. 1–10.

Ghose R, Tian Y, Dave S, et al. (2016) Digital disruption: How FinTech is forcing banking to a tipping point. Citi GPS: Global Perspectives & Solutions, 1–111.

Gilbert AS. (2017) Crowdfunding nostalgia: Kickstarter and the revival of classic PC game genres. *The Computer Games Journal* 6: 17–32.

Giles J. (2012) Like it? pay for it. *Nature* 7381: 252–253.

Giudici G, Guerini M and Rossi-Lamastra C. (2013) *Why crowdfunding projects can succeed: The role of proponents' individual and territorial social capital, retrieved from http://ssrn.com/abstract=2255944, accessed on 10 October 2017.*

Giudici G, Guerini M and Rossi-Lamastra C. (in press) Reward-based crowdfunding of entrepreneurial projects: The effect of local altruism and localized social capital on proponents' success. *Small Business Economics* DOI 10.1007/s11187-016-9830-x.

Glaser F and Risius M. (in press) Effects of transparency: analyzing social biases on trader performance in social trading. *Journal of Information Technology.*

Glasgow Chamber of Commerce, LendingCrowd, Harper Macleod LLP, et al. (2016) The Scottish crowdfunding report. 1–87.

Gleasure R and Feller J. (2016) Emerging technologies and the democratisation of financial services: A metatriangulation of crowdfunding research. *Information and Organization* 26: 101–115.

Gleasure R and Morgan L. (in press) The pastoral crowd: Exploring self-hosted crowdfunding using activity theory and social capital. *Information Systems* DOI: 10.1111/isj.12143: 1–27.

Glover J. (2014) Equity crowdfunding in the UK: Evidence from the equity tracker. Foundry House, 3 Millsands, Sheffield S3 8NH: British Business Bank, 1–28.

Glover J. (2017) How to...crowdfunding for local authorities. Third Floor, 251 Pentonville Road, London N1 9NG: LGiu: the Local Democracy Think Tank & Spacehive, pp. 1–16.

Gobble MAM. (2012) Everyone is a venture capitalist: The new age of crowdfunding. *Research Technology Management, July–August.* 4–7.

Gomber P, Koch J-A and Siering M. (2017) Digital finance and FinTech: Current research and future research directions. *Journal of Business Economics* 87: 537–580.

Gorbati A and Nelson L. (2015) Gender and the language of crowdfunding. *Academy of Management Proceedings* 2015: 15785.

Government Office for Science. (2015) FinTech futures: The UK as a world leader in financial technologies. 1 Victoria Street London SW1H 0ET: Government Office for Science, 1–67.

GP.Bullhound. (2017) FinTech: anything but alternatives. GP Bullhound LLP, 1–44.

Gramm P, Leach J and Bliley T. (1999) *The gramm leach bliley act, retrieved from http://www.gpo.gov/fdsys/pkg/PLAW-106publ102/content-detail.html, accessed on 06 September 2017.*

Green CH. (2014) *Banker's guide to new small business finance: Venture deals, crowdfunding, private equity and technology,* Hoboken, New Jersey: John Wiley & Sons, Inc., ISBN 978-1-118-83787-0.

Griffiths H. (2017) Civic crowdfunding: A guidebook for local authorities. Future Cities Catapult, 1–31.

Gutierrez DD. (2014) *Big data for finance, retrieved from https://whitepapers.em360tech.com/wp-content/files_mf/1427803213insideBIGDATAGuidetoBigDataforFinance.pdf, accessed on 05 September 2017.*

Hörisch J. (2015) Crowdfunding for environmental ventures: An empirical analysis of the influence of environmental orientation on the success of crowdfunding initiatives. *Journal of Cleaner Production* 107: 636–645.

Hamerton-Stove R. (2016) Artificial intelligence: Can it bring productivity an deconomic health back to financial services? *Infosys Insights*: 58–63.

Hardie S, Wood J and Gee D. (2017) Innovation, distributed: Mapping the FinTech bridge in the open source area. MagnaCarta, 1–27.

Hartnett B and Matan R. (2015) Crowd funding: What all nonprofit organizations should know. Sobel & Co. LLC, pp. 1–17.

Hayden R and Hou G. (2015) Faster payments: Building a business, not just an infrastructure. *McKinsey on Payments.* McKinsey & Company, 23–29.

He D, Leckow R, Haksar V, et al. (2017) Fintech and financial services: Initial considerations. IMF, SDN/17/05, 1–49.

Helmer J. (2014) (8 ways to) cut through the crowdfunding clutter. *Entrepreneur* June 2014: 86–90.

Heminway JM. (2014) Investor and market protection in the crowdfunding era: Disclosing to and for the "crowd". *Vermont Law Review* 38: 827–848.

Hernando JR. (2017) Crowdfunding: The collaborative economy for channelling institutional and household savings. *The Spanish Review of Financial Economics* 15: 12–20.

HK Financial Services Development Council. (2016) Introducing a regulatory framework for equity crowdfunding in Hong Kong. Units 3104-06, 31/F, Sunlight Tower, 248 Queen's Road East, Wan Chai, Hong Kong: HK Financial Services Development Council (FSDC), Paper No. 21, 1–56.

HK Financial Services Development Council. (2017) The future of FinTech in Hong Kong. Units 3104-06, 31/F, Sunlight Tower, 248 Queen's Road East, Wan Chai, Hong Kong: HK Financial Services Development Council (FSDC), Paper No. 29, 1–69.

Hobbs J, Grigore G and Molesworth M. (2016) Success in the management of crowdfunding projects in the creative industries. *Internet Research* 26: 146–166.

Hogg S. (2014) Mutually beneficial: Venture capital and crowdfunding learn to play nice. *Entrepreneur, December 2014, Vol. 42 Issue 12*, p. 86.

Hornuf L and Schwienbacher A. (2017) Should securities regulation promote equity crowdfunding? *Small Business Economics* 49: 579–593.

Hornuf L and Schwienbacher A. (in press) Market mechanisms and funding dynamics in equity crowdfunding. *Journal of Corporate Finance* http://dx.doi.org/10.1016/j.jcorpfin.2017.08.009.

Houlihan C and Curneen D. (2017) The regulation of crowdfunding in Ireland. Dillon Eustace, 1–4.

Hu W-C, Lee C-w and Kou W. (2005) *Advances in security and payment methods for mobile commerce,* 701 E. Chocolate Avenue, Suite 200, Hershey PA 17033: Idea Group Inc., ISBN 1-59140-345-6.

Hu Y. (2015) Crowdfunding in China. *China Today, May 2015, Vol. 64 Issue 5*, pp. 31–33.

Huang EY and Lin C-Y. (2005) Customer-oriented financial service personalization. *Industrial Management & Data Systems* 105: 26–44.

Hughes L. (2016) *Top 20 crowdcube crowdfunding success stories, retrieved from http://www.businesscloud.co.uk/news/crowdcubes-biggest-crowdfunding-success-stories-since-2011, accessed on 30 August 2017.*

Hui JS, Gerber EM and Gergle D. (2014) Understanding and leveraging social networks for crowdfunding: Opportunities and challenges. *Proceedings of the 2014 Conference on Designing Interactive Systems (DIS), 21~25 June, Vanouver, BC, Canada*, pp. 677–680.

Human D. (2013) Colleges go social in quest for donors. *Indianapolis Business Journal* 34: 3A & 30A.

IBM. (2014) Understanding big data so you can act with confidence. Software Group, Route 100, Somers, NY 10589: IBM Corporation, 1–16.

Ibrahim N and Verliyantina. (2012) The model of crowdfunding to support small and micro businesses in Indonesia through a web-based platform. *Procedia Economics and Finance* 4: 390–397.

infoDev and The World Bank. (2013) Crowdfunding's potential for the developing world. 1818 H Street NW, Washington DC 20433: Information for Development Program (infoDev)/The World Bank, pp. 1–104

Inter-American Development Bank. (2015) Creating a crowdfunding ecosystem in Chile. 1300 New York Avenue, N.W., Washington, D.C. 20577: The Multilateral Investment Fund, 1–83.

International Association of Insurance Supervisors (IAIS). (2017) FinTech developments in the insurance industry. CH-4002 Basel, Switzerland: International Association of Insurance Supervisors (IAIS), 1–46.

IPF. (2016) Real estate crowdfunding: gimmick or game changer? IPR Research, pp. 1–35.

ITWeb. (2017) *African fintech ups the ante in payments space, retrieved from http://www.itweb.co.za/index.php?option=com_content&view=article&id=163999, accessed on 11 October 2017.*

Jenq C, Pan J and Theseira W. (2015) Beauty, weight, and skin color in charitable giving. *Journal of Economic Behavior & Organization* 119: 234–253.

Jeon D-S and DomenicoMenicucci. (2011) When is the optimal lending contract in microfinance state non-contingent? *European Economic Review* 55: 720–731.

Jian L and Usher N. (2014) Crowd-funded journalism. *Journal of Computer-Mediated Communication* 19: 155–170.

Jin J, Fan B, Dai S, et al. (2017) Beauty premium: Event-related potentials evidence of how physical attractiveness matters in online peer-to-peer lending. *Neuroscience Letters* 640: 130–135.

Joly K. (2013) Higher education crowdfunding. *University Business. July.* 48–49.

Jones MM. (2014) Defying the traditional model: Crowdfunding in science fiction and fantasy. *Publishers Weekly, 24 November 2014, Vol. 261 Issue 48*, pp. 32–34.

Jones R. (2013) The in Crowd? *Music Week.* 20–21.

Josefy M, Dean TJ, Albert LS, et al. (2017) The role of community in crowdfunding success: evidence on cultural attributes in funding campaigns to "save the local theater". *Entrepreneurship Theory and Practice* March: 161–182.

Königsheim C, Lukas M and Nöth M. (in press) Financial knowledge, risk preferences and the demand for digital financial services. *Schmalenbach Business Review.*

Kamukama N and Natamba B. (2013) Social capital: Mediator of social intermediation and financial services access. *International Journal of Commerce and Management* 23: 204–215.

Kang L, Jiang Q and Tan C-H. (2017) Remarkable advocates: An investigation of geographic distance and social capital for crowdfunding. *Information & Management* 54: 336–348.

Kantor R. (2014) Why venture capital will not be crowded out by crowdfunding. *Alternative Investment Analyst Review* 3: 59–70.

Kappel T. (2009) Ex ante crowdfunding and the recording industry: A model for the US. *Loyola of Los Angeles Entertainment Law Review* 29: 375–385.

Kapron Z and Shaughnessy H. (2015) The platform for disruption: How China's FinTech will change how the world thinks about banking? Innotribe, 1–19.

Kasper G and Marcoux J. (2015) Case studies in funding innovation. Deloitte University Press, 1–76.

Kavassalis P, Stieber H, Breymann W, et al. (in press) An innovative RegTech approach to financial risk monitoring and supervisory reporting. *The Journal of Risk Finance* https://doi.org/10.1108/JRF-07-2017-0111.

Kim H and Moor LD. (2017) The case of crowdfunding in financial inclusion: A survey. *Strategic Change* 26: 193–212.

Klapper L. (2006) The role of factoring for financing small and medium enterprises. *Journal of Banking & Finance* 30: 3111–3130.

Koku PS. (2014) *Decision making in marketing and finance: An interdisciplinary approach to solving complex organizational problems,* 175 Fifth Avenue, New York, NY 10010: Palgrave Macmillan, ISBN 978-1-349-47882-8.

Koning P. (2016) Artificial intelligence (AI) for financial services. Simularity.

KPMG. (2016a) 2016 China leading FinTech 50. KPMG, 1–77.

KPMG. (2016b) FinTech 100: Leading global FinTech innovators. H2 Ventures & KPMG 1–111.

KPMG. (2016c) FinTech in India: A global growth story. KPMG, 1–63.

KPMG. (2017) The pulse of FinTech Q1 2017: Global analysis of investment in FinTech. KPMG International, 1–69.

Kraus S, Richter C, Brem A, et al. (2016) Strategies for reward-based crowdfunding campaigns. *Journal of Innovation & Knowledge* 1: 13–23.

Krittanawong C, Zhang HJ, Aydar M, et al. (2018) Crowdfunding for cardiovascular research. *International Journal of Cardiology* 250: 268–269.

Kulkarni A and Sathe S. (2014) Healthcare applications of the Internet of things: A review. *International Journal of Computer Science and Information Technologies* 5: 6229–6232.

Kunz MM, Bretschneider U, Erler M, et al. (2017) An empirical investigation of signaling in reward-based crowdfunding. *Electronic Commerce Research* 17: 425–461.

Kuppuswamy V and Bayus BL. (2013) Crowdfunding creative ideas: The dynamics of project backers in Kickstarter. *SSRN Electronic Journal* http://ssrn.com/abstract=2234765.

Löffler M and Tschiesner A. (2013) The Internet of things and the future of manufacturing. McKinsey & Company, 1–5.

Lam PTI and Law AOK. (2016) Crowdfunding for renewable and sustainable energy projects: An exploratory case study approach. *Renewable and Sustainable Energy Reviews* 60: 11–20.

Langley P. (2016) Crowdfunding in the United Kingdom: A cultural economy. *Economic Geography* 92: 301–321.

Larrimore L, Jiang L, Larrimore J, et al. (2011) Peer-to-peer lending: The relationship between language features, trustworthiness and persuasion success. *Journal of Applied Communication Research* 39: 19–371.

Laughlin L. (2013) Less risk and more reward: Applying the crowdfunding model to local and regional theater organizations. *School of Public and Environmental Affairs.*

Laycock M. (2014) *Risk management at the top: A guide to risk and its governance in financial institutions,* John Wiley & Sons Ltd, The Atrium, Southern Gate, Chichester, West Sussex, PO19 8SQ, United Kingdom: John Wiley & Sons, Ltd., ISBN 978-1-118-49742-5.

Lee I. (2016) Fintech: ecosystem and business models. *Advanced Science and Technology Letters* 142: 57–62.

Lee S. (2015) FinTech and Korea's financial investment industry. KCMI.

Lee W and Ma Q. (2015) Whom to follow on social trading services? A system to support discovering expert traders. *Proceedings of the 10th International Conference on Digital Information Management (ICDIM),* pp. 188–193.

Lehner OM. (2013) Crowdfunding social ventures: A model and research agenda. *Venture Capital* 15: 289–311.

Leiber N and Tozzi J. (2012) A new grade of startup fuel. *Bloomberg Businessweek, 12 March 2012, Issue 4307,* p. 55.

Lem MJ. (2014) Crowdfunding for real estate development: Is it even legal? *Building, February/March 2014, Vol. 64 Issue 1,* pp. 12–13.

Lemieux V. (2013) *Financial analysis and risk management: Data governance, analytics and life cycle management*: Springer-Verlag Berlin Heidelberg, ISBN 978-3-642-32231-0.

Leong C, Tan B, Xiao X, et al. (2017) Nurturing a FinTech ecosystem: The case of a youth microloan startup in China. *International Journal of Information Management* 37: 92–97.

Levin RB, Waltz PF and Wenner RW. (2017) *Meet HAL, your new robo-adviser: SEC regulatory guidance for robo-advisers, retrieved from https://sftp.polsinelli.com/publications/fintech/resources.upd0417fin.pdf, accessed onn 08 November 2017.*

Levine ML and Feigin PA. (2014) Crowdfunding provisions under the new rule 506(c): New opportunities for real estate capital formation. *The CPA Journal, June 2014, Vol. 84 Issue 6,* pp. 46–51.

Lewis T. (2015) Crowdfunding, and how it could transform the industry. *Music Trades, February 2015, Vol. 163 Issue 1,* pp. 138–146.

Ley A and Weaven S. (2011) Exporing agency dynamics of crowdfunding in start-up capital financing. *Academy of Entrepreneurship Journal* 17: 85–110.

Li F-W and Pryer KM. (2014) Crowdfunding the Azolla fern genome project: A grassroots approach. *GigaScience* 3: 1–4.

Li G, Dai JS, Park E-M, et al. (in press) A study on the service and trend of Fintech security based on text-mining: Focused on the data of Korean online news. *Journal in Computer Virology and Hacking Techniques* DOI 10.1007/s11416-016-0288-9.

Life. SREDA VC. (2014) Money of the future: Results of 2014/trends for 2015. Life. SREDA VC, 1–78.

Lin M and Viswanathan S. (2016) Home bias in online investment: An empirical study of an online crowdfunding market. *Management Science* 62: 1393–1414.

Lipusch N, Bretschneider U and Leimeister JM. (2016) Backer empowerment in reward-based crowdfunding. How participation beyond funding influences support behavior. *Proceedings of the 37th International Conference on Information System on the Digitization of the Individual (DOTI), Dublin, Ireland,* pp. 1–11.

Liu J, Kauffman RJ and Ma D. (2015) Competition, cooperation, and regulation: Understanding the evolution of the mobile payments technology ecosystem. *Electronic Commerce Research and Applications* 14: 372–391.

Liu Y. (in press) Consumer protection in mobile payments in China: A critical analysis of Alipay's service agreement. *Computer Law & Security Review* http://dx.doi.org/10.1016/j.clsr.2015.05.009.

Lopez Research. (2014) Building smarter manufacturing with the Internet of things. 2269 Chestnut Street #202 San Francisco, CA 94123: Lopez Research LLC, 1–9.

Lord S. (2014) The new era of deal marketing: Equity crowdfunding and the advent of online deal marketing. Dealflow.com, 1–13.

Loureiro YK and Gonzalez L. (2015) Competition against common sense: Insights on peer-to-peer lending as a tool to allay financial exclusion. *International Journal of Bank Marketing* 33: 605–623.

Loveland R. (2016) How prompt was regulatory corrective action during the financial crisis? *Journal of Financial Stability* 25: 16–36.

Lu CT, Xie S, Kong X, et al. (2014) Inferring the impacts of social media on crowdfunding. *Proceedings of the 7th ACM International Conference on Web Search and Data Mining, 24–28 February, New York City, USA,* pp. 573–582.

Luzar C. (2013) *List: 20 notable niche crowdfunding sites, retrieved from https://www.crowdfundinsider.com/2013/03/12380-20-niche-crowdfunding-sites/, accessed on 01 September 2017.*

Macht SA and Weatherston J. (2015) Academic research on crowdfunders: What's been done and what's to come? *Strategic Change* 24: 191–205.

Mandelbaum R. (2014) Here comes everybody. *Inc., May 2014, Vol. 36 Issue 4,* pp. 110–120.

Manning S and Bejarano TA. (2017) Convincing the crowd: Entrepreneurial storytelling in crowdfunding campaigns. *Strategic Organization* 15: 194–219.

Marakkath N and Attuel-mendes L. (2015) Can microfinance crowdfunding reduce financial exclusion? Regulatory issues. *International Journal of Bank Marketing* 33: 624–636.

Marchand FI. (2016) Crowdfunding real estate: Institutions and markets, an institutional comparison on the growth-patterns and behaviour of crowdfunding real estate markets in the Netherlands and United States. Department of Real Estate and Housing, Faculty of Architecture, Delft University of Technology, 1–132.

Mariani A, Annunziata A, Aprile MC, et al. (2017) Crowdfunding and wine business: Some insights from Fundovino experience. *Wine Economics and Policy* 6: 60–70.

Mariani A, Cataldo A and Vastola A. (2014) Consumers' engagement in co-creation of value and crowdfunding: Naked wine as a best practice. *Quality—Access to Success* 15: 70–73.

Mariani J, Quasney E and Raynor ME. (2015) Forging links into loops: The Internet of things' potential to recast supply chain management. *Deloitte Review* 17: 118–129.

Marom D, Robb A and Sade O. (2014) Gender dynamics in crowdfunding (Kickstarter): Evidence on entrepreneurs, investors, deals and taste based discrimination. *SSRN Electronic Journal* https://ssrn.com/abstract=2442954.

Marom S. (2017) Social responsibility and crowdfunding businesses: A measurement development study. *Social Responsibility Journal* 13: 235–249.

Martínez-Cañas R, Ruiz-Palomino P and Pozo-Rubio Rd. (2012) Crowdfunding and social networks in the music industry: Implications for entrepreneurship. *International Business & Economics Research Journal* 11: 1471–1476.

Marwala T. (2013) *Economic modeling using artificial intelligence methods*, Springer London Heidelberg New York Dordrecht: Springer-Verlag London, ISBN 978-1-4471-5009-1.

Marwala T. (2014) *Artificial intelligence techniques for rational decision making*, Springer Cham Heidelberg New York Dordrecht London: Springer International Publishing Switzerland, ISBN 978-3-319-11423-1.

Marwala T. (2015) *Causality, correlation and artificial intelligence for rational decision making*, 5 Toh Tuck Link, Singapore 596224: World Scientific Publishing Co. Pte. Ltd, ISBN 978-9-81463-086-3.

Marwala T and Hurwitz E. (2017a) *Artificial intelligence and economic theory: Skynet in the market*, Gewerbestrasse 11, 6330 Cham, Switzerland: Springer International Publishing AG, ISBN 978-3-319-66103-2.

Marwala T and Hurwitz E. (2017b) Behavioral economics. In: Marwala T and Hurwitz E (eds) *Artificial Intelligence and Economic Theory: Skynet in the Market*. Gewerbestrasse 11, 6330 Cham, Switzerland: Springer International Publishing AG, ISBN 978-3-319-66103-2, Chapter 5, pp. 51–61.

Marwala T and Hurwitz E. (2017c) Information asymmetry. In: Marwala T and Hurwitz E (eds) *Artificial Intelligence and Economic Theory: Skynet in the Market*. Gewerbestrasse 11, 6330 Cham, Switzerland: Springer International Publishing AG, ISBN 978-3-319-66103-2, Chapter 6, pp. 63–74.

Massolution. (2015) Crowdfunding industry report 2015/2016. Massolution, retrieved from http://crowdexpert.com/crowdfunding-industry-statistics/, accessed on 01 September 2017.

Masssolution. (2015) 2015 CF: the crowdfunding industry report. Masssolution.

Mateescu A. (2015) Peer-to-peer lending. 36 West 20th Street, 11th Floor New York, NY 10011: Data & Society Research Institute, 1–23.

Matt C, Hess T and Benlian A. (2015) Digital transformation strategies. *Business & Information Systems Engineering* 57: 339–343.

McCarthy P. (2013) Crowdfunding of technology and bio-businesses. Caledonia General Partners.

McConnell B and Nolan M. (2017) 7 banking trends to watch in 2017. Wellington Consulting Group, 1–6.

McCracken H. (2017) Beyond crowdfunding. *Fast Company* October: 46–47.

McKinsey & Company. (2015) Global payments 2015: A healthy industry confronts disruption. McKinsey & Company, 1–34.

McKinsey & Company. (2016a) Global payments 2016: Strong fundamentals despite uncertain times. McKinsey & Company, 1–43.

McKinsey & Company. (2016b) Impact of FinTech in retail banking. McKinsey & Company, 1–22.

McKinsey & Company. (2016c) Thriving in the new abnormal: North american asset management. McKinsey & Company, 1–41.

Meer J. (2014) Effects of the price of charitable giving: Evidence from an online crowdfunding platform. *Journal of Economic Behavior & Organ* 103: 113–124.

Menkveld AJ. (2016) The economics of high-frequency trading: Taking stock. *Annual Review of Financial Economics* 8: 1–24.

Meyskens M and Bird L. (2015) Crowdfunding and value creation *Entrepreneur Research Journal* 5: 155–166.

Mitina K. (2016) Business opportunites for Swiss FinTech companies in Russia. In: Bächtold J and Perova E (eds). Switzerland Global Enterprise, 1–62.

Mitra T and Gilbert E. (2014) The language that gets people to give: Phrases that predict success on Kickstarter. *Proceedings of the 17th ACM Conference on Computer Supported Cooperative Work & Social Computing, 15–19 February, Baltimore, USA*, pp. 49–61.

Mittal S and Lloyd J. (2016) The rise of FinTech in China: Redefining financial services. Asian Insights Office, DBS Group Research & EYGM Limited, 1–48.

Mixon FG, Asarta CJ and Caudill SB. (2017) Patreonomics: Public goods pedagogy for economics principles. *International Review of Economics Education* 25: 1–7.

Moisseyev A. (2013) *Crowdfunding as a marketing tool, retrieved from http://www.socialmediatoday.com/content/crowdfunding-marketing-tool, accessed on 31 March 2016.*

Mollick E. (2014) The dynamics of crowdfunding: An exploratory study. *Journal of Business Venturing* 29: 1–16.

Mollick E. (2016) The unique value of crowdfunding is not money—it's community. *Harvard Business Review.*

Mollick E and Nanda R. (in press) Wisdom or madness: Comparing crowds with expert evaluation in funding the arts. *Management Science* doi:10.1287/mnsc.2015.2207.

Mollick ER and Kuppuswamy V. (2014) After the campaign: Outcomes of crowdfunding. *SSRN Electronic Journal* https://ssrn.com/abstract=2376997.

Moss TW, Neubaum DO and Meyskens M. (2015) The effect of virtuous and entrepreneurial orientations on microfinance lending and repayment: A signaling theory perspective. *Entrepreneurship: Theory and Practice* 39: 27–52.

Mourtada W. (2014) The most important non-events of venture capital in the Middle East 2013: the recap. *Entrepreneur Middle East Edition, February 2014*, pp. 82–83.

Mourtzis D, Vlachou E and Milas N. (2016) Industrial big data as a result of IoT adoption in manufacturing. *Procedia CIRP* 55: 290–295.

Moutinho N and Leite PV. (2013) Critical success factors in crowdfunding: The case of Kickstarter. *Cadernos do Mercado de Valores Mobiliários* 45: 8–32.

Mullard A. (2015) Crowdfunding clinical trials. *Nat. Rev. Drug Discov.* 14: 593.

Multiateral Investment Fund. (2015) Creating a crowdfunding ecosystem in Chile. 1300 New York Avenue NW, Washington, DC 20005: Inter-American Development Bank, pp. 1–84.

Murray J. (2015) Equity crowdfunding and peer-to-peer lending in New Zealand: The first year. *JASSA The Finsia Journal of Applied Finance* 2: 5–10.

Mustafa SE and Adnan HM. (2017) Crowdsourcing: A platform for crowd engagement in the publishing industry. *Publishing Research Quarterly* 33: 283–296.

Nabarro. (2015) Where are they now? A report into the status of companies that have raised finance using equity crowdfunding in the UK. AltFiData, pp. 1–32.

Nakamoto S. (2008) *Bitcoin: a peer-to-peer electronic cash system, retrieved from www.bitcoin.org/bitcoin.pdf, accessed on 08 September 2017.*

Nesta. (2012) An introduction to crowdfunding. 1 Plough Place, London EC4A 1DE Nesta, pp. 1–4.

Next BE. (2013) Crowdfunding websites for college financing. *Black Enterprise.* 34.

Nicoletti B. (2017) *The future of FinTech: Integrating finance and technology in financial services*: Palgrave Macmillan, ISBN 978-3-319-51414-7.

Niu AX. (2016) Will FinTech upend the banking sector? From China's experience. *An Asia Pacific Foundatoin of Canada Research Series.* Asia Pacific Foundation of Canada, 1–13.

Nordicity. (2012) Crowdfunding in a canadian context: Exploring the potential of crowdfunding in the creative content industries. Canada Media Fund, pp. 1–35.

Nucciarelli A, Li F, Fernandes KJ, et al. (2017) From value chains to technological platforms: the effects of crowdfunding in the digital game industry. *Journal of Business Research* 78: 341–352.

O'Hara M. (2015) High frequency market microstructure. *Journal of Financial Economics* 116: 257–270.

Oehler A, Horn M and Wendt S. (2016) Benefits from social trading? Empirical evidence for certificates on Wikifolios. *International Review of Financial Analysis* 46: 202–210.

Oliver Wyman. (2017) Transforming for future value. Oliver Wyman, 1–36.

Oracle. (2015) Big data in financial services and banking: Architect's guide and reference architecture introduction. 500 Oracle Parkway, Redwood Shores, CA 94065, USA: Oracle.

Ordanini A, Miceli L, Pizzetti M, et al. (2011) Crowd-funding: Transforming customers into investors through innovative service platforms. *Journal of Service Management* 22: 443–470.

Ortiz-Yepes D. (2014) A critical review of the EMV payment tokenisation specification. *Computer Fraud & Security* October: 5–12.

Otero P. (2015) Crowdfunding: A new option for funding health projects. *Arch Argent Pediatr* 113: 154–157.

Oxera. (2015) Crowdfunding from an investor perspective. Park Central, 40/41 Park End Street, Oxford, OX1 1JD, UK.: Oxera Consulting LLP, ISBN: 978-92-79-46659-5.

Oxera. (2016) The economics of peer-to-peer lending. Oxera, retrieved from http://www.oxera.com/Latest-Thinking/Publications/Reports/2016/economics-of-peer-to-peer-lending.aspx, accessed on 23 October 2007.

Özdemir V, Faris J and Srivastava S. (2015) Crowdfunding 2.0: The next-generation philanthropy. *European Molecular Biology Organization (EMBO) Reports* 16: 267–271.

Pan W, Altshuler Y and Pentland A. (2012) Decoding social influence and the wisdom of the crowd in financial trading network *Proceedings of the International Conference on Privacy, Security, Risk and Trust (PASSAT), Amsterdam, Netherlands,* pp. 203–209.

Parhankangas A and Renko M. (2017) Linguistic style and crowdfunding success among social and commercial entrepreneurs. *Journal of Business Venturing* 32: 215–236.

Park I. (2013) Potential of crowdfunding in arts & cultural sector: A case of arts council Korea and Tumblebug. *Korean Society of Arts and Cultural Management* 6: 131–156.

Paschou M, Sakkopoulos E, Sourla E, et al. (2013) Health Internet of things: Metrics and methods for efficient data transfer. *Simulation Modelling Practice and Theory* 34: 186–199.

Paulet E and Relano F. (2017) Exploring the determinants of crowdfunding: The influence of the banking system. *Strategic Change* 26: 175–191.

Pazowski P and Czudec W. (2014) *Economic prospects and condition of crowdfunding, retrieved from http://www.toknowpress.net/ISBN/978-961-6914-09-3/papers/ML14-685.pdf, accessed on 18 February 2015.*

Pegueros V. (2012) Security of mobile banking and payments. SANC Institute, 1–27.

Pitschner S and Pitschner-Finn S. (2014) Non-profit differentials in crowd-based financing: Evidence from 50,000 campaigns. *Economics Letters* 123: 391–394.

Polzin F, Toxopeus H and Stam E. (in press) The wisdom of the crowd in funding: Information heterogeneity and social networks of crowdfunders. *Small Business Economics* DOI 10.1007/s11187-016-9829-3.

Pousttchi K. (2008) A modeling approach and reference models for the analysis of mobile payment use cases. *Electronic Commerce Research and Applications* 7: 182–201.

Preker AS, Harding A and Travis P. (2000) "Make or buy" decisions in the production of health care goods and services: New insights from institutional economics and organizational theory. *Bulletin of the World Health Organization* 78: 779–790.

Puschmann T. (2017) Fintech. *Business & Information Systems Engineering* 59: 69–76.

PwC. (2015) Money is no object: Understanding the evolving cryptocurrency market. PwC's Financial Services Institute, 1–18.
PwC. (2016a) Beyond automated advice: How FinTech is shaping asset & wealth management. PwC, 1–12.
PwC. (2016b) Blurred lines: How FinTech is shaping financial services—Netherlands. PwC, 1–13.
PwC. (2016c) Blurred lines: How FinTech is shaping financial services in Luxembourg. PwC, 1–20.
PwC. (2016d) Blurred lines: How FinTech is shaping financial services, views from Ireland. PwC, 1–19.
PwC. (2016e) Catching the FinTech wave: A survey on FinTech in Malaysia. AICB & PwC, 1–65.
PwC. (2016f) InsurTech: A golden opportunity for insurers to innovate. PwC, 1–11.
PwC. (2016g) Opportunities await: How InsurTech is reshaping insurance. PwC, 1–19.
PwC. (2016h) The rise of China's silicon dragon. PwC, HK-20160229-4-C1, 1–20.
PwC. (2016i) The un(der) banked is FinTech's largest opportunity. DeNovo & PwC, 1–47.
PwC. (2017a) FinTech 2.0: Beyond blurred lines, how financial services institutions misread the innovation landscape? PricewaterhouseCoopers New Zealand.
PwC. (2017b) FinTech trends report: India. PwC & Startupbootcamp, 1–37.
PwC. (2017c) Global FinTech survey China summary 2017. PricewaterhouseCoopers Zhong Tian LLP, HK-20170329-C1, 1–26.
PwC. (2017d) Insurance's new normal: Driving innovation with InsurTech. PwC, 1–11.
PwC. (2017e) PwC Hong Kong FinTech survey PricewaterhouseCoopers Limited, HK-20170320-C1, 1–20.
PwC. (2017f) Redrawing the lines: FinTech's growing influence on financial services. PwC, 1–20.
PwC. (2017g) The state of FinTech. PwC & Startupbootcamp FinTech, 1–88.
Quah JTS and Sriganesh M. (2008) Real-time credit card fraud detection using computational intelligence. *Expert Systems with Applications* 35: 1721–1732.
Quawasmi S. (2015) The crowdfunder. *Entrepreneur Middle East, July 2015*, p. 30.
Rana S. (2013) Philanthropic innovation and creative capitalism: A historical and comparative perspective on social entrepreneurship and corporate social responsibility. *Shruti Alabama Law Review* 64: 1121–1174.
Ravina E. (2012) *Love & loans: The effect of beauty and personal characteristics in credit markets, retrieved from https://ssrn.com/abstract=1107307, accessed on 22 November 2017.*
Rechtman Y and O'Callaghan S. (2014) Understanding the basics of crowdfunding. *Certified Public Accountants (CPA) Journal* 84: 30–33.
Reid C. (2014) Crowdfunding: publishing thrived on Kickstarter in 2014. *Publishers Weekly, 12 January 2015, Vol. 262 Issue 2*, p. 10.
Remus PC and Mendoza KA. (2012) Raising equity through crowdfunding. *New Hampshire Business Review.* 25.
Richards T. (2014) *Investing psychology: The effects of behavioral finance on investment choice and bias,* Hoboken, New Jersey: John Wiley & Sons, Inc., ISBN 978-1-118-72219-0.
Riedl J. (2013) Crowdfunding technology innovation. *Computer* 46: 100–103.
Riley-Huff DA, Herrera K, Ivey S, et al. (2016) Crowdfunding in libraries, archives and museums. *The Bottom Line* 29: 67–85.
Ring P. (2016) The retail distribution review: Retail financial services; regulation; financial advice market review. *Journal of Financial Regulation and Compliance* 24: 140–153.
Roche-Saunders G and Hunt O. (2017) Crowdfunding for charities and social enterprises. *Charity and Social Enterprise Update* Summer: 7–9.
Romãnova I and Kudinska M. (2016) Banking and Fintech: A challenge or opportunity? *Contemporary Issues in Finance: Current Challenges from Across Europe* 98: 21–35.
Roma P, Petruzzelli AM and Perrone G. (2017) From the crowd to the market: The role of reward-based crowdfunding performance in attracting professional investors. *Research Policy* 46: 1606–1628.

Rona-Tas A and Guseva A. (2014) *Plastic money: Constructing markets for credit cards in eight postcommunist countries,* Stanford, California: Stanford University Press, ISBN 978-0-8047-6857-3.

Rosenberg B. (2011) *Handbook of financial cryptography and security,* 6000 Broken Sound Parkway NW, Suite 300, Boca Raton, FL 33487-2742: Taylor and Francis Group, LLC, ISBN 978-1-4200-5982-3.

Rossi SPS. (2017) *Access to bank credit and SME financing*: Palgrave Macmillan, ISBN 978-3-319-41362-4.

Roy SK and Balaji MS. (2015) Measurement and validation of online financial service quality (OFSQ). *Marketing Intelligence & Planning* 33: 1004–1026.

Royal C and Windsor GSS. (2014) Microfinance, crowdfunding and sustainability: A case study of telecenters in a South Asian developing country. *Strategic Change* 23: 425–438.

Rubinton BJ. (2011) Crowdfunding: Disintermediated investment banking. *SSRN Electronic Journal* https://ssrn.com/abstract=1807204.

Sørensen IE. (2012) Crowdsourcing and outsourcing: The impact of online funding and distribution on the documentary film industry in the UK. *Media, Culture & Society* 34: 726–743.

Sage. (2015) Sage pay payments landscape report. Sage Pay, pp. 1–40.

Salahuddin MA, Al-Fuqaha A, Guizani M, et al. (2017) Softwarization of internet of things infrastructure for secure and smart healthcare. *Computer* July: 74–79.

Salazar J, Moore R and Verity A. (2015) Crowdfunding for emergencies. *OCHA Policy and Studies Series.* OCHA Field Information Services with the support of the OCHA Policy Development and Studies Branch (PDSB), 1–17.

Saler TD. (2014) Equity crowdfund surfing—is it for the careful investor? *Better Investing.* 9–10.

Sannajust A, Roux F and Chaibi A. (2014) Crowdfunding in France: A new revolution? *The Journal of Applied Business Research* 30: 1919–1928.

Santos A, Macedo J, Costa A, et al. (2014) Internet of things and smart objects for M-health monitoring and control. *Procedia Technology* 16: 1351–1360.

Santos DFS, Almeida HO and Perkusich A. (in press) A personal connected health system for the Internet of things based on the constrained application protocol. *Computers and Electrical Engineering.*

Sara B. (2013) Crowdfunding culture. *Journal of Mobile Media* 7: 1–30.

Savarese C. (2015) Crowdfunding and P2P lending: Which opportunities for microfinance? Rue de I'Industie 10–1000 Brussels, Belgium: European Microfinance Network (EMN) aisbl, pp. 1–34.

Savio LD. (2017) The place of crowdfunding in the discovery of scientific and social value of medical research. *Bioethics* 31: 384–392.

Scardovi C. (2017) *Digital transformation in financial services*: Springer International Publishing AG, ISBN 978-3-319-66944-1.

Schaarschmidt M, Walsh G and Evanschitzky H. (in press) Customer interaction and innovation in hybrid offerings: Investigating moderation and mediation effects for goods and services innovation. *Journal of Service Research* DOI: 10.1177/1094670517711586.

Segarra M. (2013) Will VC firms join the crowd? How crowdfunding and venture captial can coexist. *CFO, May.* 25.

Seltzer E and Mahmoudi D. (2012) Citizen participation, open innovation, and crowdsourcing: Challenges and opportunities for planning. *Journal of Planning Literature* 28: 3–18.

Seo J-H and Park E-M. (in press) A study on financing security for smartphones using text mining. *Wireless Personal Communications* DOI 10.1007/s11277-017-4121-7.

Sharf S, Shin L, Gensler L, et al. (2015) The FinTech 50: The future of your money. *Forbes* USA, 50–64.

Sharf S, Shin L, Gara A, et al. (2016) The FinTech 50: The future of your money. *Forbes* USA, 90–108.

Sheldon RC and Kupp M. (2017) A market testing method based on crowd funding. *Strategy & Leadership* 45: 19–23.

Shen S. (2017) FinTech: payments. The Henry Fund Research, University of Iowa's Tippie School of Management, 1–16.

Sheng C, Leonard M, Gangu P, et al. (2016) China InsurTech. Oliver Wyman & ZhongAn, 1–35.

Sheng C, Yip J and Cheng J. (2017) Fintech in China: Hitting the moving target. Oliver Wyman, 1–31.

Shim Y and Shin D-H. (2016) Analyzing China's Fintech industry from the perspective of actor-network theory. *Telecommunications Policy* 40: 168–181.

Sieck DR. (2012) Crowdfunding: Reaching a new class of angel investors. *Business NH Magazine, Nov 2012, Vol. 29 Issue 11*, pp. 34–35.

Sironi P. (2016) *FinTech innovation: From robo-advisors to goal based investing and gamification*: John Wiley & Sons, Inc., ISBN 978-1-119-22698-7.

Sisler J. (2012) Crowdfunding for medical expenses. *Canadian Medical Association Journal (CMAJ)* 184: E123–124.

Skan J, Lumb R, Masood S, et al. (2014) The boom in global FinTech investment: A new growth opportunity for London. Accenture, 1–16.

Skan J, Dickerson J and Masood S. (2015) The future of FinTech and banking: Digitally disrupted or reimagined? Accenture, 1–11.

Skirnevskiy V, Bendig D and Brettel M. (2017) The influence of internal social capital on serial creators' success in crowdfunding. *Entrepreneurship Theory and Practice* March: 209–236.

Snyder J. (2016) Crowdfunding for medical care: Ethical issues in an emerging health care funding practice. *The Hastings Center Report* 46: 36–42.

Snyder J, Crooks VA, Mathers A, et al. (in press) Appealing to the crowd: Ethical justifications in Canadian medical crowdfunding campaigns. *Journal of Medical Ethics* doi:10.1136/medethics-2016-103933.

Snyder J, Mathers A and Crooks VA. (2016) Fund my treatment! A call for ethics-focused social science research into the use of crowdfunding for medical care. *Social Science & Medicine* 169: 27–30.

Somerford P. (2011) A little help from my friends. *The Strad.* 56–60.

Sommerfeld B, Schaffner J and Dehaine A. (2016) To what extent will a 'robo adviser' replace your financial adviser? *Inside Magazine.* 26–35.

Soufani K. (2000) Factoring and UK small business. *Journal of Small Business & Entrepreneurship* 15: 78–89.

Stagars M. (2015) *University startups and spin-offs: Guide for entrepreneurs in academia,* 233 Spring Street, 6th Floor, New York, NY 10013: Apress, Springer Science+Business Media, ISBN 978-1-4842-0624-9.

Stanko MA and Henard DH. (2016) How crowdfunding influences innovation. *MIT Sloan Management Review* 57: 15.

Stanko MA and Henard DH. (2017) Toward a better understanding of crowdfunding, openness and the consequences for innovation. *Research Policy* 46: 784–798.

Starling Bank. (2017) Revolution or evolution? Predictions and challenges for UK FinTech in 2017. Starling Bank, 1–44.

Steigenberger N. (2017) Why supporters contribute to reward-based crowdfunding. *International Journal of Entrepreneurial Behavior & Research* 23: 336–353.

Stemler AR. (2013) The JOBS Act and crowdfunding: Harnessing the power–and money–of the masses. *Business Horizons* 56: 271–275.

Stern C, Makinen M and Qian Z. (2017) FinTechs in China—with a special focus on peer to peer lending. *Journal of Chinese Economic and Foreign Trade Studies* 10: 215–228.

Stiver A, Barroca L, Minocha S, et al. (2015) Civic crowdfunding research: Challenges, opportunities, and future agenda. *New Media & Society* 17: 249–271.

Šubelj L, Furlan Š and Bajec M. (2011) An expert system for detecting automobile insurance fraud using social network analysis. *Expert Systems with Applications* 38: 1039–1052.

Sweeting P. (2017) *Financial enterprise risk management*, University Printing House, Cambridge CB2 8BS, United Kingdom: Cambridge University Press, ISBN 978-1-107-18461-9.

Tablas M. (2013) Equity-based crowdfunding. *Black Enterprise*. 22.
Tapscott D, Tapscott A and Kirkland R. (2016) How blockchains could change the world. McKinsey & Company, 1–5.
Taylor R. (2015) Equity-based crowdfunding: Potential implications for small business capital. *Issue Brief, 14 April 2015, Number 5*, pp. 1–8.
The Board of the International Organization of Securities Commissions (IOSCO). (2017) IOSCO research report on financial technologies (FinTech). The Board of the International Organization of Securities Commissions (IOSCO), 1–76.
The Economist. (2017a) The battle in AI: Giant advantage. *The Economist* 425: 16–17.
The Economist. (2017b) Robo-advisers: Silicon speculators. *The Economist* 425: 66–67.
The Economist Intelligence Unit. (2012) Beyond branches: Innovations in emerging-market banking. The Economist Intelligence Unit, 1–46.
The Economist Intelligence Unit. (2015) Banks and big data: Risk and compliance executives weigh in. The Economist Intelligence Unit Limited, 1–4.
The Economist Intelligence Unit. (2017) Symbiosis: Your bank has your trust, can FinTech make you love it? 20 Cabot Square, London, E14 4QW, United Kingdom: The Economist Intelligence Unit Limited, 1–36.
The European Consumer Organisation (BEUC). (2017) FinTech: A more competitive and innovative european financial sector. Rue d'Arlon 80, B-1040 Brussels, Ref: BEUC-X-2017-073-15/06/2017, 1–15.
The National Crowdfunding Association (NCFA) of Canada. (2016) Alternative finance crowdfunding in Canada: unlocking real value through FinTech and crowd innovation. 1240 Bay Street, Suite 501, Toronto, ON M5R 2A7: The National Crowdfunding Association (NCFA) of Canada, pp. 1–20.
Thuerridl C and Kamleitner B. (2016) What goes around comes around? Rewards as strategic assets in crowdfunding. *California Management Review* 58: 88–110.
Thurston B. (2013) Secrets of the funded: Three lessons from early winners of the Kickstarter phenomenon. *Fast Company* November: 132.
Tian W. (2017) *Commercial banking risk management: Regulation in the wake of the financial crisis*: Palgrave Macmillan, ISBN 978-1-137-59441-9.
Tomczak A and Brem A. (2013) A conceptualized investment model of crowdfunding. *Venture Capital* 15: 335–359.
Toren M. (2014) *10 crowdfunding success stories to love, retrieved from http://www.entrepreneur.com/article/232234, accessed on 30 August 2017.*
Trelewicz JQ. (2017) Big data and big money: The role of data in the financial sector. *IT Professional* 19: 8–10.
Vachelard J, Gambarra-Soares T, Augustini G, et al. (2016) A guide to scientific crowdfunding. *PLoS Computational Biology* 14: 1–9.
Valančienė L and Jegelevičiūtė S. (2013) Valuation of crowdfunding: Benefit and drawbacks. *Economics and Management* 18: 39–48.
Valančienė L and Jegelevičiūtė S. (2014) Crowdfunding for creating value: Stakeholder approach. *Procedia-Social and Behavioral Sciences* 156: 599–604.
Vasileiadou E, Huijben JCCM and Raven RPJM. (2016) Three is a crowd? Exploring the potential of crowdfunding for renewable energy in the Netherlands. *Journal of Cleaner Production* 128: 142–155.
Verdouw CN, Beulens AJM and Vorst JGAJvd. (2013) Virtualisation of floricultural supply chains: A review from an Internet of things perspective. *Computers and Electronics in Agriculture* 99: 160–175.
Vishnu S and Tatseos P. (2016) *The bigger the better? Big data in financial services, retrieved from https://www.capco.com/insights/capco-thoughts/~/media/Capco/uploads/articlefiles/file_0_1480953589.pdf, accessed on 05 September 2017.*
Vismara S. (2016a) Equity retention and social network theory in equity crowdfunding. *Small Business Economics* 46: 579–590.

Vismara S. (2016b) Information cascades among investors in equity crowdfunding. *Entrepreneurship Theory and Practice* November: 1–31.

Voldere ID and Zeqo K. (2017) Crowdfunding: Reshaping the crowd's engagement in culture. European Union, ISBN 978-92-79-67975-9, 1–203.

Wallace B. (2014) Crowdfunding lets 'regular people' invest. *The Enterprise - Utah's Business Journal, 7–13 April 2014, Vol. 43 Issue 31*, p. 44.

Walterus B and Williams S. (2014) Crowdfunding in Belgium. KPMG, 1–24.

Wang H, Chen K, Zhu W, et al. (2015) A process model on P2P lending. *Financial Innovation* 1: 1–8.

Wang H, Chen K and Xu D. (2016) A maturity model for blockchain adoption. *Financial Innovation* 2: 1–5.

Wang JG and Yang J. (2016) *Financing without bank loans: New alternatives for funding SMEs in China*: Springer Science+Business Media Singapore, ISBN 978-981-10-0900-6.

Wang W, Zhu K, Wang H, et al. (2017a) The Impact of sentiment orientations on successful crowdfunding campaigns through text analytics. *The Institution of Engineering and Technology (IET) Software* 11: 229–238.

Wang Z, Li H and Law R. (2017b) Determinants of tourism crowdfunding performance: An empirical study. *Tourism Analysis* 22: 323–336.

Warner K, Bouwer LM and Ammann W. (2007) Financial services and disaster risk finance: examples from the community level. *Environmental Hazards* 7: 32–39.

Wechsler J. (2013) Know your crowd: The drivers of success in reward-based crowdfunding. *Department of Mass Media and Communication Research.* University of Fribourg.

Wieck E, Bretschneider U and Leimeister JM. (2013) Funding from the crowd: An Internet-based crowdfunding platform to support business set-ups from universities. *International Journal of Cooperative Information Systems* 22: 1–12.

WilmerHale. (2016) 2016 Venture capital report. 60 State Street, Boston, Massachusetts 02109: Wilmer Cutler Pickering Hale and Dorr LLP, 1–26.

Wohlgemuth V, Berger ES and Wenzel M. (2016) More than just financial performance: Trusting investors in social trading. *Journal of Business Research* 69: 4970–4974.

World Economic Forum. (2015) The future of financial services: How disruptive innovations are reshaping the way financial services are structured, provisioned and consumed. World Economic Forum & Deloitte, 1–178.

World Economic Forum. (2016) The future of financial infrastructure: An ambitiouis look at how blockchain can reshape financial services. World Economic Forum & Deloitte, 1–129.

Xing B, Nelwamondo FV, Battle K, et al. (2009) Application of artificial intelligence (AI) methods for designing and analysis of reconfigurable cellular manufacturing system (RCMS). *Proceedings of the 2nd International Conference on Adaptive Science & Technology (ICAST), 14–16 December, Accra, Ghana*, pp. 402–409. IEEE.

Xing B, Gao W-J, Nelwamondo FV, et al. (2010a) Ant colony optimization for automated storage and retrieval system. *Proceedings of The Annual IEEE Congress on Evolutionary Computation (IEEE CEC), 18–23 July, CCIB, Barcelona, Spain*, pp. 1133–1139. IEEE.

Xing B, Gao W-J, Nelwamondo FV, et al. (2010b) Artificial intelligence in reverse supply chain management: the state of the art. *Proceedings of the Twenty-First Annual Symposium of the Pattern Recognition Association of South Africa (PRASA), 22–23 November, Stellenbosch, South Africa*, pp. 305–310.

Xing B, Gao W-J, Nelwamondo FV, et al. (2010c) Can ant algorithms make automated guided vehicle system more intelligent? A viewpoint from manufacturing environment. *Proceedings of IEEE International Conference on Systems, Man, and Cybernetics (IEEE SMC), 10–13 October, Istanbul, Turkey*, pp. 3226–3234. IEEE.

Xing B, Gao W-J, Nelwamondo FV, et al. (2010d) Part-machine clustering: The comparison between adaptive resonance theory neural network and ant colony system. In: Zeng Z and Wang J (eds) *Advances in Neural Network Research & Applications, LNEE 67*, pp. 747–755. Berlin Heidelberg: Springer-Verlag.

Xing B, Gao W-J, Nelwamondo FV, et al. (2010e) Two-stage inter-cell layout design for cellular manufacturing by using ant colony optimization algorithms. In: Tan Y, Shi Y and Tan KC (eds) *Advances in Swarm Intelligence, Part I, LNCS 6145*, pp. 281–289. Berlin Heidelberg: Springer-Verlag.

Xing B, Gao W-J, Nelwamondo FV, et al. (2011) e-RL: The Internet of things supported reverse logistics for remanufacture-to-order. In: Omatu S and Fabri SG (eds) *Fifth International Conference on Advanced Engineering Computing and Applications in Sciences (ADVCOMP), 20–25 November, Lisbon, Portugal*, pp. 84–87. IARIA.

Xing B, Gao W-J and Marwala T. (2012) The applications of computational intelligence in radio frequency identification research. *IEEE International Conference on Systems, Man, and Cybernetics (IEEE SMC), 14–17 October, Seoul, Korea*, pp. 2067–2072. IEEE.

Xing B, Gao W-J and Marwala T. (2013a) The applications of computational intelligence in system reliability optimization. *IEEE Symposium Series on Computational Intelligence (IEEE SSCI), 15–19 April, Singapore*, pp. 7–14. IEEE.

Xing B, Gao W-J and Marwala T. (2013b) Intelligent data processing using emerging computational intelligence techniques. *Research Notes in Information Sciences* 12: 10–15.

Xing B, Gao W-J and Marwala T. (2013c) An overview of cuckoo-inspired intelligent algorithms and their applications. *IEEE Symposium Series on Computational Intelligence (IEEE SSCI), 15–19 April, Singapore*, pp. 85–89. IEEE.

Xing B. (2014) Computational intelligence in cross docking. *International Journal of Software Innovation* 4: 78–124.

Xing B and Gao W-J. (2014a) *Computational intelligence in remanufacturing*, 701 E. Chocolate Avenue, Suite 200, Hershey PA 17033: IGI Global, ISBN 978-1-4666-4908-8.

Xing B and Gao W-J. (2014aa) Chemical-reaction optimization algorithm. In: Xing B and Gao W-J (eds) *Innovative Computational Intelligence: A Rough Guide to 134 Clever Algorithms.* Cham Heidelberg New York Dordrecht London: Springer International Publishing Switzerland, ISBN: 978-3-319-03403-4, Chapter 25, pp. 417–428.

Xing B and Gao W-J. (2014b) *Innovative computational intelligence: A rough guide to 134 clever algorithms,* Cham Heidelberg New York Dordrecht London: Springer International Publishing Switzerland, ISBN: 978-3-319-03403-4.

Xing B and Gao W-J. (2014bb) Emerging chemistry-based CI algorithms. In: Xing B and Gao W-J (eds) *Innovative Computational Intelligence: A Rough Guide to 134 Clever Algorithms.* Cham Heidelberg New York Dordrecht London: Springer International Publishing Switzerland, ISBN: 978-3-319-03403-4, Chapter 26, pp. 429–437.

Xing B and Gao W-J. (2014c) Introduction to computational intelligence. In: Xing B and Gao W-J (eds) *Innovative Computational Intelligence: A Rough Guide to 134 Clever Algorithms.* Cham Heidelberg New York Dordrecht London: Springer International Publishing Switzerland, ISBN: 978-3-319-03403-4, Chapter 1, pp. 3–17.

Xing B and Gao W-J. (2014cc) Base optimization algorithm. In: Xing B and Gao W-J (eds) *Innovative Computational Intelligence: A Rough Guide to 134 Clever Algorithms.* Cham Heidelberg New York Dordrecht London: Springer International Publishing Switzerland, ISBN: 978-3-319-03403-4, Chapter 27, pp. 441–444.

Xing B and Gao W-J. (2014d) Bacteria inspired algorithms. In: Xing B and Gao W-J (eds) *Innovative Computational Intelligence: A Rough Guide to 134 Clever Algorithms.* Cham Heidelberg New York Dordrecht London: Springer International Publishing Switzerland, ISBN: 978-3-319-03403-4, Chapter 2, pp. 21–38.

Xing B and Gao W-J. (2014dd) Emerging mathematics-based CI algorithms. In: Xing B and Gao W-J (eds) *Innovative Computational Intelligence: A Rough Guide to 134 Clever Algorithms.* Cham Heidelberg New York Dordrecht London: Springer International Publishing Switzerland, ISBN: 978-3-319-03403-4, Chapter 28, pp. 445–448.

Xing B and Gao W-J. (2014e) Bee inspired algorithms. In: Xing B and Gao W-J (eds) *Innovative Computational Intelligence: A Rough Guide to 134 Clever Algorithms.* Cham Heidelberg New York Dordrecht London: Springer International Publishing Switzerland, ISBN: 978-3-319-03403-4, Chapter 4, pp. 45–80.

Xing B and Gao W-J. (2014ee) Introduction to remanufacturing and reverse logistics. In: Xing B and Gao W-J (eds) *Computational Intelligence in Remanufacturing.* 701 E. Chocolate Avenue, Suite 200, Hershey PA 17033: IGI Global, ISBN 978-1-4666-4908-8, Chapter 1, pp. 1–17.

Xing B and Gao W-J. (2014f) Bat inspired algorithms. In: Xing B and Gao W-J (eds) *Innovative Computational Intelligence: A Rough Guide to 134 Clever Algorithms.* Cham Heidelberg New York Dordrecht London: Springer International Publishing Switzerland, ISBN: 978-3-319-03403-4, Chapter 3, pp. 39–44.

Xing B and Gao W-J. (2014ff) Overview of computational intelligence. In: Xing B and Gao W-J (eds) *Computational Intelligence in Remanufacturing.* 701 E. Chocolate Avenue, Suite 200, Hershey PA 17033: IGI Global, ISBN 978-1-4666-4908-8, Chapter 2, pp. 18–36.

Xing B and Gao W-J. (2014g) Biogeography-based optimization algorithm. In: Xing B and Gao W-J (eds) *Innovative Computational Intelligence: A Rough Guide to 134 Clever Algorithms.* Cham Heidelberg New York Dordrecht London: Springer International Publishing Switzerland, ISBN: 978-3-319-03403-4, Chapter 5, pp. 81–91.

Xing B and Gao W-J. (2014gg) Used products return pattern analysis using agent-based modelling and simulation. In: Xing B and Gao W-J (eds) *Computational Intelligence in Remanufacturing.* 701 E. Chocolate Avenue, Suite 200, Hershey PA 17033: IGI Global, ISBN 978-1-4666-4908-8, Chapter 3, pp. 38–58.

Xing B and Gao W-J. (2014h) Cat swarm optimization algorithm. In: Xing B and Gao W-J (eds) *Innovative Computational Intelligence: A Rough Guide to 134 Clever Algorithms.* Cham Heidelberg New York Dordrecht London: Springer International Publishing Switzerland, ISBN: 978-3-319-03403-4, Chapter 6, pp. 93–104.

Xing B and Gao W-J. (2014hh) Used product collection optimization using genetic algorithms. In: Xing B and Gao W-J (eds) *Computational Intelligence in Remanufacturing.* 701 E. Chocolate Avenue, Suite 200, Hershey PA 17033: IGI Global, ISBN 978-1-4666-4908-8, Chapter 4, pp. 59–74.

Xing B and Gao W-J. (2014i) Cuckoo inspired algorithms. In: Xing B and Gao W-J (eds) *Innovative Computational Intelligence: A Rough Guide to 134 Clever Algorithms.* Cham Heidelberg New York Dordrecht London: Springer International Publishing Switzerland, ISBN: 978-3-319-03403-4, Chapter 7, pp. 105–121.

Xing B and Gao W-J. (2014ii) Used product remanufacturability evaluation using fuzzy logic. In: Xing B and Gao W-J (eds) *Computational Intelligence in Remanufacturing.* 701 E. Chocolate Avenue, Suite 200, Hershey PA 17033: IGI Global, ISBN 978-1-4666-4908-8, Chapter 5, pp. 75–94.

Xing B and Gao W-J. (2014j) Luminous insect inspired algorithms. In: Xing B and Gao W-J (eds) *Innovative Computational Intelligence: A Rough Guide to 134 Clever Algorithms.* Cham Heidelberg New York Dordrecht London: Springer International Publishing Switzerland, ISBN: 978-3-319-03403-4, Chapter 8, pp. 123–137.

Xing B and Gao W-J. (2014jj) Used product pre–sorting system optimization using teaching–learning-based optimization. In: Xing B and Gao W-J (eds) *Computational Intelligence in Remanufacturing.* 701 E. Chocolate Avenue, Suite 200, Hershey PA 17033: IGI Global, ISBN 978-1-4666-4908-8, Chapter 6, pp. 95–112.

Xing B and Gao W-J. (2014k) Fish inspired algorithms. In: Xing B and Gao W-J (eds) *Innovative Computational Intelligence: A Rough Guide to 134 Clever Algorithms.* Cham Heidelberg New York Dordrecht London: Springer International Publishing Switzerland, ISBN: 978-3-319-03403-4, Chapter 9, pp. 139–155.

Xing B and Gao W-J. (2014kk) Used product delivery optimization using agent-based modelling and simulation. In: Xing B and Gao W-J (eds) *Computational Intelligence in Remanufacturing.* 701 E. Chocolate Avenue, Suite 200, Hershey PA 17033: IGI Global, ISBN 978-1-4666-4908-8, Chapter 7, pp. 113–133.

Xing B and Gao W-J. (2014l) Frog inspired algorithms. In: Xing B and Gao W-J (eds) *Innovative Computational Intelligence: A Rough Guide to 134 Clever Algorithms.* Cham Heidelberg New York Dordrecht London: Springer International Publishing Switzerland, ISBN: 978-3-319-03403-4, Chapter 10, pp. 157–165.

Xing B and Gao W-J. (2014ll) Post–disassembly part–machine clustering using artificial neural networks and ant colony systems. In: Xing B and Gao W-J (eds) *Computational Intelligence in Remanufacturing.* 701 E. Chocolate Avenue, Suite 200, Hershey PA 17033: IGI Global, ISBN 978-1-4666-4908-8, Chapter 8, pp. 135–150.

Xing B and Gao W-J. (2014m) Fruit fly optimization algorithm. In: Xing B and Gao W-J (eds) *Innovative Computational Intelligence: A Rough Guide to 134 Clever Algorithms.* Cham Heidelberg New York Dordrecht London: Springer International Publishing Switzerland, ISBN: 978-3-319-03403-4, Chapter 11, pp. 167–170.

Xing B and Gao W-J. (2014mm) Reprocessing operations scheduling using fuzzy logic and fuzzy MAX–MIN ant systems. In: Xing B and Gao W-J (eds) *Computational Intelligence in Remanufacturing.* 701 E. Chocolate Avenue, Suite 200, Hershey PA 17033: IGI Global, ISBN 978-1-4666-4908-8, Chapter 9, pp. 151–170.

Xing B and Gao W-J. (2014n) Group search optimization algorithm. In: Xing B and Gao W-J (eds) *Innovative Computational Intelligence: A Rough Guide to 134 Clever Algorithms.* Cham Heidelberg New York Dordrecht London: Springer International Publishing Switzerland, ISBN: 978-3-319-03403-4, Chapter 12, pp. 171–176.

Xing B and Gao W-J. (2014nn) Reprocessing cell layout optimization using hybrid ant systems. In: Xing B and Gao W-J (eds) *Computational Intelligence in Remanufacturing.* 701 E. Chocolate Avenue, Suite 200, Hershey PA 17033: IGI Global, ISBN 978-1-4666-4908-8, Chapter 10, pp. 171–185.

Xing B and Gao W-J. (2014o) Invasive weed optimization algorithm. In: Xing B and Gao W-J (eds) *Innovative Computational Intelligence: A Rough Guide to 134 Clever Algorithms.* Cham Heidelberg New York Dordrecht London: Springer International Publishing Switzerland, ISBN: 978-3-319-03403-4, Chapter 13, pp. 177–181.

Xing B and Gao W-J. (2014oo) Re–machining parameter optimization using firefly algorithms. In: Xing B and Gao W-J (eds) *Computational Intelligence in Remanufacturing.* 701 E. Chocolate Avenue, Suite 200, Hershey PA 17033: IGI Global, ISBN 978-1-4666-4908-8, Chapter 11, pp. 186–202.

Xing B and Gao W-J. (2014p) Music inspired algorithms. In: Xing B and Gao W-J (eds) *Innovative Computational Intelligence: A Rough Guide to 134 Clever Algorithms.* Cham Heidelberg New York Dordrecht London: Springer International Publishing Switzerland, ISBN: 978-3-319-03403-4, Chapter 14, pp. 183–201.

Xing B and Gao W-J. (2014pp) Batch order picking optimization using ant system. In: Xing B and Gao W-J (eds) *Computational Intelligence in Remanufacturing.* 701 E. Chocolate Avenue, Suite 200, Hershey PA 17033: IGI Global, ISBN 978-1-4666-4908-8, Chapter 12, pp. 204–222.

Xing B and Gao W-J. (2014q) Imperialist competitive algorithm. In: Xing B and Gao W-J (eds) *Innovative Computational Intelligence: A Rough Guide to 134 Clever Algorithms.* Cham Heidelberg New York Dordrecht London: Springer International Publishing Switzerland, ISBN: 978-3-319-03403-4, Chapter 15, pp. 203–209.

Xing B and Gao W-J. (2014qq) Complex adaptive logistics system optimization using agent-based modelling and simulation. In: Xing B and Gao W-J (eds) *Computational Intelligence in Remanufacturing.* 701 E. Chocolate Avenue, Suite 200, Hershey PA 17033: IGI Global, ISBN 978-1-4666-4908-8, Chapter 13, pp. 223–236.

Xing B and Gao W-J. (2014r) Teaching–learning-based optimization algorithm. In: Xing B and Gao W-J (eds) *Innovative Computational Intelligence: A Rough Guide to 134 Clever Algorithms.* Cham Heidelberg New York Dordrecht London: Springer International Publishing Switzerland, ISBN: 978-3-319-03403-4, Chapter 16, pp. 211–216.

Xing B and Gao W-J. (2014rr) Conclusions and emerging topics. In: Xing B and Gao W-J (eds) *Computational Intelligence in Remanufacturing.* 701 E. Chocolate Avenue, Suite 200, Hershey PA 17033: IGI Global, ISBN 978-1-4666-4908-8, Chapter 14, pp. 238–265.

Xing B and Gao W-J. (2014s) Emerging biology-based CI algorithms. In: Xing B and Gao W-J (eds) *Innovative Computational Intelligence: A Rough Guide to 134 Clever Algorithms.* Cham Heidelberg New York Dordrecht London: Springer International Publishing Switzerland, ISBN: 978-3-319-03403-4, Chapter 17, pp. 217–317.

Xing B and Gao W-J. (2014t) Big bang–big crunch algorithm. In: Xing B and Gao W-J (eds) *Innovative Computational Intelligence: A Rough Guide to 134 Clever Algorithms.* Cham Heidelberg New York Dordrecht London: Springer International Publishing Switzerland, ISBN: 978-3-319-03403-4, Chapter 18, pp. 321–331.

Xing B and Gao W-J. (2014u) Central force optimization algorithm. In: Xing B and Gao W-J (eds) *Innovative Computational Intelligence: A Rough Guide to 134 Clever Algorithms.* Cham Heidelberg New York Dordrecht London: Springer International Publishing Switzerland, ISBN: 978-3-319-03403-4, Chapter 19, pp. 333–337.

Xing B and Gao W-J. (2014v) Charged system search algorithm. In: Xing B and Gao W-J (eds) *Innovative Computational Intelligence: A Rough Guide to 134 Clever Algorithms.* Cham Heidelberg New York Dordrecht London: Springer International Publishing Switzerland, ISBN: 978-3-319-03403-4, Chapter 20, pp. 339–346.

Xing B and Gao W-J. (2014w) Electromagnetism-like mechanism algorithm. In: Xing B and Gao W-J (eds) *Innovative Computational Intelligence: A Rough Guide to 134 Clever Algorithms.* Cham Heidelberg New York Dordrecht London: Springer International Publishing Switzerland, ISBN: 978-3-319-03403-4, Chapter 21, pp. 347–354.

Xing B and Gao W-J. (2014x) Gravitational search algorithm. In: Xing B and Gao W-J (eds) *Innovative Computational Intelligence: A Rough Guide to 134 Clever Algorithms.* Cham Heidelberg New York Dordrecht London: Springer International Publishing Switzerland, ISBN: 978-3-319-03403-4, Chapter 22, pp. 355–364.

Xing B and Gao W-J. (2014y) Intelligent water drops algorithm. In: Xing B and Gao W-J (eds) *Innovative Computational Intelligence: A Rough Guide to 134 Clever Algorithms.* Cham Heidelberg New York Dordrecht London: Springer International Publishing Switzerland, ISBN: 978-3-319-03403-4, Chapter 23, pp. 365–373.

Xing B and Gao W-J. (2014z) Emerging physics-based CI algorithms. In: Xing B and Gao W-J (eds) *Innovative Computational Intelligence: A Rough Guide to 134 Clever Algorithms.* Cham Heidelberg New York Dordrecht London: Springer International Publishing Switzerland, ISBN: 978-3-319-03403-4, Chapter 24, pp. 375–414.

Xing B. (2015a) Novel nature-derived intelligent algorithms and their applications in antenna optimization. In: Matin MA (ed) *Wideband, Multiband, and Smart Reconfigurable Antennas for Modern Wireless Communications.* 701 E. Chocolate Avenue, Hershey PA 17033: IGI Global, ISBN 978-1-4666-8645-8, Chapter 10, pp. 296–339.

Xing B. (2015b) Optimization in production management: Economic load dispatch of cyber physical power system using artificial bee colony. In: Kahraman C and Onar SÇ (eds) *Intelligent Techniques in Engineering Management: Theory and Applications.* Cham Heidelberg New York Dordrecht London: Springer International Publishing Switzerland, ISBN 978-3-319-17905-6, Chapter 12, 275–293.

Xing B and Gao W-J. (2015a) The applications of swarm intelligence in remanufacturing: A focus on retrieval. In: Khosrow-Pour M (ed) *Encyclopedia of Information Science and Technology.* 3rd ed. New York, USA: Information Science Ref. – IGI Global, ISBN 978-1-4666-5888-2, Chapter 7, pp. 66–74.

Xing B and Gao W-J. (2015b) Offshore remanufacturing. In: Khosrow-Pour M (ed) *Encyclopedia of Information Science and Technology.* 3rd ed. New York, USA: Information Science Ref. – IGI Global, ISBN 978-1-4666-5888-2, Chapter 374, pp. 3795–3804.

Xing B. (2016a) Agent-based machine-to-machine connectivity analysis for the Internet of things environment. In: Mahmood Z (ed) *Connectivity Frameworks for Smart Devices: The Internet of Things from a Distributed Computing Perspective.* Switzerland: Springer International Publishing, ISBN 978-3-319-33122-5, Chapter 3, pp. 43–61.

Xing B. (2016b) An investigation of the use of innovative biology-based computational intelligence in ubiquitous robotics systems: Data mining perspective. In: Ravulakollu KK, Khan MA and Abraham A (eds) *Trends in Ambient Intelligent Systems.* Switzerland: Springer International Publishing Switzerland, ISBN 978-3-319-30184-6, Chapter 6, pp. 139–172.

Xing B. (2016c) Smart robot control via novel computational intelligence methods for ambient assisted living. In: Ravulakollu KK, Khan MA and Abraham A (eds) *Trends in Ambient Intelligent Systems*. Switzerland: Springer International Publishing Switzerland, ISBN 978-3-319-30184-6, Chapter 2, pp. 29–55.

Xing B. (2016d) The spread of innovatory nature originated metaheuristics in robot swarm control for smart living environments. In: Espinosa HEP (ed) *Nature-Inspired Computing for Control Systems*. Cham Heidelberg New York Dordrecht London: Springer International Publishing Switzerland, ISBN 978-3-319-26228-4, Chapter 3, pp. 39–70.

Xing B. (2017) Protecting mobile payments security: A case study. In: Meng W, Luo X, Furnell S, et al. (eds) *Protecting Mobile Networks and Devices: Challenges and Solutions*. 6000 Broken Sound Parkway NW, Suite 300, Boca Raton, FL 33487-2742: CRC Press, Taylor & Francis Group, LLC, ISBN 978-1-4987-3583-4, Chapter 11, pp. 261–289.

Xing B and Marwala T. (2017) Implications of the fourth industrial age for higher education. *The Thinker: For the Thought Leaders (www.thethinker.co.za)*. South Africa: Vusizwe Media, 10–15.

Xing B and Marwala T. (2018a) Conclusion. In: Xing B and Marwala T (eds) *Smart Maintenance for Human–Robot Interaction: An Intelligent Search Algorithmic Perspective*. Gewerbestrasse 11, 6330 Cham, Switzerland: Springer International Publishing AG, ISBN 978-3-319-67479-7, Chapter 13, pp. 299–305.

Xing B and Marwala T. (2018b) Cyberware capacity–applications layer perspective. In: Xing B and Marwala T (eds) *Smart Maintenance for Human–Robot Interaction: An Intelligent Search Algorithmic Perspective*. Gewerbestrasse 11, 6330 Cham, Switzerland: Springer International Publishing AG, ISBN 978-3-319-67479-7, Chapter 8, pp. 173–191.

Xing B and Marwala T. (2018c) Cyberware capacity–energy autonomy perspective. In: Xing B and Marwala T (eds) *Smart Maintenance for Human–Robot Interaction: An Intelligent Search Algorithmic Perspective*. Gewerbestrasse 11, 6330 Cham, Switzerland: Springer International Publishing AG, ISBN 978-3-319-67479-7, Chapter 9, pp. 193–216.

Xing B and Marwala T. (2018d) Cyberware capacity–platform and middleware layers perspective. In: Xing B and Marwala T (eds) *Smart Maintenance for Human–Robot Interaction: An Intelligent Search Algorithmic Perspective*. Gewerbestrasse 11, 6330 Cham, Switzerland: Springer International Publishing AG, ISBN 978-3-319-67479-7, Chapter 7, pp. 143–171.

Xing B and Marwala T. (2018e) Hardware capacity–beginning of life perspective. In: Xing B and Marwala T (eds) *Smart Maintenance for Human–Robot Interaction: An Intelligent Search Algorithmic Perspective*. Gewerbestrasse 11, 6330 Cham, Switzerland: Springer International Publishing AG, ISBN 978-3-319-67479-7, Chapter 4, pp. 67–91.

Xing B and Marwala T. (2018f) Hardware capacity–end of life perspective. In: Xing B and Marwala T (eds) *Smart Maintenance for Human–Robot Interaction: An Intelligent Search Algorithmic Perspective*. Gewerbestrasse 11, 6330 Cham, Switzerland: Springer International Publishing AG, ISBN 978-3-319-67479-7, Chapter 6, pp. 111–139.

Xing B and Marwala T. (2018g) Hardware capacity–middle of life perspective. In: Xing B and Marwala T (eds) *Smart Maintenance for Human–Robot Interaction: An Intelligent Search Algorithmic Perspective*. Gewerbestrasse 11, 6330 Cham, Switzerland: Springer International Publishing AG, ISBN 978-3-319-67479-7, Chapter 5, pp. 93–110.

Xing B and Marwala T. (2018h) Human capacity–biopsychosocial perspective. In: Xing B and Marwala T (eds) *Smart Maintenance for Human–Robot Interaction: An Intelligent Search Algorithmic Perspective*. Gewerbestrasse 11, 6330 Cham, Switzerland: Springer International Publishing AG, ISBN 978-3-319-67479-7, Chapter 11, pp. 249–270.

Xing B and Marwala T. (2018i) Human capacity–exposome perspective. In: Xing B and Marwala T (eds) *Smart Maintenance for Human–Robot Interaction: An Intelligent Search Algorithmic Perspective*. Gewerbestrasse 11, 6330 Cham, Switzerland: Springer International Publishing AG, ISBN 978-3-319-67479-7, Chapter 12, pp. 271–295.

Xing B and Marwala T. (2018j) Human capacity–physiology perspective. In: Xing B and Marwala T (eds) *Smart Maintenance for Human–Robot Interaction: An Intelligent*

Search Algorithmic Perspective. Gewerbestrasse 11, 6330 Cham, Switzerland: Springer International Publishing AG, ISBN 978-3-319-67479-7, Chapter 10, pp. 219–247.

Xing B and Marwala T. (2018k) Introduction to human robot interaction. In: Xing B and Marwala T (eds) *Smart Maintenance for Human–Robot Interaction: An Intelligent Search Algorithmic Perspective.* Gewerbestrasse 11, 6330 Cham, Switzerland: Springer International Publishing AG, ISBN 978-3-319-67479-7, Chapter 1, pp. 3–19.

Xing B and Marwala T. (2018l) Introduction to intelligent search algorithms. In: Xing B and Marwala T (eds) *Smart Maintenance for Human–Robot Interaction: An Intelligent Search Algorithmic Perspective.* Gewerbestrasse 11, 6330 Cham, Switzerland: Springer International Publishing AG, ISBN 978-3-319-67479-7, Chapter 3, pp. 33–64.

Xing B and Marwala T. (2018m) Introduction to smart maintenance. In: Xing B and Marwala T (eds) *Smart Maintenance for Human–Robot Interaction: An Intelligent Search Algorithmic Perspective.* Gewerbestrasse 11, 6330 Cham, Switzerland: Springer International Publishing AG, ISBN 978-3-319-67479-7, Chapter 2, pp. 21–31.

Xing B and Marwala T. (2018n) *Smart maintenance for human–robot interaction: an intelligent search algorithmic perspective,* Gewerbestrasse 11, 6330 Cham, Switzerland: Springer International Publishing AG, ISBN 978-3-319-67479-7.

Xu A, Yang X, Rao H, et al. (2014) Show me the money! An analysis of project updates during crowdfunding campaigns. *Proceedings of the 32nd Annual ACM CHI Conference on Human Factors in Computing Systems, 26 April ~ 01 May, Toronto, Canada,* pp. 591–600.

Xu LD. (2011) Information architecture for supply chain quality management. *International Journal of Production Research* 49: 183–198.

Yan J, Wang K, Liu Y, et al. (in press) Mining social lending motivations for loan project recommendations. *Expert Systems with Applications* doi: 10.1016/j.eswa.2017.11.010.

Yan TC, Schulte P and Chuen DLK. (2018) InsurTech and FinTech: Banking and insurance enablement. In: Chuen DLK and Deng R (eds) *Handbook of Blockchain, Digital Finance, and Inclusion: Cryptocurrency, FinTech, InsurTech, and Regulation.* 125 London Wall, London EC2Y 5AS, United Kingdom: Elsevier Inc., ISBN 978-0-12-810441-5, Volume 1, Chapter 11, pp. 249–281.

Yang D, Chen P, Shi F, et al. (in press) Internet finance: its uncertain legal foundations and the role of big data in its development. *Emerging Markets Finance and Trade* DOI: 10.1080/1540496X.2016.1278528.

Yang H and Zhang Y. (2014) Research on influence factors of crowdfunding. *International Business Management* 9: 27–31.

Yeandle M. (2017) Trends and innovations in financial services. Toronto Financial Services Alliance (TFSA) & The Z/Yen Group, 1–34.

Yu D and Deng L. (2015) *Automatic speeech recognition: A deep learning approach,* Springer London Heidelberg New York Do: Springer-Verlag London, ISBN 978-1-4471-5778-6.

Zavolokina L, Dolata M and Schwabe G. (2016) The FinTech phenomenon: Antecedents of financial innovation perceived by the popular press. *Financial Innovation* 2: 1–16.

Zhang B, Baeck P, Ziegler T, et al. (2016) Pushing boundaries: The 2015 finance industry report. Cambridge Centre for Alternative Finance & Nesta, 1–56.

Zhang S, Xiong W, Ni W, et al. (2015) Value of big data to finance: Observations on an internet credit service company in China. *Financial Innovation* 1: 1–18.

Zhang T, Yip C, Wang G, et al. (2014) China crowdfunding report. China Impact fund (CIF) of Dao Ventures, pp. 1–64.

Zhao Q, Chen C-D, Wang J-L, et al. (2017) Determinants of backers' funding intention in crowdfunding: social exchange theory and regulatory focus. *Telematics and Informatics* 34: 370–384.

Zheng H, Li D, Wu J, et al. (2014) The role of multidimensional social capital in crowdfunding: a comparative study in China and US. *Information & Management* 51: 488–496.

Zheng H, Hung J-L, Qi Z, et al. (2016) The role of trust management in reward-based crowdfunding. *Online Information Review* 40: 97–118.

Zhong ZJ and Lin S. (2017) The antecedents and consequences of charitable donation heterogeneity on social media. *International Journal of Nonprofit and Voluntary Sector Marketing* e1585: 1–11.

Zhu L, Lu H, Zhang Q, et al. (2016) Application of crowdfunding on the financing of EV's charging piles. *Energy Procedia* 104: 336–341.

Zhu L, Zhang Q, Lu H, et al. (2017) Study on crowdfunding's promoting effect on the expansion of electric vehicle charging piles based on game theory analysis. *Applied Energy* 196: 238–248.

Zon Hv. (2001) The digital economy: Challenges for central European industry. *AI & Society* 15: 216–232.

Zvilichovsky D, Yael I and Barzilay O. (2015) Playing both sides of the market: Success and reciprocity on crowdfunding. *SSRN Electronic Journal* https://ssrn.com/abstract=2304101.

Part II

Regulator + Smart Computing = Healthier Market

CHAPTER 5

Crowdfunding Regulator

Technology Foresight and Type-2 Fuzzy Inference System

5.1 Introduction

About ten years ago, an innovative financial service form, crowdfunding, swept through the traditional financial system. Since 2013, a new buzzword has become extremely popular in both crowdfunding and crypto areas—the so-called initial coin offering (ICO), or sometimes, ITO (initial token offering), resembling the concept of initial public offering (IPO) (Leach and Melicher, 2012; Abor, 2017; Frederick et al., 2016). The marriage of ICO and crowdfunding represents a new height of disruptive funding method via the issuance of cryptocurrency or token (Zetzsche et al., 2017; Telpner and Ahmadifar, 2017). To date, statistics have shown that the capitalization of the ICO market is in the billions of dollars. In 2017 the ICO exploded with some staggering fundraising results. For example, Tezos coins (persuading 'a small nation-state') closed its ICO earlier in July and raised a record $232 m (The Economist, 2017a), while Filecoin raised $257 m (The Economist, 2017b) by September with the benefits of allowing token-holders to purchase and vend digital storage space on each other's computers. Because ICOs are open to everybody, investors can trade tokens freely after simply signing in and financing an account. Other inherent advantages of ICO also include system decentralization and automation (e.g., via Etherum's smart contract). This cryptocurrency and crowdfunding blended style of fundraising has rapidly become a new norm for startups and enthusiasts.

5.1.1 *Initial Coin Offering (ICO)*

Initial coin offering (ICO), also known as token sale or crowdsale, is a new model of crowdfunding in which a firm, organization or start-up declares its intent to pursue a blockchain-powered project and, thus, issues tokens in return for funds. This is achieved typically in the form of Bitcoin, Ether, and so on, under the terms of when those tokens are going to be sold for investment (Telpner and Ahmadifar, 2017). According to Coinschedule (an ICOs listing portal with over 200 ICOs), the funds raised during the past twelve months had reached $3.3 bn (The Economist, 2017c), dwarfing the $70 m funds raised a year ago. Compared with IPOs, ICOs seem much more attractive. Interested readers should refer to an in-depth comparative study conducted by Stellar Development Foundation & The Luxembourg House of Financial Technology (2017). The first project using ICO mechanism was probably Mastercoin in 2013, which at that time managed to raise $5 m worth of Bitcoins. Since then, a wide range of ICO projects have been initiated.

Though ICO is becoming increasingly popular, it is by no means flawless. A preliminary SWOT (strengths, weaknesses, opportunities and threats) analysis of ICO is provided as follows:

- **Strengths:** Regarding ICO's strengths, we can have the following observations:
  1) Quicker and easier fundraising;
  2) Lower transaction costs and much less documentation than IPO;
  3) Community building;
  4) Create 'skin in the game' (Taleb, 2014; Taleb and Sandis, 2016) with early adopters; and
  5) Open for business 24/7.
- **Weaknesses:** The associated weaknesses of ICO include:
  1) Price volatility;
  2) Potential fraudulence; and
  3) Network lag effect during large ICO.
- **Opportunities:** As a game changing technology, blockchain's potential for the near future is huge. The early participants involved in this race tend to be the rule-setters and enjoy the privilege of being able to harvest low-hanging fruits.
- **Threats:** However, like any other disruptive technologies, societal scale adoption and adaption to blockchain will most likely be gradual. Though future rewards might be fruitful, early game players have to bear many risks (e.g., a highly unregulated marketplace). At present, regulations or constrains on ICOs are underdeveloped. Additionally,

there is very little due diligence being performed on various offerings. Other threats also include tax evasion and potential money laundering.

5.1.2 *Crypto Crowdfunding Platform: ICO Platform*

Crypto crowdfunding platform (or called decentralizing crowdfunding platform), a new phrase that combines crowdfunding style with cryptocurrency (e.g., Bitcoin), provides a novel way to invest in new projects and open borders for the whole world (Jacynycz et al., 2016). Interestingly, both were created simultaneously at the beginning of the global financial crisis of 2008. However, this combination has only been realized recently, when the backbone of Bitcoin, i.e., blockchain technology became widely acknowledged. More specifically, during the process of developing innovative products and services that are based on blockchain, startups decided to launch ICOs and thus crowdsourced the buy and use of their companies' tokens to the cryptocurrency community (Groshoff, 2014).

At the first blush, it is just a fancy term for capital raising purpose. However, over the last year, especially in the first half of 2017, ICOs have quickly surpassed venture capital to become the major funding source in blockchain-based startups. Therefore, using blockchain technology as an activator, this new funding mechanism has kicked off a wave of quick development and innovation on decentralized customer services and business models.

Technically, we can treat the so-called ICO platform as a software workbench on which one is able to conduct an ICO and validate the associated token transactions. Using Ethereum blockchain as an example, because of the accompanied smart contract functionality of this ICO platform, token issuers can effectively program their desired tokens' behaviour into the blockchain (De Rose, 2015). The remaining part of this section briefly outlines some details about the evolution of blockchain technology and smart contract.

5.1.2.1 *Blockchain 1.0: Decentralized Digital Ledger*

In essence, blockchain 1.0, originally proposed by the mysterious Nakamoto (2008), represents a digital ledger of transactions that keeps Bitcoin (i.e., a virtual currency) secure and records the proof of who owns what at any given moment (Zohar, 2015). Therefore, in some cases, it also referred to as the Bitcoin-blockchain, which is entirely digitalized, chronologically updated, cryptographically sealed and systematically distributed (Deloitte, 2016). More information regarding Blockchain 1.0 can be found in (Li et al., in press; Prybila et al., in press; Vranken, in press; Gramoli, in press; Gupta, 2017; The Economist, 2015b; The Economist, 2015a; Scott, 2015).

5.1.2.2 Blockchain 2.0: Smart Contract

Soon after the original version of blockchain was released, clever minds quickly saw that such an establishment can not only be used for the purpose of payment transactions, but is also extendable and expandable, opening the door to the possibility of offering operations that are far more complex. In light of that, Blockchain 2.0, serves as a more generalized technological protocol and a platform. It was developed to accommodate numerous innovative derivatives (Li et al., in press; Swanson, 2016). Among others, the most popular and promising implementation is the so-called smart contract, a disruptive concept that essentially uses a piece of computer programming to verify and execute a set of embedded terms automatically whenever a predetermined event occurs (Giancaspro, 2017; Cuccuru, 2017; De Ridder et al., 2017; Mik, 2017). In fact, the term 'smart contracts' was first proposed in 1993 by Nick Szabo, a legal scholar and cryptographer (Szabo, 1997). However, at that time, the economic environment and communications infrastructure were not ready to support such an avant-garde protocol. Recently, with the maturity of blockchain technology, the concept of smart contracts resurfaced. Fundamentally, a smart contract includes the following three components (Blockchain Technology, 2016):

1) *Component 1*: Coding for determining what is programmed into a smart contract;
2) *Component 2*: Distributed ledgers for defining how the smart contract is sent out; and
3) *Component 3*: Execution for controlling how it is processed.

Once a smart contract is coded and stored into the blockchain, the contract cannot be changed and everything must be executed in accordance with its pre-programmed instructions. In other words, it is an automatable and enforceable agreement (Omohundro, 2014). Further technical background about smart contracts can be found in (Magazzeni et al., 2017; Fairfield, 2014; BBVA, 2015; Li et al., in press; Burelli et al., 2015; Tapscott and Tapscott, 2017).

The archetypal implementation platform for creating smart contracts is Ethereum (Buterin, 2014; Czepluch et al., 2015; Nordström, 2015). Like the Bitcoin-blockchain, Ethereum not only boasts its own cryptocurrency which is called 'ether', but also focuses on supporting smart, programmable contracts which are capable of providing much richer possibilities. Other competitive smart contract platforms and companies include R3's Corda platform (R3, 2017), Rootstock (Lerner, 2015), Blockstream (Back et al., 2014) and SmartContract (Ellis et al., 2017). Overall, different platforms tend to implement distinct smart contracts and transaction properties. In practice, smart contracts can be applied to multiple contexts with highly varied goals. Broadly speaking, we can divide these applications into the following two classes (Stark, 2016):

- **Smart Contract Code:** Smart contracts in this class behave more like software agents and represent counterparties in order to meet obligations or exercise rights. An immediate example may include automatically recharging a self-driving vehicle. On the one hand, the embedded smart contract code can help such a vehicle pay a pre-agreed electricity price. On the other hand, the same smart contract code can also facilitate the charging station to do the agreement check and finalize the payment as soon as the charging service is delivered properly. Other applications also include Internet of things (Salahuddin et al., 2017; Cha et al., 2017; Christidis and Devetsikiotis, 2016; Zhang and Wen, 2017; Kshetri, 2017; Pureswaran et al., 2015), energy sector (Peck and Wagman, 2017; Khaqqi et al., 2018), construction (Turk and Klinc, 2017), healthcare (Gietl et al., 2016; IBM, 2016; Yue et al., 2016) and finance (Treleaven et al., 2017; Eyal, 2017; Ferenzy et al., 2015).
- **Smart Legal Contract:** For the second class, the significance is placed on legal contracts' expression and implementation in software form. A representative example is property deeds transactions. Imagine we have a blockchain that is monitored and governed by the government. When a transaction occurs between a buyer and a seller, the associated record will automatically appear on such a blockchain. After checking the deed's correctness and the deed and buyer's affordability, an instant approval is issued. As soon as the real transaction is accomplished, the official record is kept by the blockchain. Since the whole process is fully automated, even government employees are not allowed to delay it (Nordrum, 2017). It may sound a bit futuristic, but similar projects are currently under test in several countries, e.g., Sweden and the Republic of Georgia. Other applications in this category also include voting (Borgstrup, 2014; Boucher, 2016; Higgins, 2016; Mattila, 2016), asset tracking/inventory (Bradbury, 2016; Shanley, 2017; Lu and Xu, 2017; Lee and Pilkington, 2017), escrow (Smart Contracts Alliance, 2016), insurance (Mainelli and Manson, 2017; PwC, 2016), contract law (Giancaspro, 2017; Lauslahti et al., 2017) and patents (Meitinger, 2017). Further discussions in this regard can be found in (Savelyev, 2017; Butler et al., 2017; Ølnes et al., 2017; Giancaspro, 2017; Silverberg et al., 2016; ISDA, 2017; Smart Contracts Alliance, 2016).

5.1.3 *How does an ICO Work?*

Before jumping onto the ICO bandwagon, it is worth taking a closer look at a general ICO process from start to finish (Stellar Development Foundation & The Luxembourg House of Financial Technology, 2017).

- **Stage 1–Token Creation:** The tokens are often created after the token issuer deploys a smart contract on a blockchain-based platform (e.g., Ethereum). In principle, every token has an intrinsic price (pre-designated by the issuing company) that will not change during the period of ICO. Once the ICO is completed, the price of tokens is influenced by market conditions and users' perceptions.
- **Stage 2–Announcement of An Intended ICO:** Like the preparation of any other crowdfunding campaign, the issuer needs to do the prep-work in order to test the demand for the project and get support from the community.
- **Stage 3–Whitepaper Publication:** A whitepaper is mandatory for every single ICO project. Typically, it covers important aspects of an ICO. These include issues such as what the project is about, team members, market size, what need(s) the project will fulfil upon completion, how long the ICO campaign will run for, how much money is needed to undertake the venture (i.e., token pricing), what type of money is accepted (i.e., token economy), token deployment plans, business models and revenue streams.
- **Stage 4–Legal Expertise of the Development Team:** In order to avoid the potential heaviest regulation, issuers must have competent lawyers at their disposal for solving issues like investment syndicates, taxes, securities and many more.
- **Stage 5–Pre-Sale (or called Pre-ICO):** Since askers (i.e., issuers) may need to raise funds to cover logistics (e.g., legal, operational, and developmental) costs while they prepare themselves for the full ICO, they can sell a pre-determined number of tokens (often less than 10% of the overall volume) at a discount. Notes: not all projects carry out pre-ICO.
- **Stage 6–Marketing Campaign:** This is a pivotal stage of the complete ICO, in particular, when encountering hundreds of thousands of ICO in the market. Similar to a typical crowdfunding campaign, askers should find a way to stand out among all the noise. Moreover, it is important to communicate with the target audience. Here, it is worth noting that a way to promote an ICO is to include several well-known cryptocurrency communities, e.g., Redddit and Github.
- **Stage 7–Token Sales (real ICO starts):** In an ICO, a certain portion of tokens will be sold to the public investors through a formal event in exchange for crypto funds (e.g., Bitcoin or ether). In general, there are the three types of models that can be employed to facilitate token sales (Stellar Development Foundation & The Luxembourg House of Financial Technology, 2017):

  1) Fixed price (or exchange rate) ICO, e.g., Tezos;

2) Dutch auction ICO, e.g., Gnosis; and
3) Hybrid capped sales, e.g., Mysterium.

- **Stage 8–Product Development:** Once the token sales' proceeds are received, token issuers can cash some of these investors' contributions in order to fund the proceeding product development-related matters, e.g., team, technology and business.
- **Stage 9–Product Launch:** Finally, the launching stage releases the product to the marketplace.

5.1.4 Token Types and Characteristics

Technically, a token refers to an encoded digital enclosure representing a set of rights (Massey et al., 2017). During the course of ICOs, askers sell their newly minted token to obtain public capital (in cryptocurrency form, e.g., Bitcoin and ether) to fund their business projects or community initiatives. Recently, with more and more tokens in the Ethereum ecosystem, a token standard called ERC-20 was developed by Fabian Vogelsteller (Dhillon et al., 2017). More specifically, ERC-20 describes six fundamental functions for each token, which is equivalent to creating a template. With the existence of this construct, the interchangeability among other ERC-20 compliant tokens becomes relatively straightforward and the all-round token interoperability across the entire Ethereum network becomes possible (Massey et al., 2017; Dhillon et al., 2017).

In general, tokens can be divided into three classes (Yao, 2017; PwC, 2017): (1) Cryptocurrency-like token, (2) Utility-like token and (3) Tokenized securities.

- **Token Functions as Cryptocurrency:** It is safe to say that this type of token functions like a digital currency, but with its own unique features. Meanwhile, they may be traded on a secondary market, virtual currency exchanges or other similar platforms.
- **Token Functions as Utility:** Some issuers prefer their tokens to be 'utility tokens', which means that investors can change their tokens for services or products on the sites. Token issuers themselves can keep the raised Bitcoin or ether and use the cashed funds to develop the relevant projects specified in the published whitepaper. In fact, this type of tokens are more like paid API (i.e., application program interface) keys that provide access to a service (Dhillon et al., 2017). The core of this utility is based on a DAO (or called decentralized autonomous organization) entity. It is theoretically a virtual organization, which is governed by various rules that are contained in the 'smart contracts' and are stored and implemented in a blockchain, say, Ethereum (Shermin, 2017). In general, it works through the following procedure (Castor, 2017): when one intends to create

utility-like tokens in Ethereum, he/she actually programs a smart contract, which in turn governs each individual token. This operation enables a token to have more functions (e.g., asset representation, or customized functionality) other than merely an exchange of value. Because of this merit, utility-like tokens are more likely to be utilized by crowdfunding projects or businesses during their token offerings.

- **Token Functions as Security:** Regarding this type, the United States Securities and Exchange Commission (SEC) has recently released a report offering some guidance. It stated that when a token is regarded as a security (under certain conditions), its associated ICO must be regulated under the 1934 Security Exchange Act (The Economist, 2017b). In the light of this instruction, this type of token is analogous to equity-based crowdfunding.

In practice, tokens may fuse one or more of the characteristics mentioned above. More importantly, though the primary draw of all these tokens is rather technical, the true breakthrough is the potential societal impact brought about by these tokens in terms of the formation of new business models and the construction of novel fundraising channels (BlockChannel, 2017). Indeed, this property has empowered tokens to become a 'break thing' that represents the next phase of innovation in blockchain technology.

5.1.5 *The Similarities and Differences between General Crowdfunding and Crypto Crowdfunding*

The key players involved in crypto crowdfunding include high-tech startups (askers), platform operators (e.g., Ethereum) and investors/donors/backers, just like the general crowdfunding form. On the one hand, askers choose ICO because it is not only a technically (e.g., based on open source platforms), but economically (e.g., tokens are saleable to a global crowd) feasible way for tech startups to churn out disruption innovations. On the other hand, through the lens of investors/donors/backers, they can use tokens either as a means to store wealth, with the hope of selling them at some time for speculation and capital gains (i.e., token's speculative value), or as an exchange medium within the interested project's ecosystem (i.e., token's functional value).

Generally speaking, the crypto crowdfunding platform is mainly built on the blockchain-based smart contract infrastructure (Bracamonte and Okada, 2017). Accordingly, under the generic procedure of ICOs, askers organize a campaign on the crypto crowdfunding platform and issue digital 'tokens' to raise funds for producing a new 'protocol', instead of offering a copy of a product on Kickstarter (i.e., a reward-based crowdfunding platform) or shares through Crowdfunder (i.e., an equity-based crowdfunding platform). Meanwhile, financial backers will transfer a certain amount of cryptocurrency (e.g., Bitcoin or ether) to a blockchain-generated address supplied by the askers. Similar to the general crowdfunding form, if the amount of cryptocurrency raised by a firm does not meet the original goal, investors will then be refunded and this particular ICO is regarded as a failure. On the contrary, if the funding goal are achieved within a predetermined timeframe, the accumulated funds are utilized for team construction, project development and launching, as well as new scheme initiation.

In fact, most projects on the crypto crowdfunding platform are not just for funds either. Like normal crowdfunding projects, the introduction of tokens is to create a win-win deal, i.e., technological-social. Crypto-funded new services/products have great potential in leveraging blockchain technology and incorporating token-based business models, which in turn can pave the way for a broader scale of token utilization and better community engagement. For example, RootProject (Judge et al., 2017) focuses on charity-work projects, Notary (Notary, 2017) on the way we do business, Tezos (Tezos, 2017) on security through formal verification by government rules, and Status (Status, 2017) on a mobile browser that transfers payments and smart contracts to friends within chats. Additionally, Golem (Golem, 2017) studied networks that rent or lease computing power, and Gnosis (Shin, 2017) focussed on a cryptofinance prediction market that enables collaborative information sharing and rewards participants for contributing their opinions.

By comparison, perhaps the most important aspect of difference is regulation(s). Unlike normal crowdfunding projects, the ICOs are largely unregulated and, thus, may involve a variety of risks. For example, there is no definitive agreement on what tokens actually are. This means there is no formal process for auditing askers who are conducting ICOs. Furthermore, tokens are not like equities encountered in a typical equity-based crowdfunding. Therefore, the price of tokens is not based on their fundamental value and the 'rights' offered to investors are, therefore, very limited. To summarize, the following list of similarities and dissimilarities offers readers a holistic view of ICO and general crowdfunding.

- **Similarities:** Crypto crowdfunding and general crowdfunding do share a list of similarities which are outlined as follows:
  1) *Similarity 1*: Both types of crowdfunding raise funds from crowds through an online campaign;
  2) *Similarity 2*: The major players of both ecosystems are almost the same;
  3) *Similarity 3*: In general, both forms impose an all-or-nothing policy on the projects and also set a minimum contributing price to begin with;

4) *Similarity 4*: In practice, two kinds of crowdfunding can both be used as a tool to create awareness of the project, igniting a positive network effect;
5) *Similarity 5*: In some cases, token functions as equity and thus possesses the same trait as general equity crowdfunding;
6) *Similarity 6*: In some other cases, token can also show up as rewards/donation, enabling backers to access, use or consume the interested product(s) or service(s); and
7) *Similarity 7*: Both constructs are more or less subject to the platform's own regulations, e.g., Kickstarter or Ethereum.

- **Dissimilarities:** Although the two types of crowdfunding are similar in many senses, the following dissimilarities tend to distinguish them from each other.
  1) *Dissimilarity 1*: In a general crowdfunding project, backers usually get a future product, while in an ICO backers are encouraged to use a virtual digital service;
  2) *Dissimilarity 2*: In an ICO, tokens are digital assets and can be traded in a secondary market so that their liquidity can be enhanced. However, with classical equity crowdfunding, shares are held privately and the secondary market for trading them is not available;
  3) *Dissimilarity 3*: Compared with normal crowdfunding, the mechanisms of ICO and token are very flexible in nature. More specifically, ICOs can happen in two forms (capped or uncapped), and the tokens' price and supply are either fixed or dynamic;
  4) *Dissimilarity 4*: Askers issuing equities via traditional crowdfunding platforms are required to keep the essential information open to investors. In an ICO, askers just need to issue white papers describing their team members, plans and missions, as well as the technical characteristics of the token. Since not every technological proposal is feasible, the risk here is that the intrinsic impracticability can only be discovered after investors have poured a large amount of funds into the project.
  5) *Dissimilarity 5*: Most ICO tokens currently do not offer similar properties often embedded in a typical equity, such as voting right, ownership right and right to share future earnings;
  6) *Dissimilarity 6*: During the sale of ICO tokens, askers cannot discriminate between the investors, which means they often know very little or almost nothing about their token buyers; and
  7) *Dissimilarity 7*: In essence, the next developmental step of general crowdfunding is the ICO, which is characterized by its openness, distributed nature and liquidity.

5.1.6 *A Brief Overview of Global ICO Regulatory Treatment*

Nowadays, the good news is that many countries have started to look into ICOs and have expressed the corresponding basic positions on ICOs. However, the reality is that very few of them have actually taken practical actions and ICOs are still largely under the community's self-regulation. Therefore, the legal status of ICOs still remains unclear. Some representative regulatory treatments are listed as follows and interested readers should refer to (Stellar Development Foundation & The Luxembourg House of Financial Technology, 2017; PwC, 2017; Zetzsche et al., 2017) for further discussions:

- **Canada:** On the 24th August 2017, the Canadian Securities Regulatory Authorities (or CSA) issued a staff notice on cryptocurrency offerings, indicating, "*products or other assets that are tracked and traded as part of a distributed ledger may be securities, even if they do not represent shares of a company or ownership of an entity.*" Meanwhile, a 'regulatory sandbox' was also set up by the CSA in order to offer a set of speed-elevated and flexibility-enhanced supporting services, e.g., allowing companies to register and acquire exemptions under the umbrella of securities law (Luu and Welsh, 2017; Higgins, 2017).
- **China:** On the 4th September 2017, China's authority—the People's Bank of China (or PBoC) declared ICOs illegal and ordered that all organizations that have participated in ICOs refund their token contributors (Acheson, 2017; Stellar Development Foundation & The Luxembourg House of Financial Technology, 2017; CoinDesk, 2017). In Hong Kong, the authority pointed out that the digital tokens makes the ICOs fall within the category of securities (Lim and Law, 2017).
- **European Union (EU):** On the 13th November 2017, the European Securities and Markets Authority (or ESMA) published a statement that alerts companies involved in ICOs to the necessity of meeting relevant EU regulatory requirements, e.g., prospectus directive, the markets in financial instruments directive, alternative investment fund managers directive and the Fourth anti-money laundering (AML) directive (European Securities and Markets Authority (ESMA), 2017). Additionally, with the implementation of the Fifth AML directive, a European Union-wide regulatory framework will soon be established. Given the fact that each member country tends to have its own understanding in terms of regulating cryptocurrencies, the focus of such a united network (Commission, 2016) is to mitigate money laundering risks associated with all virtual currency businesses across the European Union region.
- **Japan:** Regulations on ICO have not yet been settled in Japan. Instead, there is a rule called 'Virtual Currency (VC) Act' which was put into

effect in April 2017. In this regard, interested readers should refer to Saito (2017) for detailed information.

- **New Zealand:** The New Zealand's Financial Markets Conduct Act (FMCA) has issued a report clarifying the obligations for firms to make a 'regulated offer' of a 'financial product', regardless of where the issue occurs or where the issuer is based (Russell McVeagh, 2017).
- **Russia:** Overall, Russia has welcomed cryptocurrencies. The Central Bank of Russia considered cryptocurrencies as financial instruments, not currencies, and is actively interested in monitoring transactions and taxing the capital gains via ICOs. Apart from all this, cryptocurrency organizations must implement either AML rule or KYC (i.e., know your customer) process (Stellar Development Foundation & The Luxembourg House of Financial Technology, 2017).
- **Singapore:** On the 1st August 2017, the Monetary Authority of Singapore published a brief statement which also confirmed that certain tokens would be classified as securities under national law. In addition, the Singapore authority also developed a regulatory construct which is called 'Proposed Payment Framework'. This framework was designed to provide licensing, regulation and the supervision of certain payments and remittance businesses, e.g., cryptocurrency intermediaries (Orion, 2017).
- **South Korea:** Like China, ICOs have been completely banned in South Korea (The Economist, 2017c).
- **Switzerland:** In Switzerland, regulators informally voiced support for blockchain developments and is looking into ICOs as well. They published a report on the 29th September 2017 and indicated that Swiss Financial Market Supervisory Authority (or FINMA) will distinguish ICOs depending on the respective token functionalities. But the official announcement in terms of regulations or investor protections around ICOs was yet unavailable (Swiss Financial Market Supervisory Authority (FINMA), 2017). More details can also be found in (Hess et al., 2017).
- **United Kingdom (UK):** In the UK, though the Financial Conduct Authority considers cryptocurrencies as private currencies, in order to obtain access to cryptocurrency exchange platforms, formal examination under the AML rule or KYC process is necessary. Meanwhile, the FCA also made a regulatory sandbox available for firms involved in the development of blockchain and other forms of distributed ledgers. This service removes the need for complying with the whole set of regulatory requirements right from the beginning and, thus, enables these companies to build, test and verify all sorts of

relevant product/service concepts (Stellar Development Foundation & The Luxembourg House of Financial Technology, 2017).

- **United States (US):** On the 25th July 2017, the US SEC has issued its first public statement regarding the burgeoning market of ICOs. In its report, the SEC stated that *"determining whether a coin or token constitutes a security relates to specific factors, i.e., an investment of money in a common enterprise and with the expectation of profits that are expected to arise substantially from the efforts of a third party"* (Obie et al., 2017). However, many questions remain unanswered by the report, hence, it received mixed feedback from the community. Further discussion in this regard can be found in (Heller, 2017; Telpner and Ahmadifar, 2017).

5.2 Problem Statement

Human development can be generalized into three stages, namely, physical development, cognitive development and psychosocial development (Sigelman and Rider, 2015; Xing and Marwala, 2018a). Similarly, the process of technology development is characterized by its dynamic and cumulative nature (Volti, 2014):

- **Dynamic:** We describe it as being dynamic because no technology is too good to be improved. Perfect technology does not exist as one can always find a way to improve it.
- **Cumulative:** We define it as being cumulative since it is often observed that one advancement paves the way for another one, another advancement paves the way for the next one, and so on. In other words, through working with an existing technology, we can obtain a vast amount of experience (in terms of materials, tools, knowledge base, etc.), which in turn forms a solid foundation for the next phase of technological development.

Take human–robot interaction (Xing and Marwala, 2018k; Xing and Marwala, 2018n), its development has left conventional interaction forms (e.g., human computer or human machine) by entering a stage where the existence of entanglement has to be addressed, e.g., via smart maintenance strategy (Xing and Marwala, 2018m; Xing and Marwala, 2018e; Xing and Marwala, 2018g; Xing and Marwala, 2018f; Xing and Marwala, 2018d; Xing and Marwala, 2018b; Xing and Marwala, 2018c; Xing and Marwala, 2018j; Xing and Marwala, 2018h; Xing and Marwala, 2018i). Other exemplary technologies also include the following:

- **Reconfigurable Manufacturing System:** (Xing et al., 2006a; Xing et al., 2010a; Xing et al., 2010f; Xing et al., 2009; Xing et al., 2006b);

- **Remanufacturing:** (Xing et al., 2010b; Xing et al., 2010c; Xing et al., 2010e; Xing and Gao, 2015d; Xing and Gao, 2015b; Xing and Gao, 2015c; Xing and Gao, 2015a; Xing, 2014a; Xing and Gao, 2014ee; Xing et al., 2014; Marwala and Xing, 2011; Gao et al., 2014; Gao et al., 2013a; Gao et al., 2013b; Xing et al., 2012b; Xing et al., 2012c; Xing et al., 2013a; Xing et al., 2013e; Xing and Gao, 2014pp; Xing and Gao, 2014qq; Xing and Gao, 2014a; Xing and Gao, 2014rr; Xing and Gao, 2014ll; Xing and Gao, 2014oo; Xing and Gao, 2014nn; Xing and Gao, 2014mm; Xing and Gao, 2014hh; Xing and Gao, 2014kk; Xing and Gao, 2014jj; Xing and Gao, 2014ii; Xing and Gao, 2014gg; Xing et al., 2011a);
- **Visible Light Communication:** (Xing, 2017c; Xing and Marwala, 2018j);
- **Artificial Intelligence:** (Xing and Marwala, 2018l; Xing et al., 2013d; Xing et al., 2013b; Xing et al., 2013c; Xing and Gao, 2014d; Xing and Gao, 2014cc; Xing and Gao, 2014f; Xing and Gao, 2014e; Xing and Gao, 2014t; Xing and Gao, 2014g; Xing and Gao, 2014h; Xing and Gao, 2014u; Xing and Gao, 2014v; Xing and Gao, 2014aa; Xing and Gao, 2014i; Xing and Gao, 2014w; Xing and Gao, 2014s; Xing and Gao, 2014bb; Xing and Gao, 2014dd; Xing and Gao, 2014z; Xing and Gao, 2014k; Xing and Gao, 2014l; Xing and Gao, 2014m; Xing and Gao, 2014x; Xing and Gao, 2014n; Xing and Gao, 2014q; Xing and Gao, 2014b; Xing and Gao, 2014y; Xing and Gao, 2014c; Xing and Gao, 2014o; Xing and Gao, 2014j; Xing and Gao, 2014p; Xing and Gao, 2014r; Xing and Gao, 2014ff; Marwala, 2010a; Marwala, 2009; Marwala, 2012; Marwala and Lagazio, 2011; Marwala, 2013; Marwala, 2015; Marwala, 2010b);
- **Adaptive System Development:** (Xing, 2017a; Patel and Marwala, 2006; Marwala et al., 2001);
- **The Fourth Industrial Revolution:** (Xing and Marwala, 2017);
- **Mobile Payment:** (Xing, 2017b);
- **Internet of Things and Cyber Physical System:** (Xing, 2016a; Xing, 2016d; Xing, 2015d; Xing, 2014b; Xing, 2014c; Xing et al., 2011b; Xing et al., 2012a; Marwala, 2010a);
- **Ubiquitous Robotics:** (Xing, 2016b; Xing, 2015b);
- **Ambient Assisted Living:** (Xing, 2016e; Xing, 2016f);
- **Maintainomics:** (Xing, 2015a); and
- **Massive Open Online Course:** (Xing, 2015c).

In general, the trajectory of an individual technology development process can be illustrated on a graph, as in Fig. 5.1.

Initially, radical innovation occurs in a very humble form. As the technology is gradually accepted by the market, a series of reformative innovations will happen, though still rather slow, by following the

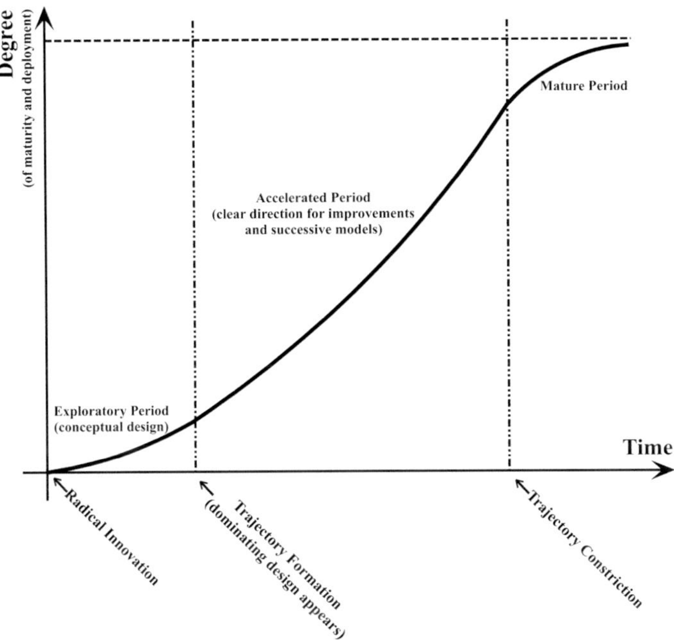

Figure 5.1: Trajectory of an individual technology development process.

trajectory of a logistic curve. After experiencing a period of rapid and intensive developmental acceleration, the process slows down again and reaches the maturity stage.

The main revolutionary gene underpinning a crowdfunding-facilitated ICO is that the token value grows as long as the crowdfunding project is succeeding, much like its counterpart, the IPO. However, unlike IPOs, ICOs are currently not governed by any specific regulation, and they are, therefore, vulnerable to the risk of fraud or money laundering (Stellar Development Foundation & The Luxembourg House of Financial Technology, 2017). Indeed, on the regulatory and legal side, many issues have been raised in terms of privacy, security and risk. Accordingly, regulators from all over the world are now starting to oversee ICOs. However, some organizations thought that it is probably too rash for regulators to interfere with ICOs, since ICOs may also give rise to viable mechanisms for companies to get a novel ecosystem established (The Economist, 2017b; Massey et al., 2017; Scott, 2015).

5.2.1 Question 5.1

From the scientific perspective, the probability of future technological development has become a key factor that long-term planning and policy-making must take into account (Miles et al., 2017). Based on the above-observed disputes about ICOs between the advantages and disadvantages, Question 5.1 is proposed as follows:

- **Question 5.1:** *How do we develop a technology evaluation framework which can address various uncertainties effectively, and among numerous existing technology foresight methodologies, how could we select the most appropriate one that reflects most characteristics of the current situation?*

By investigating Question 5.1, our expectations are two-fold: (1) be able to verify the usefulness of fuzzy systems in the area of technology evaluation, and (2) be able to develop a generic model for selecting the most relevant technology foresight approaches.

5.3 Type-2 Fuzzy Sets for Question 5.1

As we are moving from not only modelling 'numbers', but 'words' or even 'perceptions', the inherent uncertainty degree increases accordingly (John and Coupland, 2008). Conventional mathematical modelling mechanisms were designed to address crisp data (i.e., numbers) enriched problems. Nevertheless, our surroundings are characterized by pervasive uncertainties, which often prevent traditional methodologies from attaining full functionality. Under this circumstance, the concept of 'fuzzy' (Zadeh, 1965; Xing et al., 2010d) was further introduced to tackle various unsure scenarios. However, one of the main difficulties of applying the original fuzzy sets theory (the so-called type-1 fuzzy sets) in real-world applications, as commented by Zadeh (1971), is determining each set element's membership degree. The idea of type-2 fuzzy sets was then nurtured during, roughly, that period in which the uncertainty associated with each expert is taken into account. Put simply: Type-1 fuzzy sets can be used for modelling 'words', while type-2 fuzzy sets are responsible for modelling 'perceptions' (Zadeh, 1975a; Zadeh, 1975b; Zadeh, 1975c; Mendel and Wu, 2010; Bargiela and Pedrycz, 2009). More formally: In type-2 fuzzy sets, a referential set [0,1] is defined for every element's membership degree (Bustince et al., 2015; Zadeh, 1971).

5.3.1 *Type-2 Fuzzy Sets*

Type-2 fuzzy sets are essentially still fuzzy sets, but with conventional truth values expressed in the form of fuzzy sets (Nguyen and Walker, 2006). In other words, if an uncertain quantity's exact value could not be decided, then how a fuzzy set's membership would be ascertained precisely then it would be difficult to ascertain its corresponding fuzzy set's membership precisely. Such criticism applies to type-1 fuzzy sets undoubtedly, but also still applies to type-2 fuzzy sets unfortunately. The most secure means is to introduce type-∞ fuzzy sets, so that uncertainty could be properly represented. Due to the impracticality of such an expectation, type-1 and type-2 are still among the most popular fuzzy sets. Theoretically, type > 2 fuzzy sets could be employed, but the accompanied computational complexity issue is a major obstacle (Mendel, 2017).

5.3.1.1 *Basic Concept*

Over a universe of discourse (indicated by X), a type-2 fuzzy set can be represented by $\tilde{A}$. Its main feature lies in that it contains a set of pairs, denoted by $\{x, \mu_{\tilde{A}}(x)\}$, where $x \in X$, and $\mu_{\tilde{A}}(x)$ stands for the degree of the membership, which is defined via Eq. 5.1 (Dereli and Altun, 2013):

$$\tilde{A} = \int_{x \in X} \mu_{\tilde{A}}(x) x = \int_{x \in X} \left(\frac{\int_{u \in J_x} \frac{f_x(u)}{u}}{x} \right), \quad J_x \subseteq [0,1]. \qquad 5.1$$

where the second membership function is represented by $f_x(u)$, the primary membership of x is denoted by J_x, and the symbol ∫ stands for the function's continuous definition over the universe of discourse.

5.3.1.2 *Operators*

In (Dereli and Altun, 2013), t-conorm and t-norm (Klir and Yuan, 1995; Klir and Folger, 1988) operators are used to perform union/intersection-like operations on type-2 fuzzy sets (Zarandi et al., 2009). More specifically, according to (Karnik and Mendel, 1999), the following definitions can be given:

- **Union**—The union of two type-2 fuzzy sets can be defined using Eq. 5.2 (Dereli and Altun, 2013; Karnik and Mendel, 1999):

$$\tilde{A} \cup \tilde{B} = \int_{x \in X} \left(\mu_{\tilde{A}}(x) \cup \mu_{\tilde{B}}(x) \right) = \int_{x \in X} \left(\frac{\int_{u \in \left[\underline{J}_{x(\tilde{A})} \vee \underline{J}_{x(\tilde{B})}, \overline{J}_{x(\tilde{A})} \vee \overline{J}_{x(\tilde{B})} \right]} \left(\frac{1}{u} \right)}{x} \right). \qquad 5.2$$

- **Intersection**—The intersection of two type-2 fuzzy sets can be defined using Eq. 5.3 (Dereli and Altun, 2013; Karnik and Mendel, 1999):

$$\tilde{A} \cap \tilde{B} = \int_{x \in X} \left(\mu_{\tilde{A}}(x) \cap \mu_{\tilde{B}}(x) \right) = \int_{x \in X} \left(\frac{\int_{u \in \left[\underline{J}_{x(\tilde{A})} \wedge \underline{J}_{x(\tilde{B})}, \overline{J}_{x(\tilde{A})} \wedge \overline{J}_{x(\tilde{B})} \right]} \left(\frac{1}{u} \right)}{x} \right). \qquad 5.3$$

- **Complement**—The complement of a type-2 fuzzy set can be defined using Eq. 5.4 (Dereli and Altun, 2013; Karnik and Mendel, 1999):

$$\bar{\tilde{A}} \Leftrightarrow \mu_{\bar{\tilde{A}}}(x) = \neg\mu_{\tilde{A}}(x) = \int_{x\in X}\left(\frac{\int_{u\in\left[1-\bar{J}_x,\,1-\underline{J}_x\right]}\left(\frac{1}{u}\right)}{x}\right). \qquad 5.4$$

5.3.1.3 Type-2 Fuzzy Inference System

In general, a rule-based fuzzy system consists of four modules: Module of rules, module of fuzzifier, module of inference engine and module of output processor. The relationship of interconnection between these four modules is depicted in Fig. 5.2.

Once the rules module is created, one can treat the generic fuzzy system as a function that maps the outputs from the inputs. The quantitative expression of this mapping function can be given by Eq. 5.5 (Mendel, 2017):

$$y = f(\mathbf{x}). \qquad 5.5$$

In the literature, a generic fuzzy system such as this tends to have many names, e.g., fuzzy logic systems, fuzzy inference system, fuzzy-rule-based system, fuzzy expert system, fuzzy model and fuzzy logic controller, to name just a few (Castillo and Melin, 2009; Bělohlávek et al., 2017; Xing, 2016c; Xing and Gao, 2014a; Celikyilmaz and Türksen, 2009; Novák et al., 2016; Jiménez-Losada, 2017).

As the heart of a fuzzy system, rules can be extracted either from scenario-specific data or collected from different field experts. In terms of the fuzzification module, when type-1 fuzzy sets are utilized, we often call it a type-1 fuzzy inference system. If one or more interval type-2 fuzzy sets are employed, we can term it as the interval type-2 fuzzy inference system. In the situation that one or more general type-2 fuzzy sets are engaged, we

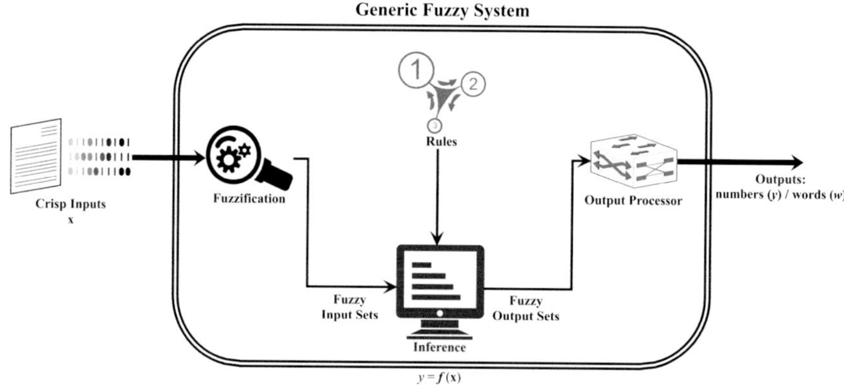

Figure 5.2: Flowchart of a typical fuzzy system.

can call it a general type-2 fuzzy system. The differences between these two kinds (i.e., interval and general) of type-2 fuzzy sets can be found in (Mendel, 2017). In practice, no matter what types of fuzzy sets are chosen, the rules are unchanged.

- **Type-1 Fuzzy Inference System (T1FIS)**—T1FIS system is essentially the same as generic fuzzy inference system, except a defuzzification module is introduced to substitute the output processor module. As depicted in Fig. 5.3, a T1FIS system is also comprised of four modules: Fuzzification module, inference engine module, rules module and defuzzification module.

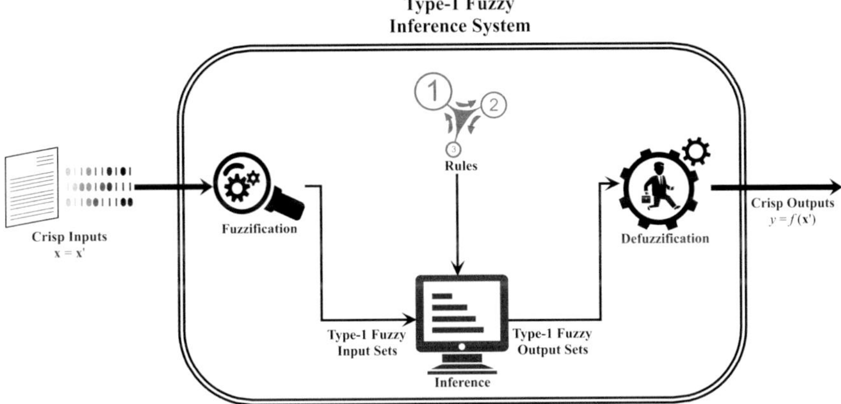

Figure 5.3: Flowchart of type-1 fuzzy inference system.

Please note that a specific value of **x** (denoted by **x′**) is added here to stimulate the entire T1FIS, and the crisp outputs will depend on it, i.e., $y = f(\mathbf{x}')$. Interested readers should refer to (Mendel, 2017) for further discussions about different settings of rules, fuzzifiers, inference engine and defuzzifiers under the T1FIS architecture.

- **Interval Type-2 Fuzzy Inference System (IT2FIS)**—Transforming a fuzzy set into a number is a key step for any rule-based fuzzy inference system. For previous T1FIS, this function is fulfilled via defuzzification module as illustrated in Fig. 5.2. When it comes to type-2 fuzzy inference system, we can have two alternatives: (1) Direct defuzzification – get a T2FIS mapped directly into a number, or (2) Type Reduction + Defuzzification – get a T2FIS first transformed into a T1FIS via type reduction, and then defuzzify the simplified fuzzy sets into an output number. These two options are applicable to both interval and general T2FIS. Regarding IT2FIS (i.e., only interval type-2 fuzzy sets involved), the process is illustrated in Fig. 5.4.

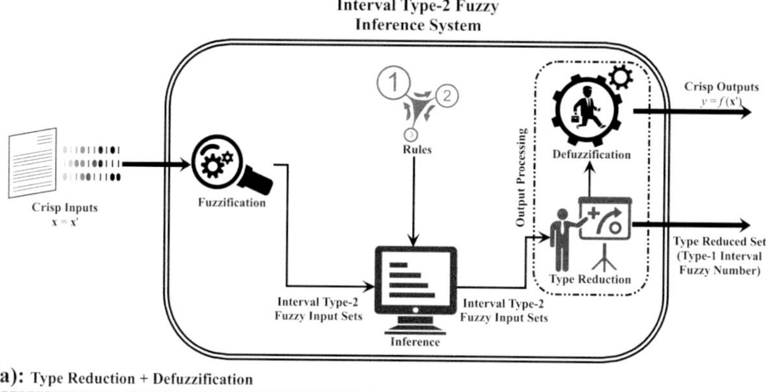

(a): Type Reduction + Defuzzification

(b): Direct Defuzzification

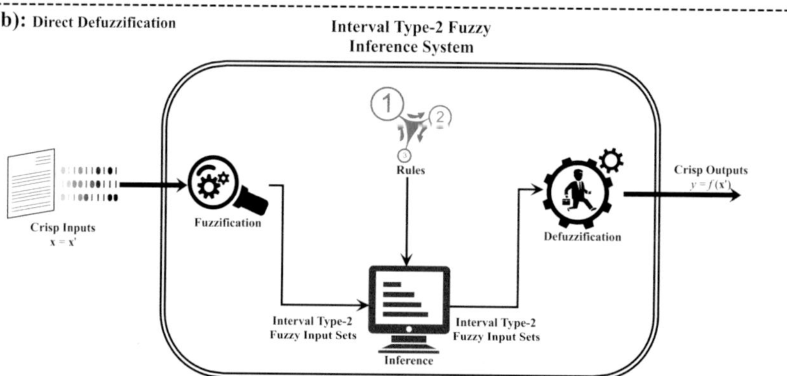

Figure 5.4: Flowchart of an interval type-2 fuzzy inference system.

Typically, an IT2FSs can be defined using Eq. 5.6 (Dereli and Altun, 2013):

$$\tilde{A} = \int_{x\in X} \left(\frac{\int_{u\in[\underline{J}_x,\overline{J}_x]} \frac{f_x(u)}{u}}{x} \right), \quad [\underline{J}_x, \overline{J}_x] \subseteq [0,1]. \qquad 5.6$$

Since the associated computational complexity (particularly when facing a large number of variables) for full T2FIS is considerably high, IT2FIS enjoys a wide popularity (Karnik and Mendel, 1999; Mendel et al., 2006; Kazemzadeh et al., 2008; Celikyilmaz and Türksen, 2009). Interested readers should refer to (Mendel, 2017) for further discussions about different settings of rules, fuzzifiers, inference engine and defuzzifiers under the IT2FIS architecture.

- **General Type-2 Fuzzy Inference System (GT2FIS)**—As the name implies, the fuzzy sets involved in GT2FIS are all (strictly speaking, one is more than enough) general type-2 fuzzy sets (Mendel, 2014). Similar to IT2FIS, there are also two kinds (i.e., direct defuzzification and type reduction + defuzzification) of GT2FIS, as depicted in Fig. 5.5.

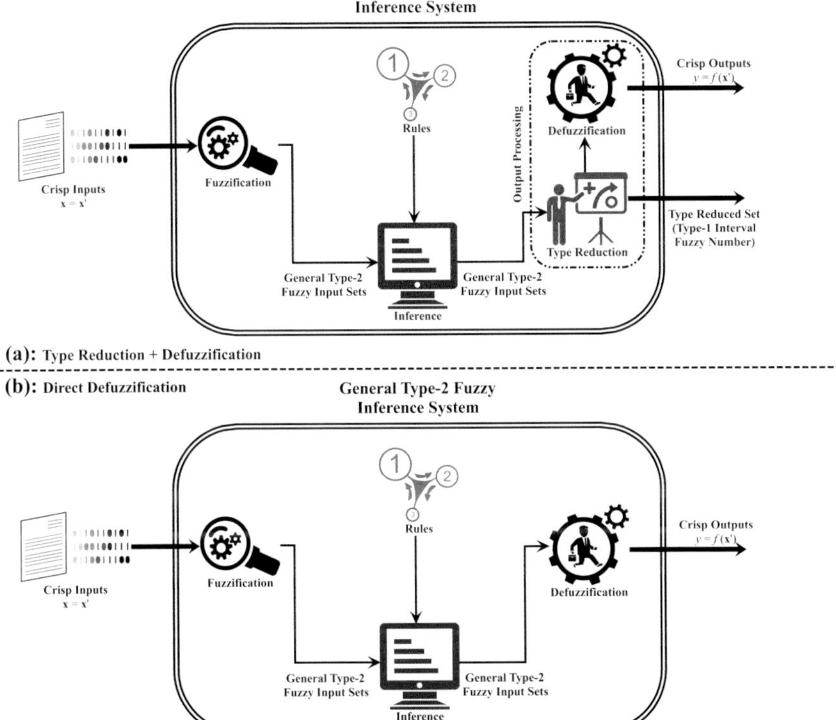

Figure 5.5: Flowchart of a general type-2 fuzzy inference system.

The underlying reasons for the importance of GT2FIS can be explained as follows (Mendel, 2017):

1) *Logical development*: When the system performance of both T1FIS and IT2FIS cannot meet our needs, the introduction of GT2FIS is an intuitive and logical resolution;
2) *Flexibility*: Because uncertainties are weighted non-uniformly by GT2FIS, its associated modelling flexibility is high.
3) *Resolvability*: Upon encountering semantically incompatible statements, the resolvability of GT2FIS is stronger than IT2FIS. Suppose a statement is given by Eq. 5.7 (Mendel, 2017; Greenfield and John, 2009):

Statement = {Initial Coin Offering via crowdfunding is good!}. 5.7

In crisp logic model, Eq. 5.7 is equivalent to Eq. 5.8 (Mendel, 2017):

$$Statement_{crisp} = \{\text{'Initial Coin Offering via crowdfunding is good!' is true.}\}. \quad 5.8$$

While for T1FIS, we tend have a group of alternatives as given by Eq. 5.9 (Mendel, 2017):

$$\begin{aligned} Statement^{(1)}_{\text{T1FIS}} &= \begin{Bmatrix} \text{'Initial Coin Offering via crowdfunding is good!'} \\ \text{has a truth value of 0.8.} \end{Bmatrix} \\ Statement^{(2)}_{\text{T1FIS}} &= \begin{Bmatrix} \text{'Initial Coin Offering via crowdfunding is good!'} \\ \text{has a truth value of 0.5.} \end{Bmatrix} \\ &\vdots \\ Statement^{(n)}_{\text{T1FIS}} &= \begin{Bmatrix} \text{'Initial Coin Offering via crowdfunding is good!'} \\ \text{has a truth value of 0.2.} \end{Bmatrix} \end{aligned} . \quad 5.9$$

By moving further into IT2FIS, the alternatives of Eq. 5.8 can be reformulated using Eq. 5.10 (Mendel, 2017; Greenfield and John, 2009):

$$\begin{aligned} Statement^{(1)}_{\text{IT2FIS}} &= \begin{Bmatrix} \text{The truth value of the statement of} \\ \begin{Bmatrix} \text{'Initial Coin Offering via crowdfunding is good!'} \\ \text{has a truth value of 0.8.} \end{Bmatrix} \text{is 1.} \end{Bmatrix} \\ Statement^{(2)}_{\text{IT2FIS}} &= \begin{Bmatrix} \text{The truth value of the statement of} \\ \begin{Bmatrix} \text{'Initial Coin Offering via crowdfunding is good!'} \\ \text{has a truth value of 0.5.} \end{Bmatrix} \text{is 1.} \end{Bmatrix} \\ &\vdots \\ Statement^{(n)}_{\text{IT2FIS}} &= \begin{Bmatrix} \text{The truth value of the statement of} \\ \begin{Bmatrix} \text{'Initial Coin Offering via crowdfunding is good!'} \\ \text{has a truth value of 0.2.} \end{Bmatrix} \text{is 1.} \end{Bmatrix} \\ &\vdots \end{aligned} . \quad 5.10$$

According to Greenfield and John (2009), the incompatibility observed among the statements of $Statement^{(1)}_{\text{IT2FIS}}$, $Statement^{(2)}_{\text{IT2FIS}}$, $Statement^{(n)}_{\text{IT2FIS}}$, etc. is attributed to assigning the same truth value to every individual T1FIS statement. By further introducing GT2FIS, different truth values, as given by Eq. 5.11 (Mendel, 2017; Greenfield and John, 2009), can be assigned to the respective T1FIS statement in order to enhance compatibility.

$$\begin{aligned}
Statement^{(1)}_{\mathrm{GT2FIS}} &= \left\{ \begin{array}{c} \text{The truth value of the statement of} \\ \left\{ \begin{array}{c} \text{'Initial Coin Offering via crowdfunding is good!'} \\ \text{has a truth value of 0.8.} \end{array} \right\} \text{ is 1.} \end{array} \right\} \\
Statement^{(2)}_{\mathrm{GT2FIS}} &= \left\{ \begin{array}{c} \text{The truth value of the statement of} \\ \left\{ \begin{array}{c} \text{'Initial Coin Offering via crowdfunding is good!'} \\ \text{has a truth value of 0.5.} \end{array} \right\} \text{ is 0.6.} \end{array} \right\} \\
&\vdots \\
Statement^{(n)}_{\mathrm{GT2FIS}} &= \left\{ \begin{array}{c} \text{The truth value of the statement of} \\ \left\{ \begin{array}{c} \text{'Initial Coin Offering via crowdfunding is good!'} \\ \text{has a truth value of 0.2.} \end{array} \right\} \text{ is 0.1.} \end{array} \right\} \\
&\vdots
\end{aligned} \quad . \; 5.11$$

Interested readers should refer to (Mendel, 2017) for further discussions about different settings of rules, fuzzifiers, inference engine and defuzzifiers under the GT2FIS architecture.

5.3.2 *Technology Evolution via Interval Type-2 Fuzzy Inference System*

In (Dereli and Altun, 2013), an innovative framework (built on IT2FIS) for evaluating technology was proposed. The general architecture of this technology evaluation framework is simplified and illustrated in Fig. 5.6.

5.3.2.1 *Framework Construction*

The rest of this section provides a brief description of the constructs of this framework. More information can be found in (Dereli and Altun, 2013).

- **Input Pre-processing:** In order to acquire the uncontaminated input information resources, filed patents and published research papers were utilized in (Dereli and Altun, 2013). More specifically, the European Patent Office (EPO) online database was used in order to retrieve relevant patent data, while the Web of Science/Knowledge (WoS/K for short) was used in order to obtain the desired publication data.
- **Fuzzy Sets Generation:** When it comes to producing input fuzzy sets, bias could be generated if only "amount" is taken into account since the importance of this factor may vary across different technology classes. Therefore, the factor of "hotness" was considered instead, when assessing a technology's trendiness, since it is more relevant to a technology's growth rate. Unlike the other practice (Arman et al., 2009) found in the literature, the hotness values measured in (Dereli and Altun, 2013) and IT2FSs (i.e., interval type-2 fuzzy sets) were utilized to deal with the associated uncertainty.

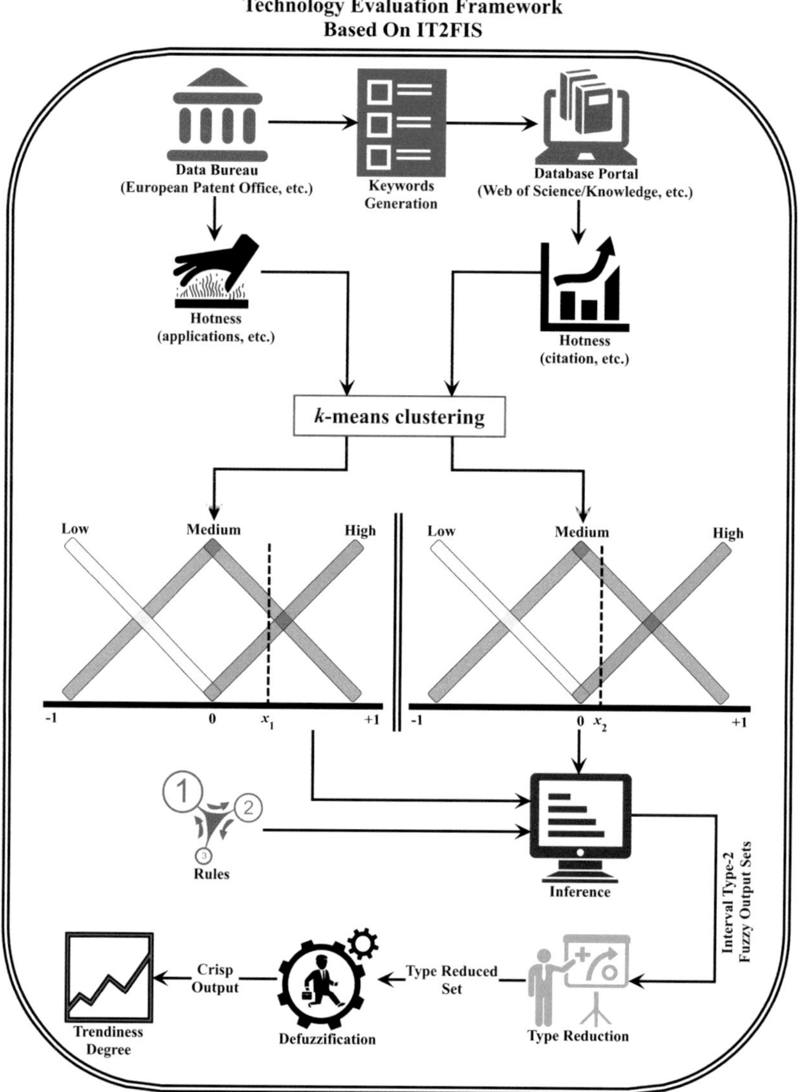

Figure 5.6: Structure of IT2FIS-enabled technology evaluation framework.

- **Membership Functions Formulation:** By classifying both datasets into three clusters, namely low, medium, and high, *k*-means clustering approach (MacQueen, 1967) was employed (Dereli and Altun, 2013) in order to search each cluster's centroids. In *k*-means, data is typically divided into disjoint sets (denoted by $Cluster_1,...,Cluster_k$), in which a centroid (indicated by μ_i) is used to represent each $Cluster_i$. The objective function of *k*-means examines the squared distance between every data point in the input set and its affiliated cluster's centroid. In general μ_i can be defined using Eq. 5.12 (Shalev-Shwartz and Ben-David, 2014):

$$\mu_i(Cluster_i) = \arg\min_{\mu \in X'} \sum_{x \in Cluster_i} d(x,\mu)^2 . \qquad 5.12$$

where the two assumptions are generally made: (1) The condition of $X \subseteq X'$ holds, that is, the X (i.e., input set) is incorporated in (X', d), i.e., some larger metric space, and (2) centroids belong to X'. Accordingly, the objective of *k*-means is given using Eq. 5.13 (Shalev-Shwartz and Ben-David, 2014):

$$G_{k\text{-}means}((X,d),(Cluster_1,\ldots,Cluster_k)) = \sum_{i=1}^{k} \sum_{x \in Cluster_i} d(x,\mu_i(Cluster_i))^2 . \qquad 5.13$$

The objective of *k*-means can also be written in the form of Eq. 5.14 (Shalev-Shwartz and Ben-David, 2014):

$$G_{k\text{-}means}((X,d),(Cluster_1,\ldots,Cluster_k)) = \min_{\mu_1,\ldots,\mu_k \in X'} \sum_{i=1}^{k} \sum_{x \in Cluster_i} d(x,\mu_i)^2 . \qquad 5.14$$

Although *k*-means objective function enjoys tremendous popularity in real-world clustering applications, the actual process of seeking the optimum *k*-means solution is proved to be NP-hard (Shalev-Shwartz and Ben-David, 2014). Therefore, a simple iterative mechanism is often employed. In other words, the outcome of this simple algorithm is often referred to as *k*-means clustering instead of the true clustering that can get the *k*-means objective cost minimized. In (Dereli and Altun, 2013), the following two steps are run alternately until no changes are found for any mean. Meanwhile, triangular membership functions are utilized in order to construct the input fuzzy sets.

1) *Rough guess*: Firstly, preliminary estimations are made towards the means of three clusters and the obtained data points serve as the initial centroid of the clusters.
2) *Data-to-cluster allocation*: Secondly, all data points are allocated to the suitable clusters.

- **Rule Base Creation:** When the fuzzy sets for all of the input and output parameters are developed, we need to create a rule-based model. In (Dereli and Altun, 2013), IT2FIS was utilized in order to fuse the data from patents and publications, respectively. Though the possession of prior experience is not compulsory in terms of producing a rule base, the importance degree of each data source has to be considered.

- **Process of Inference:** Since the input information has already been processed, the process of inference can be done as follows (Dereli and Altun, 2013).
  1) *A rule-base with N rules*: The condition given in Eq. 5.15 (Dereli and Altun, 2013) holds.

  $$\text{Rule}(n): \text{ If } x_1 \text{ is } \tilde{X}_1^n \text{ and } x_2 \text{ is } \tilde{X}_2^n \text{ then } y \text{ is } Y^n \quad n=1,2,\ldots,N \,. \tag{5.15}$$

  where $\tilde{X}_1^n$ and $\tilde{X}_2^n$ represent IT2FSs, which can be formed by patent data and publication data, respectively. The values of hotness degree of targeted technologies are denoted by x_1 and x_2, respectively, and the interval $[\underline{y}^n, \bar{y}^n]$ associated with Y^n stands for the trendiness.
  2) *Membership calculation for x_1*: On each $\tilde{X}_1^n$ as given by Eq. 5.16 (Dereli and Altun, 2013):

  $$\left[\mu_{\underline{X}_1^n}(x_1), \mu_{\bar{X}_1^n}(x_1)\right] \quad n = 1,2,\ldots,N \,. \tag{5.16}$$

  3) *Membership calculation for x_2*: On each $\tilde{X}_2^n$ as given by Eq. 5.17 (Dereli and Altun, 2013):

  $$\left[\mu_{\underline{X}_2^n}(x_2), \mu_{\bar{X}_2^n}(x_2)\right] \quad n = 1,2,\ldots,N \,. \tag{5.17}$$

  4) *Firing interval calculation for the nth rule*: $F^n(x_1, x_2)$ can be computed via Eq. 5.18 (Dereli and Altun, 2013):

  $$\begin{aligned} F^n(x_1,x_2) &= \left[\mu_{\underline{X}_1^n}(x_1)\times\mu_{\underline{X}_1^n}(x_1), \mu_{\bar{X}_1^n}(x_1)\times\mu_{\bar{X}_2^n}(x_2)\right] \\ &\equiv \left[\underline{f}^n, \bar{f}^n\right] \end{aligned} \quad n = 1,2,\ldots,N \,. \tag{5.18}$$

- **Process of Type-Reduction:** The goal of type-reduction is to transform T2FSs into T1FSs for the purpose of defuzzification. In (Dereli and Altun, 2013), one of the most popular type reducer, centre of sets (denoted by Y_{CoS}), is utilized as expressed by Eq. 5.19 (Dereli and Altun, 2013):

  $$Y_{CoS}(x) = \bigcup_{\substack{f^n \in F^n(x) \\ y^n \in Y^n}} \frac{\sum_{n=1}^{N} f^n y^n}{\sum_{n=1}^{N} f^n} = \left[y_{left}, y_{right}\right]. \tag{5.19}$$

  where the interval set's end points are denoted by y_{left} and y_{right}, respectively, which can be formulated as in Eqs. 5.20 and 5.21 (Dereli and Altun, 2013):

$$y_{left} = \frac{\sum_{n=1}^{L} \overline{f}^n \underline{y}^n + \sum_{n=L+1}^{N} \underline{f}^n \underline{y}^n}{\sum_{n=1}^{L} \overline{f}^n + \sum_{n=L+1}^{N} \underline{f}^n}. \quad 5.20$$

$$y_{right} = \frac{\sum_{n=1}^{R} \underline{f}^n \overline{y}^n + \sum_{n=R+1}^{N} \overline{f}^n \overline{y}^n}{\sum_{n=1}^{R} \underline{f}^n + \sum_{n=R+1}^{N} \overline{f}^n}. \quad 5.21$$

where L and R (representing switch points) can be specified using Eqs. 5.22 and 5.23 (Dereli and Altun, 2013):

$$\underline{y}^L \leq y_{left} \leq \underline{y}^{L+1}. \quad 5.22$$

$$\overline{y}^R \leq y_{right} \leq \overline{y}^{R+1}. \quad 5.23$$

In (Dereli and Altun, 2013), a popular method, the Karnik-Mendel algorithm (Mendel and Wu, 2010), was employed in order to identify the desired switch points. A flowchart, illustrating the calculation process suggested by Dereli and Altun (2013), is depicted in Fig. 5.7.

- **Process of Defuzzification:** The values of the interval sets' switch points are obtainable after performing the Karnik-Mendel algorithm. Then the output can be defuzzified via Eq. 5.24 (Dereli and Altun, 2013):

$$y = \frac{y_{left} + y_{right}}{2}. \quad 5.24$$

As soon as the process of defuzzification is completed, a set of crisp values is acquired, in which each value represents the associated technology's trendiness.

5.3.2.2 *Summary*

Though utilizing 'hotness indices' has certain limitations in evaluating trendiness, in particular when the difference between the amount of filed patents and the number of published candidate technologies-related publications is relatively high, the usefulness of the presented framework is well demonstrated through a case study. Dereli and Altun (2013) claimed that the realistic degree of the experimental results (obtained via the proposed framework) was higher than its benchmark counterpart.

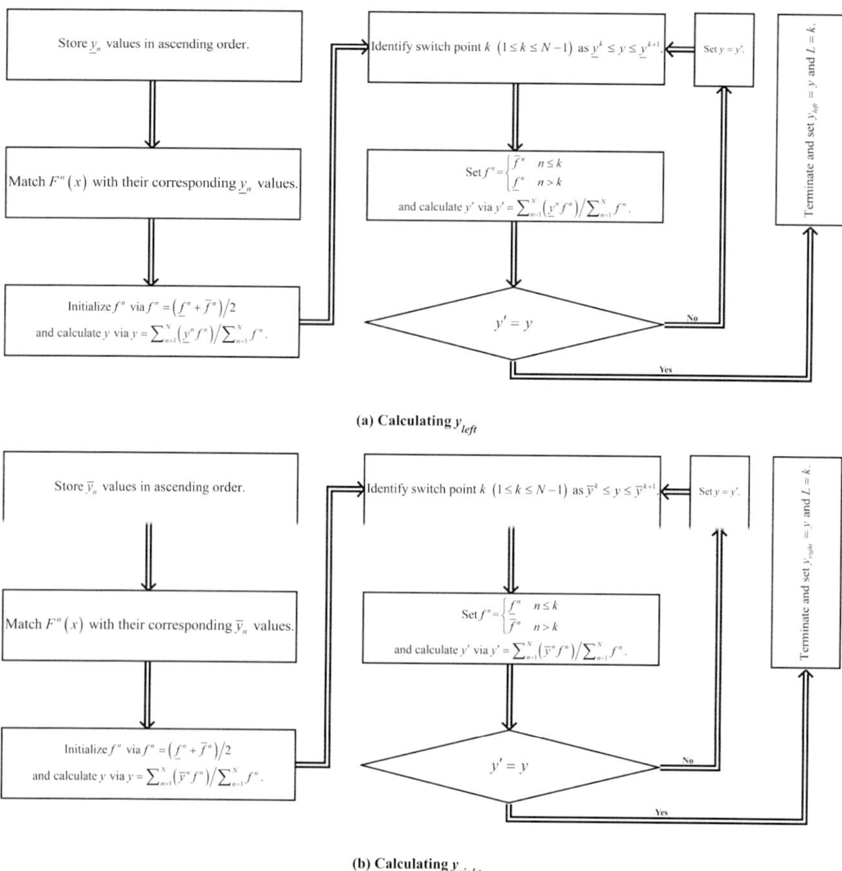

Figure 5.7: Flowchart of Karnik-Mendel algorithm for calculating y_{left} and y_{right}.

5.3.3 *Generic Technology Foresight Methods (TFM) Evaluation Procedure*

Apart from the above elaborated type-2 fuzzy inference system facilitated TFM evaluation case, an intuitionist, fuzzy (Angelov and Sotirov, 2016) group decision-making methodology was also proposed in (İntepe et al., 2013) for selecting an appropriate TFM. However, in order to provide more insightful information, required during the process of pinpointing the most favourite TFM, an attempt to offer a generic solution was made in (Esmaelian et al., 2017). In brief, the proposed model contains two phases: Qualitative analysis phase and quantitative analysis phase. Meanwhile, these two phases spread across five consecutive stages as illustrated in Fig. 5.8.

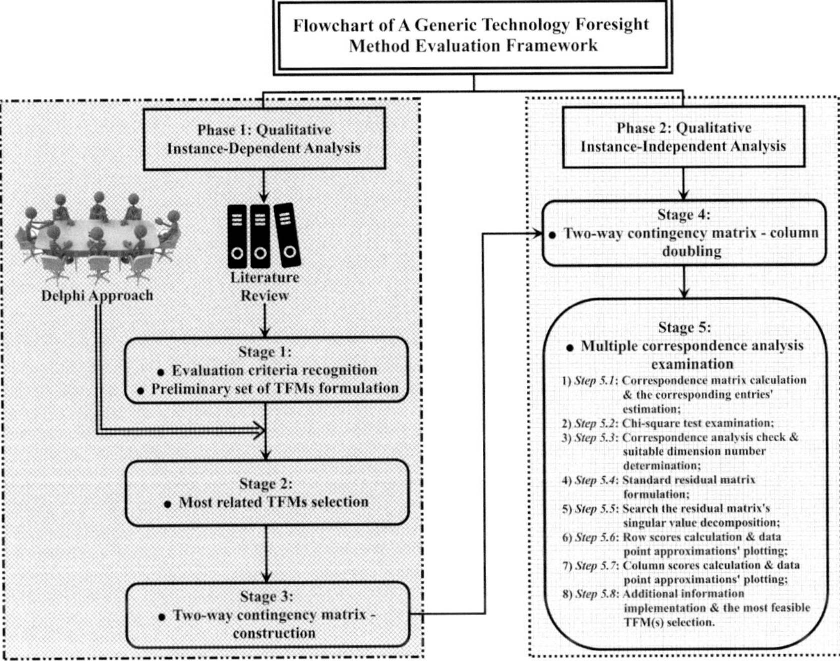

Figure 5.8: Flowchart of a generic TFM evaluation framework.

5.3.3.1 Phase 1—Qualitative Instance-Dependent Analysis

- **Stage 1–Evaluation Criteria Recognition and Preliminary Set of TFMs (i.e., technology foresight methodologies) Formulation:** Based on a properly conducted literature review, a number of evaluation criteria (denoted by m) can be recognized and the preliminary set of TFMs (represented by N) can then be formulated.
- **Stage 2–Most Related TFMs Selection:** This process typically involves experts from various backgrounds. The number of selected TFMs is denoted by n. The first two stages play a crucial role for decision-makers to unveil the complicated interconnections among various factors (Andersen and Rasmussen, 2014; Andersen and Andersen, 2014; Neij et al., 2004; Junginger et al., 2005).
- **Stage 3–Two-way Contingency Matrix—Construction:** The rows of this newly constructed matrix correspond to the selected TFMs at Stage 2, while the columns are associated with the recognized criteria at Stage 1. Expert team members are requested to grade each TFM against the pool of evaluation criteria. In (Esmaelian et al., 2017), Likert-type scores (Likert, 1932; L. Andries van der Ark et al., 2017)

were utilized as a rating scale. Accordingly, a matrix with the size of can be acquired via Eq. 25 (Esmaelian et al., 2017):

$$D = \left[x_{ij}\right]_{\substack{i=1,\ldots,n \\ j=1,\ldots,m}}. \tag{5.25}$$

where x_{ij} denotes the allocated rating value for the ith method based on the jth criterion.

5.3.3.2 *Phase 2—Quantitative Instance-Independent Analysis*

- **Stage 4–Two-way Contingency Matrix—Column Doubling:** At this stage, doubling approach is utilized in order to get a new matrix defined from D. In essence, doubling refers to the redefinition of each rating value (represented by x_{ij}) as a pair of values which are complementary to each other, that is, c_{ij}^{+} (indicating the positive or high pole) and c_{ij}^{-} (indicating the negative or low pole). In practice, it is often preferred to equip the rating scales with a lower endpoint of zero before the doubling operation is performed (Greenacre, 2007). Accordingly, a matrix with the size of $n \times 2m$ is obtained, as given by Eq. 26 (Esmaelian et al., 2017):

$$F\left[C_j^{-}, C_j^{+}\right]_{j=1,\ldots,m}. \tag{5.26}$$

where C_j^{+} and C_j^{-} represent n-dimensional column vectors as defined by Eqs. 5.27 and 5.28 (Esmaelian et al., 2017) for every $j = 1,\ldots,m$:

$$C_j^{+} = \left[c_{ij}^{+}\right]_{i=1,\ldots,n} \text{ where } c_{ij}^{+} = x_{ij} - 1. \tag{5.27}$$

$$C_j^{-} = \left[c_{ij}^{-}\right]_{i=1,\ldots,n} \text{ where } c_{ij}^{-} = M - 1 - c_{ij}^{+}. \tag{5.28}$$

- **Stage 5–Multiple Correspondence Analysis (MCA) Examination:** For the purpose of reducing the problem's dimensionality and offering meaningful graphical representation, the MCA procedure proposed in (Esmaelian et al., 2017) utilizes a similar mechanism (row principal scoring) found in the literature (Hayano et al., 2015; Yelland, 2010). For simplicity, matrix F can be defined by Eq. 29 (Esmaelian et al., 2017):

$$F = \left[f_{ik}\right]_{\substack{i=1,\ldots,n \\ j=1,\ldots,2m}}. \tag{5.29}$$

where f_{ik} denotes the F's generic element, and for every $i = 1,\ldots,n$, the following conditions hold: $f_{i1} = c_{i1}^{-}, f_{i2} = c_{i1}^{+}, f_{i3} = c_{i2}^{-}, f_{i4} = c_{i2}^{+}$, etc.

1) *Stage 5.1*: Correspondence matrix calculation and the corresponding entries' estimation. The correspondence matrix (denoted by P) is a matrix with the size of $n \times 2m$ which can be acquired from F using Eq. 5.30 (Esmaelian et al., 2017):

$$P = \left[p_{ik}\right]_{\substack{i=1,\ldots,n \\ j=1,\ldots,2m}} . \qquad 5.30$$

where p_{ik} equals to $\frac{f_{ik}}{N}$, and N equals to $\sum_{i=1}^{n} \sum_{k=1}^{2m} f_{ik}$.

Then, for every $i = 1,\ldots,n$ and $k = 1,\ldots,2m$, we can compute P_{i+}, P_{+k}, and μ_{ik}, respectively, using Eqs. 5.31–5.33 (Esmaelian et al., 2017):

$$\text{Summation of the elements in } i\text{th row of } P: \quad P_{i+} = \sum_{k=1}^{2m} p_{ik} . \qquad 5.31$$

$$\text{Summation of the elements in } k\text{th column of } P: \quad P_{+k} = \sum_{k=1}^{n} p_{ik} . \qquad 5.32$$

$$\text{Estimation of } p_{ik} \text{ under the independence assumption}: \quad \mu_{ik} = P_{i+} \cdot P_{+k} . \qquad 5.33$$

2) *Stage 5.2*: Chi-square test examination. The correspondence matrix's associated χ^2 statistics can be defined using Eq. 5.34 (Esmaelian et al., 2017):

$$\chi^2 = \sum_{i=1}^{n} \sum_{k=1}^{2m} \frac{\left(p_{ik} - \mu_{ik}\right)^2}{\mu_{ik}} . \qquad 5.34$$

More specifically, if the standard value is much smaller than chi-square statistics, then we have *P-value* ≤ 0.05, and the rows and columns of the correspondence matrix are dependent on each other.

3) *Stage 5.3*: Correspondence analysis (CA) check and suitable dimension number determination. In particular, if $\frac{\text{dimension 1's inertia + dimension 2's inertia}}{\text{total inertia}} > 50\%$, then we can plot the data points as points on a 2-dimensional perceptual map.

4) *Stage 5.4*: Standard residual matrix formulation. This matrix's elements are defined using Eq. 5.35 (Esmaelian et al., 2017):

$$\Omega = \left[\frac{p_{ik} - \mu_{ik}}{\sqrt{\mu_{ik}}}\right] . \qquad 5.35$$

5) *Stage 5.5*: Search the residual matrix's SVD, i.e., singular value decomposition. In general, SVD method (Adachi, 2016) can divide

a matrix into three sub-matrices where their product produces the original matrix as given by Eq. 5.36 (Esmaelian et al., 2017):

$$\Omega = V \Lambda W^T . \tag{5.36}$$

6) *Stage 5.6*: Plotting of row scores calculation and data point approximations. Matrix R, defined in Eq. 5.37 (Esmaelian et al., 2017), can be used to compute row scores:

$$R = \sigma_r V \Lambda . \tag{5.37}$$

where σ_r stands for the diagonal matrix as given by Eq. 5.38 (Esmaelian et al., 2017):

$$\sigma_r = \begin{bmatrix} \frac{1}{\sqrt{P_{1+}}} & 0 & 0 \\ 0 & \ddots & 0 \\ 0 & 0 & \frac{1}{\sqrt{P_{n+}}} \end{bmatrix} . \tag{5.38}$$

7) *Stage 5.7*: Plotting of column scores calculation and data point approximations. Unlike the matrix R used in Stage 5.6, the matrix C introduced at this stage is slightly different. Accordingly, a set of operations have to be performed on standard residual matrix (defined at Stage 5.4) as defined by Eqs. 5.39–5.41 (Esmaelian et al., 2017):

$$\Omega' = \Omega^T . \tag{5.39}$$

$$\Omega' = V' \Lambda' W'^T . \tag{5.40}$$

$$V'V'^T = W'W'^T = I. \tag{5.41}$$

Based on the above operations, matrix C can be defined using Eq. 5.42 (Esmaelian et al., 2017):

$$C = \sigma_c V' \Lambda' . \tag{5.42}$$

where σ_c represents the diagonal matrix as given by Eq. 5.43 (Esmaelian et al., 2017):

$$\sigma_c = \begin{bmatrix} \frac{1}{\sqrt{P_{+1}}} & 0 & 0 \\ 0 & \ddots & 0 \\ 0 & 0 & \frac{1}{\sqrt{P_{+2m}}} \end{bmatrix} . \tag{5.43}$$

8) *Stage 5.8*: Additional information implementation and the most feasible TFM(s) selection. In (Esmaelian et al., 2017), for the purpose of reflecting an appropriate relationship between the row

scores and column scores, Eq. 5.44, borrowed from (Yelland, 2010) was introduced:

$$R_{new} = \sigma^2_{r,new} \times P_{new} \times C_{new}. \tag{5.44}$$

5.3.3.3 *Summary*

Though there are still certain limitations (particularly in the qualitative part) of the generic TFM evaluation framework, we can see several associated advantages in its quantitative stage (Esmaelian et al., 2017).

- In fact, its main merit lies in the fact that it allows us to equalize the intrinsic shortcomings of relying only on qualitative analysis by performing a quantitative analysis as well. The former analysis tends to be influenced by experts' incorrect choice or inexact judgements, while the later analysis is good at using the data points format in order to synthesize criteria and alternatives and project them into a low dimensional space according to the total inertia level.
- Conventionally, the integrated assessment solutions often tend to combine consensus-based approaches with AHP (i.e., analytic hierarchy process) and/or ANP (analytic network process) models. Another merit of the proposed evaluation procedure is, thus, offering practitioners a useful alternative for performing similar tasks. The attempt at substituting AHP with MCA (i.e., multiple correspondence analysis) for fulfilling quantitative analysis can effectively mitigate the subjectivity issues.

5.4 Conclusions

Technology foresight focuses on developing proper extrapolation procedures so that the most probable technology development situations in the future can be predicted (Esmaelian et al., 2017). According to Miles et al. (2017), since the mid-1990s, technology foresight has been widely utilized by many policymakers, in particular in the STI area, i.e., science, technology and innovation (Meissner et al., 2013).

Given the controversial role of ICO technology, in this chapter, we presented an innovative technology evaluation framework (Dereli and Altun, 2013) that is built on IT2FIS (i.e., interval type-2 fuzzy inference system). This framework utilized two main data sources: Patent data extracted from the European Patent Office and publication data obtained from the Web of Science/Knowledge. With the aid of IT2FIS, two data streams were further filtered and matched according to each technology's respective degree of trendiness. The experimental results demonstrated that it could help policy-makers/decision-makers form corresponding strategic evaluations (in particular in handling uncertainties and improving outcomes) about the target technology. Considering the fact that there are also many technology foresight methodologies available in the literature, this chapter further introduced a generic model (Esmaelian et al., 2017) in order to deal with the selection of the most scenario-fit technology foresight methods. This model incorporated multiple correspondence analysis and the results verified its feasibility in adapting to a broad range of applications, including business entities and government agencies.

Technology evaluation and foresight literature is rather rich in the following forms:

- **Books:** (Inzelt and Coenen, 1996; Garud et al., 1997; Meissner et al., 2013; Gokhberg et al., 2016; Miles et al., 2016);
- **Reviews:** (Miles, 2010; Linstone, 2011; Iden et al., 2017); and
- **Articles:** (Bañuls and Salmeron, 2008; Barnard-Wills, 2017; Bildosola et al., 2017; Boe-Lillegraven and Monterde, 2015; Brandes, 2009; Chan and Daim, 2012; Chen and Ma, 2014; Chen et al., 2012; Dufva et al., 2015; Xing, 2016c; Featherston and O'Sullivan, 2017; Feige and Vonortas, 2017; Förster, 2015; Gokhberg et al., 2016; Gokhberg and Sokolov, 2017; Von der Gracht et al., 2015; Haegeman et al., 2017; Hussain et al., 2017; Jørgensen et al., 2009; Kayser and Bierwisch, 2016; Kayser and Blind, 2017; Kolominsky-Rabas et al., 2015; Lee and Geum, 2017; Li et al., 2017; Martin, 2010; Mulholland et al., 2017; Öner and Kunday, 2016; Pietrobelli and Puppato, 2016; Piirainen and Gonzalez, 2015; Proskuryakova, 2017; Quiroga and Martin, 2017; Rhisiart et al., 2017; Shah et al., 2013; Sokolov and Chulok, 2016; Stelzer et al., 2015; Vishnevskiy et al., 2017; Phaal et al., 2001; Strauss and Radnor, 2004; Amer, 2013; Moehrle et al., 2013; Marin, 2014).

However, their applications in evaluating and forecasting ICO technology are still very rare. Therefore, the framework and model presented in this chapter has not yet been tried in ICO scenarios, and is, thus, still subject to future research activities.

References

Abor JY. (2017) *Entrepreneurial finance for MSMEs: A managerial approach for developing markets*, Gewerbestrasse 11, 6330 Cham, Switzerland: Palgrave Macmillan, ISBN 978-3-319-34020-3.

Acheson N. (2017) *China's ICO ban: Understandable, reasonable and (probably) temporary, retrieved from https://www.coindesk.com/chinas-ico-ban-understandable-reasonable-probably-temporary/, accessed on 30 November 2017.*

Adachi K. (2016) *Matrix-based introduction to multivariate data analysis*, 152 Beach Road, #22-06/08 Gateway East, Singapore 189721, Singapore: Springer Nature Singapore Pte Ltd., ISBN 978-981-10-2340-8.

Amer M. (2013) Extending technology roadmap through fuzzy cognitive map-based scenarios: The case of the wind energy sector of Pakistan. Portland State University.

Andersen AD and Andersen PD. (2014) Innovation system foresight. *Technological Forecasting and Social Change* 88: 276–286.

Andersen PD and Rasmussen B. (2014) Introduction to foresight and foresight processes in practice: Note for the PhD course Strategic Foresight in Engineering. Denmark: Department of Management Engineering, Technical University of Denmark, retrieved from http://orbit.dtu.dk/files/96941116/Introduction_to_foresight.pdf, accessed on 06 December 2017.

Angelov P and Sotirov S. (2016) *Imprecision and uncertainty in information representation and processing: New tools based on intuitionistic fuzzy sets and generalized nets,* Gewerbestrasse 11, 6330 Cham, Switzerland: Springer International Publishing Switzerland, ISBN 978-3-319-26301-4.

Arman H, Hodgson A and Gindy N. (2009) Technologies watch exercise: Foresight approach enhance with scientific publications and patents analysis. *International Journal of Technology Intelligence and Planning* 5: 305–321.

Bañuls VA and Salmeron JL. (2008) Foresighting key areas in the Information Technology industry. *Technovation* 28: 103–111.

Back A, Corallo M, Dashjr L, et al. (2014) *Enabling blockchain innovations with pegged sidechains, retrieved from https://www.blockstream.com/technology/sidechains.pdf, accessed on 06 December 2017.*

Bargiela A and Pedrycz W. (2009) *Human-centric information processing through granular modelling,* Berlin Heidelberg: Springer-Verlag, ISBN 978-3-540-92915-4.

Barnard-Wills D. (2017) The technology foresight activities of European Union data protection authorities. *Technological Forecasting & Social Change* 116: 142–150.

BBVA. (2015) Smart contracts: The ultimate automation of trust? Azul Street, 4, La Vela Building - 4 and 5 floor, 28050 Madrid (Spain): BBVA Research Department.

Bělohlávek R, Dauben JW and Klir GJ. (2017) *Fuzzy logic and mathematics: A historical perspective,* 198 Madison Avenue, New York, NY 10016, United States of America: Oxford University Press, ISBN 978-0-19020-001-5.

Bildosola I, Río-Bélver RM, Garechana G, et al. (2017) TeknoRoadmap, an approach for depicting emerging technologies. *Technological Forecasting & Social Change* 117: 25–37.

Blockchain Technology. (2016) *Smart contracts explained, retrieved from http://www.blockchaintechnologies.com/blockchain-smart-contracts, accessed on 06 December 2017.*

BlockChannel. (2017) *Understanding the Ethereum ICO token hype, retrieved from https://medium.com/blockchannel/understanding-the-ethereum-ico-token-hype-429481278f45, accessed on 06 December 2017.*

Boe-Lillegraven S and Monterde S. (2015) Exploring the cognitive value of technology foresight: the case of the Cisco Technology Radar. *Technological Forecasting & Social Change* 101: 62–82.

Borgstrup J. (2014) Private, trustless and decentralized message consensus and voting schemes. *Faculty of Science.* University of Copenhagen.

Boucher P. (2016) *What if blockchain technology revolutionised voting?, Retrieved from http://www.europarl.europa.eu/RegData/etudes/ATAG/2016/581918/EPRS_ATA(2016)581918_EN.pdf, accessed on 09 November 2017.*

Bracamonte V and Okada H. (2017) An exploratory study on the influence of guidelines on crowdfunding projects in the Ethereum blockchain platform. In: Ciampaglia GL, Mashhadi A and Yasseri T (eds) *Proceedings of the 9th International Conference (SocInfo), Oxford, UK, September 13~15, 2017, Part II*, pp. 347–354.

Bradbury D. (2016) Blockchain's big deal. *Engineering & Technology, November.* 44–48.

Brandes F. (2009) The UK technology foresight programme: An assessment of expert estimates. *Technological Forecasting & Social Change* 76: 869–879.

Burelli F, John M, Cenci E, et al. (2015) Blockchain and financial services: Industry snapshot and possible future developments. 3 More London Riverside London, SE1 2RE, UK INNOVALUE Management Advisors Ltd.

Bustince H, Barrenechea E, Fernández J, et al. (2015) The origin of fuzzy extensions. In: Kacprzyk J and Pedrycz W (eds) *Springer Handbook of Computational Intelligence.*

Dordrecht Heidelberg London New York: Springer-Verlag Berlin Heidelberg, ISBN 978-3-662-43504-5, Part A Foundations, Chapter 6, pp. 89–112.

Buterin V. (2014) Ethereum white paper: A next generation smart contract & decentralized appplication platform. Ethereum.

Butler T, Khalil FA, Ceci M, et al. (2017) Smart contracts and distributed ledger technologies in financial services: Keeping lawyers in the loop. *Banking & Financial Services Policy Report* 36: 1–11.

Castillo O and Melin P. (2009) Interval type-2 fuzzy logic applications. In: Bargiela A and Pedrycz W (eds) *Human-Centric Information Processing, SCI 182.* Berlin Heidelberg: Springer-Verlag, ISBN 978-3-540-92915-4, pp. 203–231.

Castor A. (2017) *Ethereum 'tokens' are all the rage, but what are they anyway? Retrieved from http://www.coindesk.com/ethereums-erc-20-tokens-rage-anyway/, accessed on 21 June 2017.*

Celikyilmaz A and Türksen IB. (2009) *Modeling uncertainty with fuzzy logic with recent theory and applications,* Berlin Heidelberg: Springer-Verlag, ISBN 978-3-540-89923-5.

Cha S-C, Yeh K-H and Chen J-F. (2017) Toward a robust security paradigm for bluetooth low energy-based smart objects in the Internet-of-things. *Sensors* 17: 1–19.

Chan L and Daim T. (2012) Exploring the impact of technology foresight studies on innovation: Case of BRIC countries. *Futures* 44: 618–630.

Chen H, Wakeland W and Yu J. (2012) A two-stage technology foresight model with system dynamics simulation and its application in the Chinese ICT industry. *Technological Forecasting & Social Change* 79: 1254–1267.

Chen H and Ma T. (2014) Technology adoption with limited foresight and uncertain technological learning. *European Journal of Operational Research* 239: 266–275.

Christidis K and Devetsikiotis M. (2016) Blockchains and smart contracts for the Internet of things. *IEEE: Special Section on the Plethora of Research in Internet of Things (IoT)* 4: 2292–2303.

CoinDesk. (2017) *China's ban: A full translation of regulator remarks, retrieved from https://www.coindesk.com/chinas-ico-ban-a-full-translation-of-regulator-remarks/, accessed on 06 September 2017.*

Commission E. (2016) *Directive of the European parliament and of the council, retrieved from http://ec.europa.eu/justice/criminal/document/files/amldirective_en.pdf, accessed on 13 September 2017.*

Cuccuru P. (2017) Beyond bitcoin: An early overview on smart contracts. *International Journal of Law and Information Technology* 25: 179–195.

Czepluch JS, Lollike NZ and Malone SO. (2015) The use of blockchain technology in different application domains. The IT University of Copenhagen.

De Ridder CA, Tunstall MK and Prescott N. (2017) Recognition of smart contracts in the United States. *Intellectual Property & Technology Law Journal* 29: 17–19.

De Rose C. (2015) Why the Bitcoin blockchain beats out competitors. *American Banker, Vol. 180 Issue 100, retrieved from http://search.ebscohost.com/login.aspx?direct=true&db=bth&AN=103540491&site=edslive, accessed on 7 October 2015.*

Deloitte. (2016) Israel: A hotspot for blockchain innovation. Brightman Almagor Zohar & Co. Member of Deloitte Touche Tohmatsu Limited, 1–33.

Dereli T and Altun K. (2013) Technology evaluation through the use of interval type-2 fuzzy sets and systems. *Computers & Industrial Engineering* 65: 624–633.

Dhillon V, Metcalf D and Hooper M. (2017) *Blockchain enabled applications: Understand the blockchain ecosystem and how to make it work for you*: APress, ISBN 978-1-4842-3080-0.

Dufva M, Könnölä T and Koivisto R. (2015) Multi-layered foresight: Lessons from regional foresight in Chile. *Futures* 73: 100–111.

Ellis S, Juels A and Nazarov S. (2017) *ChainLink: A decentralized oracle network, retrieved from https://link.smartcontract.com/whitepaper, accessed on 06 December 2017.*

Esmaelian M, Tavana M, Caprio DD, et al. (2017) A multiple correspondence analysis model for evaluating technology foresight methods. *Technological Forecasting & Social Change* 125: 188–205.

European Securities and Markets Authority (ESMA). (2017) Statement: ESMA alerts firms involved in initial coin offerings (ICOs) to the need to meet relevant regulatory requirements. European Securities and Markets Authority (ESMA), ESMA50-157-828.

Eyal I. (2017) Blockchain technology: Transforming libertarian cryptocurrency dreams to finance and banking realities. *Computer, September*: 38–49.

Förster B. (2015) Technology foresight for sustainable production in the German automotive supplier industry. *Technological Forecasting & Social Change* 92: 237–248.

Fairfield JAT. (2014) Smart contracts, Bitcoin bots and consumer protection. *Washington and Lee Law Review Online* 71: 35–50.

Featherston CR and O'Sullivan E. (2017) Enabling technologies, lifecycle transitions and industrial systems in technology foresight: Insights from advanced materials FTA. *Technological Forecasting & Social Change* 115: 261–277.

Feige D and Vonortas NS. (2017) Context appropriate technologies for development: Choosing for the future. *Technological Forecasting & Social Change* 119: 219–226.

Ferenzy D, Tran H, French C, et al. (2015) The Internet of finance: Unleashing the potential of blockchain technology. Institute of International Finance (IIF), 1–4.

Frederick H, O'Connor A and Kuratko DF. (2016) *Entrepreneurship: Theory, process, practice,* Level 7, 80 Dorcas Street, South Melbourne, Victoria Australia 3205: Cengage Learning Australia, ISBN 978-0-17-035255-0.

Gao W-J, Xing B and Marwala T. (2013a) Computational intelligence in used products retrieval and reproduction. *International Journal of Swarm Intelligence Research* 4: 78–125.

Gao W-J, Xing B and Marwala T. (2013b) Teaching—learning-based optimization approach for enhancing remanufacturability pre-evaluation system's reliability. *IEEE Symposium Series on Computational Intelligence (IEEE SSCI), 15–19 April, Singapore,* pp. 235–239. IEEE.

Gao W-J, Xing B and Marwala T. (2014) Used products return service based on ambient recommender systems to promote sustainable choices. In: Memon QA (ed) *Distributed Network Intelligence, Security and Applications.* 6000 Broken Sound Parkway NW, Suite 300, Boca Raton, FL 33487-2742: CRC Press, Taylor & Francis Group, LLC, ISBN 978-1-4665-5958-5, Chapter 15, pp. 359–378.

Garud R, Nayyar PR and Shapira ZB. (1997) *Technological innovation: Oversights and foresights,* The Edinburgh Building, Cambridge CB2 8RU, UK: Cambridge University Press, ISBN 0-521-55299-0.

Giancaspro M. (2017) Is a 'smart contract' really a smart idea? Insights from a legal perspective. *Computer Law & Security Review* 33: 825–835.

Gietl D, Brody P, Crespigny ACd, et al. (2016) Blockchain in health: How distributed ledgers can improve provider data management and support interoperability: Ernst & Young LLP & EY, 1–12.

Gokhberg L, Meissner D and Sokolov A. (2016) *Deploying foresight for policy and strategy makers: Creating opportunities through public policies and corporate strategies in science, technology and innovation,* Switzerland: Springer International Publishing Switzerland, ISBN 978-3-319-25626-9.

Gokhberg L and Sokolov A. (2017) Technology foresight in Russia in historical evolutionary perspective. *Technological Forecasting & Social Change* 119: 256–267.

Golem. (2017) *Golem: First global market for idle computer power, retrieved from https://golem.network, accessed on 30 November 2017.*

Gramoli V. (in press) From blockchain consensus back to Byzantine consensus. *Future Generation Computer Systems* http://dx.doi.org/10.1016/j.future.2017.09.023.

Greenacre M. (2007) *Correspondence analysis in practice,* 6000 Broken Sound Parkway NW, Suite 300, Boca Raton, FL 33487-2742: Chapman and Hall/CRC, ISBN 978-1-58488-616-7.

Greenfield S and John RI. (2009) The uncertainty associated with a type-2 fuzzy set. In: Seising R (ed) *Views on Fuzzy Sets and Systems from Different Perspectives: Philosophy and Logic, Criticisms and Applications.* Berlin, Heidelberg: Springer, ISBN 978-3-540-93801-9, Chapter 23, pp. 471–483.

Groshoff D. (2014) Kickstarter my heart: Extraordinary popular delusions and the madness of crowdfunding constraints and Bitcoin bubbles. *William & Mary Business Law Review* 5: 489–557.

Gupta M. (2017) *Blockchain for dummies,* 111 River St., Hoboken, NJ 07030-5774: John Wiley & Sons, Inc., ISBN 978-1-119-37123-6.

Haegeman K, Spiesberger M and Könnölä T. (2017) Evaluating foresight in transnational research programming. *Technological Forecasting & Social Change* 115: 313–326.

Hayano RS, Tsubokura M, Miyazaki M, et al. (2015) Whole-body counter surveys of Miharu-town school children for four consecutive years after the Fukushima NPP accident. *Proceedings of the Japan Academy, Series B* 91: 92–98.

Heller M. (2017) SEC jolts Initial coin offerings. *CFO.* 16.

Hess M, Mosimann M and Lienhard S. (2017) Initial Coin Offering (ICO) & Co. - A swiss law perspective. *Wenger & Vieli Attoneys at Law: Sportlight.* Dufourstrasse 56, P.O. Box, CH-8034 Zurich: Wenger & Vieli Ltd.

Higgins S. (2016) *Abu Dhabi stock exchange launches blockchain voting, retrieved from http://www.coindesk.com/abu-dhabi-exchange-blockchain-voting/, accessed on 09 November 2017.*

Higgins S. (2017) Canaidan regulators: 'many' ICO tokens meet securities definition, retrieved from https://www.coindesk.com/canadian-regulators-many-ico-tokens-meed-securities-definition/, accessed on 30 November 2017.

Hussain M, Tapinos E and Knight L. (2017) Scenario-driven roadmapping for technology foresight. *Technological Forecasting & Social Change* 124: 160–177.

IBM. (2016) Blockchain: The chain of trust and its potential to transform healthcare—our point of view. IBM Global Business Services Public Sector Team, 1–12.

Iden J, Methlie LB and Christensen GE. (2017) The nature of strategic foresight research: A systematic literature review. *Technological Forecasting & Social Change* 116: 87–97.

İntepe G, Bozdag E and Koc T. (2013) The selection of technology forecasting method using a multi-criteria interval-valued intuitionistic fuzzy group decision making approach. *Computers and Industrial Engineering* 65: 277–285.

Inzelt A and Coenen R. (1996) *Knowledge, technology transfer and foresight,* P.O. Box 17,3300 AA Dordrecht, The Netherlands: Kluwer Academic Publishers, ISBN 978-0-7923-4274-8.

ISDA. (2017) Smart contracts and distributed ledger: a legal perspective. ISDA.

Jørgensen MS, Jørgensen U and Clausen C. (2009) The social shaping approach to technology foresight. *Futures* 41: 80–86.

Jacynycz V, Calvo A, Hassan S, et al. (2016) Betfunding: A distributed bounty-based crowdfunding platform over Ethereum. In: Omatu S, Selamat A, Bocewicz G, et al. (eds) *Proceedings of the 13th International Conference on Distributed Computing and Artificial Intelligence (DCAI), 1~3 June, Sevilla, Spain*, pp. 403–411.

Jiménez-Losada A. (2017) *Models for cooperative games with fuzzy relations among the agents: Fuzzy communications, proximity relation and fuzzy permission,* Gewerbestrasse 11, 6330 Cham, Switzerland: Springer International Publishing AG, ISBN 978-3-319-56471-5.

John R and Coupland S. (2008) Type-2 fuzzy logic and the modelling of uncertainty in applications. In: Bargiela A and Pedrycz W (eds) *Human-Centric Information Processing, SCI 182.* Berlin Heidelberg: Springer-Verlag, ISBN 978-3-540-92915-4, Chapter 8, pp. 185–201.

Judge NA, Place C, Hooper A, et al. (2017) Root project: A cryptocurrency to change the world, now. RootProject, 1–27.

Junginger M, Faaij A and Turkenburg WC. (2005) Global experience curves for wind farms. *Energy Policy* 33: 133–150.

Karnik NN and Mendel JM. (1999) Applications of type-2 fuzzy logic systems to forecasting of time-series. *Information Sciences* 120: 89–111.

Kayser V and Bierwisch A. (2016) Using Twitter for foresight: An opportunity? *Futures* 84: 50–63.

Kayser V and Blind K. (2017) Extending the knowledge base of foresight: The contribution of text mining. *Technological Forecasting & Social Change* 116: 208–215.

Kazemzadeh A, Lee S and Narayanan S. (2008) An interval type-2 fuzzy logic system to translate between emotion-related vocabularies. *Proceedings of the 9th Annual Conference of the International Speech Communication Association (INTERSPEECH 2008), Brisbane, Australia, 22–26 September*, pp. 1–4.

Khaqqi KN, Sikorski JJ, Hadinoto K, et al. (2018) Incorporating seller/buyer reputation-based system in blockchain-enabled emission trading application. *Applied Energy* 209: 8–19.

Klir GJ and Folger TA. (1988) *Fuzzy sets, uncertainty and information,* Englewood Cliffs, New Jersey 07632: Prentice Hall, ISBN 0-13-345984-5.

Klir GJ and Yuan B. (1995) *Fuzzy sets and fuzzy logic: Theory and applications,* Upper Saddle River, NJ 07458: Prentice Hall PTR, ISBN 0-13-101171-5.

Kolominsky-Rabas PL, Djanatliev A, Wahlster P, et al. (2015) Technology foresight for medical device development through hybrid simulation: The ProHTA project. *Technological Forecasting & Social Change* 97: 105–114.

Kshetri N. (2017) Can blockchain strengthen the Internet of things? *IT Professional* July/August: 68–72.

L. Andries van der Ark, Wiberg M, Culpepper SA, et al. (2017) *Quantitative psychology: The 81st Annual Meeting of the Psychometric Society, Asheville, North Carolina, 2016,* Gewerbestrasse 11, 6330 Cham, Switzerland: Springer International Publishing AG, ISBN 978-3-319-56293-3.

Lauslahti K, Mattila J and Seppälä T. (2017) Smart contracts: How will blockchain technology affect contractual practices? ETLA: The Research Institute of the Finnish Economy, No. 68. ISSN 2323–2455.

Leach JC and Melicher RW. (2012) *Entrepreneurial finance,* 5191 Natorp Boulevard, Mason, OH 45040, USA: South-Western Cengage Learning, ISBN 978-0-538-47815-1.

Lee H and Geum Y. (2017) Development of the scenario-based technology roadmap considering layer heterogeneity: An approach using CIA and AHP. *Technological Forecasting & Social Change* 117: 12–24.

Lee J-H and Pilkington M. (2017) How the blockchain revolution will reshape the consumer electronics industry. *IEEE Consumer Electronics Magazine, July.* IEEE, 19–23.

Lerner SD. (2015) *RSK: White paper overview, retrieved from https://uploads.strikinglycdn.com/files/ec5278f8-218c-407a-af3c-ab71a910246d/RSK%20White%20Paper%20-%20Overview.pdf, accessed on 06 December 2017.*

Li N, Chen K and Kou M. (2017) Technology foresight in China: Academic studies, governmental practices and policy applications. *Technological Forecasting & Social Change* 119: 246–255.

Li X, Jiang P, Chen T, et al. (in press) A survey on the security of blockchain systems. *Future Generation Computer Systems* http://dx.doi.org/10.1016/j.future.2017.08.020.

Likert R. (1932) A technique for the measurement of attitudes. *Archives of Psychology* 22: 5–55.

Lim G and Law D. (2017) Announcement clarifies regulatory position on initial coin offerings in Hong Kong. 51 Louisiana Avenue, N.W., Washington D.C. 200012113: Jones Day.

Linstone HA. (2011) Three eras of technology foresight. *Technovation* 31: 69–76.

Lu Q and Xu X. (2017) Adaptable blockchain-based systems: A case study for product traceability. *IEEE Software*: 21–27.

Luu J and Welsh S. (2017) A brave new world: Regulation of initial coin offerings. Burnet, Duckworth & Palmer LLP.

MacQueen J. (1967) Some methods for classification and analysis of multivariate observations. *Proceedings of the 5th Berkeley Symposium on Mathematical Statistics and Probability, Berkeley, CA, USA, Volume 1*, pp. 281–297. University of California Press.

Magazzeni D, McBurney P and Nash W. (2017) Validation and verification of smart contracts: A research agenda. *Computer, September.* 50–57.

Mainelli M and Manson B. (2017) A wholesale insurance executive's guide to smart contracts. 41 Lothbury, London EC2R 7HG, United Kingdom: Z/Yen Group.

Marin JC. (2014) A technology roadmap for providing predictive analytics services. *Faculty of Technology, Policy and Management*. Delft University of Technology.

Martin BR. (2010) The origins of the concept of 'foresight' in science and technology: An insider's perspective. *Technological Forecasting & Social Change* 77: 1438–1447.

Marwala T, Wilde Pd, Correia L, et al. (2001) Scalability and optimisation of a committee of agents using genetic algorithm. *Fourth International ICSC Symposium Soft computing and Intelligent Systems for Industry (SOCO/ISFI 2001), 26–29 June, Paisley, Scotland, United Kingdom*, pp. 1–7.

Marwala T. (2009) *Computational intelligence for missing data imputation, estimation and management: Knowledge optimization techniques,* New York, USA: IGI Global, ISBN 978-1-60566-336-4.

Marwala T. (2010a) *Finite-element-model updating using computational intelligence techniques: Applications to structural dynamics,* London, UK: Springer-Verlag, ISBN 978-1-84996-322-0.

Marwala T. (2010b) Finite-element-model updating using particle-swarm optimization. In: Marwala T (ed) *Finite-element-model Updating Using computational Intelligence Techniques: Applications to Structural Dynamics.* London, UK: Springer-Verlag, ISBN 978-1-84996-322-0, Chapter 4, pp. 67–84.

Marwala T and Lagazio M. (2011) *Militarized conflict modeling using computational intelligence,* London, UK: Springer-Verlag, ISBN 978-0-85729-789-1.

Marwala T and Xing B. (2011) The role of remanufacturing in building a developmental state. *The Thinker: For the Thought Leaders (www.thethinker.co.za).* South Africa: Vusizwe Media, 18–20.

Marwala T. (2012) *Condition monitoring using computational intelligence methods: Applications in mechanical and electrical systems,* London: Springer-Verlag, ISBN 978-1-4471-2379-8.

Marwala T. (2013) *Economic modeling using artificial intelligence methods,* Springer London Heidelberg New York Dordrecht: Springer-Verlag London, ISBN 978-1-4471-5009-1.

Marwala T. (2015) *Causality, correlation and artificial intelligence for rational decision making,* 5 Toh Tuck Link, Singapore 596224: World Scientific Publishing Co. Pte. Ltd, ISBN 978-9-81463-086-3.

Massey R, Dalal D and Dakshinamoorthy A. (2017) Initial coin offering: A new paradigm. Deloitte, 1–11.

Mattila J. (2016) The blockchain phenomenon: The disruptive potential of distributed consensus architectures. 2234 Piedmont Avenue, Berkeley, CA 94720-2322: Berkeley Roundtable on the International Economy (BRIE), University of California, Berkeley, 1–25.

Meissner D, Gokhberg L and Sokolov A. (2013) *Science, technology and innovation policy for the future: Potentials and limits of foresight studies,* Berlin Heidelberg: Springer-Verlag, ISBN 978-3-642-31826-9.

Meitinger TH. (2017) Smart contracts. *Informatik_Spektrum* 40: 371–375.

Mendel JM, John RI and Liu F. (2006) Interval type-2 fuzzy logic systems made simple. *IEEE Transactions on Fuzzy Systems* 14: 808–821.

Mendel JM and Wu D. (2010) *Perceptual computing: Aiding people in making subjective judgments,* 111 River Street, Hoboken, NJ 07030: IEEE Press and John Wiley & Sons, Inc., ISBN 978-0-470-47876-9.

Mendel JM. (2014) General type-2 fuzzy logic systems made simple: A tutorial. *IEEE Transactions on Fuzzy Systems* 22: 1162–1182.

Mendel JM. (2017) *Uncertain rule-based fuzzy systems: Introduction and new directions,* Gewerbestrasse 11, 6330 Cham, Switzerland: Springer International Publishing AG, ISBN 978-3-319-51369-0.

Mik E. (2017) Smart contracts: terminology, technical limitations and real world complexity. *Law, Innovation and Technology* 9: 269–300.

Miles I. (2010) The development of technology foresight: A review. *Technological Forecasting & Social Change* 77: 1448–1456.

Miles I, Saritas O and Sokolov A. (2016) *Foresight for science, technology and innovation*: Springer International Publishing Switzerland, ISBN 978-3-319-32572-9.

Miles I, Meissner D, Vonortas NS, et al. (2017) Technology foresight in transition. *Technological Forecasting & Social Change* 119: 211–218.

Moehrle MG, Isenmann R and Phaal R. (2013) *Technology roadmapping for strategy and innovation: charting the route to success,* Springer Heidelberg New York Dordrecht London: Springer-Verlag Berlin Heidelberg, ISBN 978-3-642-33922-6.

Mulholland E, Rogan F and Gallachóir BPÓ. (2017) From technology pathways to policy roadmaps to enabling measures – a multi-model approach. *Energy* 138: 1030–1041.

Nakamoto S. (2008) *Bitcoin: A peer-to-peer electronic cash system, retrieved from www.bitcoin.org/bitcoin.pdf, accessed on 08 September 2017.*

Neij L, Andersen PD and Durstewitz M. (2004) Experience curves for wind power. *International Journal of Energy Technology and Policy* 2: 15–32.

Nguyen HT and Walker EA. (2006) *A first course in fuzzy logic,* 6000 Broken Sound Parkway NW, Suite 300, Boca Raton, FL 33487-2742: CRC Press, Taylor & Francis Group, LLC, ISBN 978-1-4200-5710-2.

Nordrum A. (2017) Govern by blockchain: Dubai wants one platform to rule them all, while Illinois will try anything. *IEEE Spectrum* October: 54–55.

Nordström E. (2015) Personal clouds - Concedo. *Department of Computer Science, Electrical and Space Engineering.* Luleå University of Technology.

Notary. (2017) The Notary platform project: A crowdfunding white paper.

Novák V, Perfilieva I and Dvořák A. (2016) *Insight into fuzzy modeling,* John Wiley & Sons, Inc., 111 River Street, Hoboken, NJ 07030: John Wiley & Sons, Inc., ISBN 978-1-119-19318-0.

Obie SJ, McKown JE, Gendzier AA, et al. (2017) SEC's investigative report raises difficult questions fro ICO issuers. 51 Louisiana Avenue, N.W., Washington D.C. 20001-2113: Jones Day.

Ølnes S, Ubacht J and Janssen M. (2017) Blockchain in government: Benefits and implications of distributed ledger technology for information sharing. *Government Information Quarterly* 34: 355–364.

Omohundro S. (2014) Cryptocurrencies, smart contracts, and artificial intelligence. *AI Matters* 1: 19–21.

Öner MA and Kunday Ö. (2016) Linking technology foresight and entrepreneurship. *Technological Forecasting & Social Change* 102: 1.

Orion W. (2017) *Singapore proposes changes to its payments framework, retrieved from http://www.orionw.com/blog/news/fintech/singapore-proposes-changes-to-its-payments-framework, accessed on 09 September 2017.*

Patel PB and Marwala T. (2006) Neural networks, fuzzy inference systems and adaptive-neuro fuzzy inference systems for financial decision making. *Lecture Notes in Computer Science, vol. 4234.* Berlin Heidelberg: Springer-Verlag, pp. 430–439.

Peck ME and Wagman D. (2017) Energy trading for fun and profit: Buy your neighbor's rooftop solar power or sell your own—it'll all be on a blockchain. *IEEE Spectrum, October*: 56–57, & 61.

Phaal R, Farrukh CJ and Probert DR. (2001) Characterisation of technology roadmaps: Purpose and format. *Portland International Conference on Management of Engineering and Technology (PICMET)*, pp. 367–374.

Pietrobelli C and Puppato F. (2016) Technology foresight and industrial strategy. *Technological Forecasting & Social Change* 110: 117–125.

Piirainen KA and Gonzalez RA. (2015) Theory of and within foresight—"What does a theory of foresight even mean?". *Technological Forecasting & Social Change* 96: 191–201.

Proskuryakova L. (2017) Energy technology foresight in emerging economies. *Technological Forecasting & Social Change* 119: 205–210.

Prybila C, Schulte S, Hochreiner C, et al. (in press) Runtime verification for business processes utilizing the Bitcoin blockchain. *Future Generation Computer Systems* http://dx.doi.org/10.1016/j.future.2017.08.024.

Pureswaran V, Panikkar S, Nair S, et al. (2015) Empowering the edge: Practical insights on a decentralized Internet of things. Route 100, Somers, NY 10589: IBM Institute for Business Value, 1–24.

PwC. (2016) Blockchain in the insurance sector. PricewaterhouseCoopers (PwC) LLP, 1–2.

PwC. (2017) Initial coin offerings: A strategic perspective on ICOs. PwC, 1–5.

Quiroga MC and Martin DP. (2017) Technology foresight in traditional Bolivian sectors: Innovation traps and temporal unfit between ecosystems and institutions. *Technological Forecasting & Social Change* 119: 280–293.

R3. (2017) *Corda solution guide, retrieved from https://www.corda.net/wp-content/uploads/2017/10/Corda-Solution-Guide.pdf, accessed on 06 December 2017.*

Rhisiart M, Störmer E and Daheim C. (2017) From foresight to impact? The 2030 Future of Work scenarios. *Technological Forecasting & Social Change* 124: 203–213.
Russell McVeagh. (2017) Initial coin offerings. Vero Centre, 48 Shortland Street, PO Box 8, Auckland 1140, New Zealand: Russell McVeagh, 1–14.
Saito S. (2017) Initial coin offerings (ICO) under japanese laws. So Law Office, 1–5.
Salahuddin MA, Al-Fuqaha A, Guizani M, et al. (2017) Softwarization of internet of things infrastructure for secure and smart healthcare. *Computer* July: 74–79.
Savelyev A. (2017) Contract law 2.0: 'Smart' contracts as the beginning of the end of classic contract law. *Information & Communications Technology Law* 26: 116–134.
Scott B. (2015) Visions of a techno-leviathan: The politics of the Bitcoin blockchain. *Policy Paper, February 2015*, pp. 1–4.
Shah AN, Palacios M and Ruiz F. (2013) Strategic rigidity and foresight for technology adotion among electric utilities. *Energy Policy* 63: 1233–1239.
Shalev-Shwartz S and Ben-David S. (2014) *Understanding machine learning: From theory to algorithms,* 32 Avenue of the Americas, New York, NY 10013-2473, USA: Cambridge University Press, ISBN 978-1-107-05713-5.
Shanley A. (2017) Real-time logistics. *Pharmaceutical Technology Europe, October.* 46–47.
Shermin V. (2017) Disrupting governance with blockchains and smart contracts. *Strategic Change* 26: 499–509.
Shin L. (2017) The emperor's new coins. *Forbes Africa, September.* 18–30.
Sigelman CK and Rider EA. (2015) *Life span: Human development,* 200 First Stamford Place, 4th Floor, Stamford, CT 06902, USA: Cengage Learning, ISBN 978-1-285-45431-3.
Silverberg K, French C, Ferenzy D, et al. (2016) Getting smart: Contracts on the blockchain. Institute of International Finance (IIF).
Smart Contracts Alliance. (2016) Smart contracts: 12 use cases for business & beyond. Smart Contracts Alliance & Deloitte, 1–56.
Sokolov A and Chulok A. (2016) Priorities for future innovation: Russian S&T Foresight 2030. *Futures* 80: 17–32.
Stark J. (2016) *Making sense of blockchain smart contracts, retrieved from https://www.coindesk.com/making-sense-smart-contracts/, accessed on 06 December 2017.*
Status. (2017) *Status, retrieved from https://status.im, accessed on 30 November 2017.*
Stellar Development Foundation & The Luxembourg House of Financial Technology. (2017) Understanding initial coin offerings: Technology, benefits, risks, and regulations. The Luxembourg House of Financial Technology Foundation (LHoFT) & Stellar Development Foundation, 1–40.
Stelzer B, Meyer-Brötz F, Schiebel E, et al. (2015) Combining the scenario technique with bibliometrics for technology foresight: The case of personalized medicine. *Technological Forecasting & Social Change* 98: 137–156.
Strauss JD and Radnor M. (2004) Roadmapping for dynamic and uncertain environments. *Research Technology Management* 47: 51–57.
Swanson T. (2016) *Blockchain 2.0: Let a thousand chains blossom, retrieved from https://goo.gl/IBcLUR, accessed on 08 March 2017.*
Swiss Financial Market Supervisory Authority (FINMA). (2017) FINMA guidance: Regulatory treatment of initial coin offerings. Laupenstrasse 27, 3003 Bern: Swiss Financial Market Supervisory Authority.
Szabo N. (1997) Formalizing and securing relationships on public networks. *First Monday* 2: 1–2.
Taleb NN. (2014) The skin in the game heuristic for protection against tail events. *Review of Behavioral Economics* 1: 115–135.
Taleb NN and Sandis C. (2016) The skin-in-the-game heuristic for protection against tail events. In: DeMartino G and McCloskey D (eds) *The Oxford Handbook of Professional Economic Ethics.* 198 Madison Avenue, New York, NY 10016, USA: Oxford University Press, ISBN 978-0-19-976663-5, Chapter 2.
Tapscott D and Tapscott A. (2017) Realizing the potential of blockchain: A multistakeholder aproach to the stewardship of blockchain and cryptocurrencies. 91-93 route de la Capite, CH-1223 Cologny/Geneva, Switzerland: World Economic Forum.

Telpner JS and Ahmadifar TM. (2017) ICOs, the DAO and the investment company act of 1940. *The Investment Lawyer* 24: 16–33.
Tezos. (2017) Tezos white paper: The self-amending cryptographic ledger. Tezos.
The Economist. (2015a) Bitcoin and blockchains. *The Economist* 417: 23–26.
The Economist. (2015b) Blockchain: The next big thing. *The Economist Special Report—International Banking: Slings and Arrows* 415: 16–19.
The Economist. (2017a) Bitcoin's civil war: Breaking the chains. *The Economist* 424: 55–57.
The Economist. (2017b) Initial coin offerings: Scam or substance? *The Economist* 425: 12.
The Economist. (2017c) Initial coin offerings: Token resistance. *The Economist* 425: 59–61.
Treleaven P, Brown RG and Yang D. (2017) Blockchain technology in finance. *Computer, September*: 14–17.
Turk Ž and Klinc R. (2017) Potentials of blockchain technology for construction management. *Procedia Engineering* 196: 638–645.
Vishnevskiy K, Karasev O, Meissner D, et al. (2017) Technology foresight in asset intensive industries: the case of Russian shipbuilding. *Technological Forecasting & Social Change* 119: 194–204.
Volti R. (2014) *Society and technological change*, 41 Madison Avenue, New York, NY 10010: Worth Publishers, ISBN 978-1-4292-7897-3.
Von der Gracht HA, Bañuls VA, Turoff M, et al. (2015) Foresight support systems: The future role of ICT for foresight. *Technological Forecasting & Social Change* 97: 1–6.
Vranken H. (in press) Sustainability of bitcoin and blockchains. *Current Opinion in Environmental Sustainability* http://dx.doi.org/10.1016/j.cosust.2017.04.011.
Xing B, Bright G, Tlale NS, et al. (2006a) Reconfigurable manufacturing systems for agile mass customization manufacturing. *Proceedings of the 22nd ISPE International Conference on CAD/CAM, Robotics and Factories of the Future (CARs&FOF 2006), Vellore, India, July 2006*. pp. 473–482.
Xing B, Eganza J, Bright G, et al. (2006b) Reconfigurable manufacturing system for agile manufacturing. *Proceedings of the 12th IFAC Symposium on Information Control Problems in Manufacturing, May 2006, Saint-Etienne, France, pp. on CD.*
Xing B, Nelwamondo FV, Battle K, et al. (2009) Application of artificial intelligence (AI) methods for designing and analysis of reconfigurable cellular manufacturing system (RCMS). *Proceedings of the 2nd International Conference on Adaptive Science & Technology (ICAST), 14–16 December, Accra, Ghana*, pp. 402–409. IEEE.
Xing B, Gao W-J, Nelwamondo FV, et al. (2010a) Ant colony optimization for automated storage and retrieval system. *Proceedings of The Annual IEEE Congress on Evolutionary Computation (IEEE CEC), 18–23 July, CCIB, Barcelona, Spain*, pp. 1133–1139. IEEE.
Xing B, Gao W-J, Nelwamondo FV, et al. (2010b) Artificial intelligence in reverse supply chain management: The state of the art. *Proceedings of the Twenty-First Annual Symposium of the Pattern Recognition Association of South Africa (PRASA), 22–23 November, Stellenbosch, South Africa*, pp. 305–310.
Xing B, Gao W-J, Nelwamondo FV, et al. (2010c) Can ant algorithms make automated guided vehicle system more intelligent? A viewpoint from manufacturing environment. *Proceedings of IEEE International Conference on Systems, Man, and Cybernetics (IEEE SMC), 10–13 October, Istanbul, Turkey*, pp. 3226–3234. IEEE.
Xing B, Gao W-J, Nelwamondo FV, et al. (2010d) Cellular manufacturing system scheduling under fuzzy constraints: A group technology perspective. *Annual IEEE International Conference on Fuzzy Systems (FUZZ-IEEE), 18–23 July, CCIB, Barcelona, Spain*, pp. 887–894. IEEE.
Xing B, Gao W-J, Nelwamondo FV, et al. (2010e) Part-machine clustering: The comparison between adaptive resonance theory neural network and ant colony system. In: Zeng Z and Wang J (eds) *Advances in Neural Network Research & Applications, LNEE 67*, pp. 747–755. Berlin Heidelberg: Springer-Verlag.
Xing B, Gao W-J, Nelwamondo FV, et al. (2010f) Two-stage inter-cell layout design for cellular manufacturing by using ant colony optimization algorithms. In: Tan Y, Shi Y

and Tan KC (eds) *Advances in Swarm Intelligence, Part I, LNCS 6145*, pp. 281–289. Berlin Heidelberg: Springer-Verlag.

Xing B, Gao W-J, Nelwamondo FV, et al. (2011a) e-Reverse logistics for remanufacture-to-order: an online auction-based and multi-agent system supported solution. In: Omatu S and Fabri SG (eds) *Fifth International Conference on Advanced Engineering Computing and Applications in Sciences (ADVCOMP), 20–25 November, Lisbon, Portugal*, pp. 78–83. IARIA.

Xing B, Gao W-J, Nelwamondo FV, et al. (2011b) e-RL: the Internet of things supported reverse logistics for remanufacture-to-order. In: Omatu S and Fabri SG (eds) *Fifth International Conference on Advanced Engineering Computing and Applications in Sciences (ADVCOMP), 20–25 November, Lisbon, Portugal*, pp. 84–87. IARIA.

Xing B, Gao W-J and Marwala T. (2012a) The applications of computational intelligence in radio frequency identification research. *IEEE International Conference on Systems, Man, and Cybernetics (IEEE SMC), 14–17 October, Seoul, Korea*, pp. 2067–2072. IEEE.

Xing B, Gao W-J, Nelwamondo FV, et al. (2012b) The effects of customer perceived disposal hardship on post-consumer product remanufacturing: A multi-agent perspective. In: Tan Y, Shi Y and Ji Z (eds) *ICSI 2012, Part I, LNCS 7332*, pp. 209–216. Berlin Heidelberg: Springer-Verlag.

Xing B, Gao W-J, Nelwamondo FV, et al. (2012c) TAC-RMTO: trading agent competition in remanufacture-to-order. In: Tan Y, Shi Y and Ji Z (eds) *ICSI 2012, Part I, LNCS 7332*, pp. 519–526. Berlin Heidelberg: Springer-Verlag.

Xing B, Gao W-J and Marwala T. (2013a) The applications of computational intelligence in system reliability optimization. *IEEE Symposium Series on Computational Intelligence (IEEE SSCI), 15–19 Aprial, Singapore, pp. 7–14.* IEEE.

Xing B, Gao W-J and Marwala T. (2013b) Intelligent data processing using emerging computational intelligence techniques. *Research Notes in Information Sciences* 12: 10–15.

Xing B, Gao W-J and Marwala T. (2013c) An overview of cuckoo-inspired intelligent algorithms and their applications. *IEEE Symposium Series on Computational Intelligence (IEEE SSCI), 15–19 Aprial, Singapore, pp. 85–89.* IEEE.

Xing B, Gao W-J and Marwala T. (2013d) Taking a new look at traveling salesman problem through the lens of innovative computational intelligence. *Research Notes in Information Sciences* 12: 20–25.

Xing B, Gao W-J and Marwala T. (2013e) When vehicle routing problem becomes dynamic: can innovative computational intelligence give us a hand? *Research Notes in Information Sciences* 12: 26–31.

Xing B, Gao W-J and Marwala T. (2014) Multi-agent framework for distributed leasing based injection mould remanufacturing. In: Memon QA (ed) *Distributed Network Intelligence, Security and Applications.* 6000 Broken Sound Parkway NW, Suite 300, Boca Raton, FL 33487-2742: CRC Press, Taylor & Francis Group, LLC, ISBN 978-1-4665-5958-5, Chapter 11, pp. 267–289.

Xing B. (2014a) Computational intelligence in cross docking. *International Journal of Software Innovation* 4: 78–124.

Xing B. (2014b) Novel computational intelligence for optimizing cyber physical pre-evaluation system. In: Khan ZH, Ali ABMS and Riaz Z (eds) *Computational Intelligence for Decision Support in Cyber-Physical Systems.* Singapore Heidelberg New York Dordrecht London: Springer Science+Business Media Singapore, ISBN 978-981-4585-35-4, Chapter 15, pp. 449–464.

Xing B. (2014c) The optimization of computational stock market model-based complex adaptive cyber physical logistics system. In: Khan ZH, Ali ABMS and Riaz Z (eds) *Computational Intelligence for Decision Support in Cyber-Physical Systems.* Singapore Heidelberg New York Dordrecht London: Springer Science+Business Media Singapore, ISBN 978-981-4585-35-4, Chapter 12, pp. 357–380.

Xing B and Gao W-J. (2014a) *Computational intelligence in remanufacturing*, 701 E. Chocolate Avenue, Suite 200, Hershey PA 17033: IGI Global, ISBN 978-1-4666-4908-8.

Xing B and Gao W-J. (2014aa) Chemical-reaction optimization algorithm. In: Xing B and Gao W-J (eds) *Innovative Computational Intelligence: A Rough Guide to 134 Clever Algorithms.* Cham Heidelberg New York Dordrecht London: Springer International Publishing Switzerland, ISBN: 978-3-319-03403-4, Chapter 25, pp. 417–428.

Xing B and Gao W-J. (2014b) *Innovative computational intelligence: a rough guide to 134 clever algorithms,* Cham Heidelberg New York Dordrecht London: Springer International Publishing Switzerland, ISBN: 978-3-319-03403-4.

Xing B and Gao W-J. (2014bb) Emerging chemistry-based CI algorithms. In: Xing B and Gao W-J (eds) *Innovative Computational Intelligence: A Rough Guide to 134 Clever Algorithms.* Cham Heidelberg New York Dordrecht London: Springer International Publishing Switzerland, ISBN: 978-3-319-03403-4, Chapter 26, pp. 429–437.

Xing B and Gao W-J. (2014c) Introduction to computational intelligence. In: Xing B and Gao W-J (eds) *Innovative Computational Intelligence: A Rough Guide to 134 Clever Algorithms.* Cham Heidelberg New York Dordrecht London: Springer International Publishing Switzerland, ISBN: 978-3-319-03403-4, Chapter 1, pp. 3–17.

Xing B and Gao W-J. (2014cc) Base optimization algorithm. In: Xing B and Gao W-J (eds) *Innovative Computational Intelligence: A Rough Guide to 134 Clever Algorithms.* Cham Heidelberg New York Dordrecht London: Springer International Publishing Switzerland, ISBN: 978-3-319-03403-4, Chapter 27, pp. 441–444.

Xing B and Gao W-J. (2014d) Bacteria inspired algorithms. In: Xing B and Gao W-J (eds) *Innovative Computational Intelligence: A Rough Guide to 134 Clever Algorithms.* Cham Heidelberg New York Dordrecht London: Springer International Publishing Switzerland, ISBN: 978-3-319-03403-4, Chapter 2, pp. 21–38.

Xing B and Gao W-J. (2014dd) Emerging mathematics-based CI algorithms. In: Xing B and Gao W-J (eds) *Innovative Computational Intelligence: A Rough Guide to 134 Clever Algorithms.* Cham Heidelberg New York Dordrecht London: Springer International Publishing Switzerland, ISBN: 978-3-319-03403-4, Chapter 28, pp. 445–448.

Xing B and Gao W-J. (2014e) Bee inspired algorithms. In: Xing B and Gao W-J (eds) *Innovative Computational Intelligence: A Rough Guide to 134 Clever Algorithms.* Cham Heidelberg New York Dordrecht London: Springer International Publishing Switzerland, ISBN: 978-3-319-03403-4, Chapter 4, pp. 45–80.

Xing B and Gao W-J. (2014ee) Introduction to remanufacturing and reverse logistics. In: Xing B and Gao W-J (eds) *Computational Intelligence in Remanufacturing.* 701 E. Chocolate Avenue, Suite 200, Hershey PA 17033: IGI Global, ISBN 978-1-4666-4908-8, Chapter 1, pp. 1–17.

Xing B and Gao W-J. (2014f) Bat inspired algorithms. In: Xing B and Gao W-J (eds) *Innovative Computational Intelligence: A Rough Guide to 134 Clever Algorithms.* Cham Heidelberg New York Dordrecht London: Springer International Publishing Switzerland, ISBN: 978-3-319-03403-4, Chapter 3, pp. 39–44.

Xing B and Gao W-J. (2014ff) Overview of computational intelligence. In: Xing B and Gao W-J (eds) *Computational Intelligence in Remanufacturing.* 701 E. Chocolate Avenue, Suite 200, Hershey PA 17033: IGI Global, ISBN 978-1-4666-4908-8, Chapter 2, pp. 18–36.

Xing B and Gao W-J. (2014g) Biogeography-based optimization algorithm. In: Xing B and Gao W-J (eds) *Innovative Computational Intelligence: A Rough Guide to 134 Clever Algorithms.* Cham Heidelberg New York Dordrecht London: Springer International Publishing Switzerland, ISBN: 978-3-319-03403-4, Chapter 5, pp. 81–91.

Xing B and Gao W-J. (2014gg) Used products return pattern analysis using agent-based modelling and simulation. In: Xing B and Gao W-J (eds) *Computational Intelligence in Remanufacturing.* 701 E. Chocolate Avenue, Suite 200, Hershey PA 17033: IGI Global, ISBN 978-1-4666-4908-8, Chapter 3, pp. 38–58.

Xing B and Gao W-J. (2014h) Cat swarm optimization algorithm. In: Xing B and Gao W-J (eds) *Innovative Computational Intelligence: A Rough Guide to 134 Clever Algorithms.* Cham Heidelberg New York Dordrecht London: Springer International Publishing Switzerland, ISBN: 978-3-319-03403-4, Chapter 6, pp. 93–104.

Xing B and Gao W-J. (2014hh) Used product collection optimization using genetic algorithms. In: Xing B and Gao W-J (eds) *Computational Intelligence in Remanufacturing.* 701 E. Chocolate Avenue, Suite 200, Hershey PA 17033: IGI Global, ISBN 978-1-4666-4908-8, Chapter 4, pp. 59–74.

Xing B and Gao W-J. (2014i) Cuckoo inspired algorithms. In: Xing B and Gao W-J (eds) *Innovative Computational Intelligence: A Rough Guide to 134 Clever Algorithms.* Cham Heidelberg New York Dordrecht London: Springer International Publishing Switzerland, ISBN: 978-3-319-03403-4, Chapter 7, pp. 105–121.

Xing B and Gao W-J. (2014ii) Used product remanufacturability evaluation using fuzzy logic. In: Xing B and Gao W-J (eds) *Computational Intelligence in Remanufacturing.* 701 E. Chocolate Avenue, Suite 200, Hershey PA 17033: IGI Global, ISBN 978-1-4666-4908-8, Chapter 5, pp. 75–94.

Xing B and Gao W-J. (2014j) Luminous insect inspired algorithms. In: Xing B and Gao W-J (eds) *Innovative Computational Intelligence: A Rough Guide to 134 Clever Algorithms.* Cham Heidelberg New York Dordrecht London: Springer International Publishing Switzerland, ISBN: 978-3-319-03403-4, Chapter 8, pp. 123–137.

Xing B and Gao W-J. (2014jj) Used product pre–sorting system optimization using teaching–learning-based optimization. In: Xing B and Gao W-J (eds) *Computational Intelligence in Remanufacturing.* 701 E. Chocolate Avenue, Suite 200, Hershey PA 17033: IGI Global, ISBN 978-1-4666-4908-8, Chapter 6, pp. 95–112.

Xing B and Gao W-J. (2014k) Fish inspired algorithms. In: Xing B and Gao W-J (eds) *Innovative Computational Intelligence: A Rough Guide to 134 Clever Algorithms.* Cham Heidelberg New York Dordrecht London: Springer International Publishing Switzerland, ISBN: 978-3-319-03403-4, Chapter 9, pp. 139–155.

Xing B and Gao W-J. (2014kk) Used product delivery optimization using agent-based modelling and simulation. In: Xing B and Gao W-J (eds) *Computational Intelligence in Remanufacturing.* 701 E. Chocolate Avenue, Suite 200, Hershey PA 17033: IGI Global, ISBN 978-1-4666-4908-8, Chapter 7, pp. 113–133.

Xing B and Gao W-J. (2014l) Frog inspired algorithms. In: Xing B and Gao W-J (eds) *Innovative Computational Intelligence: A Rough Guide to 134 Clever Algorithms.* Cham Heidelberg New York Dordrecht London: Springer International Publishing Switzerland, ISBN: 978-3-319-03403-4, Chapter 10, pp. 157–165.

Xing B and Gao W-J. (2014ll) Post–disassembly part–machine clustering using artificial neural networks and ant colony systems. In: Xing B and Gao W-J (eds) *Computational Intelligence in Remanufacturing.* 701 E. Chocolate Avenue, Suite 200, Hershey PA 17033: IGI Global, ISBN 978-1-4666-4908-8, Chapter 8, pp. 135–150.

Xing B and Gao W-J. (2014m) Fruit fly optimization algorithm. In: Xing B and Gao W-J (eds) *Innovative Computational Intelligence: A Rough Guide to 134 Clever Algorithms.* Cham Heidelberg New York Dordrecht London: Springer International Publishing Switzerland, ISBN: 978-3-319-03403-4, Chapter 11, pp. 167–170.

Xing B and Gao W-J. (2014mm) Reprocessing operations scheduling using fuzzy logic and fuzzy MAX–MIN ant systems. In: Xing B and Gao W-J (eds) *Computational Intelligence in Remanufacturing.* 701 E. Chocolate Avenue, Suite 200, Hershey PA 17033: IGI Global, ISBN 978-1-4666-4908-8, Chapter 9, pp. 151–170.

Xing B and Gao W-J. (2014n) Group search optimization algorithm. In: Xing B and Gao W-J (eds) *Innovative Computational Intelligence: A Rough Guide to 134 Clever Algorithms.* Cham Heidelberg New York Dordrecht London: Springer International Publishing Switzerland, ISBN: 978-3-319-03403-4, Chapter 12, pp. 171–176.

Xing B and Gao W-J. (2014nn) Reprocessing cell layout optimization using hybrid ant systems. In: Xing B and Gao W-J (eds) *Computational Intelligence in Remanufacturing.* 701 E. Chocolate Avenue, Suite 200, Hershey PA 17033: IGI Global, ISBN 978-1-4666-4908-8, Chapter 10, pp. 171–185.

Xing B and Gao W-J. (2014o) Invasive weed optimization algorithm. In: Xing B and Gao W-J (eds) *Innovative Computational Intelligence: A Rough Guide to 134 Clever Algorithms.*

Cham Heidelberg New York Dordrecht London: Springer International Publishing Switzerland, ISBN: 978-3-319-03403-4, Chapter 13, pp. 177–181.

Xing B and Gao W-J. (2014oo) Re–machining parameter optimization using firefly algorithms. In: Xing B and Gao W-J (eds) *Computational Intelligence in Remanufacturing*. 701 E. Chocolate Avenue, Suite 200, Hershey PA 17033: IGI Global, ISBN 978-1-4666-4908-8, Chapter 11, pp. 186–202.

Xing B and Gao W-J. (2014p) Music inspired algorithms. In: Xing B and Gao W-J (eds) *Innovative Computational Intelligence: A Rough Guide to 134 Clever Algorithms*. Cham Heidelberg New York Dordrecht London: Springer International Publishing Switzerland, ISBN: 978-3-319-03403-4, Chapter 14, pp. 183–201.

Xing B and Gao W-J. (2014pp) Batch order picking optimization using ant system. In: Xing B and Gao W-J (eds) *Computational Intelligence in Remanufacturing*. 701 E. Chocolate Avenue, Suite 200, Hershey PA 17033: IGI Global, ISBN 978-1-4666-4908-8, Chapter 12, pp. 204–222.

Xing B and Gao W-J. (2014q) Imperialist competitive algorithm. In: Xing B and Gao W-J (eds) *Innovative Computational Intelligence: A Rough Guide to 134 Clever Algorithms*. Cham Heidelberg New York Dordrecht London: Springer International Publishing Switzerland, ISBN: 978-3-319-03403-4, Chapter 15, pp. 203–209.

Xing B and Gao W-J. (2014qq) Complex adaptive logistics system optimization using agent-based modelling and simulation. In: Xing B and Gao W-J (eds) *Computational Intelligence in Remanufacturing*. 701 E. Chocolate Avenue, Suite 200, Hershey PA 17033: IGI Global, ISBN 978-1-4666-4908-8, Chapter 13, pp. 223–236.

Xing B and Gao W-J. (2014r) Teaching–learning-based optimization algorithm. In: Xing B and Gao W-J (eds) *Innovative Computational Intelligence: A Rough Guide to 134 Clever Algorithms*. Cham Heidelberg New York Dordrecht London: Springer International Publishing Switzerland, ISBN: 978-3-319-03403-4, Chapter 16, pp. 211–216.

Xing B and Gao W-J. (2014rr) Conclusions and emerging topics. In: Xing B and Gao W-J (eds) *Computational Intelligence in Remanufacturing*. 701 E. Chocolate Avenue, Suite 200, Hershey PA 17033: IGI Global, ISBN 978-1-4666-4908-8, Chapter 14, pp. 238–265.

Xing B and Gao W-J. (2014s) Emerging biology-based CI algorithms. In: Xing B and Gao W-J (eds) *Innovative Computational Intelligence: A Rough Guide to 134 Clever Algorithms*. Cham Heidelberg New York Dordrecht London: Springer International Publishing Switzerland, ISBN: 978-3-319-03403-4, Chapter 17, pp. 217–317.

Xing B and Gao W-J. (2014t) Big bang–big crunch algorithm. In: Xing B and Gao W-J (eds) *Innovative Computational Intelligence: A Rough Guide to 134 Clever Algorithms*. Cham Heidelberg New York Dordrecht London: Springer International Publishing Switzerland, ISBN: 978-3-319-03403-4, Chapter 18, pp. 321–331.

Xing B and Gao W-J. (2014u) Central force optimization algorithm. In: Xing B and Gao W-J (eds) *Innovative Computational Intelligence: A Rough Guide to 134 Clever Algorithms*. Cham Heidelberg New York Dordrecht London: Springer International Publishing Switzerland, ISBN: 978-3-319-03403-4, Chapter 19, pp. 333–337.

Xing B and Gao W-J. (2014v) Charged system search algorithm. In: Xing B and Gao W-J (eds) *Innovative Computational Intelligence: A Rough Guide to 134 Clever Algorithms*. Cham Heidelberg New York Dordrecht London: Springer International Publishing Switzerland, ISBN: 978-3-319-03403-4, Chapter 20, pp. 339–346.

Xing B and Gao W-J. (2014w) Electromagnetism-like mechanism algorithm. In: Xing B and Gao W-J (eds) *Innovative Computational Intelligence: A Rough Guide to 134 Clever Algorithms*. Cham Heidelberg New York Dordrecht London: Springer International Publishing Switzerland, ISBN: 978-3-319-03403-4, Chapter 21, pp. 347–354.

Xing B and Gao W-J. (2014x) Gravitational search algorithm. In: Xing B and Gao W-J (eds) *Innovative Computational Intelligence: A Rough Guide to 134 Clever Algorithms*. Cham Heidelberg New York Dordrecht London: Springer International Publishing Switzerland, ISBN: 978-3-319-03403-4, Chapter 22, pp. 355–364.

Xing B and Gao W-J. (2014y) Intelligent water drops algorithm. In: Xing B and Gao W-J (eds) *Innovative Computational Intelligence: A Rough Guide to 134 Clever Algorithms*.

Cham Heidelberg New York Dordrecht London: Springer International Publishing Switzerland, ISBN: 978-3-319-03403-4, Chapter 23, pp. 365–373.

Xing B and Gao W-J. (2014z) Emerging physics-based CI algorithms. In: Xing B and Gao W-J (eds) *Innovative Computational Intelligence: A Rough Guide to 134 Clever Algorithms*. Cham Heidelberg New York Dordrecht London: Springer International Publishing Switzerland, ISBN: 978-3-319-03403-4, Chapter 24, pp. 375–414.

Xing B. (2015a) Graph-based framework for evaluating the feasibility of transition to maintainomics. In: Pedrycz W and Chen S-M (eds) *Information Granularity, Big Data, and Computational Intelligence*. Cham Heidelberg New York Dordrecht London: Springer International Publishing Switzerland, ISBN 978-3-319-08253-0, Chapter 5, pp. 89–119.

Xing B. (2015b) Knowledge management: Intelligent in-pipe inspection robot conceptual design for pipeline infrastructure management. In: Kahraman C and Onar SÇ (eds) *Intelligent Techniques in Engineering Management: Theory and Applications*. Cham Heidelberg New York Dordrecht London: Springer International Publishing Switzerland, ISBN 978-3-319-17905-6, Chapter 6, 129–146.

Xing B. (2015c) Massive online open course assisted mechatronics learning: A hybrid approach. In: Mesquita A and Peres P (eds) *Furthering Higher Education Possibilities through Massive Open Online Courses*. 701 E. Chocolate Avenue, Hershey PA, USA 17033: IGI Global, ISBN 978-1-4666-8279-5, Chapter 12, pp. 245–268.

Xing B. (2015d) Optimization in production management: Economic load dispatch of cyber physical power system using artificial bee colony. In: Kahraman C and Onar SÇ (eds) *Intelligent Techniques in Engineering Management: Theory and Applications*. Cham Heidelberg New York Dordrecht London: Springer International Publishing Switzerland, ISBN 978-3-319-17905-6, Chapter 12, 275–293.

Xing B and Gao W-J. (2015a) The applications of swarm intelligence in remanufacturing: A focus on retrieval. In: Khosrow-Pour M (ed) *Encyclopedia of Information Science and Technology*. 3rd ed. New York, USA: Information Science Ref. – IGI Global, ISBN 978-1-4666-5888-2, Chapter 7, pp. 66–74.

Xing B and Gao W-J. (2015b) An exploration of designing e-remanufacturing course. In: Khosrow-Pour M (ed) *Encyclopedia of Information Science and Technology*. 3rd ed. New York, USA: Information Science Ref. – IGI Global, ISBN 978-1-4666-5888-2, Chapter 66, pp. 688–698.

Xing B and Gao W-J. (2015c) Offshore remanufacturing. In: Khosrow-Pour M (ed) *Encyclopedia of Information Science and Technology*. 3rd ed. New York, USA: Information Science Ref. – IGI Global, ISBN 978-1-4666-5888-2, Chapter 374, pp. 3795–3804.

Xing B and Gao W-J. (2015d) A SWOT analysis of intelligent product enabled complex adaptive logistics systems. In: Khosrow-Pour M (ed) *Encyclopedia of Information Science and Technology*. 3rd ed. New York, USA: Information Science Ref. – IGI Global, ISBN 978-1-4666-5888-2, Chapter 490, pp. 4970–4979.

Xing B. (2016a) Agent-based machine-to-machine connectivity analysis for the Internet of things environment. In: Mahmood Z (ed) *Connectivity Frameworks for Smart Devices: The Internet of Things from a Distributed Computing Perspective*. Switzerland: Springer International Publishing, ISBN 978-3-319-33122-5, Chapter 3, pp. 43–61.

Xing B. (2016b) An investigation of the use of innovative biology-based computational intelligence in ubiquitous robotics systems: Data mining perspective. In: Ravulakollu KK, Khan MA and Abraham A (eds) *Trends in Ambient Intelligent Systems*. Switzerland: Springer International Publishing Switzerland, ISBN 978-3-319-30184-6, Chapter 6, pp. 139–172.

Xing B. (2016c) Network neutrality debate in the internet of things era: A fuzzy cognitive map extend technology roadmap perspective. In: Mahmood Z (ed) *Connectivity Frameworks for Smart Devices: The IoT Distributed Computing Perspective*. Cham Heidelberg New York Dordrecht London: Springer International Publishing Switzerland, ISBN 978-3-319-33122-5, Chapter 10, pp. 235–257.

Xing B. (2016d) Ontological framework–assisted embedded system design with security consideration. In: Pathan A-SK (ed) *Securing Cyber-Physical Systems*. 6000 Broken Sound

Parkway NW, Suite 300, Boca Raton, FL 33487-2742: CRC Press, Taylor & Francis Group, LLC, ISBN 978-1-4987-0099-3, Chapter 4, pp. 91–118.

Xing B. (2016e) Smart robot control via novel computational intelligence methods for ambient assisted living. In: Ravulakollu KK, Khan MA and Abraham A (eds) *Trends in Ambient Intelligent Systems*. Switzerland: Springer International Publishing Switzerland, ISBN 978-3-319-30184-6, Chapter 2, pp. 29–55.

Xing B. (2016f) The spread of innovatory nature originated metaheuristics in robot swarm control for smart living environments. In: Espinosa HEP (ed) *Nature-Inspired Computing for Control Systems*. Cham Heidelberg New York Dordrecht London: Springer International Publishing Switzerland, ISBN 978-3-319-26228-4, Chapter 3, pp. 39–70.

Xing B and Marwala T. (2017) Implications of the fourth industrial age for higher education. *The Thinker: For the Thought Leaders (www.thethinker.co.za)*. South Africa: Vusizwe Media, 10–15.

Xing B. (2017a) Component-based hybrid reference architecture for managing adaptable embedded software development. In: Mahmood Z (ed) *Software Project Management for Distributed Computing: Life-Cycle Methods for Developing Scalable and Reliable Tools*. Gewerbestrasse 11, 6330 Cham, Switzerland: Springer International Publishing AG, ISBN 978-3-319-54324-6, Chapter 6, pp 119–141.

Xing B. (2017b) Protecting mobile payments security: A case study. In: Meng W, Luo X, Furnell S, et al. (eds) *Protecting Mobile Networks and Devices: Challenges and Solutions*. 6000 Broken Sound Parkway NW, Suite 300, Boca Raton, FL 33487-2742: CRC Press, Taylor & Francis Group, LLC, ISBN 978-1-4987-3583-4, Chapter 11, pp. 261–289.

Xing B. (2017c) Visible light based throughput downlink connectivity for the cognitive radio networks. In: Matin MA (ed) *Spectrum Access and Management for Cognitive Radio Networks*. Singapore: Springer Science+Business Media, ISBN 978-981-10-2253-1, Chapter 8, pp. 211–232.

Xing B and Marwala T. (2018a) Conclusion. In: Xing B and Marwala T (eds) *Smart Maintenance for Human–Robot Interaction: An Intelligent Search Algorithmic Perspective*. Gewerbestrasse 11, 6330 Cham, Switzerland: Springer International Publishing AG, ISBN 978-3-319-67479-7, Chapter 13, pp. 299–305.

Xing B and Marwala T. (2018b) Cyberware capacity–applications layer perspective. In: Xing B and Marwala T (eds) *Smart Maintenance for Human–Robot Interaction: An Intelligent Search Algorithmic Perspective*. Gewerbestrasse 11, 6330 Cham, Switzerland: Springer International Publishing AG, ISBN 978-3-319-67479-7, Chapter 8, pp. 173–191.

Xing B and Marwala T. (2018c) Cyberware capacity–energy autonomy perspective. In: Xing B and Marwala T (eds) *Smart Maintenance for Human–Robot Interaction: An Intelligent Search Algorithmic Perspective*. Gewerbestrasse 11, 6330 Cham, Switzerland: Springer International Publishing AG, ISBN 978-3-319-67479-7, Chapter 9, pp. 193–216.

Xing B and Marwala T. (2018d) Cyberware capacity–platform and middleware layers perspective. In: Xing B and Marwala T (eds) *Smart Maintenance for Human–Robot Interaction: An Intelligent Search Algorithmic Perspective*. Gewerbestrasse 11, 6330 Cham, Switzerland: Springer International Publishing AG, ISBN 978-3-319-67479-7, Chapter 7, pp. 143–171.

Xing B and Marwala T. (2018e) Hardware capacity–beginning of life perspective. In: Xing B and Marwala T (eds) *Smart Maintenance for Human–Robot Interaction: An Intelligent Search Algorithmic Perspective*. Gewerbestrasse 11, 6330 Cham, Switzerland: Springer International Publishing AG, ISBN 978-3-319-67479-7, Chapter 4, pp. 67–91.

Xing B and Marwala T. (2018f) Hardware capacity–end of life perspective. In: Xing B and Marwala T (eds) *Smart Maintenance for Human–Robot Interaction: An Intelligent Search Algorithmic Perspective*. Gewerbestrasse 11, 6330 Cham, Switzerland: Springer International Publishing AG, ISBN 978-3-319-67479-7, Chapter 6, pp. 111–139.

Xing B and Marwala T. (2018g) Hardware capacity–middle of life perspective. In: Xing B and Marwala T (eds) *Smart Maintenance for Human–Robot Interaction: An Intelligent Search Algorithmic Perspective*. Gewerbestrasse 11, 6330 Cham, Switzerland: Springer International Publishing AG, ISBN 978-3-319-67479-7, Chapter 5, pp. 93–110.

Xing B and Marwala T. (2018h) Human capacity–biopsychosocial perspective. In: Xing B and Marwala T (eds) *Smart Maintenance for Human–Robot Interaction: An Intelligent Search Algorithmic Perspective.* Gewerbestrasse 11, 6330 Cham, Switzerland: Springer International Publishing AG, ISBN 978-3-319-67479-7, Chapter 11, pp. 249–270.

Xing B and Marwala T. (2018i) Human capacity–exposome perspective. In: Xing B and Marwala T (eds) *Smart Maintenance for Human–Robot Interaction: An Intelligent Search Algorithmic Perspective.* Gewerbestrasse 11, 6330 Cham, Switzerland: Springer International Publishing AG, ISBN 978-3-319-67479-7, Chapter 12, pp. 271–295.

Xing B and Marwala T. (2018j) Human capacity–physiology perspective. In: Xing B and Marwala T (eds) *Smart Maintenance for Human–Robot Interaction: An Intelligent Search Algorithmic Perspective.* Gewerbestrasse 11, 6330 Cham, Switzerland: Springer International Publishing AG, ISBN 978-3-319-67479-7, Chapter 10, pp. 219–247.

Xing B and Marwala T. (2018k) Introduction to human robot interaction. In: Xing B and Marwala T (eds) *Smart Maintenance for Human–Robot Interaction: An Intelligent Search Algorithmic Perspective.* Gewerbestrasse 11, 6330 Cham, Switzerland: Springer International Publishing AG, ISBN 978-3-319-67479-7, Chapter 1, pp. 3–19.

Xing B and Marwala T. (2018l) Introduction to intelligent search algorithms. In: Xing B and Marwala T (eds) *Smart Maintenance for Human–Robot Interaction: An Intelligent Search Algorithmic Perspective.* Gewerbestrasse 11, 6330 Cham, Switzerland: Springer International Publishing AG, ISBN 978-3-319-67479-7, Chapter 3, pp. 33–64.

Xing B and Marwala T. (2018m) Introduction to smart maintenance. In: Xing B and Marwala T (eds) *Smart Maintenance for Human–Robot Interaction: An Intelligent Search Algorithmic Perspective.* Gewerbestrasse 11, 6330 Cham, Switzerland: Springer International Publishing AG, ISBN 978-3-319-67479-7, Chapter 2, pp. 21–31.

Xing B and Marwala T. (2018n) *Smart maintenance for human–robot interaction: An intelligent search algorithmic perspective,* Gewerbestrasse 11, 6330 Cham, Switzerland: Springer International Publishing AG, ISBN 978-3-319-67479-7.

Yao M. (2017) *Demystifying ICOs: the good, the bad and the ugly, retrieved from https://www.forbes.com/sites/mariyayao/2017/10/09/demystifying-icos-the-good-the-bad-and-the-ugly/#53c1b23627e9, accessed on 30 November 2017.*

Yelland PM. (2010) An introduction to correspondence analysis. *The Mathematica Journal* 12: 1–23, retrieved from http://www.mathematica-journal.com/data/uploads/2010/2009/Yelland.pdf, accessed on 2014 December 2017.

Yue X, Wang H, Jin D, et al. (2016) Healthcare data gateways: Found healthcare intelligence on blockchain with novel privacy risk control. *Journal of Medical Systems* 40: 1–8.

Zadeh LA. (1965) Fuzzy sets. *Information and Control* 8: 338–353.

Zadeh LA. (1971) Quantitative fuzzy semantics. *Information Sciences* 3: 159–176.

Zadeh LA. (1975a) The concept of a linguistic variable and its application to approximate reasoning—I. *Information Sciences* 8: 199–249.

Zadeh LA. (1975b) The concept of a linguistic variable and its application to approximate reasoning—II. *Information Sciences* 8: 301–357.

Zadeh LA. (1975c) The concept of a linguistic variable and its application to approximate reasoning—III. *Information Sciences* 9: 43–80.

Zarandi MHF, Rezaee B, Türksen IB, et al. (2009) A type-2 fuzzy rule-based expert system model for stock price analysis. *Expert Systems with Applications* 36: 139–154.

Zetzsche D, Buckley RP, Arner DW, et al. (2017) The ICO gold rush: It's a scam, it's a bubble, it's a supper challenge for regulators. *SSRN Electronic Journal* https://ssrn.com/abstract=3072298.

Zhang Y and Wen J. (2017) The IoT electric business model: Using blockchain technology for the internet of things. *Peer-to-Peer Networking and Applications* 10: 983–994.

Zohar A. (2015) Bitcoin: Under the hood. *ACM Communication* 58: 104–113.

Part III

Asker + Smart Computing = Better Feedback

CHAPTER 6

Crowdfunding Asker Campaign Prediction and Ensemble Learning

6.1 Introduction

In our time, the worldwide crowdfunding market has grown exponentially from a $2 billion market in 2012, to $6 billion in 2013, $16 billion in 2014, and $34 billion in 2015 (Massolution, 2015). According to World Bank's estimate, it is expected to reach $93 billion in 2025 (infoDev and The World Bank, 2013). Regarding its global spread, crowdfunding has not only emerged in many developed countries but it also shares the same popularity in various developing countries as well, such as Vietnam (Bui, 2017), Zimbabwe (Munyanyi and Mapfumo, 2016) and India (Angle, 2017). According to one collection of statistics revealed by a Chinese crowdfunding website, Zhongchou, there have been 15,073 campaigns backed by 802,308 backers in which the total amount of contributions was equivalent to about $171,753,514 by 2014 (Yuan et al., 2016).

The campaign's success has a great deal of practical meanings in the realm of crowdfunding. However, in contrast to the rapid growth of the global crowdfunding market, the real success rate of many crowdfunding related campaigns is decreasing (Hu et al., 2015). According to one study (Forbes and Schaefer, 2017), among the vast majority of failed crowdfunding campaigns, nearly 81% of them reach only less than 20% of their intended fundraising target. One key underlying reason is that some individuals rush campaigns into existence based on nothing but the success stories of previous campaigns, but ignore their own preparedness in terms of initiating a project. Another reason is that a project with many fans that initially appears legitimate is often falsely assumed to be a good campaign by many askers (Hobbs et al., 2016).

In this chapter, we define askers as those who are, through the launch of one or more crowdfunding campaigns, either intending to implement their ideas, or would like to get their businesses promoted in exchange for something like rewards, interests or a share of equity securities. Although the setting-up stages for all types of crowdfunding campaigns are usually similar, to succeed, each one has its own unique tune in terms of the success and failure factors.

6.1.1 *Determinants of Success and Failure: Reward-Based Crowdfunding Campaign*

In the literature, many studies focused on reward-based crowdfunding and discussed different determinants that can influence a campaign's performance. Based on the time's continuum, a campaign can generally be divided into three phases (Beaulieu et al., 2015; Kunz et al., 2017): (1) Upcoming phase (i.e., pre-fundraising), (2) Ongoing phase (i.e., during-fundraising), and (3) Outgoing phase (i.e., post-fundraising).

- **Pre-Fundraising Phase:** By beginning with the success/failure factor in the preparation time, researchers and practitioners found that the following five steps can often help an asker launch a successful crowdfunding campaign (Steinberg and DeMaria, 2015).
    1) *Step 1*: The first thing is to prepare a novel project idea, a new product, or a good reason;
    2) *Step 2*: Next, carefully choose a relevant platform;
    3) *Step 3*: The third thing is to be ready to tell a compelling story;
    4) *Step 4*: Then, carefully compile your fundraising information (e.g., how long a campaign will run), and set the reward levels;
    5) *Step 5*: The last, but not least, is to implement a strategic social media plan, e.g., informing the appropriate people before the official launch.

Generally speaking, to successfully crowdfund, an asker must offer both a product that is appropriate for crowdfunding and an online campaign organized in the form of text, images and videos that matches the product (Giles, 2012).

Yet, Hobbs et al. (2016) pointed out that these general preparations do not actually address what might impact each specific campaign the most. Following this logic, Yuan et al. (2016) suggested that project goal-setting is one key factor in terms of leveraging a crowdfunding campaign's success because the 'All-or-Nothing' rule is the most adopted policy (Giudici et al., 2013) in a real crowdfunding environment. The implication of such a principle is that the fundraising campaign is forced to terminate and the project is regarded as a failure if an insufficient amount of funds (typically asker pre-determined) is collected before the due date. To put it simply, most projects fail because their original fundraising goals are not achieved.

In addition, Parhankangas and Renko (2017), Yuan et al. (2016), Wang et al. (2017) and Mitra and Gilbert (2014) focused on the descriptions of projects (e.g., linguistic style and textual features) that may affect a fundraising campaign's success. Other features that predict a crowdfunding project's success include the campaign website (Lacan and Desmet, in press), project categories (Cordova et al., 2015; Hobbs et al., 2016; Roma et al., 2017; Renwick and Mossialos, 2017; Bao and Huang, 2017; Cha, in press; Gamble et al., 2017; Cumming et al., 2017), type of rewards (Bao and Huang, 2017; Bi et al., 2017; Chertkow and Feehan, 2014), quality of the product (Mollick, 2014; Zheng et al., 2014), fundraising amount (Chung, 2014; Buff and Alhadeff, 2013; Frydrych et al., 2014) and askers' characteristics, such as gender (Marom et al., 2014), geography (Mollick, 2014), personal information (Boeuf et al., 2014; Frydrych et al., 2014) and photograph (Colombo et al., 2015).

Moreover, the effects of whether a project is connected with other social networks, such as Twitter or Facebook (Giudici et al., 2013; Belleflamme et al., 2014; Schwienbacher and Larralde, 2010; Lu et al., 2014; Bechter et al., 2011), or whether a project involves a video (Wheat et al., 2013) have been investigated as well.

- **During-Fundraising Phase:** During the actual campaign running period, several researchers drew on a variety of theories and models in order to explain how to promote the fundraising campaigns, including social capital theory (Zheng et al., 2014; Bao and Huang, 2017; Kromidha and Robson, 2016; Giudici et al., in press), signalling theory (Kunz et al., 2017; Courtney et al., 2017), communication theory (Wu et al., 2015), language expectancy theory (Parhankangas and Renko, 2017) and persuasion theory (Bi et al., 2017; Allison et al., 2017). In addition, similar to selling a product via an online platform, there is a clear link between management endeavours and marketing efforts devoted to a project promotion and its final success. In light of this, Hobbs et al. (2016) focused on how to manage campaigns in order to meet the target financial goal. Additionally, Kraus et al. (2016) discussed the strategies for reward-based crowdfunding projects, whereas Mollick (2014), Kuppuswamy and Bayus (2013), as well as Xu et al. (2014), observed that projects with frequently updated information are more likely to attract funding from the crowd than other unattended projects. Agrawal et al. (2014), Colombo et al. (2015) and Ordanini et al. (2011) pointed out that early contributions played a crucial role in determining crowdfunding projects' eventual successes. Xu et al. (2016) observed that the backers' satisfaction degree is also very useful in measuring projects' success.

- **Post-Fundraising Phase:** As the name implies, a crowdfunding campaign is terminated at this phase, however, it does not necessarily mean that an asker's journey is over as well. In fact, the follow-up stage is on the same scale as the other two parts of a campaign (i.e., activating and engaging). As crowdfunding includes an open innovation aspect, delivering feedback to the askers is important (Kuppuswamy and Bayus, 2013; Ordanini et al., 2011; Stanko and Henard, 2017). However, crowdfunding literature remains largely unclear in this regard, in spite of the fact that the post-fundraising phase means a great deal to a business, particularly for pre-selling and idea-validating projects (Cao, 2014; Sheldon and Kupp, 2017; Huizingh, 2011). In general, once a campaign is complete, project creators need to take care of the following two things:
  1) *First*, askers should review the feasibility of their delivery plan. According to Mollick (2014), the majority of Kickstarter-funded projects (roughly 75%) deliver only after considerable delays, and some (approximately 4%) even fail to deliver at all;
  2) *Second*, clear and open ongoing communication is necessary, since backers often prefer experiencing the entire product development process (Agrawal et al., 2014). In addition, Roma et al. (2017) also highlighted that patents and entrepreneur social capital are important in accessing further funding from professional investors.

To summarize, what are the characteristics that distinguish a truly successful crowdfunding campaign from the rest? One answer is, perhaps, the fact that only having million-dollar funds is not enough (Brown et al., 2017; Roma et al., 2017). What carries more weight is whether a product has been delivered and how the company's next steps of growth are planned towards profitability.

6.1.2 *Determinants of Success and Failure: Donation-Based Crowdfunding Campaign*

In practice, donation-based crowdfunding is also called charitable crowdfunding. This refers to an alternative way to raise money for a charitable project by asking a large number of backers (or named donors) to donate a small amount to it (Australian Charities and Not-for-Profits Commission, 2017; Belleflamme et al., 2015). Examples include films and performing arts (Mishra, 2014), environment protection (Hörisch, 2015; Ings, 2015), medical care (Berliner and Kenworthy, 2017) and emergency fundraising (Sisler, 2012). Unlike reward-based crowdfunding, research on donation-based crowdfunding is relatively weak, which means we still only have a limited understanding of which factors influence donors' behaviour the most.

In fact, individual donations to charities are not new. Previous studies have suggested numerous factors that determine whether backers attract charity appeal (giving some money). A non-exhaustive list of these variables is outlined as follows:

- Intrinsic motives (e.g., altruism and public recognition) underlying various charitable donation activities (Webb et al., 2000; Teah et al., 2014; Smith and McSweeney, 2007);
- Extrinsic factors which include backers' demographic (Andreoni et al., 2003; Chang and Lee, 2011; Dvorak and Toubman, 2013);
- Cultural factors (Stojcic et al., 2016);
- Socio-economic factors (Burgoyne et al., 2005; Roberts and Roberts, 2012);
- Situational characteristics factors (Andreoni and Payne, 2011; Rai et al., 2017); and
- Organisational motivation factors (Hankinson, 2001; Michel and Rieunier, 2012).

More details about motivational factors underpinning traditional charitable giving are found in (Nguyen, 2012; Bekkers and Wiepking, 2011; Bock et al., in press).

Compared with traditional fundraising behaviours, online donation-based crowdfunding opens another door to observing a new norm around how backers behave towards a donation. More specifically, Gleasure and Feller (2016) summarized the following three main differences:

- **Anonymity:** Most charitable donations made using crowdfunding platforms are anonymous;
- **Social Connection:** Though some crowdfunding donations can be predicted by considering geography and social proximity factors, the social connections between donors and fund-seekers for most donation cases are hard to identify; and
- **Proactivity:** Charitable donors on crowdfunding platforms tend to be more proactive, rather than waiting passively for donation requests to emerge.

In addition, online donation-based crowdfunding also provides a speedy and easy way to offer immediate financial assistance, and generates instant awareness for the mission (BBB WGA, 2013; Nuwer, 2015). Further discussions about the differences between donation-based crowdfunding and traditional fundraising can be found in (Ania and Charlesworth, 2015).

Askers' behaviour can also make a significant contribution to the success of donation-based fundraising. As claimed by Sanders (2017), a big part of donation-based fundraising's failure stems from a dysfunctional relationship between askers and backers. Grace Garey, Watsi's co-founder, has also indicated that "people don't just want to know where their money goes—they expect to meet the people who benefit" (Nuwer, 2015). Unfortunately, Corazzini et al. (2015) pointed out that many askers tend to offer the same fundraising motivation in resembling tones and manners alike. Thus, it is no surprise to see that many of these askers are struggling to get sufficient funds.

To resolve these issues, various attempts were made in the literature: Dragojlovic and Lynd (2016) conducted a study in which an online survey of prospective backers was performed in order to find out which kind of biomedical research projects are most expected to appeal to backers. In a similar vein, Krittanawong et al. (2018) focused on cardiovascular research and pointed out that crowdfunding campaigns are more suitable for young research scholars who have small pilot projects and want to raise a small sum of funds in order to support their project during training. Meanwhile, Meer (2014) used the data from DonorsChoose. DonorsChoose is an online platform that examines the impact of price of giving. In addition, Dune (2013) investigated how museums cultivate their online millennial backers. The research findings suggested that social media (e.g., Twitter and Facebook) is a key tool in creating a successful campaign. Furthermore, Agerström et al. (2016) focused on the feasibility of utilizing descriptive norms in order to increase charitable giving. More in-depth relevant studies have been done by Zhong and Lin (2017), Gao (2016) and Korolov et al. (2016), in which they concluded that social pressure can be used as a key ingredient in predicting online donation strategy. Finally, a book authored by Miller (2013) gives a comprehensive view on how to make a good story, engage with one's community and raise more money.

6.1.3 *Determinants of Success and Failure: Peer-to-Peer (P2P)-Based Crowdfunding Campaign*

In the literature, peer-to-peer (P2P) lending usually includes two main forms (other alternatives can be found in Chapter 9) (Groshoff, 2014): (1) repayment of loan with interest, i.e., marketplace lending, and (2) socially motivated lending that is interest free, i.e., microfinance/microcredit or social lending. Although both forms can be viewed as a means of fundraising, the key factors that influence the campaign's fate differ from each other (Bruton et al., 2014).

- **Marketplace Lending:** In the marketplace-lending model, borrowers (askers) are project owners and backers are lenders. On platforms driven by this model, askers (individuals or small businesses) have to set up their loans (e.g., student loan, personal loan, mortgages, car finance, invoice, equipment, etc.), often called listings. Furthermore, backers decide under what conditions they are willing to finance these listings. The winning lenders are required to fund the target loan(s) upon the auction-determined interest rate. Thanks to various advanced technologies (e.g., artificial intelligence and big data analytics) and social media, marketplace lending has grown quickly in recent years. In the US alone (i.e., Prosper and Lending Club), loans originating from marketplace lenders have grown from $161 million in 2010 to $2,660 million on 30 September 2014 (KPMG, 2015).

However, empirical studies suggest that the return on investment of this model has a poor record. For example, of over 293,976 launched listings in Prosper from May 2006 to August 2008, Puro et al. (2010) observed that only a handful survived, i.e., 8.4% listings have actually been funded. Even the Lending Club, the most hyped 'success', the theoretical predicted interest rates are significantly higher than the platform's actually charged interest rate (Emekter et al., 2015). In other words, investment is not compensated adequately for the risks taken. In addition to failing investment, askers also find funding elusive. Backers remain skittish about the listings, and attracting their attentions is onerous. Therefore, different sources of data need to be collected and thoroughly analyzed in order to find out what the determinants of success are.

In general, from an asker's point of view, we can identify the following four general hurdles:

1) *First*, askers are relatively unprofessional and, thus, have difficulties in setting their loan requests properly, such as starting interest rate, loan period and loan amount (Puro et al., 2010);
2) *Second*, askers often suffer from a trust deficit (Zhang et al., 2014; Chen et al., 2014);
3) *Third*, the asymmetric information effect is particularly amplified in the marketplace lending domain (Lin et al., 2013); and
4) *Finally*, the local financial environment where askers reside may be jointly impaired as well (Hortaçsu et al., 2009; Ordanini et al., 2011; Agrawal et al., 2011; Lin and Viswanathan, 2016).

As a result, in order for askers to get ideal loans, they must optimize their strategic options (Puro et al., 2010), make the loan process more efficient (Wu and Xu, 2011) and maintain their credit scores (Emekter et al., 2015). To this end, a set of studies were conducted by Lin et al. (2017), Gonzalez and Loureiro (2014), Harkness (2016) and Lin et al. (2011), respectively, in which they examined which borrower's personal characteristic affects the default rate most; another group of studies performed by Dorfleitner et al. (2016), Larrimore et al. (2011), Herzenstein et al. (2011) and Michels (2012), respectively, indicated that borrowers' narratives (e.g., loan's description) have a significant effect on the funding's success. Moreover, Durguner (in press) investigated the importance of various borrower-lender relationship variables, whereas Lin et al. (2013) studied whether borrowers' social networks (i.e., the association that borrowers maintain) influence the success rate of their loan requests. Similar studies can also be found in (Chen et al., 2016; Freedman and Jin, 2014; Ge et al., 2016; Ge et al., 2017; Freedman and Jin, 2017; Castillo et al., 2014). In addition, Khang and Byers (2017) took loan types into account in order to verify the hypothesis that the efficacy of finance firms is higher than banks in terms of evaluating lower quality borrowers and securing their much needed loans. Meanwhile, several studies evaluated borrowers' pictures and argued that there is a significant racial discrimination phenomenon in existence. Some of them even claimed that facial features (e.g., trustworthy) also influence the funding's success (Gonzalez and Loureiro, 2014; Ravina, 2012; Duarte et al., 2012; Pope and Sydnor, 2011).

Overall, the factors that affect the funding success of marketplace-type lending can be grouped into four categories (Han et al., 2018): (1) loan characteristics (e.g., loan rate, amount and duration), (2) asker's personal information (e.g., credit level, age and gender), (3) asker's voluntary information (e.g., photograph), and (4) asker's soft information (e.g., social network).

- **Microfinance:** More recently, a microfinance mechanism, devised by Yunus (2010) in the mid-1970s, has proved its effectiveness in expanding formal credit to the poor population. In the literature, the usefulness of increasing economic opportunities via credit in the context of developing marketplaces is broadly recognized (B. A. de Aghion et al., 2007). In real-world situations, low-income families tend to lack access to official banking systems and the associated services (Asgary and McNulty, 2017). Therefore, the concept of microfinance is to provide the financial services to these households. Notable applicable developing nations include Brazil (Agier and Szafarz, 2013), India (Parwez et al., in press; Ashta et al., in press; Maitra et al., 2017), China (Hsu, 2016; Hsu, 2014) and South Africa (Baiyegunhi and Fraser, 2014; Chisasa, 2014; Gininda et al., 2014; Kim et al., 2007; Pronyk et al., 2008; Ward and Oladele, 2013). Some developed countries, e.g., Canada (Emery and Ferrer, 2015), are also actively exploring the feasibility of using the microfinance method in order to finance immigrants who are often less competitive in obtaining credit. The key difference here (Morduch and Haley, 2002) is that developing countries focus more on financing small business setups or equipment purchases, while developed countries pay more attention to training migrant workers. Further discussion in this regard can be found in (Milana and Ashta, 2012).

However, no system is perfect. Asymmetric information (or called adverse selection) (Marwala and Hurwitz, 2017a) is also the main obstacle encountered by microcredit institutions (Ahlin and BrianWaters, 2016; Yum et al., 2012). Recently, findings from the empirical literature concluded that individual lending should consider this factor as well (Jeon and DomenicoMenicucci, 2011; Ahlin and BrianWaters, 2016; Widiarto et al., 2017). To tackle this problem, some available solutions are listed below:

1) One option is to take a more precise measurement of the borrower's risk profile (Hernandez and Torero, 2014). In this regard, Dorfleitner and Oswald (2016) studied the determinants of the repayment behaviours of borrowers while Jenq et al. (2015) focused on the implications of borrowers' appearances.
2) Another is to disburse loans to groups of borrowers (often women) with peer monitoring (Stiglitz, 1990; Besley and Coate, 1995) and joint liability (Duggan, 2016; Ghatak and Guinnane, 1999; Chowdhury et al., 2014). In this regard, Schaaf (2013) examined the effects of team membership and Chen et al. (2017) investigated the impact of team competition on P2P microfinance activity.
3) The third option is to introduce a higher level of transparency. For example, Burtch et al. (2014) considered the distance of geography and the differences in culture in their study. They also tried to check the dual roles of these two factors on affecting lenders' decisions when selecting borrowers. Similarly, the transaction costs and social distance factors were covered by Meer and Rigbi (2013) in order to examine the hidden causations.

In the domain of P2P lending, microfinance has grown from a traditional philanthropic movement into a more innovative and globalized industry. Famous examples include Kiva (2017), Lendwithcare (2017) and Babyloan (2017). Moreover, with the popularity of various online P2P lending platforms, microfinance has evolved into several alternatives, such as online social investment which offers a 1% to 3% return on the funds involved (e.g., MicroPlace), and 'direct' social lending which facilitates loans between people who already know each other (e.g., Virgin Money). Through a comparison between traditional and online P2P-type microfinances, we can identify the following differences:

1) Borrowers and lenders are directly connected to an online P2P microfinance environment.
2) Meanwhile, they also differ in terms of lending strategies (e.g., directly matching) and methods of linking people (e.g., web 2.0).
3) Furthermore, one remarkable difference is the borrowers' characteristics. Borrowers in conventional microfinance markets tend to be poor and self-employed (Khavul, 2010); while in online microfinance scenarios, disadvantaged customers and poor entrepreneurs are dominating borrowers (Freedman, 2000; Savarese, 2015).

4) Another significant difference is social trust. Traditional microfinance institutions typically rely on the group relationships (e.g., peer monitoring and social collaterals) in order to overcome the asymmetric information problem, while new online askers are often tied to their social networking (e.g., blogs, peer' reviews, and peer' communities), which is helpful for reducing costs of default. Therefore, online askers can acquire loans with lower interest rates, fewer transaction costs and a larger operation of scale (Hartley, 2010) than ever before.

6.1.4 *Determinants of Success and Failure: Equity-Based Crowdfunding Campaign*

In recent years, equity-based crowdfunding has received academic attention because of its rapid development. Nevertheless, Lukkarinen et al. (2016) noticed that most campaigns within this category end up failing. To explain why this type of crowdfunding often fails, scholarly publications cover the following aspects:

- Belleflamme et al. (2014) and Cholakova and Clarysse (2015) argued that equity-based crowdfunding differs from other crowdfunding forms, therefore, the participated investors are also essentially different.
- Vulkan et al. (2016) and Krupa and Żołądkiewicz (2017) explained that equity-based crowdfunding tends to be much more technically complicated and expensive (e.g., restrictive legal requirements and due diligence processes) than other forms of funding.
- Ahlers et al. (2015) suggested that start-ups and small and medium enterprises (SMEs) within in this category should find more solid ways to signal their value clearly to potential investors.
- Regulation was regarded by Collins and Pierrakis (2012), and Zhu and Zhou (2016) as one of the main obstacles hindering the success of this model.
- Other interpretations include the existence of information asymmetry (Moritz et al., 2015; Agrawal et al., 2014), herding effect (Vismara, 2016b; Mohammadi and Shafi, in press) and fraud detection (Kantor, 2014; Huang and Zhao, in press).

Nevertheless, entrepreneurs are eager to adopt equity-based crowdfunding despite the low success rate and various unresolved difficulties. The underlying reason is that the equity stake is far more flexible (e.g., low barriers to entry) and convenient (e.g., high speed) for early-growth or new product development than any other means of investment (Collins and Pierrakis, 2012; Zhu and Zhou, 2016). As

Lukkarinen et al. (2016) put it in their study, the key to understanding failure is finding critical success signals for growth.

To this end, many authors tried to analyze the determinants of success in this domain. Among them, Medziausyte and Neugebauer (2017) found that the determinants could include previous crowdfunding history, equity participation, field of business, companies' life cycle phase and social media presence. Block et al. (in press) studied the impact of information updating during an equity-based crowdfunding campaign. Vismara (2016a) showed that equity retention influences the success of this crowdfunding category. Mohammadi and Shafi (in press) attempted to understand whether or not gender-related differences exist in the behaviour of investors. Mamonov et al. (2017) analysed the Title II equity (Freedman and Nutting, 2015) crowdfunding success. Hornuf and Neuenkirch (in press) examined the auction mechanism in the equity-based crowdfunding in order to analyse backers' preparedness to pay for cash flow rights. Vulkan et al. (2016) investigated the data from Seedrs, one of the leading UK equity crowdfunding platforms, and uncovered that starting strong and having many backers, in particular at least one backer who provided a large pledge, are the key factors leading to final success. Moreover, Vismara (2016b) confirmed a similar hypothesis, i.e., the earlier the more successful. The study concluded that in the early days of an equity-based crowdfunding campaign, the involvement of public profile investors is crucial for attracting other types of investors. In addition, Ahlers et al. (2015) suggested that a campaign's success in this category is related to a couple of well-tuned start-up features, such as the structure of the board, risk elements and the plotted exit strategies. Lukkarinen et al. (2016) thoroughly examined various campaign factors associated with equity crowdfunding. These included the number of investors, amount of raised funds, team grading, markets grading, concept ranking, terms grading, state grading, average grading, fundraising target, minimum amount of investment, the duration of the campaign, availability of financials, early funding raised from private networks and linkage to social media networks. Their results suggested that the role of conventionally used investment decision criteria (in the context of venture capital and angel investing) becomes somewhat irrelevant in equity crowdfunding scenarios. Instead, the factors carrying more weight are pre-determined campaign features and the usage of social networks (both private and public).

6.2 Problem Statement

A summary of the previous introductory section gives a rather complicated crowdfunding landscape (see Fig. 6.1).

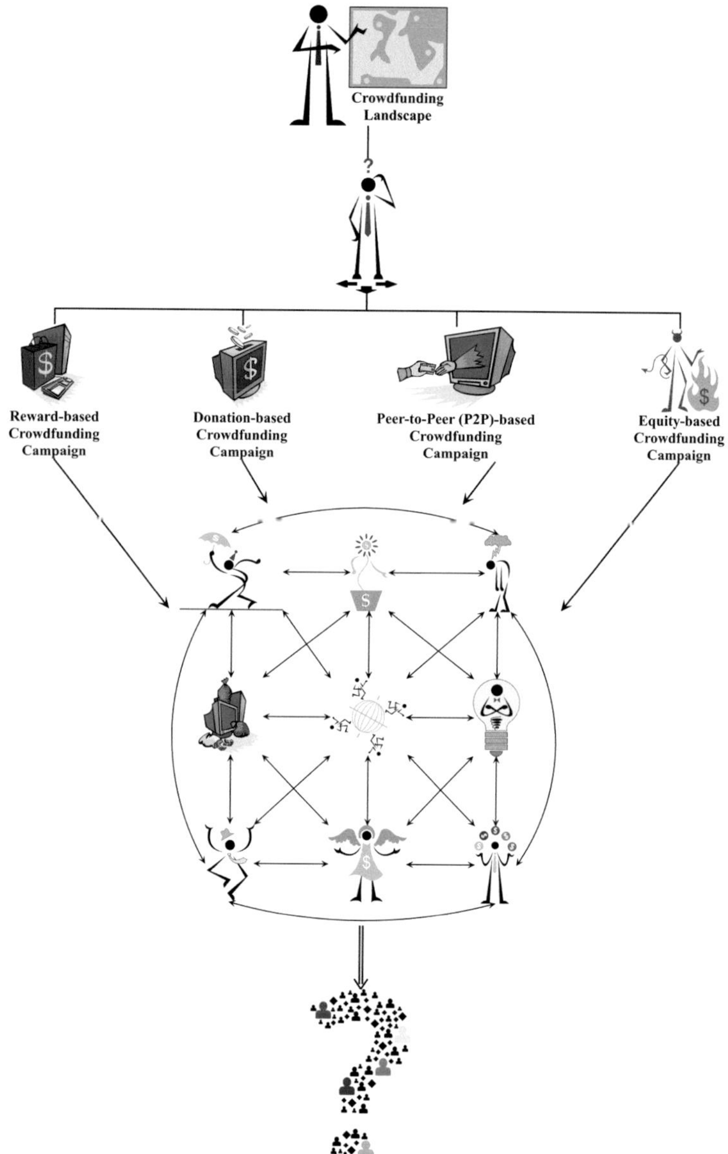

Figure 6.1: Landscape of crowdfunding from an asker's perspective.

Using human robot interaction scenario (Xing and Marwala, 2018n; Marwala and Hurwitz, 2017b) as an example, in the context of the Fourth Industrial Revolution (Xing and Marwala, 2017), robots are rapidly marching towards us (Xing and Marwala, 2018a), by not only forming ubiquitous robotic systems (Xing, 2016b; Xing and Marwala, 2018k) and realizing pervasive intelligence (Xing and Marwala, 2018i), but also diversifying and spreading into many areas, such as ambient assisted living (Xing, 2016e; Xing, 2016f), human physiological sensing (Xing and Marwala, 2018j), extra-personal communication (Xing and Marwala, 2018h), energy autonomy (Xing and Marwala, 2018c), adaptive control (Xing and Marwala, 2018b), assistive technology devices (Xing and Marwala, 2018d), remanufacturing (Xing and Marwala, 2018f; Xing and Gao, 2014a; Marwala and Xing, 2011; Xing and Gao, 2015; Xing, 2014), design automation (Xing and Marwala, 2018e; Xing, 2016d), human robot collaboration (Xing and Marwala, 2018g) and smart maintenance (Xing and Marwala, 2018m; Xing, 2015b; Xing, 2015a).

Let us now borrow the blue ocean philosophy (Kim and Mauborgne, 2015; Kim and Mauborgne, 2005) and assume that the robot market universe consists of two types of oceans: the red one and the blue one. Essentially, red ocean represents all the relevant industries that currently exist today, i.e., the known marketplace. Blue ocean denotes all the relevant industries that do not yet exist, i.e., the unknown marketplace. To this end, in the 'red' robot market, industry standards and borders (e.g., industrial robot, field robot, etc.) are often determined and followed, and the competition rules and regulations of the game are typically acknowledged. Under this circumstance, in order to grab a bigger robot market share, firms have to try very hard to outperform their competitors. When the marketplace becomes more crowded, competition turns the 'red' side into a free-for-all, while on the 'blue' robot market side, the marketplace is mostly untapped. The creation of blue oceans may be either well beyond the existing confinement, or within red oceans themselves but with the existing walls expanded. The utmost good news for blue robot markets is that competition is inessential since the rules and regulations of the game are largely unsettled.

When faced with such a complicated terrain but deciding to resort to crowdfunding, this red and blue principle is applicable as well. Other similar domains also include mobile payment (Xing, 2017b), visible light communication (Xing, 2017c), advanced manufacturing system (Xing et al., 2006), software development (Xing, 2017a) and Internet of things (Xing, 2016a; Xing, 2016c). The asker's effective management capability is essential in order to achieve positive differentiation in the minds of backers. To achieve this, an asker needs to overcome some of the following challenges (Michaelidou and Micevski, 2015; Bennett and Barkensjo, 2005; Venable et al., 2005):

- Differentiate the existing market and identify a target market;
- Communicate his/her value proposition;
- Establish a sense of attraction or urgency;

- Offer a higher potential incentive for backers to pitch in;
- Secure and simplify the transaction; and
- Keep backers that continue to contribute (i.e., backer retention).

In this regard, Aido campaign (by Ingen Dynamics Inc.) has done a great job. As the next generation social robot, Aido is family friendly and uniquely mobile. The smartness and interactivity make Aido the first-ever social robot that goes around one's home/office environment in order to improve the owner's lifestyle. The function of Aido robot includes playing with children, helping with household chores, planning schedules, patrolling the house, and so on. Equipped with all this uniqueness, this project was launched on the Indiegogo crowdfunding platform and closed with a total raised funds of $888,881 in pre orders, that is 594% funded on 24 April 2016 (Ingen Dynamics Inc., 2016).

6.2.1 *Question 6.1*

Based on the above observations, Question 6.1 is proposed as follows:

- **Question 6.1:** *How do we build a predictive model using smart computing techniques that can not only predict a campaign's success rate, but also approximate its fundraising range?*

By investigating Question 6.1, our expectations are threefold: (1) Be able to follow the evolution trajectory of both projects and backers, (2) Be able to demonstrate suitable technique(s) for predicting the result of a project and the optimal sought-after funding levels, and (3) Be able to offer intelligent search and ascertaining strategy according to a project's time series patterns.

6.3 Ensemble Learning for Question 6.1

In practice, there is a legion of intelligent algorithms (Xing and Marwala, 2018l; Marwala, 2009; Xing and Gao, 2014b; Xing and Gao, 2014aa; Xing and Gao, 2014bb; Xing and Gao, 2014c; Xing and Gao, 2014cc; Xing and Gao, 2014d; Xing and Gao, 2014dd; Xing and Gao, 2014e; Xing and Gao, 2014f; Xing and Gao, 2014g; Xing and Gao, 2014h; Xing and Gao, 2014i; Xing and Gao, 2014j; Xing and Gao, 2014k; Xing and Gao, 2014l; Xing and Gao, 2014m; Xing and Gao, 2014n; Xing and Gao, 2014o; Xing and Gao, 2014p; Xing and Gao, 2014q; Xing and Gao, 2014r; Xing and Gao, 2014s; Xing and Gao, 2014t; Xing and Gao, 2014u; Xing and Gao, 2014v; Xing and Gao, 2014w; Xing and Gao, 2014x; Xing and Gao, 2014y; Xing and Gao, 2014z; Xing and Gao, 2014ff; Xing et al., 2010a; Xing et al., 2013) that can be used for classification tasks. Among them, the simplest and the most popular data classification approaches is Naive Bayes (NB). In NB, a joint model can be constructed by incorporating a D-dimensional input vector (denoted by $\mathbf{x}$) and the associated class label (represented by c) as given by Eq. 6.1 (Barker, 2012):

$$\Pr(\mathbf{x},c)=\Pr(c)\prod_{i=1}^{D}\Pr(x_i|c). \quad 6.1$$

With the aid of a careful selection for each conditional distribution, denoted by $\Pr(x_i|c)$, a classifier can be formed based on Bayes' rule for an input vector (denoted by $\mathbf{x}^*$), as is given by Eq. 6.2 (Barker, 2012):

$$\Pr(c|\mathbf{x}^*)=\frac{\Pr(\mathbf{x}^*|c)\Pr(c)}{\Pr(\mathbf{x}^*)}=\frac{\Pr(\mathbf{x}^*|c)\Pr(c)}{\sum_c \Pr(\mathbf{x}^*|c)\Pr(c)}. \quad 6.2$$

where only two classes are considered, i.e., domain(c) = {0,1}.

Although NB is indeed a powerful tool for classification, it can sometimes become too zealous to deal with small data counts scenarios. If an individual attribute (denoted by i) has no counts for a class (represented by c), then NB classifier will always yield an output of $\mathbf{x}$ not belonging to class c, irrespective of the remaining attributes. The underlying reason lies in the fact that a product between zero and anything else is unchanged. In order to tackle the so-called overconfidence effect, a simple Bayesian approach can be introduced. Let $\mathcal{D}=\{(\mathbf{x}^n,c^n),n=1,\dots,N\}$ represent a dataset, then an input's class can be predicted using Eq. 6.3 (Barker, 2012):

$$\Pr(c|\mathbf{x},\mathcal{D})\propto\Pr(\mathbf{x},\mathcal{D},c)\Pr(c|\mathcal{D})\propto\Pr(\mathbf{x}|\mathcal{D},c)\Pr(c|\mathcal{D}). \quad 6.3$$

For simplicity, we can use maximum likelihood method to set $\Pr(c|\mathcal{D})$ using Eq. 6.4 (Barker, 2012):

$$\Pr(c|\mathcal{D})=\frac{1}{N}\sum_n \mathbb{I}\left[c^n=c\right]. \quad 6.4$$

An in-depth discussion regarding NB and well-known variants, say, Bayesian NB and tree-augmented NB, can be found in (Barker, 2012). One of the latest versions is called rough Gaussian NB classifier (Babu et al., 2017), in which the concept drift issue found in a data streaming scenario was addressed using rough set theory, and then Gaussian NB was modified in order to deal with data dynamics.

In general, ensemble learning denotes learning a weighted combination of base models, as defined by Eq. 6.5 (Murphy, 2012):

$$f(y|\mathbf{x},\pi)=\sum_{m\in\mathcal{M}} w_m f_m(y|\mathbf{x}). \quad 6.5$$

where w_m represents tuneable parameters. Since each based model f_m can get a weighted opportunity to vote, ensemble learning is essentially a committee method. To some extent, we can also claim that artificial neural network (ANN) belongs to ensemble learning category as well and can be used in different domains (Xing et al., 2010b), where the mth hidden layer is denoted by f_m, and the corresponding weights of the output layer are represented by w_m. In other words, a simple ANN is associated to linear combinations of various learners as given by Eq. 6.6 (Alpaydın, 2010; Alpaydın, 2014):

$$y_i = \sum_j w_j d_{ji} \quad \text{where } w_j \geq 0, \ \sum_j w_j = 1. \qquad 6.6$$

A simplified illustration corresponding to this formulation is provided in Fig. 6.2, in which d_j denotes various basic learners, some function $f(\cdot)$ is employed to combine basic learners' outputs. Please note that only a single input and output are considered in this graph.

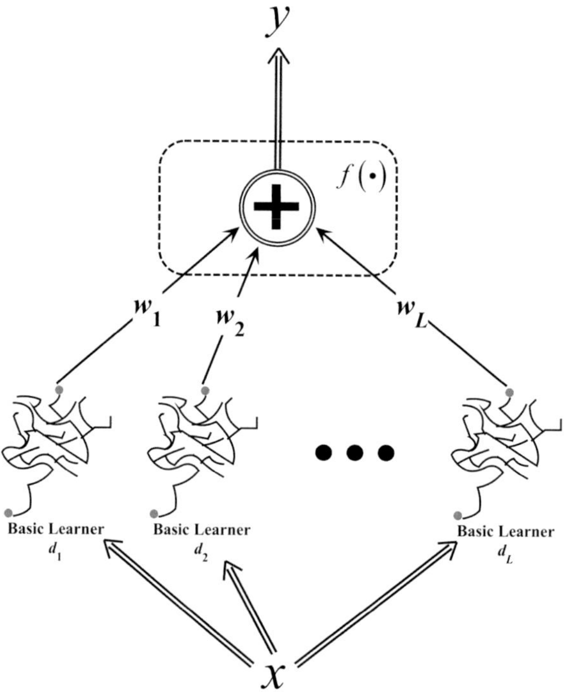

Figure 6.2: Schematic representation of linear combination of multiple learners.

In practice, there are distinct means to combine the multiple learners for the purpose of generating the final output (Alpaydın, 2010; Alpaydın, 2014):

- **Combination via Multiexpert:** In this class, basic learners can work together in parallel according to two strategies: (1) global strategy (i.e., learner fusion), given an input, a set of outputs is delivered by all involving learners and all these outputs are used to generate the final output. Voting and stacking schemes are representative examples. (2) local strategy (i.e., learner selection), in which a gating mechanism is introduced in order to select a limited number of learners to take charge of output generation.
- **Combination via Multistage:** Its serial nature, in which the subsequent learners are only deployed on cases where the previous learners do not perform well enough, often features this category. The core concept is sorting the learners into ascending order based on their complexities, and a complex learner is not employed until simpler learners are diffident. Cascading scheme is a typical example in this regard.

Although these combined models often share the same drawback of being quite difficult to analyse, consisting of many (dozens or hundreds) individual learners and, thus, making it rather hard to separate factors in terms of their corresponding contributions, some advancements have been made recently. Some forms of ensemble learning are described in detail below (Marwala, 2012; Marwala, 2013a; Marwala, 2013b).

- **Bagging:** In order to combine the decisions associated to different models, various outputs have to be mixed together in order to form a single prediction. The easiest way to finalize this is to take a vote, in the case of classification, or to compute the average if numeric prediction is involved. Bagging adopts this strategy by assigning equal weight to different models. In a cost-sensitive classification domain, bagging is a competitive option due to its accurate probability estimations. Unfortunately, bagged classifiers are notoriously difficult to analyse (Rokach and Maimon, 2015; Alpaydın, 2010; Alpaydın, 2014; Witten and Frank, 2005; Witten et al., 2011).
- **Wagging:** As a variant of bagging (Bauer and Kohavi, 1999), though each classifier in wagging is trained against the whole training dataset, a weight is stochastically assigned to each instance. In fact bagging can be regarded as wagging when weights are allocated from the Poisson distribution (Rokach and Maimon, 2015).
- **Randomization:** In general, a diverse ensemble of classifiers can be created with the bagging mechanism (via the introduction of randomness into the input of the learning algorithm) which is often coupled with superior performance. However, there are alternative options for creating diversity, say, using a built-in randomization module. The backpropagation algorithm, through assigning small randomly selected values to the network weights, facilitates

multilayer perceptron learning. Therefore, one way to improve the classifier's outcome stability is to first run the learner a couple of times under distinct random number choices and then fuse the predictions of classifiers using voting or averaging operation. Although similar results can often be obtained through either bagging or randomization, it is worth paying attention to the following issues: More randomness induces more variety in the learner but keeps the data underutilized, this can decrease each individual model's accuracy; the way we introduce randomness by two mechanisms may be differentiated or complemented. More pros and cons regarding randomization versus bagging can be found in (Witten et al., 2011).

- **Boosting:** Intuitively, the spirit of combining multiple models is best realized when the participating models share different strengths and weaknesses. Ideally, they can complement each other. Based on such insight, boosting methods attempt to seek models that meet this need (Witten et al., 2011).
  1) *Similarities*: Like bagging, boosting also makes use of voting or averaging operations in order to combine each individual model's output. Meanwhile, boosting also combines the same category of models as bagging does.
  2) *Dissimilarities*: Unlike bagging, where each model is constructed independently, every newly introduced model is impacted by the incumbent models' performance. In other words, boosting encourages newcomers to focus on existing inappropriately handled instances by allotting heavier weights. Finally, a model's contribution in boosting is weighted by its confidence rather than by equally distributed weight.
- **Additive Regression:** There has been a long tradition of using additive models in statistics. Briefly, any means of producing predictions by summing up other models' contributions is termed as additive regression. In practice, boosting is often reformulated as a greedy algorithm by implementing a forward stepwise additive modelling scheme. At the beginning, these variants often consist of an empty ensemble. They then gradually and sequentially absorb new model members. At each step, only the model that can maximise the whole ensemble's predictive performance is accepted under the condition of keeping the ensemble's existing models unaltered (Witten et al., 2011).
- **Stacking:** Another way of combining multiple models is called stacked generalization (or 'stacking'). In general, an intuitive way of calculating the weights defined in Eq. $f(y|\mathbf{x}, \boldsymbol{\pi}) = \sum_{m \in \mathcal{M}} w_m f_m(y|\mathbf{x})$ is through Eq. 6.7 (Murphy, 2012):

$$\hat{\mathbf{w}} = \arg\min_{\mathbf{w}} \sum_{i=1}^{N} L\left[y_i, \sum_{m=1}^{M} w_m f_m(\mathbf{x})\right]. \quad 6.7$$

Unfortunately, this operation will lead to the overfitting problem, i.e., for most complicated models, the value of w_m might be too large. The easiest solution to address this issue is to introduce cross-validation as given by Eq. 6.8 (Murphy, 2012):

$$\hat{\mathbf{w}} = \arg\min_{\mathbf{w}} \sum_{i=1}^{N} L\left[y_i, \sum_{m=1}^{M} w_m \hat{f}_m^{-i}(\mathbf{x})\right]. \quad 6.8$$

where $\hat{f}_m^{-i}(\mathbf{x})$ represents the predictor acquired from training against dataset after eliminating $(\mathbf{x}_i, y_i)$. This process is the so-called stacking, i.e., stacked generalization (Wolpert, 1992). Unlike its peer strategies, e.g., bagging and boosting, stacking is applied in order to combine models formed by varied learning algorithms. However, stacking is less popular in practice than other strategies due to the inherent difficulty in terms of theoretical analysis. Further explanations in this regard can be found in (Witten et al., 2011). However, one noteworthy application is the Netflix team that used this approach to tie the winning team's submission with respect to accuracy (Sill et al., 2009).

- **Error-Correcting Output Codes:** Another interesting construct of ensemble learning is ECOC (i.e., error-correcting output codes) (Dietterich and Bakiri, 1995), which is often employed in order to address the multi-class classification issue. Suppose that a symbol (i.e., the class label) with C possible state needs to be decoded, a bit vector with the length of $B = \lceil \log_2 C \rceil$ can be introduced to encode the class label, then we can use different binary classifiers to train B for predicting each bit. The decoding rule can be formulated as Eq. 6.9 (Murphy, 2012):

$$\hat{c}(\mathbf{x}) = \min_{c} \sum_{b=1}^{B} \left| Codeword_{cb} - \hat{p}_b(\mathbf{x}) \right|. \quad 6.9$$

  where a Codeword's bth bit for class c is denoted by $Codeword_{cb}$.

- **Ensemble Learning ≠ Bayes Model Averaging:** Another way of selecting the best model and then making predictions based on this is to perform a weighted averaging option for each model's predictions, as given by Eq. 6.10 (Murphy, 2012):

$$\Pr(y \mid \mathbf{x}, \mathcal{D}) = \sum_{m \in \mathcal{M}} \Pr(y \mid \mathbf{x}, m, \mathcal{D}) \Pr(m \mid \mathcal{D}). \quad 6.10$$

This process is termed as Bayes model averaging (or BMA for short). Since it is often impractical to perform an averaging operation over all models, an easy solution is to sample a few models from the posterior (Murphy, 2012). Meanwhile, it is crucial to bear in mind the difference between BMA and ensemble learning (Minka, 2002). In brief, ensemble learning tries to enlarge the model space by defining a single new model, which is essentially a convex combination of different base models, as given by Eq. 6.11 (Murphy, 2012):

$$\Pr(y|\mathbf{x},\boldsymbol{\pi}) = \sum_{m \in \mathcal{M}} \pi_m \Pr(y|\mathbf{x},m). \tag{6.11}$$

The value of $\Pr(\boldsymbol{\pi}|\mathcal{D})$ can then be calculated using Bayesian inference which is followed by a decision based on $\Pr(y|\mathbf{x},\mathcal{D}) = \int \Pr(y|\mathbf{x},\boldsymbol{\pi}) \Pr(\boldsymbol{\pi}|\mathcal{D}) d\boldsymbol{\pi}$ (Murphy, 2012).

6.3.1 *Boosting*

Boosting (Schapire, 2012) is essentially a greedy algorithm in which the weak learner is sequentially applied to a dataset's weighted versions, that is, previously misclassified examples are allocated more weight. Boosting was initially proposed by computational learning theorists (Schapire, 1990; Freund and Schapire, 1996) with a focus on binary classification. Since boosting is less prone to the overfitting problem, statisticians conducted a great deal of work in order to examine how a variety of loss functions are handled by boosting. The core concept is to build a strong learner based on a weak learning algorithm. Based on some insights provided in (Bühlmann and Hothorn, 2007; Hastie et al., 2009), this section presents a glimpse of the statistical interpretation of boosting.

6.3.1.1 Forward Stepwise Additive Modelling

The aim of boosting is to find the answer for an optimization problem, as in Eq. 6.12 (Murphy, 2012):

$$\min_f \sum_i^N L\left[y_i, f(\mathbf{x}_i)\right]. \tag{6.12}$$

where $L(y, \hat{y})$ is used to denote a loss function, and $f(\cdot)$ is often assumed to be an adaptive basis-function model (ABM) as defined by Eq. 6.13 (Murphy, 2012):

$$f(\mathbf{x}) = w_0 + \sum_{m=1}^{M} w_m \phi_m(\mathbf{x}). \tag{6.13}$$

where the mth basis function is denoted by $\phi_m(\mathbf{x})$ which is learnable from dataset. In general, the basis functions are featured by their parametric characteristics, that is, we can formulate it in the form of Eq. 6.14 (Murphy, 2012):

$$\phi_m(\mathbf{x}) = \phi_m(\mathbf{x}; \mathbf{v}_m). \tag{6.14}$$

where the basis function's parameters are represented by $\mathbf{v}_m$. Typically, we can use $\theta = (w_0, \mathbf{w}_{1:M}, \{\mathbf{v}_m\}_{m=1}^{M})$ to stand for the whole parameter set.

In practice, the most commonly used loss functions are squared error, absolute error, exponential loss and logistic loss. Take squared error, the optimal approximation can be given by Eq. 6.15 (Murphy, 2012):

$$f^*(\mathbf{x}) = \underset{f(\mathbf{x})}{\arg\min} = \mathbb{E}_{y|\mathbf{x}}\left[\left(Y - f(\mathbf{x})\right)^2\right] = \mathbb{E}\left[Y|\mathbf{x}\right]. \quad 6.15$$

However, since the real conditional distribution, i.e., $\Pr(y|\mathbf{x})$, has to be known, the direct computation of Eq. 6.15 is often not practicable. Therefore, for binary classification case, one can see that the optimal approximation is obtainable using Eq. 6.16 (Murphy, 2012):

$$f^*(\mathbf{x}) = \frac{1}{2}\log\frac{\Pr(\tilde{y}=1|\mathbf{x})}{\Pr(\tilde{y}=-1|\mathbf{x})}. \quad 6.16$$

Given the fact that searching the optimal $f(\cdot)$ is difficult, one can solve it sequentially, that is, (1) Initialize using Eq. 6.17 (Murphy, 2012):

$$f_0(\mathbf{x}) = \arg\min_{\gamma}\sum_{i=1}^{N} L\left[y_i, f(\mathbf{x}_i;\gamma)\right]. \quad 6.17$$

Here, if squared error loss is employed, we can let $f_0(\mathbf{x})$ equal to $\bar{y}$; (2) Compute Eq. 6.18 (Murphy, 2012) at the mth iteration:

$$(\beta_m, \gamma_m) = \underset{\beta,\gamma}{\arg\min}\sum_{i=1}^{N} L\left[y_i, f_{m-1}(\mathbf{x}_i) + \beta\phi(\mathbf{x}_i;\gamma)\right]. \quad 6.18$$

and (3) Formulate $f_m(\mathbf{x})$ using Eq. 6.19 (Murphy, 2012):

$$f_m(\mathbf{x}) = f_{m-1}(\mathbf{x}) + \beta_m\phi(\mathbf{x}_i;\gamma_m). \quad 6.19$$

The implications drawn from the above three steps are that no backward operation and earlier parameters' adjustment are performed. This is where the name "forward stepwise additive modelling" comes from.

6.3.1.2 Boosting Variants

In order to resolve Eq. 6.18, we can introduce different forms of loss functions in order to construct distinct boosting variants.

- **L_2 boosting:** If the squared error loss function is utilized, then the loss can be formulated at the mth stage as Eq. 6.20 (Murphy, 2012):

$$L\left[y_i, f_{m-1}(\mathbf{x}_i) + \beta\phi(\mathbf{x}_i;\gamma)\right] = \left[r_{im} - \phi(\mathbf{x}_i;\gamma)\right]^2. \quad 6.20$$

where the present residual is denoted by $r_{im} \triangleq y_i - f_{m-1}(\mathbf{x}_i)$, and β is set to be equal to one without losing generality. This boosting version is, thus, termed as least squares boosting (or L_2 boosting for short) (Bühlmann and Yu, 2003).

- **AdaBoost:** Suppose an exponential loss function is employed in a binary classification case, then at the mth stage we need to minimize Eq. 6.21 (Murphy, 2012):

$$L_m(\phi) = \sum_{i=1}^{N} \exp\left\{-\tilde{y}_i\left[f_{m-1}(\mathbf{x}_i) + \beta\phi(\mathbf{x}_i)\right]\right\} = \sum_{i=1}^{N} w_{i,m} \exp\left[-\beta\tilde{y}_i\phi(\mathbf{x}_i)\right]. \quad 6.21$$

where $w_{i,m} \exp[-\beta\tilde{y}_i\,\phi(\mathbf{x}_i)]$ is a weight factor imposed on the ith dataset, and $\tilde{y}_i$ belongs to $\{-1,+1\}$. Accordingly, the objective function can be written in the form of Eq. 6.22 (Murphy, 2012):

$$\begin{aligned} L_m &= e^{-\beta} \sum_{\tilde{y}_i = \phi(\mathbf{x}_i)} w_{i,m} + e^{\beta} \sum_{\tilde{y}_i \neq \phi(\mathbf{x}_i)} w_{i,m} \\ &= \left(e^{\beta} - e^{-\beta}\right) \sum_{i=1}^{N} w_{i,m} \mathbb{I}\left[\tilde{y}_i \neq \phi(\mathbf{x}_i)\right] + e^{-\beta} \sum_{i=1}^{N} w_{i,m} \end{aligned}. \quad 6.22$$

Subsequently one can introduce the optimal function as given by Eq. 6.23 (Murphy, 2012):

$$\phi_m = \arg\min_{\phi} w_{i,m} \mathbb{I}\left[\tilde{y}_i \neq \phi(\mathbf{x}_i)\right]. \quad 6.23$$

By substituting Eq. 6.23 into Eq. 6.22, we can get Eq. 6.24 (Murphy, 2012):

$$\beta_m = \frac{1}{2}\log\frac{1 - err_m}{err_m}. \quad 6.24$$

where err_m equals to $\frac{\sum_{i=1}^{N} w_{i,m}\mathbb{I}\left[\tilde{y}_i \neq \phi(\mathbf{x}_i)\right]}{\sum_{i=1}^{N} w_{i,m}}$. Therefore, the total update can be given by Eq. 6.25 (Murphy, 2012):

$$f_m(\mathbf{x}) = f_{m-1}(\mathbf{x}) + \beta_m\phi(\mathbf{x}). \quad 6.25$$

Based on this, the weights the next iteration can be given by Eq. 6.26 (Murphy, 2012):

$$\begin{aligned} w_{i,m+1} &= w_{i,m} e^{-\beta_m \tilde{y}_i \phi_m(\mathbf{x}_i)} = w_{i,m} e^{\beta_m\left\{2\mathbb{I}\left[\tilde{y}_i \neq \phi(\mathbf{x}_i)\right] - 1\right\}} \\ &= w_{i,m} e^{2\beta_m \mathbb{I}\left[\tilde{y}_i \neq \phi(\mathbf{x}_i)\right]} e^{-\beta_m} \end{aligned}. \quad 6.26$$

where the values of $-\tilde{y}_i\,\phi_m(\mathbf{x}_i)$ are given by Eq. 6.27 (Murphy, 2012):

$$\begin{cases} -\tilde{y}_i\phi_m(\mathbf{x}_i) = -1 & \text{if } \tilde{y}_i = \phi_m(\mathbf{x}_i) \\ -\tilde{y}_i\phi_m(\mathbf{x}_i) = +1 & \text{otherwise} \end{cases}. \qquad 6.27$$

The result of this process is the algorithm called adaptive boosting (or AdaBoost for short) (Freund and Schapire, 1997).

- **LogitBoost:** One disadvantage of using exponential loss lies in the fact that it places a great deal of weight on misclassified cases, which in turn makes the approach particularly sensitive to outlier data. Therefore an intuitive alternative solution is to introduce a logistic loss function, which makes it possible to extract probability from the learned function using Eq. 6.28 (Murphy, 2012):

$$\Pr(y = 1|\mathbf{x}) = \frac{e^{f(\mathbf{x})}}{e^{-f(\mathbf{x})} + e^{f(\mathbf{x})}} = \frac{1}{1+e^{-2f(\mathbf{x})}}. \qquad 6.28$$

Here the aim is to minimize the expected log-loss, given by Eq. 6.29 (Murphy, 2012). This algorithm is commonly known as LogitBoost (Friedman et al., 2000).

$$L_m(\phi) = \sum_{i=1}^{N} \log\left[1+\exp\left(-2\tilde{y}_i\left(f_{m-1}(\mathbf{x}) + \phi(\mathbf{x}_i)\right)\right)\right]. \qquad 6.29$$

- **Gradient Boosting:** In practice, it is often useful to develop a generic boosting formulation, i.e., gradient boosting (Friedman, 2001), instead of proposing many independent versions for each loss function. Given a minimization task, as in Eq. 6.30 (Murphy, 2012):

$$\hat{\mathbf{f}} = \underset{\mathbf{f}}{\arg\min}\, L(\mathbf{f}). \qquad 6.30$$

where $\mathbf{f} = (f(\mathbf{x}_i),\ldots,f(\mathbf{x}_N))$ represents a set of parameters. Here, a gradient descent approach can be introduced in order to solve Eq. $\hat{\mathbf{f}} = \underset{\mathbf{f}}{\arg\min}\, L(\mathbf{f})$ stepwise. At the mth step, let the gradient of $L(\mathbf{f})$ be $\mathbf{g}_m$ which can be evaluated at $\mathbf{f} = \mathbf{f}_{m-1}$, as given by Eq. 6.31 (Murphy, 2012):

$$g_{im} = \left[\frac{\partial L(y_i, f(\mathbf{x}_i))}{\partial f(\mathbf{x}_i)}\right]_{f=f_{m-1}}. \qquad 6.31$$

Then the update can be made according to Eq. 6.32 (Murphy, 2012):

$$\mathbf{f}_m = \mathbf{f}_{m-1} - \rho_m \mathbf{g}_m. \qquad 6.32$$

where ρ_m denotes the length of each step and is determined by Eq. 6.33 (Murphy, 2012):

$$\rho_m = \arg\min_{\rho} L(\mathbf{f}_{m-1} - \rho\,\mathbf{g}_m). \qquad 6.33$$

This process is usually termed as functional gradient descent. Once we have this, a modification can be made in the algorithm by introducing a weak learner in order to get the negative gradient signal approximated, as defined by Eq. 6.34 (Murphy, 2012):

$$\gamma_m = \arg\min_{\gamma} \sum_{i=1}^{N} \left[-g_{im} - \phi(\mathbf{x}_i;\gamma)\right]^2. \qquad 6.34$$

- **Sparse Boosting:** When we would like to explore all possible variables (denoted by $j = 1{:}\ D$), and identify one variable, represented by $j(m)$, that enables the residual vector to be best predicted, as defined by Eqs. 6.35–6.37 (Murphy, 2012):

$$j(m) = \arg\min_{j} \sum_{i=1}^{N} \left(r_{im} - \hat{\beta}_{jm} x_{ij}\right)^2. \qquad 6.35$$

$$\hat{\beta}_{jm} = \frac{\sum_{i=1}^{N} x_{ij} r_{im}}{\sum_{i=1}^{N} x_{ij}^2}. \qquad 6.36$$

$$\phi_m(\mathbf{x}) = \hat{\beta}_{j(m),m}\, x_{j(m)}. \qquad 6.37$$

This process is called sparse boosting (Bühlmann and Yu, 2006).

6.3.1.3 *Why does Boosting Perform Better?*

It is widely acknowledged that the boosting method performs extremely well, in particular for classification tasks. The underlying reasons are two-fold (Murphy, 2012): (1) Through the viewpoint of a form of ℓ_1 regularization, boosting is good at preventing overfitting by removing irrelevant features. Take ℓ_1-Adaboost (Duchi and Singer, 2009), the best features (i.e., weak learners) are greedily added using boosting strategy and all irrelevant ones are then trimmed off with the aid of ℓ_1 regularization. (2) The second explanation goes to "margin" concept, that is, the training dataset's margin is maximized by boosting scheme.

Recently, a learning approach which was built on AdaBoost was proposed in (Baig et al., 2017) for the purpose of learning a non-linear feed-forward ANN with a sole hidden layer and output neuron. In brief, the proposed boosting-enabled ANN learning contains three key components: (1) Basic Boostron (denoting a boosting-bolstered perceptron learning mechanism) which is capable of learning a perceptron with no hidden layer of neurons involved; (2) Extended Boostron that learns linear neurons' single output feed-forward network; and (3) Novel scheme for neurons to possess non-linearity by appropriate activation functions' representation.

6.3.2 *Decision Trees*

Although in operations research domain, a decision tree often incorporates a hierarchical collection of decisions and the associated consequences, in data mining area, a decision tree is typically regarded as a predictive model that represents either classifiers or regression models. Accordingly, when a decision tree is involved in classification or regression tasks, we tend to have a corresponding classification tree or regression tree. A classification tree is often employed in order to divide an object or an instance into a predetermined set of classes, according to the corresponding value of attributes. In principle, a classification tree is often helpful in terms of exploring task space.

6.3.2.1 Classification and Regression Trees (CART)

Classification and regression tress (or CART for short) are a kind of decision trees, which can perform partition operations recursively within the input space and then get a local mode defined for each of the input space's divided sub-spaces. This process is interpretable using a tree structure, in which each piece of leaf is used to represent a portion of sub-region.

- **Fundamentals:** Take the regression tree illustrated in Fig. 6.3, the first node examines the relationship between x_1 and some threshold value of t_1.
  1) *On the one hand*, if the result is $x_1 \leq t_1$, the relationship between x_2 and another threshold value of t_2 is then checked. If the output of this node is $x_2 \leq t_2$, the first sub-region (denoted by $Subregion_1$) is reached; otherwise, the relationship of $x_1 \leq t_3$ is evaluated and two more sub-regions, namely $Subregion_4$ and $Subregion_5$, can be obtained respectively.
  2) *On the other hand*, if $x_1 \leq t_1$ does not hold, the relationship between x_2 and another threshold value of t_4 is then measured. If the outcome of this node is $x_2 \leq t_4$, the second sub-region (indicated by $Subregion_2$) is acquired; otherwise, $Subregion_3$ is earned.

The aim of these 'axis parallel splits' operations is to divide a two-dimensional space into five distinct sub-regions. Mathematically, CART model can be written in the general form as Eq. 6.38 (Murphy, 2012):

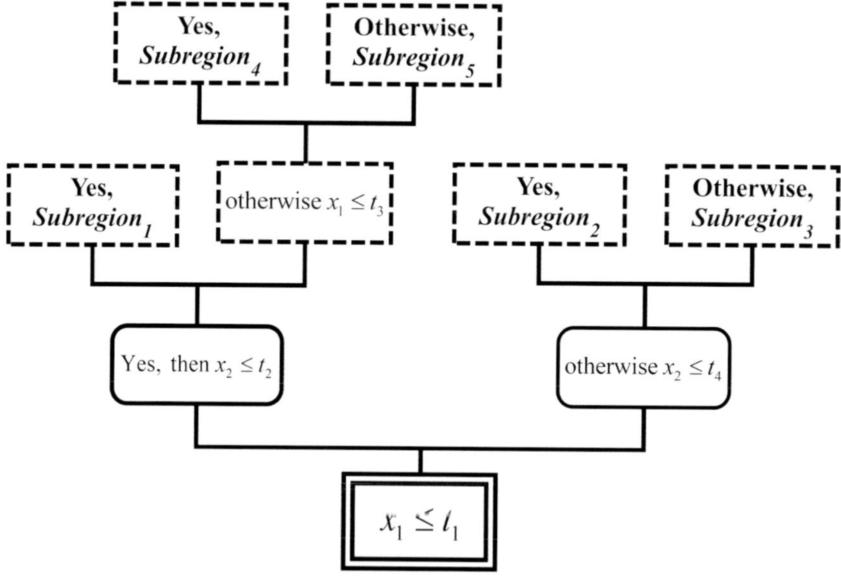

Figure 6.3: A simplified regression tree with two input values.

$$f(\mathbf{x}) = \mathbb{E}\left[y|\mathbf{x}\right] = \sum_{m=1}^{M} w_m \mathbb{I}\left(\mathbf{x} \in Subregion_m\right) = \sum_{m=1}^{M} w_m \phi\left(\mathbf{x}; \mathbf{v}_m\right). \quad 6.38$$

where the mth sub-region is represented by $Subregion_m$, the mean response of the mth sub-region is denoted by w_m, and $\mathbf{v}_m$ incorporates two elements, namely the selection of parameter to be split on, and the dedicated threshold value spreading from the root to the mth leaf. More discussions regarding how the CART model can be treated from an ABM (i.e., adaptive basis-function model) perspective can be found in (Murphy, 2012).

- **Tree Growing:** Since the search involved in optimal data partition is NP-complete (Hyafil and Rivest, 1976), it is therefore a common practice to employ a greedy procedure in order to calculate a locally optimal maximum likelihood estimate. Briefly, the split function selects the best feature and the corresponding best value using Eq. 6.39 (Murphy, 2012):

$$\left(j^*, t^*\right) = \arg \min_{j \in \{1,\ldots,D\}} \min_{t \in T_j} \text{cost}\left(\left\{\mathbf{x}_i, y_i : x_{ij} \le t\right\}\right) + \text{cost}\left(\left\{\mathbf{x}_i, y_i : x_{ij} > t\right\}\right). \quad 6.39$$

For simplicity, suppose we only have real-valued inputs, it is therefore possible to make a comparison between a feature (represented by x_{ij}) and a numeric value (denoted by t). For the jth feature, we can perform a sorting operation with respect to all x_{ij}'s unique values in order to acquire a set of associated threshold values (indicated by $\mathcal{T}_j$). For instance, if feature one contains the values of {6.5,–15,80,–15}, then $\mathcal{T}_1$ is equal to {–15,6.5,80}. In terms of categorical-type inputs, a usual way is to consider the separation in the form of $x_{ij} = c_l$ and $x_{ij} \neq c_l$, for each potential class labelled by c_l. When it comes to the examination of whether a node is worth splitting, one can employ different stopping heuristics such as (Murphy, 2012): (1) Whether the cost reduction is too small. In general, the gain of utilizing a feature can be defined as a normalized measure of cost reduction, as given by Eq. 6.40 (Murphy, 2012):

$$\Delta \triangleq \text{cost}(\mathcal{D}) - \left[\frac{|\mathcal{D}_{Left}|}{|\mathcal{D}|}\text{cost}(\mathcal{D}_{Left}) + \frac{|\mathcal{D}_{Right}|}{|\mathcal{D}|}\text{cost}(\mathcal{D}_{Right})\right]. \tag{6.40}$$

(2) Whether the maximal desired depth has been reached by the tree; (3) Whether the homogeneous degree for the distribution of the response is adequate in either $\mathcal{D}_{Left}$ or $\mathcal{D}_{Right}$; and (4) Whether the example number in either $\mathcal{D}_{Left}$ or $\mathcal{D}_{Right}$ is too small. The remaining task after all this is to determine the cost measure for evaluating a proposed split's feasibility, and this is dependent on our initial goal, that is, classification or regression.

1) *Regression cost*: In the context of regression, the cost can be defined using Eq. 6.41 (Murphy, 2012):

$$\text{cost}(\mathcal{D}) = \sum_{i \in \mathcal{D}} (y_i - \bar{y})^2. \tag{6.41}$$

where $\bar{y}$ represents the mean of the response parameter within the specified dataset which is equal to $\frac{1}{|\mathcal{D}|}\sum_{i\in\mathcal{D}} y_i$.

2) *Classification cost*: In terms of classification context, for data in the leaf that satisfies the measure of $X_j < t$, one can get a multi-noulli model fit to them by estimating the class-conditional probabilities given by Eq. 6.42 (Murphy, 2012):

$$\hat{\pi}_c = \frac{1}{|\mathcal{D}|} \sum_{i \in \mathcal{D}} \mathbb{I}(y_i = c). \tag{6.42}$$

where $\mathcal{D}$ denotes the data in the leaf.

- **Tree Pruning:** To avoid the issue of overfitting, the tree's growth can be halted if the error reduction is not enough to compensate for the additional complexity caused by introducing an extra sub-tree. Nevertheless, this intuitive attempt is often proved to be overly myopic. Accordingly, the widely acknowledged flowchart is first to get a tree fully grown and then to cut off the branches that lead the least error increase using some scheme. Please refer to (Breima et al., 1984) for more details.

It is often advisable to employ a shallow tree for lowering variance. For instance, we have trees with *J* leaves: (1) If *J* is equal to two, we encounter a stump, i.e., only a single parameter can be used to split it; (2) If *J* is equal to three, two-parameter interaction is possible, and so on. In practice, it is often suggested that one should take the value of $J \approx 6$ (Hastie et al., 2009; Caruana and Niculescu-Mizil, 2006). Since a shallow tree is prone to deviation, the gradient boosting scheme can be combined with shallow regression trees in order to form the MART model (i.e., multivariate adaptive regression trees). The variables at the tree's leaves can be re-estimated in order to minimize the loss function which is given by Eq. 6.43 (Murphy, 2012):

$$\gamma_{jm} = \arg\min_{\gamma} \sum_{\gamma_i \in R_{jm}} L\left(y_i, f_{m-1}(\mathbf{x}_i) + \gamma\right). \tag{6.43}$$

where R_{jm} represents the region for the *j*th leaf on the *m*th tree, and γ_{jm} denotes the corresponding variable in regression and classification problems, respectively.

6.3.2.2 Random Forests

The popularity of CART models is due to the fact that they are easily interpretable, well-suited for dealing with mixed inputs (i.e., discrete and continuous), insensitive towards the inputs' monotone transformations, capable of performing automatic parameter selection, relatively robust to outliers, able to be scaled up to large datasets and able to be modified for handling missing inputs. Nevertheless, CART models also suffer from several disadvantages. The main drawback lies in their low prediction accuracy in comparison to other types of models, partially attributed to the tree structure's inherent greedy nature. In other words, the stability of trees is low: tiny changes in the input dataset tend to have a great effect on the tree's structure.

One way to fix this issue is to perform an average operation over various estimates. However, highly correlated predictors are often caused by simply re-running the same learning algorithm on a dataset's distinct sub-sets, which in turn restrict the amount of variance that can potentially be reduced. Accordingly, the random forests (RF) technique was proposed in (Breiman, 2001) with the goal of using learning trees to de-correlate the base learners using a randomly selected sub-set of input parameters and data cases. In general, the prediction accuracy of RF models is excellent and they have been widely used in various real-world applications, e.g., body pose recognition (Shotton et al., 2011).

Meanwhile, by having a set of randomized regression trees, an assembly of multi-layered neural networks can be introduced in order to reconstruct this ensemble of trees under certain connection weights. Based on this understanding, neural RF was recently proposed in (Biau et al., 2016). By comparing neural forests with other benchmarks, the number of variables that need to be tuned is smaller, and the restrictions on the decision boundaries' geometry is looser.

6.3.3 *The Applications of Ensemble Learning in Predicting Project Success Rate and Fundraising Range*

In order to develop a predictor for estimating project success rate and fundraising range, a pioneer study was conducted in (Chung, 2014), in which datasets are carefully analysed and features are specifically selected.

6.3.3.1 *Datasets*

In order to understand the impact of introducing social media information to project success rate, two datasets were collected and analysed in (Chung, 2014).

- **Kickstarter Dataset:** As one of the most beloved crowdfunding platforms, two types of data were collected in (Chung, 2014).
  1) *Static data*, which includes all project page information and the associated use, page information. Firstly, 168,851 project pages were collected, and the information embedded in each project page includes the duration of the project, the description of the project and rewards, the funding target, etc. Secondly, the associated 146,721 different user pages were also gathered with each page covering the information regarding the user's biography, the longevity of the account, location details, the amount of supported projects, the number of established projects, etc. After a screening process, a collection of 151,608 project pages and 142,890 user pages is chosen in (Chung, 2014) for further analysis; and
  2) *Temporal data*, which includes secured funds on a daily basis and the daily fluctuation of backer's number during the course of a project. Finally, there are 74,053 projects' temporal data assembled in (Chung, 2014).
- **Twitter Dataset:** In (Chung, 2014), the author discovered that about 13.4% of users (i.e., 19,138 users) in their chosen dataset have connected their Twitter user profile pages with the corresponding Kickstarter user profile pages, and the associated number of the established projects is 22,408. In order to make use of this Twitter account information, each Twitter user's profile (based on the Kickstarter user profile at

hand) is extracted by including issues such as the total number of tweets, the number of tweets posted during the course of a campaign, the number of profiles that the user is following, and the number of followers. Through a proper filtering, the information regarding a total of 17,908 Twitter users was collected in (Chung, 2014) and then all this information was combined with the relevant 21,028 Kickstarter campaign pages in order to form Twitter dataset.

Based on the aforementioned data collection plans, two datasets were constructed. Please refer to (Chung, 2014) for further analysis regarding users and campaigns.

6.3.3.2 Features

Built on the initial analysis, the following features (49 in total) were employed in (Chung, 2014) in order to develop a predictor that is capable of forecasting project success rate and fundraising range.

- **Project Features:** From a campaign page, 11 features (i.e., *Features 1~11*) were summarized in (Chung, 2014).
  1) *Feature 1–Category;*
  2) *Feature 2–Duration;*
  3) *Feature 3–Goal;*
  4) *Feature 4–Number of pictures;*
  5) *Feature 5–Number of video-clips;*
  6) *Feature 6–Amount of frequently-asked-questions;*
  7) *Feature 7–Amount of rewards;*
  8) *Feature 8–Reward description's SMOG grade*: For measuring the readability of the text involved in reward description;
  9) *Feature 9–Main page description's SMOG grade*: Similarly, to assess the readability of the main page involved in a campaign;
  10) *Feature 10–Number of sentences contained in a reward description;* and
  11) *Feature 11–Number of sentences included in a campaign's main description.*

On the subject of SMOG grade, it was proposed in (McLaughlin, 1969) and has since been widely used in order to evaluate the year-of-education required for comprehending a paragraph of writing. In principle, a higher score implies a better comprehension with respect to campaign/reward descriptions. In (Chung, 2014), Eq. 6.44 was employed in order to calculate the relevant SMOG grade:

$$\text{SMOG grade} = 1.043\sqrt{|\text{Polysyllables}| \times \frac{30}{|\text{sentences}|}} + 3.1291. \qquad 6.44$$

where the term 'Polysyllables' is used to denote the word count with three or more syllables.

- **User Features:** Based on a user's profile page and the corresponding past experience, 28 features (i.e., *Features 12~39*) were created in (Chung, 2014).
  1) *Feature 12~26–The distribution of the supported projects belonging to one of the fifteen main categories;*
  2) *Feature 27–Number of supported projects;*
  3) *Feature 28–Number of previously established projects;*
  4) *Feature 29–Number of previously posted comments;*
  5) *Feature 30–Number of connected websites found in a user's profile;*
  6) *Feature 31–Number of Facebook friends that a user's account has;*
  7) *Feature 32~34–Is each of the user accounts from Facebook, Twitter, and YouTube, respectively, linked?*
  8) *Feature 35–User biographical description's SMOG grade;*
  9) *Feature 36–Number of sentences contained in a user biographical description;*
  10) *Feature 37–How many days observed between the following two dates, that is, an asker's Kickstarter joining date and an asker's campaign launching date;*
  11) *Feature 38–The success rate of a project given that it is supported by an asker;*
  12) *Feature 39–The success rate of previously launched campaign by an asker.*
- **Temporal Features:** Through a set of transformation operations performed on temporal data, two time-series features (i.e., *Features 40&41*) were produced in (Chung, 2014).
  1) *Feature 40–Accumulated secured funds over time;*
  2) *Feature 41–Assembled number of campaign supporters over time.*
- **Twitter Features:** Based on the established Twitter dataset, eight features (i.e., *Features 42~49*) were captured in (Chung, 2014).
  1) *Feature 42–Number of tweets;*
  2) *Feature 43–Number of profiles that a user is following;*
  3) *Feature 44–Number of followers;*
  4) *Feature 45–Number of favourites;*
  5) *Feature 46–Number of listings that a user has taken part in;*
  6) *Feature 47–Number of tweets posted during the course of a campaign;*
  7) *Feature 48–Number of tweets incorporating keyword "Kickstarter" posted during the course of a campaign;*
  8) *Feature 49–SMOG grade of combined tweets that have been posted during the course of a campaign.*

More specifically, the first five features (i.e., *Features 42~46*) were generally utilized to label any project that has been established by a user while the remaining three features were introduced to mark each project individually due to their differentiated lasting period.

6.3.3.3 *Experimental Settings*

The detailed experimental settings are as follows (Chung, 2014):

- **Three Dataset Combinations:** Three combinations were used in (Chung, 2014), which include (1) Kickstarter static dataset (consisting of 151,608 projects and 39 features), (2) Kickstarter static and Twitter combined dataset (consisting of 21,028 projects and 47 features) and (3) Kickstarter static, temporal and Twitter combined dataset (consisting of 11,675 projects and 49 features).
- **Classification Algorithms:** Given the fact that different classification algorithms perform differently against distinct datasets, three representative classification algorithms, namely NB (i.e., Naïve Bayes), RF (i.e., random forests) and AdaboostM1 (a variant of Adaboost with RF as the base learner), were employed in (Chung, 2014). Meanwhile, Weka workbench (Witten et al., 2017) was selected in (Chung, 2014) for implementing the algorithms and building predictive models.
- **Feature Selection:** In order to examine whether an effective predictor can gain positive contribution from the above proposed features, each feature's χ^2 value (Yang and Pedersen, 1997) was measured in (Chung, 2014). In terms of the χ^2 value, the higher it is, the greater the discriminative power of the associated feature.
- **Predictive Model Evaluation:** For evaluating a classifier's performance, ROC (i.e., receiver operating characteristic) curves are often employed in the data mining domain. Originating from the signal detection application area, ROC curves describe a classifier's performance. Further discussions can be found in (Witten et al., 2017; Bramer, 2016). In (Chung, 2014), 'accuracy' was selected as the main evaluation measure, while the 'area' under the ROC curves was chosen as the secondary criterion. Meanwhile, a five-fold cross-validation approach was used to assess each classifier.

6.3.4 *Summary*

- **Campaign Success Rate Predictor:** If only Kickstarter static features were included, the predictor proposed in (Chung, 2014) was able to reach an accuracy of 76.4% which is higher than previous work (Greenberg et al., 2013). By introducing features extracted from social media, an increase of 2.5% can be observed in prediction accuracy.

When the temporal features were also included, the accuracy improved consistently. Among three classification algorithms, the performance of AdaboostM1 is superior in terms of both accuracy and AUC for all three scenarios.

- **Fundraising Range Predictor:** Given the fact that the All-or-Nothing policy is utilized by many crowdfunding platforms, a number of classes were further defined in (Chung, 2014). In two classes design, only one threshold value ($5,000 in this case) was considered, where class1 ≤ $5,000 and class2 > $5,000. Similarly, when it comes to three classes, two threshold values ($100 and $10,000 in this case) were taken into account, where class1 ≤ $100, $100 < class2 ≤ $10,000 and class3 > $10,000. Based on these designs, the experiments demonstrated the possibility of pre-determining a campaign's anticipated fundraising range. Among three employed algorithms, AdaboostM1 was again outperforming the others for all three scenarios.

6.4 Conclusions

Overall, the growing prominence of crowdfunding is, in some ways, surprising. The HiveWire, a full-service crowdfunding agency in Canada, chose the following four key factors as being instrumental in influencing a campaign's success (Ania and Charlesworth, 2015):

- Idea quality and pitch;
- Campaign marketing professionalism;
- Strength of social network; and
- Rewards enticement.

As we have shown in the earlier part of this chapter, there are a myriad of similar pickups existing in the literature. In order to offer askers a much more quantitative and qualitative support system, we presented a comprehensive study conducted by Chung (2014) in which a large amount of data (collected from Kickstarter) regarding the projects and backers were analysed. The study revealed that there is an interesting overlap occurring among campaign participants, that is, the co-existence of two identities, being not only project creators (for their own projects) but also active backers (for another creator's project). By analysing the collected projects' temporal data, two peak times were successfully identified, namely beginning and closing. Moreover, through the adoption of four types of features, Chung (2014) built several predictive models using different ensemble learning techniques. The performance of these frameworks was further tested against different feature sets, in particular, the inclusion of Twitter and temporal features increases the prediction accuracy for both project success rate predictor and range of asked-for funds predictor. The predictive models and the valuable results discussed throughout this chapter will certainly be very helpful for someone who is preparing a crowdfunding project.

References

Agerström J, Carlsson R, Nicklasson L, et al. (2016) Using descriptive social norms to increase charitable giving: The power of local norms. *Journal of Economic Psychology* 52: 147–153.

Agier I and Szafarz A. (2013) Subjectivity in credit allocation to micro-entrepreneurs: Evidence from Brazil. *Small Business Economics* 41: 263–275.

Agrawal A, Catalini C and Goldfarb A. (2011) The geography of crowdfunding. National Bureau of Economic Research (NBER), Working Paper No. 16820.

Agrawal A, Catalini C and Goldfarb A. (2014) Some simple economics of crowdfunding. *Innovation Policy and the Economy* 14: 63–97.

Ahlers GKC, Cumming D, Günther C, et al. (2015) Signaling in equity crowdfunding. *Entrepreneurship: Theory and Practice* 39: 955–980.

Ahlin C and BrianWaters. (2016) Dynamic microlending under adverse selection: can it rival group lending? *Journal of Development Economics* 121: 237–257.

Allison TH, Davis BC, Webb JW, et al. (2017) Persuasion in crowdfunding: An elaboration likelihood model of crowdfunding performance. *Journal of Business Venturing* 32: 707–725.

Alpaydın E. (2010) *Introduction to machine learning,* 55 Hayward Street, Cambridge, MA 02142: The MIT Press, ISBN 978-0-262-01243-0.

Alpaydın E. (2014) *Introduction to machine learning,* 55 Hayward Street, Cambridge, MA 02142: The MIT Press, ISBN 978-0-262-02818-9.

Andreoni J, Brown E and Rischall I. (2003) Charitable giving by married couple: Who decides and why does it matter? *The Journal of Human Rescources* XXXVIII: 111–133.

Andreoni J and Payne AA. (2011) Is crowding out due entirely to fundraising? Evidence from a panel of charities. *Journal of Public Economics* 95: 334–343.

Angle P. (2017) Crowd power. *Indian Management* 56: 81–87.

Ania A and Charlesworth C. (2015) Crowdfunding guide: For nonprofits, charities and social impact projects. 215 Spadina Ave, Suite 400, Toronto, ON, M5T 2C7: HiveWire & Centre for Social Innovation, Document Version: 1701 - D.

Asgary NH and McNulty RE. (2017) Contributions and challenges in the struggle to end poverty: The case of Kiva. *Information Technology for Development* 23: 367–387.

Ashta A, Ghosh C, Guha S, et al. (in press) Knowledge in microsocial milieus: The case of microfinance practices among women in India. *Journal of the Knowledge Economy* DOI 10.1007/s13132-016-0372-x.

Australian Charities and Not-for-Profits Commission. (2017) Crowdfunding and charities: Information for charities, donors, and fundraisers about the use of crowdfunding. Australian Government, 1–9.

Bühlmann P and Yu B. (2003) Boosting with the L_2 loss: Regression and classification. *Journal of the American Statistical Association* 98: 324–339.

Bühlmann P and Yu B. (2006) Sparse boosting. *Journal of Machine Learning Research* 7: 1001–1024.

Bühlmann P and Hothorn T. (2007) Boosting algorithms: Regularization, prediction and model fitting. *Statistical Science* 22: 477–505.

B. A. de Aghion, Armendáriz B and Morduch J. (2007) *The economics of microfinance,* Cambridge: MIT Press, ISBN 9780262512015.

Babu DK, Ramadevi Y and Ramana KV. (2017) RGNBC: Rough Gaussian naïve Bayes classifier for data stream classification with recurring concept drift. *Arabian Journal for Science and Engineering* 42: 705–714.

Babyloan. (2017) *Babyloan, retrieved from https://www.babyloan.org/en, accessed on 25 July 2017.*

Baig MM, Awais MM and El-Alfy E-SM. (2017) AdaBoost-based artificial neural network learning. *Neurocomputing* 248: 120–126.

Baiyegunhi LJS and Fraser GCG. (2014) Smallholder farmers' access to credit in the Amathold district municipality, eastern cape province, South Africa. *Journal of Agriculture and Rural Development in the Tropic and Subtropics* 115: 79–89.

Bao Z and Huang T. (2017) External supports in reward-based crowdfunding campaigns: a comparative study focused on cultural and creative projects. *Online Information Review* 41: 626–642.

Barker D. (2012) *Bayesian reasoning and machine learning,* Cambridge CB2 8BS, United Kingdom: Cambridge University Press, ISBN 978-0-521-51814-7.

Bauer E and Kohavi R. (1999) An empirical comparison of voting classification algorithms: bagging, boosting, and variants. *Machine Learning* 35: 1–38.

BBB WGA. (2013) Crowdfunding for charitable causes. *BBB Wise Giving Alliance, Summer/Fall.* 2–9.

Beaulieu T, Sarker S and Sarker S. (2015) A conceptual framework for understanding crowdfunding. *Communications of the Association for Information Systems* 37: 1–31.

Bechter C, Jentzsch S and Michael F. (2011) From wisdom to wisdom of the crowd and crowdfunding. *Journal of Communication and Computer* 8: 951–957.

Bekkers R and Wiepking P. (2011) A literature review of empirical studies of philanthropy: eight mechanisms that drive charitable giving. *Nonprofit and Voluntary Sector Quarterly* 40: 924–973.

Belleflamme P, Lambert T and Schwienbacher A. (2014) Crowdfunding: Tapping the right crowd. *Journal of Business Venturing* 29: 585–609.

Belleflamme P, Omrani N and Peitz M. (2015) The economics of crowdfunding platforms. *Information Economics and Policy* 33: 11–28.

Bennett R and Barkensjo A. (2005) Causes and consequences of donor perceptions of the quality of the relationship marketing activities of charitable organizations. *Journal of Targeting, Measurement and Analysis for Marketing* 13: 122–139.

Berliner LS and Kenworthy NJ. (2017) Producing a worthy illness: Personal crowdfunding amidst financial crisis. *Social Science & Medicine* 187: 233–242.

Besley T and Coate S. (1995) Group lending, repayment incentives and social collateral. *Journal of Development Economics* 46: 1–18.

Bi S, Liu Z and Usman K. (2017) The influence of online information on investing decisions of reward-based crowdfunding. *Journal of Business Research* 71: 10–18.

Biau G, Scornet E and Welbl J. (2016) Neural random forests. retrieved from https://arxiv.org/pdf/1604.07143.pdf, accessed on 30 November 2017.

Block J, Hornuf L and Moritz A. (in press) Which updates during an equity crowdfunding campaign increase crowd participation? *Small Business Economics* DOI 10.1007/s11187-017-9876-4.

Bock DE, Eastman JK and Eastman KL. (in press) Encouraging consumer charitable behavior: The impact of charitable motivations, gratitude, and materialism. *Journal of Business Ethics* DOI 10.1007/s10551-016-3203-x.

Boeuf B, Darveau J and Legoux R. (2014) Financing creativity: Crowdfunding as a new approach for theatre projects. *International Journal of Arts Management* 16: 33–48.

Bramer M. (2016) *Principles of data mining,* 236 Gray's Inn Road, London WC1X 8HB, United Kingdom: Springer-Verlag London Ltd., ISBN 978-1-4471-7306-9.

Breima L, Friedman J, Stone CJ, et al. (1984) *Classification and regression trees*: Chapman and Hall/CRC, ISBN 978-0-412-0484-1.

Breiman L. (2001) Random forests. *Machine Learning* 45: 5–32.

Brown TE, Boon E and Pitt LF. (2017) Seeking funding in order to sell: Crowdfunding as a marketing tool. *Business Horizons* 60: 189–195.

Bruton G, Khavul S, Siegel D, et al. (2014) New financial alternatives in seeding entrepreneurship: Microfinance, crowdfunding and peer-to-peer innovations. *Entrepreneurship: Theory and Practice* 39: 9–26.

Buff LA and Alhadeff P. (2013) Budgeting for crowdfunding rewards. *MEIEA Journal* 13: 27–44.

Bui HH. (2017) Creating a robust crowdfunding ecosystem in Vietnam. *Bachelor of Business Administration | International Business.* Turku University of Applied Science.

Burgoyne CB, Young B and Walker CM. (2005) Deciding to give to charity: A focus group study in the context of the household economy. *Journal of Community and Applied Social Psychology* 15: 383–405.

Burtch G, Ghose A and Wattal S. (2014) Cultural differences and geography as determinants of online pro-social lending. *MIS Quarterly* 38: 773–794.

Cao J. (2014) *How VCs Use Kickstarter to Kick the Tires on Hardware Startups, retrieved from http://www.bloomberg.com/news/2014-08-08/how-vcs-use-kickstarter-to-kick-the-tires-on-hardware-startups.html, accessed on 30 November 2017.*

Caruana R and Niculescu-Mizil A. (2006) An empirical comparison of supervised learning algorithms. *Proceedings of the 23rd International Conference on Machine Learning (ICML '06), Pittsburgh, Pennsylvania, USA, 25–29 June*, pp. 161–168.

Castillo M, Petrie R and Wardell C. (2014) Fundraising through online social networks: A field experiment on peer-to-peer solicitation. *Journal of Public Economics* 114: 29–35.

Cha J. (in press) Crowdfunding for video games: Factors that influence the success of and capital pledged for campaigns. *International Journal of Media Management.*

Chang CT and Lee YK. (2011) The 'I' of the beholder: How gender differences and self-referencing influence charity advertising. *International Journal of Advertising* 30: 447–478.

Chen D, Lai F and Lin Z. (2014) A trust model for online peer-to-peer lending: A lender's perspective. *Information Technology and Management* 15: 239–254.

Chen R, Chen Y, Liu Y, et al. (2017) Does team competition increase pro-social lending? Evidence from online microfinance. *Games and Economic Behavior* 101: 311–333.

Chen X, Zhou L and Wan D. (2016) Group social capital and lending outcomes in the financial credit market: An empirical study of online peer-to-peer lending. *Electronic Commerce Research and Applications* 15: 1–13.

Chertkow R and Feehan J. (2014) Setting your crowdfunding rewards. *Electronic Musician.* NewBay Media, LLC, 84–85.

Chisasa J. (2014) A diagnosis of rural agricultural credit markets in South Africa: empirical evidence from north west and mpumalanga provinces. *Banks and Bank Systems* 9: 100–111.

Cholakova M and Clarysse B. (2015) Does the possibility to make equity investments in crowdfunding projects crowd out reward-based investments? *Entrepreneurship Theory and Practice* 39: 145–172.

Chowdhury S, Chowdhury PR and Sengupta K. (2014) Sequential lending with dynamic joint liability in micro-finance. *Journal of Development Economics* 111: 167–180.

Chung J. (2014) Long-term study of crowdfunding platform: Predicting project success and fundraising amount. Logan, Utah: Utah State University.

Collins L and Pierrakis Y. (2012) The venture crowd: Crowdfunding equity invesment into business. 1 Plough Place, London EC4A 1DE, UK: Nesta, ISBN 978-1-84875-138-5, 1–36.

Colombo MG, Franzoni C and Rossi-Lamastra C. (2015) Internal social capital and the attraction of early contribution in crowdfunding. *Entrepreneurship Theory and Practice* 39: 75–100.

Corazzini L, Cotton C and Valbonesi P. (2015) Donor coordination in project funding: Evidence from a threshold public goods experiment. *Journal of Public Economics* 128: 16–29.

Cordova A, Dolci J and Gianfrate G. (2015) The determinants of crowdfunding success: Evidence from technology projects. *Procedia-Social and Behavioral Sciences* 181: 115–124.

Courtney C, Dutta S and Li Y. (2017) Resolving information asymmetry: Signaling, endorsement, and crowdfunding success. *Entrepreneurship Theory and Practice* March: 265–290.

Cumming DJ, Leboeuf G and Schwienbacher A. (2017) Crowdfunding cleantech. *Energy Economics* 65: 292–303.

Dietterich TG and Bakiri G. (1995) Solving multiclass learning problems via error-correcting output codes. *Journal of Artificial Intelligence Research* 2: 263–286.

Dorfleitner G and Oswald E-M. (2016) Repayment behavior in peer-to-peer microfinancing: Empirical evidence from Kiva. *Review of Financial Economics* 30: 45–59.

Dorfleitner G, Priberny C, Schuster S, et al. (2016) Description-text related soft information in peer-to-peer lending—evidence from two leading European platforms. *Journal of Banking & Finance* 64: 169–187.

Dragojlovic N and Lynd LD. (2016) What will the crowd fund? Preferences of prospective donors for drug development fundraising campaigns. *Drug Discovery Today* 21: 1863–1868.

Duarte J, Siegel S and Young L. (2012) Trust and credit: The role of appearance in peer-to-peer lending. *The Review of Financial Studies* 25: 2455–2484.

Duchi J and Singer Y. (2009) Boosting with structural sparsity. *Proceedings of the 26th International Conference on Machine Learning (ICML '09), Montreal, Quebec, Canada, 14–18 June*, pp. 297–304.

Duggan CSM. (2016) Doing bad by doing good? Theft and abuse by lenders in the microfinance markets of Uganda. *Studies in Comparative International Development* 51: 189–208.

Dune K. (2013) Crowdfunding: Why museums should cultivate the millennial online donor. *Department of Museum Studies.* Philadelphia, Pennsylvania: University of the Arts.

Durguner S. (in press) Do borrower-lender relationships still matter for small business loans? *Journal of International Financial Markets, Institutions & Money* http://dx.doi.org/10.1016/j.intfin.2017.09.007.

Dvorak T and Toubman SR. (2013) Are women more generous than men? Evidence from alumni donations. *Eastern Economic Journal* 39: 121–131.

Emekter R, Tu Y, Jirasakuldech B, et al. (2015) Evaluating credit risk and loan performance in online peer-to-peer (P2P) lending. *Applied Economics* 47: 54–70.

Emery JCH and Ferrer A. (2015) The social rate of return to investing in character: An economic evaluation of Alberta's immigrant access fund microloan program. *Journal of International Migration and Integration* 16: 205–224.

Forbes H and Schaefer D. (2017) Guidelines for successful crowdfunding. *Procedia CIRP* 60: 398–403.

Freedman DM and Nutting MR. (2015) *Equity crowdfunding for investors: a guide to risks, returns, regulations, funding portals, due diligence, and deal terms,* 111 River Street, Hoboken, NJ 07030: John Wiley & Sons, Inc., ISBN 978-1-1188-5356-6.

Freedman MP. (2000) Challenges to launching grassroots microlending programs: A case study. *Journal of Developmental Entrepreneurship* 5: 235–248.

Freedman S and Jin GZ. (2014) The signaling value of online social networks: Lessons from peer-to-peer lending. Indiana University & University of Maryland, pp. 1–53.

Freedman S and Jin GZ. (2017) The information value of online social networks: Lessons from peer-to-peer lending. *International Journal of Industrial Organization* 51: 185–222.

Freund Y and Schapire RE. (1996) Experiments with a new boosting algorithm. *Proceedings of the Thirteenth International Conference on International Conference on Machine Learning (ICML'96), Bari, Italy, 03–06 July*, pp. 148–156.

Freund Y and Schapire RE. (1997) A decision-theoretic generalization of on-line learning and an application to boosting. *Journal of Computer and System Sciences* 55: 119–139.

Friedman J, Hastie T and Tibshirani R. (2000) Additive logistic regression: A statistical view of boosting. *Annals of Statistics* 28: 337–374.

Friedman J. (2001) Greedy function approximation: A gradient boosting machine. *Annals of Statistics* 29: 1189–1232.

Frydrych D, Bock AJ, Kinder T, et al. (2014) Exploring entrepreneurial legitimacy in reward-based crowdfunding. *Venture Capital* 16: 247–269.

Gamble JR, Brennan M and McAdam R. (2017) A rewarding experience? Exploring how crowdfunding is affecting music industry business models. *Journal of Business Research* 70: 25–36.

Gao F. (2016) Social media as a communication strategy: Content analysis of top nonprofit foundations' micro-blogs in China. *International Journal of Strategic Communication* 10: 255–271.

Ge R, Feng J and Gu B. (2016) Borrower's default and self-disclosure of social media information in P2P lending. *Financial Innovation* 2: 1–6.

Ge R, Feng J, Gu B, et al. (2017) Predicting and deterring default with social media information in peer-to-peer lending. *Journal of Management Information Systems* 34: 401–424.

Ghatak M and Guinnane TW. (1999) The economics of lending with joint liability: Theory and practice. *Journal of Developing Economics* 60: 195–228.

Giles J. (2012) Like it? pay for it. *Nature* 7381: 252–253.

Gininda PS, Antwi MA and Oladele OI. (2014) Smallholder sugarcane farmers' perception of the effect of micro agricultural finance institution of South Africa on livelihood outcomes in nkomazi local municipality, mpumalanga province. *Mediterranean Journal of Social Sciences* 5: 1032–1042.

Giudici G, Guerini M and Rossi-Lamastra C. (2013) *Why crowdfunding projects can succeed: the role of proponents' individual and territorial social capital, retrieved from http://ssrn.com/abstract=2255944, accessed on 10 October 2017.*

Giudici G, Guerini M and Rossi-Lamastra C. (in press) Reward-based crowdfunding of entrepreneurial projects: the effect of local altruism and localized social capital on proponents' success. *Small Business Economics.*

Gleasure R and Feller J. (2016) Does heart or head rule donor behaviors in charitable crowdfunding markets? *International Journal of Electronic Commerce* 20: 499–524.

Gonzalez L and Loureiro YK. (2014) When can a photo increase credit? the impact of lender and borrower profiles on online peer-to-peer loans. *Journal of Behavioral and Experimental Finance* 2: 44–58.

Greenberg MD, Pardo B, Hariharan K, et al. (2013) Crowdfunding support tools: Predicting success & failure. *Proceedings of the Extended Abstracts on Human Factors in Computing Systems (CHI EA '13), Paris, France, 27 April–02 May*, pp. 1815–1820.

Groshoff D. (2014) Kickstarter my heart: Extraordinary popular delusions and the madness of crowdfunding constraints and Bitcoin bubbles. *William & Mary Business Law Review* 5: 489–557.

Hörisch J. (2015) Crowdfunding for environmental ventures: An empirical analysis of the influence... crowdfunding initiatives. *Journal of Cleaner Production* 107: 636–645.

Han J-T, Chen Q, Liu J-G, et al. (2018) The persuasion of borrowers' voluntary information in peer-to-peer lending: An empirical study based on elaboration likelihood model. *Computers in Human Behavior* 78: 200–214.

Hankinson P. (2001) Brand orientation in the charity sector: a framework for discussion and research. *International Journal of Nonprofit and Voluntary Sector Marketing* 6: 231–242.

Harkness SK. (2016) Discrimination in lending markets: Status and the intersections of gender and race. *Social Psychology Quarterly* 79: 81–93.

Hartley SE. (2010) *Crowd-sourced microfinance and cooperation in group lending, retrieved from http://ssrn.com/abstract=1572182, accessed on 26 November 2017.*

Hastie T, Tibshirani R and Friedman J. (2009) *The elements of statistical learning: Data mining, inference, and prediction*, Berlin Heidelberg: Springer-Verlag, ISBN 978-0-3878-4857-0.

Hernandez MA and Torero M. (2014) Parametric versus non-parametric methods in risk scoring: An application to microcredit. *Empirical Economics* 46: 1057–1079.

Herzenstein M, Sonenshein S and Dholakia UM. (2011) Tell me a good story and I may lend you money: The role of narratives in peer-to-peer lending decisions. *Journal of Marketing Research* 48: S138–S149.

Hobbs J, Grigore G and Molesworth M. (2016) Success in the management of crowdfunding projects in the creative industries. *Internet Research* 26: 146–166.

Hornuf L and Neuenkirch M. (in press) Pricing shares in equity crowdfunding. *Small Business Economics* DOI 10.1007/s11187-016-9807-9.

Hortaçsu A, Martínez-Jerez FA and Douglas J. (2009) The geography of trade in online transactions: Evidence from eBay and mercadolibre. *American Economic Journal: Microeconomics* 1: 53–74.

Hsu B. (2016) The 'impossible' default: Qualitative data on borrower responses to two types of social-collateral microfinance structures in rural China. *The Journal of Development Studies* 52: 147–159.

Hsu BY. (2014) Alleviating poverty or reinforcing inequality? Interpreting micro-finance in practice, with illustrations from rural China. *The British Journal of Sociology* 65: 245–265.

Hu M, Li X and Shi M. (2015) Product and pricing decisions in crowdfunding. *Marketing Science* 34: 331–345.

Huang T and Zhao Y. (in press) Revolution of securities law in the Internet age: A review on equity crowd-funding. *Computer Law & Security Review.*

Huizingh EK. (2011) Open innovation: State of the art and future perspectives. *Technovation* 31: 2–9.

Hyafil L and Rivest RL. (1976) Constructing optimal binary decision trees is NP-complete. *Information Processing Letters* 5: 15–17.

infoDev and The World Bank. (2013) Crowdfunding's potential for the developing world. 1818 H Street NW, Washington DC 20433: Information for Development Program (infoDev)/The World Bank, pp. 1–104.

Ingen Dynamics Inc. (2016) *Aido: Next gen home robot, retrieved from www.indiegogo.com/projects/aido-next-gen-home-robot--2#/, accessed on 18 December 2017.*

Ings S. (2015) Design out of disaster. *New Scientist* 226: 44–45.

Jenq C, Pan J and Theseira W. (2015) Beauty, weight, and skin color in charitable giving. *Journal of Economic Behavior & Organization* 119: 234–253.

Jeon D-S and DomenicoMenicucci. (2011) When is the optimal lending contract in microfinance state non-contingent? *European Economic Review* 55: 720–731.

Kantor R. (2014) Why venture capital will not be crowded out by crowdfunding. *Alternative Investment Analyst Review* 3: 59–70.

Khang K and Byers S. (2017) The effect of lender and loan type on a borrowing firm's equity return. *Applied Economics* 49: 4099–4115.

Khavul S. (2010) Microfinance: Creating opportunities for the poor? *Academy of Management Perspectives* 24: 57–71.

Kim JC, Watts CH, Hargreaves JR, et al. (2007) Understanding the impact of a microfinance-based intervention on women's empowerment and the reduction of intimate partner violence in South Africa. *American Journal of Public Health* 97: 1794–1802.

Kim WC and Mauborgne R. (2005) Blue ocean strategy: From theory to practice. *California Management Review* 47: 105–121.

Kim WC and Mauborgne R. (2015) *Blue ocean strategy: How to create uncontested market space and make the competition irrelevant,* 60 Harvard Way, Boston, Massachusetts 02163: Harvard Business School Publishing Corporation, ISBN 978-1-62527-450-2.

Kiva. (2017) *Kiva, retrieved from https://www.kiva.org/, accessed on 25 July 2017.*

Korolov R, Peabody J, Lavoie A, et al. (2016) Predicting charitable donations using social media. *Social Network Analysis and Mining* 6: 1–10.

KPMG. (2015) Value-based compliance: A marketplace lending call to action. KPMG, 1–16.

Kraus S, Richter C, Brem A, et al. (2016) Strategies for reward-based crowdfunding campaigns. *Journal of Innovation & Knowledge* 1: 13–23.

Krittanawong C, Zhang HJ, Aydar M, et al. (2018) Crowdfunding for cardiovascular research. *International Journal of Cardiology* 250: 268–269.

Kromidha E and Robson P. (2016) Social identity and signalling success factors in online crowdfunding. *Entrepreneurship & Regional Development* 28: 605–629.

Krupa D and Żołądkiewicz A. (2017) Equity crowdfunding as a form of financing projects in Poland. *Entrepreneurship and Management* XVIII.

Kunz MM, Bretschneider U, Erler M, et al. (2017) An empirical investigation of signaling in reward-based crowdfunding. *Electronic Commerce Research* 17: 425–461.

Kuppuswamy V and Bayus BL. (2013) Crowdfunding creative ideas: The dynamics of project backers in Kickstarter. *SSRN Electronic Journal* http://ssrn.com/abstract=2234765.

Lacan C and Desmet P. (in press) Does the crowdfunding platform matter? Risks of negative attitudes in two-sided markets. *Journal of Consumer Marketing* https://doi.org/10.1108/JCM-03-2017-2126.

Larrimore L, Jiang L, Larrimore J, et al. (2011) Peer-to-peer lending: The relationship between language features, trustworthiness, and persuasion success. *Journal of Applied Communication Research* 39: 19–371.

Lendwithcare. (2017) *Lendwithcare, retrieved from https://www.lendwithcare.org/, accessed on 25 July 2017.*

Lin M, Prabhala NR and Viswanathan S. (2013) Judging borrowers by the company they keep: Friendship networks and information asymmetry in online peer-to-peer lending. *Management Science* 59: 17–35.

Lin M and Viswanathan S. (2016) Home bias in online investment: An empirical study of an online crowdfunding market. *Management Science* 62: 1393–1414.

Lin TT, Lee C-C and Chen C-H. (2011) Impacts of the borrower's attributes, loan contract contents, and collateral characteristics on mortgage loan default. *The Service Industries Journal* 31: 1385–1404.

Lin X, Li X and Zheng Z. (2017) Evaluating borrower's default risk in peer-to-peer lending: Evidence from a lending platform in China. *Applied Economics* 49: 3538–3545.

Lu CT, Xie S, Kong X, et al. (2014) Inferring the impacts of social media on crowdfunding. *Proceedings of the 7th ACM International Conference on Web Search and Data Mining, 24–28 February, New York City, USA*, pp. 573–582.

Lukkarinen A, Teich JE, Wallenius H, et al. (2016) Success drivers of online equity crowdfunding campaigns. *Decision Support Systems* 87: 26–38.

Maitra P, Mitra S, Mookherjee D, et al. (2017) Financing smallholder agriculture: An experiment with agent-intermediated microloans in India. *Journal of Development Economics* 127: 306–337.

Mamonov S, Malaga R and Rosenblum J. (2017) An exploratory analysis of Title II equity crowdfunding success. *Venture Capital* 19: 239–256.

Marom D, Robb A and Sade O. (2014) Gender dynamics in crowdfunding (Kickstarter): Evidence on entrepreneurs, investors, deals and taste based discrimination. *SSRN Electronic Journal* https://ssrn.com/abstract=2442954.

Marwala T. (2009) *Computational intelligence for missing data imputation, estimation and management: knowledge optimization techniques,* New York, USA: IGI Global, ISBN 978-1-60566-336-4.

Marwala T and Xing B. (2011) The role of remanufacturing in building a developmental state. *The Thinker: For the Thought Leaders (www.thethinker.co.za).* South Africa: Vusizwe Media, 18–20.

Marwala T. (2012) On-line condition monitoring using ensemble learning. In: Marwala T (ed) *Condition Monitoring Using Computational Intelligence Methods: Applications in Mechanical and Electrical Systems.* London: Springer-Verlag, ISBN 978-1-4471-2379-8, Chapter 11, pp. 211–226.

Marwala T. (2013a) Multi-agent approaches to economic modeling: Game theory, ensembles, evolution and the stock market. In: Marwala T (ed) *Economic Modeling Using Artificial Intelligence Methods.* Springer London Heidelberg New York Dordrecht: Springer-Verlag London, ISBN 978-1-4471-5009-1, Chapter 11, pp. 195–213.

Marwala T. (2013b) Real-time approaches to computational economics: Self adaptive economic systems. In: Marwala T (ed) *Economic Modeling Using Artificial Intelligence Methods.* Springer London Heidelberg New York Dordrecht: Springer-Verlag London, ISBN 978-1-4471-5009-1, Chapter 10, pp. 173–193.

Marwala T and Hurwitz E. (2017a) Information asymmetry. In: Marwala T and Hurwitz E (eds) *Artificial Intelligence and Economic Theory: Skynet in the Market.* Gewerbestrasse 11, 6330 Cham, Switzerland: Springer International Publishing AG, ISBN 978-3-319-66103-2, Chapter 6, pp. 63–74.

Marwala T and Hurwitz E. (2017b) Introduction to man and machines. In: Marwala T and Hurwitz E (eds) *Artificial Intelligence and Economic Theory: Skynet in the Market*. Gewerbestrasse 11, 6330 Cham, Switzerland: Springer International Publishing AG, ISBN 978-3-319-66103-2, Chapter 1, pp. 1–14.

Massolution. (2015) *Crowdfunding industry report 2015/2016, retrieved from http://crowdexpert.com/crowdfunding-industry-statistics/, accessed on 01 September 2017.*

McLaughlin GH. (1969) SMOG grading—a new readability formula. *Journal of Reading* 12: 639–646.

Medziausyte J and Neugebauer P. (2017) Financing success through equity crowdfunding: The case of start-ups and SMEs funded on European crowdfunding platform. *International Business School*. Jönköping University.

Meer J and Rigbi O. (2013) The effects of transactions costs and social distance: Evidence from a field experiment. *The B.E. Journal of Economic Analysis & Policy* 13: 271–296.

Meer J. (2014) Effects of the price of charitable giving: Evidence from an online crowdfunding platform. *Journal of Economic Behavior & Organ* 103: 113–124.

Michaelidou N and Micevski M. (2015) Consumers' intention to donate to two children's charity banrds: A comparison of Barnardo's and BBC children in need. *Journal of Product & Brand Management* 24: 134–146.

Michel G and Rieunier S. (2012) Nonprofit brand image and typicality influences on charitable giving. *Journal of Business Research* 65: 701–707.

Michels J. (2012) Do unverifiable disclosures matter? Evidence from peer-to-peer lending. *The Accounting Review* 87: 1385–1413.

Milana C and Ashta A. (2012) Developing microfinance: A survey of the literature. *Strategic Change* 21: 299–320.

Miller KL. (2013) *Content marketing for nonprofits: A communications map for engaging your community, becoming a favorite cause and raising more money*, One Montgomery Street, Suite 1200, San Francisco, CA 94104-4594: Jossey-Bass, John Wiley & Sons, Inc., ISBN 978-1-118-44402-3.

Minka TP. (2002) Bayesian model averaging is not model combination. retrieved from https://tminka.github.io/papers/minka-bma-isnt-mc.pdf, accessed on 30 November 2017.

Mishra A. (2014) Million-donor movies: More and more independent regional filmmakers are looking at crowdfunding to make movies. *Business Today, 2 February 2014, Vol. 23 Issue 2*, pp. 98–101.

Mitra T and Gilbert E. (2014) The language that gets people to give: Phrases that predict success on Kickstarter. *Proceedings of the 17th ACM Conference on Computer Supported Cooperative Work & Social Computing, 15–19 February, Baltimore, USA*, pp. 49–61.

Mohammadi A and Shafi K. (in press) Gender differences in the contribution patterns of equity-crowdfunding investors. *Small Business Economics* DOI 10.1007/s11187-016-9825-7.

Mollick E. (2014) The dynamics of crowdfunding: An exploratory study. *Journal of Business Venturing* 29: 1–16.

Morduch J and Haley B. (2002) Analysis of the effects of microfinance on poverty reduction. NYU Wagner Working Paper No. 1014.

Moritz A, Block J and Lutz E. (2015) Investor communication in equity-based crowdfunding: A qualitative-empirical study. *Qualitative Research in Financial Markets* 7: 309–342.

Munyanyi W and Mapfumo A. (2016) Factors influencing crowdfunding plausibility in post hyperinflationary Zimbabwe. *Journal of Entrepreneurship and Business Innovation* 3: 18–28.

Murphy KP. (2012) *Machine learning: a probabilistic perspective*: The MIT Press, ISBN: 978-0-262-01802-9.

Nguyen NH. (2012) Women's motivation for sponsoring children with a charity: An exploratory study. *Faculty of Business and Law*. Auckland University of Technology.

Nuwer R. (2015) Time to donate direct. *New Scientist, 22 November 2015, Vol. 224 Issue 2996*, p. 26.

Ordanini A, Miceli L, Pizzetti M, et al. (2011) Crowd-funding: Transforming customers into investors through innovative service platforms. *Journal of Service Management* 22: 443–470.

Parhankangas A and Renko M. (2017) Linguistic style and crowdfunding success among social and commercial entrepreneurs. *Journal of Business Venturing* 32: 215–236.

Parwez S, Patel R and Shekar KC. (in press) A review of microfinance-led development: evidence from Gujarat. *Global Social Welfare* DOI 10.1007/s40609-017-0095-3.

Pope DG and Sydnor JR. (2011) What's in a picture? Evidence of discrimination from prosper.com. *Journal of Human Resources* 46: 53–92.

Pronyk PM, Harpham T, Busza J, et al. (2008) Can social capital be intentionally generated? A randomized trial from rural South Africa. *Social Science & Medicine* 67: 1559–1570.

Puro L, Teich JE, Wallenius H, et al. (2010) Borrower decision aid for people-to-people lending. *Decision Support Systems* 49: 52–60.

Rai D, Lin C-W and Yang C-M. (2017) The effects of temperature cues on charitable donation. *Journal of Consumer Marketing* 34: 20–28.

Ravina E. (2012) *Love & loans: The effect of beauty and personal characteristics in credit markets, retrieved from https://ssrn.com/abstract=1107307, accessed on 22 November 2017.*

Renwick MJ and Mossialos E. (2017) Crowdfunding our health: Economic risks and benefits. *Social Science & Medicine* 191: 48–56.

Roberts JA and Roberts CR. (2012) Money matters: Does the symbolic presence of money affect charitable giving and attitudes among adolescents? *Young Consumers: Insight and Ideas for Responsible Marketers* 13: 329–336.

Rokach L and Maimon O. (2015) *Data mining with decision trees: Theory and applications*, 5 Toh Tuck Link, Singapore 596224: World Scientific Publishing Co. Pte. Ltd., ISBN 978-9-8145-9007-5.

Roma P, Petruzzelli AM and Perrone G. (2017) From the crowd to the market: The role of reward-based crowdfunding performance in attracting professional investors. *Research Policy* 46: 1606–1628.

Sanders M. (2017) Social influences on charitable giving in the workplace. *Journal of Behavioral and Experimental Economics* 66: 129–136.

Savarese C. (2015) Crowdfunding and P2P lending: Which opportunities for microfinance? Rue de l'Industie 10–1000 Brussels, Belgium: European Microfinance Network (EMN) aisbl, pp. 1–34.

Schaaf P. (2013) Effect of teams on lending behavior on Kiva.org. *Sociology.* Erasmus University Rotterdam.

Schapire RE. (1990) The strength of weak learnability. *Machine Learning* 5: 197–227.

Schapire RE. (2012) *Boosting: foundations and algorithms*, Cambridge, Massachusetts and London, England: The MIT Press, ISBN 978-0-262-01718-3.

Schwienbacher A and Larralde B. (2010) Crowdfunding of small entrepeneurial ventures. *SSRN Electronic Journal* http://dx.doi.org/10.2139/ssrn.1699183.

Sheldon RC and Kupp M. (2017) A market testing method based on crowd funding. *Strategy & Leadership* 45: 19–23.

Shotton J, Fitzgibbon A, Cook M, et al. (2011) Real-time human pose recognition in parts from single depth images. *IEEE Conference on Computer Vision and Pattern Recognition (CVPR), Colorado Springs, CO, USA, 20–25 June*, pp. 1–8.

Sill J, Takacs G, Mackey L, et al. (2009) Feature-weighted linear stacking. retrieved from https://arxiv.org/pdf/0911.0460.pdf, accessed on 30 November 2017.

Sisler J. (2012) Crowdfunding for medical expenses. *Canadian Medical Association Journal (CMAJ)* 184: E123–124.

Smith JR and McSweeney A. (2007) Charitable giving: The effectiveness of a revised theory of planned behavior model in predicting donating intentions and behavior. *Journal of Community & Applied Social Psychology* 17: 363–386.

Stanko MA and Henard DH. (2017) Toward a better understanding of crowdfunding, openness and the consequences for innovation. *Research Policy* 46: 784–798.

Steinberg S and DeMaria R. (2015) *The crowdfunding bible: How to raise money for any startup, video game, or project*, www.asmallbusinessexpert.com: Overload Entertainment, LLC, ISBN 978-1-105-72628-6.

Stiglitz J. (1990) Peer monitoring and credit markets. *World Bank Economic Review* 4: 351–366.

Stojcic I, Kewen L and Xiaopeng R. (2016) Does uncertainty avoidance keep charity away? Comparative research between charitable behavior and 79 national cultures. *Culture and Brain* 4: 1–20.

Teah M, Lwin M and Cheah I. (2014) Moderating role of religious beliefs on attitudes towards charities and motivation to donate. *Asia Pacific Journal of Marketing and Logistics* 26: 738–760.

Venable BT, Rose GM, Bush VD, et al. (2005) The role of brand personality in charitable giving: an assessment and validation. *Journal of the Academy of Marketing Science* 33: 295–312.

Vismara S. (2016a) Equity retention and social network theory in equity crowdfunding. *Small Business Economics* 46: 579–590.

Vismara S. (2016b) Information cascades among investors in equity crowdfunding. *Entrepreneurship Theory and Practice* November: 1–31.

Vulkan N, Åstebro T and Sierra MF. (2016) Equity crowdfunding: A new phenomena. *Journal of Business Venturing Insights* 5: 37–49.

Wang W, Zhu K, Wang H, et al. (2017) The Impact of sentiment orientations on successful crowdfunding campaigns through text analytics. *The Institution of Engineering and Technology (IET) Software* 11: 229–238.

Ward L and Oladele OI. (2013) Factors influencing farmers' attitude towards formal and informal financial markets in the northern cape, South Africa. *Life Science Journal* 10: 2997–3001.

Webb DJ, Green CL and Brashear TG. (2000) Development and validation of scales to measure attitudes influencing monetary donations to charitable organizations. *Journal of Academy of Marketing Science* 28: 299–309.

Wheat RE, Wang Y, Byrnes JE, et al. (2013) Raising money for scientific research through crowdfunding. *Trends in Ecology & Evolution* 28: 71–72.

Widiarto I, Emrouznejad A and Anastasakis L. (2017) Observing choice of loan methods in not-for-profit microfinance using data envelopment analysis. *Expert Systems with Applications* 82: 278–290.

Witten IH and Frank E. (2005) *Data mining: Practical machine learning tools and techniques*, 500 Sansome Street, Suite 400, San Francisco, CA 94111: Morgan Kaufmann, Elsevier Inc., ISBN 0-12-088407-0.

Witten IH, Frank E and Hall MA. (2011) *Data mining: Practical machine learning tools and techniques*, 30 Corporate Drive, Suite 400, Burlington, MA 01803, USA: Morgan Kaufmann, Elsevier Inc., ISBN 978-0-12-374856-0.

Witten IH, Frank E, Hall MA, et al. (2017) *Data mining: Practical machine learning tools and techniques*, 30 Corporate Drive, Suite 400, Burlington, MA 01803, USA: Morgan Kaufmann, Elsevier Inc., ISBN 978-0-12-804291-5.

Wolpert D. (1992) Stacked generalization. *Neural Networks* 5: 241–259.

Wu J and Xu Y. (2011) A decision support system for borrower's loan in P2P lending. *Journal of Computers* 6: 1183–1190.

Wu S, Wang G and Li Y. (2015) How to attract the crowd in crowdfunding. *International Journal of Entrepreneurship and Small Business* 24: 322–334.

Xing B, Eganza J, Bright G, et al. (2006) Reconfigurable manufacturing system for agile manufacturing. *Proceedings of the 12th IFAC Symposium on Information Control Problems in Manufacturing, May 2006, Saint-Etienne, France, pp. on CD.*

Xing B, Gao W-J, Nelwamondo FV, et al. (2010a) Artificial intelligence in reverse supply chain management: The state of the art. *Proceedings of the Twenty-First Annual Symposium of the Pattern Recognition Association of South Africa (PRASA), 22–23 November, Stellenbosch, South Africa*, pp. 305–310.

Xing B, Gao W-J, Nelwamondo FV, et al. (2010b) Part-machine clustering: The comparison between adaptive resonance theory neural network and ant colony system. In: Zeng Z and Wang J (eds) *Advances in Neural Network Research & Applications, LNEE 67*, pp. 747–755. Berlin Heidelberg: Springer-Verlag.

Xing B, Gao W-J and Marwala T. (2013) An overview of cuckoo-inspired intelligent algorithms and their applications. *IEEE Symposium Series on Computational Intelligence (IEEE SSCI), 15–19 April, Singapore*, pp. 85–89. IEEE.

Xing B. (2014) Computational intelligence in cross docking. *International Journal of Software Innovation* 4: 78–124.

Xing B and Gao W-J. (2014a) *Computational intelligence in remanufacturing*, 701 E. Chocolate Avenue, Suite 200, Hershey PA 17033: IGI Global, ISBN 978-1-4666-4908-8.

Xing B and Gao W-J. (2014aa) Chemical-reaction optimization algorithm. In: Xing B and Gao W-J (eds) *Innovative Computational Intelligence: A Rough Guide to 134 Clever Algorithms*. Cham Heidelberg New York Dordrecht London: Springer International Publishing Switzerland, ISBN: 978-3-319-03403-4, Chapter 25, pp. 417–428.

Xing B and Gao W-J. (2014b) *Innovative computational intelligence: a rough guide to 134 clever algorithms*, Cham Heidelberg New York Dordrecht London: Springer International Publishing Switzerland, ISBN: 978-3-319-03403-4.

Xing B and Gao W-J. (2014bb) Emerging chemistry-based CI algorithms. In: Xing B and Gao W-J (eds) *Innovative Computational Intelligence: A Rough Guide to 134 Clever Algorithms*. Cham Heidelberg New York Dordrecht London: Springer International Publishing Switzerland, ISBN: 978-3-319-03403-4, Chapter 26, pp. 429–437.

Xing B and Gao W-J. (2014c) Introduction to computational intelligence. In: Xing B and Gao W-J (eds) *Innovative Computational Intelligence: A Rough Guide to 134 Clever Algorithms*. Cham Heidelberg New York Dordrecht London: Springer International Publishing Switzerland, ISBN: 978-3-319-03403-4, Chapter 1, pp. 3–17.

Xing B and Gao W-J. (2014cc) Base optimization algorithm. In: Xing B and Gao W-J (eds) *Innovative Computational Intelligence: A Rough Guide to 134 Clever Algorithms*. Cham Heidelberg New York Dordrecht London: Springer International Publishing Switzerland, ISBN: 978-3-319-03403-4, Chapter 27, pp. 441–444.

Xing B and Gao W-J. (2014d) Bacteria inspired algorithms. In: Xing B and Gao W-J (eds) *Innovative Computational Intelligence: A Rough Guide to 134 Clever Algorithms*. Cham Heidelberg New York Dordrecht London: Springer International Publishing Switzerland, ISBN: 978-3-319-03403-4, Chapter 2, pp. 21–38.

Xing B and Gao W-J. (2014dd) Emerging mathematics-based CI algorithms. In: Xing B and Gao W-J (eds) *Innovative Computational Intelligence: A Rough Guide to 134 Clever Algorithms*. Cham Heidelberg New York Dordrecht London: Springer International Publishing Switzerland, ISBN: 978-3-319-03403-4, Chapter 28, pp. 445–448.

Xing B and Gao W-J. (2014e) Bee inspired algorithms. In: Xing B and Gao W-J (eds) *Innovative Computational Intelligence: A Rough Guide to 134 Clever Algorithms*. Cham Heidelberg New York Dordrecht London: Springer International Publishing Switzerland, ISBN: 978-3-319-03403-4, Chapter 4, pp. 45–80.

Xing B and Gao W-J. (2014f) Bat inspired algorithms. In: Xing B and Gao W-J (eds) *Innovative Computational Intelligence: A Rough Guide to 134 Clever Algorithms*. Cham Heidelberg New York Dordrecht London: Springer International Publishing Switzerland, ISBN: 978-3-319-03403-4, Chapter 3, pp. 39–44.

Xing B and Gao W-J. (2014ff) Overview of computational intelligence. In: Xing B and Gao W-J (eds) *Computational Intelligence in Remanufacturing*. 701 E. Chocolate Avenue, Suite 200, Hershey PA 17033: IGI Global, ISBN 978-1-4666-4908-8, Chapter 2, pp. 18–36.

Xing B and Gao W-J. (2014g) Biogeography-based optimization algorithm. In: Xing B and Gao W-J (eds) *Innovative Computational Intelligence: A Rough Guide to 134 Clever Algorithms*. Cham Heidelberg New York Dordrecht London: Springer International Publishing Switzerland, ISBN: 978-3-319-03403-4, Chapter 5, pp. 81–91.

Xing B and Gao W-J. (2014h) Cat swarm optimization algorithm. In: Xing B and Gao W-J (eds) *Innovative Computational Intelligence: A Rough Guide to 134 Clever Algorithms*.

Cham Heidelberg New York Dordrecht London: Springer International Publishing Switzerland, ISBN: 978-3-319-03403-4, Chapter 6, pp. 93–104.

Xing B and Gao W-J. (2014i) Cuckoo inspired algorithms. In: Xing B and Gao W-J (eds) *Innovative Computational Intelligence: A Rough Guide to 134 Clever Algorithms.* Cham Heidelberg New York Dordrecht London: Springer International Publishing Switzerland, ISBN: 978-3-319-03403-4, Chapter 7, pp. 105–121.

Xing B and Gao W-J. (2014j) Luminous insect inspired algorithms. In: Xing B and Gao W-J (eds) *Innovative Computational Intelligence: A Rough Guide to 134 Clever Algorithms.* Cham Heidelberg New York Dordrecht London: Springer International Publishing Switzerland, ISBN: 978-3-319-03403-4, Chapter 8, pp. 123–137.

Xing B and Gao W-J. (2014k) Fish inspired algorithms. In: Xing B and Gao W-J (eds) *Innovative Computational Intelligence: A Rough Guide to 134 Clever Algorithms.* Cham Heidelberg New York Dordrecht London: Springer International Publishing Switzerland, ISBN: 978-3-319-03403-4, Chapter 9, pp. 139–155.

Xing B and Gao W-J. (2014l) Frog inspired algorithms. In: Xing B and Gao W-J (eds) *Innovative Computational Intelligence: A Rough Guide to 134 Clever Algorithms.* Cham Heidelberg New York Dordrecht London: Springer International Publishing Switzerland, ISBN: 978-3-319-03403-4, Chapter 10, pp. 157–165.

Xing B and Gao W-J. (2014m) Fruit fly optimization algorithm. In: Xing B and Gao W-J (eds) *Innovative Computational Intelligence: A Rough Guide to 134 Clever Algorithms.* Cham Heidelberg New York Dordrecht London: Springer International Publishing Switzerland, ISBN: 978-3-319-03403-4, Chapter 11, pp. 167–170.

Xing B and Gao W-J. (2014n) Group search optimization algorithm. In: Xing B and Gao W-J (eds) *Innovative Computational Intelligence: A Rough Guide to 134 Clever Algorithms.* Cham Heidelberg New York Dordrecht London: Springer International Publishing Switzerland, ISBN: 978-3-319-03403-4, Chapter 12, pp. 171–176.

Xing B and Gao W-J. (2014o) Invasive weed optimization algorithm. In: Xing B and Gao W-J (eds) *Innovative Computational Intelligence: A Rough Guide to 134 Clever Algorithms.* Cham Heidelberg New York Dordrecht London: Springer International Publishing Switzerland, ISBN: 978-3-319-03403-4, Chapter 13, pp. 177–181.

Xing B and Gao W-J. (2014p) Music inspired algorithms. In: Xing B and Gao W-J (eds) *Innovative Computational Intelligence: A Rough Guide to 134 Clever Algorithms.* Cham Heidelberg New York Dordrecht London: Springer International Publishing Switzerland, ISBN: 978-3-319-03403-4, Chapter 14, pp. 183–201.

Xing B and Gao W-J. (2014q) Imperialist competitive algorithm. In: Xing B and Gao W-J (eds) *Innovative Computational Intelligence: A Rough Guide to 134 Clever Algorithms.* Cham Heidelberg New York Dordrecht London: Springer International Publishing Switzerland, ISBN: 978-3-319-03403-4, Chapter 15, pp. 203–209.

Xing B and Gao W-J. (2014r) Teaching–learning-based optimization algorithm. In: Xing B and Gao W-J (eds) *Innovative Computational Intelligence: A Rough Guide to 134 Clever Algorithms.* Cham Heidelberg New York Dordrecht London: Springer International Publishing Switzerland, ISBN: 978-3-319-03403-4, Chapter 16, pp. 211–216.

Xing B and Gao W-J. (2014s) Emerging biology-based CI algorithms. In: Xing B and Gao W-J (eds) *Innovative Computational Intelligence: A Rough Guide to 134 Clever Algorithms.* Cham Heidelberg New York Dordrecht London: Springer International Publishing Switzerland, ISBN: 978-3-319-03403-4, Chapter 17, pp. 217–317.

Xing B and Gao W-J. (2014t) Big bang–big crunch algorithm. In: Xing B and Gao W-J (eds) *Innovative Computational Intelligence: A Rough Guide to 134 Clever Algorithms.* Cham Heidelberg New York Dordrecht London: Springer International Publishing Switzerland, ISBN: 978-3-319-03403-4, Chapter 18, pp. 321–331.

Xing B and Gao W-J. (2014u) Central force optimization algorithm. In: Xing B and Gao W-J (eds) *Innovative Computational Intelligence: A Rough Guide to 134 Clever Algorithms.* Cham Heidelberg New York Dordrecht London: Springer International Publishing Switzerland, ISBN: 978-3-319-03403-4, Chapter 19, pp. 333–337.

Xing B and Gao W-J. (2014v) Charged system search algorithm. In: Xing B and Gao W-J (eds) *Innovative Computational Intelligence: A Rough Guide to 134 Clever Algorithms*. Cham Heidelberg New York Dordrecht London: Springer International Publishing Switzerland, ISBN: 978-3-319-03403-4, Chapter 20, pp. 339–346.

Xing B and Gao W-J. (2014w) Electromagnetism-like mechanism algorithm. In: Xing B and Gao W-J (eds) *Innovative Computational Intelligence: A Rough Guide to 134 Clever Algorithms*. Cham Heidelberg New York Dordrecht London: Springer International Publishing Switzerland, ISBN: 978-3-319-03403-4, Chapter 21, pp. 347–354.

Xing B and Gao W-J. (2014x) Gravitational search algorithm. In: Xing B and Gao W-J (eds) *Innovative Computational Intelligence: A Rough Guide to 134 Clever Algorithms*. Cham Heidelberg New York Dordrecht London: Springer International Publishing Switzerland, ISBN: 978-3-319-03403-4, Chapter 22, pp. 355–364.

Xing B and Gao W-J. (2014y) Intelligent water drops algorithm. In: Xing B and Gao W-J (eds) *Innovative Computational Intelligence: A Rough Guide to 134 Clever Algorithms*. Cham Heidelberg New York Dordrecht London: Springer International Publishing Switzerland, ISBN: 978-3-319-03403-4, Chapter 23, pp. 365–373.

Xing B and Gao W-J. (2014z) Emerging physics-based CI algorithms. In: Xing B and Gao W-J (eds) *Innovative Computational Intelligence: A Rough Guide to 134 Clever Algorithms*. Cham Heidelberg New York Dordrecht London: Springer International Publishing Switzerland, ISBN: 978-3-319-03403-4, Chapter 24, pp. 375–414.

Xing B and Gao W-J. (2015) Offshore remanufacturing. In: Khosrow-Pour M (ed) *Encyclopedia of Information Science and Technology*. 3rd ed. New York, USA: Information Science Ref. – IGI Global, ISBN 978-1-4666-5888-2, Chapter 374, pp. 3795–3804.

Xing B. (2015a) Graph-based framework for evaluating the feasibility of transition to maintainomics. In: Pedrycz W and Chen S-M (eds) *Information Granularity, Big Data, and Computational Intelligence*. Cham Heidelberg New York Dordrecht London: Springer International Publishing Switzerland, ISBN 978-3-319-08253-0, Chapter 5, pp. 89–119.

Xing B. (2015b) Knowledge management: Intelligent in-pipe inspection robot conceptual design for pipeline infrastructure management. In: Kahraman C and Onar SÇ (eds) *Intelligent Techniques in Engineering Management: Theory and Applications*. Cham Heidelberg New York Dordrecht London: Springer International Publishing Switzerland, ISBN 978-3-319-17905-6, Chapter 6, 129–146.

Xing B. (2016a) Agent-based machine-to-machine connectivity analysis for the Internet of things environment. In: Mahmood Z (ed) *Connectivity Frameworks for Smart Devices: The Internet of Things from a Distributed Computing Perspective*. Switzerland: Springer International Publishing, ISBN 978-3-319-33122-5, Chapter 3, pp. 43–61.

Xing B. (2016b) An investigation of the use of innovative biology-based computational intelligence in ubiquitous robotics systems: Data mining perspective. In: Ravulakollu KK, Khan MA and Abraham A (eds) *Trends in Ambient Intelligent Systems*. Switzerland: Springer International Publishing, ISBN 978-3-319-30184-6, Chapter 6, pp. 139–172.

Xing B. (2016c) Network neutrality debate in the internet of things era: A fuzzy cognitive map extend technology roadmap perspective. In: Mahmood Z (ed) *Connectivity Frameworks for Smart Devices: The IoT Distributed Computing Perspective*. Cham Heidelberg New York Dordrecht London: Springer International Publishing Switzerland, ISBN 978-3-319-33122-5, Chapter 10, pp. 235–257.

Xing B. (2016d) Ontological framework–assisted embedded system design with security consideration. In: Pathan A-SK (ed) *Securing Cyber-Physical Systems*. 6000 Broken Sound Parkway NW, Suite 300, Boca Raton, FL 33487-2742: CRC Press, Taylor & Francis Group, LLC, ISBN 978-1-4987-0099-3, Chapter 4, pp. 91–118.

Xing B. (2016e) Smart robot control via novel computational intelligence methods for ambient assisted living. In: Ravulakollu KK, Khan MA and Abraham A (eds) *Trends in Ambient Intelligent Systems*. Switzerland: Springer International Publishing Switzerland, ISBN 978-3-319-30184-6, Chapter 2, pp. 29–55.

Xing B. (2016f) The spread of innovatory nature originated metaheuristics in robot swarm control for smart living environments. In: Espinosa HEP (ed) *Nature-Inspired Computing for Control Systems*. Cham Heidelberg New York Dordrecht London: Springer International Publishing Switzerland, ISBN 978-3-319-26228-4, Chapter 3, pp. 39–70.

Xing B and Marwala T. (2017) Implications of the fourth industrial age for higher education. *The Thinker: For the Thought Leaders (www.thethinker.co.za)*. South Africa: Vusizwe Media, 10–15.

Xing B. (2017a) Component-based hybrid reference architecture for managing adaptable embedded software development. In: Mahmood Z (ed) *Software Project Management for Distributed Computing: Life-Cycle Methods for Developing Scalable and Reliable Tools*. Gewerbestrasse 11, 6330 Cham, Switzerland: Springer International Publishing AG, ISBN 978-3-319-54324-6, Chapter 6, pp 119–141.

Xing B. (2017b) Protecting mobile payments security: A case study. In: Meng W, Luo X, Furnell S, et al. (eds) *Protecting Mobile Networks and Devices: Challenges and Solutions*. 6000 Broken Sound Parkway NW, Suite 300, Boca Raton, FL 33487-2742: CRC Press, Taylor & Francis Group, LLC, ISBN 978-1-4987-3583-4, Chapter 11, pp. 261–289.

Xing B. (2017c) Visible light based throughput downlink connectivity for the cognitive radio networks. In: Matin MA (ed) *Spectrum Access and Management for Cognitive Radio Networks*. Singapore: Springer Science+Business Media, ISBN 978-981-10-2253-1, Chapter 8, pp. 211–232.

Xing B and Marwala T. (2018a) Conclusion. In: Xing B and Marwala T (eds) *Smart Maintenance for Human–Robot Interaction: An Intelligent Search Algorithmic Perspective*. Gewerbestrasse 11, 6330 Cham, Switzerland: Springer International Publishing AG, ISBN 978-3-319-67479-7, Chapter 13, pp. 299–305.

Xing B and Marwala T. (2018b) Cyberware capacity–applications layer perspective. In: Xing B and Marwala T (eds) *Smart Maintenance for Human–Robot Interaction: An Intelligent Search Algorithmic Perspective*. Gewerbestrasse 11, 6330 Cham, Switzerland: Springer International Publishing AG, ISBN 978-3-319-67479-7, Chapter 8, pp. 173–191.

Xing B and Marwala T. (2018c) Cyberware capacity–energy autonomy perspective. In: Xing B and Marwala T (eds) *Smart Maintenance for Human–Robot Interaction: An Intelligent Search Algorithmic Perspective*. Gewerbestrasse 11, 6330 Cham, Switzerland: Springer International Publishing AG, ISBN 978-3-319-67479-7, Chapter 9, pp. 193–216.

Xing B and Marwala T. (2018d) Cyberware capacity–platform and middleware layers perspective. In: Xing B and Marwala T (eds) *Smart Maintenance for Human–Robot Interaction: An Intelligent Search Algorithmic Perspective*. Gewerbestrasse 11, 6330 Cham, Switzerland: Springer International Publishing AG, ISBN 978-3-319-67479-7, Chapter 7, pp. 143–171.

Xing B and Marwala T. (2018e) Hardware capacity–beginning of life perspective. In: Xing B and Marwala T (eds) *Smart Maintenance for Human–Robot Interaction: An Intelligent Search Algorithmic Perspective*. Gewerbestrasse 11, 6330 Cham, Switzerland: Springer International Publishing AG, ISBN 978-3-319-67479-7, Chapter 4, pp. 67–91.

Xing B and Marwala T. (2018f) Hardware capacity–end of life perspective. In: Xing B and Marwala T (eds) *Smart Maintenance for Human–Robot Interaction: An Intelligent Search Algorithmic Perspective*. Gewerbestrasse 11, 6330 Cham, Switzerland: Springer International Publishing AG, ISBN 978-3-319-67479-7, Chapter 6, pp. 111–139.

Xing B and Marwala T. (2018g) Hardware capacity–middle of life perspective. In: Xing B and Marwala T (eds) *Smart Maintenance for Human–Robot Interaction: An Intelligent Search Algorithmic Perspective*. Gewerbestrasse 11, 6330 Cham, Switzerland: Springer International Publishing AG, ISBN 978-3-319-67479-7, Chapter 5, pp. 93–110.

Xing B and Marwala T. (2018h) Human capacity–biopsychosocial perspective. In: Xing B and Marwala T (eds) *Smart Maintenance for Human–Robot Interaction: An Intelligent Search Algorithmic Perspective*. Gewerbestrasse 11, 6330 Cham, Switzerland: Springer International Publishing AG, ISBN 978-3-319-67479-7, Chapter 11, pp. 249–270.

Xing B and Marwala T. (2018i) Human capacity–exposome perspective. In: Xing B and Marwala T (eds) *Smart Maintenance for Human–Robot Interaction: An Intelligent

Search Algorithmic Perspective. Gewerbestrasse 11, 6330 Cham, Switzerland: Springer International Publishing AG, ISBN 978-3-319-67479-7, Chapter 12, pp. 271–295.

Xing B and Marwala T. (2018j) Human capacity–physiology perspective. In: Xing B and Marwala T (eds) *Smart Maintenance for Human–Robot Interaction: An Intelligent Search Algorithmic Perspective.* Gewerbestrasse 11, 6330 Cham, Switzerland: Springer International Publishing AG, ISBN 978-3-319-67479-7, Chapter 10, pp. 219–247.

Xing B and Marwala T. (2018k) Introduction to human robot interaction. In: Xing B and Marwala T (eds) *Smart Maintenance for Human–Robot Interaction: An Intelligent Search Algorithmic Perspective.* Gewerbestrasse 11, 6330 Cham, Switzerland: Springer International Publishing AG, ISBN 978-3-319-67479-7, Chapter 1, pp. 3–19.

Xing B and Marwala T. (2018l) Introduction to intelligent search algorithms. In: Xing B and Marwala T (eds) *Smart Maintenance for Human–Robot Interaction: An Intelligent Search Algorithmic Perspective.* Gewerbestrasse 11, 6330 Cham, Switzerland: Springer International Publishing AG, ISBN 978-3-319-67479-7, Chapter 3, pp. 33–64.

Xing B and Marwala T. (2018m) Introduction to smart maintenance. In: Xing B and Marwala T (eds) *Smart Maintenance for Human–Robot Interaction: An Intelligent Search Algorithmic Perspective.* Gewerbestrasse 11, 6330 Cham, Switzerland: Springer International Publishing AG, ISBN 978-3-319-67479-7, Chapter 2, pp. 21–31.

Xing B and Marwala T. (2018n) *Smart maintenance for human–robot interaction: An intelligent search algorithmic perspective,* Gewerbestrasse 11, 6330 Cham, Switzerland: Springer International Publishing AG, ISBN 978-3-319-67479-7.

Xu A, Yang X, Rao H, et al. (2014) Show me the money! An analysis of project updates during crowdfunding campaigns. *Proceedings of the 32nd Annual ACM CHI Conference on Human Factors in Computing Systems, 26 April ~ 01 May, Toronto, Canada,* pp. 591–600.

Xu B, Zheng H, Xu Y, et al. (2016) Configurational paths to sponsor satisfaction in crowdfunding. *Journal of Business Research* 69: 915–927.

Yang Y and Pedersen JO. (1997) A comparative study on feature selection in text categorization. *Proceedings of the Fourteenth International Conference on Machine Learning (ICML '97), 08–12 July,* pp. 412–420.

Yuan H, Lau RYK and Xu W. (2016) The determinants of crowdfunding success: A semantic text analytics approach. *Decision Support Systems* 91: 67–76.

Yum H, Lee B and Chae M. (2012) From the wisdom of crowds to my own judgment in microfinance through online peer-to-peer lending platforms. *Electronic Commerce Research and Applications* 11: 469–483.

Yunus M. (2010) *Banker to the poor: Micro-lending and the battle against world poverty,* New York: ReadHowYouWant, ISBN 9781458780287.

Zhang T, Tang M, Lu Y, et al. (2014) Trust building in online peer-to-peer lending. *Journal of Global Information Technology Management* 17: 250–266.

Zheng H, Li D, Wu J, et al. (2014) The role of multidimensional social capital in crowdfunding: A comparative study in China and US. *Information & Management* 51: 488–496.

Zhong ZJ and Lin S. (2017) The antecedents and consequences of charitable donation heterogeneity on social media. *International Journal of Nonprofit and Voluntary Sector Marketing* e1585: 1–11.

Zhu H and Zhou ZZ. (2016) Analysis and outlook of applications of blockchain technology to equity crowdfunding in China. *Financial Innovation* 2: 1–11.

Part IV

Backer + Smart Computing = Firmer Support

CHAPTER 7

Crowdfunding Backer Sentiment Analysis and Fuzzy Product Ontology

7.1 Introduction

In the crowdfunding domain, individuals are typically allowed to post one or more fundraising projects via Kickstarter and Indiegogo-like online platforms. The requests can include causes such as new product/service development (Belleflamme et al., 2014; Mollick, 2014; Sheldon and Kupp, 2017), artistic works (Boeuf et al., 2014; Figliomeni, 2014a; Figliomeni, 2014b; Mixon et al., 2017), journalistic concepts (Jian and Usher, 2014; Carvajal et al., 2012; Sanchez-González and Palomo-Torres, 2014), scientific endeavours (Weigmann, 2013; Özdemir et al., 2015; Wood et al., 2013; Otero, 2015) and philanthropic activities (Saxton and Wang, 2014). Therefore, anyone who is interested in becoming a backer or supporter for a particular project can pledge a small amount of contributions.

It is, therefore, understandable that backers' behaviour is crucial for any project's success (Stanko and Henard, 2017; Mollick, 2016; Xu et al., 2016). There are, however, some obstacles that may hinder backers' devotion.

- **Differentiation:** Backers are often offered multiple projects before making their final funding decisions (Kuppuswamy and Bayus, 2013), while the essential differences among many of these project options are usually hard to identify (Corazzini et al., 2015; Meer, 2014). Because of this, several scholars pointed out that the chances for prospective backers to fund crowdfunding campaigns are particularly low under certain conditions, such as when different projects' qualities are hardly distinguishable (Zhao and Vinig, 2017; Ward and Ramachandran, 2010; Bi et al., 2017) and when some project has already received a lot of support (Kuppuswamy and Bayus, 2013).

- **Stimulation:** In addition, Mollick (2014), Gerber and Hui (2013) and Moysidou (2016) noticed that the backers' true motivations are extremely heterogeneous. For instance, Kuppuswamy and Bayus (2013) showed that the deadline plea (via continuous project corresponded updates) can often assist project askers in achieving their fundraising goals. Meanwhile, different response patterns towards the project associated narratives were also observed by Allison et al. (2014), that is, if the expected contributions/donations are highlighted as an opportunity for helping others in a project's narratives, backers are more likely to respond to it actively; in contrary to this, backers behave less actively when the narratives emphasize the business opportunity aspect.
- **Population:** Furthermore, Lipusch et al. (2016) pointed out that with the mushrooming of crowdfunding sites and the exponential growth of list projects, backers are quickly turning to a scarce resource. To this end, Chen et al. (2009) and Mollick (2013) focused on the role of preparedness in bolstering a project.

To summarize, all this suggests that the efforts involved in interpreting backers' sentiment are more likely to pay askers back than other fundraising strategies (Colistra and Duvall, 2017; Xu et al., 2016). Indeed, similar to the role of consumer satisfaction in the online shopping context, backers' sentiment is not only a key predictor in forecasting backers' contribution, re-contribution and e-'word-of-mouth (WoM)' behaviour (Van Montfort et al., 2000; Posselt and Gerstner, 2005; Kuo and Wu, 2012), but a crucial indicator of whether or not a project can be successful (Lipusch et al., 2016; Bi et al., 2017).

7.1.1 *Key Factors Influencing Backers' Donating Intention/ Motivation in Non-Profit Campaigns*

The non-profit campaign category (i.e., backers voluntarily made their contributions and the attachment of definite rewards is usually not necessary) often incorporates crowdfunding and charity in healthcare (Renwick and Mossialos, 2017; Granville, 2016), civic (Davies, 2014; Stiver et al., 2015; Griffiths, 2017; Feinberg, 2014), education (Meer, 2014; Next, 2013) and social causes (Nuwer, 2015; Mano, 2014).

The factors that drive non-profit giving include but are not limited to: geographic dispersion of backers studied in (Mendes-Da-Silva et al., 2016; Lin and Viswanathan, 2016; Giudici et al., in press), dynamic behaviour patterns of backers examined by Kuppuswamy and Bayus (2013) and Mollick (2014) via Kickstarter; social identity and social status considered by Gerber and Hui (2013); online social media effect investigated by Mano (2014) with respect to engaging voluntarily, giving behaviour, and contributing funds; the relationship between crowdfunding and diaspora philanthropy outlined by Flanigan (2017); trust issue covered by MacMillan et al. (2005) when donating to non-profit institutions; and mutual visibility analysed by Saxton and Wang (2014) when tapping the crowd. Although non-profit campaigns were more successful to some extent, Pitschner and Pitschner-Finn (2014) argued that the total number of backers and the overall funding amount were comparatively low. Therefore, in another study, Zhong and Lin (2017) claimed that donating is a rational behaviour that balances altruism and self-interest. They suggested that a backer's motivation could be influenced by differentiating donation strategies (e.g., frequency and magnitude of giving). In a similar vein, the studies on microfinance (Kasper and Marcoux, 2015; Asgary and McNulty, 2017; McKinnon et al., 2013; Staats et al., 2013; Ly and Mason, 2012) found that the backer may be more interested in supporting social good promoted venture (e.g., helping poor people and lifting them out of poverty), other than any other returns generated by loan (i.e., at an interest rate of zero). The best-known example that implements these findings is the Kiva platform, which includes non-profit elements.

The full list of factors might be too formidable to outline. For simplicity, a good review and summarization which covers eight dimensions, offered by Bekkers and Wiepking (2011), is provided below:

1) *Dimension 1*: Awareness of need;
2) *Dimension 2*: Costs and benefits;
3) *Dimension 3*: Solicitation;
4) *Dimension 4*: Values;
5) *Dimension 5*: Altruism;
6) *Dimension 6*: Reputations;
7) *Dimension 7*: Psychological gains; and
8) *Dimension 8*: Efficacy.

Among them, reputations, psychological benefits and efficacy are considered to be associated with backers' self-interests.

7.1.2 *Key Factors Influencing Backers' Donating Intention/ Motivation in Incentive-Based Campaigns*

At the same time, incentive-based crowdfunding has become the subject of many academic studies as well. According to (Bretschneider and Leimeister, in press), incentive-based campaigns include reward-, equity-

and peer-to-peer (P2P)-based crowdfunding. Because of this, several scholars pointed out that rewards (either intrinsic or extrinsic) serve the key factors for backers participating in incentive-based crowdfunding campaigns (Schwienbacher and Larralde, 2010; Kleemann et al., 2008; Belleflamme et al., 2014; Hobbs et al., 2016; Collins and Pierrakis, 2012).

- **Intrinsic Rewards:** Regarding this type of reward (e.g., for pleasure or simply for joy-of-doing something), Colistra and Duvall (2017) found that backers were willing to fund projects not only because they are much more likely to have the products, but also because they feel involved in the process of testing the ideas and/or facilitating the projects. In this stream, Zhao and Vinig (2017) investigated the impact of backers' hedonic and utilitarian values on projects' results, while some others focused on whether the projects have inherent social meaning, say, importance to a community (Mollick, 2016; Gerber and Hui, 2013) or to the broader human society (Kuppuswamy and Bayus, 2017; Gerber et al., 2012). Generally speaking, the main incentive that drives backers' engagement under this branch can be termed as 'experience participation'.
- **Extrinsic Rewards:** Regarding this type of reward (e.g., pre-ordering goods or monetary returns), backers are mainly motivated by the idea of receiving favoured products in the first place (Gerber et al., 2012) and harvesting returns on investment (Ordanini et al., 2011; Cholakova and Clarysse, 2015). For example, entrepreneurs usually like using 'positive signal' (e.g., how many rewards will be offered and the number of granted patents) to indicate the potential value of the new venture (Ahlers et al., 2015; Kunz et al., 2017; Courtney et al., 2017; Hsu and Ziedonis, 2013; Colombo et al., 2015). In light of this, Kunz et al. (2017) proposed the idea of applying signalling theory to incentive-based crowdfunding campaigns in order to help backers evaluate a campaign. Meanwhile, Cholakova and Clarysse (2015) based their study on self-determination theory and cognitive evaluation theory in order to examine different types of causations between financial and non-financial incentives. In addition, Zhao et al. (2017) explored the factors that could potentially influence backers' funding intention via social exchange theory.

  To be sure, this group of backers is inherently different from the former experience backer group. However, the backers' attitudes towards the funded projects do share some common characteristics, e.g., empowerment and well-documented profile. For instance, Ortega and Bell (2008) observed that Zopa (i.e., a P2P lending platform) motivated its backers by allowing them to make complex financial decisions. A similar conclusion was also drawn in (Lipusch et al., 2016). Regarding the projects' information, Koufaris and Kambil (2001) explained that such adequacy can help potential backers save a huge amount of time. The funding process is, therefore, much more controllable from the backers' perspective.

In addition, since extrinsic reward-driven crowdfunding bears some financial risks, existing literature (Zhang et al., 2014; Zheng et al., 2016; Greiner and Wang, 2010; Oosterhoff, 2015; Reimink, 2014) also raised the issue of trust and commitment. Unlike angel investors or venture capital funds, Belleflamme et al. (2014) stressed that backers might not have any special knowledge about the interested crowdfunding projects, let alone the underlying industry ecosystem. Gerber and Hui (2013) underlined another important factor that deters potential backers from becoming true donors is the distrust regarding askers' projected usage of funds. Likewise, Colistra and Duvall (2017) concluded that one of the backers' major concern was on the part of crowdfunding platforms, i.e., their accountability in administering askers to complete the promised projects. Other similar studies were also carried out in (Zheng et al., 2016; Zhao et al., 2017).

Overall, for the spectrum of incentive-based crowdfunding campaigns, the backers' motivation can be portrayed by referring to the following axes (Steigenberger, 2017; Cholakova and Clarysse, 2015; Belleflamme et al., 2014; Gerber and Hui, 2013; Colombo et al., 2015; Bretschneider and Leimeister, in press; Gómez-Diago, 2015; De Buysere et al., 2012; Harms, 2007):

1) *Axis 1*: Needs and likes (e.g., consumption);
2) *Axis 2*: Altruism;
3) *Axis 3*: Social belonging (e.g., image or recognition stimulus);
4) *Axis 4*: Quality of the projects; and
5) *Axis 5*: Return on investment.

7.1.3 How to Pitch a Project?

An online crowdfunding platform typically serves as a link between the askers and the backers. Therefore, when resorting to crowdfunding for whatever reasons, e.g., project/loan, some assistant add-ons can make a big difference between a successful and a failed campaign.

- **A Narrative and A Short Video:** Briefly, a narrative describes the meaning of the narrator's ideas, business plans, development timelines, use of funds, etc., while a project video (often less than three minutes) contains an essay (or a story) outlined by the askers and some further information about the projects. According to Kickstarter's data (Yonata, 2017), the success funded rate for projects without/with videos is quite significant –50% vs. 30% (given the overall average success rate is around 40%). Other benefits of using a narrative or a short video to pitch a project include (Burtch, 2013; Stern, 2013; Beier and Wagner, 2014; Wheat et al., 2013): (1) Bringing leaps of joy to backers, (2) Engaging and leveraging an amount of trust between backers and askers and (3) Boosting backers' retention to new heights.

In addition, crowdfunding platforms also enable real-time updating public comments, e.g., the expression of best wishes for the campaign, the joint discussion with other backers about the project and the questions and answers regarding the project. In (Bretschneider and Leimeister, in press; Zheng et al., 2014; Boeuf et al., 2014), the authors observed that askers and backers all use this functionality very often in exchanging thanks, praises and acknowledgements for received supports and deliverables.

- **Social Media Messaging:** Indeed, from a backer's point of view, a campaigns' homepage information and the related social media (e.g., Facebook and twitters) messages draw a big picture in order to convince backers to provide funding for the project (Wang et al., 2017; Nambisan and Zahra, 2016; Moisseyev, 2013; Zheng et al., 2014). For example, several authors (Huntley, 2006; Liu et al., 2011; Colombo et al., 2015) emphasized that relational communications (directly or indirectly) could influence not only the customers' purchasing decision, but the potential backers' willingness to donate. In recent years, we are witnessing a growing scholarly interest, e.g. (Ward and Ramachandran, 2010; Mollick and Nanda, 2014), regarding whether and how pseudo-personal (e.g., videos or social media messaging) communications can reduce information asymmetries and actuate backers' donating decisions.

7.1.4 *Sentiment Analysis*

In reality, with the ever-expanding information tree, backers begin to face the difficulty of locating the right comments that are in their best interests (Zhao and Vinig, 2017; Hamilton and Thompson, 2007). In addition, text and video information are among the fuzziest and least understood information sources (Gao and Lin, 2016; Schau and Gilly, 2003). Therefore it is not easy to obtain a full set of interpretations (Sonenshein, 2010), and the fraudulent implications are not uncommon (Schlenker and Weigold, 1992; Snyder et al., 2016). Because of this, there is a need to assist backers when massive information is not truly helpful.

In the literature, examining the influence of linguistic features mined from textual or video descriptions is often called sentiment (content) analysis (Pang and Lee, 2008; Liu, 2015). A more formal definition (Wang et al., 2017; Bryman and Bell, 2015) is that sentiment analysis refers to the use of any technique, e.g., natural language processing (NLP), opinion mining, text analysis and computational linguistics, to systematically and objectively extract subjective meanings, implications and characteristics from various sources of materials, such as political texts (Haselmayer and Jenny, 2017), public opinions (Ali et al., 2017), brand evaluations (Hornikx and Hendriks, 2015) and financial decisions (Li et al., 2017; Bollen et al., 2010).

Among others, one means of enhancing the sentiment (or feeling) is to resort to social capital (Beier and Wagner, 2014; Mano, 2014). According to Putnam (1993), individuals tend to use their social capital (including networks, norms, trusts, etc.) in order to facilitate the cooperation with others for common goals and mutual benefits. In fact, previous studies in financial and marketing fields also agreed with the usefulness of incorporating social capital, which includes helping businesses build powerful ties with customers (Walther, 1992), creating a pool of creditworthy borrowers (Everett, 2010), increasing awareness (Gandia, 2011), reducing information asymmetry (Zhang et al., 2016; Collier and Hampshire, 2010; Lin et al., 2013), promoting customers' willingness-to-pay (Kim and Crompton, 2001) and predicting charitable donations (Korolov et al., 2016; Huntley, 2006; Bekkers and Wiepking, 2011; Auger, 2013), to name just a few. For more details, one should refer to several recently published books (Cambria et al., 2017b; Liu, 2015; Zhang, 2014; Lischka, 2016; Cambria and Hussain, 2015; Peterson, 2016).

7.1.4.1 The Role of Economic Sentiment

According to Lischka (2016), economic sentiment deals with how economic and business news and opinions (i.e., usually posted on social media and blogs) affect the collective economic expectations and behaviours, such as making investments and private purchases. Therefore, economic sentiment becomes one of the pillars of behavioural finance, i.e., studying human fallibility in competitive marketplaces (Shleifer, 2000; Marwala and Hurwitz, 2017b). In the domain of economic sentiment, researchers focus mainly on the following two branches (Kearney and Liu, 2013):

- **The Narrow One (e.g., investor sentiment):** This branch concerns the manner in which individuals/institutional investors build their beliefs about the market and future economic trends. Following this narrow angle, in recent years, economic sentiment applications have focused mainly on stock market predictions. For example, Saade (2015) examined, at the time of a technological company's IPO (i.e., initial public offering), the effect of investor sentiment (both individual and institutional) towards the entire market on the aftermarket performance of such company IPO shares; while Yao et al. (2017) aimed to capture tail dependence between investor sentiment and stock market index. Other examples are included in (Burghardt, 2010; Zhang, 2014).

- **The Broad One (e.g., natural language- and text-based sentiment):** This branch involves the classification of polarities (e.g., positive or negative, strong or weak, active or passive, or simply neutral), or features (e.g., happy, angry or sad) of a given text piece or video clip. Regarding this broad angle, the popularity of entrepreneurial finances has prompted researchers to focus on their needs, since entrepreneurs have not only evoked a variety of novel investment motivations (e.g., non-financial) but have also brought a basket of opinions about the new investment approaches (e.g., crowdfunding), business models (e.g., provision of advice) and valuation measures (e.g., social return on investment). All this naturally leads to the problem of sentiment analysis. Some typical questions currently under consideration (Block et al., in press; Kuppuswamy and Bayus, 2013; Brüntje and Gajda, 2016) include: (1) How can young innovative startups find alternative finance channels for company growth, innovation, and internationalization? (2) How can the foundation of entrepreneurial culture (e.g., values and beliefs) be strengthened? (3) What can be implemented in order to enhance individual entrepreneurial spirit?

7.1.4.2 Where can Sentiment Analysis Fit into Crowdfunding?

In the past few years, a popular new player has gradually entered into entrepreneurial finance arena. Thanks to the growing use of crowdfunding, a solid foundation for sentiment analysis has been laid in this field. Some of the promising directions are outlined as follows:

- **Social Networks:** The function of social networks in crowdfunding is well recognized and broadly researched by many scholars (Moisseyev, 2013; Zheng et al., 2014; Agrawal et al., 2015; Hui et al., 2014; Aprilia and Wibowo, 2017; Skirnevskiy et al., 2017; Horvát et al., 2015; Lin et al., 2013).
- **Campaign Profiles:** Every day, backers are flooding into crowdfunding platforms with all types of data such as story texts, pictures and videos. Therefore, digging valuable information from people's current and/or past sentiment has become extremely important for a project to survive. In this regard, Wang et al. (2017) studied the impact of sentiment orientations on crowdfunding campaigns through text analytics. Parhankangas and Renko (2017) concentrated on linguistic styles that appeared in crowdfunding pitches and tried to answer how a successful fundraising is potentially related to the recognized styles. Herzenstein et al. (2011b) examined how backers' decisions are

influenced by the identity claims existing in narratives made by the askers. Siering et al. (2016) used sentiment analysis in order to detect fraudulent behaviour on crowdfunding platforms. Nambisan and Zahra (2016) explored the demand-side narratives, whereas Allison et al. (2014) conducted their study in the context of micro-lending markets and covered the significance of entrepreneurial narratives' intrinsic and extrinsic cues. Gao and Lin (2014) analysed correlations between the linguistic styles of borrower phrased texts and the quality of received loans and most recently, a set of survey data from the Netherlands was utilized by Polzin et al. (in press) to recognize in-crowd and out-crowd funders. Their study intended to examine the inherent heterogeneity associated with their information usage.

- **People's Mood:** In the stock market, people tend to be overoptimistic about the growth prospects of a technologically advanced company. In the crowdfunding domain, the similar sentiment emerges when people meet technologically advanced projects. In this regard, some studies discovered that the decisions made by backers/investors are not always the outcome of referring to solid financial data, but often the product of emotional behaviours that are unnoticeably controlled by different decision biases such as herding effect (Berkovich, 2011; Herzenstein et al., 2011a; Kuppuswamy and Bayus, 2013; Lee and Lee, 2012; Zhang and Liu, 2012; Yum et al., 2012), free riding behaviour (Burtch et al., 2013), home bias (Agrawal et al., 2011; Guenther et al., in press; Agrawal et al., 2015; Lin and Viswanathan, 2016; Giudici et al., in press), technology bias (Cordova et al., 2015; Fernandes, 2013; Samanci and Kiss, 2014) and some taste-based discriminations, e.g., gender (Mohammadi and Shafi, in press; Chen et al., in press; Piper and Schnepf, 2008; Agier and Szafarz, 2013; Moodie, 2013; Riggins and Weber, 2017), appearance (Pope and Sydnor, 2011) and race (Harkness, 2016; Walter, 2008).
- **Emerging Areas:** Some niche but interesting topics regarding entrepreneurs' psychology and philosophy are worth attentions as well. Examples of these include: how to tell a good story (Tugend, 2015; Bluestein, 2014; Cavanagh, 2014); When are entrepreneurs' lying behaviours forgivable by investors? (Pollack and Bosse, 2014); Is lending discrimination always expensive? (Ferguson and Peters, 2000); What are the differences between online (i.e., web-based platforms) and offline (i.e., face-to-face interactions) crowdfunding? (Gras et al., 2017); How is the relationship between crowdfunding and social responsibility (Marom, 2017), and why the askers who has received an over-requested amount of funds are very unlikely to complete their projects in time or on time? (Mollick, 2014).

7.1.4.3 Methods and Models for Sentiment Analysis

At first glance, people might get the false impression that sentiment analysis is just a matter of scrutinizing a document (or a sentence) to see if it contains any positive or negative sentiment (or opinion). Yet, in practice, the involvement of different aspects and features tend to complicate the analysis. Previously, in order to measure the sentiment, two methods were intensively used, i.e., survey-based and market data-based methods. However, both suffer notable disadvantages, that is, the former is too costly to generate while the latter is too difficult to interpret (Burghardt, 2010).

With the explosive growth of social media applications since the early 2000s, the field of sentiment analysis again becomes very active, particularly thanks to the rapid advancement of data mining techniques. During this period, several new text-/web-mining approaches and models were developed. Generally speaking, one can cluster them into three groups: statistical approaches, knowledge-based techniques and hybrid methods (Cambria et al., 2017a):

- **Statistical Approaches:** In this group, the most popular approaches are deep learning and support vector machines. Several commercial and academic tools (including IBM, SAS, Oracle, Luminoso, etc.) are built upon statistical properties of texts/words in order to offer comments and predict trends. The main drawback of these statistical approaches lies in the fact that they are semantically weak, therefore, their applications are often limited to a polarity evaluation or a mood classification. In addition, the training data used to learn patterns for this group of approaches is in notoriously short supply. As a result, they are just survey-based methods but enhanced by advanced algorithms (Cambria et al., 2017b).
- **Knowledge-Based Techniques:** In general, knowledge-based techniques include information entropy, rough entropy and knowledge granulation. However, when linguistic principles are implemented, these techniques often perform poorly when attempting to recognize affection. Meanwhile, flexibility is another weakness of these techniques since knowledge representations are usually strictly defined (Cambria et al., 2017a).
- **Hybrid Methods:** In fact, the next-generation sentiment mining systems are more likely to be hybridized, that is, offer a good integration between the structured machine treatable data (via machine learning algorithms) and unstructured multimodal linguistic knowledge (Cambria et al., 2017b). One noteworthy method that represents this trend is sentic computing (Cambria and Hussain, 2012; Cambria and Hussain, 2015) framework.

7.2 Problem Statement

Typically, there are two types of social networks involved in crowdfunding (Young, 2013), namely (1) The internal one provided by crowdfunding platforms for the purpose of publishing project relevant information (e.g., descriptions and videos) and engaging with potential backers; and (2) The external one developed by askers themselves via other third-party social network websites (e.g., Facebook, Twitter, LinkedIn and YouTube). This social network can also be utilized for promoting the campaign and interacting with backers.

According to some existing studies (Bechter et al., 2011; Agrawal et al., 2015; Zheng et al., 2014; Baeck et al., 2014; Lu et al., 2014; Giudici et al., 2013; Giudici et al., in press), social networks are dynamically linked to the overall ratio of successful project. Indeed some other studies (Mollick, 2014; Lehner, 2013; Greiner and Wang, 2010; Vismara, 2016; Colombo et al., 2015; Ordanini et al., 2011; Lin et al., 2013) even suggested that the amount of capital assembled via the crowdfunding channel is closely related to the range of an asker's social networks. Other benefits of social networks also include obtaining backers' feedback (Gerber and Hui, 2013), collecting creative business ideas and solutions (Schwienbacher and Larralde, 2010; Ordanini et al., 2011) and accumulating community benefits (Gerber et al., 2012; Belleflamme et al., 2014).

However, the valuation of social networks' credibility is subjective and problematic as well. As Podolny (2001) pointed out that social networks can serve as 'pipes', yet they can also function as 'prisms' as well. By pipes we mean a direct means for information transfer; while the term 'prisms' implies an indirect way to show the relationship between the third parties' perceptions (regarding the goods and services supplied by an asker) and this particular asker's social connections. In this regard, recent studies disclosed some looming problems as a by-product of the popular use of social networks during crowdfunding campaigns. For example, Siering et al. (2016) found that the risk of fraud has gone up, while Bretschneider and Leimeister (in press) observed some kind of moderating effect which is caused by backers' irrational herding behaviour and has a remarkable impact on their reward motivations. A further investigation was conducted by Kuppuswamy and Bayus (2013) who discover some interesting differences in backers' herding (i.e., bathtub shaped pattern) phenomenon between reward-based crowdfunding and other crowdfunding alternatives, e.g., donation (Burtch et al., 2013), equity or peer-to-peer (P2P) lending crowdfunding (Zhang and Liu, 2012; Liu et al., 2015; Lee and Lee, 2012; Herzenstein et al., 2011a; Zhang and Chen, 2017; Chen and Lin, 2014).

7.2.1 *Question 7.1*

Faced with those uncertainties, it is necessary to leverage social networks for crowdfunding campaigns in order to receive askers' understanding and, thus, improve their confidence. In other words, the question is no longer whether backers' sentiment (e.g., opinion, evaluation, attitude, emotion and mood) is important, but rather to move onto the next stage. In view of this, Question 7.1 is proposed as follows:

- **Question 7.1:** *How do we measure and verify existing backers' sentiment (e.g., opinion, evaluation, attitude, emotion and mood) so that the newcomers can interpret these subjective feelings and beliefs correctly?*

In investigating Question 7.1, our expectations are numerous: (1) To explore an advanced social analytics methodology, which can effectively extract social intelligence from consumer comments published on social media websites; (2) To examine an innovative fuzzy algorithm for mining product ontology with various product aspects (both explicit and implicit) and the associated aspect-related sentiments captured; (3) To test a semi-supervised statistical learning approach with autonomous context-sensitive sentiments extraction capability.

7.3 Sentic Computing for Question 7.1

In order to address Question 7.1, a set of explorative studies were conducted by Cambria and Hussain (2012), Lau et al. (2014) and Fernandes (2013). In this section, we first introduce some background knowledge about the employed methodologies, which is then followed by a case study of crowdfunding.

7.3.1 *Sentic Computing*

As a pioneering multi-disciplinary toolkit, sentic computing (Cambria and Hussain, 2015) blends research results from other fields (e.g., affective computing, information extracting, social science, computer science and common-sense reasoning) in order to process and understand natural language. Sentic was termed by referring the Latin '*sentire*' (origin of sentiment, sentience, etc.) and 'sensus' (meaning common-sense) (Cambria and Hussain, 2015). In general, sentic computing accelerates the following three novel shifts (Cambria and Hussain, 2015):

- **Methodological Shift:** In shifting from mono-disciplinary to the desired multi-disciplinary, sentic computing promotes the ensemble utilization of the following techniques.

1) *Artificial intelligence and semantic Web techniques*: For representing and inferring knowledge;
2) *Mathematics*: For mining graph and reducing multi-dimensionality alike tasks;
3) *Linguistics*: For analysing discourse and exploring pragmatics;
4) *Psychology*: For modelling cognition and affection;
5) *Sociology*: For following social network dynamics and valuing social influence;
6) *Ethics*: For comprehending the nature of the mind and nurturing emotional machines.

For more details one should refer to (Xing and Gao, 2014b; Xing and Gao, 2014x; Xing and Gao, 2014q; Xing and Gao, 2014m; Xing and Gao, 2014s; Xing and Gao, 2014z; Xing and Gao, 2014bb; Xing and Gao, 2014dd; Xing and Gao, 2014c).

- **Semantical Shift:** In shifting from syntax to semantics, sentic computing focuses more on the bag-of-concepts model rather than simply getting the co-occurrence frequency of words counted.
- **Linguistical Shift:** In shifting from statistics to linguistics, sentiments are allowed to move freely between concepts according to the dependency relationship between clauses. Using any technological crowdfunding campaign as an example, the expression "The HubDock video is good but boring" equates to "The HubDock video is boring but good" on the surface through the lens of bag-of-words. However, the underlying polarities of these two sentences might be completely opposite: the person in the first scenario does not seem to be interested in the introduced product, though the video itself is not bad at all; while for the second scenario, the person seems not interested in the video itself, but has a good feeling about the introduced product.

7.3.2 *SenticNet*

As a publicly accessible semantic repository for assisting concept-level sentiment analysis, SenticNet (Cambria et al., 2014) intends to fill the gap (both conceptual and affective) between the explicit word-level natural language data and the associated implicit concept-level attitudes and sentiments, through a toolkit that combines graph mining and multi-dimensional scaling techniques. SenticeNet serves as a knowledge base which can be very helpful for developing diverse applications in various areas, e.g., socially assistive robot (Xing and Marwala, 2018h). At present, there are totally 30,000 common-sense concepts associated semantics and sentics accessible via SenticNet (Cambria and Hussain, 2015).

In contrast to other sentiment-analysis counterparts, which are often constructed via labelling bits and pieces of knowledge (often extracted from general natural language processing resources) manually, SenticNet is built autonomously through the following stages (Cambria and Hussain, 2012; Cambria and Hussain, 2015). A simplified illustration is provided in Fig. 7.1.

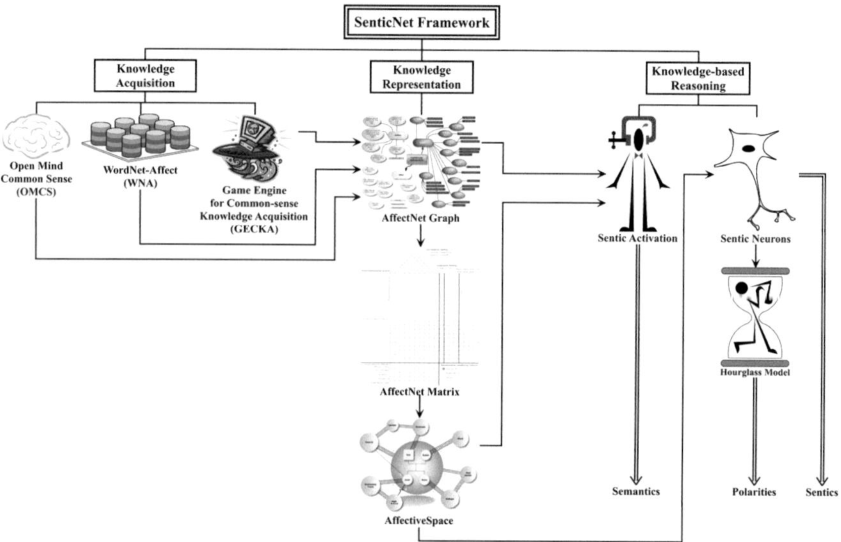

Figure 7.1: Schematic representation of SenticNet framework establishment.

7.3.2.1 Knowledge Acquisition

The main knowledge bases and sources underpinning SenticNet cover (Cambria and Hussain, 2012; Cambria and Hussain, 2015): (1) The Open Mind Common Sense for extracting general common sense knowledge, (2) WordNet-Affect for delivering affective knowledge and (3) GECKA (i.e., game engine for common-sense knowledge acquisition) for crowdsourcing practical common sense knowledge.

7.3.2.2 Knowledge Representation

When knowledge is successfully collected from different bases and sources, we need to figure out a means to represent it. In (Cambria and Hussain, 2012; Cambria and Hussain, 2015), three levels were proposed: Network level, matrix level and vector space level.

- **AffectNet Graph:** At Level One, a triple form <concept-relationship-concept> is employed in order to integrate the bits and pieces of knowledge (obtained via accumulating or crowdsourcing) into a semantic network. In the context of SenticNet, the output of this process is often termed as AffectNet Graph (Cambria and Hussain, 2012; Cambria and Hussain, 2015).

- **AffectNet Matrix:** At Level Two, the matrix form is introduced in order to interpret the AffectNet Graph. In general, rows of the matrix consist of various concepts; while the columns of the matrix incorporate the combination of <relationship-concept>. By doing so, we can see the entire knowledge-related data repository through the lens of a large matrix. One of the key advantages of such a transformation is the potential of performing cumulative analogy, as proposed in (Chklovski, 2003; Turney, 2002). The core of cumulative analogy is the use of similarity, which represents the most intuitive way for humans to categorize things. However, one drawback of time approach lies in its time- and resource-consuming nature. The underlying reason is the formation of a fat matrix, which is the result of introducing a couple of thousand rows and columns (for indicating different features) (Cambria and Hussain, 2012; Cambria and Hussain, 2015).
- **AffectiveSpace:** One possible way to address the matrix issue arising at Level Two is to introduce a vector space at Level Three. With the aid of multi-dimensionality reducing methodologies, we can reasonably reduce the amount of concept-linked semantic features, i.e., the <relationship-concept> combination, without sacrificing too much knowledge representation. In practice, intuition is often used by us in order to tackle a situation that we have never encountered before. As put by Cambria and Hussain (2015), intuition can be loosely defined as a process of identifying analogies between the problem at hand and other previously solved problems in the hope of finding a favourable solution. Minsky (1986) termed this property as 'difference-engines'. Accordingly, AffectiveSpace was proposed in (Cambria, 2015), aiming at transforming conceptual structures (featured by their diffusion and distention characteristics) into more concentrated versions. In order to represent the matrix form of AffectNet, the principal component analysis (PCA) (Marwala, 2015k; Marwala, 2009; Marwala, 2012) approach was introduced, while the truncated singular value decomposition technique was utilized for reducing dimensionality (Cambria and Hussain, 2015). When such operations are performed, we can get a low-rank approximation of the original matrix, as given by Eq. 7.1 (Cambria and Hussain, 2015):

$$\min_{\tilde{\mathbf{A}}|rank(\tilde{\mathbf{A}})=k} \left|\mathbf{A}-\tilde{\mathbf{A}}\right| = \min_{\tilde{\mathbf{A}}|rank(\tilde{\mathbf{A}})=k} \left|\Sigma-\mathbf{U}^{*}\tilde{\mathbf{A}}\mathbf{V}\right| = \min_{\tilde{\mathbf{A}}|rank(\tilde{\mathbf{A}})=k} \left|\Sigma-\mathbf{S}\right|. \qquad 7.1$$

where **S** represents a diagonal matrix, and $\tilde{\mathbf{A}}$ is assumed to equal to $\mathbf{USV}^*$. According to (Eckart and Young, 1936), we can see this as matrix **A**'s best approximation from the least-square perspective as given by Eqs. 7.2 and 7.3 (Cambria and Hussain, 2015):

$$\min_{\tilde{A}|rank(\tilde{A})=k} |\Sigma - S| = \min_{s_i} \sqrt{\sum_{i=1}^{n} (\sigma_i - s_i)^2} . \qquad 7.2$$

$$\min_{s_i} \sqrt{\sum_{i=1}^{n} (\sigma_i - s_i)^2} = \min_{s_i} \sqrt{\sum_{i=1}^{k} (\sigma_i - s_i)^2 + \sum_{i=k+1}^{n} \sigma_i^2} = \sqrt{\sum_{i=k+1}^{n} \sigma_i^2} . \qquad 7.3$$

To further address the ever-increasing amount of concepts and semantic features, random projection (Balduzzi, 2013) was employed in order to replace the above matrix operation. The mathematical soundness of this approach was provided in (Achlioptas, 2003), as given by Eq. 7.4 (Cambria and Hussain, 2015):

$$\sqrt{\frac{m}{d}} \|x - y\|_2 (1 - \varepsilon) \le \|\mathbf{\Phi} x - \mathbf{\Phi} y\|_2 \le \sqrt{\frac{m}{d}} \|x - y\|_2 (1 + \varepsilon) . \qquad 7.4$$

where d represents the original search space's dimensionality, m stands for our desired search space's dimensionality, ε indicates a tolerance variable and $\mathbf{\Phi}$ denotes a random matrix. Initially, a Gaussian **N**(0,1) matrix was utilized in random projection. A sparse version of random projection intends to replace the Gaussian matrix with independent and identically distributed entries, as given by Eq. 7.5 (Cambria and Hussain, 2015):

$$\phi_{ij} = \sqrt{s} \begin{cases} 1 & \Pr = \frac{1}{2s} \\ 0 & \Pr = 1 - \frac{1}{s} . \\ -1 & \Pr = \frac{1}{2s} \end{cases} \qquad 7.5$$

Nevertheless, the intrinsic sparse nature of AffectNet makes this sparse random projection inadvisable. Instead, when the training samples' number is significantly smaller than the features' number (i.e., $n \ll d$), a sub-sampled randomized Hadamard transform, as defined by Eq. 7.6 (Lu et al., 2013; Tropp, 2011) for $d = 2^p$ where p represents any positive integer, is advisable:

$$\mathbf{\Phi} = \sqrt{\frac{d}{m}} \mathbf{RHD} . \qquad 7.6$$

where:

1) m: The number that we would like to random sub-sample from d features;
2) **R**: A random matrix with the size of $m \times d$;
3) $\mathbf{H} \in \mathbb{R}^{d \times d}$: A normalized Walsh-Hadamard matrix which is formed recursively by $H_d = \begin{bmatrix} H_{\frac{d}{2}} & H_{\frac{d}{2}} \\ H_{\frac{d}{2}} & H_{\frac{d}{2}} \end{bmatrix}$ with $H_2 = \begin{bmatrix} +1 & +1 \\ +1 & +1 \end{bmatrix}$; and
4) **D**: A diagonal matrix with the size of $d \times d$.

In (Cambria and Hussain, 2015), k number of singular values were designed to be included in the construction of AffectiveSpace. For the purpose of finding a suitable k, a technique called CF-IOF (i.e., concept frequency—inverse opinion frequency) (Cambria et al., 2015), as given by Eq. 7.7 (Cambria and Hussain, 2015), was employed:

$$\left(\textit{Concept Frequency-Inverse Opinion Frequency}\right)_{concept,d} = \frac{n_{concept,d}}{\sum_k n_{k,d}} \log \sum_k \frac{n_k}{n_{concept}} \quad .7.7$$

More discussions regarding the AffectiveSpace transformation can be found in (Cambria and Hussain, 2015). To summarize, two mapping functions were finally obtained as expressed by Eqs. 7.8 and 7.9 (Cambria and Hussain, 2015):

$$x_{ij}^{*} = \tanh\left(\frac{x_{ij} - \mu_i}{a \cdot \sigma_i}\right). \quad 7.8$$

$$x_{ij}^{*} = \frac{x_{ij} - \mu_i}{a \cdot \sigma_i + \left|x_{ij} - \mu_i\right|}. \quad 7.9$$

7.3.2.3 Knowledge-Based Reasoning

In order to generate semantics and sentics from the aforementioned three types of knowledge representations, several techniques can be employed. Among them, the inference of semantics can be performed via the spreading activation; while the creation of sentics can be accomplished via a categorized emotion model and a group of neural networks, collectively. A thorough explanation regarding these approaches can be found in (Cambria and Hussain, 2012; Cambria and Hussain, 2015). In this section, we offer readers only a brief introduction.

- **Sentic Activation:** A major difference between human intelligence (Sigelman and Rider, 2015; Csikszentmihalyi, 2014; Sternberg, 1990) and artificial intelligence (AI) (Russell and Norvig, 2010; Xing et al., 2009; Xing et al., 2010a; Marwala and Hurwitz, 2017a; Marwala, 2013b; Marwala, 2014; Marwala, 2013c; Xing and Gao, 2014a; Xing and Gao, 2014ff; Xing and Gao, 2014aa; Xing and Gao, 2014b; Xing and Gao, 2014bb; Xing and Gao, 2014c; Xing and Gao, 2014cc; Xing and Gao, 2014d; Xing and Gao, 2014dd; Xing and Gao, 2014e; Xing and Gao, 2014f; Xing and Gao, 2014g; Xing and Gao, 2014h; Xing and Gao, 2014i; Xing and Gao, 2014j; Xing and Gao, 2014k; Xing and Gao, 2014l; Xing and Gao, 2014m; Xing and Gao, 2014n; Xing and Gao, 2014o; Xing and Gao, 2014p; Xing and Gao, 2014q; Xing and Gao, 2014r; Xing and Gao, 2014s; Xing and Gao, 2014t; Xing and Gao, 2014u; Xing and Gao, 2014v; Xing and Gao, 2014w; Xing and Gao, 2014x; Xing and Gao, 2014y; Xing and Gao, 2014z; Xing et al., 2013) is the human capability of adapting to novel or unknown situations where AI fails to prove itself, at least for now. Therefore, in order to maintain a harmonic relationship between man and machines (Xing and Marwala, 2018n; Xing and Marwala, 2018k; Xing and Marwala, 2018m; Xing and Marwala, 2018l; Xing and Marwala, 2018e; Xing and Marwala, 2018g; Xing and Marwala, 2018f; Xing and Marwala, 2018d; Xing and Marwala, 2018b; Xing and Marwala, 2018c; Xing and Marwala, 2018j; Xing and Marwala, 2018i; Xing and Marwala, 2018a), machines need to be endowed with human-like reasoning strategies (Marwala and Hurwitz, 2017c; Xing, 2017). Amongst them, sentic activation (Cambria et al., 2012b) is one such solution which broadly incorporates unconscious and conscious reasoning mechanisms. Further explanation can be found in (Cambria and Hussain, 2012; Cambria and Hussain, 2015).

- **Hourglass Model:** Emotion is often regarded as one of the most difficult (and still largely unexplored) areas of psychology (Orsay, 2014; Ekman and Davidson, 1994). The origin of 'emotion' is the Latin root word *mot*, which means 'to move'. Emotion can be loosely defined (over 90 definitions can be found in the literature) as the feeling aspect of consciousness, which is often characterized by three key ingredients (Ciccarelli and White, 2018), namely some kind of physical arousal, some behaviour that stimulates feeling with respect to the outer environment and an internal sentience in terms of feeling. In (Cambria and Hussain, 2012; Cambria and Hussain, 2015), the Hourglass model of emotions (Cambria et al., 2012a) was constructed, which is based on some results proposed in (Plutchik, 2001). As a model that draws inspirations from biological studies and is motivated by psychological findings, the underlying concept of the Hourglass model is that the brain's selective activation/deactivation when met with different stimuli can lead to distinct emotional states.

Apart from detecting the emotional situation, the Hourglass model can also be employed for detecting polarity. Mathematically, polarity can be formulated by Eq. 7.10 (Cambria and Hussain, 2012; Cambria and Hussain, 2015), with respect to the found affective dimensions:

$$polarity = \sum_{i=1}^{N}\left[\frac{Pleasantness(concept_i)+|Attention(concept_i)|-|Sensitivity(concept_i)|+Aptitude(concept_i)}{3N}\right]. \quad 7.10$$

where an input concept is denoted by $concept_i$, N stands for the overall number of concepts, the normalization coefficient is set as 3, and the dimensions of the Hourglass model are defined as floating numbers in the interval of [–1,+1]. More detailed information regarding the Hourglass model can be found in (Cambria and Hussain, 2012; Cambria and Hussain, 2015).

- **Sentic Neurons:** The processing of the information (both cognitive and affective) corresponding to natural language concepts often stays at the centre of affective analogical reasoning (Cambria and Hussain, 2012; Cambria and Hussain, 2015). Essentially, affective analogical reasoning resembles the brain's capacity to forming semantic patterns through association. If a new notion is comparable to something that the brain has already known, the likelihood that the brain to absorb that piece of new information is high. In order to make this semantic association, we often need to take the following factors into account (Cambria and Hussain, 2012; Cambria and Hussain, 2015):
  1) *Superior performance in terms of generalization*: For a better matchmaking between conceptual and affective patterns;
  2) *Quick speed in terms of learning*: For recalculating concept associations whenever a new expression is accepted by AffectNet; and
  3) *Low complexity in terms of computation*: For coping with big social data scenarios.
- **Extreme Learning Machine:** All these requirements serve as good motivation for us to resort to a powerful tool—extreme learning machine (ELM) (Huang et al., 2011)—as recommended in (Cambria and Hussain, 2012; Cambria and Hussain, 2015). Originally, it was proposed in order to address some issues pertaining to back-propagation network training (Marwala, 2007). The basic setting of ELM learning (Huang et al., 2012; Huang, 2014) consists of a training set (represented by X) and a number of N tagged pairs, i.e., $(\mathbf{x}_i, y_i)$ where $\mathbf{x}_i$ (an element of R^m) represents the ith input vector and y_i (an element of R) stands for the associated expectation value. Similar to artificial neural networks (Marwala, 2000; Marwala, 2015h; Xing et al., 2010b; Marwala, 2015j), the input layer of an ELM contains m

neurons which are connected with the 'hidden' layer (comprising N_h neurons) via a collection of weights $\{\hat{\mathbf{w}}_j \in \mathrm{R}^m; j = 1,..., N_h\}$. A bias term (denoted by $\hat{b}_j$) and a nonlinear activation function, represented by $\varphi(\cdot)$, are integrated into the jth hidden neuron. Therefore, when an input stimulus $\mathbf{x}$ emerges, the feedback generated by neurons can be formulated as Eq. 7.11 (Huang et al., 2011; Huang et al., 2012; Huang, 2014; Cambria and Huang, 2013):

$$a_j(\mathbf{x}) = \varphi\left(\hat{\mathbf{w}}_j \cdot \mathbf{x} + \hat{b}_j\right). \tag{7.11}$$

Meanwhile, a vector of weighted links $\bar{\mathbf{w}}_j$ (an element of R^{N_h}) connects the hidden neurons and the output neuron with no additional bias introduced (Huang, 2014). Accordingly, the entire network's output function can be expressed as Eq. 7.12 (Huang et al., 2011; Huang et al., 2012; Huang, 2014; Cambria and Huang, 2013):

$$f(\mathbf{x}) = \sum_{j=1}^{N_H} \bar{\mathbf{w}}_j a_j(\mathbf{x}). \tag{7.12}$$

For convenience, an activation matrix (denoted by $\mathbf{H}$) can also be defined with the entry h_{ij} (an element of $\mathbf{H}$ where $i = 1,...,N$ and $j = 1,...,N_h$) representing the jth hidden neuron's activation value for the ith input pattern. The expression of $\mathbf{H}$ is given by Eq. 7.13 (Huang et al., 2011; Huang et al., 2012; Huang, 2014; Cambria and Huang, 2013):

$$\mathbf{H} \equiv \begin{bmatrix} \varphi\left(\hat{\mathbf{w}}_1 \cdot \mathbf{x}_1 + \hat{b}_1\right) & \cdots & \varphi\left(\hat{\mathbf{w}}_{N_h} \cdot \mathbf{x}_1 + \hat{b}_{N_h}\right) \\ \vdots & \ddots & \vdots \\ \varphi\left(\hat{\mathbf{w}}_1 \cdot \mathbf{x}_N + \hat{b}_1\right) & \cdots & \varphi\left(\hat{\mathbf{w}}_{N_h} \cdot \mathbf{x}_N + \hat{b}_{N_h}\right) \end{bmatrix}. \tag{7.13}$$

In addition, the quantities of $\{\hat{\mathbf{w}}_j, \hat{b}_j\}$ (see Eq. 7.11) are set in a random manner, and the quantities of $\{\bar{\mathbf{w}}_j, \bar{b}\}$ (see Eq. 7.12) stand only for degrees of freedom. Therefore, the training problem can be reduced to the convex cost minimization problem, as given by Eq. 7.14 (Huang et al., 2011; Huang et al., 2012; Huang, 2014; Cambria and Huang, 2013):

$$\min_{\{\bar{\mathbf{w}}, \bar{b}\}} \left\| \mathbf{H}\bar{\mathbf{w}} - \mathbf{y} \right\|. \tag{7.14}$$

A unique L_2 solution can be produced with a matrix pseudo-inversion operation, as given by Eq. 7.15 (Huang et al., 2011; Huang et al., 2012; Huang, 2014; Cambria and Huang, 2013):

$$\bar{\mathbf{w}} = \mathbf{H}^{+}\mathbf{y}. \tag{7.15}$$

Furthermore, as recommended by Huang et al. (2012), the ELM's generalization performance can be enhanced by implementing different regularization strategies. Therefore, the cost function (see Eq. 7.14) is augmentable by an L_2 regularization factor, as reformulated in Eq. 7.16 (Huang et al., 2012; Cambria and Hussain, 2015):

$$\min_{\bar{\mathbf{w}}} \left\{ \left\| \mathbf{H}\bar{\mathbf{w}} - \mathbf{y} \right\|^2 + \lambda \left\| \bar{\mathbf{w}} \right\|^2 \right\}. \quad 7.16$$

Regarding the construction of an emotion categorization framework, readers should refer to (Cambria and Hussain, 2012; Cambria and Hussain, 2015) for further discussions.

7.3.3 *Fuzzy Product Ontology*

In (Lau et al., 2014), the definition of fuzzy product ontology was given in tuple form, as shown in Eq. 7.17 (Lau et al., 2014):

$$Ontology := \left\langle Concepts, R_{taxonomic}, R_{non-toxonomic} \right\rangle. \quad 7.17$$

where the concept set is denoted by term *Concepts*, which can include but is not limited to product, aspect and sentiment. Meanwhile, the strengths of taxonomic (i.e., sub- or super-class) or non-taxonomic relationships among different concepts within the *Concepts* set are defined by membership functions given by Eqs. 7.18 and 7.19, respectively (Lau et al., 2014):

$$\mu_{R_{taxonomic}} : Concepts \times Concepts \mapsto [0,1]. \quad 7.18$$

$$\mu_{R_{non-taxonomic}} : Concepts \times Concepts \mapsto [0,1]. \quad 7.19$$

Suppose we know the meanings of two concepts (denoted by $Concept_i$ and $Concept_j$), if every attribute of $Concept_i$ also belongs to $Concept_j$, as given by Eq. 7.20 (Lau et al., 2014):

$$\begin{aligned} &\left\{ attribute_1^{concept_i}, attribute_2^{concept_i}, \ldots, attribute_n^{concept_i} \right\} \\ &\subset \left\{ attribute_1^{concept_j}, attribute_2^{concept_j}, \ldots, attribute_n^{concept_j} \right\}. \end{aligned} \quad 7.20$$

then $concept_i$ is regarded as subsuming $concept_j$ (Cimiano et al., 2005).

7.3.4 *Latent Topic Modelling for Product Aspects Mining*

7.3.4.1 Notations

In this chapter, the following notations are employed in order to depict the descriptions or comments utilized for the purpose of training or evaluation (Crain et al., 2012).

- **Descriptions:** *Descriptions* represent a corpus of M descriptions (indexed by *description*). Meanwhile, there are a number of W different terms included in the vocabulary (indexed by v). The term-description matrix X is thus a $W \times M$ matrix that encodes each term's occurrence in every piece of description. In the latent Dirichlet allocation (LDA) (Blei et al., 2003; Moro et al., 2015) probabilistic generative model, the number of topics is typically denoted by K and indexed by i, while N is used to indicate the token number in any set (a subscript is introduced to stand for the specific set), that is, N_i represents the allocated token number for the ith topic. The complement of a set is indicated by a bar symbol: $\overline{z}_{ndescription} \equiv \{\overline{z}_{ndescription} : description' \neq description$ or $n' \neq n\}$.
- **Multinomial Distribution:** In the text mining area, a commonly used model is multinomial distribution, as given by Eq. 7.21 (Mendenhall et al., 2013; Crain et al., 2012; Johnson, 2018):

$$M(\mathbf{X} \mid \mathbf{\Psi}) \propto \prod_{v=1}^{W} \psi_v^{x_v} \tag{7.21}$$

where the terms' relative frequency in a description can be captured.

- **Dirichlet Distribution:** As the multinomial distribution's conjugate version, Dirichlet distribution can be defined by Eq. 7.22 (Ross, 2014; Crain et al., 2012):

$$Dir(\mathbf{\Psi} \mid \Xi) = \frac{\Gamma\left(\sum_{i=1}^{K} \xi_i\right)}{\prod_{i=1}^{K} \Gamma(\xi_i)} \prod_{i=1}^{K} \psi_i^{\xi_i - 1}. \tag{7.22}$$

Dirichlet distribution supports imbalanced multinomial distributions, in particular, when a small number of values determines most of the probability mass. This characteristic makes it particularly suitable for modelling human language, which is characterized by power law distributions.

- **Generative Process:** In general, a generative process stands for an algorithm that depicts how an output was actually generated. Take die-rolling, the generative process can be described as selecting one side from a multinomial distribution with $\frac{1}{6}$ probability on every single side (six sides in total). In the topic modelling domain, it is often useful to introduce a random generative process for the sake of capturing real statistical correlations between topics and terms, even if the terms' selection in a description does not show randomness.

7.3.4.2 Latent Dirichlet Allocation Topic Modelling

In the text analysis area, though classical methods often offer a good basis, there are two drawbacks that need to be further addressed: (1) A great number of parameters included in conventional approaches often tend to grow linearly with the increase of description's number, which in turn can cause over-fitting of the training data. (2) In the case of a description not contained in the training data, it is often hard to find a natural way to compute its probability. In this regard, latent Dirichlet allocation (LDA) was introduced as an alternative solution. Since a process of generating each description's topics is included in LDA, the parameter number that needs to be learned is greatly reduced. This merit makes the calculation of an arbitrary description's probability become more defined. In essence, LDA is built on a hypothetical generative process for a corpus. The relationships among different random parameters are illustrated in Fig. 7.2.

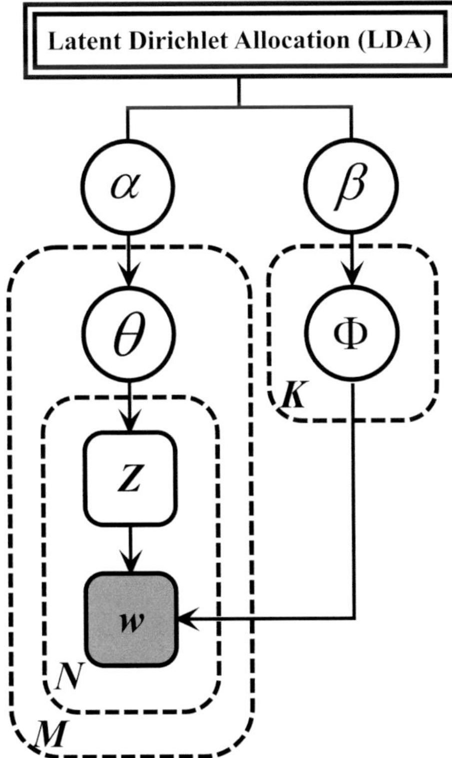

Figure 7.2: Schematic representation of LDA graphical model.

- **Term Probability Selection for Each Topic:** The terms' distribution for the *i*th topic is represented by a multinomial distribution (denoted by $\mathbf{\Phi}_i$) which is derived from a symmetric Dirichlet distribution with variable β, as given by Eq. 7.23 (Aggarwal and Zhai, 2012):

$$\mathbf{\Phi}_i \sim Dir(\beta);\ \Pr(\mathbf{\Phi}_i|\beta) = \frac{\Gamma(W\beta)}{[\Gamma(\beta)]^W}\prod_{v=1}^{W}\phi_{iv}^{\beta-1}. \quad 7.23$$

- **Topic Selection for Each Description:** The topic distribution for description is formulated as a multinomial distribution (denoted by $\theta_{description}$), which is also derived from a Dirichlet distribution with variable α, as given by Eq. 7.24 (Aggarwal and Zhai, 2012):

$$\boldsymbol{\theta}_{description} \sim Dir(\alpha);\ \Pr(\boldsymbol{\theta}_{description}|\alpha) = \frac{\Gamma\left(\prod_{i=1}^{K}\alpha_i\right)}{\prod_{i=1}^{K}\Gamma(\alpha_i)}\prod_{v=1}^{W}\theta_{idescription}^{\alpha_i-1}. \quad 7.24$$

- **Topic Selection for Each Token:** The topic (denoted by $z_{ndescription}$) for each token index (indicated by n) is selected from the distribution of description topic, as given by Eq. 7.25 (Aggarwal and Zhai, 2012):

$$z_{ndescription} \sim M(\boldsymbol{\theta}_{description});\ \Pr(z_{ndescription} = i|\boldsymbol{\theta}_{description}) = \boldsymbol{\theta}_{idescription}. \quad 7.25$$

- **Token Selection:** Each token w at each index is selected from the multinomial distribution corresponding to the selected topic, as given by Eq. 7.26 (Aggarwal and Zhai, 2012):

$$w_{ndescription} \sim M(\phi_{z_{ndescription}});\ \Pr(w_{ndescription} = v|z_{ndescription} = i, \phi_i) = \phi_{iv}. \quad 7.26$$

7.3.4.3 LDA-Based Topic Modelling for Product Aspects Mining

Since product aspects often involve two levels of information (high- and elementary-level) that are related to both product descriptions and consumer comments, LDA-based topic modelling methodology was utilized in (Lau et al., 2014) in order to extract the relevant product aspects from both implicit and explicit perspectives, respectively. The objective of their study was to calculate the conditional probability of $\Pr(t_i|z_i)$, which indicates a product aspect (i.e., a latent topic). More specifically, the goal of training an LDA model is to search for a group of optimal parameters which can maximize the probability of creating the training description set. Typically, the probability of the training descriptions under a given

LDA model is named as the empirical likelihood (denoted by *likelihood*), as given by Eq. 7.27 (Aggarwal and Zhai, 2012):

$$\begin{aligned} likelihood &= \prod_{description=1}^{M} \prod_{n=1}^{N} \left[\begin{array}{l} \Pr\left(w_{ndescription} \middle| z_{ndescription}, \mathbf{\Phi}\right) \Pr\left(z_{ndescription} \middle| \boldsymbol{\theta}_{description}\right) \\ \Pr\left(\boldsymbol{\theta}_{description} \middle| \alpha\right) \Pr\left(\mathbf{\Phi} \middle| \beta\right) \end{array} \right] \\ &= \phi_{zw} \theta_{zdescription} \frac{\Gamma\left(\sum_{i=1}^{K} \alpha_i\right)}{\prod_{i=1}^{K} \Gamma(\alpha_i)} \prod_{i=1}^{K} \theta_{idescription}^{\alpha_i - 1} \frac{\Gamma(W\beta)}{\left|\Gamma(\beta)\right|^{W}} \prod_{v=1}^{W} \phi_v^{\beta-1} \end{aligned} \quad . \ 7.27$$

However, the straight computation of the likelihood is difficult since the direct observation of the $z_{ndescription}$ (i.e., topic assignments) is often not possible, even for a single description case. To cope with this issue, Gibbs sampling (Steyvers et al., 2004; Rosen-Zvi et al., 2010) was proposed in the literature for the purpose of calculating the approximations of ϕ and θ. In (Lau et al., 2014), the MCMC (i.e., Markov chain Monte Carlo) algorithm was introduced in order to calibrate the Gibbs sampling approach so as to obtain the estimated conditional probabilities of ϕ and θ. In other words, for each observable term, a Markov chain was constructed based on its conditional probability against other parameters by repeatedly selecting a latent topic (Steyvers et al., 2004). Built on this formulation, the approximated multinomial distributions ϕ and θ (denoted by $\bar{\phi}$ and $\bar{\theta}$, respectively) are given by Eqs. 7.28 and 7.29 (Blei et al., 2003; Steyvers et al., 2004):

$$\bar{\theta} = \frac{Concepts_{np}^{ZDescriptions} + \alpha}{\sum_{n' \in Z} Concepts_{n'p}^{ZDescriptions} + |Z|\alpha} . \quad 7.28$$

$$\bar{\phi} = \frac{Concepts_{mn}^{VZ} + \beta}{\sum_{m' \in V} Concepts_{m'n}^{VZ} + |V|\beta} . \quad 7.29$$

where a count matrix is represented by $Concepts_{mn}^{VZ}$ which is used to capture how many times a term (indicated by $t_i = m$) has been allocated to a latent topic (denoted by $z_i = n$), not including the current word position. In Eq. 7.29, the role of term $\frac{Concepts_{mn}^{VZ} + \beta}{\sum_{m' \in V} Concepts_{m'n}^{VZ} + |V|\beta}$ is to get the estimated probability of the t_i term, given the z_i latent topic, while the term $\frac{Concepts_{np}^{ZDescriptions} + \alpha}{\sum_{n' \in Z} Concepts_{n'p}^{ZDescriptions} + |Z|\alpha}$ is introduced in order to acquire the approximated probability of the z_i

latent topic, given the $description_i$. More details regarding the underlying computational complexity can be found in (Lau et al., 2014; Aggarwal and Zhai, 2012).

In the information retrieval domain (Kraft and Colvin, 2017; Melucci, 2015), language models are often utilized, together with the probabilistic method, for the purpose of determining the upcoming part of speech. Since the representative speech patterns/terms can be found together, and the associated next term prediction task is simpler than searching for a term in a corpus that is full of random data, language models typically outperform traditional vector models.

Based on this understanding, one can further apply either probability functions or possibility functions (i.e., fuzziness) to the targeted language model. In this regard, the authors of (Lau et al., 2014) proposed a probabilistic language model by combining previously developed unigram language models (Ponte and Croft, 1998; Zhai and Lafferty, 2004) for the purpose of evaluating two concepts' (denoted by $concept_i$ and $concept_j$, respectively) mutual subsumption relationship, as given by Eqs. 7.30–7.33 (Lau et al., 2014):

$$\Pr\left(\phi_{concept_i} \middle| \phi_{concept_j}\right) = \prod_{t_i \in concept_i} \Pr\left(t_i \middle| \phi_{concept_j}\right). \tag{7.30}$$

$$\Pr\left(t_i \middle| \phi_{concept_j}\right) = (1-\lambda)\Pr_{ML}\left(t_i \middle| \phi_{concept_j}\right) + \lambda \Pr_{ML}\left(t_i \middle| \phi_{Descriptions}\right). \tag{7.31}$$

$$\Pr_{MLE}\left(t_i \middle| \phi_{concept_j}\right) = \frac{\Pr\left(t_i \middle| z_j\right)}{\sum_{l=1}^{top_t} \Pr\left(t_l \middle| z_j\right)}. \tag{7.32}$$

$$\Pr_{MLE}\left(t_i \middle| \phi_{Descriptions}\right) = \frac{tf\left(t_i\right)}{\left|Descriptions\right|}. \tag{7.33}$$

where $\phi_{concept_i}$ denotes the $concept_i$'s language model, which is acquired through the process of LDA-based topic modelling. One shortcoming here is the potential zero probability issue which will lead to an infinite perplexity situation. A possible solution is to utilize smoothing techniques in order to make sure that every possible event can be allocated with a probability mass, even when very small in nature. One representation methodology in this category is the Jelinek-Mercer interpolated smoothing approach (Zhai and Lafferty, 2004; Goldberg, 2017; Chen and Goodman, 1996; Jelinek and Mercer, 1980), as given by Eq. 7.34 (Goldberg, 2017):

$$\hat{p}_{\text{int}}\left(w_{i+1}=m \middle| w_{i-k:i}\right)=\lambda_{w_{i-k:i}}\frac{\#\left(w_{i-k:i+1}\right)}{\#\left(w_{i-k:i}\right)}+\left(1-\lambda_{w_{i-k:i}}\right)\hat{p}_{\text{int}}\left(w_{i+1}=m \middle| w_{i-(k-1):i}\right). \quad 7.34$$

where $\hat{p}$ stands for good estimates, $\#(w_{i:j})$ indicates the count of words' (denoted by $w_{i:j}$) sequence in a corpus and the smoothing factor is denoted by λ which falls within the range of [0.1,0.7] (Zhai and Lafferty, 2004). More alternative methods can be found in (Goldberg, 2017).

With the aid of the aforementioned smoothing technique, the probability of one language model (denoted by $\phi_{concept_i}$) being generated by another language model ($\phi_{concept_j}$) can be approximated via two maximum likelihood estimates, that is, $\Pr_{MLE}(t_i|\phi_{concept_j})$ and $\Pr_{MLE}(t_i|\phi_{Descriptions})$, as given in Eqs. 7.32 and 7.33 (Lau et al., 2014). Further parameter settings can be found in (Lau et al., 2014). In addition to all this, the membership function, denoted by $\mu_{R_{TAX}}$ $(concept_i, concept_j)$ and corresponded to the fuzzy taxonomic relationship between $concept_i$ and $concept_j$, is defined by Eq. 7.35 (Lau et al., 2014):

$$\mu_{R_{TAX}}\left(concept_i, concept_j\right)=\text{normal}\left[\frac{\Pr\left(\phi_{concept_i}\middle|\phi_{concept_j}\right)-\Pr\left(\phi_{concept_j}\middle|\phi_{concept_i}\right)}{\Pr\left(\phi_{concept_j}\middle|\phi_{concept_i}\right)}\right]. \quad 7.35$$

where $\text{normal}(x)=\frac{x-\min}{\max-\min}$ represents a normalization function in linear form which is used to normalize the minimum and the maximum values among a collection of other values. Thorough explanations regarding Eqs. 7.30–7.33 and Eq. 7.35 can be found in (Lau et al., 2014).

7.3.4.4 Context-Sensitive Sentiments Learning

In (Lau et al., 2014), a group of consumer reviews (typically user-graded) is utilized in order to formulate the non-taxonomic relationships (denoted by $R_{non\text{-}taxonomic}$) between product aspects and sentiments through a process of offline learning. As suggested in (Subrahmanian and Recupero, 2008), a virtual text window of size (denoted by a weight factor of ω_{window}) is also introduced in order to measure the product aspects' relevant adjectives or adverbs within in a review description, which are then extracted and serve as the sentiment candidates. In particular, if ω_{window} is equal to one, the identified adjectives or adverbs will be extracted immediately. In (Lau et al., 2014), the value of ω_{window} was set to six, indicating that the boundary of a sentence has been taken into account. In other words, the authors of (Lau et al., 2014) only extract product aspect-related adjectives or adverbs when they co-exist within the same sentence. Meanwhile, an association measure, denoted by Association($sentiment_i$,$aspect_j$), was developed in (Lau et al., 2014) for representing the correlation between sentiment and product aspect. The inspiration for this formulation comes from (Lau et al., 2009), in which a measure representing mutual information was successfully utilized in order to address the fuzzy domain ontology mining issue. More specifically, the measure of Association$(sentiment_i, aspect_j)$ is defined by Eq. 7.36 (Lau et al., 2014):

$$\text{Association}\left(sentiment_i, aspect_j\right) = \omega_{association} \times \left\{ \begin{array}{l} \Pr\left(t_i, t_j\right)\log_2\left[\frac{\Pr\left(t_i,t_j\right)+1}{\Pr\left(t_i\right)\Pr\left(t_j\right)}\right] + \\ \Pr\left(\neg t_i, \neg t_j\right)\log_2\left[\frac{\Pr\left(\neg t_i,\neg t_j\right)+1}{\Pr\left(\neg t_i\right)\Pr\left(\neg t_j\right)}\right] \end{array} \right\} - \left(1-\omega_{association}\right) \times \left\{ \begin{array}{l} \Pr\left(t_i, \neg t_j\right)\log_2\left[\frac{\Pr\left(t_i,\neg t_j\right)+1}{\Pr\left(t_i\right)\Pr\left(\neg t_j\right)}\right] + \\ \Pr\left(\neg t_i, t_j\right)\log_2\left[\frac{\Pr\left(\neg t_i,t_j\right)+1}{\Pr\left(\neg t_i\right)\Pr\left(t_j\right)}\right] \end{array} \right\}. \quad 7.36$$

where the correlation degree between a sentiment and a product aspect (denoted by $sentiment_i$ and $aspect_j$, respectively) is indicated by Association$(sentiment_i, aspect_j)$. The weight factor $\omega_{association}$, typically falling within the range of $[0.5,0.7]$, is also introduced in order to control the relative importance of two classes of evidence and create an associated relationship. Interested readers please refer to (Lau et al., 2014) for a more in-depth discussion.

In (Lau et al., 2008), a word divergence measure, denoted by Word-Divergence and built on Kullback-Leibler (KL) divergence measure (Kullback and Leibler, 1951; Amari and Nagaoka, 2000), has been successfully used for mining various keywords (e.g., positive, negative, and neutral) according to the user's needs. Unlike the original formulation of KL divergence, the Word-Divergence measure performs a subtraction operation between the positive and negative events' conditional probabilities (Lau et al., 2014). Accordingly, a sentiment-aspect pair's polarity score, denoted by $sentiment\text{-}aspect := (sentiment_i, aspect_j)$, can be defined by Eqs. 7.37 and 7.38 (Lau et al., 2014):

$$\text{Word-Divergence}\left(sentiment\text{-}aspect\right) = \tanh\left\{ \begin{array}{l} \frac{df\left(sentiment\text{-}aspect\right)}{\omega_{negative}} \times \Pr\left(positive \middle| sentiment\text{-}aspect\right) \times \\ \log_2\left[\frac{\Pr\left(positive \middle| sentiment\text{-}aspect\right)}{\Pr\left(positive\right)}\right] - \frac{df\left(sentiment\text{-}aspect\right)}{\omega_{negative}} \times \\ \Pr\left(negative \middle| sentiment\text{-}aspect\right) \times \log_2\left[\frac{\Pr\left(negative \middle| sentiment\text{-}aspect\right)}{\Pr\left(negative\right)}\right] \end{array} \right\}. \quad 7.37$$

$$polarity_{\text{Ontology}}(sentiment\text{-}aspect) = \begin{cases} \frac{\text{Word-Divergence}(sentiment\text{-}aspect) - \omega_{word\text{-}divergence}}{1-\omega_{word\text{-}divergence}} & \text{if Word-Divergence}(sentiment\text{-}aspect) > \omega_{word\text{-}divergence} \\ -\frac{|\text{Word-Divergence}(sentiment\text{-}aspect)| - \omega_{word\text{-}divergence}}{1-\omega_{word\text{-}divergence}} & \text{if WordDivergence}(sentiment\text{-}aspect) < \omega_{word\text{-}divergence} \\ 0 & \text{otherwise} \end{cases} \quad 7.38$$

The detailed explanation regarding this formulation and the corresponding pseudo code of the complete automated product ontology learning algorithm, built on product description and customer feedback corpora, can be found in (Lau et al., 2014).

7.3.4.5 *Aspect-Driven Sentiment Analysis*

Given the fact that a group of sentiment-aspect pairs has been extracted from a collection of *Descriptions* regarding a product (denoted by $product_i$), the corresponding aspect score of a particular product and its different aspects, represented by Aspect-Score($product_i$,$aspect_i$), is equal to the weighted average of the sentiment-aspect pair set's polarity scores as given by Eq. 7.39 (Lau et al., 2014):

$$\text{Aspect-Score}(product_i, aspect_i) = \frac{\sum_{sentiment\text{-}aspect \in Sentiment\text{-}Aspect} \omega_{source} \times polarity(sentiment\text{-}aspect)}{|Sentiment\text{-}Aspect|} \quad .7.39$$

where the value of weight factor ω_{source} can equal either 1, given the definition of a pair of sentiment-aspects is available in a product ontology, or 0.5, when the definition of the sentiment is only available in a generic sentiment lexicon form.

Meanwhile, when the constituent pairs of sentiment-aspect are pinpointed in a particular description (denoted by $description_{sentiment\text{–}aspect}$), their weighted polarity scores can be used to estimate the description's polarity score, as given by Eq. 7.40 (Lau et al., 2014):

$$polarity(description) = \frac{\sum_{sentiment\text{-}aspect \in description} \omega_{source} \times polarity(sentiment\text{-}aspect)}{|description_{sentiment\text{-}aspect}|} \quad .7.40$$

Apart from this, when all pairs of sentiment-aspect have been pinpointed from the collection of descriptions with respect to a specific product (denoted by $Descriptions_{product_i}$), the weighted average of their collective polarity scores represents a product's polarity score, as given by Eq. 7.41 (Lau et al., 2014):

$$polarity(product_i) = \frac{\sum_{sentiment\text{-}aspect \in Descriptions_{product_i}} \omega_{source} \times polarity(sentiment\text{-}aspect)}{\left|Descriptions_{sentiment\text{-}aspect}\right|} \quad 7.41$$

where $Descriptions_{sentiment\text{–}aspect}$ stands for every individual pair of sentiment-aspects that has been identified in the set of $Descriptions_{product_i}$.

7.3.4.6 *Summary*

The performance evaluation of information retrieval results is critical, in particular when it comes to concluding whether a newly minted approach is outperforming its competitors or at least showing improvements when compared to the previous version. Two very often used measures are precision and recall, which stay at the centre of any evaluation metric. Briefly, the role of 'precision' is to check how many descriptions were mistakenly collected together with the correctly captured ones, while the task of examining how many descriptions were accidentally omitted is the main responsibility of 'recall'. The values of precision and recall, in general, fall within the range of [0,1], where the best performance is denoted by one. This fundamental rule can, thus, be formulated by Eqs. 7.42 and 7.43 (Lupu et al., 2017a; Lupu et al., 2017b):

$$Precision = \frac{number\ of\ related\ terms\ identified}{number\ of\ terms\ identified}. \quad 7.42$$

$$Recall = \frac{number\ of\ related\ terms\ identified}{number\ of\ related\ terms}. \quad 7.43$$

Since it is typically not so straightforward to understand what the analysing results mean, evaluation metrics are often used in order to perform this comparative task. However, the answer to the question of what makes a good metric varies a lot, some suggestions are given in the literature as follows (Lupu et al., 2017a; Lupu et al., 2017b):

- **Suitability:** In practice, some retrieval tasks emphasize the precision factor, while others focus more on recall. Therefore, the right balance for the targeted task must be achieveable with a new evaluation metric.

- **Stability:** The criteria used to conclude that one system is superior to another, based on two similar test sets, have to be consistent.
- **Distinguishability:** The evaluation metric has to be capable of quantifying all the differences between two ranked lists. A common situation that must be avoided is that although the metric itself shows both precision and recall, they are incapable of differentiating between the same document set's distinct rankings.

In line with the above requirements, the evaluation metrics employed in (Lau et al., 2014) are given by Eqs. 7.44–7.47 (Van Rijsbergen, 1979; Kraft and Colvin, 2017):

$$Precision = \frac{a}{a+b}. \quad 7.44$$

$$Recall = \frac{a}{a+c}. \quad 7.45$$

$$Accuracy = \frac{a+d}{a+b+c+d}. \quad 7.46$$

$$F_{\beta=1} = \frac{(1+\beta^2)\times Precision \times Recall}{\beta^2 \times Precision + Recall} = \frac{2a}{2a+b+c}. \quad 7.47$$

where β denotes a user defined parameter, a represents correctly excavated concepts' number, b stands for wrongly excavated concepts' number, c indicates missed related concepts' number, and d means missed irrelevant concepts' number. In the context of consumer descriptions' polarity forecasting, variables a, b, c, and d can be used to indicate several numbers as well, i.e., categorized positive/negative descriptions' number, and non-categorized positive/negative descriptions' number.

Built on this structure, a series of experiments were run on the basis of a well-selected dataset (Lau et al., 2014) in order to examine the usefulness of the proposed methodology. By comparing the performance among several benchmark algorithms, the experimental algorithm achieved the best scores in terms of precision, recall, $F_{\beta=1}$ and accuracy.

7.3.5 *Analysis of Crowdfunding Project Videos*

In (Fernandes, 2013), a pioneer study was carried out in order to examine the importance of crowdfunding project videos, with a particular interest on technological projects. Each project video was deliberately decomposed according to each associated attribute/characteristic (labelled as X_i).

7.3.5.1 Project Video Selection

The technology domain was selected by Fernandes (2013) as the focal study area. Meanwhile, the project fundraising goal was constrained within the interval of [$30000, $500000]. Finally, eight project videos were selected from the Kickstarter platform, which cover four funded projects and four unfunded projects (see Fig. 7.3 for illustration). In (Fernandes, 2013), the success was defined according to whether or not the project achieved its fundraising goal within the allotted time period.

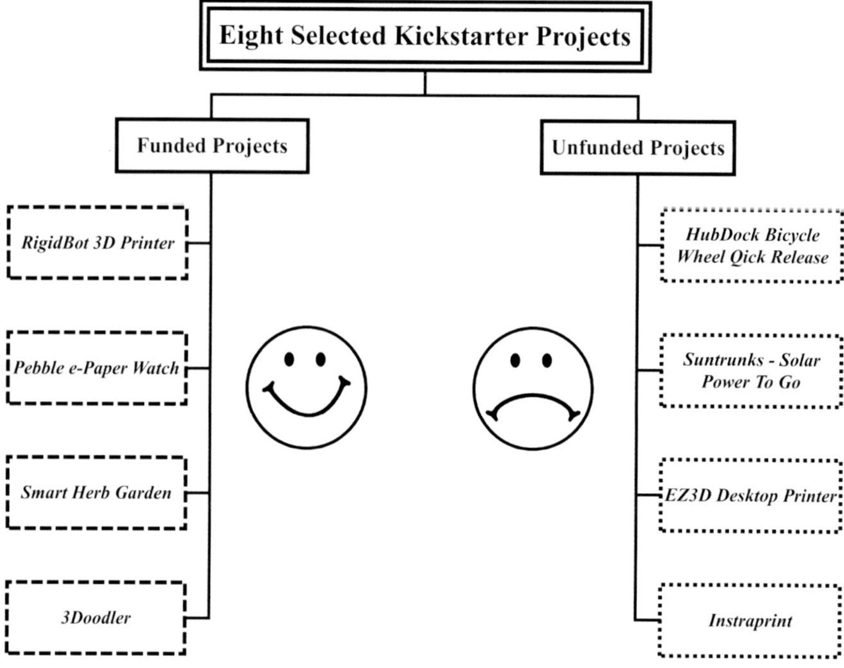

Figure 7.3: An illustration of eight selected kickstarter projects.

7.3.5.2 Video Watcher Survey

Fernandes (2013) surveyed twelve undergraduate students in order to check how each one of eight project videos are rated by watchers, given some predetermined attributes. The detailed ranking grades are illustrated in Fig. 7.4.

Meanwhile, Fernandes (2013) also introduced a comment section so that participants can outline their feelings, suggestions and so on. In addition, both technological and non-technological disciplinary undergraduate students (e.g., bioengineering, literature, mechanical engineering, political science, etc.) at Massachusetts Institute of Technology were chosen in

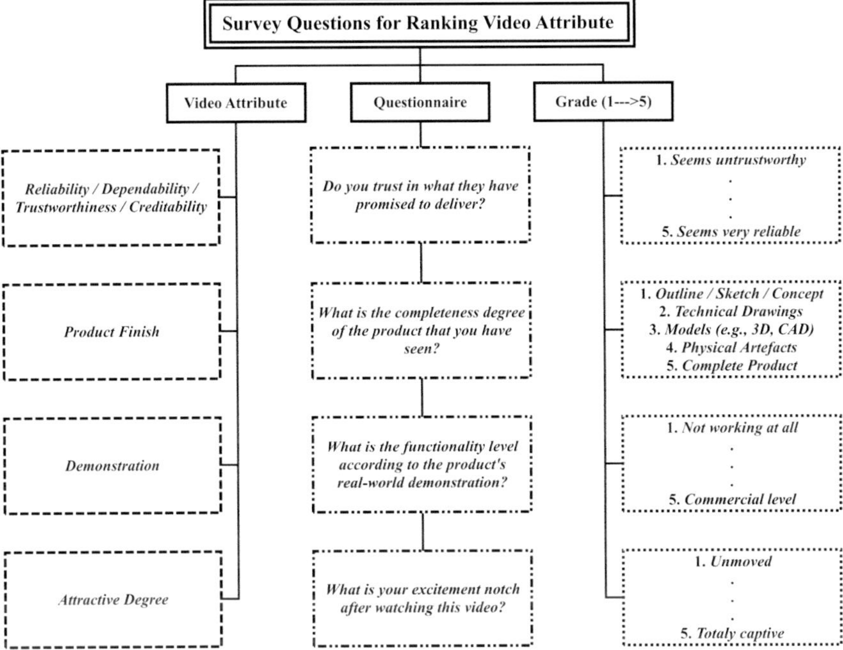

Figure 7.4: An illustration of ranking grades used to rate the watched video.

order to constitute a diverse group, though the survey was performed individually to avoid group persuasion (Aronson et al., 2016). Other settings regarding video watching can be found in (Fernandes, 2013).

7.3.5.3 Parameter Definition and Correlation Analysis

- **Success Coefficient:** Once the relevant information was well collected, Fernandes (2013) further defined a parameter named "success coefficient" for each project, as given by Eq. 7.48 (Fernandes, 2013):

$$\Psi = \frac{\text{Funds Accumulated}}{\text{Funds Begged}}. \qquad 7.48$$

- **Causation and Correlation:** The role of causation, correlation and artificial intelligence in making rational decisions is widely recognized (Marwala, 2015b; Marwala, 2015f; Marwala, 2015k; Marwala, 2015j; Marwala, 2015d; Marwala, 2015h; Marwala, 2015i; Marwala, 2015a; Marwala, 2015e; Marwala, 2015g; Marwala, 2015c; Marwala, 2013c; Marwala, 2013a). Based on this common agreement, Fernandes (2013) introduced the Spearman rank correlation variable ρ, as given by Eq. 7.49 (Johnson, 2018; Brase and Brase, 2013; Sullivan, 2018):

$$\rho = 1 - \frac{6\sum d_i^2}{n\left(n^2 - 1\right)}. \tag{7.49}$$

where the number of videos (i.e., eight in this case) is denoted by n, and the calculations of d_i^2 can be found in (Johnson, 2018; Brase and Brase, 2013; Sullivan, 2018). Basically, the use of this Spearman rank coefficient is to measure the association strength between two graded variables, namely each video attribute's average rating (denoted by X_{ij}), and each video's 'success coefficient' (represented by Ψ_j).

7.3.5.4 *Summary*

Through the calculation of the Spearman rank correlation, Fernandes (2013) confirmed the existence of a correlation between a crowdfunding project's success rate and certain corresponding project video's qualities, at least in the technological domain. More specifically, Fernandes (2013) observed that when a project video's attractiveness degree increases, not only does the project initiator's trustworthiness improve from the video watchers' perspective, but the associated project's success rate also increases.

7.4 Conclusions

Within the realm of crowdfunding, as an open call is often made through the Internet, the factors that can make a contribution to a project's success include (non-exhaustive list) the specified project fundraising goal, the success rate of competitive projects, the backer's incentives, the profile of the project, the type of information available, the project initiator's credentials and creditability, etc. Among these factors, social capital tends to be widely used by all parties. From a backer's point of view, social media can offer real-time information flow relevant to the projects.

In this regard, Beier and Wagner (2014) and Mollick (2014) have confirmed that videos provide the richest form of media communications and can, therefore, leverage the success of a project. However, despite the importance of project videos for fundraising campaigns, many of their basic aspects are still not well understood. Some micro-level questions remain open: Which kind of information affects backers' decisions most? What can we infer about the backer's expectations from their opinions/feelings? On the macro-level side, the creditworthiness of these inference, illation or reasoning tactics is largely unknown, in part due to the lack of effective tools needed to measure the sentiment expressed through online social media.

In this chapter, we elaborated the sentic computing framework and fuzzy product ontology principle in order to address the aforementioned concerns. On one hand, four attributes (i.e., creator's trustworthiness, product demonstration degree, finish of the product and video's attractiveness) were selected by Fernandes (2013) in order to analyse how a project video is related to a project's success. Meanwhile, a survey was conducted in which each individual participant was requested to rank these attributes for videos of both successful and failed projects. On the other hand, through the utilization of the LDA (i.e., latent Dirichlet allocation) facilitated topic modelling technique and automatically established product ontologies, Lau et al. (2014) introduced a system that can precisely predict aspect-level sentiments' polarities with no need for the expensive manual tagging process for training examples. Such design artefacts are particularly useful for consumers who wish to perform comparison tasks when sentiments are involved. Though the studies considered in this chapter are far from extensive, they do convince us, to some extent, that both hard facts (e.g., narratives and videos provided by askers) and soft facts (e.g., comments provided by peers) all play an important role in influencing backers' final decisions, as claimed by Moritz et al. (2015).

References

Achlioptas D. (2003) Database-friendly random projections: Johnson-Lindenstrauss with binary coins. *Journal of Computer and System Sciences* 66: 671–687.

Aggarwal CC and Zhai C. (2012) *Mining text data,* 233 Spring Street, New York, NY 10013, USA: Springer Science+Business Media, LLC, ISBN 978-1-4614-3222-7.

Agier I and Szafarz A. (2013) Microfinance and gender: Is there a glass ceiling on loan size? *World Development* 42: 165–181.

Agrawal A, Catalini C and Goldfarb A. (2011) The geography of crowdfunding. National Bureau of Economic Research (NBER), Working Paper No. 16820.

Agrawal A, Catalini C and Goldfarb A. (2015) Crowdfunding: geography, social networks, and the timing of investment decisions. *Journal of Economics & Management Strategy* 24: 253–274.

Ahlers GKC, Cumming D, Günther C, et al. (2015) Signaling in equity crowdfunding. *Entrepreneurship: Theory and Practice* 39: 955–980.

Ali F, Kwak D, Khan P, et al. (2017) Fuzzy ontology-based sentiment analysis of transportation and city feature reviews for safe traveling. *Transportation Research Part C* 77: 33–48.

Allison TH, Davis BC, Short JC, et al. (2014) Crowdfunding in a prosocial microlending environment: Examining the role of intrinsic versus extrinsic cues. *Entrepreneurship: Theory and Practice* 39: 53–73.

Amari S-i and Nagaoka H. (2000) *Methods of information geometry,* P.O.Box 6248, Providence, Rhode Island 02940-6248: Originally published in Japanese by Iwanami Shoten in 1993. English version was translated by Daishi Harada and published by American Mathematical Society and Oxford University Press, ISBN 0-8218-0531-2.

Aprilia L and Wibowo SS. (2017) The impact of social captial on crowdfunding performance. *The South East Asian Journal of Management* 11: 44–57.

Aronson E, Wilson TD, Akert RM, et al. (2016) *Social Psychology*: Pearson Education, Inc., ISBN 978-0-13-393654-4.

Asgary NH and McNulty RE. (2017) Contributions and challenges in the struggle to end poverty: The case of Kiva. *Information Technology for Development* 23: 367–387.

Auger GA. (2013) Fostering democracy through social media: Evaluating diametrically opposed nonprofit advocacy organizations' use of Facebook, Twitter and YouTube. *Public Relations Review* 39: 369–376.

Baeck P, Collins L and Zhang B. (2014) Understanding alternative finance: The UK alternative finance industry report 2014. 1 Plough Place, London, EC4A 1DE Nesta, pp. 1–95.

Balduzzi D. (2013) Randomized co-training: From cortical neurons to machine learning and back again. Retrieved from https://arxiv.org/abs/1310.6536, accessed on 14 December 2017.

Bechter C, Jentzsch S and Michael F. (2011) From wisdom to wisdom of the crowd and crowdfunding. *Journal of Communication and Computer* 8: 951–957.

Beier M and Wagner K. (2014) Crowdfunding between social media and e-commerce: Online communication, online relationships and fundraising success on crowdfunding platforms. Switzerland: Hochschule fuer Technik and Wirtschaft University of Applied Sciences, Chur.

Bekkers R and Wiepking P. (2011) A literature review of empirical studies of philanthropy: Eight mechanisms that drive charitable giving. *Nonprofit and Voluntary Sector Quarterly* 40: 924–973.

Belleflamme P, Lambert T and Schwienbacher A. (2014) Crowdfunding: Tapping the right crowd. *Journal of Business Venturing* 29: 585–609.

Berkovich E. (2011) Search and herding effects in peer-to-peer leanding: Evidence from prosper.com. *Annals of Finance* 7: 389–405.

Bi S, Liu Z and Usman K. (2017) The influence of online information on investing decisions of reward-based crowdfunding. *Journal of Business Research* 71: 10–18.

Blei DM, Ng AY and Jordan MI. (2003) Latent Dirichlet allocation. *Journal of Machine Learning Research* 3: 993–1022.

Block JH, Colombo MG, Cumming DJ, et al. (in press) New players in entrepreneurial finance and why they are there. *Small Business Economics* DOI 10.1007/s11187-016-9826-6.

Bluestein A. (2014) How I got started in the beginning ... : A well-honed founding story can help you connect with investors, employees, and consumers—and, with any luck, keep them listening. *Inc., February 2014*, pp. 28–40.

Boeuf B, Darveau J and Legoux R. (2014) Financing creativity: Crowdfunding as a new approach for theatre projects. *International Journal of Arts Management* 16: 33–48.

Bollen J, Mao H and Zeng X. (2010) Twitter mood predicts the stock market. *Journal of Computer Science* 2: 1–8.

Brüntje D and Gajda O. (2016) *Crowdfunding in Europe: state of the art in theory and practice*, Cham Heidelberg New York Dordrecht London: Springer International Publishing Switzerland, ISBN 978-3-319-18016-8.

Brase CH and Brase CP. (2013) *Understandable statistics: Concepts and methods*, 20 Channel Center Street, Boston, MA 02210, USA: Brooks/Cole, Cengage Learning, ISBN 978-0-8400-4838-7.

Bretschneider U and Leimeister JM. (in press) Not just an ego-trip: Exploring backers' motivation for funding in incentive-based crowdfunding. *Journal of Strategic Information Systems*.

Bryman A and Bell E. (2015) *Business research methods*, New York: Oxford University Press, ISBN 9780199668649.

Burghardt M. (2010) Retail investor sentiment and behavior: An empirical analysis. *Faculty of Economic.* Karlsruhe Institute of Technology.

Burtch G. (2013) An empirical examination of factors influencing participant behavior in crowdfunded markets. Temple University.

Burtch G, Ghose A and Wattal S. (2013) An empirical examination of the antecedents and consequences of investment patterns in crowd-funded markets. *Information Systems Research* 24: 499–519.

Cambria E and Hussain A. (2012) *Sentic computing: Techniques, tools, and applications*, Dordrecht Heidelberg New York London: Springer, ISBN 978-94-007-5069-2.

Cambria E, Livingstone A and Hussain A. (2012a) The Hourglass of emotions. In: Esposito A, Esposito EM, Vinciarelli A, et al. (eds) *Cognitive Behavioural Systems, Lecture Notes in Computer Science, Vol. 7403*. Berlin, Heidelberg: Springer, ISBN 978-3-642-34583-8, Chapter 11, pp. 144–157.

Cambria E, Olsher D and Kwok K. (2012b) Sentic activation: A two-level affective common sense reasoning framework. *Proceedings of the Twenty-Sixth AAAI Conference on Artificial Intelligence (AAAI-12), 22–26 July, Sheraton Centre, Toronto, Canada*, pp. 186–192.

Cambria E and Huang G-B. (2013) Extreme learning machines. *IEEE Intelligent Systems* 28: 30–31.

Cambria E, Olsher D and Rajagopal D. (2014) SenticNet 3: A common and common-sense knowledge based for cognition-driven sentiment analysis. *Proceedings of the Twenty-Eighth AAAI Conference on Artificial Intelligence (AAAI-14), 27–31 July, Québec City, Québec, Canada*, pp. 1515–1521.

Cambria E and Hussain A. (2015) *Sentic computing: A common-sense-based framework for concept-level sentiment analysis*: Springer International Publishing Switzerland, ISBN 978-3-319-23653-7.

Cambria E, Fu J, Bisio F, et al. (2015) AffectiveSpace 2: Enabling affective intuition for concept-level sentiment analysis. *Proceedings of the Twenty-Ninth AAAI Conference on Artificial Intelligence (AAAI-15), 25–30 January, Austin, Texas, USA*, pp. 508–514.

Cambria E, Das D, Bandyopadhyay S, et al. (2017a) Affective computing and sentiment analysis. In: Cambria E, Das D, Bandyopadhyay S, et al. (eds) *A Practical Guide to Sentiment Analysis.* Springer International Publishing AG, ISBN 978-3-319-55392-4, Chapter 1, pp. 1–10.

Cambria E, Das D, Bandyopadhyay S, et al. (2017b) *A practical guide to sentiment analysis*: Springer International Publishing AG, ISBN 978-3-319-55392-4.

Carvajal M, García-Avilés JA and González JL. (2012) Crowdfunding and non-profit media: the emergence of new models for public interest journalism. *Journalism Practice* 6: 638–647.

Cavanagh S. (2014) Choosing the right words to lure financial backers. *Education Week, 17 September 2014, Vol. 34 Issue 4*, p. 8.

Chen D, Li X and Lai F. (in press) Gender discrimination in online peer-to-peer credit lending: evidence from a lending platform in China. *Electronic Commerce Research* DOI 10.1007/s10660-016-9247-2.

Chen D and Lin Z. (2014) Rational or irrational herding in online microloan markets: Evidence from China. *SSRN Electronic Journal* https://ssrn.com/abstract=2425047.

Chen SF and Goodman J. (1996) An empirical study of smoothing techniques for language modeling. *Proceedings of the 34th Annual Meeting of the Association for Computational Linguistics, accessed on 14 November 2017, retrieved from http://aclweb.org/anthology/P96-1041.*

Chen X, Yao X and Kotha S. (2009) Entrepreneurial passion and preparedness in business plan presentations: A persuasion analysis of venture capitalists' funding decisions. *Academy of Management Journal* 52: 199–214.

Chklovski T. (2003) Learner: A system for acquiring commonsense knowledge by analogy. *Proceedings of the 2nd International Conference on Knowledge Capture (K-CAP '03), 23–25 October, Sanibel Island, FL, USA*, pp. 4–12.

Cholakova M and Clarysse B. (2015) Does the possibility to make equity investments in crowdfunding projects crowd out reward-based investments? *Entrepreneurship Theory and Practice* 39: 145–172.

Ciccarelli SK and White JN. (2018) *Psychology*, Edinburgh Gate, Harlow, Essex CM20 2JE, England: Pearson Education Limited, ISBN 978-1-292-15971-3.

Cimiano P, Hotho A and Staab S. (2005) Learning concept hierarchies from text corpora using formal concept analysis. *Journal of Artificial Intelligence Research* 24: 305–339.

Colistra R and Duvall K. (2017) Show me the money: Importance of crowdfunding factors on backers' decisions to financially support Kickstarter campaigns. *Social Media + Society* October–December: 1–12.

Collier B and Hampshire R. (2010) Sending mixed signals: Multilevel reputation effects in peer-to-peer lending markets. *Proceedings of the 2010 ACM Conference on Computer Supported Cooperative Work (CSCW), 6–10 Februrary, Savannah, Georhia, USA*, pp. 197–206.

Collins L and Pierrakis Y. (2012) The venture crowd: Crowdfunding equity invesment into business. 1 Plough Place, London EC4A 1DE, UK: Nesta, ISBN 978-1-84875-138-5, 1–36.

Colombo MG, Franzoni C and Rossi-Lamastra C. (2015) Internal social capital and the attraction of early contribution in crowdfunding. *Entrepreneurship Theory and Practice* 39: 75–100.

Corazzini L, Cotton C and Valbonesi P. (2015) Donor coordination in project funding: Evidence from a threshold public goods experiment. *Journal of Public Economics* 128: 16–29.

Cordova A, Dolci J and Gianfrate G. (2015) The determinants of crowdfunding success: Evidence from technology projects. *Procedia-Social and Behavioral Sciences* 181: 115–124.

Courtney C, Dutta S and Li Y. (2017) Resolving information asymmetry: Signaling, endorsement, and crowdfunding success. *Entrepreneurship Theory and Practice* March: 265–290.

Crain SP, Zhou K, Yang S-H, et al. (2012) Dimensionality reduction and topic modeling. In: Aggarwal CC and Zhai C (eds) *Mining Text Data.* 233 Spring Street, New York, NY 10013, USA: Springer Science+Business Media, LLC, ISBN 978-1-4614-3222-7, Chapter 5, pp. 129–161.

Csikszentmihalyi M. (2014) *Applications of flow in human development and education: The collected works of Mihaly Csikszentmihalyi*: Springer Science+Business Media, ISBN 978-94-017-9093-2.

Davies R. (2014) Three provocations for civic crowdfunding. *The 15th Annual meeting of the association of Internet Researchers (AoIR), Daegu, Korea, 22–24 October*, pp. 1–3.

De Buysere K, Gajda O, Kleverlaan R, et al. (2012) *A framework for European crowdfunding*: European Crowdfunding Network, ISBN 978-3-00-040193-0.

Eckart C and Young G. (1936) The approximation of one matrix by another of lower rank. *Psychometrika* 1: 211–218.

Ekman P and Davidson RJ. (1994) *The nature of emotion: Fundamental questions,* 198 Madison Avenue, New York, 10016-4314: Oxford Unviersity Press, Inc., ISBN 0-19-508943-X.

Everett GR. (2010) Group membership, relationship banking and default risk: The case of online social lending. *SSRN Electronic Journal* http://ssrn.com/abstract=1114428/.

Feinberg M. (2014) Crowdfunding for municipal projects. *Government Procurement, December/January 2014, Vol. 21 Issue 6*, p. 14.

Ferguson MF and Peters SR. (2000) Is lending discrimination always costly? *Journal of Real Estate Finance and Economics* 21: 23–44.

Fernandes R. (2013) Analysis of crowdfunding descriptions for technology projects. Massachusetts Institute of Technology.

Figliomeni M. (2014a) The entrepreneurial artist: How crowdfunding can jump-start your project Part 1. *Canadian Musician, July/August 2014, Vol. 36 Issue 4*, p. 62.

Figliomeni M. (2014b) The entrepreneurial artist: What can equity crowdfunding do for me? Part 2. *Canadian Musician, September/October 2014, Vol. 36 Issue 5*, p. 62.

Flanigan ST. (2017) Crowdfunding and diaspora philanthropy: An integration of the literature and major concepts. *Voluntas* 28: 492–509.

Gómez-Diago G. (2015) Communication in crowdfunding online platforms. In: Zagalo N and Branco P (eds) *Creativity in the Digital Age.* London, UK: Springer-Verlag, ISBN 978-1-4471-6680-1, Part III, Chapter 10, pp. 171–190.

Gandia JL. (2011) Internet disclosure by non-profit organizations: Empirical evidence of nongovernmental organizations of development in Spain. *Nonprofit and Voluntary Sector Quarterly* 321: 115–127.

Gao Q and Lin M. (2014) Linguistic features and peer-to-peer loan quality: A machine learning approach. *SSRN Electronic Journal* http://papers.ssrn.com/sol3/papers.cfm?abstract_id?446114.

Gao Q and Lin M. (2016) Economic value of texts: Evidence from online debt crowdfunding. *SSRN Electronic Journal* http://dx.doi.org/10.2139/ssrn.2446114.

Gerber E and Hui J. (2013) Crowdfunding: motivations and deterrents for participation. *ACM Transactions on Computer-Human Interaction* 20: 1–32.

Gerber EM, Hui JS and Kuo PY. (2012) Crowdfunding: Why people are motivated to post and fund projects on crowdfunding platforms. *Proceedings of the International Workshop on Design, Influence, and Social Techniques, 11–15 February, 2012, Seattle, Washington, USA*, pp. 1–10.

Giudici G, Guerini M and Rossi-Lamastra C. (2013) *Why crowdfunding projects can succeed: The role of proponents' individual and territorial social capital, retrieved from http://ssrn.com/abstract=2255944, accessed on 10 October 2017.*

Giudici G, Guerini M and Rossi-Lamastra C. (in press) Reward-based crowdfunding of entrepreneurial projects: The effect of local altruism and localized social capital on proponents' success. *Small Business Economics.*

Goldberg Y. (2017) *Neural network methods in natural language processing*, www.morganclaypool.com: Morgan & Claypool, ISBN 978-1-62705-298-6.

Granville VJ. (2016) Peer-to-peer fundraising and crowdfunding in health care philanthropy. *Healthcare Philanthropy Journal* Spring: 30–35.

Gras D, Nason RS, Lerman M, et al. (2017) Going offline: Broadening crowdfunding research beyond the online context. *Venture Capital* 19: 217–237.

Greiner ME and Wang H. (2010) Building consumer-to-consumer trust in e-finance marketplaces: An empirical analysis. *International Journal of Electronic Commerce* 15: 105–136.

Griffiths H. (2017) Civic crowdfunding: A guidebook for local authorities. Future Cities Catapult., 1–31.

Guenther C, Johan S and Schweizer D. (in press) Is the crowd sensitive to distance? How investment decisions differ by investor type. *Small Business Economics* DOI 10.1007/s11187-016-9834-6.

Hamilton RW and Thompson DV. (2007) Is there a substitute for direct experience? Comparing consumer's preferences after direct and indirect product experiences. *Journal of Consumer Research* 34: 546–555.

Harkness SK. (2016) Discrimination in lending markets: Status and the intersections of gender and race. *Social Psychology Quarterly* 79: 81–93.

Harms M. (2007) What drives motivation to participate financially in a crowdfunding community? Amsterdam: Vrije Universitaet Amsterdam.

Haselmayer M and Jenny M. (2017) Sentiment analysis of political communication: Combining a dictionary approach with crowdcoding. *Quality & Quantity* 51: 2623–2646.

Herzenstein M, Dholakia UM and Andrews RL. (2011a) Strategic herding behavior in peer-to-peer loan auctions. *Journal of Interactive Marketing* 25: 27–36.

Herzenstein M, Sonenshein S and Dholakia UM. (2011b) Tell me a good story and I may lend you money: The role of narratives in peer-to-peer lending decisions. *Journal of Marketing Research* 48: S138–S149.

Hobbs J, Grigore G and Molesworth M. (2016) Success in the management of crowdfunding projects in the creative industries. *Internet Research* 26: 146–166.

Hornikx J and Hendriks B. (2015) Consumer tweets about brands: A content analysis of sentiment tweets about goods and services. *Journal of Creative Communications* 10: 176–185.

Horvát EÁ, Uparna J and Uzzi B. (2015) Network vs market relations: The effect of friends in crowdfunding. *Proceedings of the 2015 IEEE/ACM International Conference on Advances in Social Networks Analysis and Mining (ASONAM), 25~28 August, Paris, France*, pp. 226–233.

Hsu DH and Ziedonis RH. (2013) Resources as dual sources of advantage: Implications for valuing entrepreneurial-firm patents. *Strategic Management Journal* 34: 761–781.

Huang G-B, Wang DH and Lan Y. (2011) Extreme learning machines: A survey. *International Journal of Machine Learning and Cybernetics* 2: 107–122.

Huang G-B, Zhou H, Ding X, et al. (2012) Extreme learning machine for regression and multiclass classification. *IEEE Transactions on Systems, Man, and Cybernetics—Part B: Cybernetics* 42: 513–529.

Huang G-B. (2014) An insight into extreme learning machines: Random neurons, random features and kernels. *Cognitive Computation* 6: 376–390.

Hui JS, Gerber EM and Gergle D. (2014) Understanding and leveraging social networks for crowdfunding: Opportunities and challenges. *Proceedings of the 2014 Conference on Designing Interactive Systems (DIS), 21~25 June, Vanouver, BC, Canada*, pp. 677–680.

Huntley JK. (2006) Conceptualization and measurement of relationship quality: Linking relationship quality to actual sales and recommendation intention. *Industrial Marketing Management* 35: 703–714.

Jelinek F and Mercer R. (1980) Interpolated estimation of Markov source parameters from sparse data. *Proceedings of the Workshop on Pattern Recognition in Practice.*

Jian L and Usher N. (2014) Crowd-Funded Journalism. *Journal of Computer-Mediated Communication* 19: 155–170.

Johnson RA. (2018) *Miller & Freund's probability and statistics for engineers,* Edinburgh Gate, Harlow, Essex CM20 2JE, England: Pearson Education Limited, ISBN 978-1-292-17601-7.

Kasper G and Marcoux J. (2015) Case studies in funding innovation. Deloitte University Press, 1–76.

Kearney C and Liu S. (2013) Textual sentiment in finance: A survey of methods and models. *SSRN Electronic Journal* http://ssrn.com/abstract=2213801.

Kim SS and Crompton JL. (2001) The effects of different types of information messages on perceptions of price and stated willingness-to-pay. *Journal of Leisure Research* 33: 299–318.

Kleemann F, Voß GG and Rieder K. (2008) Un(der) paid innovators: The commercial utilization of consumer work through crowdsourcing. *Science, Technology & Innovation Studies* 4: 5–26.

Korolov R, Peabody J, Lavoie A, et al. (2016) Predicting charitable donations using social media. *Social Network Analysis and Mining* 6: 1–10.

Koufaris M and Kambil PALA. (2001) Consumer behavior in web-based commerce: An empirical study. *International Journal of Electronic Commerce* 6: 115–138.

Kraft DH and Colvin E. (2017) *Fuzzy information retrieval,* www.morganclaypool.com: Morgan and Claypool, ISBN 978-1-62705-952-7.

Kullback S and Leibler RA. (1951) On information and sufficiency. *Annals of Mathematical Statistics* 22: 79–86.

Kunz MM, Bretschneider U, Erler M, et al. (2017) An empirical investigation of signaling in reward-based crowdfunding. *Electronic Commerce Research* 17: 425–461.

Kuo Y-F and Wu C-M. (2012) Satisfaction and post-purchase intentions with service recovery of online shopping websites: Perspectives on perceived justice and emotions. *International Journal of Information Management* 32: 127–138.

Kuppuswamy V and Bayus BL. (2013) Crowdfunding creative ideas: The dynamics of project backers in Kickstarter. *SSRN Electronic Journal* http://ssrn.com/abstract=2234765.

Kuppuswamy V and Bayus BL. (2017) Does my contribution to your crowdfunding project matter? *Journal of Business Venturing* 32: 72–89.

Lau RYK, Bruza P and Song D. (2008) Towards a belief revision based adaptive and context-sensitive information retrieval system. *ACM Transactions on Information Systems* 26: Article 8.

Lau RYK, Song D, Li Y, et al. (2009) Towards a fuzzy domain ontology extraction method for adaptive e-learning. *IEEE Transactions on Knowledge and Data Engineering* 21: 800–813.

Lau RYK, Li C and Liao SSY. (2014) Social analytics: Learning fuzzy product ontologies for aspect-oriented sentiment analysis. *Decision Support Systems* 65: 80–94.

Lee E and Lee B. (2012) Herding behavior in online P2P lending: An empirical investigation. *Electronic Commerce Research and Applications* 11: 495–503.

Lehner OM. (2013) Crowdfunding social ventures: A model and research agenda. *Venture Capital* 15: 289–311.

Li B, Chan KCC, Ou C, et al. (2017) Discovering public sentiment in social media for predicting stock movement of publicly listed companies. *Information Systems* 69C: 81–92.

Lin M, Prabhala NR and Viswanathan S. (2013) Judging borrowers by the company they keep: Friendship networks and information asymmetry in online peer-to-peer lending. *Management Science* 59: 17–35.

Lin M and Viswanathan S. (2016) Home bias in online investment: An empirical study of an online crowdfunding market. *Management Science* 62: 1393–1414.

Lipusch N, Bretschneider U and Leimeister JM. (2016) Backer empowerment in reward-based crowdfunding: How participation beyond funding influences support behavior. *Proceedings of the 37th International Conference on Information System on the Digitization of the Individual (DOTI), Dublin, Ireland*, pp. 1–11.

Lischka JA. (2016) *Economic news, sentiment, and behavior: How economic and business news affects the economy*: Springer Fachmedien Wiesbaden, ISBN 978-3-658-11540-1.

Liu B. (2015) *Sentiment analysis: Mining opinions, sentiments and emotions,* 32 Avenue of the Americas, New York, NY 10013-2473, USA: Cambridge University Press, ISBN 978-1-107-01789-4.

Liu CT, Gao YM and Lee CH. (2011) The effects of relationship quality and switching barriers on customer loyalty. *International Journal of Information Management* 31: 71–79.

Liu D, Brass DJ, Lu Y, et al. (2015) Friendships in online peer-to-peer lending: Pipes, prisms and relational herding. *MIS Quarterly* 39: 729–742.

Lu CT, Xie S, Kong X, et al. (2014) Inferring the impacts of social media on crowdfunding. *Proceedings of the 7th ACM International Conference on Web Search and Data Mining, 24–28 February, New York City, USA*, pp. 573–582.

Lu Y, Dhillon P, Foster DP, et al. (2013) Faster ridge regression via the sub-sampled randomized hadamard transform. In: Burges CJC, Bottou L, Welling M, et al. (eds) *Advances in Neural Information Processing Systems*, pp. 369–377. New York: Neural Information Processing Systems Foundation, Inc.

Lupu M, Mayer K, Kando N, et al. (2017a) *Current challenges in patent information retrieval,* Heidelberger Platz 3, 14197 Berlin, Germany: Springer-Verlag GmbH Germany, ISBN 978-3-662-53816-6.

Lupu M, Piroi F and Stefanov V. (2017b) An introduction to contemporary search technology. In: Lupu M, Mayer K, Kando N, et al. (eds) *Current Challenges in Patent Information Retrieval.* 2nd ed. Heidelberger Platz 3, 14197 Berlin, Germany: Springer-Verlag GmbH Germany, ISBN 978-3-662-53816-6, Chapter 2, pp. 47–73.

Ly P and Mason G. (2012) Individual preferences over development projects: Evidence from microlending on Kiva. *International Society for Third-Sector Research* 23: 1036–1055.

MacMillan K, Money K, Money A, et al. (2005) Relationship marketing in the not-for-profit sector: an extension and application of the commitment-trust theory. *Journal of Business Research* 58: 806–818.

Mano RS. (2014) Social media, social causes, giving behavior and money contributions. *Computers in Human Behavior* 31: 287–293.

Marom S. (2017) Social responsibility and crowdfunding businesses: A measurement development study. *Social Responsibility Journal* 13: 235–249.

Marwala T. (2000) Fault identification using neural networks and vibration data. *St. John's College.* University of Cambridge.

Marwala T. (2007) Bayesian training of neural networks using genetic programming. *Pattern Recognition Letters* 28: 1452–1458.

Marwala T. (2009) *Computational intelligence for missing data imputation, estimation and management: knowledge optimization techniques,* New York, USA: IGI Global, ISBN 978-1-60566-336-4.

Marwala T. (2012) Data processing techniques for condition monitoring. In: Marwala T (ed) *Condition Monitoring Using Computational Intelligence Methods: Applications in Mechanical and Electrical Systems.* London: Springer-Verlag, ISBN 978-1-4471-2379-8, Chapter 2, pp. 27–51.

Marwala T. (2013a) Correlations versus causality approaches to economic modeling. In: Marwala T (ed) *Economic Modeling Using Artificial Intelligence Methods.* Springer London Heidelberg New York Dordrecht: Springer-Verlag London, ISBN 978-1-4471-5009-1, Chapter 8, pp. 137–154.

Marwala T. (2013b) *Economic modeling using artificial intelligence methods,* Springer London Heidelberg New York Dordrecht: Springer-Verlag London, ISBN 978-1-4471-5009-1.

Marwala T. (2013c) Introduction to economic modeling. In: Marwala T (ed) *Economic Modeling Using Artificial Intelligence Methods.* Springer London Heidelberg New York Dordrecht: Springer-Verlag London, ISBN 978-1-4471-5009-1, Chapter 1, pp. 1–21.

Marwala T. (2014) *Artificial intelligence techniques for rational decision making,* Springer Cham Heidelberg New York Dordrecht London: Springer International Publishing Switzerland, ISBN 978-3-319-11423-1.

Marwala T. (2015a) Causal, correlation and automatic relevance determination machines for Granger causality. In: Marwala T (ed) *Causality, Correlation and Artificial Intelligence for Rational Decision Making.* 5 Toh Tuck Link, Singapore 596224: World Scientific Publishing Co. Pte. Ltd, ISBN 978-9-81463-086-3, Chapter 7, pp. 125–145.

Marwala T. (2015b) *Causality, correlation and artificial intelligence for rational decision making,* 5 Toh Tuck Link, Singapore 596224: World Scientific Publishing Co. Pte. Ltd, ISBN 978-9-81463-086-3.

Marwala T. (2015c) Conclusions and further work. In: Marwala T (ed) *Causality, Correlation and Artificial Intelligence for Rational Decision Making.* 5 Toh Tuck Link, Singapore 596224: World Scientific Publishing Co. Pte. Ltd, ISBN 978-9-81463-086-3, Chapter 10, pp. 187–189.

Marwala T. (2015d) Correlation machines using optimization methods. In: Marwala T (ed) *Causality, Correlation and Artificial Intelligence for Rational Decision Making.* 5 Toh Tuck Link, Singapore 596224: World Scientific Publishing Co. Pte. Ltd, ISBN 978-9-81463-086-3, Chapter 4, pp. 65–86.

Marwala T. (2015e) Flexibly-bounded rationality. In: Marwala T (ed) *Causality, Correlation and Artificial Intelligence for Rational Decision Making.* 5 Toh Tuck Link, Singapore 596224: World Scientific Publishing Co. Pte. Ltd, ISBN 978-9-81463-086-3, Chapter 8, pp. 147–166.

Marwala T. (2015f) Introduction to artificial intelligence based decision making. In: Marwala T (ed) *Causality, Correlation and Artificial Intelligence for Rational Decision Making.* 5 Toh Tuck Link, Singapore 596224: World Scientific Publishing Co. Pte. Ltd, ISBN 978-9-81463-086-3, Chapter 1, pp. 1–21.

Marwala T. (2015g) Marginalization of irrationality in decision making. In: Marwala T (ed) *Causality, Correlation and Artificial Intelligence for Rational Decision Making.* 5 Toh Tuck Link, Singapore 596224: World Scientific Publishing Co. Pte. Ltd, ISBN 978-9-81463-086-3, Chapter 9, pp. 167–185.

Marwala T. (2015h) Neural networks for modeling granger causality. In: Marwala T (ed) *Causality, Correlation and Artificial Intelligence for Rational Decision Making.* 5 Toh Tuck Link, Singapore 596224: World Scientific Publishing Co. Pte. Ltd, ISBN 978-9-81463-086-3, Chapter 5, pp. 87–103.

Marwala T. (2015i) Rubin, Pearl and Granger causality models: A unified view. In: Marwala T (ed) *Causality, Correlation and Artificial Intelligence for Rational Decision Making.* 5 Toh Tuck Link, Singapore 596224: World Scientific Publishing Co. Pte. Ltd, ISBN 978-9-81463-086-3, Chapter 6, pp. 105–124.

Marwala T. (2015j) What is a causal machine? In: Marwala T (ed) *Causality, Correlation and Artificial Intelligence for Rational Decision Making*. 5 Toh Tuck Link, Singapore 596224: World Scientific Publishing Co. Pte. Ltd, ISBN 978-9-81463-086-3, Chapter 3, pp. 43–63.

Marwala T. (2015k) What is a correlation machine? In: Marwala T (ed) *Causality, Correlation and Artificial Intelligence for Rational Decision Making*. 5 Toh Tuck Link, Singapore 596224: World Scientific Publishing Co. Pte. Ltd, ISBN 978-9-81463-086-3, Chapter 2, pp. 23–42.

Marwala T and Hurwitz E. (2017a) *Artificial intelligence and economic theory: Skynet in the market*, Gewerbestrasse 11, 6330 Cham, Switzerland: Springer International Publishing AG, ISBN 978-3-319-66103-2.

Marwala T and Hurwitz E. (2017b) Behavioral economics. In: Marwala T and Hurwitz E (eds) *Artificial Intelligence and Economic Theory: Skynet in the Market*. Gewerbestrasse 11, 6330 Cham, Switzerland: Springer International Publishing AG, ISBN 978-3-319-66103-2, Chapter 5, pp. 51–61.

Marwala T and Hurwitz E. (2017c) Introduction to man and machines. In: Marwala T and Hurwitz E (eds) *Artificial Intelligence and Economic Theory: Skynet in the Market*. Gewerbestrasse 11, 6330 Cham, Switzerland: Springer International Publishing AG, ISBN 978-3-319-66103-2, Chapter 1, pp. 1–14.

McKinnon SL, Dickinson E, Carr JN, et al. (2013) Kiva.org, person-to-person lending and the conditions of intercultural contact. *Howard Journal of Communications* 24: 327–347.

Meer J. (2014) Effects of the price of charitable giving: Evidence from an online crowdfunding platform. *Journal of Economic Behavior & Organ* 103: 113–124.

Melucci M. (2015) *Introduction to information retrieval and quantum mechanics*, Heidelberg New York Dordrecht London: Springer-Verlag Berlin Heidelberg, ISBN 978-3-662-48312-1.

Mendenhall W, Beaver RJ and Beaver BM. (2013) *Introduction to probability and statistics*, 20 Channel Center Street, Boston, MA 02210, USA: Brooks/Cole, Cengage Learning, ISBN 978-1-133-10375-2.

Mendes-Da-Silva W, Rossoni L, Conte BS, et al. (2016) The impacts of fundraising periods and geographic distance on financing music production via crowdfunding in Brazil. *Journal of Cultural Economics* 40: 75–99.

Minsky M. (1986) *The society of mind*, Simon & Schuster Building, Rockefeller Center, 1230 Avenue of the Americas, New York, NY 10020: Simon & Schuster, Inc., ISBN 0-671-60740-5.

Mixon FG, Asarta CJ and Caudill SB. (2017) Patreonomics: Public goods pedagogy for economics principles. *International Review of Economics Education* 25: 1–7.

Mohammadi A and Shafi K. (in press) Gender differences in the contribution patterns of equity-crowdfunding investors. *Small Business Economics* DOI 10.1007/s11187-016-9825-7.

Moisseyev A. (2013) Effect of social media on crowdfunding project results. Lincoln, Nebraska: University of Nebraska.

Mollick E. (2013) Swept away by the crowd? Crowdfunding, venture capital and the selection of entrepreneurs. *SSRN Electronic Journal* http://ssrn.com/abstract=2239204.

Mollick E. (2014) The dynamics of crowdfunding: An exploratory study. *Journal of Business Venturing* 29: 1–16.

Mollick E and Nanda R. (2014) *Wisdom or madness? Comparing crowds with expert evaluation in funding the arts, retrieved from http://papers.ssrn.com/sol3/papers.cfm?abstract_id2443114, accessed on 12 June 2014.*

Mollick E. (2016) The unique value of crowdfunding is not money—it's community. *Harvard Business Review*.

Moodie M. (2013) Microfinance and the gender of risk: The case of Kiva.org. *Signs* 38: 279–302.

Moritz A, Block J and Lutz E. (2015) Investor communication in equity-based crowdfunding: A qualitative-empirical study. *Qualitative Research in Financial Markets* 7: 309–342.

Moro S, Cortez P and Rita P. (2015) Business intelligence in banking: A literature analysis from 2002 to 2013 using text mining and latent Dirichlet allocation. *Expert Systems with Applications* 42: 1314–1324.

Moysidou K. (2016) Motivations to contribute financially to crowdfunding projects. In: Salampasis D and Mention AL (eds) *Open Innovation: Unveiling the Power of the Human Element.* World Scientific Publishing, ISBN 9789813140868.

Nambisan S and Zahra SA. (2016) The role of demand-side narratives in opportunity formation and enactment. *Journal of Business Venturing Insights* 5: 70–75.

Next BE. (2013) Crowdfunding websites for college financing. *Black Enterprise.* 34.

Nuwer R. (2015) Time to donate direct. *New Scientist, 22 November 2015, Vol. 224 Issue 2996,* p. 26.

Oosterhoff A. (2015) Crowdfunding and the law of trusts. *Proceedings of the 18th Annual Estates and Trusts Summit, Law Society of Upper Canada,* pp. 1–24.

Ordanini A, Miceli L, Pizzetti M, et al. (2011) Crowd-funding: Transforming customers into investors through innovative service platforms. *Journal of Service Management* 22: 443–470.

Orsay J. (2014) *Psychology & sociology,* Osote Publishing, New Jersey: Examkrackers, Inc., ISBN 978-1-893858-70-1.

Ortega ACB and Bell F. (2008) Online social lending: Borrower-generated content. *Proceedings of Americas Conference on Information Systems (AMCIS), Toronto, Canada,* pp. 1–10.

Otero P. (2015) Crowdfunding: A new option for funding health projects. *Arch Argent Pediatr* 113: 154–157.

Özdemir V, Faris J and Srivastava S. (2015) Crowdfunding 2.0: The next-generation philanthropy. *European Molecular Biology Organization (EMBO) Reports* 16: 267–271.

Pang B and Lee L. (2008) Opinion mining and sentiment analysis. *Foundations and Trends® in Information Retrieval* 2: 1–135.

Parhankangas A and Renko M. (2017) Linguistic style and crowdfunding success among social and commercial entrepreneurs. *Journal of Business Venturing* 32: 215–236.

Peterson RL. (2016) *Trading on sentiment: The power of minds over markets,* 111 River Street, Hoboken, NJ 07030: John Wiley & Sons, Inc., ISBN 978-1-119-12276-0.

Piper G and Schnepf SV. (2008) Gender differences in charitable giving in Great Britain. *Voluntas* 19: 103–124.

Pitschner S and Pitschner-Finn S. (2014) Non-profit differentials in crowd-based financing: Evidence from 50,000 campaigns. *Economics Letters* 123: 391–394.

Plutchik R. (2001) The nature of emotions. *American Scientist* 89: 344–350.

Podolny JM. (2001) Networks as the pipes and prisms of the market. *American Journal of Sociology* 107: 33–60.

Pollack JM and Bosse DA. (2014) When do investors forgive entrepreneurs for lying? *Journal of Business Venturing* 29: 741–754.

Polzin F, Toxopeus H and Stam E. (in press) The wisdom of the crowd in funding: Information heterogeneity and social networks of crowdfunders. *Small Business Economics* DOI 10.1007/s11187-016-9829-3.

Ponte JM and Croft WB. (1998) A language modelling approach to information retrieval. *Proceedings of the 21st Annual International ACM SIGIR Conference on Research and Development in Information Retrieval, Melbourne, Australia,* pp. 275–281.

Pope DG and Sydnor JR. (2011) What's in a picture? Evidence of discrimination from prosper.com. *Journal of Human Resources* 46: 53–92.

Posselt T and Gerstner E. (2005) Pre-sale vs. post-sale e-satisfaction: Impact on repurchase intention and overall satisfaction. *Journal of Interactive Marketing* 19: 35–47.

Putnam R. (1993) Bowling alone: America's declining soical captial *Journal of Democracy* 6: 65–68.

Reimink M. (2014) *Crowdfunding in Dutch small and medium enterprises: an Empirical analysis of factors influencing the intention to invest in a crowdfunding initiative, retrieved from http://essay.utwente.nl/65168/1/Reimink_MA_MB.pdf, accessed on 23 July 2017.*

Renwick MJ and Mossialos E. (2017) Crowdfunding our health: Economic risks and benefits. *Social Science & Medicine* 191: 48–56.

Riggins FJ and Weber DM. (2017) Information asymmetries and identification bias in P2P social microlending. *Information Technology for Development* 23: 107–126.

Rosen-Zvi M, Chemudugunta C, Griffiths TL, et al. (2010) Learning author-topic models from text corpora. *ACM Transactions on Information Systems* 28: Article 4.

Ross SM. (2014) *Introduction to probability models,* The Boulevard, Langford Lane, Kidlington, Oxford OX5 1GB, UK: Elsevier Inc., ISBN 978-0-12-407948-9.

Russell SJ and Norvig P. (2010) *Artificial intelligence: a modern approach,* 1 Lake Street, Upper Saddle River, NJ 07458: Pearson Education, Inc., ISBN 978-0-13-604259-4.

Saade S. (2015) Investor sentiment and the underperformance of technology firms initial public offerings. *Research in International Business and Finance* 34: 205–232.

Samanci M and Kiss G. (2014) Exploratory study on technology related successfully crowdfunding projects' post online market presence. *School of Economics and Management.* Lund University.

Sanchez-González M and Palomo-Torres M-B. (2014) Knowledge and assessment of crowdfunding in communication: the view of journalists and future journalists. *Media Education Reseacrh Journal* 43: 101–110.

Saxton GD and Wang LL. (2014) The social network effect: the determinants of giving through social media. *Nonprofit and Voluntary Sector Quarterly* 43: 850–868.

Schau HJ and Gilly MC. (2003) We are what we post? Self-presentation in personal web space. *Journal of Consumer Research* 30: 385–404.

Schlenker BR and Weigold MF. (1992) Interpersonal processes involving impression regulation and management. *Annual Review of Psychology* 43: 133–168.

Schwienbacher A and Larralde B. (2010) Crowdfunding of small entrepeneurial ventures. *SSRN Electronic Journal* http://dx.doi.org/10.2139/ssrn.1699183.

Sheldon RC and Kupp M. (2017) A market testing method based on crowd funding. *Strategy & Leadership* 45: 19–23.

Shleifer A. (2000) *Inefficient markets: An introduction to behavioral finance,* Great Clarendon Street, Oxford OX2 6DP: Oxford University Press, ISBN 9780191606892.

Siering M, Koch J-A and Deokar AV. (2016) Detecting fraudulent behavior on crowdfunding platforms: The role of linguistic and content-based cues in static and dynamic contexts. *Journal of Management Information Systems* 33: 421–455.

Sigelman CK and Rider EA. (2015) *Life-span: Human development,* 200 First Stamford Place, 4th Floor, Stamford, CT 06902, USA: Cengage Learning, ISBN 978-1-285-45431-3.

Skirnevskiy V, Bendig D and Brettel M. (2017) The influence of internal social capital on serial creators success in crowdfunding. *Entrepreneurship Theory and Practice* March: 209–236.

Snyder J, Mathers A and Crooks VA. (2016) Fund my treatment!: A call for ethics-focused social science research into the use of crowdfunding for medical care. *Social Science & Medicine* 169: 27–30.

Sonenshein S. (2010) We're changing, or are we? Untangling the role of progressive, regressive and stability narratives during strategic change implementation. *Academy of Management Journal* 53: 477–512.

Staats S, Sintjago A and Fitzpatrick R. (2013) Kiva microloans in a learning community: An assignment for interdisciplinary synthesis. *Innovative Higher Education* 38: 173–187.

Stanko MA and Henard DH. (2017) Toward a better understanding of crowdfunding, openness and the consequences for innovation. *Research Policy* 46: 784–798.

Steigenberger N. (2017) Why supporters contribute to reward-based crowdfunding. *International Journal of Entrepreneurial Behavior & Research* 23: 336–353.

Stern JS. (2013) Characteristics of content and social spread strategy on the IndieGoGo crowdfunding platform. The University of Texas at Austin.

Sternberg RJ. (1990) Wisdom and its relations to intelligence and creativity. In: Sternberg RJ (ed) *Wisdom: Its Nature, Originds, and Development.* The Pitt Building, Trumpington Street, Cambridge CB2 1RP: Cambridge University Press, ISBN 0-521-36453-1, Chapter 7, pp. 142–159.

Steyvers M, Smyth P, Rosen-Zvi M, et al. (2004) Probabilistic author-topic models for information discovery. In: Kim W, Kohavi R, Gehrke J, et al. (eds) *Proceedings of the Tenth ACM SIGKDD International Conference on Knowledge Discovery and Data Mining, 22–25 August, Seattle, Washington,* pp. 306–315. ACM.

Stiver A, Barroca L, Minocha S, et al. (2015) Civic crowdfunding research: Challenges, opportunities, and future agenda. *New Media & Society* 17: 249–271.

Subrahmanian VS and Recupero DR. (2008) AVA: Adjective-verb-adverb combinations for sentiment analysis. *IEEE Intelligent Systems* 23: 43–50.

Sullivan M. (2018) *Statistics: Informed decisions using data,* Edinburgh Gate, Harlow, Essex CM20 2JE, England: Pearson Education Limited, ISBN 978-1-292-15711-5.

Tropp JA. (2011) Improved analysis of the sub-sampled randomized Hadamard transform. *Advances in Adaptive Data Analysis* 3: 115–126.

Tugend A. (2015) The tales we tell. *Entrepreneur* 43: 46–48.

Turney PD. (2002) Thumbs up or thumbs down? Semantic orientation applied to unsupervised classification of reviews. *Proceedings of the 40th Annual Meeting on Association for Computational Linguistics (ACL '02), 7–12 July, Philadelphia, Pennsylvania, USA*, pp. 417–424.

Van Montfort K, Masurel E and Van Rijn I. (2000) Service satisfaction: An empirical analysis of consumer satisfaction in financial services. *The Service Industries Journal* 20: 80–94.

Van Rijsbergen CJ. (1979) *Information retrieval,* Newton, MA, USA: Butterworth-Heinemann, ISBN 0-4087-0929-4.

Vismara S. (2016) Equity retention and social network theory in equity crowdfunding. *Small Business Economics* 46: 579–590.

Walter T. (2008) Competition to default: Racial discrimination in the market for online peer-to-peer lending. *Business*: 1–44.

Walther JB. (1992) Interpersonal effects in computer-mediated interaction: A relational perspective. *Communication Research* 19: 52–90.

Wang W, Zhu K, Wang H, et al. (2017) The Impact of sentiment orientations on successful crowdfunding campaigns through text analytics. *The Institution of Engineering and Technology (IET) Software* 11: 229–238.

Ward C and Ramachandran V. (2010) Crowdfunding the next hit: microfunding online experience goods. *Workshop on Computational Social Science and the Wisdom of Crowds at Neural Informaiton Processing Systems Conference (NIPS), 6–9 December, 2010, Vancouver, Canada*, pp. 1–5.

Weigmann K. (2013) Tapping the crowds for research funding. *European Molecular Biology Organization (EMBO) Reports* 14: 1043–1046.

Wheat RE, Wang Y, Byrnes JE, et al. (2013) Raising money for scientific research through crowdfunding. *Trends in Ecology & Evolution* 28: 71–72.

Wood J, Sames L, Moore A, et al. (2013) Multifaceted roles of ultra-rare and rare disease patients/parents in drug discovery. *Drug Discovery Today* 18: 1043–1051.

Xing B, Nelwamondo FV, Battle K, et al. (2009) Application of artificial intelligence (AI) methods for designing and analysis of reconfigurable cellular manufacturing system (RCMS). *Proceedings of the 2nd International Conference on Adaptive Science & Technology (ICAST), 14–16 December, Accra, Ghana*, pp. 402–409. IEEE.

Xing B, Gao W-J, Nelwamondo FV, et al. (2010a) Artificial intelligence in reverse supply chain management: the state of the art. *Proceedings of the Twenty-First Annual Symposium of the Pattern Recognition Association of South Africa (PRASA), 22–23 November, Stellenbosch, South Africa*, pp. 305–310.

Xing B, Gao W-J, Nelwamondo FV, et al. (2010b) Part-machine clustering: the comparison between adaptive resonance theory neural network and ant colony system. In: Zeng Z and Wang J (eds) *Advances in Neural Network Research & Applications, LNEE 67*, pp. 747–755. Berlin Heidelberg: Springer-Verlag.

Xing B, Gao W-J and Marwala T. (2013) An overview of cuckoo-inspired intelligent algorithms and their applications. *IEEE Symposium Series on Computational Intelligence (IEEE SSCI), 15–19 April, Singapore*, pp. 85–89. IEEE.

Xing B and Gao W-J. (2014a) *Computational intelligence in remanufacturing,* 701 E. Chocolate Avenue, Suite 200, Hershey PA 17033: IGI Global, ISBN 978-1-4666-4908-8.

Xing B and Gao W-J. (2014aa) Chemical-reaction optimization algorithm. In: Xing B and Gao W-J (eds) *Innovative Computational Intelligence: A Rough Guide to 134 Clever Algorithms.*

Cham Heidelberg New York Dordrecht London: Springer International Publishing Switzerland, ISBN: 978-3-319-03403-4, Chapter 25, pp. 417–428.

Xing B and Gao W-J. (2014b) *Innovative computational intelligence: A rough guide to 134 clever algorithms,* Cham Heidelberg New York Dordrecht London: Springer International Publishing Switzerland, ISBN: 978-3-319-03403-4.

Xing B and Gao W-J. (2014bb) Emerging chemistry-based CI algorithms. In: Xing B and Gao W-J (eds) *Innovative Computational Intelligence: A Rough Guide to 134 Clever Algorithms.* Cham Heidelberg New York Dordrecht London: Springer International Publishing Switzerland, ISBN: 978-3-319-03403-4, Chapter 26, pp. 429–437.

Xing B and Gao W-J. (2014c) Introduction to computational intelligence. In: Xing B and Gao W-J (eds) *Innovative Computational Intelligence: A Rough Guide to 134 Clever Algorithms.* Cham Heidelberg New York Dordrecht London: Springer International Publishing Switzerland, ISBN: 978-3-319-03403-4, Chapter 1, pp. 3–17.

Xing B and Gao W-J. (2014cc) Base optimization algorithm. In: Xing B and Gao W-J (eds) *Innovative Computational Intelligence: A Rough Guide to 134 Clever Algorithms.* Cham Heidelberg New York Dordrecht London: Springer International Publishing Switzerland, ISBN: 978-3-319-03403-4, Chapter 27, pp. 441–444.

Xing B and Gao W-J. (2014d) Bacteria inspired algorithms. In: Xing B and Gao W-J (eds) *Innovative Computational Intelligence: A Rough Guide to 134 Clever Algorithms.* Cham Heidelberg New York Dordrecht London: Springer International Publishing Switzerland, ISBN: 978-3-319-03403-4, Chapter 2, pp. 21–38.

Xing B and Gao W-J. (2014dd) Emerging mathematics-based CI algorithms. In: Xing B and Gao W-J (eds) *Innovative Computational Intelligence: A Rough Guide to 134 Clever Algorithms.* Cham Heidelberg New York Dordrecht London: Springer International Publishing Switzerland, ISBN: 978-3-319-03403-4, Chapter 28, pp. 445–448.

Xing B and Gao W-J. (2014e) Bee inspired algorithms. In: Xing B and Gao W-J (eds) *Innovative Computational Intelligence: A Rough Guide to 134 Clever Algorithms.* Cham Heidelberg New York Dordrecht London: Springer International Publishing Switzerland, ISBN: 978-3-319-03403-4, Chapter 4, pp. 45–80.

Xing B and Gao W-J. (2014f) Bat inspired algorithms. In: Xing B and Gao W-J (eds) *Innovative Computational Intelligence: A Rough Guide to 134 Clever Algorithms.* Cham Heidelberg New York Dordrecht London: Springer International Publishing Switzerland, ISBN: 978-3-319-03403-4, Chapter 3, pp. 39–44.

Xing B and Gao W-J. (2014ff) Overview of computational intelligence. In: Xing B and Gao W-J (eds) *Computational Intelligence in Remanufacturing.* 701 E. Chocolate Avenue, Suite 200, Hershey PA 17033: IGI Global, ISBN 978-1-4666-4908-8, Chapter 2, pp. 18–36.

Xing B and Gao W-J. (2014g) Biogeography-based optimization algorithm. In: Xing B and Gao W-J (eds) *Innovative Computational Intelligence: A Rough Guide to 134 Clever Algorithms.* Cham Heidelberg New York Dordrecht London: Springer International Publishing Switzerland, ISBN: 978-3-319-03403-4, Chapter 5, pp. 81–91.

Xing B and Gao W-J. (2014h) Cat swarm optimization algorithm. In: Xing B and Gao W-J (eds) *Innovative Computational Intelligence: A Rough Guide to 134 Clever Algorithms.* Cham Heidelberg New York Dordrecht London: Springer International Publishing Switzerland, ISBN: 978-3-319-03403-4, Chapter 6, pp. 93–104.

Xing B and Gao W-J. (2014i) Cuckoo inspired algorithms. In: Xing B and Gao W-J (eds) *Innovative Computational Intelligence: A Rough Guide to 134 Clever Algorithms.* Cham Heidelberg New York Dordrecht London: Springer International Publishing Switzerland, ISBN: 978-3-319-03403-4, Chapter 7, pp. 105–121.

Xing B and Gao W-J. (2014j) Luminous insect inspired algorithms. In: Xing B and Gao W-J (eds) *Innovative Computational Intelligence: A Rough Guide to 134 Clever Algorithms.* Cham Heidelberg New York Dordrecht London: Springer International Publishing Switzerland, ISBN: 978-3-319-03403-4, Chapter 8, pp. 123–137.

Xing B and Gao W-J. (2014k) Fish inspired algorithms. In: Xing B and Gao W-J (eds) *Innovative Computational Intelligence: A Rough Guide to 134 Clever Algorithms.* Cham Heidelberg

New York Dordrecht London: Springer International Publishing Switzerland, ISBN: 978-3-319-03403-4, Chapter 9, pp. 139–155.

Xing B and Gao W-J. (2014l) Frog inspired algorithms. In: Xing B and Gao W-J (eds) *Innovative Computational Intelligence: A Rough Guide to 134 Clever Algorithms.* Cham Heidelberg New York Dordrecht London: Springer International Publishing Switzerland, ISBN: 978-3-319-03403-4, Chapter 10, pp. 157–165.

Xing B and Gao W-J. (2014m) Fruit fly optimization algorithm. In: Xing B and Gao W-J (eds) *Innovative Computational Intelligence: A Rough Guide to 134 Clever Algorithms.* Cham Heidelberg New York Dordrecht London: Springer International Publishing Switzerland, ISBN: 978-3-319-03403-4, Chapter 11, pp. 167–170.

Xing B and Gao W-J. (2014n) Group search optimization algorithm. In: Xing B and Gao W-J (eds) *Innovative Computational Intelligence: A Rough Guide to 134 Clever Algorithms.* Cham Heidelberg New York Dordrecht London: Springer International Publishing Switzerland, ISBN: 978-3-319-03403-4, Chapter 12, pp. 171–176.

Xing B and Gao W-J. (2014o) Invasive weed optimization algorithm. In: Xing B and Gao W-J (eds) *Innovative Computational Intelligence: A Rough Guide to 134 Clever Algorithms.* Cham Heidelberg New York Dordrecht London: Springer International Publishing Switzerland, ISBN: 978-3-319-03403-4, Chapter 13, pp. 177–181.

Xing B and Gao W-J. (2014p) Music inspired algorithms. In: Xing B and Gao W-J (eds) *Innovative Computational Intelligence: A Rough Guide to 134 Clever Algorithms.* Cham Heidelberg New York Dordrecht London: Springer International Publishing Switzerland, ISBN: 978-3-319-03403-4, Chapter 14, pp. 183–201.

Xing B and Gao W-J. (2014q) Imperialist competitive algorithm. In: Xing B and Gao W-J (eds) *Innovative Computational Intelligence: A Rough Guide to 134 Clever Algorithms.* Cham Heidelberg New York Dordrecht London: Springer International Publishing Switzerland, ISBN: 978-3-319-03403-4, Chapter 15, pp. 203–209.

Xing B and Gao W-J. (2014r) Teaching–learning-based optimization algorithm. In: Xing B and Gao W-J (eds) *Innovative Computational Intelligence: A Rough Guide to 134 Clever Algorithms.* Cham Heidelberg New York Dordrecht London: Springer International Publishing Switzerland, ISBN: 978-3-319-03403-4, Chapter 16, pp. 211–216.

Xing B and Gao W-J. (2014s) Emerging biology-based CI algorithms. In: Xing B and Gao W-J (eds) *Innovative Computational Intelligence: A Rough Guide to 134 Clever Algorithms.* Cham Heidelberg New York Dordrecht London: Springer International Publishing Switzerland, ISBN: 978-3-319-03403-4, Chapter 17, pp. 217–317.

Xing B and Gao W-J. (2014t) Big bang–big crunch algorithm. In: Xing B and Gao W-J (eds) *Innovative Computational Intelligence: A Rough Guide to 134 Clever Algorithms.* Cham Heidelberg New York Dordrecht London: Springer International Publishing Switzerland, ISBN: 978-3-319-03403-4, Chapter 18, pp. 321–331.

Xing B and Gao W-J. (2014u) Central force optimization algorithm. In: Xing B and Gao W-J (eds) *Innovative Computational Intelligence: A Rough Guide to 134 Clever Algorithms.* Cham Heidelberg New York Dordrecht London: Springer International Publishing Switzerland, ISBN: 978-3-319-03403-4, Chapter 19, pp. 333–337.

Xing B and Gao W-J. (2014v) Charged system search algorithm. In: Xing B and Gao W-J (eds) *Innovative Computational Intelligence: A Rough Guide to 134 Clever Algorithms.* Cham Heidelberg New York Dordrecht London: Springer International Publishing Switzerland, ISBN: 978-3-319-03403-4, Chapter 20, pp. 339–346.

Xing B and Gao W-J. (2014w) Electromagnetism-like mechanism algorithm. In: Xing B and Gao W-J (eds) *Innovative Computational Intelligence: A Rough Guide to 134 Clever Algorithms.* Cham Heidelberg New York Dordrecht London: Springer International Publishing Switzerland, ISBN: 978-3-319-03403-4, Chapter 21, pp. 347–354.

Xing B and Gao W-J. (2014x) Gravitational search algorithm. In: Xing B and Gao W-J (eds) *Innovative Computational Intelligence: A Rough Guide to 134 Clever Algorithms.* Cham Heidelberg New York Dordrecht London: Springer International Publishing Switzerland, ISBN: 978-3-319-03403-4, Chapter 22, pp. 355–364.

Xing B and Gao W-J. (2014y) Intelligent water drops algorithm. In: Xing B and Gao W-J (eds) *Innovative Computational Intelligence: A Rough Guide to 134 Clever Algorithms*. Cham Heidelberg New York Dordrecht London: Springer International Publishing Switzerland, ISBN: 978-3-319-03403-4, Chapter 23, pp. 365–373.

Xing B and Gao W-J. (2014z) Emerging physics-based CI algorithms. In: Xing B and Gao W-J (eds) *Innovative Computational Intelligence: A Rough Guide to 134 Clever Algorithms*. Cham Heidelberg New York Dordrecht London: Springer International Publishing Switzerland, ISBN: 978-3-319-03403-4, Chapter 24, pp. 375–414.

Xing B. (2017) Visible light based throughput downlink connectivity for the cognitive radio networks. In: Matin MA (ed) *Spectrum Access and Management for Cognitive Radio Networks*. Singapore: Springer Science+Business Media, ISBN 978-981-10-2253-1, Chapter 8, pp. 211–232.

Xing B and Marwala T. (2018a) Conclusion. In: Xing B and Marwala T (eds) *Smart Maintenance for Human–Robot Interaction: An Intelligent Search Algorithmic Perspective*. Gewerbestrasse 11, 6330 Cham, Switzerland: Springer International Publishing AG, ISBN 978-3-319-67479-7, Chapter 13, pp. 299–305.

Xing B and Marwala T. (2018b) Cyberware capacity–applications layer perspective. In: Xing B and Marwala T (eds) *Smart Maintenance for Human–Robot Interaction: An Intelligent Search Algorithmic Perspective*. Gewerbestrasse 11, 6330 Cham, Switzerland: Springer International Publishing AG, ISBN 978-3-319-67479-7, Chapter 8, pp. 173–191.

Xing B and Marwala T. (2018c) Cyberware capacity–energy autonomy perspective. In: Xing B and Marwala T (eds) *Smart Maintenance for Human–Robot Interaction: An Intelligent Search Algorithmic Perspective*. Gewerbestrasse 11, 6330 Cham, Switzerland: Springer International Publishing AG, ISBN 978-3-319-67479-7, Chapter 9, pp. 193–216.

Xing B and Marwala T. (2018d) Cyberware capacity–platform and middleware layers perspective. In: Xing B and Marwala T (eds) *Smart Maintenance for Human–Robot Interaction: An Intelligent Search Algorithmic Perspective*. Gewerbestrasse 11, 6330 Cham, Switzerland: Springer International Publishing AG, ISBN 978-3-319-67479-7, Chapter 7, pp. 143–171.

Xing B and Marwala T. (2018e) Hardware capacity–beginning of life perspective. In: Xing B and Marwala T (eds) *Smart Maintenance for Human–Robot Interaction: An Intelligent Search Algorithmic Perspective*. Gewerbestrasse 11, 6330 Cham, Switzerland: Springer International Publishing AG, ISBN 978-3-319-67479-7, Chapter 4, pp. 67–91.

Xing B and Marwala T. (2018f) Hardware capacity–end of life perspective. In: Xing B and Marwala T (eds) *Smart Maintenance for Human–Robot Interaction: An Intelligent Search Algorithmic Perspective*. Gewerbestrasse 11, 6330 Cham, Switzerland: Springer International Publishing AG, ISBN 978-3-319-67479-7, Chapter 6, pp. 111–139.

Xing B and Marwala T. (2018g) Hardware capacity–middle of life perspective. In: Xing B and Marwala T (eds) *Smart Maintenance for Human–Robot Interaction: An Intelligent Search Algorithmic Perspective*. Gewerbestrasse 11, 6330 Cham, Switzerland: Springer International Publishing AG, ISBN 978-3-319-67479-7, Chapter 5, pp. 93–110.

Xing B and Marwala T. (2018h) Human capacity–biopsychosocial perspective. In: Xing B and Marwala T (eds) *Smart Maintenance for Human–Robot Interaction: An Intelligent Search Algorithmic Perspective*. Gewerbestrasse 11, 6330 Cham, Switzerland: Springer International Publishing AG, ISBN 978-3-319-67479-7, Chapter 11, pp. 249–270.

Xing B and Marwala T. (2018i) Human capacity–exposome perspective. In: Xing B and Marwala T (eds) *Smart Maintenance for Human–Robot Interaction: An Intelligent Search Algorithmic Perspective*. Gewerbestrasse 11, 6330 Cham, Switzerland: Springer International Publishing AG, ISBN 978-3-319-67479-7, Chapter 12, pp. 271–295.

Xing B and Marwala T. (2018j) Human capacity–physiology perspective. In: Xing B and Marwala T (eds) *Smart Maintenance for Human–Robot Interaction: An Intelligent Search Algorithmic Perspective*. Gewerbestrasse 11, 6330 Cham, Switzerland: Springer International Publishing AG, ISBN 978-3-319-67479-7, Chapter 10, pp. 219–247.

Xing B and Marwala T. (2018k) Introduction to human robot interaction. In: Xing B and Marwala T (eds) *Smart Maintenance for Human–Robot Interaction: An Intelligent

Search Algorithmic Perspective. Gewerbestrasse 11, 6330 Cham, Switzerland: Springer International Publishing AG, ISBN 978-3-319-67479-7, Chapter 1, pp. 3–19.

Xing B and Marwala T. (2018l) Introduction to intelligent search algorithms. In: Xing B and Marwala T (eds) *Smart Maintenance for Human–Robot Interaction: An Intelligent Search Algorithmic Perspective.* Gewerbestrasse 11, 6330 Cham, Switzerland: Springer International Publishing AG, ISBN 978-3-319-67479-7, Chapter 3, pp. 33–64.

Xing B and Marwala T. (2018m) Introduction to smart maintenance. In: Xing B and Marwala T (eds) *Smart Maintenance for Human–Robot Interaction: An Intelligent Search Algorithmic Perspective.* Gewerbestrasse 11, 6330 Cham, Switzerland: Springer International Publishing AG, ISBN 978-3-319-67479-7, Chapter 2, pp. 21–31.

Xing B and Marwala T. (2018n) *Smart maintenance for human–robot interaction: an intelligent search algorithmic perspective,* Gewerbestrasse 11, 6330 Cham, Switzerland: Springer International Publishing AG, ISBN 978-3-319-67479-7.

Xu B, Zheng H, Xu Y, et al. (2016) Configurational paths to sponsor satisfaction in crowdfunding. *Journal of Business Research* 69: 915–927.

Yao CZ, Sun BY and Lin JN. (2017) A study of correlation between investor sentiment and stock market based on Copula model. *Kybernetes* 46: 550–571.

Yonata J. (2017) *4 important components of successful animated crowdfunding videos, retrieved from https://breadnbeyond.com/articles/animated-crowdfunding-video-tips/, accessed on 11 November 2017.*

Young TE. (2013) *The everything guide to crowdfunding: Learn how to use social media for small-business funding,* 57 Littlefield Street, Avon, MA 02322 U.S.A.: Adams Media, ISBN 978-1-4405-5033-1.

Yum H, Lee B and Chae M. (2012) From the wisdom of crowds to my own judgment in microfinance through online peer-to-peer lending platforms. *Electronic Commerce Research and Applications* 11: 469–483.

Zhai C and Lafferty J. (2004) A sutdy of smoothing methods for language models applied to information retrieval. *ACM Transactions on Information Systems* 22: 179–214.

Zhang J and Liu P. (2012) Rational herding in microloan markets. *Management Science* 58: 892–912.

Zhang K and Chen X. (2017) Herding in a P2P lending market: rational inference or irrational trust? *Electronic Commerce Research and Applications* 23: 45–53.

Zhang T, Tang M, Lu Y, et al. (2014) Trust building in online peer-to-peer lending. *Journal of Global Information Technology Management* 17: 250–266.

Zhang Y. (2014) *Stock message boards: A quantitative approach to measuring investor sentiment*: Palgrave Macmillan, ISBN 978-1-137-37417-2.

Zhang Y, Jia H, Diao Y, et al. (2016) Research on credit scoring by fusing social media information in online peer-to-peer lending. *Procedia Computer Science* 91: 168–174.

Zhao L and Vinig T. (2017) Hedonic value and crowdfunding project performance: a propensity score matching-based analysis. *Review of Behavioral Finance* 9: 169–186.

Zhao Q, Chen C-D, Wang J-L, et al. (2017) Determinants of backers' funding intention in crowdfunding: Social exchange theory and regulatory focus. *Telematics and Informatics* 34: 370–384.

Zheng H, Li D, Wu J, et al. (2014) The role of multidimensional social capital in crowdfunding: A comparative study in China and US. *Information & Management* 51: 488–496.

Zheng H, Hung J-L, Qi Z, et al. (2016) The role of trust management in reward-based crowdfunding. *Online Information Review* 40: 97–118.

Zhong ZJ and Lin S. (2017) The antecedents and consequences of charitable donation heterogeneity on social media. *International Journal of Nonprofit and Voluntary Sector Marketing* e1585: 1–11.

Part V

Investor + Smart Computing = Fatter Return

CHAPTER 8

Crowdfunding Investor Credit Scoring and Support Vector Machine

8.1 Introduction

In recent years, we have witnessed considerable worldwide success in peer-to-peer (P2P) lending, and there is a remarkable trend for provisioning credit via P2P lending platforms. For example, Zopa, the first company in the P2P lending space, was initially launched in the UK in 2005 and has since lent over £900 million (Akkizidis and Stagars, 2016). According to another study (Huang et al., 2016), the number of Chinese P2P platforms reached 2589 by the end of 2015, and the annual turnover for 2015 was totalled at 982.3 billion RMB. In addition, in the USA, P2P consumer lending accounts for a lion's share of market volume (reached a record of $21 billion in 2016) (Ziegler et al., 2017). Meanwhile, more and more governments, e.g., UK (Nesta, 2012; Baeck et al., 2014; Baeck and Collins, 2013; Zhang et al., 2016a; White Label Crowdfunding Limited, 2016; Collins and Pierrakis, 2012), USA (Gajda and Mason, 2013) and China (Deer et al., 2015; Stern et al., 2017; Wang et al., 2016), are actively embracing P2P lending platforms due to their unique ability to make loans available to poor people and opaque SMEs.

In the literature, P2P lending is often termed as crowd-lending and is, thus, characterized by inclusiveness, i.e., including unsecured personal/ small and medium-size enterprises (SMEs) loans, which are considered high risk by banks (Liang et al., in press; DeZoort et al., 2017). In other words, poor people and startups with low revenues are the main customers that need access to P2P lending platforms in order to get external finance. For instance, by comparing data from Germany's largest and oldest P2P lending platform (i.e., Auxmoney) and traditional bank lending (i.e.,

Deutsche Bundesbank), we can learn the following (Roure et al., 2016; Akkizidis and Stagars, 2016):

- **Borrowers (Askers):** In P2P lending context, one can typically find two kinds of borrowers: Individuals who are looking for debt refinancing options at suitable rates, and small businesses that are struggling to get low-value funding from conventional banks.
- **Lenders (Investors):** The lenders involved in P2P lending are mainly individual investors who are searching for a higher rate of return on investment, rather than solely relying on their interest-bearing accounts.

Since most platforms offering P2P lending and the participating investors are profit-driven, and default is often a risk, there is the question of what they must avoid in the first place. Therefore, in order to offer loan opportunities outside traditional lending institutions, identifying and determining a borrower's creditworthiness is a mandatory task for any P2P lending operator (Kibbe, 2013). For simplicity, we generalize the working process of a generic P2P platform as follows (Kibbe, 2013; Emekter et al., 2015; Wang et al., 2015):

- **Application Initiation:** The borrowers often initiate the lending procedure.
- **Request Listing:** Upon the approval of an application, the institution gives the borrower a risk grade and the associated interest rate. The approved loan request (coupled with some necessary information) is then listed on the platform's virtual marketplace.
- **Investment:** Next, potential investors can peruse the listed loan requests (borrower's identity information is typically omitted or limited at this stage) and pick the favourite ones for further investment via either a normal auction mechanism or some form of a fixed rate auction. For a normal auction, investors usually compete with each other to fund the loan at the lowest interest rate. Generally speaking, investors have their control in both cases, and can decide which request to buy into by scrutinizing loans one by one.

8.1.1 *Credit Risk in Online Peer-to-Peer (P2P) Lending*

Today, credit is a typical feature of all global economies. The most popular form of credit is credit cards, other credit domains include home mortgages, student loans, auto loans, consumer durable goods and other consumption expenditure (Joseph, 2013; Davis and Kim, 2017). In fact, credit is good, if used wisely. However, there are notable cases where both lenders and borrowers suffered as a result of credit.

According to one definition from the Basel Committee on Banking Supervision (2000), one can regard credit risk as "the potential a bank borrower will fail to meet its obligations in accordance with agreed terms". Indeed, credit risk, caused by information asymmetry (Marwala and Hurwitz, 2017b), is one of the major risks that is often encountered by investors (He et al., 2008; Zhou and Pham, 2004; Cumming et al., 2017; Loureiro and Gonzalez, 2015; Claesson and Tengvall, 2015). Moreover, according to other studies (Lin et al., 2013; Guo et al., 2016; Emekter et al., 2015; Ma and Wang, 2016), the credit risk found in a P2P lending environment is even more pronounced and pervasive than in its counterpart—traditional financial market.

Of course, credit risk is by no means insurmountable. One intuitive solution, as pointed out by Freedman and Jin (2011), is 'learning by doing' which can be used by investors as an effective tool for minimizing hidden risks. Meanwhile, online P2P lending platforms often also implement one or more of the following strategies (Emekter et al., 2015; Guo et al., 2016) to address credit risk issue: (1) Work closely with credit agency via mutually sharing borrowers' credit information, (2) Produce risk ratings for each approved loan request, (3) Put a restriction on the loan amount so as to balance the overall risk level, (4) Provide additional services (e.g., credit matchmaking systems and portfolio recommendation systems) in order to help investors avoid default risks and (5) In the case of a borrower failing to fulfil the repayment, the credit agency will be noticed and a collection agency will be hired to represent the investor(s) for further funds collection.

Furthermore, with the widespread adoption and rapid advancement of big data analytics (Xing, 2016b; Xing, 2015; Xing et al., 2013a; Chen et al., 2012; Ince and Aktan, 2009), innovative computational intelligence algorithms (Xing and Gao, 2014b; Xing and Marwala, 2018l; Xing and Gao, 2014c; Xing and Gao, 2014q; Xing and Gao, 2014m; Xing, 2016d; Xing and Gao, 2014x; Xing and Gao, 2014aa; Xing and Gao, 2014bb; Xing and Gao, 2014cc; Xing and Gao, 2014d; Xing and Gao, 2014dd; Xing and Gao, 2014e; Xing and Gao, 2014f; Xing and Gao, 2014g; Xing and Gao, 2014h; Xing and Gao, 2014i; Xing and Gao, 2014j; Xing and Gao, 2014k; Xing and Gao, 2014l; Xing and Gao, 2014n; Xing and Gao, 2014o; Xing and Gao, 2014p; Xing and Gao, 2014r; Xing and Gao, 2014s; Xing and Gao, 2014t; Xing and Gao, 2014u; Xing and Gao, 2014v; Xing and Gao, 2014w; Xing and Gao, 2014y; Xing and Gao, 2014z; Xing and Gao, 2014ff; Xing and Gao, 2014a), Internet of things (Xing, 2016a; Xing et al., 2013b) and cloud computing techniques (Xing et al., 2011), one can explore and analyse different multi-objective tasks in order to reduce investors' search costs, mitigate the difficulties involved in evaluating potential borrowers' creditworthiness and facilitate credit scoring evaluation (Akkizidis and Stagars, 2016; Yan et al., 2015).

8.1.2 Background of Credit Scoring

To assess the creditworthiness of loan applicants, thereby avoiding credit risk, credit scoring (a credit guarantee scheme) has become the most widely used method of justifying whether or not a line of credit can be granted (Marron, 2007; Dryver and Sukkasem, 2009; Anderson, 2007). This scheme is based on a set of decision models and technologies that specialise in calculating the default risk (i.e., creditworthiness) of each individual customer from his/her past payment history. In general, the benefits of introducing credit scoring are threefold (Mirtalaei et al., 2012) and consist of: (1) Lowering the cost in analysing creditworthiness, (2) Speeding up the decision-making process and (3) Better decision-making consistence with the aid of a set of pre-defined benchmarks.

Credit scoring began in the late 1800s and the early 1900s, as modern accounting and finance were developed. At that time, financial institutions used it to increase the efficiency of idle economic resources and, thus, the main focus was on the cooperation's balance sheet. In the 1950s, and particularly from the 1970s onwards, the emphasis began to shift towards the consumer market due to competitive pressures, and a set of probabilistic conceptions of risk in consumer lending were defined (Marron, 2007; Burton, 2012). Since then, the problem of credit scoring has been transformed into the binary or multiclass classification domain (Lee et al., 2002), and the methods used for deciding whether the loans could be granted include linear discriminate analysis (Rosenberg and Gleit, 1994), Kolmogorov-Smirnov statistic method (Zeng, in press; Dryver and Sukkasem, 2009), logistic regression (Wiginton, 1980; Dryver and Sukkasem, 2009), Markov chains (Frydman et al., 1985) and classification trees (Feldman and Gross, 2005). Typical borrowers' information flows (Chandler and Parker, 1989) include applicant's age, residence address (current and previous), home ownership status, employment status (current and previous), income, banking relationship, phone number, debt ratio, credit references, etc.

Later, after the adoption of statistical modelling, several researchers (Burton, 2012; Gutiérrez-Nieto et al., 2016) pointed out that a deep analysis and interpretation of social, cultural and moral constructed nature can also play a vital role in averting risk. For example, some studies (Gonzalez and Loureiro, 2014; Chen et al., in press; Duarte et al., 2012; Mirtalaei et al., 2012) examined the effects of both lender and/or borrower personal characteristics (e.g., gender, and perceived trustworthiness and attractiveness) on lending decisions. Golden et al. (2016) investigated the relationship between credit scores and insurance losses. In addition, Lin et al. (2013) and Singh (2015) examined the role of social connections that affect the lending performance, while Moro et al. (in press) investigated the impact of the legal system on credit risk. In the meantime, the methods that are used to analyse both structured and unstructured data have been upgraded as well. Popular examples include artificial neural networks (Angelini et al., 2008; Mirtalaei et al., 2012; Zhao et al., 2015; Dilsha, 2015), genetic programming (Abdou, 2009; Ong et al., 2005), fuzzy system (Alam et al., 2000; Ignatius et al., in press), granular computing (Saberi et al., 2013) and some hybrid variations, e.g., neural fuzzy (Akkoç, 2012).

8.1.3 Models and Mechanisms for Credit Scoring Analysis

In the literature, many advanced models have been developed in order to enhance credit scoring's predictive accuracy and, in general, one can divide these models into the following two classes (Sohn and Kim, 2013):

- **The Technological Credit Scoring Model:** Traditionally, credit-scoring models were mainly built on statistical modelling, operational research and mathematical programming techniques (Kumar and Ravi, 2007; Crook et al., 2007; Angilella and Mazzù, 2015). For example, Sohn et al. (2005) developed a scoring model based on logistic regression technique; Xia et al. (2017) proposed a sequential ensemble credit scoring model on the basis of a gradient boosting machine's variant; and Guo et al. (2016) employed an instance-based model via a logistic regression of borrower's credit attributes in order to identify credit rationing. Further discussions on this subject can be found in (Ince and Aktan, 2009; Xiao et al., 2006).
- **The Behavioural Credit Scoring Model:** As computational intelligence technologies and innovation in data mining analytics are rapidly advancing, P2P platform operators have recently begun to offer loan monitoring services, in which behavioural factors are included. For example, Kreditech, a German P2P lending platform, examined potential borrowers' behavioural data (e.g., the frequency of utilizing capital letters, or how fast the mouse is moved) in order to evaluate his/her credit risks (Yan et al., 2015). In addition, Hsieh (2004) developed a self-organizing neural network map-based credit scoring model which takes the customers' repayment behaviour into consideration; Zhang et al. (2016b) created a credit scoring model by fusing social media information. Other examples can also be found in (Bastos, 2010; Zhang et al., 2007; Huang et al., 2004; Patel and Marwala, 2006; Marwala and Hurwitz, 2017a; Sohn and Kim, 2013). One of the main benefits of introducing behavioural factors into credit scoring models lies in the fact that it may assist investors in understanding the reasons associated with a particular behaviour, rather than just gambling on how they should respond to different occurrences.

Meanwhile, Ma and Wang (2016) also pointed out that under the assistance of big data analytics, three mechanisms that can support an automatic default testing should be considered (1) The audit mechanism for examining and verifying the feasibility of making loans; (2) The credit rating mechanism for determining the loan amount and rate; and (3) The information disclosure mechanism for sharing critical information, previous experiences and necessary resources with each other.

8.1.4 Multi-Dimensional Information for Credit Scoring Analysis

Typically, there are four main sources of data used for credit scoring analysis (Han et al., 2018): (1) Loan characteristics (e.g., loan rate, loan amount and loan duration), (2) Borrower's personal information (e.g., bank accounts, assets, credit grades from credit-reporting agency, income tax returns and even working environment), (3) Voluntary information (e.g., photograph) and (4) Soft information (e.g., social networks and group allocation).

- **Loan Characteristics:** Qiu et al. (2012) claimed that loan characteristics play a significant role in influencing the outcomes of a loan. This source of information often includes loan title (or the so-called listing name), loan purpose, loan amount, loan period, loan quality, loan duration, and so on. Under this branch, some kind of decision support system has been investigated by several scholars (Puro et al., 2010; Wu and Xu, 2011) for helping borrowers evaluate their listing characteristics, such as the starting interest rate, loan amount and duration.
- **Borrower's Personal Information (or called Hard Information):** Lin et al. (2013) concluded that the lower the credit score a borrower gets, the less chance his/her loan succeeds with higher borrowing rates. The information source in this domain can range from credit profile to different borrowers' demographic-related attributes. For instance, Collier and Hampshire (2010) explored borrowers' community reputation systems in terms of structural and behavioural signals and found that borrowers' financial situations, such as their incomes, greatly influenced the likelihood of a successful loan request. Michels (2012) also showed that borrowers' income and expense data can indeed sway lenders' decisions, e.g., decreasing the charged interest rate. Similar implications can be drawn from other studies (Berger and Gleisner, 2009; Everett, 2010; Kumar, 2007; Loureiro and Gonzalez, 2015) as well. In addition, on the basis of the elaboration likelihood model, the influence that borrowers' economic status (e.g., credit grade, debt-to-income ratio, home ownership and previous successful loan history) has on lenders in terms of their assessment, was systematically investigated by Greiner and Wang (2010). Several authors went even further by exploring the relationship between the geographic and/or cultural distance and the lender's final decisions

(Burtch et al., 2014; Kang et al., 2017; Agrawal et al., 2015; Mendes-Da-Silva et al., 2016).

- **Voluntary Information:** In the literature, several authors (Michels, 2012; Han et al., 2018; Ge et al., 2017) discovered that this type of information can also be influential. In general, the most useful voluntary information is often obtained from the photograph, though this might cause the following disputes:
  1) *Discrimination*: For example, Iyer et al. (2016) confirmed that the inclusion of borrowers' personal pictures in the application can have a very positive effect on a loan request's final fate. More specifically, through an in-depth investigation on the pictures in terms of gender, skin colour, appearance and age, they learned that investors do exhibit taste-powered discrimination when making their own lending decisions. Indeed, the debate about whether there is a discrimination phenomenon in lending abounds in the traditional credit market. Some (Labiea et al., 2015; Dietrich, 2005; Boehe and Cruz, 2013) confirmed the existence of discrimination; while others (Han, 2011; Han, 2004) strongly denied their judgements. Therefore, as suggested in (Blake, 2006; Saliyaa and Jayasinghe, 2016), the effect of discrimination must be specifically investigated, case by case. In the Chinese P2P lending arena, Chen et al. (in press) also investigated the conflicting gender discrimination issue. The results suggested that though the reasons for the discrimination might be different, profit-based statistical discrimination (i.e., rational choice) and taste-based discrimination (i.e., irrational choice) do exist in their experiments. However, another similar study, conducted by Barasinska and Schäfer (2014) under distinct scenario settings, showed different results. They suggested that gender discrimination is more likely to be a platform-specific phenomenon instead of a normal attribute belonging to a P2P lending market. Likewise, Moodie (2013) examined the effect of gender discrimination on a popular P2P platform. Kiva; Piper and Schnepf (2008) focused on gender differences in the charitable giving context in the United Kingdom. Walter (2008) focused on the incidents and causes of racial discrimination that emerged in the P2P lending markets and concluded that the connection between disparate treatment and racial prejudice (an often accused potential driving force) is weak, and that the degree of statistical discrimination that the 5 market seems to exhibit is inefficient.
  2) *Appearance*: Duarte et al. (2012) and Ravina (2012) showed separately that borrowers who have a trustworthy look do enjoy some advantages in terms of being funded. Similar results were

also reported by Jin et al. (2017), Jenq et al. (2015) and Gonzalez and Loureiro (2014), respectively.

3) *Age*: Some studies, e.g. (Pope and Sydnor, 2011), also mentioned the effect of the age issue. They found that elderly people tend to encounter a relatively low probability of receiving funding. To make things worse, even if their loans are sometimes accidentally granted, the appended repayment terms are rather strict.

- **Soft Information:** According to some studies (Petersen and Rajan, 1994; Diamond, 1989; Fama, 1985), soft information often denotes the information that is transmitted as a result of repetitive interactions and observations between senders (e.g., loan officer) and receivers (e.g., firm's manager), mainly consisting of non-financial information or technical specifications (Moro and Fink, 2013; Huang, in press; Grunert and Norden, 2011). In fact, digging useful and helpful data from soft information sources is not new in traditional financial markets. Several scholars (McCann and McIndoe-Calder, 2015; Cole et al., 2004; Berger et al., 2005; D'Aurizio et al., 2015) even pointed out that investing in soft information's collection, monitoring, and producing is not an optional but a compulsory activity, required in order to supplement the aggregated financial dataset which credit scoring models are based on. More recently, 'soft information' sources were intensively studied by various researchers (Grunert et al., 2005; Berger and Frame, 2007; Chen et al., 2015) with a hope of further enhancing the power of default prediction models. Some of them are briefly discussed as follows:

1) *Friendship*: In one study (Lin et al., 2013), the authors defined borrowers' friendships as one type of soft information. They examined the data extracted from Prosper (a leading P2P loan site) and observed a way of linking credit quality with the borrowers' online friendships. In a similar vein, Iyer et al. (2016) showed that by referring soft information like friend endorsement numbers and self-described loan purpose, investors are capable of capturing a potential borrower's creditworthiness. Freedman and Jin (2014) also performed a relevant study by checking the importance of social ties, particularly in cases where loan requests have support from borrower's friends. Apart from some commonly agreed advantages, they also found evidence of risks associated with borrowers' participation in social networking. Interested readers please refer to (Huang, in press) for further discussions. To cope with this issue, a four-stage framework was proposed by Sherchan et al. (2013) in order to increase online social networks' trust.

2) *Text Description*: In practice, the soft information obtained from interpreting the loan description texts written by borrowers

themselves also play an important role, as suggested by Dorfleitner et al. (2016), Larrimore et al. (2011), and Jiang et al. (in press), respectively. Factors like text length, orthography and the keywords (both social and emotional) are covered by their studies.

3) *Association*: Regarding this soft information source, some author claimed that the investors' community (e.g., online discussion forums in the social network) is a very powerful organization that can help investors make their final decisions (Herzenstein et al., 2011). In another study conducted by Sonenshein et al. (2010), the authors investigated the effects of social accounts' possessiveness in influencing investors' assessment about whether the money should be lent out or not. Similar studies in this regard include (Moritz et al., 2015).

8.2 Problem Statement

In reality, things often become somewhat more complicated than logical analysis. Although a range of critical information from the borrower's side (e.g., credit grade and the intended use of the loan) has been released to investors, it is still quite challenging to make a proper investment decision, at least for the following five reasons:

- **First**, little is known about whether platform operator-determined borrowers' credit grades affect investors' assessments and the subsequent lending decisions. Iyer et al. (2009) highlighted that investors on Prosper can only see an aggregated credit category. However, if every borrower's exact credit score could be known in advance, it may help investors infer which borrower is more creditworthy than his/her peers within the same credit category.
- **Second**, it is tempting to believe that the default risk problem can be solved with borrowers' high credit ratings. Nevertheless, Malekipirbazari and Aksakalli (2015) suggested that the high credit ratings obtained from borrowers' financial information cannot always guarantee that their default probability is low. This problem is particularly severe in certain countries. According to a survey conducted by the Association of Chartered Certified Accountants (ACCA) (Deer et al., 2015), among the borrowers that they have surveyed, over half of which having a limited prior borrowing record, in particular from the traditional banking system, it was observed that these borrowers' available credit information is poor. This inevitably raises a call for heightened vigilance.
- **Third**, different P2P lending platforms tend to focus on different types of borrowers and, consequently, diverse lending strategies tend

to be recommended to investors. The consequence of this practice often sparks a debate about the tension between helping some and confusing others.

- **Fourth**, there is strong evidence that shows that both irrational and rational herding behaviours of investors can emerge in a P2P lending marketplace (Zhang and Liu, 2012; Mohammadi and Shafi, in press; Moritz et al., 2015; Zhang, 2013; Lee and Lee, 2012; Herzenstein et al., 2011).
- **Finally**, the problem is especially acute among investors themselves, due to their inadequate financial knowledge and screening expertise (Oxera, 2015; Iyer et al., 2009; Greiner and Wang, 2010). As a result, there is a lack a 'true' insight into default risks.

As a result, though P2P lending has enjoyed a tremendous success in many places around the world, the loans involved in P2P are still extremely vulnerable since they often involve a substantial default risk. Therefore, from the investor's viewpoint, it is good that the P2P platform provides a credit graded lending marketplace, but the risk is not fully diluted. Nowadays, as the environment becomes even more uncertain (Barth and Kaufman, 2016; Rossi and Malavasi, 2016; Beccalli and Poli, 2015; Elson, 2017), P2P lending investors are eagerly looking for more accurate credit scoring schemes. In practice, P2P's credit scoring is made not only at two levels of abstractions (i.e., either to lend or not), but for a range of solutions. For example, the Lending Club assigns a loan credit grade to each loan, ranging from A1 to G5 in descending credit ranks (Emekter et al., 2015), while Prosper.com uses a seven-level risk rating (i.e., AA, A, B, C, D, E, HR), among which AA represents borrowers' lowest loan default rate (Guo et al., 2016; Luo and Lin, 2013). Although investors can establish their own investment strategy by referring to the available credit scores at hand, the P2P lending environment tends to exhibit unique dynamics. According to (Emekter et al., 2015), the following findings can be observed:

- **Higher Credit Scores:** In general, borrowers with high-income and high credit scores do not usually resort to P2P lending for solutions. More specifically, the top one third of qualified borrowers do not establish any loan listings.
- **High Interest Rates:** More often than not, the high interest rates charged for low credit-scored borrowers cannot overcome the corresponding default risk.

To this end, investors may still need to invest a great deal of effort in digging deeper and establishing their own credit scoring system, incorporating borrowers' 'hard information' which is mainly provided by the P2P platforms (e.g., ratings or debt-to-income ratios), as well as borrowers' 'soft information' that is often distilled from other channels, such as social network and blogs.

8.2.1 Question 8.1

In the lending domain, 'good borrowers' often refers to those who can finalize their repayment in full within a predetermined due time (Malekipirbazari and Aksakalli, 2015). Therefore, from a profitability viewpoint, the implications drawn from the above observations imply that the effort related to identifying potential good borrowers is of huge importance to investors taking part in P2P lending. The favourable situation of P2P investors has a trickle-down influence on the overall sustainability of a P2P lending market. Based on this observation, Question 8.1 is proposed as follows:

- **Question 8.1:** *How do we improve credit-scoring models via smart computing techniques so that interested investors are able to identify good borrowers?*

By investigating Question 8.1, we expect to be able to offer alternative quantitative methodologies that, given the present industrial common practice of employing logistic regression, have the possibility of increasing a P2P lending market's profitability, transparency, and sustainability through improved credit scoring functionality.

8.3 Support Vector Machine for Question 8.1

A 'support vector machine' (SVM) often refers to an abstract machine with learning capabilities, i.e., a training data set can be used to assist in learning activities with a goal of generalizing and making proper predictions (Marwala, 2014b). The applications of SVM and other smart computing techniques in the credit scoring area are not short of evidence (Yu et al., 2008; Marwala, 2013a; Marwala, 2014a). In general, the training data set often consists of a group of input vectors (denoted by $\mathbf{x}_i$) and each of these input vectors is composed of several component features. Meanwhile, a bunch of corresponding labels (denoted by y_i) are also introduced in order to pair with these input vectors, which in turn forms an amount of $i = 1,2,...,m$ pairs.

8.3.1 Support Vector Machine (SVM)

In the literature, a common way of explaining SVM is to employ a binary classification task as an example, since predicting over two classes is often found in various real-world scenarios. As shown in Fig. 8.1, the collective tagged datapoints residing in an input space can be viewed as the training data set.

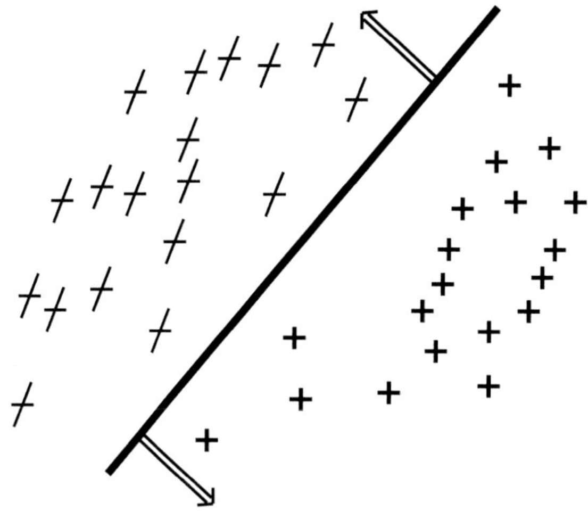

Figure 8.1: Illustration of binary classification

Take the well separable two data classes, the goal of the learning task is to identify a directed hyperplane. In other words, the datapoints can be separated by an oriented hyperplane such that we can label one side as y_i equal to +1 and the other side as y_i equal to –1. The main feature of the identified hyperplane lies in that the distance from any of the two data classes to the hyperplane is maximized. Therefore, as the name implies, the support vectors refer to the closest data points on both sides, which affect the position of the desired hyperplane the most. Typically, $\mathbf{w} \cdot \mathbf{x} + b$ is used to define a hyperplane, where the offset of the hyperplane from the input space's origin is represented by b, the points residing within the hyperplane are indicated by $\mathbf{x}$, and the normal to the hyperplane are the weights (denoted by $\mathbf{w}$) which in turn influence the orientation of the hyperplane.

8.3.1.1 Maximum Margin Hyperplane

From the statistical learning theory viewpoint, the motivation of employing a binary classifier SVM stems from the concept of theoretical upper bound in terms of the generalization error. In other words, the theoretical prediction error generated after applying the classifier to novel, unseen cases. Such generalization errors mostly contain the following two critical characteristics (Campbell and Ying, 2011):

- **Feature 1:** The bound is minimized via the margin (denoted by γ) maximization. This refers, as shown in Fig. 8.2, to the minimal distance between the hyperplane (which separates the two data point classes) and the nearest data point(s) to the hyperplane.

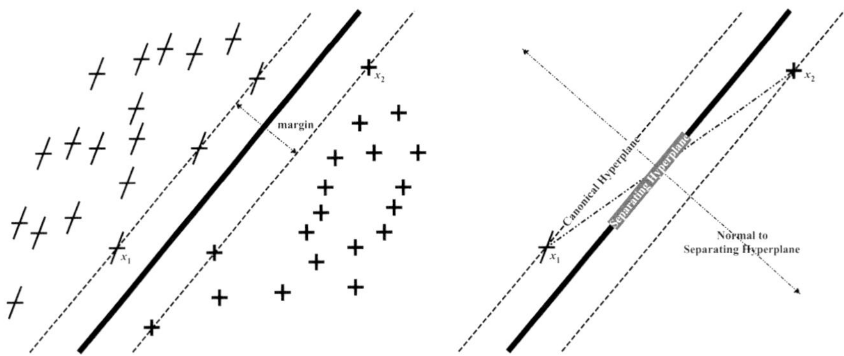

Figure 8.2: Illustration of feature 1.

- **Feature 2:** The bound is independent to the space's dimensionality.

Suppose we need to perform a binary classification over $\mathbf{x}_i (i = 1,2,\ldots,m)$ data points, the decision function can be defined via Eq. 8.1 (Campbell and Ying, 2011):

$$f(\mathbf{x}) = \text{sign}(\mathbf{w}\cdot\mathbf{x} + b). \tag{8.1}$$

where the scalar or inner product is denoted by $(\cdot)$, that is, $\mathbf{w}\cdot\mathbf{x} \equiv \mathbf{w}^T\cdot\mathbf{x}$.

As we can see from Eq. 8.1, one can successfully classify the datapoints under the condition given via Eq. 8.2 (Campbell and Ying, 2011):

$$y_i(\mathbf{w}\cdot\mathbf{x}_i + b) \geq 1 \quad \forall i. \tag{8.2}$$

where $\mathbf{w}\cdot\mathbf{x} + b$ is positive when the corresponding label $y_i = +1$; otherwise, $\mathbf{w}\cdot\mathbf{x} + b$ is negative when the corresponding label $y_i = -1$. To eliminate some potential ambiguities, we can further introduce a scale for $(\mathbf{w}, b)$, as given via Eq. 8.3 (Campbell and Ying, 2011):

$$\begin{cases} \mathbf{w}\cdot\mathbf{x}_i + b = 1 & \text{for the nearest datapoints on one side} \\ \mathbf{w}\cdot\mathbf{x}_i + b = -1 & \text{for the nearest datapoints on the other side} \end{cases}. \tag{8.3}$$

Here, the canonical hyperplanes are often used to denote those hyperplanes that pass through $\mathbf{w}\cdot\mathbf{x} + b = 1$ and $\mathbf{w}\cdot\mathbf{x} + b = -1$, and the region between these canonical hyperplanes is termed as the margin band.

As illustrated in Fig. 8.3, let $\mathbf{x}_1$ and $\mathbf{x}_2$ be two sets of datapoints that reside within the canonical hyperplanes on two sides, respectively.

If the conditions of $\mathbf{w}\cdot\mathbf{x} + b = 1$ and $\mathbf{w}\cdot\mathbf{x} + b = -1$ hold, we can get $\mathbf{w}\cdot(\mathbf{x}_1 - \mathbf{x}_2) = 2$. In terms of the separating hyperplane (i.e., $\mathbf{w}\cdot\mathbf{x} + b = 0$), we have the normal vector as $\frac{\mathbf{w}}{\|\mathbf{w}\|_2}$ (where the square root of $\mathbf{w}^T\mathbf{w}$ is represented by $\|\mathbf{w}\|_2$). Accordingly, the distance between the two hyperplanes that pass through $\mathbf{w}\cdot\mathbf{x} + b = 1$ and $\mathbf{w}\cdot\mathbf{x} + b = -1$, respectively, is equal to $(\mathbf{x}_1$

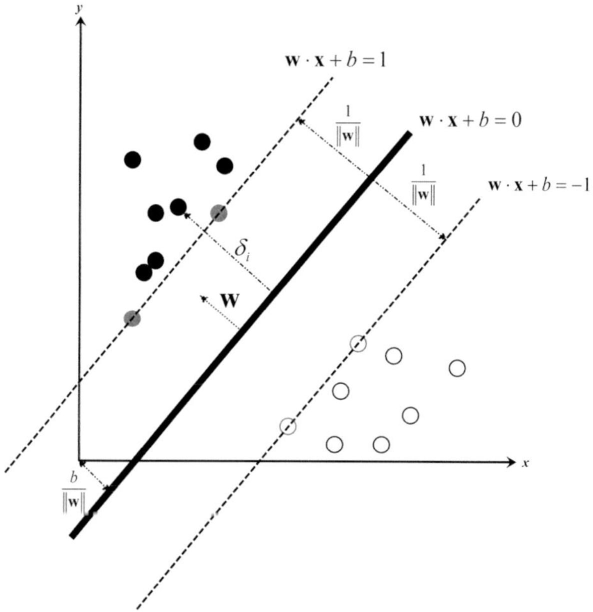

Figure 8.3: Geometric illustration of margin and maximum margin hyperplane.

$-\ \mathbf{x}_2$)'s projection onto $\frac{\mathbf{w}}{\|\mathbf{w}\|_2}$, which in turn gives us $\frac{\mathbf{w}\cdot(\mathbf{x}_1-\mathbf{x}_2)}{\|\mathbf{w}\|_2}=\frac{2}{\|\mathbf{w}\|_2}$. Thus, the corresponding margin is $\gamma=\frac{1}{\|\mathbf{w}\|_2}$. Finally, the margin maximization can be reached by minimizing Eq. 8.4 (Campbell and Ying, 2011):

$$\begin{aligned} &\frac{1}{2}\|\mathbf{w}\|_2^2 \\ &\text{subject to} \quad y_i(\mathbf{w}\cdot\mathbf{x}_i+b)\geq 1 \quad \forall i \end{aligned} \tag{8.4}$$

8.3.1.2 Lagrangian Methods for Constrained Optimization

A typical constrained optimization problem is given by Eq. 8.4, in which the goal is to minimize the objective function (denoted by $\frac{1}{2}\|\mathbf{w}\|_2^2$) under the constraints defined by $y_i(\mathbf{w}\cdot\mathbf{x}_i+b)\geq 1 \quad \forall i$.

- **Primal Formulation:** In practice, such formulation can also be reduced to another form, that is, minimizing Lagrange function (Morgan, 2015), which is composed of the objective function's sum and a number of m constraints (multiplied by each constraint's corresponding Lagrange multiplier). This transformation is often called primal formulation, as given by Eq. 8.5 (Campbell and Ying, 2011):

$$L(\mathbf{w},b)=\frac{1}{2}(\mathbf{w}\cdot\mathbf{w})-\sum_{i=1}^{m}\alpha_i\left[y_i(\mathbf{w}\cdot\mathbf{x}_i+b)-1\right]. \quad 8.5$$

where α_i represents the Lagrange multiplier (i.e., $\alpha_i \geq 0$). By taking the partial derivatives at the minimal place in terms of **w** and b and setting them to zero, we can get Eqs. 8.6 and 8.7, respectively (Campbell and Ying, 2011):

$$\frac{\partial L}{\partial b}=-\sum_{i=1}^{m}\alpha_i y_i=0. \quad 8.6$$

$$\frac{\partial L}{\partial \mathbf{w}}=\mathbf{w}-\sum_{i=1}^{m}\alpha_i y_i \mathbf{x}_i=0. \quad 8.7$$

- **Dual Formulation:** By working on Eqs. 8.5 and 8.7, we can get the dual formulation (i.e., Wolfe dual), as given by Eq. 8.8 (Campbell and Ying, 2011):

$$W(\alpha)=\sum_{i=1}^{m}\alpha_i-\frac{1}{2}\sum_{i,j=1}^{m}\alpha_i\alpha_j y_i y_j(\mathbf{x}_i\cdot\mathbf{x}_j). \quad 8.8$$

Here, the goal is to maximize the Wolfe dual in terms of α_i, under the conditions given by Eq. 8.9 (Campbell and Ying, 2011):

$$\alpha_i \geq 0; \quad \sum_{i=1}^{m}\alpha_i y_i=0. \quad 8.9$$

To summarize, minimization is involved in the primal formulation, while the corresponding dual formulation consists of maximization. When both formulations' solutions have been obtained, one can see that both objective functions tend to share the same value. From the dual objective given by Eq. 8.8, we can discover that the data points (denoted by $\mathbf{x}_i$) only occur within an inner product. In order to get an alternative data representation, the data points can be mapped into a feature space (i.e., a dimensionality differed space) via a replacement given by Eq. 8.10 (Campbell and Ying, 2011):

$$\mathbf{x}_i\cdot\mathbf{x}_j\rightarrow\Phi(\mathbf{x}_i)\cdot\Phi(\mathbf{x}_j). \quad 8.10$$

where $\Phi(\cdot)$ stands for the mapping function. An intuitive reason for executing a mapping transformation lies in the fact that it may not always be possible to separate the presented data linearly within the input space. In other words, when separating two data classes by a directed hyperplane in input space is not practicable, it is often possible to separate these data in a higher space of dimensionality. This observation also matches the aforementioned 'feature 2'.

8.3.1.3 Kernel Functions

The exact form of the mapping function $\Phi(\cdot)$ does not have to be known, since it is possible to define such functions via the choice of inner product in feature space, or Kernelization concept (Kung, 2014; Marwala, 2009; Marwala, 2013a; Schölkopf and Smola, 2002) as given by Eq. 8.11 (Campbell and Ying, 2011):

$$K(\mathbf{x}_i, \mathbf{x}_j) = \Phi(\mathbf{x}_i) \cdot \Phi(\mathbf{x}_j). \tag{8.11}$$

For instance, we have a set of two-class datapoints whose linear separability is unknown. Initially, we can begin with a linear kernel, denoted by $K(\mathbf{x}_i, \mathbf{x}_j) = \mathbf{x}_i \cdot \mathbf{x}_j$, with no mapping transformation imposed. If the data is linearly inseparable, it is then impossible to reach zero training error situation by employing Eqs. 8.8 and 8.9, that is, the misclassification of training datapoints is often the case. However, if we introduce Gaussian kernel, given via Eq. 8.12 (Campbell and Ying, 2011), in order to construct a space with higher dimensionality, this problem may be solved.

$$K\left(\mathbf{x}_i, \mathbf{x}_j\right) = e^{\frac{-\left(\mathbf{x}_i - \mathbf{x}_j\right)^2}{2\sigma^2}}. \tag{8.12}$$

The process of employing a kernel together with its introduced feature space is often called kernel substitution. Other exemplary kernel functions are given by Eq. 8.13 (Campbell and Ying, 2011):

$$K\left(\mathbf{x}_i, \mathbf{x}_j\right) = \left(\mathbf{x}_i \cdot \mathbf{x}_j + 1\right)^d; \quad K\left(\mathbf{x}_i, \mathbf{x}_j\right) = \tanh\left(\beta \mathbf{x}_i \cdot \mathbf{x}_j + b\right). \tag{8.13}$$

where a polynomial classifier and a feed-forward neural network classifier are defined, respectively.

For a typical binary classification task with a pre-determined kernel function, the learning task can be simplified by maximizing Eq. 8.14 (Campbell and Ying, 2011):

$$W(\alpha) = \sum_{i=1}^{m} \alpha_i - \frac{1}{2} \sum_{i,j=1}^{m} \alpha_i \alpha_j y_i y_j K\left(\mathbf{x}_i, \mathbf{x}_j\right). \tag{8.14}$$

where the constraints are still defined via Eq. 8.9, i.e., $\alpha_i \geq 0$; $\sum_1^m \alpha_i y_i = 0$. In terms of the bias (denoted by b), for data points tagged as $y_i = +1$, we can get Eq. 8.15 (Campbell and Ying, 2011):

$$\min_{\{i|y_i=+1\}} \left[\mathbf{w} \cdot \mathbf{x}_i + b\right] = \min_{\{i|y_i=+1\}} \left[\sum_{j=1}^{m} \alpha_j y_j K\left(\mathbf{x}_i, \mathbf{x}_j\right)\right] + b = 1. \tag{8.15}$$

A similar formulation can also be obtained for data points labelled by $y_i = -1$. Accordingly, we arrive at Eq. 8.16 (Campbell and Ying, 2011):

$$b = -\frac{1}{2}\left\{ \max_{\{i|y_i=-1\}} \left[\sum_{j=1}^{m} \alpha_j y_j K(\mathbf{x}_i, \mathbf{x}_j) \right] + \min_{\{i|y_i=+1\}} \left[\sum_{j=1}^{m} \alpha_j y_j K(\mathbf{x}_i, \mathbf{x}_j) \right] \right\}. \quad 8.16$$

In order to create an SVM-enabled binary classifier, one can first insert the data points $(\mathbf{x}_i, y_i)$ into Eq. 8.14 and then maximize $W(\alpha)$ under the constrained conditions defined by Eq. 8.9, i.e., $\alpha_i \geq 0$; $\sum_1^m \alpha_i y_i = 0$. Upon the obtained values of α_i (denoted by α_i^*), the bias b can be computed via Eq. 8.16. Accordingly, for a novel input vector $\mathbf{z}$, the class can be predicted via Eq. 8.17 (Campbell and Ying, 2011):

$$\phi(\mathbf{z}) = \sum_{i=1}^{m} \alpha_i^* y_i K(\mathbf{x}_i, \mathbf{z}) + b^*. \quad 8.17$$

where b^* stands for the bias value at the optimal place. This formulation is constructed by substituting $\mathbf{w}^* = \sum_{i=1}^{m} \alpha_i^* y_i \Phi(\mathbf{x}_i)$ (obtained from Eq. 8.7) into the decision function defined via Eq. 8.1, i.e., $f(\mathbf{z}) = \text{sign}[\mathbf{w}^* \cdot \Phi(\mathbf{z}) + b^*]$ $= \text{sign}[\sum_i \alpha_i^* y_i K(\mathbf{x}_i, \mathbf{z}) + b^*]$. Based on this establishment, a solution denoted by (α^*, b^*) is often regarded as a hypothetical modelling of data.

Once we identify the maximal margin hyperplane in the feature space, the data points with the property of $\alpha_i^* > 0$ (i.e., stay nearest to the hyperplane) can be pinpointed and labelled as support vectors. Alternatively, all other data points are characterized by $\alpha_i^* = 0$ and treated as non-support vectors. If some of these data points are removed, the identified separating hyperplane remains unchanged.

From the optimization theory perspective, one can also interpret this via Karush-Kuhn-Tucker (KKT) conditions, that is, at the point where a constrained optimization problem's optimum is reached, the whole condition set has to be satisfied. When a projection to feature space is absent, one of the KKT conditions can be written via Eq. 8.18 (Campbell and Ying, 2011):

$$\alpha_i [y_i (\mathbf{w} \cdot \mathbf{x}_i + b) - 1] = 0. \quad 8.18$$

Based on Eq. 8.18, the following two cases can be deduced: (1) $y_i (\mathbf{w} \cdot \mathbf{x}_i + b) > 1$ for non-support vectors and, thus, $\alpha_i = 0$; and (2) $y_i (\mathbf{w} \cdot \mathbf{x}_i + b) = 1$ for support vectors and, thus, $\alpha_i > 0$.

8.3.1.4 Soft Margin Classifiers

In practice, most datasets incorporate noise, therefore, poor generalization can occur when applying the SVM to these data sets, since outlier data points impose undue effect on hyperplane separation. To cope with this issue, a soft margin concept is introduced, which typically has the following two popular schemes (Campbell and Ying, 2011; Abe, 2010; Hamel, 2009):

- ℓ_1 **Error Norm:** In this case, the learning task remains almost the same as given by Eqs. 8.14 and 8.9, apart from introducing a new box constraint as defined via Eq. 8.19 (Campbell and Ying, 2011):

$$0 \leq \alpha_i \leq C. \tag{8.19}$$

Meanwhile, a non-negative slack variable (denoted by ξ_i) is also introduced into Eq. 8.2 to form Eq. 8.20 (Campbell and Ying, 2011):

$$y_i(\mathbf{w} \cdot \mathbf{x}_i + b) \geq 1 - \xi_i \quad \forall i. \tag{8.20}$$

For now, the focal task is to minimize the sum of two elements, i.e., $\|\mathbf{w}\|^2$ and errors $\sum_{i=1}^{m} \xi_i$, as given by Eq. 8.21 (Campbell and Ying, 2011):

$$\min\left(\frac{1}{2}\mathbf{w} \cdot \mathbf{w} + C\sum_{i=1}^{m} \xi_i\right). \tag{8.21}$$

If ξ_i is greater than zero, then a margin error is obtained. A reformulation can be performed on Eq. 8.21 in order to construct a primal Lagrange, leading us to Eq. 8.22 (Campbell and Ying, 2011):

$$L(\mathbf{w}, b, \alpha, \xi) = \frac{1}{2}(\mathbf{w} \cdot \mathbf{w}) + C\sum_{i=1}^{m} \xi_i - \sum_{i=1}^{m} \alpha_i\left[yi(\mathbf{w} \cdot \mathbf{x}_i + b) - 1 + \xi_i\right] - \sum_{i=1}^{m} r_i \xi_i. \tag{8.22}$$

where Eq. 8.20 is addressed by using Eq. 8.22 under the condition of Lagrange multipliers $\alpha_i \geq 0$, and the requirement of $\xi_i \geq 0$ can be met under the situation of $r_i \geq 0$. By calculating the partial derivatives at the optimal place with respect to three parameters, **w**, b, and ξ, respectively, we can get Eqs. 8.23–8.25 (Campbell and Ying, 2011):

$$\frac{\partial L}{\partial \mathbf{w}} = \mathbf{w} - \sum_{i=1}^{m} \alpha_i y_i \mathbf{x}_i = 0. \tag{8.23}$$

$$\frac{\partial L}{\partial b} = \sum_{i=1}^{m} \alpha_i y_i = 0. \tag{8.24}$$

$$\frac{\partial L}{\partial \xi_i} = C - \alpha_i - r_i = 0. \tag{8.25}$$

After eliminating **w** from $L(\mathbf{w}, b, \alpha, \xi)$, we can get the same dual objective function as given by Eq. 8.14. For an ℓ_1 error norm, one can identify the bias in the decision function given by Eq. 8.17, in which the following Eqs. 8.26 and 8.27 are included (Campbell and Ying, 2011):

$$r_i \xi_i = 0. \qquad 8.26$$

$$\alpha_i \left[y_i (\mathbf{w} \cdot \mathbf{x}_i + b) - 1 + \xi_i \right] = 0. \qquad 8.27$$

Accordingly, the bias can be acquired by performing an averaging operation, as given by Eq. 8.28 (Campbell and Ying, 2011):

$$b = y_k - \sum_{i=1}^{m} \alpha_i^* y_i \left(\mathbf{x}_i \cdot \mathbf{x}_k \right). \qquad 8.28$$

- **ℓ_2 Error Norm:** The objective function of this scenario can be given via Eq. 8.29 (Campbell and Ying, 2011):

$$\begin{aligned} &\min \left(\frac{1}{2} \mathbf{w} \cdot \mathbf{w} + C \sum_{i=1}^{m} \xi_i^2 \right) \\ &\text{subject to} \quad y_i \left(\mathbf{w} \cdot \mathbf{x}_i + b \right) \geq 1 - \xi_i \qquad \xi_i \geq 0 \end{aligned} . \qquad 8.29$$

After the Lagrange multipliers are introduced in order to incorporate the constraints, a primal Lagrange function can be obtained via Eq. 8.30 (Campbell and Ying, 2011) with $\alpha_i \geq 0$ and $r_i \geq 0$:

$$L\left(\mathbf{w}, b, \alpha, \xi\right) = \frac{1}{2}\left(\mathbf{w} \cdot \mathbf{w}\right) + C \sum_{i=1}^{m} \xi_i^2 - \sum_{i=1}^{m} \alpha_i \left[y_i \left(\mathbf{w} \cdot \mathbf{x}_i + b \right) - 1 + \xi_i \right] - \sum_{i=1}^{m} r_i \xi_i. \qquad 8.30$$

By performing kernel substitution, we can get a dual objective function, as given by Eq. 8.31 (Campbell and Ying, 2011):

$$W\left(\alpha\right) = \sum_{i=1}^{m} \alpha_i - \frac{1}{2} \sum_{i,j=1}^{m} \alpha_i \alpha_j y_i y_j K\left(\mathbf{x}_i, \mathbf{x}_j\right) - \frac{1}{4C} \sum_{i=1}^{m} \alpha_i^2. \qquad 8.31$$

- **v-SVM:** Since it is often the case that there is no clear interpretation that can be related to C and λ under L_1 and L_2 cases, an alternative v-SVM is often employed. In essence, v-SVM is a variant of L_1 case, with the primal formulation given by Eq. 8.32 (Campbell and Ying, 2011):

$$\begin{aligned} &\min_{\mathbf{w}, \xi, \rho} \left[L\left(\mathbf{w}, \xi, \rho\right) = \frac{1}{2}\left(\mathbf{w} \cdot \mathbf{w}\right) - v\rho + \frac{1}{m} \sum_{i=1}^{m} \xi_i \right]. \\ &\text{subject to} \quad y_i \left(\mathbf{w} \cdot \mathbf{x}_i + b \right) \geq \rho - \xi_i, \ \xi_i \geq 0, \text{ and } \rho \geq 0 \end{aligned} \qquad 8.32$$

In a similar way, the primal Lagrange function can be given by Eq. 8.33 (Campbell and Ying, 2011) with $\alpha_i \geq 0, \beta_i \geq 0$ and $\delta \geq 0$:

$$L(\mathbf{w}, b, \alpha, \xi) = \frac{1}{2}(\mathbf{w} \cdot \mathbf{w}) - v\rho + \frac{1}{m}\sum_{i=1}^{m} \xi_i - \sum_{i=1}^{m}\left\{\alpha_i\left[y_i(\mathbf{w} \cdot \mathbf{x}_i + b) - \rho + \xi_i\right] + \beta_i \xi_i\right\} - \delta\rho \quad 8.33$$

The dual formulation of v-SVM can then be given by Eq. 8.34 (Campbell and Ying, 2011):

$$\max_{\alpha}\left[\mathbf{w}(\alpha) = -\frac{1}{2}\sum_{i,j=1}^{m} \alpha_i \alpha_j y_i y_j K(\mathbf{x}_i, \mathbf{x}_j)\right]$$
$$\text{subject to} \quad 0 \le \alpha_i \le \frac{1}{m}, \ \sum_{i=1}^{m} \alpha_i y_i = 0, \text{ and } \sum_{i=1}^{m} \alpha_i \ge v \quad 8.34$$

8.3.2 Fuzzy Support Vector Machine (FSVM)

Though standard SVM is a powerful method, there are still several limitations to this approach. In various cases, training data points may no longer belong to one individual class. This is to say that there is some kind of fuzziness associated with each training data point. To this end, the fuzzy version of standard SVM was introduced in the literature (Lin and Wang, 2002).

8.3.2.1 Standard Fuzzy Support Vector Machine (FSVM)

Similar to standard SVM, the optimization problem under the FSVM can be given by Eq. 8.35 (Yu et al., 2008; Lin and Wang, 2002):

$$\min_{\mathbf{w}, b, \xi_i, s_i}\left[\Psi(\mathbf{w}, b, \xi_i, s_i) = \frac{1}{2}\|\mathbf{w}\|^2 + C\sum_{i=1}^{l} s_i \xi_i\right]$$
$$\text{subject to} \quad y_i\left[\mathbf{w}^T \Phi(\mathbf{x}_i) + b\right] \ge 1 - \xi_i, \quad \xi_i \ge 0, \quad i = 1, 2, \ldots, l \quad 8.35$$

It is evident that the membership value s_i is introduced to scale the error term ξ_i. In general, the fuzzy membership values are employed in order to get the soft penalty terms weighed, which in turn can give us the relative training samples' confidence degree during the process of training. In other words, the training samples with larger membership values tend to have a greater influence on training FSVM than those with smaller membership values. The dual formulation of FSVM is in many ways similar to standard SVM, as given by Eq. 8.36 (Yu et al., 2008; Lin and Wang, 2002):

$$\max_{\alpha}\left[W(\alpha)=-\frac{1}{2}\sum_{i,j=1}^{m}\alpha_i\alpha_j y_i y_j K(\mathbf{x}_i,\mathbf{x}_j)+\sum_{i=1}^{m}\alpha_i\right]$$

$$\text{subject to}\quad \sum_{i=1}^{m} y_i\alpha_i = 0 \qquad i=1,2,\ldots,l \qquad .8.36$$

$$0 \le \alpha_i \le s_i C$$

With different s_i values, the trade-off between two factors, that is, the maximum margin and the number of constraints violations, can be controlled. Since s_i with a smaller value can reduce the importance degree of the corresponding data point $\mathbf{x}_i$ during the process of training, a proper selection of the fuzzy membership function plays a crucial role in utilizing FSVM. In (Lin and Wang, 2002), a linear fuzzy membership was proposed. Take a training sample $(x_1, y_1, s_1),\ldots,(x_l, y_l, s_l)$, the mean of "+1" labelled class is denoted by $\mathbf{x}_+$ while the mean of "–1" labelled class is represented by $\mathbf{x}_-$. The radius of two classes can then be given via Eqs. 8.37 and 8.38, respectively (Shi and Xu, 2016):

$$r_+ = \max|\mathbf{x}_+ - \mathbf{x}_i| \quad \text{where } y_i = 1. \qquad 8.37$$

$$r_- = \max|\mathbf{x}_- - \mathbf{x}_i| \quad \text{where } y_i = -1. \qquad 8.38$$

Accordingly, each sample's fuzzy membership is given via Eq. 8.39 (Shi and Xu, 2016) with $\delta > 0$:

$$s_i = \begin{cases} 1-\dfrac{|\mathbf{x}_+ - \mathbf{x}_i|}{r_+ + \delta} & \text{where } y_i = 1 \\ 1-\dfrac{|\mathbf{x}_- - \mathbf{x}_i|}{r_- + \delta} & \text{where } y_i = -1 \end{cases}. \qquad 8.39$$

To further improve FSVM's performance, a nonlinear membership model was proposed by (Tang, 2011), in which the mapping function $\varphi(x)$ was introduced. In general, the formulation of φ_+ and φ_- can be given via Eq. 8.40 (Shi and Xu, 2016):

$$\begin{aligned}\phi_+ &= \frac{1}{n_+}\sum_{y_i=1}\phi(\mathbf{x}_i) \\ \phi_- &= \frac{1}{n_-}\sum_{y_i=-1}\phi(\mathbf{x}_i)\end{aligned}. \qquad 8.40$$

where n_+ and n_- stand for the sample number in each class. Similarly, the radius can also be defined via Eq. 8.41 (Shi and Xu, 2016):

$$\begin{aligned} r_+ &= \max|\phi_+ - \phi(\mathbf{x}_i)| \quad \text{where } y_i = 1 \\ r_- &= \max|\phi_- - \phi(\mathbf{x}_i)| \quad \text{where } y_i = -1 \end{aligned}. \qquad 8.41$$

Accordingly, within the feature space, the square of distance can be computed via Eq. 8.42 (Shi and Xu, 2016):

$$d_{i+}^2 = \left\|\phi_- - \phi(\mathbf{x}_i)\right\|^2 = K(\mathbf{x}_i, \mathbf{x}_i) - \frac{2}{n_+}\sum_{y_j=1} K(\mathbf{x}_j, \mathbf{x}_i) + \frac{1}{n_+^2}\sum_i \sum_{y_j=1} K(\mathbf{x}_i, \mathbf{x}_j)$$
$$d_{i-}^2 = \left\|\phi_+ - \phi(\mathbf{x}_i)\right\|^2 = K(\mathbf{x}_i, \mathbf{x}_i) - \frac{2}{n_+}\sum_{y_j=1} K(\mathbf{x}_j, \mathbf{x}_i) + \frac{1}{n_+^2}\sum_i \sum_{y_j=1} K(\mathbf{x}_i, \mathbf{x}_j) \quad . \quad 8.42$$

Finally, each sample's fuzzy membership can be obtained via Eq. 8.43 (Shi and Xu, 2016) with $\delta > 0$:

$$s_i = \begin{cases} 1 - \sqrt{\dfrac{d_{i+}^2}{r_+^2 + \delta}} & \text{where } y_i = 1 \\ 1 - \sqrt{\dfrac{d_{i-}^2}{r_-^2 + \delta}} & \text{where } y_i = -1 \end{cases} . \quad 8.43$$

8.3.2.2 Least Squares Fuzzy Support Vector Machine (FSVM)

One of the main disadvantages of standard FSVM is its inadequacy in terms of handling various large-scale real-world scenarios. To address this issue, the least squares version of FSVM (i.e., LS-FSVM) was proposed and formulated via Eq. 8.44 (Yu et al., 2008; Suykens and Vandewalle, 1999; Yu, 2014; Lai et al., 2006):

$$\min_{\mathbf{w},b,\xi_i,s_i} \left[\Psi(\mathbf{w}, b, \xi_i, s_i) = \frac{1}{2}\|\mathbf{w}\|^2 + \frac{C}{2}\sum_{i=1}^{l} s_i \xi_i^2 \right]$$
$$\text{subject to} \quad y_i\left[\mathbf{w}^T \Phi(\mathbf{x}_i) + b\right] = 1 - \xi_i \qquad i = 1, 2, \ldots, l \quad . \quad 8.44$$

The Lagrangian formulation can then be given by Eq. 8.45 (Yu et al., 2008; Suykens and Vandewalle, 1999; Yu, 2014; Lai et al., 2006):

$$\max_{\alpha_i} \min_{\mathbf{w},b,\xi_i} \left\{ \begin{aligned} L(\mathbf{w}, b, \xi_i; \alpha_i) = \frac{1}{2}\|\mathbf{w}\|^2 + \frac{C}{2}\sum_{i=1}^{l} s_i \xi_i \\ -\sum_{i=1}^{l} \alpha_i \left[y_i \left(\mathbf{w} \cdot \phi(\mathbf{x}_i) + b \right) - 1 + \xi_i \right] \end{aligned} \right\} . \quad 8.45$$

where the *i*th Lagrangian multiplier is represented by α_i. After performing a differentiation operation on Eq. 8.45, we can get optimal conditions, as given by Eqs. 8.46–8.49 (Yu et al., 2008; Suykens and Vandewalle, 1999; Yu, 2014; Lai et al., 2006):

$$\frac{\partial L}{\partial \mathbf{w}} = \mathbf{w} - \sum_{i=1}^{l} \alpha_i y_i \phi(\mathbf{x}_i) = 0 . \quad 8.46$$

$$\frac{\partial L}{\partial b} = -\sum_{i=1}^{l} \alpha_i y_i = 0 . \qquad 8.47$$

$$\frac{\partial L}{\partial \xi_i} = s_i C \xi_i - \alpha_i = 0 . \qquad 8.48$$

$$\frac{\partial L}{\partial \alpha_i} = y_i \left[\mathbf{w} \cdot \phi(\mathbf{x}_i) + b \right] - 1 + \xi_i = 0 . \qquad 8.49$$

Accordingly, one can further get the following optimal conditions as given by Eqs. 8.50–8.53 (Yu et al., 2008; Suykens and Vandewalle, 1999; Yu, 2014; Lai et al., 2006):

$$\mathbf{w} = \sum_{i=1}^{l} \alpha_i y_i \phi(\mathbf{x}_i) = 0 . \qquad 8.50$$

$$\sum_{i=1}^{l} \alpha_i y_i = 0 . \qquad 8.51$$

$$s_i C \xi_i = \alpha_i . \qquad 8.52$$

$$y_i \left[\mathbf{w} \cdot \phi(\mathbf{x}_i) + b \right] - 1 + \xi_i = 0. \qquad 8.53$$

By introducing matrix form, the above representations can also be expressed via Eq. 8.54 (Yu et al., 2008; Suykens and Vandewalle, 1999; Yu, 2014; Lai et al., 2006):

$$\begin{bmatrix} \mathbf{\Omega} & \mathbf{Y} \\ \mathbf{Y}^T & 0 \end{bmatrix} \begin{bmatrix} \boldsymbol{\alpha} \\ 0 \end{bmatrix} = \begin{bmatrix} 1 \\ 0 \end{bmatrix} . \qquad 8.54$$

where $\mathbf{\Omega}$, $\mathbf{Y}$, and $\mathbf{1}$ are defined by Eq. 8.55 (Yu et al., 2008; Suykens and Vandewalle, 1999; Yu, 2014; Lai et al., 2006):

$$\begin{aligned} \mathbf{\Omega}_{ij} &= y_i y_j \phi(\mathbf{x}_i) + (s_i C)^{-1} I \\ \mathbf{Y} &= (y_1, y_2, \ldots, y_l)^T \\ \mathbf{1} &= (1, 1, \ldots, 1)^T \end{aligned} . \qquad 8.55$$

Based on Eqs. 8.54 and 8.55, $\boldsymbol{\alpha}$ and b can be calculated via Eqs. 8.56 and 8.57 (Yu et al., 2008; Suykens and Vandewalle, 1999; Yu, 2014; Lai et al., 2006), respectively:

$$\boldsymbol{\alpha} = \mathbf{\Omega}^{-1} (\mathbf{1} - b\mathbf{Y}). \qquad 8.56$$

$$b = \frac{\mathbf{Y}^T \mathbf{\Omega}^{-1} \mathbf{1}}{\mathbf{Y}^T \mathbf{\Omega}^{-1} \mathbf{Y}} . \qquad 8.57$$

8.3.3 The Application of Support Vector Machine (SVM) in Credit Scoring

8.3.3.1 Algorithm Implementation Environment Selection

In (Haltuf, 2014), a trial study was conducted with the aid of LIBSVM (Chang and Lin, 2013) MATLAB (Sizemore and Mueller, 2015) library. Among different SVM libraries, LIBSVM was selected due to its longest development history, continuous updating and rich application scenarios, to name just a few.

8.3.3.2 Data Description and Pre-Processing

The data set was acquired from an Estonian P2P lending platform called IsePankur.ee/Bondora.com (Bondora Capital OÜ, 2017). In particular, it included 6818 loan records from the beginning of 2009 to the middle of 2014. Since the goal of (Haltuf, 2014) was to establish a model which is able to predict the default probability over a one year horizon, only 2932 data points were actually selected, i.e., loan records between the 28th February 2009 and the 16th April 2013.

Within each loan record, one can find a range of information such as the borrower's personal information, borrower's behavioral information, loan's performance and the default loan's recovery process. The full list of dataset values is depicted in Fig. 8.4.

- **Categorical Variables:** In SVM, it is often assumed that a vector of real numbers is employed in order to represent each input observation. However, when it comes to credit evaluation scenarios, many features are categorical variables (e.g., gender, education background, marital status, home ownership, etc.) rather than real numbers. Therefore, a transformation operation needs to be performed on these variables. One popular method in this regard is to make use of the weight of evidence (WoE), which is defined via Eq. 8.58 (Witzany, 2010) for each particular value of one categorical variable *c*.

$$WoE(c) = \ln P(c|nondefault) - \ln P(c|default). \quad 8.58$$

  The advantage of the WoE technique lies in the fact that it leaves the number of dimensions unchanged. The details of alternative conversion approaches are outside the scope of this chapter but interested readers should refer to (Haltuf, 2014; Hsu et al., 2010) for further discussion.

- **Scaling and Standardizing:** Data scaling is another crucial pre-processing step which can eliminate the possibility of the attributes in smaller numeric ranges being dominated by those in greater ranges.

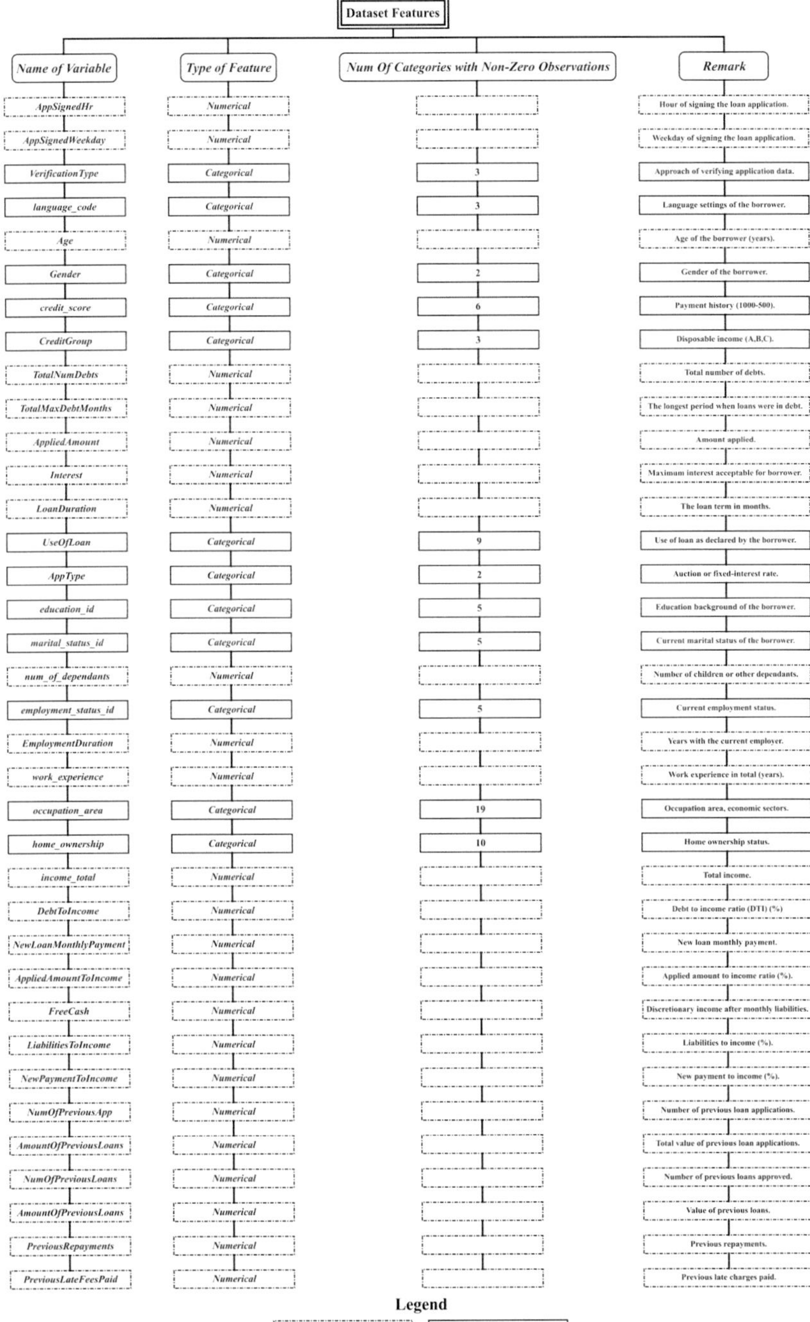

Figure 8.4: Complete list of dataset features.

In general, the range of [–1,+1] or [0,1] is often employed for numerical variables to be linearly scaled to. Meanwhile, the operation of mean subtraction and divided by the standard deviation (i.e., standardizing) is often used in data pre-processing. A thorough explanation can be retrieved from (Sarle, 2002).

- **Missing Values:** Missing value problem (Marwala, 2009) is often an issue encountered in real-world datasets. The underlying reasons are numerous, as are the possible solutions. However, a rule of thumb practice would be: if a certain piece of information is missed, use the most common value instead.

8.3.3.3 Feature Selection

Although many real-life datasets consist of high input feature numbers, credit scoring fortunately does not suffer from these high-dimensional problems. In general, reducing the target problem's dimensionality is beneficial in many ways, such as computational cost reduction and model performance improvement.

- ***F-Score***: As one of the best theoretically justified and quantitatively tested feature selection techniques, the application of F-score in SVM was proposed in (Chen and Lin, 2006). Suppose we have a set of training vectors (denoted by $\mathbf{x}_k$, $k = 1,\dots,m$), the number of positive and negative classes (indicated by m^+ and m^-, respectively), and the averages of the ith feature (represented by $\overline{x}_i$, $\overline{x}_i^+$, and $\overline{x}_i^-$, respectively), then the ith input feature's F-score can be defined via Eq. 8.59 (Chen and Lin, 2006):

$$F(i) = \frac{\left(\overline{x}_i^+ - \overline{x}_i\right)^2 + \left(\overline{x}_i^- - \overline{x}_i\right)^2}{\frac{1}{m^+ - 1}\sum_{k=1}^{n^+}\left(\overline{x}_{k,i}^+ - \overline{x}_i^+\right)^2 + \frac{1}{m^- - 1}\sum_{k=1}^{n^-}\left(\overline{x}_{k,i}^- - \overline{x}_i^-\right)^2}. \tag{8.59}$$

where the specific attribute's discriminative power is in linear relationship with the value of F-score.

Other feature selection approaches also include forward/backward selection, soft computing techniques (e.g., genetic algorithm). An in-depth comparison can be found in (Haltuf, 2014).

8.3.3.4 Model Selection

In the context of SVM, one often needs to select an appropriate kernel function and identify its optimal parameters. Although linear or Gaussian kernels are often employed in credit scoring applications (Haltuf, 2014), a generic methodology in terms of selecting an optimal model is still required. In practice, there are three approaches, varying in precision and computing resource requirements, that can be utilized, i.e., hold-out validation, k-fold cross validation and leave-one-out validation.

8.3.3.5 Model Evaluation

In order to select the most suitable model from a pool of candidates, one has to establish a criterion in order to compare these alternatives' performance. In this regard, a simplified confusion matrix (Witten et al., 2011; Van Gestel et al., 2003; Lessmann et al., 2015) for addressing the classification problem is illustrated in Fig. 8.5.

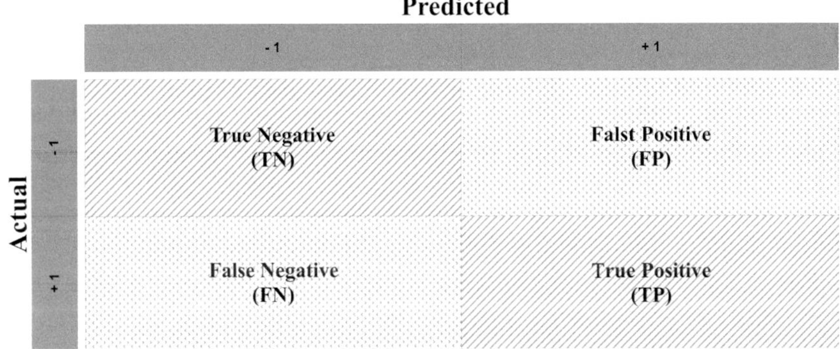

Figure 8.5: Confusion matrix of the classification model.

As illustrated in Fig. 8.5, several simple criteria can be deduced, as given via Eqs. 8.60–8.63 (Haltuf, 2014):

$$\textbf{Accuracy} = \frac{\text{True Positive} + \text{True Negative}}{\text{True Positive} + \text{True Negative} + \text{False Negative} + \text{False Positive}}. \quad 8.60$$

$$\textbf{Classification Error} = \frac{\text{False Positive} + \text{False Negative}}{\text{True Positive} + \text{True Negative} + \text{False Negative} + \text{False Positive}}. \quad 8.61$$

$$\textbf{Sensitivity} = \frac{\text{True Positive}}{\text{True Positive} + \text{False Negative}}. \quad 8.62$$

$$\textbf{Specificity} = \frac{\text{True Negative}}{\text{True Negative} + \text{False Positive}}. \quad 8.63$$

Interested readers can find further details in (Haltuf, 2014).

8.3.3.6 Experimental Study

In (Haltuf, 2014), the author conducted a set of experiments in order to address the credit scoring issue encountered in the P2P lending environment. The models selected for comparison purposes include logistic regression, SVM with linear kernel and SVM with Gaussian kernel.

8.3.4 *Summary*

The real power of SVM tends to be exhibited in cases where some non-linear behavior features are shown by the input data. One possible explanation could be that SVM separates the hyperplane of the shapes. According to (Haltuf, 2014), by applying chosen models to the Bondora credit scoring dataset, SVM with Gaussian kernel performed better than other models.

To further verify the potential benefits that the tailored models have against other native investment practices (e.g., random diversification or intuitive selection), the author of (Haltuf, 2014) considered three types of investors, namely, RANDOM, NAIVE and FORMAL. Based on the results of his simulations, he argued that with the virtual marketplace within the crowdfunding platform becoming more and more efficient, it is, therefore, reasonable to expect that the average return on investment will gradually decrease since the crowdfunding platform's operators tend to wipe off the so-called "early adopters premium" (Haltuf, 2014) phenomenon (often observed in early years of some crowdfunding platforms, such as Bondora and Lending Club).

8.4 Conclusions

In the simplest context, 'credit' means 'complete the purchase now, and finalize the payment later', no matter what the buying activity is for, either instant consumption, tangible products or other intangible services that meet consumers' needs, or productive companies. The word originated from the Latin word *'credo'* meaning 'trust in' or 'rely on' (Anderson, 2007). Therefore, borrowers have to pay the cost of (1) establishing their own impression of trust, (2) finalizing the repayment based on the agreed terms and (3) accepting a certain risk premium in order to account for default situations. In the modern era, data sets about borrowers' financial and other corresponding situations give lenders a great opportunity to enhance the desired trust. On the other hand, 'scoring' means ranking order instances (e.g., people, firms, countries, etc.) via some numerical tool based on various real or recognized qualities (e.g., productivity, profitability, saleability, etc.) (Anderson, 2007). The ultimate goal of scoring is, thus, to be able to distinguish between these cases. In many scenarios where deterministic realities are rare but only with stochastic probabilities, scoring can also find its existence. So, when it comes to 'credit scoring', it is simply the process of utilizing statistical models in order to interpret related data information in the form of numerical values, which in turn can assist lenders in making credit decisions.

P2P lending is characterized by the rife presence of amateur investors whose investment policy is rather empirical and incorporates only one or two factors. Under such conditions, an objective and quantitative methodology-assisted credit scoring strategy is much needed in order for a concerned investor to achieve excess returns. As quoted in (Anderson, 2007), D.N. Chorafas made the following recommendations in 1990 regarding the drivers that will generate returns from the initial information technology investments: (1) artificial intelligence (AI) and networking in contrast to stand-alone data computing and processing tools; and (2) staying ahead of competition rather than conventional return-on-investment measures. Here, the automated credit scoring belongs to the former camp.

Nowadays, the benefits of introducing credit scoring to different business sectors are numerous. However, the drawbacks of credit scoring are also extensive, and here we only look at two representative issues: Process automation and data quality.

- **Process Automation:** In terms of process automation, as suggested by (Thomas, 2000), not everyone is happy about being automatically and periodically scored. Major concerns are (Anderson, 2007):
    1) *Impersonal*: Many people do not like the fact that faceless machines control their lives. In this regard, a continuous development and maintenance of a smart and reliable human machine interactive system (Xing and Marwala, 2018e; Xing and Marwala, 2018g; Xing and Marwala, 2018f; Xing and Marwala, 2018d; Xing and Marwala, 2018b; Xing and Marwala, 2018c; Xing and Marwala, 2018j; Xing and Marwala, 2018h; Xing and Marwala, 2018i; Xing, 2016d; Xing, 2016c; Xing and Marwala, 2018m; Xing and Marwala, 2018k; Xing and Marwala, 2018n; Xing and Marwala, 2018a) will certainly reduce scepticism.
    2) *Causation*: Since credit scoring models usually pay more attention to correlations, the underlying causes can hardly be revealed (Marwala, 2015b; Marwala, 2015f; Marwala, 2015k; Marwala, 2015j; Marwala, 2015d; Marwala, 2015h; Marwala, 2015i; Marwala, 2015a; Marwala, 2015e; Marwala, 2015g; Marwala, 2015c).

    Other issues in this category also include inadequate disputes handling mechanisms, blacklisting and privacy (Anderson, 2007).
- **Data Quality:** The success of predictive models largely depends on the quality of input data; if the data is inferior, the quality of the final scoring results will inevitable be poor. Regarding data quality, major issues often arise from missing data, misrepresentation and mis-

capture (Anderson, 2007). Take missing data, for example. More often than not, we find that for one or more record entries, no information is enclosed; either missing at random, missing completely at random, missing not at random or missing by design (Marwala, 2013b; Marwala, 2014c; Marwala, 2009).

Since credit scoring is more like a process of industrializing trust, the mass-produced products often come with some 'prices' that are absent from the mass-customized ones. Accordingly, in some certain cases, borrowers with the highest credit scores do not necessarily qualify as "good". This implies that conventional credit grading metrics cannot fully cope with the unique dynamics of a P2P lending market. Therefore, in this chapter, we utilized a small investor-welcome P2P platform as a representative case study. The models built on the SVM smart computing technique were subsequently elaborated in order to address the focal issue. As illustrated in (Haltuf, 2014), through a thorough comparison between the chosen models and the industry-standard logistic regression, the developed models could lead to a profitable outcome for those who are investing or planning to invest higher amount of funds in absolute terms. Prior studies also showed that smart computing, as a decision-making support technique, has the capacity to reduce information asymmetry in P2P lending (Yang et al., in press) and get the most applicable results with the least time and cost (Yan et al., 2015). Similar attempts can also be found in (Guo et al., 2016; Malekipirbazari and Aksakalli, 2015).

References

Abdou HA. (2009) Genetic programming for credit scoring: The case of Egyptian public sector banks. *Expert Systems with Applications* 36: 11402–11417.

Abe S. (2010) *Support vector machines for pattern classification,* London Dordrecht Heidelberg New York: Springer-Verlag London Limited, ISBN 978-1-84996-097-7.

Agrawal A, Catalini C and Goldfarb A. (2015) Crowdfunding: Geography, social networks, and the timing of investment decisions. *Journal of Economics & Management Strategy* 24: 253–274.

Akkizidis I and Stagars M. (2016) *Marketplace lending, financial analysis, and the future of credit,* The Atrium, Southern Gate, Chichester, West Sussex, PO19 8SQ, United Kingdom: John Wiley & Sons Ltd, ISBN 978-1-119-09916-1.

Akkoç S. (2012) An empirical comparison of conventional techniques, neural networks and the three stage hybrid Adaptive Neuro Fuzzy Inference System (ANFIS) model for credit scoring analysis: The case of Turkish credit card data. *European Journal of Operational Research* 222: 168–178.

Alam P, Booth D, Lee K, et al. (2000) The use of fuzzy clustering algorithm and self-organizing neural network for identifying potentially failing banks: An experiment study. *Expert Systems with Applications* 18: 185–199.

Anderson R. (2007) *The credit scoring toolkit: Theory and practice for retail credit risk management and decision automation,* Great Clarendon Street, Oxford OX2 6DP: Oxford University Press, ISBN 978-0-19-922640-5.

Angelini E, Tollo G and Roli A. (2008) A neural network approach for credit risk evaluation. *Quarterly Review of Economics and Finance* 48: 733–755.

Angilella S and Mazzù S. (2015) The financing of innovative SMEs: A multicriteria credit rating model. *European Journal of Operational Research* 244: 540–554.

Baeck P and Collins L. (2013) Working the crowd: A short guide to crowdfunding and how it can work for you. 1 Plough Place, London EC4A 1DE, UK: Nesta, 1–19.

Baeck P, Collins L and Zhang B. (2014) Understanding alternative finance: The UK alternative finance industry report 2014. 1 Plough Place, London, EC4A 1DE Nesta, pp. 1–95.

Barasinska N and Schäfer D. (2014) Is crowdfunding different? Evidence on the relation between gender and funding success from a German peer-to-peer lending platform. *German Economic Review* 15: 436–452.

Barth JR and Kaufman GG. (2016) *The first great financial crisis of the 21st century: A retrospective,* 5 Toh Tuck Link, Singapore 596224 and PO Box 1024, Hanover, MA 02339, USA: World Scientific Publishing Co. Pte. Ltd. and now publishers Inc.

Basel Committee on Banking Supervision. (2000) Principles for the management of credit risk. Basel, Switzerland: Bank for International Settlements.

Bastos JA. (2010) Forecasting bank loans loss-given-default. *Journal of Banking & Finance* 34: 2510–2517.

Beccalli E and Poli F. (2015) *Lending, investments and the financial crisis,* 175 Fifth Avenue, New York, NY 10010: Palgrave Macmillan, ISBN 978-1-349-56498-9.

Berger AN, Frame WS and Miller NH. (2005) Credit scoring and the availability, price, and risk of small business credit. *Journal of Money, Credit and Banking* 37: 191–222.

Berger AN and Frame WS. (2007) Small business credit scoring and credit availability. *Journal of Small Business Management* 45: 5–22.

Berger S and Gleisner F. (2009) Emergence of financial intermediaries on electronic markets: The case of online P2P lending. *Business Research* 2: 39–65.

Blake MK. (2006) Gendered lending: Gender, context and the rules of business lending. *Venture Capital* 8: 183–201.

Boehe DM and Cruz LB. (2013) Gender and microfinance performance: Why does the institutional context matter? *World Development* 47: 121–135.

Bondora Capital OÜ. (2017) Public reports: Loan dataset. retrieved from https://www.bondora.com/public-reports, accessed on 13 December 2017.

Burtch G, Ghose A and Wattal S. (2014) Cultural differences and geography as determinants of online pro-social lending. *MIS Quarterly* 38: 773–794.

Burton D. (2012) Credit scoring, risk, and consumer lendingscapes in emerging markets. *Environment and Planning A* 44: 111–124.

Campbell C and Ying Y. (2011) *Learning with support vector machines,* www.morganclaypool.com: Morgan & Claypool, ISBN 978-1-608-45616-1.

Chandler GG and Parker LE. (1989) The predictive value of credit bureau reports. *Journal of Retail Banking* 11: 47–54.

Chang C-C and Lin C-J. (2013) LIBSVM: A library for support vector machines. Taipei, Taiwan: National Taiwan University, retrieved from www.csie.ntu.edu.tw/~cjlin/papers/libsvm.pdf, accessted on 05 November 2017.

Chen D, Li X and Lai F. (in press) Gender discrimination in online peer-to-peer credit lending: Evidence from a lending platform in China. *Electronic Commerce Research* DOI 10.1007/s10660-016-9247-2.

Chen H, Chiang RH and Storey VC. (2012) Business intelligence and analytics: From big data to big impact. *MIS Quarterly* 36: 1165–1188.

Chen Y-W and Lin C-J. (2006) Combining SVMs with various feature selection strategies. In: Guyon I, Nikravesh M, Gunn S, et al. (eds) *Feature Extraction, Studies in Fuzziness and Soft Computing, Vol. 207.* Berlin, Heidelberg: Springer, ISBN 978-3-540-35487-1, Chapter 13, pp. 315–324.

Chen Y, Huang RJ, Tsai J et al. (2015) Soft information and small business lending. *Journal of Financial Services Research* 47: 115–133.

Claesson G and Tengvall M. (2015) Peer-to-peer lending: The effects of institutional involvement in social lending. *Jönköping International Business School.* Jönköping: Jönköping University.

Cole RA, Goldberg LG and White LJ. (2004) Cookie-cutter vs. character: the microstructure of small business lending by large and small banks. *Journal of Financial and Quantitative Analysis* 39: 227–251.

Collier B and Hampshire R. (2010) Sending mixed signals: Multilevel reputation effects in peer-to-peer lending markets. *Proceedings of the 2010 ACM Conference on Computer Supported Cooperative Work (CSCW), 6–10 Februrary, Savannah, Georhia, USA*, pp. 197–206.

Collins L and Pierrakis Y. (2012) The venture crowd: Crowdfunding equity invesment into business. 1 Plough Place, London EC4A 1DE, UK: Nesta, ISBN 978-1-84875-138-5, 1–36.

Crook JN, Edelman DB and Thomas LC. (2007) Recent developments in consumer credit risk assessment. *European Journal of Operational Research* 183: 1447–1465.

Cumming DJ, Leboeuf G and Schwienbacher A. (2017) Crowdfunding cleantech. *Energy Economics* 65: 292–303.

D'Aurizio L, Oliviero T and Romano L. (2015) Family firms, soft information and bank lending in a financial crisis. *Journal of Corporate Finance* 33: 279–292.

Davis A and Kim J. (2017) Explaining changes in the US credit card market: Lenders are using more information. *Economic Modelling* 61: 76–92.

Deer L, Mi J and Yu Y. (2015) The rise of peer-to-peer lending in China: An overview and survey case study. 29 Lincoln's Inn Fields London WC2A 3EE United Kingdom: The Association of Chartered Certified Accountants (ACCA), 1–24.

DeZoort FT, Wilkins A and Justice SE. (2017) The effect of SME reporting framework and credit risk on lender's judgments and decisions. *Journal of Accounting and Public Policy* 36: 302–315.

Diamond E. (1989) Reputation acquisition in debt markets. *Journal of Political Economy* 97: 828–862.

Dietrich J. (2005) Under-specified models and detection of discrimination: A case study of mortgage lending. *The Journal of Real Estate Finance and Economics* 31: 83–105.

Dilsha KM. (2015) A neural network approach for microfinance credit scoring. *Journal of Statistics and Management Systems* 18: 121–138.

Dorfleitner G, Priberny C, Schuster S, et al. (2016) Description-text related soft information in peer-to-peer lending—evidence from two leading European platforms. *Journal of Banking & Finance* 64: 169–187.

Dryver AL and Sukkasem J. (2009) Validating risk models with a focus on credit scoring models. *Journal of Statistical Computation and Simulation* 79: 181–193.

Duarte J, Siegel S and Young L. (2012) Trust and credit: The role of appearance in peer-to-peer lending. *The Review of Financial Studies* 25: 2455–2484.

Elson A. (2017) *The global financial crisis in retrospect: evolution, resolution, and lessons for prevention*, 1 New York Plaza, New York, NY 10004, USA: Palgrave Macmillan, ISBN 978-1-137-59749-6.

Emekter R, Tu Y, Jirasakuldech B, et al. (2015) Evaluating credit risk and loan performance in online peer-to-peer (P2P) lending. *Applied Economics* 47: 54–70.

Everett GR. (2010) Group membership, relationship banking and default risk: The case of online social lending. *SSRN Electronic Journal* http://ssrn.com/abstract=1114428/.

Fama E. (1985) What's different about banks? *Journal of Monetary Economics* 15: 29–39.

Feldman D and Gross S. (2005) Mortgage default: Classification trees analysis. *The Journal of Real Estate Finance and Economics* 30: 369–396.

Freedman S and Jin GZ. (2011) Learning by doing with asymmetric information: Evidence from prosper.com. NBER Working Paper, no. 16855.

Freedman S and Jin GZ. (2014) The signaling value of online social networks: Lessons from peer-to-peer lending. Indiana University & University of Maryland, pp. 1–53.

Frydman H, Kallberg JG and Kao DL. (1985) Testing the adequacy of markov chain and mover-stayer models as representations of credit behavior. *Operations Research* 33: 1203–1214.

Gajda O and Mason N. (2013) Crowdfunding for impact in Europe and the USA. Toniic llc & Ecn, 1–24.

Ge R, Feng J, Gu B, et al. (2017) Predicting and deterring default with social media information in peer-to-peer lending. *Journal of Management Information Systems* 34: 401–424.

Golden LL, Brockett PL, Ai J, et al. (2016) Empirical evidence on the use of credit scoring for predicting insurance losses with psycho-social and biochemical explanations. *North American Actuarial Journal* 20: 233–251.

Gonzalez L and Loureiro YK. (2014) When can a photo increase credit? The impact of lender and borrower profiles on online peer-to-peer loans. *Journal of Behavioral and Experimental Finance* 2: 44–58.

Greiner ME and Wang H. (2010) Building consumer-to-consumer trust in e-finance marketplaces: An empirical analysis. *International Journal of Electronic Commerce* 15: 105–136.

Grunert J, Norden L and Weber M. (2005) The role of non-financial factors in internal credit ratings. *Journal of Banking & Finance* 29: 509–531.

Grunert J and Norden L. (2011) Soft information matters in SME lending. *RSM Insight* 2nd Quarter: 10–11.

Guo Y, Zhou W, Luo C, et al. (2016) Instance-based credit risk assessment for investment decisions in P2P lending. *European Journal of Operational Research* 249: 417–426.

Gutiérrez-Nieto B, Serrano-Cinca C and Camón-Cala J. (2016) A credit score system for socially responsible lending. *Journal of Business Ethics* 133: 691–701.

Haltuf M. (2014) Support vector machines for credit scoring. *Department of Banking and Insurance, Faculty of Finance.* Prague: University of Economics in Prague.

Hamel L. (2009) *Knowledge discovery with support vector machines,* 111 River Street, Hoboken, NJ 07030: JohnWiley & Sons, Inc., ISBN 978-0-470-37192-3.

Han J-T, Chen Q, Liu J-G, et al. (2018) The persuasion of borrowers' voluntary information in peer to peer lending: An empirical study based on elaboration likelihood model. *Computers in Human Behavior* 78: 200–214.

Han S. (2004) Discrimination in lending: Theory and evidence. *Journal of Real Estate Finance and Economics* 29: 5–46.

Han S. (2011) Creditor learning and discrimination in lending. *Journal of Financial Services Research* 40: 1–27.

He X, Inman JJ and Mittal V. (2008) Gender jeopardy in financial risk taking. *Journal of Marketing Research* 45: 414–424.

Herzenstein M, Dholakia UM and Andrews RL. (2011) Strategic herding behavior in peer-to-peer loan auctions. *Journal of Interactive Marketing* 25: 27–36.

Hsieh NC. (2004) An integrated data mining and behavioral scoring model for analyzing bank customers. *Expert Systems with Applications* 27: 623–633.

Hsu C-W, Chang C-C and Lin C-J. (2010) A practical guide to support vector classification. National Taiwan University, Taipei, Taiwan, retrieved from http://citeseerx.ist.edu/viewdoc/summary?doi=10.1.1.224.4115, accessed on 12 December 2017.

Huang A. (in press) A risk detection system of e-commerce: Researches based on soft information extracted by affective computing web texts. *Electronic Commerce Research* DOI 10.1007/s10660-017-9262-y.

Huang Z, Chen H, Hsu CJ, et al. (2004) Credit rating analysis with support vector mahines and neural networks: A market comparative study. *Decision Support Systems* 37: 543–558.

Huang Z, Lei Y and Shen S. (2016) China's personal credit reporting system in the internet finance era: Challenges and opportunities. *China Economic Journal* 9: 288–303.

Ignatius J, Hatami-Marbini A, Rahman A, et al. (in press) A fuzzy decision support system for credit scoring. *Neural Computing & Application* DOI 10.1007/s00521-016-2592-1.

Ince H and Aktan B. (2009) A comparison of data mining techniques for credit scoring in banking: A managerial perspective. *Journal of Business Economics and Management* 10: 233–240.

Iyer R, Khwaja AI, Luttmer EFP et al. (2009) Screening in new credit markets: Can individual lenders infer borrower creditworthiness in peer-to-peer lending? Cambridge, MA.: NBER Working Paper No. 15242, 1–42.

Iyer R, Khwaja AI, Luttmer EFP, et al. (2016) Screening peers softly: Inferring the quality of small borrowers. *Management Science* 62: 1554–1577.

Jenq C, Pan J and Theseira W. (2015) Beauty, weight, and skin color in charitable giving. *Journal of Economic Behavior & Organization* 119: 234–253.

Jiang C, Wang Z, Wang R, et al. (in press) Loan default prediction by combining soft information extracted from descriptive text in online peer-to-peer lending. *Annals of Operations Research* DOI 10.1007/s10479-017-2668-z.

Jin J, Fan B, Dai S, et al. (2017) Beauty premium: Event-related potentials evidence of how physicalattractiveness matters in online peer-to-peer lending. *Neuroscience Letters* 640: 130–135.

Joseph C. (2013) *Advanced credit risk analysis and management,* The Atrium, Southern Gate, Chichester, West Sussex, PO19 8SQ, United Kingdom: John Wiley & Sons, Ltd, ISBN 978-1-118-60491-5.

Kang L, Jiang Q and Tan C-H. (2017) Remarkable advocates: An investigation of geographic distance and social capital for crowdfunding. *Information & Management* 54: 336–348.

Kibbe J. (2013) Too big to disintermediate? Peer-to-peer lending takes on traditional consumer lending. 200 Liberty Street, New York, NY 10281: Richards Kibbe & Orbe LLP.

Kumar PR and Ravi V. (2007) Bankruptcy prediction in banks and firms via statistical and intelligent techniques: A review. *European Journal of Operational Research* 180: 1–28.

Kumar S. (2007) Bank of one: Empirical analysis of peer-to-peer financial marketplace. *Proceedings of 13th Americas Conference on Information System (AIS), Atlanta, US*, pp. 1–8.

Kung SY. (2014) *Kernel methods and machine learning,* University Printing House, Cambridge CB2 8BS, United Kingdom: Cambridge University Press, ISBN: 978 1 107 702496 0.

Labiea M, Méon P-G, Mersland R, et al. (2015) Discrimination by microcredit officers: Theory and evidence on disability in Uganda. *The Quarterly Review of Economics and Finance* 58: 44–55.

Lai KK, Yu L, Zhou L, et al. (2006) Credit risk evaluation with least square support vector machine. In: Wang G-Y, Peters JF, Skowron A, et al. (eds) *RSKT 2016, LNAI 4062.* Berlin Heidelberg: Springer-Verlag, pp. 490–495.

Larrimore L, Jiang L, Larrimore J, et al. (2011) Peer to peer lending: The relationship between language features, trustworthiness, and persuasion success. *Journal of Applied Communication Research* 39: 19–37l.

Lee E and Lee B. (2012) Herding behavior in online P2P lending: An empirical investigation. *Electronic Commerce Research and Applications* 11: 495–503.

Lee TS, Chiu CC, Lu CJ, et al. (2002) Credit scoring using the hybrid neural discriminant technique. *Expert Systems with Applications* 23: 245–254.

Lessmann S, Baesens B, Seow H-V, et al. (2015) Benchmarking state-of-the-art classification algorithms for credit scoring: An update of research. *European Journal of Operational Research* 247: 124–136.

Liang L, Huang B, Liao C, et al. (in press) The impact of SME's lending and credit guarantee on bank efficiency in South Korea. *Review of Development Finance* http://dx.doi.org/10.1016/j.rdf.2017.04.003.

Lin C-F and Wang S-D. (2002) Fuzzy support vector machines. *IEEE Transactions on Neural Networks* 13: 464–471.

Lin M, Prabhala NR and Viswanathan S. (2013) Judging borrowers by the company they keep: Friendship networks and information asymmetry in online peer-to-peer lending. *Management Science* 59: 17–35.

Loureiro YK and Gonzalez L. (2015) Competition against common sense: Insights on peer-to-peer lending as a tool to allay financial exclusion. *International Journal of Bank Marketing* 33: 605–623.

Luo B and Lin Z. (2013) A decision tree model for herd behavior and empirical evidence from the online P2P lending market. *Information Systems and e-Business Management* 11: 141–160.

Ma H-Z and Wang X-R. (2016) Influencing factor analysis of credit risk in P2P lending based on interpretative structural modeling. *Journal of Discrete Mathematical Sciences & Cryptography* 19: 777–786.

Malekipirbazari M and Aksakalli V. (2015) Risk assessment in social lending via random forests. *Expert Systems with Applications* 42: 4621–4631.

Marron D. (2007) 'Lending by numbers': Credit scoring and the constitution of risk within American consumer credit. *Economy and Society* 36: 103–133.

Marwala T. (2009) *Computational intelligence for missing data imputation, estimation and management: Knowledge optimization techniques,* New York, USA: IGI Global, ISBN 978-1-60566-336-4.

Marwala T. (2013a) *Economic modeling using artificial intelligence methods,* Springer London Heidelberg New York Dordrecht: Springer-Verlag London, ISBN 978-1-4471-5009-1.

Marwala T. (2013b) Missing data approaches to economic modeling: Optimization approach. In: Marwala T (ed) *Economic Modeling Using Artificial Intelligence Methods.* Springer London Heidelberg New York Dordrecht: Springer-Verlag London, ISBN 978-1-4471-5009-1, Chapter 7, pp. 119–136.

Marwala T. (2014a) *Artificial intelligence techniques for rational decision making,* Springer Cham Heidelberg New York Dordrecht London: Springer International Publishing Switzerland, ISBN 978-3-319-11423-1.

Marwala T. (2014b) Introduction to rational decision making. In: Marwala T (ed) *Artificial Intelligence Techniques for Rational Decision Making.* Springer Cham Heidelberg New York Dordrecht London: Springer International Publishing Switzerland, ISBN 978-3-319-11423-1, Chapter 1, pp. 1–17.

Marwala T. (2014c) Missing data approaches for rational decision making: application to antenatal data. In: Marwala T (ed) *Artificial Intelligence Techniques for Rational Decision Making.* Springer Cham Heidelberg New York Dordrecht London: Springer International Publishing Switzerland, ISBN 978-3-319-11423-1, Chapter 4, pp. 55–71.

Marwala T. (2015a) Causal, correlation and automatic relevance determination machines for Granger causality. In: Marwala T (ed) *Causality, Correlation and Artificial Intelligence for Rational Decision Making.* 5 Toh Tuck Link, Singapore 596224: World Scientific Publishing Co. Pte. Ltd, ISBN 978-9-81463-086-3, Chapter 7, pp. 125–145.

Marwala T. (2015b) *Causality, correlation and artificial intelligence for rational decision making,* 5 Toh Tuck Link, Singapore 596224: World Scientific Publishing Co. Pte. Ltd, ISBN 978-9-81463-086-3.

Marwala T. (2015c) Conclusions and further work. In: Marwala T (ed) *Causality, Correlation and Artificial Intelligence for Rational Decision Making.* 5 Toh Tuck Link, Singapore 596224: World Scientific Publishing Co. Pte. Ltd, ISBN 978-9-81463-086-3, Chapter 10, pp. 187–189.

Marwala T. (2015d) Correlation machines using optimization methods. In: Marwala T (ed) *Causality, Correlation and Artificial Intelligence for Rational Decision Making.* 5 Toh Tuck Link, Singapore 596224: World Scientific Publishing Co. Pte. Ltd, ISBN 978-9-81463-086-3, Chapter 4, pp. 65–86.

Marwala T. (2015e) Flexibly-bounded rationality. In: Marwala T (ed) *Causality, Correlation and Artificial Intelligence for Rational Decision Making.* 5 Toh Tuck Link, Singapore 596224: World Scientific Publishing Co. Pte. Ltd, ISBN 978-9-81463-086-3, Chapter 8, pp. 147–166.

Marwala T. (2015f) Introduction to artificial intelligence based decision making. In: Marwala T (ed) *Causality, Correlation and Artificial Intelligence for Rational Decision Making.* 5 Toh Tuck Link, Singapore 596224: World Scientific Publishing Co. Pte. Ltd, ISBN 978-9-81463-086-3, Chapter 1, pp. 1–21.

Marwala T. (2015g) Marginalization of irrationality in decision making. In: Marwala T (ed) *Causality, Correlation and Artificial Intelligence for Rational Decision Making.* 5 Toh Tuck Link, Singapore 596224: World Scientific Publishing Co. Pte. Ltd, ISBN 978-9-81463-086-3, Chapter 9, pp. 167–185.

Marwala T. (2015h) Neural networks for modeling granger causality. In: Marwala T (ed) *Causality, Correlation and Artificial Intelligence for Rational Decision Making*. 5 Toh Tuck Link, Singapore 596224: World Scientific Publishing Co. Pte. Ltd, ISBN 978-9-81463-086-3, Chapter 5, pp. 87–103.

Marwala T. (2015i) Rubin, Pearl and Granger causality models: A unified view. In: Marwala T (ed) *Causality, Correlation and Artificial Intelligence for Rational Decision Making*. 5 Toh Tuck Link, Singapore 596224: World Scientific Publishing Co. Pte. Ltd, ISBN 978-9-81463-086-3, Chapter 6, pp. 105–124.

Marwala T. (2015j) What is a causal machine? In: Marwala T (ed) *Causality, Correlation and Artificial Intelligence for Rational Decision Making*. 5 Toh Tuck Link, Singapore 596224: World Scientific Publishing Co. Pte. Ltd, ISBN 978-9-81463-086-3, Chapter 3, pp. 43–63.

Marwala T. (2015k) What is a correlation machine? In: Marwala T (ed) *Causality, Correlation and Artificial Intelligence for Rational Decision Making*. 5 Toh Tuck Link, Singapore 596224: World Scientific Publishing Co. Pte. Ltd, ISBN 978-9-81463-086-3, Chapter 2, pp. 23–42.

Marwala T and Hurwitz E. (2017a) *Artificial intelligence and economic theory: Skynet in the market*, Gewerbestrasse 11, 6330 Cham, Switzerland: Springer International Publishing AG, ISBN 978-3-319-66103-2.

Marwala T and Hurwitz E. (2017b) Information asymmetry. In: Marwala T and Hurwitz E (eds) *Artificial Intelligence and Economic Theory: Skynet in the Market*. Gewerbestrasse 11, 6330 Cham, Switzerland: Springer International Publishing AG, ISBN 978-3-319-66103-2, Chapter 6, pp. 63–74.

McCann F and McIndoe-Calder T. (2015) Firm size, credit scoring accuracy and bank's production of soft information. *Applied Economics* 47: 3594–3611.

Mendes-Da-Silva W, Rossoni L, Conte BS, et al. (2016) The impacts of fundraising periods and geographic distance on financing music production via crowdfunding in Brazil. *Journal of Cultural Economics* 40: 75–99.

Michels J. (2012) Do unverifiable disclosures matter? Evidence from peer-to-peer lending. *The Accounting Review* 87: 1385–1413.

Mirtalaei MS, Saberi M, Hussain OK, et al. (2012) A trust-based bio-inspired approach for credit lending decisions. *Computing* 94: 541–577.

Mohammadi A and Shafi K. (in press) Gender differences in the contribution patterns of equity-crowdfunding investors. *Small Business Economics* DOI 10.1007/s11187-016-9825-7.

Moodie M. (2013) Microfinance and the gender of risk: The case of Kiva.org. *Signs* 38: 279–302.

Morgan PB. (2015) *An explanation of constrained optimization for economists*, Toronto Buffalo London: University of Toronto Press, ISBN 978-1-4426-4278-2.

Moritz A, Block J and Lutz E. (2015) Investor communication in equity-based crowdfunding: A qualitative-empirical study. *Qualitative Research in Financial Markets* 7: 309–342.

Moro A and Fink M. (2013) Loan managers' trust and credit access for SMEs. *Journal of Banking & Finance* 37: 927–936.

Moro A, Maresch D and Ferrando A. (in press) Creditor protection, judicial enforcement and credit access. *The European Journal of Finance* http://dx.doi.org/10.1080/1351847X.2016.1216871.

Nesta. (2012) An introduction to crowdfunding. 1 Plough Place, London EC4A 1DE Nesta, pp. 1–4.

Ong C, Huang J and Tzeng G. (2005) Building credit scoring models using genetic programming. *Expert Systems with Applications* 29: 41–47.

Oxera. (2015) Crowdfunding from an investor perspective. Park Central, 40/41 Park End Street, Oxford, OX1 1JD, UK: Oxera Consulting LLP, ISBN 978-92-79-46659-5.

Patel PB and Marwala T. (2006) Neural networks, fuzzy inference systems and adaptive-neuro fuzzy inference systems for financial decision making. *Lecture Notes in Computer Science, Vol. 4234*. Berlin Heidelberg: Springer-Verlag, pp. 430–439.

Petersen MA and Rajan RG. (1994) The benefits of lending relationship: Evidence from small business data. *Journal of Finance* 49: 3–37.

Piper G and Schnepf SV. (2008) Gender differences in charitable giving in Great Britain. *Voluntas* 19: 103–124.

Pope DG and Sydnor JR. (2011) What's in a picture? Evidence of discrimination from prosper. com. *Journal of Human Resources* 46: 53–92.

Puro L, Teich JE, Wallenius H, et al. (2010) Borrower decision aid for people-to-people lending. *Decision Support Systems* 49: 52–60.

Qiu J, Lin Z and Luo B. (2012) Effects of borrower-defined conditions in the online peer-to-peer lending market. In: Shaw MJ, Zhang D and Yue WT (eds) *E-Life: Web-Enabled Convergence of Commerce, Work, and Social Life.* Springer-Verlag Berlin Heidelberg, ISBN 978-3-642-29872-1, Part II, pp. 167–179.

Ravina E. (2012) *Love & loans: The effect of beauty and personal characteristics in credit markets, retrieved from https://ssrn.com/abstract=1107307, accessed on 22 November 2017.*

Rosenberg E and Gleit A. (1994) Quantitative methods in credit management: A survey. *Operational Research* 42: 589–613.

Rossi SPS and Malavasi R. (2016) *Financial crisis, bank behaviour and credit crunch,* Cham Heidelberg New York Dordrecht London: Springer International Publishing Switzerland, ISBN 978-3-319-17412-9.

Roure Cd, Pelizzon L and Tasca P. (2016) How does P2P lending fit into the consumer credit market? Wilhelm-Epstein-Straße 14, 60431 Frankfurt am Main, Postfach 10 06 02, 60006 Frankfurt am Main: Deutsche Bundesbank, ISBN 978-3-95729-286-5, 1–24.

Saberi M, Mirtalaie MS, Hussain FK, et al. (2013) A granular computing-based approach to credit scoring modeling. *Neurocomputing* 122: 100–115.

Saliyaa CA and Jayasinghe K. (2016) Creating and reinforcing discrimination: The controversial role of accounting in bank lending. *Accounting Forum* 40: 235–250.

Sarle WS. (2002) Neural network FAQ, part 2 of 7: Learning. retrieved from ftp://ftp.sas.com/pub/neural/FAQ2.html, accessed on 13 December 2017.

Schölkopf B and Smola AJ. (2002) *Learning with kernels: Support vector machines, regularization, optimization and beyond,* Cambridge, Massachusetts: The MIT Press, ISBN 0-262-19475-9.

Sherchan W, Nepal S and Paris C. (2013) A survey of trust in social networks. *ACM Computing Surveys* 45: 1–33.

Shi J and Xu B. (2016) Credit scoring by fuzzy support vector machines with a novel membership function. *Journal of Risk and Financial Management* 9: Article 13.

Singh SF. (2015) Social sorting as 'social transformation': Credit scoring and the reproduction of populations as risks in South Africa. *Security Dialogue* 46: 365–383.

Sizemore J and Mueller JP. (2015) *MATLAB for dummies,* 111 River Street, Hoboken, NJ 07030-5774: John Wiley & Sons, Inc., ISBN 978-1-118-882010-0.

Sohn SY, Moon TH and Kim HS. (2005) Improved technology scoring model for credit guarantee fund. *Expert Systems with Applications* 28: 327–331.

Sohn SY and Kim YS. (2013) Behavioral credit scoring model for technology-based firms that considers uncertain financial ratios obtained from relationship banking. *Small Business Economics* 41: 931–943.

Sonenshein S, Herzenstein M and Dholakia UM. (2010) How accounts shape lending decisions through fostering perceived trustworthiness. *Organizational Behavior and Human Decision Processes* 115: 69–84.

Stern C, Makinen M and Qian Z. (2017) FinTechs in China—with a special focus on peer-to-peer lending. *Journal of Chinese Economic and Foreign Trade Studies* 10: 215–228.

Suykens JAK and Vandewalle J. (1999) Least squares support vector machine classifiers. *Neural Processing Letters* 9: 293–300.

Tang WM. (2011) Fuzzy SVM with a new fuzzy membership function to solve the two-class problems. *Neural Processing Letters* 34: 290.

Thomas LC. (2000) A survey of credit and behavioural scoring: Forecasting financial risk of lending to consumers. *International Journal of Forecasting* 16: 149–172.

Van Gestel T, Viaene S, Stepanova M, et al. (2003) Benchmarking state-of-the-art classification algorithms for credit scoring. *Journal of the Operational Research Society* 54: 627–635.

Walter T. (2008) Competition to default: Racial discrimination in the market for online peer-to-peer lending. *Business*: 1–44.

Wang H, Chen K, Zhu W, et al. (2015) A process model on P2P lending. *Financial Innovation* 1: 1–8.

Wang J, Shen Y and Huang Y. (2016) Evaluating the regulatory scheme for internet finance in China: The case of peer-to-peer lending. *China Economic Journal* 9: 272–287.

White Label Crowdfunding Limited. (2016) Differences in the regulatory systems governing peer-to-peer lending platforms between the UK and Malaysia. White Label Crowdfunding Limited, 1–35.

Wiginton JC. (1980) A note on the comparison of logit and discriminant models of consumer credit behavior. *Journal of Financial and Quantitative Analysis* 15: 757–770.

Witten IH, Frank E and Hall MA. (2011) *Data mining: Practical machine learning tools and techniques,* 30 Corporate Drive, Suite 400, Burlington, MA 01803, USA: Morgan Kaufmann, Elsevier Inc., ISBN 978-0-12-374856-0.

Witzany J. (2010) *Credit risk management and modeling,* Czech Republic: Praha: Oeconomica, ISBN 978-80-245-1682-0.

Wu J and Xu Y. (2011) A decision support system for borrower's loan in P2P lending. *Journal of Computers* 6: 1183–1190.

Xia Y, Liu C, Li Y, et al. (2017) A boosted decision tree approach using Bayesian hyper-parameter optimization for credit scoring. *Expert Systems with Applications* 78: 225–241.

Xiao W, Zhao Q and Fei Q. (2006) A comparative study of data mining methods in consumer loans credit scoring management. *Journal of Systems Science and Systems Engineering* 15: 419–435.

Xing B, Gao W-J, Nelwamondo FV, et al. (2011) e-Reverse logistics for remanufacture-to-order: an online auction-based and multi-agent system supported solution. In: Omatu S and Fabri SG (eds) *Fifth International Conference on Advanced Engineering Computing and Applications in Sciences (ADVCOMP), 20–25 November, Lisbon, Portugal*, pp. 78–83. IARIA.

Xing B, Gao W-J and Marwala T. (2013a) Intelligent data processing using emerging computational intelligence techniques. *Research Notes in Information Sciences* 12: 10–15.

Xing B, Gao W-J and Marwala T. (2013b) An overview of cuckoo-inspired intelligent algorithms and their applications. *IEEE Symposium Series on Computational Intelligence (IEEE SSCI), 15–19 April, Singapore*, pp. 85–89. IEEE.

Xing B and Gao W-J. (2014a) *Computational intelligence in remanufacturing,* 701 E. Chocolate Avenue, Suite 200, Hershey PA 17033: IGI Global, ISBN 978-1-4666-4908-8.

Xing B and Gao W-J. (2014aa) Chemical-reaction optimization algorithm. In: Xing B and Gao W-J (eds) *Innovative Computational Intelligence: A Rough Guide to 134 Clever Algorithms.* Cham Heidelberg New York Dordrecht London: Springer International Publishing Switzerland, ISBN: 978-3-319-03403-4, Chapter 25, pp. 417–428.

Xing B and Gao W-J. (2014b) *Innovative computational intelligence: A rough guide to 134 clever algorithms,* Cham Heidelberg New York Dordrecht London: Springer International Publishing Switzerland, ISBN: 978-3-319-03403-4.

Xing B and Gao W-J. (2014bb) Emerging chemistry-based CI algorithms. In: Xing B and Gao W-J (eds) *Innovative Computational Intelligence: A Rough Guide to 134 Clever Algorithms.* Cham Heidelberg New York Dordrecht London: Springer International Publishing Switzerland, ISBN: 978-3-319-03403-4, Chapter 26, pp. 429–437.

Xing B and Gao W-J. (2014c) Introduction to computational intelligence. In: Xing B and Gao W-J (eds) *Innovative Computational Intelligence: A Rough Guide to 134 Clever Algorithms.* Cham Heidelberg New York Dordrecht London: Springer International Publishing Switzerland, ISBN: 978-3-319-03403-4, Chapter 1, pp. 3–17.

Xing B and Gao W-J. (2014cc) Base optimization algorithm. In: Xing B and Gao W-J (eds) *Innovative Computational Intelligence: A Rough Guide to 134 Clever Algorithms.* Cham Heidelberg New York Dordrecht London: Springer International Publishing Switzerland, ISBN: 978-3-319-03403-4, Chapter 27, pp. 441–444.

Xing B and Gao W-J. (2014d) Bacteria inspired algorithms. In: Xing B and Gao W-J (eds) *Innovative Computational Intelligence: A Rough Guide to 134 Clever Algorithms.* Cham Heidelberg New York Dordrecht London: Springer International Publishing Switzerland, ISBN: 978-3-319-03403-4, Chapter 2, pp. 21–38.

Xing B and Gao W-J. (2014dd) Emerging mathematics-based CI algorithms. In: Xing B and Gao W-J (eds) *Innovative Computational Intelligence: A Rough Guide to 134 Clever Algorithms.* Cham Heidelberg New York Dordrecht London: Springer International Publishing Switzerland, ISBN: 978-3-319-03403-4, Chapter 28, pp. 445–448.

Xing B and Gao W-J. (2014e) Bee inspired algorithms. In: Xing B and Gao W-J (eds) *Innovative Computational Intelligence: A Rough Guide to 134 Clever Algorithms.* Cham Heidelberg New York Dordrecht London: Springer International Publishing Switzerland, ISBN: 978-3-319-03403-4, Chapter 4, pp. 45–80.

Xing B and Gao W-J. (2014f) Bat inspired algorithms. In: Xing B and Gao W-J (eds) *Innovative Computational Intelligence: A Rough Guide to 134 Clever Algorithms.* Cham Heidelberg New York Dordrecht London: Springer International Publishing Switzerland, ISBN: 978-3-319-03403-4, Chapter 3, pp. 39–44.

Xing B and Gao W-J. (2014ff) Overview of computational intelligence. In: Xing B and Gao W-J (eds) *Computational Intelligence in Remanufacturing.* 701 E. Chocolate Avenue, Suite 200, Hershey PA 17033: IGI Global, ISBN 978-1-4666-4908-8, Chapter 2, pp. 18–36.

Xing B and Gao W-J. (2014g) Biogeography-based optimization algorithm. In: Xing B and Gao W-J (eds) *Innovative Computational Intelligence: A Rough Guide to 134 Clever Algorithms.* Cham Heidelberg New York Dordrecht London: Springer International Publishing Switzerland, ISBN: 978-3-319-03403-4, Chapter 5, pp. 81–91.

Xing B and Gao W-J. (2014h) Cat swarm optimization algorithm. In: Xing B and Gao W-J (eds) *Innovative Computational Intelligence: A Rough Guide to 134 Clever Algorithms.* Cham Heidelberg New York Dordrecht London: Springer International Publishing Switzerland, ISBN: 978-3-319-03403-4, Chapter 6, pp. 93–104.

Xing B and Gao W-J. (2014i) Cuckoo inspired algorithms. In: Xing B and Gao W-J (eds) *Innovative Computational Intelligence: A Rough Guide to 134 Clever Algorithms.* Cham Heidelberg New York Dordrecht London: Springer International Publishing Switzerland, ISBN: 978-3-319-03403-4, Chapter 7, pp. 105–121.

Xing B and Gao W-J. (2014j) Luminous insect inspired algorithms. In: Xing B and Gao W-J (eds) *Innovative Computational Intelligence: A Rough Guide to 134 Clever Algorithms.* Cham Heidelberg New York Dordrecht London: Springer International Publishing Switzerland, ISBN: 978-3-319-03403-4, Chapter 8, pp. 123–137.

Xing B and Gao W-J. (2014k) Fish inspired algorithms. In: Xing B and Gao W-J (eds) *Innovative Computational Intelligence: A Rough Guide to 134 Clever Algorithms.* Cham Heidelberg New York Dordrecht London: Springer International Publishing Switzerland, ISBN: 978-3-319-03403-4, Chapter 9, pp. 139–155.

Xing B and Gao W-J. (2014l) Frog inspired algorithms. In: Xing B and Gao W-J (eds) *Innovative Computational Intelligence: A Rough Guide to 134 Clever Algorithms.* Cham Heidelberg New York Dordrecht London: Springer International Publishing Switzerland, ISBN: 978-3-319-03403-4, Chapter 10, pp. 157–165.

Xing B and Gao W-J. (2014m) Fruit fly optimization algorithm. In: Xing B and Gao W-J (eds) *Innovative Computational Intelligence: A Rough Guide to 134 Clever Algorithms.* Cham Heidelberg New York Dordrecht London: Springer International Publishing Switzerland, ISBN: 978-3-319-03403-4, Chapter 11, pp. 167–170.

Xing B and Gao W-J. (2014n) Group search optimization algorithm. In: Xing B and Gao W-J (eds) *Innovative Computational Intelligence: A Rough Guide to 134 Clever Algorithms.* Cham Heidelberg New York Dordrecht London: Springer International Publishing Switzerland, ISBN: 978-3-319-03403-4, Chapter 12, pp. 171–176.

Xing B and Gao W-J. (2014o) Invasive weed optimization algorithm. In: Xing B and Gao W-J (eds) *Innovative Computational Intelligence: A Rough Guide to 134 Clever Algorithms.* Cham Heidelberg New York Dordrecht London: Springer International Publishing Switzerland, ISBN: 978-3-319-03403-4, Chapter 13, pp. 177–181.

Xing B and Gao W-J. (2014p) Music inspired algorithms. In: Xing B and Gao W-J (eds) *Innovative Computational Intelligence: A Rough Guide to 134 Clever Algorithms.* Cham Heidelberg New York Dordrecht London: Springer International Publishing Switzerland, ISBN: 978-3-319-03403-4, Chapter 14, pp. 183–201.

Xing B and Gao W-J. (2014q) Imperialist competitive algorithm. In: Xing B and Gao W-J (eds) *Innovative Computational Intelligence: A Rough Guide to 134 Clever Algorithms.* Cham Heidelberg New York Dordrecht London: Springer International Publishing Switzerland, ISBN: 978-3-319-03403-4, Chapter 15, pp. 203–209.

Xing B and Gao W-J. (2014r) Teaching–learning-based optimization algorithm. In: Xing B and Gao W-J (eds) *Innovative Computational Intelligence: A Rough Guide to 134 Clever Algorithms.* Cham Heidelberg New York Dordrecht London: Springer International Publishing Switzerland, ISBN: 978-3-319-03403-4, Chapter 16, pp. 211–216.

Xing B and Gao W-J. (2014s) Emerging biology-based CI algorithms. In: Xing B and Gao W-J (eds) *Innovative Computational Intelligence: A Rough Guide to 134 Clever Algorithms.* Cham Heidelberg New York Dordrecht London: Springer International Publishing Switzerland, ISBN: 978-3-319-03403-4, Chapter 17, pp. 217–317.

Xing B and Gao W-J. (2014t) Big bang–big crunch algorithm. In: Xing B and Gao W-J (eds) *Innovative Computational Intelligence: A Rough Guide to 134 Clever Algorithms.* Cham Heidelberg New York Dordrecht London: Springer International Publishing Switzerland, ISBN: 978-3-319-03403-4, Chapter 18, pp. 321–331.

Xing B and Gao W-J. (2014u) Central force optimization algorithm. In: Xing B and Gao W-J (eds) *Innovative Computational Intelligence: A Rough Guide to 134 Clever Algorithms.* Cham Heidelberg New York Dordrecht London: Springer International Publishing Switzerland, ISBN: 978-3-319-03403-4, Chapter 19, pp. 333–337.

Xing B and Gao W-J. (2014v) Charged system search algorithm. In: Xing B and Gao W-J (eds) *Innovative Computational Intelligence: A Rough Guide to 134 Clever Algorithms.* Cham Heidelberg New York Dordrecht London: Springer International Publishing Switzerland, ISBN: 978-3-319-03403-4, Chapter 20, pp. 339–346.

Xing B and Gao W-J. (2014w) Electromagnetism-like mechanism algorithm. In: Xing B and Gao W-J (eds) *Innovative Computational Intelligence: A Rough Guide to 134 Clever Algorithms.* Cham Heidelberg New York Dordrecht London: Springer International Publishing Switzerland, ISBN: 978-3-319-03403-4, Chapter 21, pp. 347–354.

Xing B and Gao W-J. (2014x) Gravitational search algorithm. In: Xing B and Gao W-J (eds) *Innovative Computational Intelligence: A Rough Guide to 134 Clever Algorithms.* Cham Heidelberg New York Dordrecht London: Springer International Publishing Switzerland, ISBN: 978-3-319-03403-4, Chapter 22, pp. 355–364.

Xing B and Gao W-J. (2014y) Intelligent water drops algorithm. In: Xing B and Gao W-J (eds) *Innovative Computational Intelligence: A Rough Guide to 134 Clever Algorithms.* Cham Heidelberg New York Dordrecht London: Springer International Publishing Switzerland, ISBN: 978-3-319-03403-4, Chapter 23, pp. 365–373.

Xing B and Gao W-J. (2014z) Emerging physics-based CI algorithms. In: Xing B and Gao W-J (eds) *Innovative Computational Intelligence: A Rough Guide to 134 Clever Algorithms.* Cham Heidelberg New York Dordrecht London: Springer International Publishing Switzerland, ISBN: 978-3-319-03403-4, Chapter 24, pp. 375–414.

Xing B. (2015) Graph-based framework for evaluating the feasibility of transition to maintainomics. In: Pedrycz W and Chen S-M (eds) *Information Granularity, Big Data, and Computational Intelligence.* Cham Heidelberg New York Dordrecht London: Springer International Publishing Switzerland, ISBN 978-3-319-08253-0, Chapter 5, pp. 89–119.

Xing B. (2016a) Agent-based machine-to-machine connectivity analysis for the Internet of things environment. In: Mahmood Z (ed) *Connectivity Frameworks for Smart Devices: The Internet of Things from a Distributed Computing Perspective.* Switzerland: Springer International Publishing, ISBN 978-3-319-33122-5, Chapter 3, pp. 43–61.

Xing B. (2016b) An investigation of the use of innovative biology-based computational intelligence in ubiquitous robotics systems: Data mining perspective. In: Ravulakollu

KK, Khan MA and Abraham A (eds) *Trends in Ambient Intelligent Systems.* Switzerland: Springer International Publishing Switzerland, ISBN 978-3-319-30184-6, Chapter 6, pp. 139–172.

Xing B. (2016c) Smart robot control via novel computational intelligence methods for ambient assisted living. In: Ravulakollu KK, Khan MA and Abraham A (eds) *Trends in Ambient Intelligent Systems.* Switzerland: Springer International Publishing Switzerland, ISBN 978-3-319-30184-6, Chapter 2, pp. 29–55.

Xing B. (2016d) The spread of innovatory nature originated metaheuristics in robot swarm control for smart living environments. In: Espinosa HEP (ed) *Nature-Inspired Computing for Control Systems.* Cham Heidelberg New York Dordrecht London: Springer International Publishing Switzerland, ISBN 978-3-319-26228-4, Chapter 3, pp. 39–70.

Xing B and Marwala T. (2018a) Conclusion. In: Xing B and Marwala T (eds) *Smart Maintenance for Human–Robot Interaction: An Intelligent Search Algorithmic Perspective.* Gewerbestrasse 11, 6330 Cham, Switzerland: Springer International Publishing AG, ISBN 978-3-319-67479-7, Chapter 13, pp. 299–305.

Xing B and Marwala T. (2018b) Cyberware capacity–applications layer perspective. In: Xing B and Marwala T (eds) *Smart Maintenance for Human–Robot Interaction: An Intelligent Search Algorithmic Perspective.* Gewerbestrasse 11, 6330 Cham, Switzerland: Springer International Publishing AG, ISBN 978-3-319-67479-7, Chapter 8, pp. 173–191.

Xing B and Marwala T. (2018c) Cyberware capacity–energy autonomy perspective. In: Xing B and Marwala T (eds) *Smart Maintenance for Human–Robot Interaction: An Intelligent Search Algorithmic Perspective.* Gewerbestrasse 11, 6330 Cham, Switzerland: Springer International Publishing AG, ISBN 978-3-319-67479-7, Chapter 9, pp. 193–216.

Xing B and Marwala T. (2018d) Cyberware capacity–platform and middleware layers perspective. In: Xing B and Marwala T (eds) *Smart Maintenance for Human–Robot Interaction: An Intelligent Search Algorithmic Perspective.* Gewerbestrasse 11, 6330 Cham, Switzerland: Springer International Publishing AG, ISBN 978-3-319-67479-7, Chapter 7, pp. 143–171.

Xing B and Marwala T. (2018e) Hardware capacity–beginning of life perspective. In: Xing B and Marwala T (eds) *Smart Maintenance for Human–Robot Interaction: An Intelligent Search Algorithmic Perspective.* Gewerbestrasse 11, 6330 Cham, Switzerland: Springer International Publishing AG, ISBN 978-3-319-67479-7, Chapter 4, pp. 67–91.

Xing B and Marwala T. (2018f) Hardware capacity–end of life perspective. In: Xing B and Marwala T (eds) *Smart Maintenance for Human–Robot Interaction: An Intelligent Search Algorithmic Perspective.* Gewerbestrasse 11, 6330 Cham, Switzerland: Springer International Publishing AG, ISBN 978-3-319-67479-7, Chapter 6, pp. 111–139.

Xing B and Marwala T. (2018g) Hardware capacity–middle of life perspective. In: Xing B and Marwala T (eds) *Smart Maintenance for Human–Robot Interaction: An Intelligent Search Algorithmic Perspective.* Gewerbestrasse 11, 6330 Cham, Switzerland: Springer International Publishing AG, ISBN 978-3-319-67479-7, Chapter 5, pp. 93–110.

Xing B and Marwala T. (2018h) Human capacity–biopsychosocial perspective. In: Xing B and Marwala T (eds) *Smart Maintenance for Human–Robot Interaction: An Intelligent Search Algorithmic Perspective.* Gewerbestrasse 11, 6330 Cham, Switzerland: Springer International Publishing AG, ISBN 978-3-319-67479-7, Chapter 11, pp. 249–270.

Xing B and Marwala T. (2018i) Human capacity–exposome perspective. In: Xing B and Marwala T (eds) *Smart Maintenance for Human–Robot Interaction: An Intelligent Search Algorithmic Perspective.* Gewerbestrasse 11, 6330 Cham, Switzerland: Springer International Publishing AG, ISBN 978-3-319-67479-7, Chapter 12, pp. 271–295.

Xing B and Marwala T. (2018j) Human capacity–physiology perspective. In: Xing B and Marwala T (eds) *Smart Maintenance for Human–Robot Interaction: An Intelligent Search Algorithmic Perspective.* Gewerbestrasse 11, 6330 Cham, Switzerland: Springer International Publishing AG, ISBN 978-3-319-67479-7, Chapter 10, pp. 219–247.

Xing B and Marwala T. (2018k) Introduction to human robot interaction. In: Xing B and Marwala T (eds) *Smart Maintenance for Human–Robot Interaction: An Intelligent*

Search Algorithmic Perspective. Gewerbestrasse 11, 6330 Cham, Switzerland: Springer International Publishing AG, ISBN 978-3-319-67479-7, Chapter 1, pp. 3–19.

Xing B and Marwala T. (2018l) Introduction to intelligent search algorithms. In: Xing B and Marwala T (eds) *Smart Maintenance for Human–Robot Interaction: An Intelligent Search Algorithmic Perspective.* Gewerbestrasse 11, 6330 Cham, Switzerland: Springer International Publishing AG, ISBN 978-3-319-67479-7, Chapter 3, pp. 33–64.

Xing B and Marwala T. (2018m) Introduction to smart maintenance. In: Xing B and Marwala T (eds) *Smart Maintenance for Human–Robot Interaction: An Intelligent Search Algorithmic Perspective.* Gewerbestrasse 11, 6330 Cham, Switzerland: Springer International Publishing AG, ISBN 978-3-319-67479-7, Chapter 2, pp. 21–31.

Xing B and Marwala T. (2018n) *Smart maintenance for human–robot interaction: an intelligent search algorithmic perspective,* Gewerbestrasse 11, 6330 Cham, Switzerland: Springer International Publishing AG, ISBN 978-3-319-67479-7.

Yan J, Yu W and Zhao JL. (2015) How signaling and search costs affect information asymmetry in P2P lending: the economics of big data. *Financial Innovation* 1: 1–11.

Yang D, Chen P, Shi F, et al. (in press) Internet finance: its uncertain legal foundations and the role of big data in its development. *Emerging Markets Finance and Trade* DOI: 10.1080/1540496X.2016.1278528.

Yu L, Wang S, Lai KK, et al. (2008) *Bio-Inspired credit risk analysis: Computational intelligence with support vector machines,* Heidelberg, Germany: Springer-Verlag Berlin Heidelberg, ISBN 978-3-540-77802-8.

Yu L. (2014) Credit risk evaluation with a least squares fuzzy support vector machine classifier. *Discrete Dynamics in Nature and Society* 2014: Article ID 564213.

Zeng G. (in press) A comparison study of computational methods of Kolmogorov-Smirnov statistic in credit scoring. *Communications in Statistics—Simulation and Computation* DOI: 10.1080/03610918.2016.1249883.

Zhang B, Baeck P, Ziegler T, et al. (2016a) Pushing boundaries: The 2015 finance industry report. Cambridge Centre for Alternative Finance & Nesta, 1–56.

Zhang D, Chen Q and Wei L. (2007) Building behavior scoring model using genetic algorithm and support vector machines. *Computational Science* 4488: 482–485.

Zhang J and Liu P. (2012) Rational herding in microloan markets. *Management Science* 58: 892–912.

Zhang J. (2013) The wisdom of crowdfunding. *Communities & Banking* Winter: 30–31.

Zhang Y, Jia H, Diao Y, et al. (2016b) Research on credit scoring by fusing social media information in online peer-to-peer lending. *Procedia Computer Science* 91: 168–174.

Zhao Z, Xu S, Kang BH, et al. (2015) Investigation and improvement of multi-layer perceptron neural networks for credit scoring. *Expert Systems with Applications* 42: 3508–3516.

Zhou R and Pham MT. (2004) Promotion and prevention across mental accounts: When financial products dictate consumers' investment goals. *Journal of Consumer Research* 31: 125–135.

Ziegler T, Reedy EJ, Le A, et al. (2017) The Americas alternative finance industry report: Hitting stride. Cambridge Centre for Alternative Finance, Judge Business School, University of Cambridge, 1–77.

Part VI

Operator + Smart Computing = Wiser Service

CHAPTER 9

Crowdfunding Operator

Portfolio Selection and Metaheuristics

9.1 Introduction

Nowadays, crowdfunding platforms, in particular peer-to-peer (P2P) lending, provide a wealth-management service for individuals, which allows lenders to pursue high returns with a streamlined process, while borrowers can look for low interest rates with fast approvals. In general, there are two operation models that P2P lending can use in order to add value: (1) auction-based lending, and (2) automatic-matched lending. The former matches individual lenders and borrowers through an auction style process, while the latter helps the two sides to establish an automatic lending relationship (De Buysere et al., 2012).

Inevitably, in either model, both lenders and borrowers face multi-objective choices. From a lender's perspective, he/she is faced with the problem of choosing among an enormous number of loan proposals and must consider various issues such as loan risks (i.e., default probability of each one of those proposals), the trading efficiency (i.e., winning-bid probability and fully-funded probability) and the whole performance (i.e., optimal portfolio selection). On the other hand, from the borrower's perspective, he/she must think about many factors as well, e.g., the loan's duration, the desirable amount (even if a lender would like you to borrow more) and the rate of repayment. Accordingly, it is not surprising to see that plenty of researches have been conducted in order to develop models for relevant recommendations. Usually, these models include two kinds of roles: (1) assess loan requests and investment offers based on multiple objectives (i.e., forming the search space), and (2) portfolio selection and optimization (i.e., seeking the optimal solution).

- **Loan Requests and Investment Offers Assessment:** Regarding this task, earlier contributions were mainly focused on evaluating the credit risk of borrowers in terms of their static and/or dynamic features. For example, Lin et al. (2017) examined the default risk based on demographic characteristics of borrowers (e.g., gender, marital status, debt-to-income ratio and delinquency history), while Qiu et al. (2012) focused more on loan amount, loan period and acceptable maximum interest rate. Yet, due to the specific working mechanism of P2P lending, matching borrowers' demand directly with investors' supply is different to traditional credit risk evaluation. In a P2P lending scenario, the automatic evaluating and combining systems should not only identify the mutual matched requests and offers, but also need to maximize the platform's total amount of traded funds.

- **Portfolio Selection and Optimization:** The goal of this task is to hold the most desirable group of loans. Indeed, this is extremely crucial for the platform's profitability, since the P2P revenue model mainly depends on the investor's net returns (interest gain) and origination fees charged to borrowers (Oxera, 2016; Kibbe, 2013). To this end, most practices incorporate metaheuristics methods in order to find a trade-off between the quality of the solution space and the computational cost needed to reach a broader spectrum of this goal (Gorgulho et al., 2013).

9.1.1 The Role of Peer-to-Peer (P2P) Lending Platform Operator

In general, the P2P lending platforms are for-profit business entities that act as intermediaries between borrowers (askers) and lenders (investors) via the Internet channel (Schwienbacher and Larralde, 2010; Belleflamme et al., 2015). The term 'P2P lending' refers to a debt-based form of crowdfunding that facilitates an interaction between two parties on a user-friendly online plaza (Mateescu, 2015; PwC, 2015; Renton, 2015; Assetz Capital, 2014). Further information about P2P lending can be found in (Verstein, 2008; Fin and Murphy, 2016; Segal, 2015; Rosenblum et al., 2015).

Typically, the operator of a P2P lending platform plays the following five abstract roles (Lacan and Desmet, in press; Rochet and Tirole, 2003; Nguyen, 2014): (1) Building an information gateway, (2) Cultivating positive network effects, (3) Organizing the contractual relationships between both parties and collecting payments from the contributors depending on the lending types, (4) Implementing advanced assisting technologies, and (5) Supplying support services such as social networking features (Freedman and Jin, 2017) and moral responsibilities (Marom, 2017).

More specifically, the main responsibilities of an operator include conducting credit assessments, administering the loans on behalf of the investors, facilitating the requesting and bidding process, providing legal assurance, and, if a loan is successfully released, operators will coordinate the payment process (Larrimore et al., 2011; Kibbe, 2013; Oxera, 2016). In return, the platform charges borrowers a type of origination fee for each lending transaction, which is proportional to the perceived credit risk for each borrower.

Compared to traditional financial institutes (e.g., banks), P2P lending platforms enjoy several advantages (Milne and Parboteeah, 2016; Kibbe, 2013; Oxera, 2016; Fong, 2015): (1) Connecting individual borrowers and lenders directly, (2) Providing better investment opportunities than traditional bank deposits could offer, coupled with low costs for borrowers, (3) Provisioning credit to several types of borrowers who are otherwise incapable of accessing normal bank lending channels, (4) Behaving more responsibly and carrying greater social value than the incumbent banking system, (5) Delivering improved service quality and speed via technological innovation which benefits both borrowers and lenders, (6) Supplying some additional services, such as early repayment options, speed of funding and loan match recommendation, and (7) Protecting borrowers from embarrassment (which they have often experienced via other channels) due to some degree of anonymity.

Broadly speaking, the P2P lending landscape consists of several business models (i.e., microfinance lending, social investing, marketplace lending, peer-to-business lending and social lending) whose capacities and challenges can be very diverse (Groshoff, 2014).

- **Microfinance Lending:** The main objective of this lending type is to alleviate poverty and encourage the growth of enterprises (particularly micro and small scale) in developing countries. To this end, these platforms with international microfinance institutions and/or non-profit organizations match individual lenders (backers) with low-income entrepreneurs (askers) in developing countries and some selected cities within the developed countries. The platforms enable these askers to receive microfinance without any interest being paid to the backers (see Fig. 9.1 for illustration). For them, the challenge is how to sustain member engagement and increase contributions to the projects (Chen et al., 2017). Examples include ALSOL in Mexico (Barboza and Trejos, 2009), YiLongDai in China (Li and Wang, 2017) and Kiva in the USA (Chen et al., 2017).
- **Social Investing:** Generally speaking, this type refers to investments that focus on social purposes. In fact, it is a variant of microfinance lending, since it allows investors to invest in the bonds of security issuers rather than just loan requests (Lehner, 2013; Reiser and Dean, 2015; Roche-Saunders and Hunt, 2017), as illustrated in Fig. 9.2. Thereby, the investors can get returns varying from 1% to 3%, while the platform can earn a commission from the security issuers (Ashta and Assadi, 2008). Notable examples in this model are Microplace in the USA and Babyloan in France.

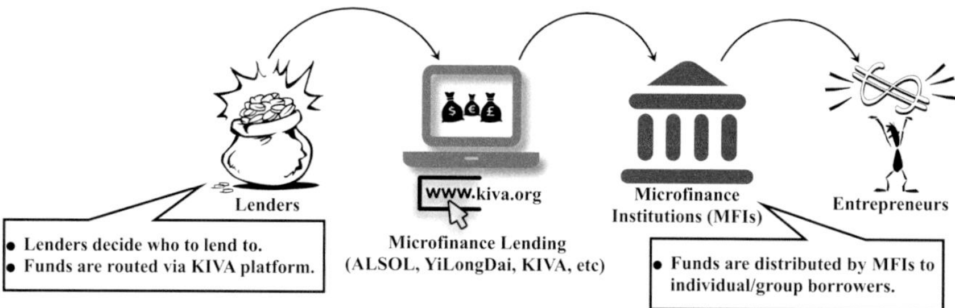

Figure 9.1: Microfinance lending platform (e.g., KIVA) serves as an intermediary role between lenders and the microfinance institutions.

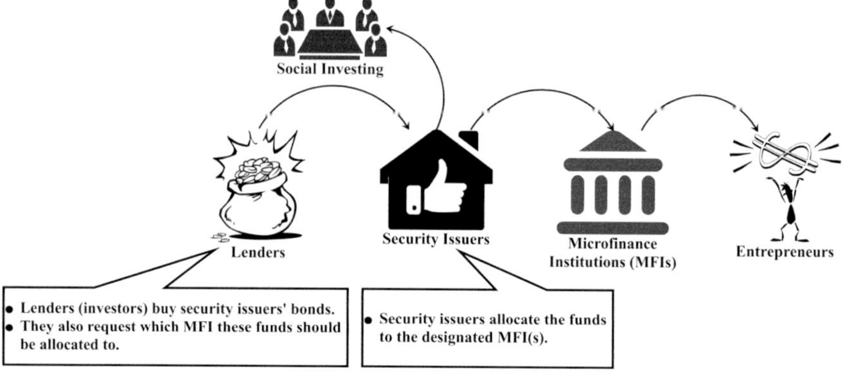

Figure 9.2: Social investing serves as a broker role between lenders and security issuers (a commission fee is charged from security issuers).

In this model, one of the main challenges lies in the ever-increasing pressure to measure and monetize social impact (Martin and Osberg, 2015; Kickul and Lyons, 2015). Another challenge is that the underlying values and rules associated with the enterprises involved in this model are commonly less agreed upon (Brown, 2006). In addition, like their counterparts—microfinance lending, how to attract potential and retain existing contributors is another key challenging issue (Yu, 2011). Interested readers please refer to (Martin and Osberg, 2015; Johnsen, 2017; Laratta et al., 2011; Hines, 2005; Roche-Saunders and Hunt, 2017; Ashta et al., 2015; Ania and Charlesworth, 2015; Calic and Mosakowski, 2016; Parhankangas and Renko, 2017; Rana, 2013) for further information.

- **Marketplace Lending:** With marketplace lending, the main motivation for the investors is not to rely on goodwill but rather a higher financial return. Therefore, these platform operators generally create two different types of value for each side within their marketplace.

For entrepreneurs or individuals (or called askers/borrowers), the operator serves as a mediator to help those who often lack access to public debt markets reach a large group of risk-prone investors; while for investors, it is a place that offers the opportunity of lending to those overlooked borrowers at a high interest rate. A simplified relationship in this regard is illustrated in Fig. 9.3. In this domain, famous platforms range from the Zopa and Funding Circle in UK, through to the Prosper and Lending Club in the USA (Freedman and Jin, 2017; Emekter et al., 2015; Malekipirbazari and Aksakalli, 2015) and to the PPDai and Yooli in China (Chen et al., 2014; Lin et al., 2017).

Accordingly, challenges are also differentiated. For lenders, due to online anonymity and most individual investors' insufficient financial expertise, although a borrower's credit history is available for perusal by all lenders, the classical information asymmetry problem (Marwala and Hurwitz, 2017a) becomes even more severe than in traditional market environments (Klafft, 2008; Lee and Lee, 2012). So, the critical work for these platforms is checking borrowers' default risk and generating adaptable interest rates; while for borrowers, the biggest challenge lies in the privacy risks (e.g., personal data disclosure), i.e., the side effect of inadequate implementation of information security systems by platform operators (Groshoff, 2014). To deal with this challenge, platforms have to protect themselves from attacks. More information regarding P2P networks' security issues can be found in (Selvaraj and Anand, 2012). In addition, P2P lending platforms also have their own management challenges, e.g., notorious herding problem (Herzenstein et al., 2011a; Lee and Lee, 2012; Zhang and Liu, 2012; Berkovich, 2011).

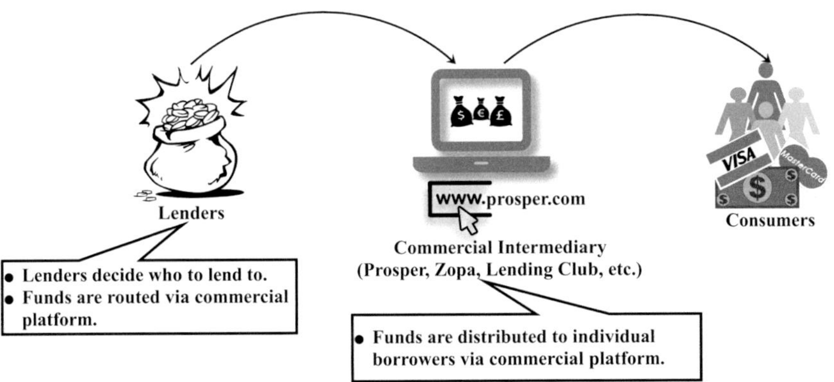

Figure 9.3: Marketplace lending platform (e.g., Prosper, Zop and Lending Club) serves as an intermediary role directly between lenders and individual borrowers.

- **Peer-to-Business Lending:** Similar to marketplace lending, the loans provided by these platforms are not targeting individual borrowers, but are specifically aimed towards small and medium enterprises (SMEs) borrowers (as illustrated in Fig. 9.4). Examples in this category include First Circle and Kabbage in UK.
- **Social Lending:** This model, also called the real P2P movement of funds, can be regarded as a special case of P2P lending, since people who use this platform often know each other (e.g., family and friends) before approaching the platform (Ashta and Assadi, 2008), as depicted in Fig. 9.5. One famous example is Virgin Money whose main objective is to facilitate the fund transactions.

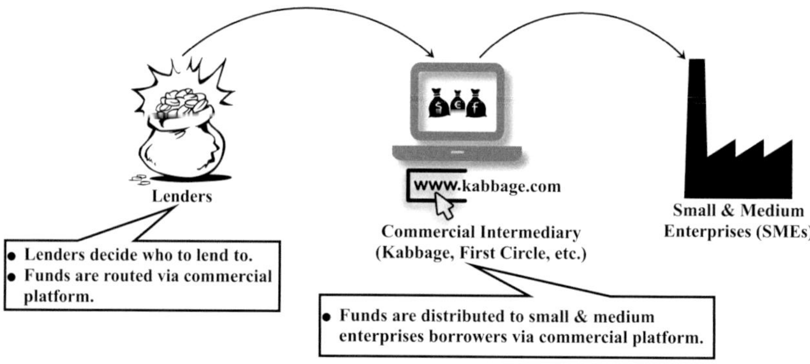

Figure 9.4: Peer-to-business lending platform (e.g., Kabbage, First Circle, etc.) serves as an direct intermediary between lenders and SMEs borrowers.

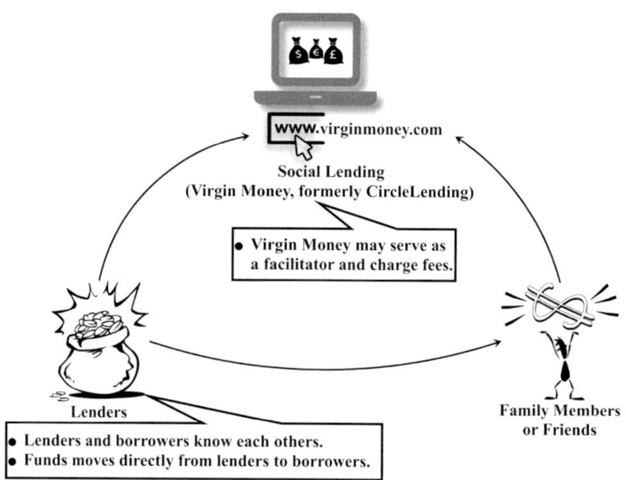

Figure 9.5: Social lending platform (e.g., Virgin Money, etc.) serves as a facilitator between lenders and individual borrowers.

9.1.2 The Peer-to-Peer (P2P) Lending Mechanisms

Nowadays, we find that P2P lending platforms consist of a diverse spectrum of activities. For example, some of the platforms act as a middleman, while others act only as match-makers (De Buysere et al., 2012). Accordingly, not all P2P operators implement the same lending model. Typically, we can divide different lending models into the following two categories (Milne and Parboteeah, 2016):

- **Auction-based Lending (e.g., Zopa, Prosper and Lending Club):** Under this lending model, the mediation of financial institutions is not required. Instead, borrowers (either households or small firms) post their loan requests (e.g., the amount and the maximum interest rate they are willing to pay) and get their current financial situation (e.g., debt-to-income ratio and credit score) information proved online. Meanwhile, lenders indicate the minimum rate they are looking for in order to obtain potential borrowers that can best meet lenders' risk-return preferences (Herzenstein et al., 2011a). When new borrowers enter into the platform, they are matched with lenders who are looking for opportunities to provide loans via the platform. Due to the nature of P2P lending, lenders can work collectively to offer a bulk amount of funds for one single borrower, which means financing a large loan request from the same borrower is often possible without demanding each individual lender to make a significant contribution. In order to achieve this goal, an automatic 'reverse auction' is often conducted by the platform, where the payable interest rate on the loan is gradually increased until the loan can be fully funded by enough bids (Freedman and Jin, 2017). The advantage of this collaborative practice not only offers an opportunity to each investor in terms of diversifying the investment portfolio by selecting distinct loan types, and investing only intended amount of funds, but also enables borrowers to get loan(s) at a lower rate without worrying too much about the collateral (Magee, 2011; Groshoff, 2014).
- **Automatic-Matched Lending (e.g., Kabbage):** Regarding this model, a borrower's loan requests are often first underwritten and completed with the platform or its designated financial partner(s). Then the platform sells the loan to potential interested lenders. The transparency between borrowers and lenders is typically low under this lending model and, thus, leaves less room for lenders to manipulate their investment options (Manbeck and Franson, 2016). In other words, the platforms will take full responsibility in ensuring that the loan and investment characteristics are best matched with each other. Theoretically, the auto-matching functionality could maximally diversify the lenders' loan portfolio by taking factors such as the investors' preferences, borrowers' credit scores, intended use

of funds and other demographic data into account. In practice, delays may occur due to the imbalanced number of lenders and borrowers. One possible solution would be to adjust the interest rate over time in order to attenuate the problem (Milne and Parboteeah, 2016).

9.2 Problem Statement

As we have mentioned earlier in this chapter, any P2P lending platform operator will encounter the following two types of problems to one degree or another:

- **Multi-Objective Loan Assessments Problem:** Typically, every P2P operator uses proprietary loan assessment algorithms, designed to evaluate a borrower's and/or lender's financial and demographic information. Though the foci may vary a lot, say, it was reported that 68% of the time, the employment information will be verified by the Lending Club; while for Prosper, only a sub-set of funded listings will routinely be selected for employment verification purposes (Kibbe, 2013), the basic need remains the same. Thus, the surety of a large set of loan requests must be confirmed in a way that satisfies certain criteria fixed by the operators. In addition, as P2P lending belongs to the class of two-sided platforms, operators should help lenders in finding the loan requests that they are interested in investing in (Choo et al., 2014; Zhao et al., 2014; Belleflamme et al., 2015; Oxera, 2016). Generally, there are the following four streams of topics within this research niche:
  1) *Borrower's characteristics*: Typically, the key factor of a loan check is to control the borrower's loan default probability. To address this, Gonzalez and McAleer (2011) analysed the differences between Prosper and Zopa from several perspectives, such as the requested loan amount, maturity status, interest rate, credit grading and previous borrower experience. Emekter et al. (2015) evaluated credit risk in terms of credit scoring, debt-to-income ratio and FICO (created by Fair Issac Corporation) score. Additionally, a classification model for identifying good borrowers which is based on random forest approach was proposed by Malekipirbazari and Aksakalli (2015). Similarly, Harris (2013) employed support vector machines in order to evaluate the credit risk. Other instances can also be found in (Lin et al., 2017; Guo et al., 2016; Serrano-Cinca and Gutiérrez-Nieto, 2016; Puro et al., 2010; Jiang et al., in press; Cai et al., 2016).
  2) *Loan's description*: In some cases, a high default probability may also be observed among high credit-rated borrowers, based on their financial background information, as suggested by

Malekipirbazari and Aksakalli (2015). Accordingly, additional factors that may potentially lead to loan default must be further introduced by P2P platforms, given the fact that a rich user data set is available now, say, descriptive loan text which is highly related to the loan's viability assessment (Nowak et al., in press). Following this understanding, Dorfleitner et al. (2016) analysed borrowers' description texts in the hope of predicting default probabilities in a P2P lending scenario. Meanwhile, the relationship between a borrower written loan description and the final investors supporting status is also examined by Nowak et al. (in press). Additionally, mining both borrowers' and lenders' motivations is also selected as the study focus of Yan et al. (in press), in which text data was utilized for pinpointing effective loan match recommendations. Further information regarding this topic can be found in (Jiang et al., in press; Wang et al., 2016; Gao and Lin, 2016; Iyer et al., 2016; Herzenstein et al., 2011b; Larrimore et al., 2011; Han et al., 2018).

3) *Lender's traits*: In the real-world, P2P operators should evaluate the availability of lenders' funds as well, in particular, for those individuals belonging to middle- to low-income class who often have relatively insufficient money to make investments (Wang and Yang, 2016). In light of this, the distribution of lender investing preferences over distinct borrower risk classes was examined by Krumme and Herrero (2009). The results suggested that lenders are often prone to overinvesting in high-risk classes, which could result in leverage risk. Meanwhile, the investment performance of SMEs, given the occurrence of loans being denied, was studied by Mueller and Reize (2013), while special attention was paid by Prystav (2016) to the associated consequences of lenders' investment behaviour. Apart from all this, a herding phenomenon in the P2P lending environment is also widely acknowledged among various researchers (Lee and Lee, 2012; Gao and Feng, 2014; Yum et al., 2012; Berkovich, 2011; Zhang and Chen, 2017).
4) *Loan's lender-borrower matching process*: In this branch, Lin and Lo (2006) designed a model that can match loan quality. Three different factors for assessing the potential credit risk were included in their model. In another study, Chen and Song (2013) developed a two-sided matching model for the loan market. With their model, how banks and firms choose each other can be understood to some extent. In addition, Rakesh et al. (2016) proposed a probabilistic model for exploring information (e.g., status of projects, lenders' preference, and the collective preference of the group) in order to solve the mismatch problem between borrowers and lenders.

- **Multi-Objective Portfolio Selection Problem:** A portfolio selection problem can be broadly defined as the process of selecting various components (e.g., loan request, stocks, bonds, projects and/or programs) that will deliver the best possible results (Gorgulho et al., 2013; Elton et al., 2014; Hünseler, 2013; Walker, 2014). Typically, it is constructed diversely due to risk-aversion (Richards Kibbe & Orbe LLP, 2013). For decades, the benefits of diversification have been widely accepted. For example, Haugen (2001) suggested that diversification is an effective way to reduce the risk within a collection of assets. In addition, given the long existing asymmetric information issue in finance markets, scholars emphasized that diversification could decrease the financial intermediation cost (Diamond, 1984) and increase the monitoring motivation (Cesari and Daltung, 2000). Furthermore, portfolio decision should be ambiguity-aversion as well. In this regard, Guidolin and Rinaldi (2013) systematically surveyed the literature and examined the implications of portfolio choice under ambiguity. More studies can be found in (Epstein and Schneider, 2010; Gilboa et al., 2008; Camerer and Weber, 1992).

  Under the umbrella of P2P lending, Guo et al. (2016) claimed that a key challenge in the P2P lending environment is how to effectively allocate investors funds across different loan requests while making accurate assessments of each loan's risk. Xia et al. (2017) pointed out that most of today's loan assessment models focus mainly on determining whether a particular loan request should be funded while leaving the reasonable funding amount, i.e., portfolio selection, mostly untouched. Indeed, as suggested by Emekter et al. (2015), the study on risk management for P2P lending is very useful for any lenders trying to optimize their investment portfolios. Nevertheless, the activities involved in identifying the most suitable asset collection and the optimal asset weightings are often characterized by numerous uncertainties. Using the findings reported by Freedman and Jin (2011) as an example, the likelihood for a lender on Prosper to fund additional loan request(s) is significantly low if most existing loans in this lender's portfolio are already late. When it comes to the portfolio's performance, they also found a great deal of heterogeneity among lenders. Given the fact that an investment decision is rather forward-looking, it would be great to have the associated assets' gains and risk degrees forecasted, rather than simply observed (Michaud and Michaud, 2008; Rachev et al., 2008).

9.2.1 *Question 9.1*

Based on the above observations, Question 9.1 is proposed as follows:

- **Question 9.1:** *How do we automatically assess loan requests and investment offers based on practical needs and help both borrowers and lenders with portfolio selection and optimization?*

In investigating Question 9.1, our expectations are twofold: (1) Since P2P lending represents a business opportunity and in many cases a means to promote profit efficiency while reducing the risk, the first goal is, thus, to explore an optimal combination result which can satisfy all involved parties, i.e., borrowers, lenders and platform operators; (2) Essentially, portfolio selection and optimization is a process that fouses on finding the optimal capital weights for a basket of investments in order to obtain the highest gain with the least amount of risk (Hsu, 2014). Therefore, our second goal in this chapter is to understand how portfolio selection and optimization can be accomplished under certain fuzzy settings.

9.3 Metaheuristics for Question 9.1

In order to address Question 9.1, a pioneer system was built by Martinho (2009) in order to combine the loan requests and investment offers in the P2P lending scenario. In this section, we first introduce some background knowledge about the employed metaheuristics (Xing et al., 2013; Xing and Gao, 2014d; Xing and Gao, 2014f; Xing and Gao, 2014aa; Xing and Gao, 2014bb; Xing and Gao, 2014cc; Xing and Gao, 2014dd; Xing and Gao, 2014g; Xing and Gao, 2014h; Xing and Gao, 2014i; Xing and Gao, 2014j; Xing and Gao, 2014k; Xing and Gao, 2014l; Xing and Gao, 2014m; Xing and Gao, 2014n; Xing and Gao, 2014o; Xing and Gao, 2014p; Xing and Gao, 2014q; Xing and Gao, 2014r; Xing and Gao, 2014s; Xing and Gao, 2014t; Xing and Gao, 2014u; Xing and Gao, 2014v; Xing and Gao, 2014w; Xing and Gao, 2014x; Xing and Gao, 2014y; Xing and Gao, 2014z) algorithm, which is then followed by a detailed description of the system construction.

9.3.1 Search and Optimization—Hill Climbing

Hill climbing (HC) is essentially a simple loop that keeps moving in the direction of increasing value, i.e., uphill (Lucci and Kopec, 2016; Marwala and Lagazio, 2011b). The algorithm stops when a peak value is reached and no higher value is found around the neighbourhood. Since there is no search tree being maintained by the algorithm, the current node's data structure only takes the state and the objective function's value into account. Accordingly, only the current state's immediate neighbours are within the scope of the algorithm. Although it is often called a greedy local search, HC often performs well. The reason for it to make a fast move towards a solution lies in the fact that the algorithm often easily moves away from an inappropriate state (Russell and Norvig, 2010; Flasiński, 2016).

A successful implementation of HC algorithm mostly depends on the state space landscape's shape: (1) If the existence of local maxima and plateaux is scarce, the algorithm which is driven by random mechanism often gets a good solution very quickly; (2) However, many real-world problems are notoriously characterized by the so-called NP-hard nature, which means the algorithm can easily get stuck on an exponential number of local maximum traps.

Let us consider a simple maze example, the candle's flame flickers faster as we get closer to the exit. Therefore, the heuristic function (denoted by h) can be defined as follows (Flasiński, 2016):

- $h(n)$ = **5:** The value of five means that the flame stays stable;
- $h(n)$ = **4:** The value of four means that the flame starts to flicker a bit;
- $h(n)$ = **3:** The value of three means that the flame is flickering at low frequency;
- $h(n)$ = **2:** The value of two means that the flame is flickering at high frequency;
- $h(n)$ = **1:** The value of one means that the flame is flickering at a very high frequency; and
- $h(n)$ = **0:** The value of zero means that the flame is blown out (i.e., the exit has been reached).

Since the process of a HC search incorporates expanding the nodes that are the best in terms of the heuristic function's value, it will take us four steps in order to find the desired solution (as illustrated in Fig. 9.6).

9.3.2 *Search and Optimization—Simulated Annealing*

In materials science and engineering (Callister and Rethwisch, 2012), annealing is a term that refers to a type of heat treatment during which a material is exposed to a raised temperature for a deliberately prolonged duration of time, this is followed by a slow cooling period. The goal of the annealing process is typically threefold (Callister and Rethwisch, 2012): (1) Stresses relief, (2) Softness, ductility or toughness improvement, and/or (3) Specific microstructure creation. In principle, any annealing process is comprised of three phases (Callister, 2001): (1) Get the temperature of the material heated up to the desired degree, (2) Keep the material steady or "soaking" at that temperature, and (3) Cool the material down to room temperature.

As the name implies, the simulation of this annealing process is termed as simulated annealing (SA) (Černý, 1985; Kirkpatrick et al., 1983; Marwala, 2014b) where the faultless crystal state corresponds to the energy configuration in its global minimum situation. The similarity between an

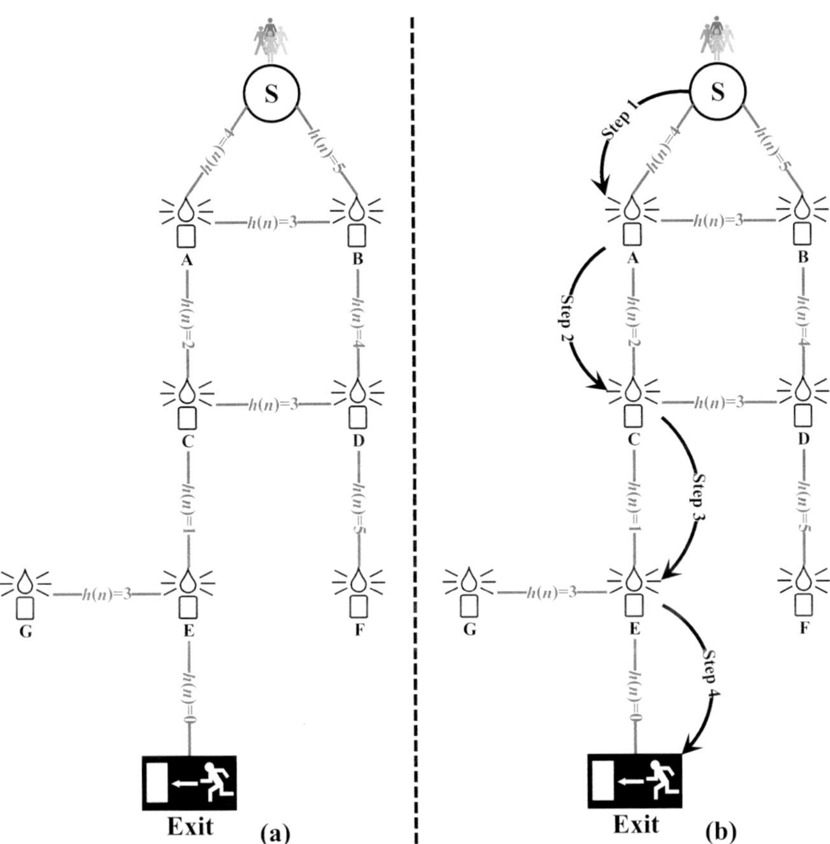

Figure 9.6: The maze problem: (a) heuristic function's values; and (b) the process of HC search.

optimization process and a SA can be stated as follows (Du and Swamy, 2016): (1) Targeted optimization problem solutions are represented by physical material states, (2) A solution's cost is indicated by a state's energy, and (3) Controlling parameter is mimicked by the manipulated temperature.

In the first instance, the Metropolis algorithm (Metropolis et al., 1953) was proposed as an approach to simulate the evolution of a solid's thermal equilibrium process under a given temperature. As one of the Metropolis algorithm's variants, the temperature setting in SA has been changed to vary from high to low (Kirkpatrick et al., 1983). In essence, SA consists of two fundamental stochastic processes (Du and Swamy, 2016), that is, one is responsible for solution generation, while the other one is taking charge of solution acceptance.

9.3.2.1 Basic Simulated Annealing

Based on statistical thermodynamics, at the absolute temperature T, the probability of a physical system being in α state and containing E_α amount of energy satisfies the Boltzmann distribution, as given by Eq. 9.1 (Hardy and Binek, 2014; Du and Swamy, 2016):

$$\Pr_a = \frac{1}{Z} e^{\left(\frac{-E_\alpha}{k_B} T\right)}. \qquad 9.1$$

where the Boltzmann's constant is denoted by k_B, and Z represents the partition function which can be given by Eq. 9.2 (Hardy and Binek, 2014; Du and Swamy, 2016):

$$Z = \sum_\beta \left(e^{\frac{-E_\alpha}{k_B} T} \right). \qquad 9.2$$

In SA, the Boltzmann's constant is eliminated. At the beginning, i.e., temperature T is high, small changes associated with the energy are typically ignored by the system, which makes it approach thermal equilibrium quickly. Under this situation, SA performs a rather rough search across the global states' space and identifies a good enough minimum. Then when temperature T is gradually lowered, the system starts responding to small energy changes by initiating a fine search around the neighbourhood of the already identified good enough minimum. During this process, a better minimum is determined. Finally, when the temperature T reaches zero, any state change of the system will no longer cause an energy increase, which can, thus, keep the system at an equilibrium situation. Accordingly, a state change's probability is decided by the Boltzmann distribution of two states' energy difference, as given by Eq. 9.3 (Du and Swamy, 2016):

$$\Pr = e^{\left(-\frac{\Delta E}{T}\right)}. \qquad 9.3$$

9.3.2.2 Cooling Scheme

Also known as Boltzmann annealing, the cooling plan for temperature T is crucial for the efficacy of the basic SA (Du and Swamy, 2016; Marwala, 2009; Marwala, 2010c). This is also the case when processing materials (Callister and Rethwisch, 2012): (1) On the one hand, if T is lowered too quickly, it may lead to a premature convergence on a local minimum (warping or cracking phenomenon observed in physical annealing scenario); (2) On the other hand, if T is reduced too slowly, the whole algorithm suffers from a low speed towards convergence. By analysing the SA process via Markov chain theory (Marwala et al., 2017; Xing and Marwala, 2018h; Revuz, 1984), an easy but necessary and sufficient setting of the cooling schedule is proposed in (Geman and Geman, 1984), that is, T has to be lowered based on the condition given by Eq. 9.4 (Du and Swamy, 2016; Geman and Geman, 1984):

$$T(t) \geq \frac{T_0}{\ln(1+t)}, \quad t = 1, 2, \ldots. \tag{9.4}$$

where T_0 represents a large enough temperature in the beginning.

9.3.3 Search and Optimization—Genetic Algorithm

Genetic algorithm (GA) (Holland, 1992; Vose, 1999; Reeves and Rowe, 2002) represents a particular category of evolutionary computation techniques that is typically utilized for searching and optimizing tasks. In brief, GA is featured by maintaining a group of search points instead of a single point. The corresponding comparisons and interactions among these points lead to the system's evolution (Wegener, 2001; Neumann and Witt, 2010; Auger and Doerr, 2011). A well-cited introduction to GA has been written by (Melanie, 1996). In the past decades, GA has proved itself in many applications, such as finite element updating, remanufacturing, economic modelling and modelling interstate conflict (Forrest and Mitchell, 2016; Xing and Marwala, 2018c; Marwala, 2010a; Xing and Gao, 2014a; Xing and Gao, 2014hh).

9.3.3.1 Genetic Algorithm (GA) Framework

One of the key mechanisms of GA is a process whereby the population is updated in discrete iteration manner. At the beginning, GA is often equipped with a randomly created population, which often serves as parents. Descendants are then generated from the parent pool. Once this process is completed, the next generation is typically produced. In general, there are two fundamental schemes for accomplishing this: The generational scheme and the steady-state scheme (Rowe, 2015).

- **Generational Scheme:** This approach tends to use the parent population in order to generate descendants repetitively until a completely new population is formed. Under this scheme, all descendants are produced within one generation. The general procedure of the generational GA is as follows (Rowe, 2015).
  1) *Initialization*: Use the points from the search space to randomly form the initial population with the size of μ.
  2) *Iteration*: Select a point within the population, modify the point via mutation and crossover mechanisms and add the resultant descendant into the new population. Repeat these steps until the termination criterion is met.
  3) *Stop*.

- **Steady-State Scheme:** In contrast to the generational scheme, this approach tends to employ the current parents to produce a single descendant, and then get it inserted into the population by substituting some other individual. Therefore, a generation under this scheme refers to the creation of a new single solution. The general flowchart of the steady-state GA is as follows (Rowe, 2015).
  1) *Initialization*: This step is the same as the generational scheme, that is, use the points from the search space to randomly form the initial population with the size of μ.
  2) *Iteration*: Select a point within the population, modify the point via mutation and crossover mechanisms, identify an existing member from the population and use the newly generated descendant to substitute such existing member. Repeat all this until the termination criterion is met.
  3) *Stop*.

As we can see, in either instance the population size remains unaltered throughout the procedure. An important parameter of the GA algorithm is a good selection of a population worth investigating. Some general principles were summarized in (Jansen and Wegener, 2001). Meanwhile, a more detailed comparison between generational and steady-state GAs can be found in (Rowe, 2015). Overall, a successful GA design depends on the following crucial factors (Rowe, 2015):

- **Selection Strategy:** This factor determines the manner in which points are selected from the current population. For general optimization problems, the selection strategy favours better solutions.
- **Mutation Strategy:** In essence, this factor will bring minor variations into the current selected solutions.
- **Crossover Strategy:** Though it is also used to modify the selected points, as mutation strategies do, this factor typically constructs a better solution by combining bits and pieces from various good solutions.
- **Replacement Strategy:** Apart from the selection, mutation and crossover operators are the common threads found in both cases; there is a special factor posed by steady-state GA in terms of identifying the way to select an individual to be replaced, i.e., replacement strategy.

9.3.3.2 *Selection Strategy*

As a determinant factor for GA in guiding its search process towards better solutions, selection strategies are often defined via fitness functions, which distribute a positive value to every point within the search space, and the solution with the maximum/minimum fitness degree is selected as the optimal one. More often than not, the targeted problem's objective function can serve as the fitness function. In the literature, a couple of selection strategies have been proposed (Rowe, 2015) but only some of them are described in this section.

- **Proportional Selection:** Under this strategy, the probability for an individual (denoted by x) to be chosen is treated as being proportional to their own fitness degree within the whole population, as given via Eq. 9.5 (Rowe, 2015; Goldberg, 1989):

$$\Pr = \frac{f(x)}{\sum_{y} f(y)}. \quad 9.5$$

  where the sum is for all population members. In practice, a roulette wheel (Marwala, 2009; Xing and Gao, 2014b; Marwala, 2013b; Goldberg, 1989) is often employed in order to implement this paradigm.

- **Stochastic Universal Sampling Selection:** Since one often has to get μ individuals chosen from the population for the purpose of finalizing one generation, a significant $O(\mu^2)$ running time is required if proportional selection strategy is employed. Under this circumstance, stochastic universal sampling (Baker, 1987) selection strategy can be used as an alternative. If the whole population's total fitness degree is denoted by T, then we can have Eq. 9.6 (Rowe, 2015):

$$E[i] = \frac{f(i)}{T}\mu. \quad 9.6$$

  where the expected number of item i's copies is denoted by $E[i]$. This selection strategy can make sure that either $\lfloor E[i] \rfloor$ or $\lceil E[i] \rceil$ copies of item i are chosen. The running time can, thus, be reduced to $O(\mu)$. Please refer to (Rawlins, 1992; Sedgewick and Wayne, 2011) for more information about notation.

- **Scaling Method Facilitated Selection:** When GA is approaching the optimum point, the population's diversity degree is considerably less than it was in its initial stage, that is, the fitness values may often stay unchanged. To address this issue, one solution is to scale the fitness function somehow via, say, sigma scaling (Melanie, 1996; Goldberg, 1989). Suppose we have the original fitness function as $f: X \rightarrow \mathbb{R}$, the scaled version can be given by Eq. 9.7 (Rowe, 2015):

$$h(x) = 1 + \frac{f(x) - \bar{f}}{2\sigma}. \quad 9.7$$

  where the standard deviation and the average of the fitness within the population are denoted by σ and $\bar{f}$, respectively.

- **Rank Selection:** In order to mitigate the drawback of using probability to select individuals, which is rather sensitive to the fitness function's relative scale, we can resort to a ranking method in order to sort the population, with the best one getting a score μ while the worst one only receives a score of one (Whitley, 1989). A common practice in this regard is to linearly scale the rank to reach a score between two numbers (a and b) as given via Eq. 9.8 (Rowe, 2015):

$$h(i) = \frac{(b-a)r(i) + \mu a - b}{\mu - 1}. \tag{9.8}$$

where $r(i)$ stands for an item i's rank within the population. Items can, thus, be selected in proportion to their h-value, as given by Eq. 9.9 (Rowe, 2015):

$$\Pr = \frac{2\left[(b-a)r(i) + \mu a - b\right]}{\mu(\mu-1)(b+a)}. \tag{9.9}$$

9.3.3.3 *Mutation Strategy*

The function of mutation is to facilitate GA in order to perform new points sampling within the search space. The underlying concept of mutation is to produce a variant with the population that might even be better than the present best member of the population. In a general manner, mutation strategy can be regarded as first selecting a representation for the search space's points, and then using a group of operators to work on that representation. Suppose we have n bits in a representation, it will in turn offer us n corresponding operators. For binary strings, one of the most common mutation practices is to flip each bit randomly and independently based on a fixed probability u, which is often called the mutation rate (Rowe, 2015). This is equivalent to flipping a sub-set of bits (size k) according to the probability given via Eq. 9.10 (Rowe, 2015):

$$u^k(1-u)^{n-k}. \tag{9.10}$$

Typically, the mutation rate is set as $\frac{1}{n}$, while for other scenario dependent settings, one can refer to (Rowe, 2015; Jansen and Wegener, 2010; Witt, 2012; Böttcher et al., 2010).

9.3.3.4 *Crossover Strategy*

The core of this strategy is to combine parts of two distinct solutions (often good parts) so as to form a third solution (hopefully better). In the literature, several paradigms exist for executing crossover strategy. Take a binary strings-type of representation, there are the following three typical options (Rowe, 2015):

- **One-Point Crossover:** This paradigm randomly selects a bit location and combines two types of bit values, that is, all the bit values below this location from one parent, and all the remaining bit values from the other parent, as described by Eq. 9.11 (Rowe, 2015):

$$\begin{array}{lcccccccc} & 0 & 1 & 0 & 0 & \boxed{1} & 1 & 0 & 1 \\ \text{Parents} & 1 & 1 & 1 & 0 & \boxed{0} & 1 & 1 & 1 \\ & & & & & \Downarrow & & & \\ \text{Descendant} & 0 & 1 & 0 & 0 & \boxed{0} & 1 & 1 & 1 \end{array} . \tag{9.11}$$

  where the fifth location is selected as our cutting point.

- **Two-Point Crossover:** Likewise, two-bit locations are randomly chosen under this paradigm. The bit values between these two cutting points are inherited from one parent, while the other parent contributes to the remaining values, as described by Eq. 9.12 (Rowe, 2015):

$$\begin{array}{lcccccccc} & 0 & \boxed{1} & 0 & 0 & 1 & \boxed{1} & 0 & 1 \\ \text{Parents} & 1 & \boxed{1} & 1 & 0 & 0 & \boxed{1} & 1 & 1 \\ & & & & & \Downarrow & & & \\ \text{Descendant} & 0 & \boxed{1} & 1 & 0 & 0 & \boxed{1} & 0 & 1 \end{array} . \tag{9.12}$$

  where the second and the sixth locations are selected as the cutting point.

- **Uniform Crossover:** Since both of the aforementioned types of crossover strategies impose bias on the ordering of bits, one can use uniform crossover which can randomly select bit values from either parent. In general, a bit string (the so-called mask) is first produced in a random manner, then the mask's values determine which parents the bit values should be taken from, as described by Eq. 9.13 (Rowe, 2015):

$$\begin{array}{lcccccccc} \text{Parents} & 0 & 1 & 0 & 0 & 1 & 1 & 0 & 1 \\ & 1 & 1 & 1 & 0 & 0 & 1 & 1 & 1 \\ \text{Mask} & 0 & 1 & 0 & 1 & 0 & 1 & 0 & 1 \\ & & & & & \Downarrow & & & \\ \text{Descendant} & 0 & 1 & 0 & 0 & 1 & 1 & 0 & 1 \end{array} . \tag{9.13}$$

9.3.3.5 Replacement Strategy

As we have mentioned earlier in this chapter, there is a need to develop an approach by which the steady-state algorithm can insert a new descendant solution into the population and in the meantime replace one of the existing population members. Though different replacement strategies enjoy distinct strengths, most of them can find their roots in matters discussed regarding selection strategy. This section briefly discusses one of them; other useful strategies can be found in (Rowe, 2015).

- **Inverse Selection:** As a replacement strategy, inverse selection is essentially based on selection methods. By establishing an inverse fitness-proportional replacement paradigm, one can use the fitness degree in order to determine the probability of a member being replaced. In one instance, the existing global optimal value can be used in order to subtract the fitness, which can guarantee that the already found optimum will never be replaced. Under this circumstance, such probability is given via Eq. 9.14 (Rowe, 2015):

$$\Pr = \frac{f^{*} - f(i)}{\mu f^{*} - \sum_{j} f(j)}. \qquad 9.14$$

where the optimal fitness value is denoted by f^{*}.

9.3.4 *Search and Optimization—Particle Swarm Optimization*

The attempt at utilizing many autonomous particle agents that can not only act collectively in simple means, but be able to generate surprisingly complex emergent behaviour, can be traced back to the computer graphics (Hughes et al., 2014) domain, in which liquids or gases (fuzzy objects), e.g., smoke/smog, clouds/fog, fire/light, and water/liquid, are often modelled as particle systems (Reeves, 1983). Typically, a particle system consists of a group of individual particle agents, each of which can be defined as a single point mass.

Based on the studies concerned with observing bird flocks, fish schools and ant colonies, PSO was proposed in (Kennedy and Eberhart, 1995; Kennedy and Eberhart, 2001; Clerc and Kennedy, 2002) in order to involve a population of particles which behave akin to social norms found in swarms. Because of its uncomplicated implementation procedure, PSO enjoys a great popularity in many domains where its potential for fast convergence to a good enough solution is well demonstrated (Marwala, 2014a; Marwala, 2009; Marwala, 2013b; Marwala and Lagazio, 2011a; Marwala, 2005; Marwala, 2010b).

9.3.4.1 Basic Particle Swarm Optimization (PSO)

For an optimization problem with n variables, one can define a population of particles (denoted by N_p). Within the targeted n-dimensional search space, a random position is allocated to each particle, forming a potential candidate solution. Each particle possesses its own trajectory, namely, position and velocity (denoted by $\mathbf{x}_i$ and $\mathbf{v}_i$, respectively), and explores the search space by continuously updating its trajectory. The entire population of particles adjust their corresponding trajectories by comparing the best locations between what they have discovered so far, i.e., personal best, denoted by $\mathbf{x}_i^*$ (for $i = 1,...,N_p$), and what other particles have visited as well, i.e., global best, denoted by $\mathbf{x}^{Gbest}$. All particles have intrinsic fitness values, which are subject to evaluation by the fitness function to be optimized. The current optimal particles will lead the remaining particles through the solution space. For every generational iteration, the best solution found by the population is referred to as the local best.

In general, at the $(t + 1)$th iteration, the particle swarm of the basic PSO is updated via Eq. 9.15 (Kennedy and Eberhart, 1995):

$$\begin{aligned} \mathbf{v}_i(t+1) &= \mathbf{v}_i(t) + cr_1\left[\mathbf{x}_i^* - \mathbf{x}_i(t)\right] + cr_2\left[\mathbf{x}^{Gbest}(t) - \mathbf{x}_i(t)\right], \\ \mathbf{x}_i(t+1) &= \mathbf{x}_i(t) + \mathbf{v}_i(t+1), \end{aligned} \quad i = 1,\ldots,N_p. \tag{9.15}$$

where the acceleration constant is indicated by c, which is greater than zero, and two uniform random numbers that both belong to [0,1] are denoted by r_1 and r_2, respectively. One drawback of this basic version of PSO is that it cannot control the velocity's magnitude, which may cause population explosion and divergence. One possible solution is to set a threshold value (denoted by v_{max}) on $\mathbf{v}_i$'s absolute values (Du and Swamy, 2016).

9.3.4.2 Neighbourhood Topology

As a key factor of PSO, the layout of the neighborhood topology determines how social information is shared within a swarm of particles. There are typically three popular neighbourhood topologies found in the literature (Du and Swamy, 2016; Kiranyaz et al., 2014), namely, von Neumann, global best (Gbest), and local best (Lbest), as illustrated in Fig. 9.7.

- **Fully Topology (or Gbest):** Among these three topologies, Gbest topology is often employed by the basic PSO, in which all particles within the group can obtain the information regarding the globally discovered best solution(s). In other words, under this layout, the entire flock makes up the neighborhood and each individual particle serves as the neighbor of every other remaining particle.
- **Ring Topology (or Lbest):** There is a ring lattice configuration in the Lbest topology, that is, each particle's neighborhood consists of itself and two (or more) of its immediate neighbors which are often selected according to the associated adjacent indices rather than other factors, i.e., physical locations.
- **von Neumann Topology:** Under this layout (two-dimensional lattice structure with all four sides wrapped), there are four neighbors

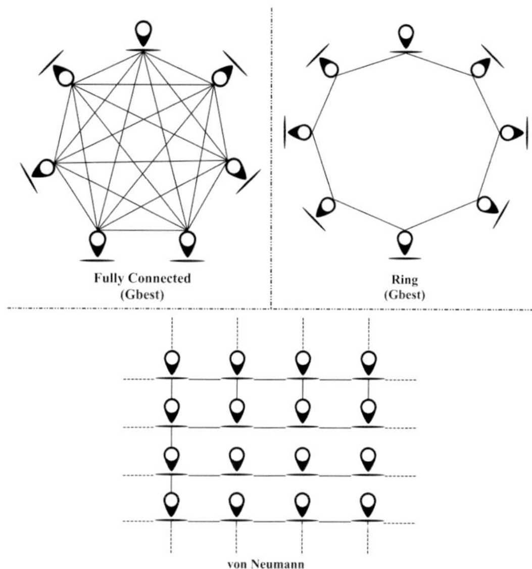

Figure 9.7: An illustration of different neighborhood topologies.

belonging to each particle which itself is located at the center of its four neighbors.

By testing PSO under different network topology configurations, the suggestions given in (Kennedy and Mendes, 2002; Kennedy, 1999) are: for complex problems, it is better to restrict PSO with a small neighbourhood number; while for relative simple problems, one can relax such constraint by incorporating a large neighbourhood number. In general, the von Neumann network layout often outperforms the other two layouts, i.e., Gbest and Lbest (Kennedy and Mendes, 2002). Regarding other PSO improvements and variations, please refer to (Marwala, 2015; Marwala, 2013a; Kiranyaz et al., 2014) for more details.

9.3.5 *Portfolio Optimization in Stock Market Scenario*

Before we jump to the explanation of the system construction for automatically combining loan requests and investment offers, let us first explore how portfolio selection can be optimized under fuzzy settings.

9.3.5.1 Markowitz's Model

By using the historical returns' mean and variance, Markowitz's model (Marwala and Hurwitz, 2017b; Markowitz, 1952; Markowitz, 1959; Markowitz, 1991; Fernholz, 2002; Elton et al., 2014) intends to measure a portfolio's expected return and associated risk within a stock market.

It can, thus, be formulated as a multi-objective problem, as given by Eqs. 9.16–9.19 (Nayebpur and Bokaei, 2017):

$$R_p = \sum_{i=1}^{i=N} r_i \cdot x_i \,. \tag{9.16}$$

$$\sigma = \sqrt{\sum_{i=1}^{i=N} \sum_{j=1}^{j=N} x_i \cdot x_j \cdot \text{cov}_{ij}} \,. \tag{9.17}$$

$$\sum_{i=1}^{i=N} x_i = 1 \quad 0 \le x_i \le 1 \quad i = 0, 1, \ldots, m \,. \tag{9.18}$$

$$\sum_{j=1}^{j=N} x_j = 1 \quad 0 \le x_j \le 1 \quad j = 0, 1, \ldots, m \,. \tag{9.19}$$

where R_p represents the portfolio's return and can be treated as the weighted average of the stocks' (contained in the portfolio) returns; the deviance of a stock's return from the mean can be defined as σ which indicates the portfolio's risk; i stands for the stock index; N denotes the total stock number; r_i and r_j indicate the ith and the jth stock's return, respectively; and x_i and x_j represent the percentage of the ith and the jth stock in the portfolio, respectively.

9.3.5.2 *Fuzzy Synthetic Evaluation and Genetic Algorithm (GA) for Portfolio Selection and Optimization*

In practice, determining the proper weights associated with invested assets plays a crucial role for every investor (Sefiane and Benbouziane, 2012). In order to address the issue of calculating criteria importance, an approach combining fuzzy synthetic evaluation (FSE) and GA was recently proposed in (Nayebpur and Bokaei, 2017).

- **FSE System:** In order to take the firm's performance into account, a model previously proposed in (Edirisinghe and Zhang, 2007) was borrowed in (Nayebpur and Bokaei, 2017). Briefly, the model (denoted by O_0) covers the following six main criteria category (Nayebpur and Bokaei, 2017): (1) Profitability O_1, (2) Utilization O_2, (3) Liquidity O_3, (4) Leverage O_4, (5) Valuation O_5 and (6) Growth O_6. Through a hierarchical decomposing process, the grand objective can be first broken down to several more specific objectives, which, in turn, form the macro layer. Furthermore, different sets of decision questions from the micro layer serve as determinants for the macro layer.
- **FSE Process:** By using fuzzy mathematics, FSE is able to transform unclear information (Kuo and Chen, 2006). In (Nayebpur and Bokaei,

2017), the FSE process was designed in order to incorporate the following six main steps:

1) *Evaluation criteria set formulation*: $U = \{u_i\},\ i = 1,2,\ldots,m$;
2) *Evaluation criteria grade set formulation*: $V = \{v_j\},\ j = 1,2,\ldots,n$ and each grade is based on a five-point Likert scale, that is, very low, low, average, high and very high.
3) *Membership function matrix calculation*: The assessment matrix can be formulated as Eq. 9.20 (Nayebpur and Bokaei, 2017):

$$\tilde{R} = \left[r_{ij}\right]_{n\times m} = \begin{bmatrix} r_{11} & r_{12} & \cdots & r_{1n} \\ r_{21} & r_{22} & \cdots & r_{2n} \\ \vdots & \vdots & \vdots & \vdots \\ r_{m1} & r_{m2} & \cdots & r_{mn} \end{bmatrix}. \qquad 9.20$$

4) *Weight of criteria determination*: At this step, GA was utilized in (Nayebpur and Bokaei, 2017) in order to determine the effective criteria's weight (denoted by $\tilde{w}$).
5) *Critical vector acquirement*: In (Nayebpur and Bokaei, 2017), a fuzzy operator was employed in order to acquire a key vector of FSE (denoted by $\tilde{B}$) as given by Eq. 9.21 (Nayebpur and Bokaei, 2017):

$$\tilde{B} = \tilde{W} \circ \tilde{R}. \qquad 9.21$$

where the symbol of $\circ$ represents the fuzzy composition operator, which often has the following four types (Feng and Xu, 1999a; Feng and Xu, 1999b; Mordeson and Nair, 2001): Type 1 operator, where only crucial factors are considered. It is, therefore, suitable for scenarios in which an individual element needs to be emphasized. The mathematical formulation of Type 1 operator is given via Eq. 9.22 (Nayebpur and Bokaei, 2017; Feng and Xu, 1999a; Feng and Xu, 1999b; Mordeson and Nair, 2001):

$$M(\Lambda, \mathrm{V}), \quad b_j = \overset{m}{\underset{i=1}{\mathrm{V}}}\left(a_k \Lambda r_{kj}\right) = \max\left\{\min\left(a_k, r_{kj}\right)\right\} \quad 1 < k < m. \qquad 9.22$$

Type 2 operator, where crucial factors are emphasized, and several non-major factors are covered as well, i.e., a finer solution than type 1 operator, is mathematically formulated in Eq. 9.23 (Nayebpur and Bokaei, 2017; Feng and Xu, 1999a; Feng and Xu, 1999b; Mordeson and Nair, 2001):

$$M(\cdot, \mathrm{V}), \quad b_j = \overset{m}{\underset{i=1}{\mathrm{V}}}\left(a_k \cdot r_{kj}\right) = \max\left\{a_k, r_{kj}\right\} \quad 1 < k < m. \qquad 9.23$$

Type 3 operator is slightly different to type 2 operator, which make it suitable for evaluating indistinguishable results acquired from Type 2.

The mathematical formulation of Type 3 operator is given via Eq. 9.24 (Nayebpur and Bokaei, 2017; Feng and Xu, 1999a; Feng and Xu, 1999b; Mordeson and Nair, 2001):

$$M(\Lambda,\oplus), \quad b_j = \oplus\left(a_k \Lambda r_{kj}\right) = \sum_{k=1}^{m} \min\left(a_k, r_{kj}\right). \tag{9.24}$$

Finally, type 4 operator takes each individual factor into account, that is, all criteria are involved according to their corresponding weight coefficients. The mathematical formulation of Type 4 operator is given via Eq. 9.25 (Nayebpur and Bokaei, 2017; Feng and Xu, 1999a; Feng and Xu, 1999b; Mordeson and Nair, 2001):

$$M(\cdot,+), \quad b_j = \sum_{k=1}^{m} \min\left(a_k, r_{kj}\right). \tag{9.25}$$

6) *Best fuzzy composition operation determination*: In (Nayebpur and Bokaei, 2017), the operator is selected based on whether it can minimize the fitness score, given by Eq. 9.26 (Nayebpur and Bokaei, 2017):

$$\text{Fitness Score} = \frac{\text{Computed overall}}{\text{performance evaluation}} - \frac{\text{Surveyed overall}}{\text{performance evaluation}}. \tag{9.26}$$

- **GA Optimizing Criteria Weight**: In (Nayebpur and Bokaei, 2017), Eq. 9.27 is employed for computing the associated weight of each criterion:

$$\begin{gathered} \min e(S) = \sqrt{\sum_{i=1}^{m} \left(d \cdot w_i - o_i\right)^2} \\ \sum_{i=1}^{m} w_i = 1, \quad 0 \le w_i \le 1 \quad i = 1, 2, \ldots, m \end{gathered}. \tag{9.27}$$

where $e(S)$ denotes the Euclidean distance between two overall performance evaluations, i.e., computed and surveyed, o_i indicates the computed overall performance evaluation, d represents the surveyed overall performance evaluation and S stands for parameters set.

9.3.5.3 *Summary*

Though having some limitations (e.g., simple fitness function, limited benchmark comparison, etc.), the work conducted in (Nayebpur and Bokaei, 2017) proved the effectiveness of employing GA and FSE to address two conflicting factors (i.e., return and risk) often encountered in portfolio selection scenarios. For their particular testing environment, Tehran stock market, the proposed method was able to rank the six main criteria in the following sequence: O_5 (valuation), O_6 (growth), O_1 (profitability), O_3 (liquidity), O_2 (utilization) and O_4 (leverage).

9.3.6 Loan Requests and Investment Offers Combination in Crowdfunding Scenario

The system proposed in (Martinho, 2009) can be briefly described as follows.

9.3.6.1 Problem Formulation

The targeted problem, $P = (S, f)$, can be formulated as a generic optimization problem by specifying the following elements (Martinho, 2009):

- **Parameters Set:** Various parameters included in this set are outlined below (Martinho, 2009):
  1) N : the number of participating lenders;
  2) M : the number of participating borrowers;
  3) $Rate_{\min_i}$: the minimum rate for the *i*th lender to give out the money;
  4) $Rate_{\max_j}$: the maximum rate for the *j*th borrower to accept the money;
  5) $Amount_{\min_i}$ and $Amount_{\max_i}$: the minimum and maximum amount of money that the *i*th lender is able to lend; and
  6) $Amount_{\min_j}$ and $Amount_{\max_j}$: the minimum and maximum amount of money that the *j*th borrower intends to borrow.
- **Decision Variables Set:** The set is defined as $X = \{rate_{11}, amount_{11}, rate_{12}, amount_{12}, \ldots, rate_{ij}, amount_{ij}, \ldots, rate_{NM}, amount_{NM}\}$, where the meanings of notations are outlined as below (Martinho, 2009):
  1) $rate_{ij}$: The rate variable stands for the rate at which the *i*th lender has decided to lend the funds to the *j*th borrower;
  2) $amount_{ij}$: The amount variable stands for the amount of money that the *i*th lender has agreed to lend to the *j*th borrower.
- **Decision Variables Domains:** In (Martinho, 2009), the domains for rate and amount were defined as follows:
  1) **dom**($rate_{ij}$) : The domain of rate falls within the interval of [0,1], where the maximum rate is only allowed to reach 100%;
  2) **dom**($amount_{ij}$) : The domain of amount falls within the interval of [0, $Amount_j$], where the maximum amount requested by the *j*th borrower is denoted by $Amount_j$.
- **Constraints:** The constraints considered in (Martinho, 2009) are as follows:
  1) Constraint 1 – The first constraint determines the relationship between the overall loan rate for the *j*th borrower and the proposed maximum rate as given by Eq. 9.28 (Martinho, 2009):

$$\frac{\sum_{i=1}^{N} amount_{ij} \cdot rate_{ij}}{\sum_{i=1}^{N} amount_{ij}} \le Rate_{\max_j}, \;\; \forall\, j \in \{1, \ldots, M\}. \tag{9.28}$$

2) Constraint 2 – The second constraint determines the relationship between the overall investment rate for the *i*th lender and the proposed minimum rate as given by Eq. 9.29 (Martinho, 2009):

$$\frac{\sum_{j=1}^{M} amount_{ij} \cdot rate_{ij}}{\sum_{j=1}^{M} amount_{ij}} \geq Rate_{\min_i}, \ \forall\ i \in \{1,\ldots,N\}. \qquad 9.29$$

3) Constraint 3 – The third constraint determines the relationship between the actual total amount of funds invested by the *i*th lender and the original amount of funds offered by the same lender as given by Eq. 9.30 (Martinho, 2009):

$$Amount_{\min_i} \leq \sum_{j=1}^{M} amount_{ij} \leq Amount_{\max_i}, \ \forall\ i \in \{1,\ldots,N\}. \qquad 9.30$$

4) Constraint 4 – The fourth constraint determines the relationship between the actual amount of funds accepted by the *j*th borrower and the initial amount of funds requested by the same borrower as given by Eq. 9.31 (Martinho, 2009):

$$Amount_{\min_j} \leq \sum_{i=1}^{N} amount_{ij} \leq Amount_{\max_j}, \ \forall\ j \in \{1,\ldots,M\}. \qquad 9.31$$

- **Function:** In (Martinho, 2009), a function was formulated via Eq. 9.32:

$$\begin{aligned} f : &\mathbf{dom}(rate_{11}) \times \mathbf{dom}(amount_{11}) \times \cdots \times \\ &\mathbf{dom}(rate_{ij}) \times \mathbf{dom}(amount_{ij}) \times \cdots \times \\ &\mathbf{dom}(rate_{NM}) \times \mathbf{dom}(amount_{NM}) \rightarrow \Re \end{aligned}. \qquad 9.32$$

The function is defined in order to map the domains of decision variables to a set of decimal values, which can then be used to calculate the maximum utility.

- **Solutions Set:** The solution space was defined as a set which is formed by all valid solutions, as given by Eq. 9.33 (Martinho, 2009):

$$S = \left\{ \begin{array}{l} s = \{\ldots,(rate_{ij}, Rate_{ij}),(amount_{ij}, Rate_{ij}),\ldots\} : \\ Rate_{ij} \in \mathbf{dom}(rate_{ij}) \wedge Amount_{ij} \in \mathbf{dom}(amount_{ij}) \wedge s \\ \text{meets all the constraints} \end{array} \right\}. \qquad 9.33$$

9.3.6.2 *Utility Functions*

The importance of defining suitable utility function(s) is well-known, though the true practice is by no means a trivial task. A rigorous comparative study regarding distinct utility functions (Aleskerov et al.,

2007) is outside the scope of this chapter. Here we only briefly present the following two utility functions that are considered in (Martinho, 2009):

- **Cost of Dissatisfaction:** This utility function tries to make sure that all parties (borrowers and lenders) are all better off, making that particular crowdfunding operator more competitive in the market. In order to achieve this goal, each member's squared margins are first summed and then minimized, as given by Eq. 9.34 (Martinho, 2009):

$$\text{Cost-of-Dissatisfaction} = \sum_{i=1}^{N}\sum_{j=1}^{M}\left[\left(rate_{ij} - Rate_{\min_i}\right)^2 + \left(Rate_{\max_j} - rate_{ij}\right)^2\right]. \quad 9.34$$

- **Tight Margin Utility:** This utility function intends to get the profitable and fair solutions rewarded. Accordingly, there are three key factors taken into account by this function: Member profit, fairness and requests satisfaction. More specifically, the mathematical formulation of this utility function is given by Eq. 9.35 (Martinho, 2009):

$$\begin{aligned}\text{Tight-Margin-Utility} = &\left(k_{\text{Member-Margin}} \cdot \text{Member-Margin}\right)\cdot \\ &\left(k_{\text{Tightness}} \cdot \text{Tightness}\right)\cdot\left(k_{\text{Completion-Rate}} \cdot \text{Completion-Rate}\right)\end{aligned}. \quad 9.35$$

  where $k_{\text{Member-Margin}}$, $k_{\text{Tightness}}$ and $k_{\text{Completion-Rate}}$ are three specific parameters for utility function fine-tuning and other variables are defined as follows (Martinho, 2009):

  1) Member-Margin: The summation of each member's margin is defined as Member-Margin, which can be calculated via Eq. 9.36 (Martinho, 2009):

$$\text{Member-Margin} = \sum_{i=1}^{N}\left(rate_{ij} - Rate_{\min_i}\right) + \sum_{j=1}^{M}\left(Rate_{\max_j} - rate_{ij}\right). \quad 9.36$$

  2) Tightness: In (Martinho, 2009), the homogeneous margins among all members was considered tight. Therefore, the measure for examining the closeness degree of each member's individual gain is defined as Tightness, which is computable via Eq. 9.37 (Martinho, 2009):

$$\text{Tightness} = \frac{1}{\sigma\left(\left\{\left(rate_{ij} - Rate_{\min_i}\right): i\in[1,N], j\in[1,M]\right\} \vee \left\{\left(Rate_{\max_j} - rate_{ij}\right): i\in[1,N], j\in[1,M]\right\}\right)}. \quad 9.37$$

  3) Completion-Rate: From the crowdfunding operator's perspective, it would be great if all the available funds are invested and all the funds requests are fulfilled as well. Keeping this in mind, the relevance of the amount matching is termed as the Completion-Rate, which is calculable via Eq. 9.38 (Martinho, 2009):

$$\text{Completion-Rate} = \frac{2 \cdot \text{Matched-Amount-In-Total}}{\text{Offered-Amount-In-Total} + \text{Asked-Amount-In-Total}}. \quad 9.38$$

9.3.6.3 Summary

In order to design a matching system for a crowdfunding platform, the solution framework proposed in (Martinho, 2009) consists of three main modules: A toolkit module for addressing constraints, a suite module for optimization purpose and an adapter module for dealing with specific issues. Further discussions regarding their respective functionalities can be found in (Martinho, 2009). Meanwhile, in order to implement the aforementioned solution framework, a detailed breakdown explanation with respect to each individual class's responsibility was also offered in (Martinho, 2009).

One determining factor for conducting a meaningful experiment is whether there is enough reliable data at hand. The solution came from (Martinho, 2009) who developed a simple data generator and then used it to generate a dataset for testing purposes. Accordingly, the following parameters were specifically defined (Martinho, 2009):

- **Numbers of Lenders and Borrowers:** Denoted by N and M, respectively;
- **Mean Lending Rate:** Indicated by $\widehat{rate}_i$;
- **Standard Deviation of the Lending Rate:** Represented by σ_{rate_i};
- **Mean Lending Amount:** Indicated by $\widehat{amount}_i$;
- **Standard Deviation of the Lending Amount:** Represented by σ_{amount_i};
- **Mean Borrowing Rate:** Indicated by $\widehat{rate}_j$;
- **Standard Deviation of the Borrowing Rate:** Represented by σ_{amount_j};
- **Mean Borrowing Amount:** Indicated by $\widehat{amount}_j$; and
- **Standard Deviation of the Borrowing Rate:** Represented by σ_{amount_j}.

Following these definitions, two scenarios were created in (Martinho, 2009) and the corresponding exact parameter settings are illustrated in Fig. 9.8:

- **Tight Market:** In this market, a lender is willing to offer low interest rates while a borrower is also happy to accept higher interests when a suitable match is identified. This market is characterized by its small deviation in terms of the rates of distribution, which keeps the market more homogeneous.
- **Loose Market:** The characteristics of this market's lenders and borrowers are essentially similar to the tight market. However, the loose market tends to have larger deviation with respect to the

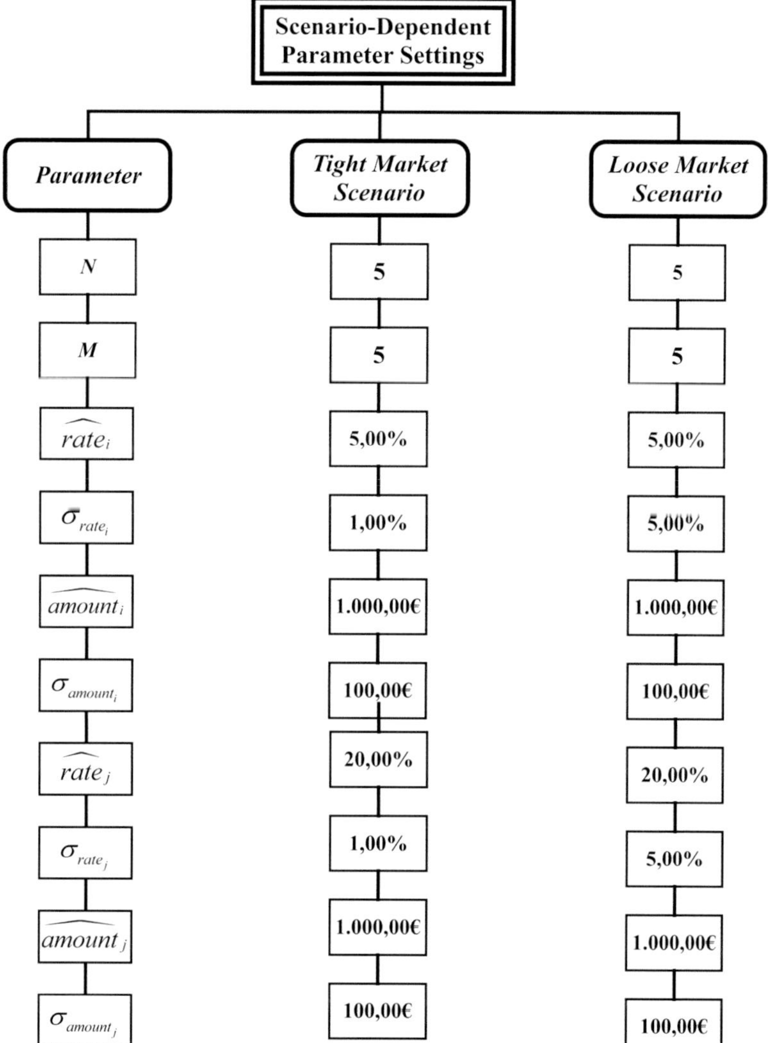

Figure 9.8: An illustration of scenario-dependent parameter settings.

rates distribution, allowing the market be more heterogeneous and diversified.

Under each generated scenario, several metaheuristics algorithms (i.e., HC, SA, GA, and PSO) were implemented in (Martinho, 2009) with a fixed budget for 1000 iterations.

- **Scenario 1–Tight Market:** GA was demonstrated to be the most efficient methodology, though its evolutionary process is somewhat irregular. The second best-performing algorithm was PSO, whose evolutionary process was rather continuous but unfortunately got stuck at a local optimum. The performance of HC (ranked third) is particularly interesting, in that it exhibited an almost linear utility score improvement progress. Finally, the baseline approach goes to SA.

- **Scenario 2–Loose Market:** Again, GA was ranked as the best algorithm. At the end of 1000 iterations, HC sightly outperformed PSO. The performance of SA remained largely unchanged.

9.4 Conclusions

In a crowdfunding environment, in particular the P2P lending community, loan requests are often determined by the actual requested amount of funds and the maximum adopted rate of repayment; while investment offers are often influenced by the true offered amount of funds and the minimum accepted rate of repayment. One or more optimal combination solutions are often of interest to not only crowdfunding platform operators from both the platform's sustainability and profitability perspectives, but also researchers from a broader portfolio optimization perspective.

In the literature, nature computing techniques (Xing and Gao, 2014b; Xing and Marwala, 2018l; Xing and Gao, 2014c; Xing and Gao, 2014q; Xing and Gao, 2014m; Xing, 2016; Xing and Gao, 2014x), powered intelligent trading machines (Marwala and Hurwitz, 2017b), trading systems (Davey, 2014) and algorithmic trading (Chan, 2013; Pole, 2007; Leshik and Cralle, 2011) have assisted people in selecting and optimizing portfolios. From a developmental perspective, any one of these is still in its infancy, which means varying incarnations of robotic traders will emerge during the next few years (Xing and Marwala, 2018a). However, humans are not always as rational as we have imagined (Camerer et al., 2004; Cartwright, 2011; Marwala and Hurwitz, 2017a). The power of minds over markets is already well recognized (Peterson, 2016). Under this circumstance, smart maintenance (Xing and Marwala, 2018m) of a healthy relationship between humans and robots (Xing and Marwala, 2018k; Xing and Marwala, 2018n) makes a lot of sense in various different ways, such as the hardware capacity angle (Xing and Marwala, 2018e; Xing and Marwala, 2018g; Xing and Marwala, 2018f), cyber-ware capacity angle (Xing and Marwala, 2018d; Xing and Marwala, 2018b; Xing and Marwala, 2018c) and human-ware capacity angle (Xing and Marwala, 2018j; Xing and Marwala, 2018h; Xing and Marwala, 2018i).

In this chapter, we employed a set of metaheuristics algorithms, namely SA (simulated annealing), GA (genetic algorithm) and PSO (particle swarm optimization), together with FSE (fuzzy synthetic evaluation) approach in order to explore the combination issue (investment offers & loan requests) and the relevant portfolio selection and optimization problem in the context of Portugal and Iran, respectively. Other similar attempts also include the Pareto ant colony optimization (Xing et al., 2010d; Xing et al., 2010a; Xing et al., 2010c; Xing et al., 2010b) algorithm employed in (Doerner et al., 2004); an integrated approach using an artificial bee colony (Xing and Gao, 2014e; Xing, 2015) and genetic programming (Marwala, 2007) proposed in (Hsu, 2014); a fuzzy hybrid portfolio selection method introduced by Tavana et al. (2015) and a novel variant of particle swarm optimization utilized by Sun et al. (2011). For more portfolio selection examples, one should refer to (Wei and Chang, 2011; Oh et al., 2012; Chen et al., 2009). Due to several simplifications, the overall results discussed in this chapter may not be of high production quality. But significant headway has already been made towards constructing crowdfunding infrastructure.

References

Aleskerov F, Bouyssou D and Monjardet B. (2007) *Utility maximization, choice and preference,* Berlin Heidelberg: Springer-Verlag, ISBN 978-3-540-34182-6.

Ania A and Charlesworth C. (2015) Crowdfunding guide: For nonprofits, charities and social impact projects. 215 Spadina Ave, Suite 400, Toronto, ON, M5T 2C7: HiveWire & Centre for Social Innovation, Document Version: 1701 - D.

Ashta A and Assadi D. (2008) Do social cause and social technology meet? Impact of web 2.0 technologies on peer-to-peer lending transactions. *SSRN Electronic Journal* http://ssrn.com/abstract=1281373.

Ashta A, Assadi D and Marakkath N. (2015) The strategic challenges of a social innovation: The case of Rang De in crowdfunding. *Strategic Change* 24: 1–14.

Assetz Capital. (2014) Peer-to-peer lending: Industry overview & understanding the marketplace. Assetz House, Newby Road, Stockport, Cheshire, SK7 5DA Assetz Capital, pp. 1–24.

Auger A and Doerr B. (2011) *Theory of randomized search heuristics,* River Edge, NJ, USA: World Scientific Publishing Co. Pte. Ltd., ISBN 978-981-4284-66-6.

Baker JE. (1987) Reducing bias and inefficiency in selection algorithm. *Proceedings of the Second International Conference on Genetic Algorithms, Cambridge, Massachusetts, USA,* pp. 14–21.

Barboza G and Trejos S. (2009) Micro Credit in Chiapas, Mexico: Poverty reduction through group lending. *Journal of Business Ethics* 88: 283–299.

Belleflamme P, Omrani N and Peitz M. (2015) The economics of crowdfunding platforms. *Information Economics and Policy* 33: 11–28.

Berkovich E. (2011) Search and herding effects in peer-to-peer leading: Evidence from prosper.com. *Annals of Finance* 7: 389–405.

Böttcher S, Doerr B and Neumann F. (2010) Optimal fixed and adaptive mutation rates for the LeadingOnes problem. In: Schaefer R, Cotta C, Kołodziej J, et al. (eds) *Parallel Problem Solving from Nature, PPSN XI, PPSN 2010, Lecture Notes in Computer Science, Vol. 6238.* Berlin, Heidelberg: Springer, ISBN 978-3-642-15843-8, pp. 1–10.

Brown J. (2006) Equity finance for social enterprises. *Social Enterprise Journal* 2: 73–81.

Cai S, Lin X, Xu D, et al. (2016) Judging online peer-to-peer lending behavior: A comparison of first-time and repeated borrowing requests. *Information & Management* 53: 857–867.

Calic G and Mosakowski E. (2016) Kicking off social entrepreneurship: How a sustainability orientation influences crowdfunding success. *Journal of Management Studies* 53: 738–767.

Callister WD. (2001) *Fundamentals of materials science and engineering: An interactive e.text,* 605 Third Avenue, New York, NY 10158-0012: John Wiley & Sons, Inc., ISBN 0-471-39551-X.

Callister WD and Rethwisch DG. (2012) *Fundamentals of materials science and engineering: An integrated approach,* 111 River Street, Hoboken, NJ 07030-5774: John Wiley & Sons, Inc., ISBN 978-1-118-06160-2.

Camerer C and Weber M. (1992) Recent developments in modeling preferences: Uncertainty and ambiguity. *Journal of Risk and Uncertainty* 5: 235–370.

Camerer CF, Loewenstein G and Rabin M. (2004) *Advances in behavioral economics,* 41 William Street, Princeton, New Jersey 08540: Princeton University Press, ISBN 0-691-11681-4.

Cartwright E. (2011) *Behavioral economics,* 2 Park Square, Milton Park, Abingdon, Oxon OX14 4RN: Routledge, ISBN 978-0-415-57309-2.

Černý V. (1985) Thermodynamical approach to the traveling salesman problem: An efficient simulation algorithm. *Journal of Optimization Theory and Applications* 45: 41–51.

Cesari V and Daltung S. (2000) The optimal size of a bank: Costs and benefits of diversification. *European Economic Review* 44: 1701–1726.

Chan EP. (2013) *Algorithmic trading: Winning strategies and their rationale,* 111 River Street, Hoboken, NJ 07030: John Wiley & Sons, Inc., ISBN 978-1-118-46014-6.

Chen D, Lai F and Lin Z. (2014) A trust model for online peer-to-peer lending: A lender's perspective. *Information Technology and Management* 15: 239–254.

Chen J-S, Hou J-L, Wu S-M, et al. (2009) Constructing investment strategy portfolios by combination genetic algorithms. *Expert Systems with Applications* 36: 3824–3828.

Chen J and Song K. (2013) Two-sided matching in the loan market. *International Journal of Industrial Organization* 31: 145–152.

Chen R, Chen Y, Liu Y, et al. (2017) Does team competition increase pro-social lending? Evidence from online microfinance. *Games and Economic Behavior* 101: 311–333.

Choo J, Lee D, Dilkina B, et al. (2014) To gather together for a better world: Understanding and leveraging communities in micro-lending recommendation. *Proceedings of the 23rd International Conference on World Wide Web, ACM,* pp. 249–260.

Clerc M and Kennedy J. (2002) The particle swarm—explosion, stability, and convergence in a multidimensional complex space. *IEEE Transactions on Evolutionary Computation* 6: 58–73.

Davey KJ. (2014) *Building winning algorithmic trading systems: A trader's journey from data mining to Monte Carlo simulation to live trading,* 111 River Street, Hoboken, NJ 07030: John Wiley & Sons, Inc., ISBN 978-1-118-77891-3.

De Buysere K, Gajda O, Kleverlaan R, et al. (2012) *A framework for European crowdfunding*: European Crowdfunding Network, ISBN 978-3-00-040193-0.

Diamond DW. (1984) Financial intermediation and delegated monitoring. *Review of Economic Studies* 51: 393–414.

Doerner K, Gutjahr WJ, Hartl RF, et al. (2004) Pareto ant colony optimization: A metaheuristic approach to multiobjective portfolio selection. *Annals of Operations Research* 131: 79–99.

Dorfleitner G, Priberny C, Schuster S, et al. (2016) Description-text related soft information in peer-to-peer lending—evidence from two leading European platforms. *Journal of Banking & Finance* 64: 169–187.

Du K-L and Swamy MNS. (2016) *Search and optimization by metaheuristics: Techniques and algorithms inspired by nature*: Springer International Publishing Switzerland, ISBN 978-3-319-41191-0.

Edirisinghe NCP and Zhang X. (2007) Portfolio selection under DEA-based relative financial strength indicators: Case of US industries. *Journal of the Operational Research Society* 59: 842–856.

Elton EJ, Gruber MJ, Brown SJ, et al. (2014) *Modern portfolio theory and investment analysis,* 111 River Street, Hoboken, NJ 07030-5774: JohnWiley & Sons, Inc., ISBN 978-1-118-46994-1.

Emekter R, Tu Y, Jirasakuldech B, et al. (2015) Evaluating credit risk and loan performance in online peer-to-peer (P2P) lending. *Applied Economics* 47: 54–70.

Epstein LG and Schneider M. (2010) Ambiguity and asset markets. *Annual Review of Financial Economics* 2: 315–346.
Feng S and Xu L. (1999a) An intelligent decision support for fuzzy comprehensive evaluation of urban development. *Expert Systems with Applications* 16: 21–32.
Feng S and Xu LD. (1999b) Decision support for fuzzy comprehensive evaluation of urban development. *Fuzzy Sets and Systems* 105: 1–12.
Fernholz ER. (2002) *Stochastic portfolio theory,* New York: Springer Science+Business Media, Inc., ISBN 978-1-4419-2987-7.
Fin KDS and Murphy J. (2016) Peer-to-peer lending: Structures, risks and regulation. *The Finsia Journal of Applied Finance* 3: 37–44.
Flasiński M. (2016) *Introduction to artificial intelligence*: Springer International Publishing Switzerland, ISBN 978-3-319-40020-4.
Fong A. (2015) Regulation of peer-to-peer lending in Hong Kong: State of play. *Law and Financial Markets Review* 9: 251–259.
Forrest S and Mitchell M. (2016) Adaptive computation: The multidisciplinary legacy of John H. Holland. *Communications of the ACM* 59: 58–63.
Freedman S and Jin GZ. (2011) Learning by doing with asymmetric information: Evidence from prosper.com. NBER Working Paper, no. 16855.
Freedman S and Jin GZ. (2017) The information value of online social networks: Lessons from peer-to-peer lending. *International Journal of Industrial Organization* 51: 185–222.
Gao Q and Lin M. (2016) Economic value of texts: Evidence from online debt crowdfunding. *SSRN Electronic Journal* http://dx.doi.org/10.2139/ssrn.2446114.
Gao R and Feng J. (2014) An overview study on P2P lending. *International Business and Management* 8: 14–18.
Geman S and Geman D. (1984) Stochastic relaxation, Gibbs distributions, and the Bayesian restoration of images. *IEEE Transactions on Pattern Analysis and Machine Intelligence* 6: 721–741.
Gilboa I, Postlewaite AW and Schmeidler D. (2008) Probability and uncertainty in economic modeling. *Journal of Economic Perspectives* 22: 173–188.
Goldberg DE. (1989) *Genetic algorithms in search, optimization, and machine learning*: Addison-Wesley Publishing Company, Inc., ISBN 0-201-15767-5.
Gonzalez L and McAleer K. (2011) Online social lending: A peak at U.S. Prosper & U.K. Zopa. *Journal of Accounting, Finance and Economics* 1: 26–41.
Gorgulho AMSBS, Horta NCG and Neves RFMF. (2013) *Intelligent financial portfolio composition based on evolutionary computation strategies,* Springer Heidelberg New York Dordrecht London: Springer, ISBN 978-3-642-32988-3.
Groshoff D. (2014) Kickstarter my heart: Extraordinary popular delusions and the madness of crowdfunding constraints and Bitcoin bubbles. *William & Mary Business Law Review* 5: 489–557.
Guidolin M and Rinaldi F. (2013) Ambiguity in asset pricing and portfolio choice: A review of the literature. *Theory and Decision* 74: 183–217.
Guo Y, Zhou W, Luo C, et al. (2016) Instance-based credit risk assessment for investment decisions in P2P lending. *European Journal of Operational Research* 249: 417–426.
Hünseler M. (2013) *Credit portfolio management: a practitioner's guide to the active management of credit risks,* 175 Fifth Avenue, New York, NY 10010: Palgrave Macmillan, ISBN 978-1-349-35162-6.
Han J-T, Chen Q, Liu J-G, et al. (2018) The persuasion of borrowers' voluntary information in peer-to-peer lending: An empirical study based on elaboration likelihood model. *Computers in Human Behavior* 78: 200–214.
Hardy RJ and Binek C. (2014) *Thermodynamics and statistical mechanics: An integrated approach,* The Atrium, Southern Gate, Chichester, West Sussex, PO19 8SQ, United Kingdom: John Wiley & Sons Ltd., ISBN 978-1-118-50101-6.
Harris T. (2013) Quantitative credit risk assessment using support vector machines: Broad versus narrow default definitions. *Expert Systems with Applications* 40: 4404–4413.

Haugen RA. (2001) *Modern investment theory,* New Jersey: Prentice Hall, ISBN 9780130191700.

Herzenstein M, Dholakia UM and Andrews RL. (2011a) Strategic herding behavior in peer-to-peer loan auctions. *Journal of Interactive Marketing* 25: 27–36.

Herzenstein M, Sonenshein S and Dholakia UM. (2011b) Tell me a good story and I may lend you money: The role of narratives in peer-to-peer lending decisions. *Journal of Marketing Research* 48: S138–S149.

Hines F. (2005) Viable social enterprise: An evaluation of business support to social enterprises. *Social Enterprise Journal* 1: 13–28.

Holland JH. (1992) *Adaptation in natural and artificial systems: An introductory analysis with applications to biology, control, and artificial intelligence,* Cambridge, Massachusetts and London, England The MIT Press, ISBN 978-0-262-08213-6.

Hsu C-M. (2014) An integrated portfolio optimisation procedure based on data envelopment analysis, artificial bee colony algorithm and genetic programming. *International Journal of System Science* 45: 2645–2664.

Hughes JF, Dam AV, McGuire M, et al. (2014) *Computer graphics: Principles and practice,* One Lake Street, Upper Saddle River, New Jersey 07458: Pearson Education, Inc., ISBN 978-0-321-39952-6.

Iyer R, Khwaja AI, Luttmer EFP, et al. (2016) Screening peers softly: Inferring the quality of small borrowers. *Management Science* 62: 1554–1577.

Jansen T and Wegener I. (2001) On the utility of populations in evolutionary algorithms. *Proceedings of the Genetic and Evolutionary Computation Conference (GECCO'2001),* pp. 1034–1041.

Jansen T and Wegener I. (2010) On the choice of the mutation probability for the (1+1) EA. In: Schoenauer M (ed) *Parallel Problem Solving from Nature, PPSN VI, PPSN 2000, Lecture Notes in Computer Science, Vol. 1917.* Berlin, Heidelberg: Springer, ISBN 978-3-540-41056-0, pp. 89–98.

Jiang C, Wang Z, Wang R, et al. (in press) Loan default prediction by combining soft information extracted from descriptive text in online peer-to-peer lending. *Annals of Operations Research* DOI 10.1007/s10479-017-2668-z.

Johnsen S. (2017) Social enterprise in the United Arab Emirates. *Social Enterprise Journal* 13: 392–409.

Kennedy J and Eberhart R. (1995) Particle swarm optimization. *IEEE International Joint Conference on Neural Networks, 27 November–01 December,* pp. 1942–1948. IEEE.

Kennedy J. (1999) Small worlds and mega-minds: Effects of neighborhood topology on particle swarm performance. *Proceedings of the Congress on Evolutionary Computation (CEC 99), 6–9 May, Washington, DC, USA,* pp. 1931–1938.

Kennedy J and Eberhart RC. (2001) *Swarm Intelligence,* 525 B Street, Suite 1900, San Diego, CA 92101-4495, USA: Academic Press, ISBN 1-55860-595-9.

Kennedy J and Mendes R. (2002) Population structure and particle swarm performance. *Proceedings of the Congress on Evolutionary Computation (CEC'02), 12–17 May, Honolulu, HI, USA,* pp. 1671–1676.

Kibbe J. (2013) Too big to disintermediate? Peer-to-peer lending takes on traditional consumer lending. 200 Liberty Street, New York, NY 10281: Richards Kibbe & Orbe LLP.

Kickul J and Lyons TS. (2015) Financing social enterprises. *Entrepreneur Research Journal* 5: 83–85.

Kiranyaz S, Ince T and Gabbouj M. (2014) *Multidimensional particle swarm optimization for machine learning and pattern recognition,* Springer Heidelberg New York Dordrecht London: Springer-Verlag Berlin Heidelberg, ISBN 978-3-642-37845-4.

Kirkpatrick S, Gelatt CD and Vecchi MP. (1983) Optimization by simulated annealing. *Science* 220: 671–680.

Klafft M. (2008) Online peer-to-peer lending: A lenders' perspective. *SSRN Electronic Journal* https://ssrn.com/abstract=1352352.

Krumme KA and Herrero S. (2009) Lending behavior and community structure in an online peer-to-peer economic network. *Proceedings of the 12th International Conference on Computational Science and Engineering, 29~31 August, Vancouver, Canada*, pp. 613–618.

Kuo YF and Chen PC. (2006) Selection of mobile value-added services for system operators using fuzzy synthetic evaluation. *Expert Systems with Applications* 30: 612–620.

Lacan C and Desmet P. (in press) Does the crowdfunding platform matter? Risks of negative attitudes in two-sided markets. *Journal of Consumer Marketing* https://doi.org/10.1108/JCM-03-2017-2126.

Laratta R, Nakagawa S and Sakurai M. (2011) Japanese social enterprises: Major contemporary issues and key challenges. *Social Enterprise Journal* 7: 50–68.

Larrimore L, Jiang L, Larrimore J, et al. (2011) Peer to peer lending: The relationship between language features, trustworthiness and persuasion success. *Journal of Applied Communication Research* 39: 19–371.

Lee E and Lee B. (2012) Herding behavior in online P2P lending: An empirical investigation. *Electronic Commerce Research and Applications* 11: 495–503.

Lehner OM. (2013) Crowdfunding social ventures: A model and research agenda. *Venture Capital* 15: 289–311.

Leshik EA and Cralle J. (2011) *An introduction to algorithmic trading: Basic to advanced strategies*, The Atrium, Southern Gate, Chichester, West Sussex, PO19 8SQ, United Kingdom: John Wiley & Sons Ltd, ISBN 978-0-470-68954-7.

Li Y and Wang C. (2017) Risk identification, future value and credit capitalization: Research on the theory and policy of poverty alleviation by Internet finance. *China Finance and Economic Review* 5: 1–12.

Lin TT and Lo I-H. (2006) Analysis of required and matching loan qualities in financial institutions. *Journal of Statistics and Management Systems* 9: 105–122.

Lin X, Li X and Zheng Z. (2017) Evaluating borrower's default risk in peer-to-peer lending: Evidence from a lending platform in China. *Applied Economics* 49: 3538–3545.

Lucci S and Kopec D. (2016) *Artificial intelligence in the 21st century: A living introduction*, 22841 Quicksilver Drive, Dulles, VA 20166: Mercury Learning and Information, ISBN 978-1-942270-00-3.

Magee J. (2011) Peer-to-peer lending in the United States: Surviving after Dodd-Frank. *North Carolina Banking Institute Journal* 15: 139–174.

Malekipirbazari M and Aksakalli V. (2015) Risk assessment in social lending via random forests. *Expert Systems with Applications* 42: 4621–4631.

Manbeck P and Franson M. (2016) The regulation of marketplace lending: A summary of the principal issues. Chapman and Cutler LLP.

Markowitz HM. (1952) Portfolio selection. *Journal of Finance* 7: 77–91.

Markowitz HM. (1959) *Portfolio selection: efficient diversification of investments*, New York: John Wiley & Sons, Inc.

Markowitz HM. (1991) Foundations of portfolio selection. *Journal of Finance* 46: 469–477.

Marom S. (2017) Social responsibility and crowdfunding businesses: A measurement development study. *Social Responsibility Journal* 13: 235–249.

Martin RL and Osberg SR. (2015) Two keys to sustainable social enterprise. *Harvard Business Review* May: 86–94.

Martinho LPdCB. (2009) Combining loan requests and investment offers. *Faculdade de Engenharia*. Universidade do Porto.

Marwala T. (2005) Finite element model updating using particle swarm optimization. *International Journal of Engineering Simulation* 6: 25–30.

Marwala T. (2007) Bayesian training of neural networks using genetic programming. *Pattern Recognition Letters* 28: 1452–1458.

Marwala T. (2009) *Computational intelligence for missing data imputation, estimation and management: Knowledge optimization techniques*, New York, USA: IGI Global, ISBN 978-1-60566-336-4.

Marwala T. (2010a) Finite-element-model updating using genetic algorithm. In: Marwala T (ed) *Finite-element-model Updating Using Computational Intelligence Techniques: Applications to Structural Dynamics.* London, UK: Springer-Verlag, ISBN 978-1-84996-322-0, Chapter 3, pp. 49–66.

Marwala T. (2010b) Finite-element-model updating using particle-swarm optimization. In: Marwala T (ed) *Finite-element-model Updating Using Computational Intelligence Techniques: Applications to Structural Dynamics.* London, UK: Springer-Verlag, ISBN 978-1-84996-322-0, Chapter 4, pp. 67–84.

Marwala T. (2010c) Finite-element-model updating using simulated annealing. In: Marwala T (ed) *Finite-element-model Updating Using Computational Intelligence Techniques: Applications to Structural Dynamics.* London, UK: Springer-Verlag, ISBN 978-1-84996-322-0, Chapter 5, pp. 85–102.

Marwala T and Lagazio M. (2011a) *Militarized conflict modeling using computational intelligence,* London, UK: Springer-Verlag, ISBN 978-0-85729-789-1.

Marwala T and Lagazio M. (2011b) Particle swarm optimization and hill-climbing optimized rough sets for modeling interstate conflict. In: Marwala T and Lagazio M (eds) *Militarized Conflict Modeling Using Computational Intelligence.* London, UK: Springer-Verlag, ISBN 978-0-85729-789-1, Chapter 8, pp. 147–164.

Marwala T. (2013a) Introduction to economic modeling. In: Marwala T (ed) *Economic Modeling Using Artificial Intelligence Methods.* Springer London Heidelberg New York Dordrecht: Springer-Verlag London, ISBN 978-1-4471-5009-1, Chapter 1, pp. 1–21.

Marwala T. (2013b) Missing data approaches to economic modeling: Optimization approach. In: Marwala T (ed) *Economic Modeling Using Artificial Intelligence Methods.* Springer London Heidelberg New York Dordrecht: Springer-Verlag London, ISBN 978-1-4471-5009-1, Chapter 7, pp. 119–136.

Marwala T. (2014a) Flexibly-bounded rationality in interstate conflict. In: Marwala T (ed) *Artificial Intelligence Techniques for Rational Decision Making.* Springer Cham Heidelberg New York Dordrecht London: Springer International Publishing Switzerland, ISBN 978-3-319-11423-1, Chapter 6, pp. 91–109.

Marwala T. (2014b) Rational counterfactuals and decision making: Application to interstate conflict. In: Marwala T (ed) *Artificial Intelligence Techniques for Rational Decision Making.* Springer Cham Heidelberg New York Dordrecht London: Springer International Publishing Switzerland, ISBN 978-3-319-11423-1, Chapter 5, pp. 73–89.

Marwala T. (2015) *Causality, correlation and artificial intelligence for rational decision making,* 5 Toh Tuck Link, Singapore 596224: World Scientific Publishing Co. Pte. Ltd, ISBN 978-9-81463-086-3.

Marwala T, Boulkaibet I and Adhikari S. (2017) *Probabilistic finite element model updating using Bayesian statistics: Applications to aeronautical and mechanical engineering,* The Atrium, Southern Gate, Chichester, West Sussex, PO19 8SQ, United Kingdom: John Wiley & Sons, Ltd, ISBN 978-1-1191-5301-6.

Marwala T and Hurwitz E. (2017a) *Artificial Intelligence and Economic Theory: Skynet in the Market,* Gewerbestrasse 11, 6330 Cham, Switzerland: Springer International Publishing AG, ISBN 978-3-319-66103-2.

Marwala T and Hurwitz E. (2017b) Portfolio theory. In: Marwala T and Hurwitz E (eds) *Artificial Intelligence and Economic Theory: Skynet in the Market.* Gewerbestrasse 11, 6330 Cham, Switzerland: Springer International Publishing AG, ISBN 978-3-319-66103-2, Chapter 11, pp. 125–136.

Mateescu A. (2015) Peer-to-peer lending. 36 West 20th Street, 11th Floor New York, NY 10011: Data & Society Research Institute, 1–23.

Melanie M. (1996) *An introduction to genetic algorithms,* Cambridge, Massachusetts: The MIT Press, ISBN 0-262-13316-4.

Metropolis N, Rosenbluth AW, Rosenbluth MN, et al. (1953) Equation of state calculations by fast computing machines. *The Journal of Chemical Physics* 21: 1087–1092.

Michaud RO and Michaud RO. (2008) *Efficient asset management: A practical guide to stock portfolio optimization and asset allocation,* New York: Oxford University Press, ISBN 978-0-19-533191-2.

Milne A and Parboteeah P. (2016) The business models and economics of peer-to-peer lending. Place du Congrès 1, B-1000 Brussels, Belgium: European Credit Research Institute, No. 17, ISBN 978-94-6138-526-0.

Mordeson JN and Nair PS. (2001) *Fuzzy mathematics: An introduction for engineers and scientists,* Berlin Heidelberg: Springer-Verlag, ISBN 978-3-7908-2494-0.

Mueller E and Reize F. (2013) Loan availability and investment: Can innovative companies better cope with loan denials? *Applied Economics* 45: 5001–5011.

Nayebpur H and Bokaei MN. (2017) Portfolio selection with fuzzy synthetic evaluation and genetic algorithm. *Engineering Computations* 34: 2422–2434.

Neumann F and Witt C. (2010) *Bioinspired computation in combinatorial—optimization algorithms and their computational complexity,* Berlin, Heidelberg: Springer-Verlag, ISBN 978-3-642-16544-3.

Nguyen GT. (2014) Exploring collaborative consumption business models—case peer-to-peer digital platforms. *School of Business.* Aalto University.

Nowak A, Ross A and Yencha C. (in press) Small business borrowing and peer-to-peer lending: Evidence from Lending Club. *Contemporary Economic Policy* doi:10.1111/coep.12252.

Oh J, Yang J and Lee S. (2012) Managing uncertainty to improve decision-making in NPD portfolio management with a fuzzy expert system. *Expert Systems with Applications* 39: 9868–9885.

Oxera. (2016) The economics of peer-to-peer lending. Oxera, retrieved from http://www.oxera.com/Latest-Thinking/Publications/Reports/2016/economics-of-peer-to-peer-lending.aspx, accessed on 23 October 2007.

Parhankangas A and Renko M. (2017) Linguistic style and crowdfunding success among social and commercial entrepreneurs. *Journal of Business Venturing* 32: 215–236.

Peterson RL. (2016) *Trading on sentiment: The power of minds over markets,* 111 River Street, Hoboken, NJ 07030: John Wiley & Sons, Inc., ISBN 978-1-119-12276-0.

Pole A. (2007) *Statistical arbitrage: Algorithmic trading insights and techniques,* 111 River Street, Hoboken, NJ 07030: John Wiley & Sons, Inc., ISBN 978-0-470-13844-1.

Prystav F. (2016) Personal information in peer-to-peer loan applications: Is less more? *Journal of Behavioral and Experimental Finance* 9: 6–19.

Puro L, Teich JE, Wallenius H, et al. (2010) Borrower decision aid for people-to-people lending. *Decision Support Systems* 49: 52–60.

PwC. (2015) Peer pressure: how P2P lending platforms are transforming the consumer lending industry. PricewaterhouseCoopers LLP, pp. 1–18.

Qiu J, Lin Z and Luo B. (2012) Effects of borrower-defined conditions in the online peer-to-peer lending market. In: Shaw MJ, Zhang D and Yue WT (eds) *E-Life: Web-Enabled Convergence of Commerce, Work, and Social Life.* Springer-Verlag Berlin Heidelberg, ISBN 978-3-642-29872-1, Part II, pp. 167–179.

Rachev ST, Stoyanov SV and Fabozzi FJ. (2008) *Advanced stochastic models, risk assessment and portfolio optimization: The ideal risk, uncertainty and performance measures,* Hoboken, NJ: John Wiley & Sons, Inc., ISBN 978-0-470-05316.

Rakesh V, Lee WC and Reddy CK. (2016) Probabilistic group recommendation model for crowdfunding domains. *Proceedings of the 9th ACM International Conference on Web Search and Data Mining (WSDM), 22~25 February, San Francisco, CA, USA*, pp. 257–266.

Rana S. (2013) Philanthropic innovation and creative capitalism: A historical and comparative perspective on social entrepreneurship and corporate social responsibility. *Shruti Alabama Law Review* 64: 1121–1174.

Rawlins GJE. (1992) *Compared to what? An introduction to the analysis of algorithms,* 41 Madison Avenue, New Yorker, NY 10010: Computer Science Press, W.H. Freeman and Company, ISBN 0-7167-8243-X.

Reeves CR and Rowe JE. (2002) *Genetic algorithms–principles and perspectives: A guide to GA theory*: Kluwer Academic Publishers, ISBN 1-4020-7240-6.

Reeves WT. (1983) Particle systems—a technique for modeling a class of fuzzy objects. *ACM Transactions on Graphics* 2: 91–108.

Reiser DB and Dean SA. (2015) SE(c)(3): A catalyst for social enterprise crowdfunding. *Indiana Law Journal* 90: 1091–1129.

Renton P. (2015) Understanding peer-to-peer lending. LendAcademy.com, pp. 1–13.

Revuz D. (1984) *Markov chains,* P.O. Box 1991, 1000 BZ Amsterdam, The Netherlands: Elsevier Science Publishers B.V., ISBN 0444-86400-8.

Richards Kibbe & Orbe LLP. (2013) Too big to disintermediate? Peer-to-peer lending takes on traditional consumer lending. Richards Kibbe & Orbe LLP, 1–9.

Roche-Saunders G and Hunt O. (2017) Crowdfunding for charities and social enterprises. *Charity and Social Enterprise Update* Summer: 7–9.

Rochet JC and Tirole J. (2003) Platform competition in two-sided markets. *Journal of the European Economic Association* 1: 990–1209.

Rosenblum RH, Gault-Brown SI and Caiazza AB. (2015) Peer-to-peer lending platforms: Securities law considerations. *Journal of Investment Compliance* 16: 15–18.

Rowe JE. (2015) Genetic algorithms. In: Kacprzyk J and Pedrycz W (eds) *Springer Handbook of Computational Intelligence.* Dordrecht Heidelberg London New York: Springer-Verlag Berlin Heidelberg, ISBN 978-3-662-43504-5, Part E Evolutionary Computation, Chapter 42, pp. 825–844.

Russell SJ and Norvig P. (2010) *Artificial intelligence: A modern approach,* 1 Lake Street, Upper Saddle River, NJ 07458: Pearson Education, Inc., ISBN 978-0-13-604259-4.

Schwienbacher A and Larralde B. (2010) Crowdfunding of small entrepeneurial ventures. *SSRN Electronic Journal* http://dx.doi.org/10.2139/ssrn.1699183.

Sedgewick R and Wayne K. (2011) *Algorithms,* Rights and Contracts Department, 501 Boylston Street, Suite 900, Boston, MA 02116: Pearson Education, Inc., ISBN 978-0-321-57351-3.

Sefiane S and Benbouziane M. (2012) Portfolio selection using genetic algorithm. *Journal of Applied Finance & Banking* 2: 143–154.

Segal M. (2015) Peer-to-peer lending: A financing alternative for small businesses. 409 Third Street, S.W., Washington, DC 20416: US Small Business Administration (SBA) Office of Economic Research of the Office of Advocacy, 1–14.

Selvaraj C and Anand S. (2012) A survey on security issues of reputation management systems for peer-to-peer networks. *Computer Science Review* 6: 145–160.

Serrano-Cinca C and Gutiérrez-Nieto B. (2016) The use of profit scoring as an alternative to credit scoring systems in peer-to-peer (P2P) lending. *Decision Support Systems* 89: 113–122.

Sun J, Fang W, Wu X, et al. (2011) Solving the multi-stage portfolio optimization problem with a novel particle swarm optimization. *Expert Systems with Applications* 38: 6727–6735.

Tavana M, Keramatpour M, Santos-Arteaga FJ, et al. (2015) A fuzzy hybrid project portfolio selection method using data envelopment analysis,TOPSIS and integer programming. *Expert Systems with Applications* 42: 8432–8444.

Verstein A. (2008) Peer-to-peer lending update and regulatory considerations. Filene Research Institute, ISBN 978-1-932795-35-6, 1–49.

Vose MD. (1999) *The simple genetic algorithm,* Cambridge, MA, USA: The MIT Press, ISBN 0-2622205-8-X.

Walker ST. (2014) *Understanding alternative investments: Creating diversified portfolios that ride the wave of investment success*: Palgrave Macmillan, ISBN 978-1-137-37018-1.

Wang JG and Yang J. (2016) *Financing without bank loans: New alternatives for funding SMEs in China*: Springer Science+Business Media Singapore, ISBN 978-981-10-0900-6.

Wang S, Qi Y, Fu B, et al. (2016) Credit risk evaluation based on text analysis. *International Journal of Cognitive Informatics & Natural Intelligence* 10: 1–11.

Wegener I. (2001) Theoretical aspects of evolutionary algorithms. In: Orejas F, Spirakis PG and van Leeuwen J (eds) *Automata, Languages and Programming, ICALP 2001, Lecture*

Notes in Computer Science, Vol. 2076. Berlin, Heidelberg: Springer, ISBN 978-3-540-42287-7, Chapter 6, pp. 64–78.

Wei C-C and Chang H-W. (2011) A new approach for selecting portfolio of new product development projects. *Expert Systems with Applications* 38: 429–434.

Whitley D. (1989) The GENITOR algorithm and selection pressure: Why ran-based allocation of reproductive trials is best. In: Schaffer JD (ed) *Proceedings of the Third International Conference on Genetic Algorithms, June, San Mateo, CA*, pp. 116–121.

Witt C. (2012) Optimizing linear functions with randomized search heuristics. In: Dürr C and Wilke T (eds) *Proceedings of the 29th Symposium on Theoretical Aspects of Computer Science (STACS'12), February, Paris, France*, pp. 420–431.

Xia Y, Liu C and Liu N. (2017) Cost-sensitive boosted tree for loan evaluation in peer-to-peer lending. *Electronic Commerce Research and Applications* 24: 30–49.

Xing B, Gao W-J, Nelwamondo FV, et al. (2010a) Ant colony optimization for automated storage and retrieval system. *Proceedings of The Annual IEEE Congress on Evolutionary Computation (IEEE CEC), 18–23 July, CCIB, Barcelona, Spain*, pp. 1133–1139. IEEE.

Xing B, Gao W-J, Nelwamondo FV, et al. (2010b) Can ant algorithms make automated guided vehicle system more intelligent? A viewpoint from manufacturing environment. *Proceedings of IEEE International Conference on Systems, Man and Cybernetics (IEEE SMC), 10–13 October, Istanbul, Turkey*, pp. 3226–3234. IEEE.

Xing B, Gao W-J, Nelwamondo FV, et al. (2010c) Part-machine clustering: The comparison between adaptive resonance theory neural network and ant colony system. In: Zeng Z and Wang J (eds) *Advances in Neural Network Research & Applications, LNEE 67*, pp. 747–755. Berlin Heidelberg: Springer-Verlag.

Xing B, Gao W-J, Nelwamondo FV, et al. (2010d) Two-stage inter-cell layout design for cellular manufacturing by using ant colony optimization algorithms. In: Tan Y, Shi Y and Tan KC (eds) *Advances in Swarm Intelligence, Part I, LNCS 6145*, pp. 281–289. Berlin Heidelberg: Springer-Verlag.

Xing B, Gao W-J and Marwala T. (2013) An overview of cuckoo-inspired intelligent algorithms and their applications. *IEEE Symposium Series on Computational Intelligence (IEEE SSCI), 15–19 April, Singapore*, pp. 85–89. IEEE.

Xing B and Gao W-J. (2014a) *Computational intelligence in remanufacturing*, 701 E. Chocolate Avenue, Suite 200, Hershey PA 17033: IGI Global, ISBN 978-1-4666-4908-8.

Xing B and Gao W-J. (2014aa) Chemical-reaction optimization algorithm. In: Xing B and Gao W-J (eds) *Innovative Computational Intelligence: A Rough Guide to 134 Clever Algorithms*. Cham Heidelberg New York Dordrecht London: Springer International Publishing Switzerland, ISBN: 978-3-319-03403-4, Chapter 25, pp. 417–428.

Xing B and Gao W-J. (2014b) *Innovative computational intelligence: A rough guide to 134 clever algorithms*, Cham Heidelberg New York Dordrecht London: Springer International Publishing Switzerland, ISBN: 978-3-319-03403-4.

Xing B and Gao W-J. (2014bb) Emerging chemistry-based CI algorithms. In: Xing B and Gao W-J (eds) *Innovative Computational Intelligence: A Rough Guide to 134 Clever Algorithms*. Cham Heidelberg New York Dordrecht London: Springer International Publishing Switzerland, ISBN: 978-3-319-03403-4, Chapter 26, pp. 429–437.

Xing B and Gao W-J. (2014c) Introduction to computational intelligence. In: Xing B and Gao W-J (eds) *Innovative Computational Intelligence: A Rough Guide to 134 Clever Algorithms*. Cham Heidelberg New York Dordrecht London: Springer International Publishing Switzerland, ISBN: 978-3-319-03403-4, Chapter 1, pp. 3–17.

Xing B and Gao W-J. (2014cc) Base optimization algorithm. In: Xing B and Gao W-J (eds) *Innovative Computational Intelligence: A Rough Guide to 134 Clever Algorithms*. Cham Heidelberg New York Dordrecht London: Springer International Publishing Switzerland, ISBN: 978-3-319-03403-4, Chapter 27, pp. 441–444.

Xing B and Gao W-J. (2014d) Bacteria inspired algorithms. In: Xing B and Gao W-J (eds) *Innovative Computational Intelligence: A Rough Guide to 134 Clever Algorithms*. Cham Heidelberg New York Dordrecht London: Springer International Publishing Switzerland, ISBN: 978-3-319-03403-4, Chapter 2, pp. 21–38.

Xing B and Gao W-J. (2014dd) Emerging mathematics-based CI algorithms. In: Xing B and Gao W-J (eds) *Innovative Computational Intelligence: A Rough Guide to 134 Clever Algorithms*. Cham Heidelberg New York Dordrecht London: Springer International Publishing Switzerland, ISBN: 978-3-319-03403-4, Chapter 28, pp. 445–448.

Xing B and Gao W-J. (2014e) Bee inspired algorithms. In: Xing B and Gao W-J (eds) *Innovative Computational Intelligence: A Rough Guide to 134 Clever Algorithms*. Cham Heidelberg New York Dordrecht London: Springer International Publishing Switzerland, ISBN: 978-3-319-03403-4, Chapter 4, pp. 45–80.

Xing B and Gao W-J. (2014f) Bat inspired algorithms. In: Xing B and Gao W-J (eds) *Innovative Computational Intelligence: A Rough Guide to 134 Clever Algorithms*. Cham Heidelberg New York Dordrecht London: Springer International Publishing Switzerland, ISBN: 978-3-319-03403-4, Chapter 3, pp. 39–44.

Xing B and Gao W-J. (2014g) Biogeography-based optimization algorithm. In: Xing B and Gao W-J (eds) *Innovative Computational Intelligence: A Rough Guide to 134 Clever Algorithms*. Cham Heidelberg New York Dordrecht London: Springer International Publishing Switzerland, ISBN: 978-3-319-03403-4, Chapter 5, pp. 81–91.

Xing B and Gao W-J. (2014h) Cat swarm optimization algorithm. In: Xing B and Gao W-J (eds) *Innovative Computational Intelligence: A Rough Guide to 134 Clever Algorithms*. Cham Heidelberg New York Dordrecht London: Springer International Publishing Switzerland, ISBN: 978-3-319-03403-4, Chapter 6, pp. 93–104.

Xing B and Gao W-J. (2014hh) Used product collection optimization using genetic algorithms. In: Xing B and Gao W-J (eds) *Computational Intelligence in Remanufacturing*. 701 E. Chocolate Avenue, Suite 200, Hershey PA 17033: IGI Global, ISBN 978-1-4666-4908-8, Chapter 4, pp. 59–74.

Xing B and Gao W-J. (2014i) Cuckoo inspired algorithms. In: Xing B and Gao W-J (eds) *Innovative Computational Intelligence: A Rough Guide to 134 Clever Algorithms*. Cham Heidelberg New York Dordrecht London: Springer International Publishing Switzerland, ISBN: 978-3-319-03403-4, Chapter 7, pp. 105–121.

Xing B and Gao W-J. (2014j) Luminous insect inspired algorithms. In: Xing B and Gao W-J (eds) *Innovative Computational Intelligence: A Rough Guide to 134 Clever Algorithms*. Cham Heidelberg New York Dordrecht London: Springer International Publishing Switzerland, ISBN: 978-3-319-03403-4, Chapter 8, pp. 123–137.

Xing B and Gao W-J. (2014k) Fish inspired algorithms. In: Xing B and Gao W-J (eds) *Innovative Computational Intelligence: A Rough Guide to 134 Clever Algorithms*. Cham Heidelberg New York Dordrecht London: Springer International Publishing Switzerland, ISBN: 978-3-319-03403-4, Chapter 9, pp. 139–155.

Xing B and Gao W-J. (2014l) Frog inspired algorithms. In: Xing B and Gao W-J (eds) *Innovative Computational Intelligence: A Rough Guide to 134 Clever Algorithms*. Cham Heidelberg New York Dordrecht London: Springer International Publishing Switzerland, ISBN: 978-3-319-03403-4, Chapter 10, pp. 157–165.

Xing B and Gao W-J. (2014m) Fruit fly optimization algorithm. In: Xing B and Gao W-J (eds) *Innovative Computational Intelligence: A Rough Guide to 134 Clever Algorithms*. Cham Heidelberg New York Dordrecht London: Springer International Publishing Switzerland, ISBN: 978-3-319-03403-4, Chapter 11, pp. 167–170.

Xing B and Gao W-J. (2014n) Group search optimization algorithm. In: Xing B and Gao W-J (eds) *Innovative Computational Intelligence: A Rough Guide to 134 Clever Algorithms*. Cham Heidelberg New York Dordrecht London: Springer International Publishing Switzerland, ISBN: 978-3-319-03403-4, Chapter 12, pp. 171–176.

Xing B and Gao W-J. (2014o) Invasive weed optimization algorithm. In: Xing B and Gao W-J (eds) *Innovative Computational Intelligence: A Rough Guide to 134 Clever Algorithms*. Cham Heidelberg New York Dordrecht London: Springer International Publishing Switzerland, ISBN: 978-3-319-03403-4, Chapter 13, pp. 177–181.

Xing B and Gao W-J. (2014p) Music inspired algorithms. In: Xing B and Gao W-J (eds) *Innovative Computational Intelligence: A Rough Guide to 134 Clever Algorithms*. Cham Heidelberg

New York Dordrecht London: Springer International Publishing Switzerland, ISBN: 978-3-319-03403-4, Chapter 14, pp. 183–201.

Xing B and Gao W-J. (2014q) Imperialist competitive algorithm. In: Xing B and Gao W-J (eds) *Innovative Computational Intelligence: A Rough Guide to 134 Clever Algorithms.* Cham Heidelberg New York Dordrecht London: Springer International Publishing Switzerland, ISBN: 978-3-319-03403-4, Chapter 15, pp. 203–209.

Xing B and Gao W-J. (2014r) Teaching–learning-based optimization algorithm. In: Xing B and Gao W-J (eds) *Innovative Computational Intelligence: A Rough Guide to 134 Clever Algorithms.* Cham Heidelberg New York Dordrecht London: Springer International Publishing Switzerland, ISBN: 978-3-319-03403-4, Chapter 16, pp. 211–216.

Xing B and Gao W-J. (2014s) Emerging biology-based CI algorithms. In: Xing B and Gao W-J (eds) *Innovative Computational Intelligence: A Rough Guide to 134 Clever Algorithms.* Cham Heidelberg New York Dordrecht London: Springer International Publishing Switzerland, ISBN: 978-3-319-03403-4, Chapter 17, pp. 217–317.

Xing B and Gao W-J. (2014t) Big bang–big crunch algorithm. In: Xing B and Gao W-J (eds) *Innovative Computational Intelligence: A Rough Guide to 134 Clever Algorithms.* Cham Heidelberg New York Dordrecht London: Springer International Publishing Switzerland, ISBN: 978-3-319-03403-4, Chapter 18, pp. 321–331.

Xing B and Gao W-J. (2014u) Central force optimization algorithm. In: Xing B and Gao W-J (eds) *Innovative Computational Intelligence: A Rough Guide to 134 Clever Algorithms.* Cham Heidelberg New York Dordrecht London: Springer International Publishing Switzerland, ISBN: 978-3-319-03403-4, Chapter 19, pp. 333–337.

Xing B and Gao W-J. (2014v) Charged system search algorithm. In: Xing B and Gao W-J (eds) *Innovative Computational Intelligence: A Rough Guide to 134 Clever Algorithms.* Cham Heidelberg New York Dordrecht London: Springer International Publishing Switzerland, ISBN: 978-3-319-03403-4, Chapter 20, pp. 339–346.

Xing B and Gao W-J. (2014w) Electromagnetism-like mechanism algorithm. In: Xing B and Gao W-J (eds) *Innovative Computational Intelligence: A Rough Guide to 134 Clever Algorithms.* Cham Heidelberg New York Dordrecht London: Springer International Publishing Switzerland, ISBN: 978-3-319-03403-4, Chapter 21, pp. 347–354.

Xing B and Gao W-J. (2014x) Gravitational search algorithm. In: Xing B and Gao W-J (eds) *Innovative Computational Intelligence: A Rough Guide to 134 Clever Algorithms.* Cham Heidelberg New York Dordrecht London: Springer International Publishing Switzerland, ISBN: 978-3-319-03403-4, Chapter 22, pp. 355–364.

Xing B and Gao W-J. (2014y) Intelligent water drops algorithm. In: Xing B and Gao W-J (eds) *Innovative Computational Intelligence: A Rough Guide to 134 Clever Algorithms.* Cham Heidelberg New York Dordrecht London: Springer International Publishing Switzerland, ISBN: 978-3-319-03403-4, Chapter 23, pp. 365–373.

Xing B and Gao W-J. (2014z) Emerging physics-based CI algorithms. In: Xing B and Gao W-J (eds) *Innovative Computational Intelligence: A Rough Guide to 134 Clever Algorithms.* Cham Heidelberg New York Dordrecht London: Springer International Publishing Switzerland, ISBN: 978-3-319-03403-4, Chapter 24, pp. 375–414.

Xing B. (2015) Optimization in production management: Economic load dispatch of cyber physical power system using artificial bee colony. In: Kahraman C and Onar SÇ (eds) *Intelligent Techniques in Engineering Management: Theory and Applications.* Cham Heidelberg New York Dordrecht London: Springer International Publishing Switzerland, ISBN 978-3-319-17905-6, Chapter 12, 275–293.

Xing B. (2016) The spread of innovatory nature originated metaheuristics in robot swarm control for smart living environments. In: Espinosa HEP (ed) *Nature-Inspired Computing for Control Systems.* Cham Heidelberg New York Dordrecht London: Springer International Publishing Switzerland, ISBN 978-3-319-26228-4, Chapter 3, pp. 39–70.

Xing B and Marwala T. (2018a) Conclusion. In: Xing B and Marwala T (eds) *Smart Maintenance for Human–Robot Interaction: An Intelligent Search Algorithmic Perspective.* Gewerbestrasse

11, 6330 Cham, Switzerland: Springer International Publishing AG, ISBN 978-3-319-67479-7, Chapter 13, pp. 299–305.

Xing B and Marwala T. (2018b) Cyberware capacity–applications layer perspective. In: Xing B and Marwala T (eds) *Smart Maintenance for Human–Robot Interaction: An Intelligent Search Algorithmic Perspective.* Gewerbestrasse 11, 6330 Cham, Switzerland: Springer International Publishing AG, ISBN 978-3-319-67479-7, Chapter 8, pp. 173–191.

Xing B and Marwala T. (2018c) Cyberware capacity–energy autonomy perspective. In: Xing B and Marwala T (eds) *Smart Maintenance for Human–Robot Interaction: An Intelligent Search Algorithmic Perspective.* Gewerbestrasse 11, 6330 Cham, Switzerland: Springer International Publishing AG, ISBN 978-3-319-67479-7, Chapter 9, pp. 193–216.

Xing B and Marwala T. (2018d) Cyberware capacity–platform and middleware layers perspective. In: Xing B and Marwala T (eds) *Smart Maintenance for Human–Robot Interaction: An Intelligent Search Algorithmic Perspective.* Gewerbestrasse 11, 6330 Cham, Switzerland: Springer International Publishing AG, ISBN 978-3-319-67479-7, Chapter 7, pp. 143–171.

Xing B and Marwala T. (2018e) Hardware capacity–beginning of life perspective. In: Xing B and Marwala T (eds) *Smart Maintenance for Human–Robot Interaction: An Intelligent Search Algorithmic Perspective.* Gewerbestrasse 11, 6330 Cham, Switzerland: Springer International Publishing AG, ISBN 978-3-319-67479-7, Chapter 4, pp. 67–91.

Xing B and Marwala T. (2018f) Hardware capacity–end of life perspective. In: Xing B and Marwala T (eds) *Smart Maintenance for Human–Robot Interaction: An Intelligent Search Algorithmic Perspective.* Gewerbestrasse 11, 6330 Cham, Switzerland: Springer International Publishing AG, ISBN 978-3-319-67479-7, Chapter 6, pp. 111–139.

Xing B and Marwala T. (2018g) Hardware capacity–middle of life perspective. In: Xing B and Marwala T (eds) *Smart Maintenance for Human–Robot Interaction: An Intelligent Search Algorithmic Perspective.* Gewerbestrasse 11, 6330 Cham, Switzerland: Springer International Publishing AG, ISBN 978-3-319-67479-7, Chapter 5, pp. 93–110.

Xing B and Marwala T. (2018h) Human capacity–biopsychosocial perspective. In: Xing B and Marwala T (eds) *Smart Maintenance for Human–Robot Interaction: An Intelligent Search Algorithmic Perspective.* Gewerbestrasse 11, 6330 Cham, Switzerland: Springer International Publishing AG, ISBN 978-3-319-67479-7, Chapter 11, pp. 249–270.

Xing B and Marwala T. (2018i) Human capacity–exposome perspective. In: Xing B and Marwala T (eds) *Smart Maintenance for Human–Robot Interaction: An Intelligent Search Algorithmic Perspective.* Gewerbestrasse 11, 6330 Cham, Switzerland: Springer International Publishing AG, ISBN 978-3-319-67479-7, Chapter 12, pp. 271–295.

Xing B and Marwala T. (2018j) Human capacity–physiology perspective. In: Xing B and Marwala T (eds) *Smart Maintenance for Human–Robot Interaction: An Intelligent Search Algorithmic Perspective.* Gewerbestrasse 11, 6330 Cham, Switzerland: Springer International Publishing AG, ISBN 978-3-319-67479-7, Chapter 10, pp. 219–247.

Xing B and Marwala T. (2018k) Introduction to human robot interaction. In: Xing B and Marwala T (eds) *Smart Maintenance for Human–Robot Interaction: An Intelligent Search Algorithmic Perspective.* Gewerbestrasse 11, 6330 Cham, Switzerland: Springer International Publishing AG, ISBN 978-3-319-67479-7, Chapter 1, pp. 3–19.

Xing B and Marwala T. (2018l) Introduction to intelligent search algorithms. In: Xing B and Marwala T (eds) *Smart Maintenance for Human–Robot Interaction: An Intelligent Search Algorithmic Perspective.* Gewerbestrasse 11, 6330 Cham, Switzerland: Springer International Publishing AG, ISBN 978-3-319-67479-7, Chapter 3, pp. 33–64.

Xing B and Marwala T. (2018m) Introduction to smart maintenance. In: Xing B and Marwala T (eds) *Smart Maintenance for Human–Robot Interaction: An Intelligent Search Algorithmic Perspective.* Gewerbestrasse 11, 6330 Cham, Switzerland: Springer International Publishing AG, ISBN 978-3-319-67479-7, Chapter 2, pp. 21–31.

Xing B and Marwala T. (2018n) *Smart maintenance for human–robot interaction: An intelligent search algorithmic perspective,* Gewerbestrasse 11, 6330 Cham, Switzerland: Springer International Publishing AG, ISBN 978-3-319-67479-7.

Yan J, Wang K, Liu Y, et al. (in press) Mining social lending motivations for loan project recommendations. *Expert Systems with Applications* doi: 10.1016/j.eswa.2017.11.010.

Yu X. (2011) Social enterprise in China: Driving forces, development patterns and legal framework. *Social Enterprise Journal* 7: 9–32.

Yum H, Lee B and Chae M. (2012) From the wisdom of crowds to my own judgment in microfinance through online peer-to-peer lending platforms. *Electronic Commerce Research and Applications* 11: 469–483.

Zhang J and Liu P. (2012) Rational herding in microloan markets. *Management Science* 58: 892–912.

Zhang K and Chen X. (2017) Herding in a P2P lending market: Rational inference or irrational trust? *Electronic Commerce Research and Applications* 23: 45–53.

Zhao H, Wu L, Liu Q, et al. (2014) Investment recommendation in P2P lending: A portfolio perspective with risk management. *Proceedings of IEEE International Conference on Data Mining (ICDM), 14~17 December, Shenzhen, China*, pp. 1109–1114.

CHAPTER 10

Crowdfunding Operator

Channel Competition, Strategic Interaction and Game Theory

10.1 Introduction

The crowdfunding channel provides an exciting way to get the sponsorship of products into the market. Take the donation/reward-based crowdfunding channel, donors typically sponsor one or more projects and they are often repaid using some rewards associated with the project's development. In practice, different platforms' foci and the accommodated project types vary greatly. For instance, Kickstarter is often regarded as a creativity-bolstering channel and its user base is often characterized by the early adoption of innovative technological and artistic artefacts. Indiegogo promotes itself as a vehicle of empowerment and passion, which makes this channel more suitable for projects proposed by not only activists and ideologists, but also non-profit campaigners. As a result, crowdfunding may not always be a one-size-fits-all solution for every business practice or product concept.

10.1.1 Successful Campaigns Using Generic Crowdfunding Channels

In general, crowdfunding channels can provide three types of service: (1) To get the funds that one needs, (2) To test one's ideas/upgrades/products before or after entering into the market and (3) To connect with consumers.

- **Individuals:** In this domain, projects are focused on the following:
    1) *Creative projects*: A good example in this category is the Oculus Rift case on Kickstarter (Kickstarter, 2014b; Chafkin, 2015).

2) *Artistic projects*: For instance, Artist Marina Abramovic has managed to raise $661,452 through Kickstarter in order to fund her dream project: Marina Abramovic Institute (Kickstarter, 2014a).
3) *Research projects*: e.g., Neurodome case on Kickstarter (Kickstarter, 2013).

- **Startups:** Take Pebble Smartwatch, three highly successful campaigns were run on the Kickstarter platform. First, in 2012, there were 68,929 backers who supported their project and the total raised funding reached around $10 million; secondly, in 2015, the number of backers reached 78,471 and the total raised funding was surprisingly $20 million, and third, in 2016, they successfully raised another $13 million (Bajarin, 2014; Chang, 2015; Alois, 2016). Other notable examples also include Scanadu Scout on Indiegogo (Indiegogo, 2013; Brooke, 2013) and Ouya on Kickstarter (Kickstarter, 2012).
- **Enterprises:** In order to absorb the creativity from grassroots and ignite the enthusiasm of the entire new community, GE's FirstBuild team (Vanderbilt, 2015) also used the crowdfunding channel and service to reach its potential customer base. For instance, the team launched two products, namely, Paragon Induction Cooktop and Opal Nugget Ice Maker. This was done using Indiegogo, which enabled these companies to acquire 10,000 new customers and secure $3 million along the way. Meanwhile, they were also ranked among the top ten of Indiegogo's most successful projects of all time (Vanderbilt, 2015). Other noteworthy cases also include a beer dispenser (called Vessi™ Fermentor) crowdfunded (over $200,000) by Whirlpool Corporation using Indiegogo (Dolan, 2016), the 'save your seat' crowdfunding campaign initiated by Heineken through Indiegogo for supporting the restoration of Miami Marine Stadium (Yeh, 2016) and the 'healthier tomorrow challenge' campaign rolled out by RB for bolstering health and well-being related startups through Indiegogo (Sustainable Brands, 2016).

10.1.2 Unsuccessful Campaigns Using Crowdfunding Channels

However, since every crowdfunding channel may have its own preferred behavioural outcomes and the associated distinct utilities, seeking an appropriate channel is not always a trivial task in practice. Take Indiegogo and Kickstarter, it is believed that virtually any kind of project is allowable through the Indiegogo channel for money raising purposes, while this is often not the case when it comes to the Kickstarter channel. Meanwhile, the Kickstarter channel follows the notorious 'all or nothing' policy, while the Indiegogo's model is a bit more gentle, that is, 'keep whatever you raise for whatever you want'.

From a customer/backer's viewpoint, most likely very little is known about, say, your motivations regarding the use of the online crowdfunding channel, or what you can do to make your ongoing project a successful one. Take the pre-ordering type of crowdfunding campaign, though it is often focused on the specialized creative products, most backers do not usually favour free gifts (e.g., a T-shirt) but prefer other forms of rewards (e.g., possible price discrimination). Sometimes campaigns are running perfectly within the first 24 hours, but unfortunately could not make it in the end (e.g., Ubuntu Edge Phone through Indiegogo channel).

- **Multiple Channel:** Some businesses tend to submit their projects to more than one crowdfunding platform at a time. For example, a company called Skarp launched their product—Laser Razor using two channels, i.e., Kickstarter and Indiegogo, but failed eventually.
- **Own Channel:** Behind some later successful campaigns, there are also different stories of failure in the beginning. However, the success finally arrived in other forms, say, using the entrepreneurs' own established crowdfunding channel. Take the Lockitron case, when Kickstarter founders initially rejected it the entrepreneurs managed to set up their own website in order to facilitate crowdfunding Lockitron, a phone-aided door-locking equipment. As soon as the project was launched, their initial $150,000 fundraising goal was reached (in pre-orders) after only 24 hours. When the scheduled 30 day campaign reached its end, their product was reserved by nearly 15,000 people and the total advance orders was valued at $2.3 million (Goodman, 2013; Bergl, 2013).
- **Specific Channel:** Take Cultivate Ignite, it allows you to request for concepts and to allocate an amount of budget to each one of your employees for evaluating and investing their favourite ideas. Another case is Patreon, which serves as a monthly subscription crowdfunding channel allowing backers or donors to make regular contributions instead of a single bulk input (Mixon et al., 2017). Other examples include PledgeMusic for musicians, Razoo and Crowdrise for 'worthy causes' beyond simply goods and profits, Fundrise for real estate and Appbacker for mobile app developers.

10.1.3 Specialized Crowdfunding Channel Trials by Incumbents

Conventionally, the customers serviced by generic crowdfunding channels are mainly individuals, startups and small companies. However, Indiegogo recently proposed a new service, the Enterprise Crowdfunding (Kastrenakes, 2016), which focuses on helping large corporations crowdfund their next key products. Meanwhile, several existing establishments have signalled their interest in setting up their own crowdfunding channels so that their employees are encouraged to devise now concepts and are supported with internal resources. Several examples are outlined as follows:

- **IBM's iFundIT:** IBM has its own crowdfunding platform for assisting projects with potential. Similar to some leading channels (e.g., Kickstarter and Indiegogo), project askers can spur their concepts within the enterprise's internal social network. The iFundIT program initially received about $300,000 in seed grants from the parent company (Luckerson, 2014).
- **P&G's 'Connect + Develop':** This program was designed for strengthening partnerships and meeting various needs across the P&G business (Procter & Gamble, 2017).
- **Dell's IdeaStorm:** Customers are able to present, vote and comment using this channel, in particular on ideas that they wish Dell to implement (Bayus, 2013; Dell, 2017).

10.2 Problem Statement

Crowdfunding is rapidly becoming a widely used channel, through which people can raise funds simply by using Internet services. In fact, it brings not only a fund-raising chance, but also comes with other huge potentials, for example, as a marketing channel (Agrawal et al., 2014; Arkrot et al., 2017; Brown et al., 2017; Gerber and Hui, 2013; Sheldon and Kupp, 2017), an effective customer engagement channel (Paykacheva, 2014; Mustafa and Adnan, 2017; Marchegiani, in press) and as a bridge between institutional or professional investors as well as start-ups or households (Hernando, 2017; Ordanini et al., 2011; Roma et al., 2017). However, despite variations in focus, the significant interdependencies between product firms and service providers (i.e., crowdfunding platforms) have not been widely investigated. Indeed, this interdependency can potentially help to create a unified channel that allows askers to appeal to a large number of potential backers/investors for much more than just financial support. In addition, more and more manufacturers have started to offer services integrated products in order to gain various competitive advantages, such as reverse logistic (Xing et al., 2010b; Xing and Gao, 2014gg; Xing and Gao, 2014hh; Xing and Gao, 2014kk; Xing and Gao, 2015c; Xing et al., 2010a) and remanufacturing (Xing and Gao, 2014ii; Xing and Gao, 2014ee; Marwala and Xing, 2011; Xing and Gao, 2014a; Xing and Gao, 2015a; Xing and Gao, 2014jj; Xing and Gao, 2015b). Under these situations, we should pay more attention to 'which' appropriate channel structure could potentially benefit the greatest number of people. Therefore, this section outlines the following generic questions encountered by platform operators within a crowdfunding ecosystem.

10.2.1 Question 10.1

The relationship between a company's performance and its servitization degree is often influenced by different factors, say, the innovation and complexity of the offered product, the targeted business areas, the structure of organization and the investment intensity. Based on this observation, Question 10.1 is proposed as follows:

- **Question 10.1:** *When can we regard servitization as a competitive strategy in terms of profit making?*

  By investigating Question 10.1, our expectations are twofold:

  1) Be able to perform a theoretical investigation on different channel situations correlated to the servitization strategy's superior performance under the equilibrium of price and quality, respectively;
  2) Be able to identify the influencing elements that have an impact on the anticipated payoffs for a producer who is considering the implementation of servitization strategy.

10.2.2 Question 10.2

The identification of what are the optimum decisions and how they can be reached is always the main theme of decision theory. Many theories related to decision-making are often normative and prescriptive under a simplified assumption that people are fully informed and behave rationally. However, the findings derived from the experimental economics and psychological studies have demonstrated the mismatches between the real evidence regarding human behaviour and the corresponding theoretical assumption made in classical decision theory (Yukalov and Sornette, 2011). In other words, people often tend to show deviations from ideal rationality when performing decision-making due to the inherent cognitive and emotional biases. In the literature, many attempts, e.g., bounded rationality theory proposed in (Simon, 1955) and artificial intelligence techniques (Marwala, 2014a; Marwala, 2015), have been made to account for various paradoxes such as the Allais paradox (Cartwright, 2011) and various other paradoxical cases (Camerer et al., 2004; Levine, 2012). Nevertheless, we are still far from understanding decision-maker's behaviour in a precise way. Based on this concern, Question 10.2 is proposed in the following way.

- **Question 10.2:** *How do we acquire a probabilistic analysis of a decision-maker's behaviour given the availability of multi-level servitization strategies?*

By investigating Question 10.1, we aim to find a way to compute a decision-maker's strategic state according to admissible strategy numbers and quantized strategy dominance.

10.3 Game Theory for Question 10.1

Game theory often finds its applications in many areas, e.g., modelling multi-agent systems (Marwala and Hurwitz, 2017; Marwala, 2013; Mathieu et al., 2006). In order to address Question 10.1, a pioneer work was conducted in (Lee et al., 2016) in order to deal with the channel competition issue emerging from the introduction of a new channel where products with full servitization features are offered.

10.3.1 Experimental Setting

The model proposed by (Lee et al., 2016) demonstrated the competition of channel between two companies in the marketplace regarding price and quality. Terms used in this model are depicted in Fig. 10.1.

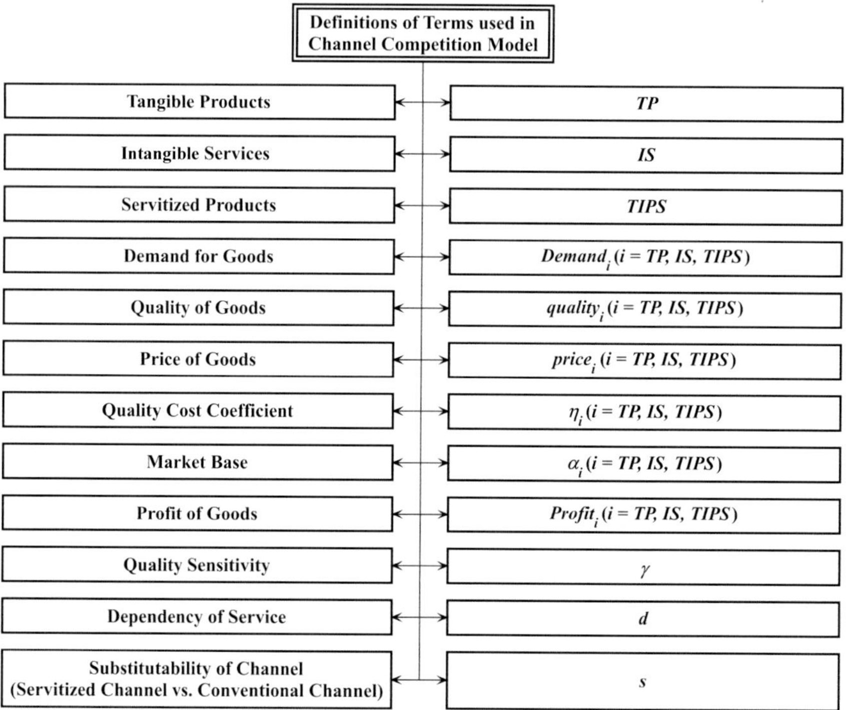

Figure 10.1: Definitions of terms used in channel competition model.

Detailed settings are outlined as follows (Lee et al., 2016):

- **Channel:** (1) In a customary channel, a producer offers tangible products (*TP*), and a service company supplies intangible services (*IS*) to customers. The producer and service company are independent of each other. (2) In a servitized channel, servitized products (*TIPS*), weaving the features of tangible goods and intangible services together, are offered by a company. Accordingly, competition happens between the union of a producer and service supplier (in customary channel) and the servitized company (in servitized channel).
- **Game:** A two-stage game is set to be played by actors involved in this competition. At stage 1, the levels of quality (*quality*) are simultaneously chosen by all players and then the corresponding price (*price*) levels are determined at stage 2. Demand (*Demand*) is therefore influenced by these factors, in particular, a linear function was employed in (Lee et al., 2016) in order to model the associated relationship by following typical practices found in the literature, e.g., (Banker et al., 1998; Bernstein and Federgruen, 2004). Since a demand function describes a relationship between the demanded quantity and diverse factors that have impacted the demand, interested readers please refer to (Huang et al., 2013) for a detailed explanation of various demand functions.
- **Parameter:** Two parameters were given special attention in (Lee et al., 2016), that is, dependency of service (d) and substitutability of channel (s). (1) The dependency of service means the degree of service needed for product's usage and, thus, d falls within the range of [0,1]. As d gets close to its right boundary, services become necessary for a proper utilization of products; when d approaches zero, the dependence on services will be reduced to a very low level. (2) The substitutability level between the customary channel and its competitive servitized one is denoted by s which also falls within the range of [0,1]. In a similar manner, the substitutability is high between the integrated (goods and services) and separated (goods and services) situations when s gets close to one; while, as s approaches zero, two channels will co-exist and enjoy the dominant position in their respective market segments.
- **Demand Function:** Due to the nature of channel structures' asymmetry, three demand functions were defined in (Lee et al., 2016) for *TP*, *IS*, and *TIPS*, respectively. For a conventional channel, α_{TP} is used to denote the intrinsic demand for goods, i.e., the market base for goods. The market base for services is indicated by α_{IS} which is transformable into $d_{\alpha_{IS}}$ according to the market service dependency. Built on these definitions, the demand functions can be formulated as Eqs. 10.1–10.3 (Lee et al., 2016):

$$Demand_{TP} = \alpha_{TP} - price_{TP} - d_{p_{IS}} + s_{p_{TIPS}} + \gamma \cdot \left(quality_{TP} + d_{p_{IS}} - s_{p_{TIPS}} \right). \quad 10.1$$

$$\begin{aligned} Demand_{TIPS} &= \alpha_{TIPS} - price_{TIPS} + s \cdot \left(price_{TP} + d_{p_{IS}} \right) \\ &+ \gamma \cdot \left[quality_{TIPS} - s \cdot \left(quality_{TP} + d_{q_{IS}} \right) \right] \end{aligned}. \quad 10.2$$

$$Demand_{IS} = \alpha_{IS} - price_{IS} + \gamma \cdot quality_{IS}. \quad 10.3$$

Here, Eq. 10.1 shows that the demand for goods is linearly affected by factors from two channels, namely, (1) Conventional channel's quality and price of tangible products and intangible services, and (2) Servitized channel's quality and price of servitized products. A similar pattern can also be observed in Eq. 10.2 where demand for servitized products is defined.

- **Profit Function:** The convex cost functions (Boyd and Vandenberghe, 2004) were employed in (Lee et al., 2016) in order to express competing companies' quality level. The cost of reaching the desired quality (i.e., the degree of difficulty) is denoted by coefficient η. Accordingly, the profit functions can be formulated as Eqs. 10.4–10.6 (Lee et al., 2016):

$$Profit_{TP} = Demand_{TP} price_{TP} - \tfrac{1}{2}\eta_{TP} \left(quality_{TP} \right)^2. \quad 10.4$$

$$Profit_{IS} = Demand_{IS} price_{IS} - \tfrac{1}{2}\eta_{IS} \left(quality_{IS} \right)^2. \quad 10.5$$

$$Profit_{TIPS} = Demand_{TIPS} price_{TIPS} - \tfrac{1}{2}\eta_{TIPS} \left(quality_{TIPS} \right)^2. \quad 10.6$$

10.3.1.1 Scenario 1: Producer and Service Supplier are Independent of Each Other

In scenario 1, we find that the tangible product producer and intangible service supplier are typically independent of each other and coexist within the conventional channel ecosystem (see Fig. 10.2).

Under this circumstance, product producers and service suppliers offer tangible products and intangible services, respectively, to customers via separately determined optimal quality and price. In the same manner, servitized goods are provided to customers by servitized firms under their own optimal quality and price strategies. In order to maximize profit, each company's profit (denoted by $Profit_{TP}$, $Profit_{IS}$, and $Profit_{TIPS}$, respectively) should have concavity conditions under the equilibrium situation of quality and price. In the optimization area, a set $X \subseteq \mathbb{R}^n$ (Euclidean n-space

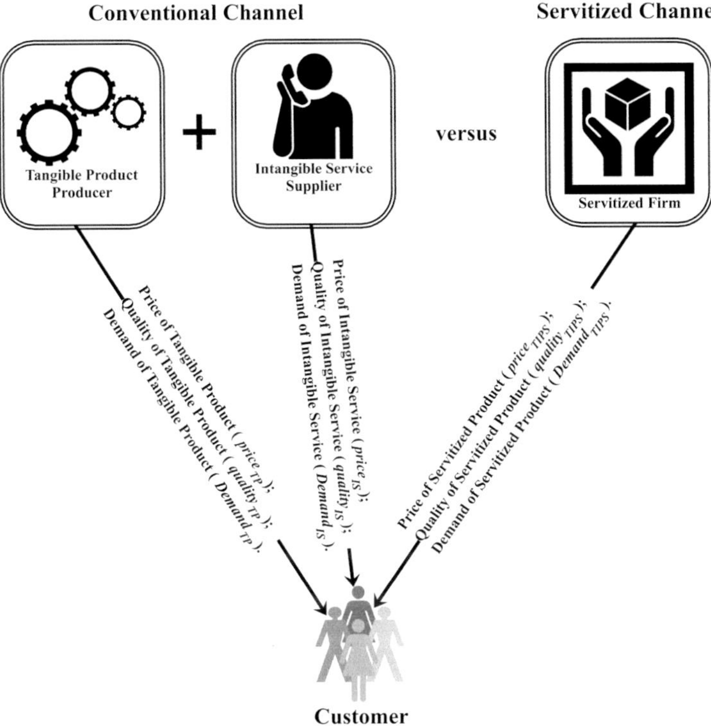

Figure 10.2: Channel competition: Scenario 1.

is denoted by $\mathbb{R}^n$) is convex if for any $x, y \in X$ and any $\theta \in (0,1)$, Eq. 10.7 holds (Butenko and Pardalos, 2014):

$$\theta x + (1 - \theta) \cdot y \in X. \tag{10.7}$$

In other words, a set can be regarded as roughly convex if every point within the set is observable by every other point using an unobstructed (i.e., lying in the set) straight path between them (Boyd and Vandenberghe, 2004). See Fig. 10.3 for illustration.

Given that the domain of a function (denoted by **dom** f) is a convex set, the function $f : \mathfrak{R}^n \rightarrow \mathfrak{R}$ is convex if for any $x, y \in$ **dom** f and $\varphi \in [0,1]$, Eq. 10.8 holds (Boyd and Vandenberghe, 2004):

$$f\left[\varphi \cdot x + (1-\varphi) \cdot y\right] \leq \varphi \cdot f(x) + (1-\varphi) \cdot f(y). \tag{10.8}$$

Geometrically, the meaning of this inequality refers to the graph of a convex function plotted over the range of $[x,y]$ which lies on or below the line segment between two points, that is, $(x,f(x))$ and $(y,f(y))$, as illustrated in Fig. 10.4.

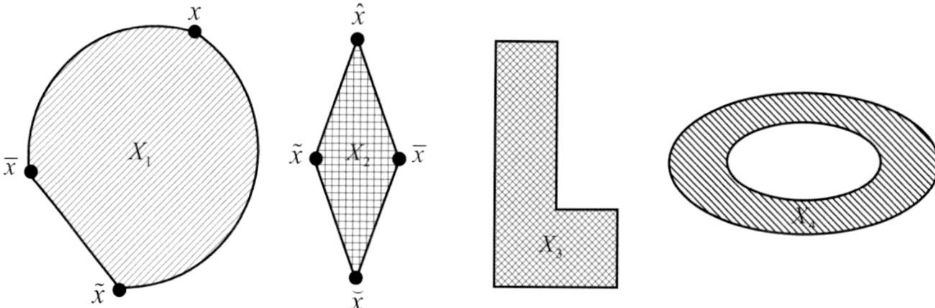

Figure 10.3: Examples: convex (X_1 and X_2) sets vs. nonconvex (X_3 and X_4) sets (numbering from left to right with X_1 on the left).

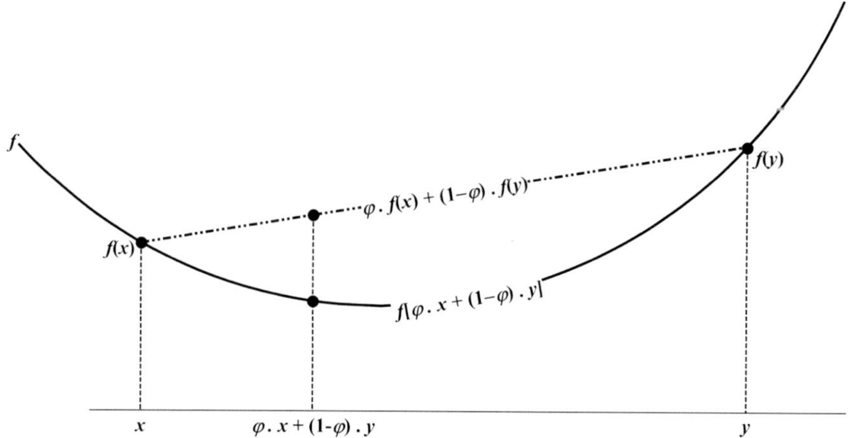

Figure 10.4: Geometric depiction of a convex function.

Meanwhile, a function (*f*) is strictly convex if strict inequality holds in Eq. $f\left[\varphi\cdot x+(1-\varphi)\cdot y\right]\leq\varphi\cdot f(x)+(1-\varphi)\cdot f(y)$ whenever $x \neq y$, that is Eq. 10.9 holds (Butenko and Pardalos, 2014):

$$f\left[\varphi\cdot x+(1-\varphi)\cdot y\right]<\varphi\cdot f(x)+(1-\varphi)\cdot f(y);\ \forall x,y\in X,\ y\neq x,\ \varphi\in(0,1). \quad 10.9$$

Accordingly, we can call a function (*f*) concave if $-f$ is convex, and strictly concave if $-f$ is strictly convex.

Built on these properties, the first proposition proposed by (Lee et al., 2016) can be concluded as follows: The company's profit can be regarded as strictly concave in price if quality of the coefficient of quality cost is adequately greater than the sensitivity of quality. In particular, the tangible product producer's profit ($Profit_{TP}$) is strictly concave in $price_{TP}$ and $quality_{TP}$ if $\eta_{TP} > \frac{\gamma^2}{2}$. The intangible service supplier's profit ($Profit_{IS}$) is strictly concave in $price_{IS}$ and $quality_{IS}$ if $\eta_{IS} > \frac{\gamma^2}{2}$. The servitized firm's profit ($Profit_{TIPS}$) is strictly concave in $price_{TIPS}$ and $quality_{TIPS}$ if $\eta_{TIPS} > \frac{\gamma^2}{2}$.

By solving the model using backward induction, we can obtain prices of each offering using Eqs. 10.10–10.12 (Lee et al., 2016):

$$price^*_{TP} = \frac{2\gamma\cdot\left(s^2-2\right)\cdot\left(2quality_{TP}+d_{qIS}\right)+\left(2d-4-s^2\cdot d\right)\cdot\alpha_{TP}+2s\cdot\left(\gamma\cdot quality_{TIPS}-\alpha_{TIPS}\right)}{2\left(s^2-4\right)}. \quad 10.10$$

$$price^*_{IS} = \frac{\gamma\cdot quality_{IS}+d\cdot\alpha_{TP}}{2}. \quad 10.11$$

$$price^*_{TIPS} = \frac{2\gamma\cdot\left(s^2-2\right)\cdot quality_{TIPS}+\gamma\cdot s\cdot\left(2quality_{TP}+d_{qIS}\right)\cdot 2quality_{TP}-\left(2+d^2\right)\cdot s\cdot\alpha_{TP}-4\alpha_{TIPS}}{2\left(s^2-4\right)}. \quad 10.12$$

Based on these equilibrium conditions, we can then verify that the quality of non-servitized (servitized) goods have a positive (negative) impact on the price of non-servitized goods. By inserting Eq. 10.10 back into Eq. 10.4, respectively, we can get the optimal quality levels under the concave situations as given by Eqs. 10.13–10.15 (Lee et al., 2016):

$$quality^*_{TP} = \frac{4\gamma^3\cdot S_2^2\cdot\left\{s\cdot\alpha_{TIPS}\cdot\Gamma_{2IS}+\alpha_{TP}\cdot\left[\Gamma_{2IS}+d^2\cdot\Gamma_{IS}\cdot(1+S_2)\right]\right\}-2\gamma\cdot\eta_{TIPS}\cdot S_2\left[\left(2\alpha_{TP}+s\cdot\alpha_{TIPS}\right)\cdot\Gamma_{2IS}+d^2\cdot\alpha_{TP}\cdot\Gamma_{IS}\cdot S_2\right]\cdot S_4}{\Gamma_{2IS}\cdot\left[4\gamma^4\cdot S_2^2\cdot(1+S_2)-2\gamma^2\cdot(\eta_{TP}+\eta_{TIPS})\cdot S_2^2\cdot S_4+\eta_{TP}\cdot\eta_{TIPS}\cdot S_4^3\right]}. \quad 10.13$$

$$quality^*_{IS} = -\frac{\gamma\cdot d\cdot\alpha_{TP}}{\Gamma_{2IS}}. \quad 10.14$$

$$quality^*_{TIPS} = \frac{4\gamma^3\cdot(d\cdot\alpha_{TP}+\alpha_{TIPS})\cdot S_2^2-2\gamma\cdot\left[2\alpha_{TIPS}\cdot\Gamma_{2IS}+s\cdot\alpha_{TP}\cdot\left(d^2\cdot\Gamma_{IS}+\Gamma_{2IS}\right)\right]\cdot\eta_{TP}\cdot S_2\cdot S_4}{\Gamma_{2IS}\cdot\left[4\gamma^4\cdot S_2^2\cdot(1+S_2)-2\gamma^2\cdot(\eta_{TP}+\eta_{TIPS})\cdot S_2^2\cdot S_4+\eta_{TP}\cdot\eta_{TIPS}\cdot S_4^3\right]}. \quad 10.15$$

where $\Gamma_{2TIPS} = \gamma^2 - 2\eta_{TIPS}$, $\Gamma_{TIPS} = \gamma^2 - \eta_{TIPS}$, $\Gamma_{2IS} = \gamma^2 - 2\eta_{IS}$, $\Gamma_{IS} = \gamma^2 - \eta_{IS}$, $\Gamma_{2TP} = \gamma^2 - 2\eta_{TP}$, $\Gamma_{TP} = \gamma^2 - \eta_{TP}$, $S_1 = s^2 - 1$, $S_2 = s^2 - 2$, and $S_4 = s^2 - 4$.

Furthermore, if we insert Eq. 10.13 back into Eq. 10.10, respectively, the optimal price levels can be acquired using Eqs. 10.16–10.18 (Lee et al., 2016):

$$price^*_{TP} = \frac{S_4\cdot\eta_{TP}\cdot\left\{\alpha_{TP}\cdot\left\{S_4\cdot\eta_{TIPS}\cdot\left[\left(d^2\cdot S_2+4\right)\cdot\eta_{IS}-\gamma^2\cdot\left(d^2\cdot S_2+2\right)\right]+2\gamma^2\cdot S_2\cdot\left[\left(d^2\cdot S_1+2\right)\cdot\Gamma_{IS}-\gamma^2\right]\right\}+s\cdot\alpha_{TIPS}\cdot\Gamma_{2IS}\cdot\left(2\gamma^2\cdot S_2-S_4\cdot\eta_{TIPS}\right)\right\}}{\Gamma_{2IS}\left[S_4\cdot\eta_{TP}\cdot\left(S_4^2\cdot\eta_{TIPS}-2\gamma^2\cdot S_2^2\right)+2\gamma^2\cdot S_2^2\cdot\left(2\gamma^2\cdot S_1-S_4\cdot\eta_{TIPS}\right)\right]}. \quad 10.16$$

$$price^*_{IS} = -\frac{d\cdot\alpha_{TP}\cdot\eta_{IS}}{\Gamma_{2IS}}. \quad 10.17$$

$$price^*_{TIPS} = \frac{S_4 \cdot \eta_{TIPS} \cdot \left\{2\alpha_{TIPS} \cdot \Gamma_{2IS} \cdot \left(\gamma^2 \cdot S_2 - S_4 \cdot \eta_{TP}\right) + d \cdot \alpha_{TP} \cdot \left\{2\gamma^2 \cdot S_2 \cdot \Gamma_{2IS} - S_4 \cdot \eta_4 \cdot \left[\left(d^2+2\right) \cdot \Gamma_{IS} - \gamma^2\right]\right\}\right\}}{\Gamma_{2IS}\left[S_4 \cdot \eta_{TP} \cdot \left(S_4^2 \cdot \eta_{TIPS} - 2\gamma^2 \cdot S_2^2\right) + 2\gamma^2 \cdot S_2^2 \cdot \left(2\gamma^2 \cdot S_1 - S_4 \cdot \eta_{TIPS}\right)\right]}. \quad 10.18$$

With the above-mentioned information at hand, we can calculate the corresponding profits. By comparing the profits of tangible product producer and servitized firm, the conditions of enjoying a competitive advantage were proposed by (Lee et al., 2016) in their second proposition. There exist threshold values for both service dependency (*d*) and channel substitutability (*s*) so that the profit of the servitized firm ($Profit_{TIPS}$) is better off than the tangible product producer's profit ($Profit_{TP}$) when (1) both *d* and *s* are greater than the threshold values, and (2) $\alpha_{TIPS} \cdot \eta > \frac{\gamma^2}{2}$ (assuming $\eta = \eta_{TP} = \eta_{IS} = \eta_{TIPS}$ given the cost structure is not a concern). The relevant proof of this proposition can be found in (Lee et al., 2016).

10.3.1.2 Scenario 2: Producer and Service Supplier are Integrated

In this scenario (as illustrated in Fig. 10.5), tangible product producer and intangible service supplier coexist as an integrated entity (the former integrates the latter, or vice versa). In line with some common practices, as mentioned in (Lee et al., 2016; Fang et al., 2008; Han et al., 2013; Suarez et al., 2013), the price and quality for products and services are determined separately by the integrated players.

Furthermore, the combined profit (denoted by $Profit_{TP+IS}$) is introduced in (Lee et al., 2016) as follows: $Profit_{TP+IS}$ is strictly concave in $price_{TP}$, $price_{IS}$, $quality_{TP}$, and $quality_{IS}$; while $Profit_{TIPS}$ is strictly concave in $price_{TIPS}$ and $quality_{TIPS}$ if $\eta_{TIPS} > \frac{\gamma^2}{2}$. By solving this model using backward induction, the equilibrium price levels can be estimated using Eqs. 10.19–10.21 (Lee et al., 2016):

$$price^*_{TP} = \frac{\left[d \cdot \left(d \cdot \alpha_{TP} - \gamma \cdot quality_{IS}\right) - 2\gamma \cdot quality_{TP}\right] \cdot \left(s^2 - 2\right) + 4\alpha_{TP} + 2s \cdot \left(\alpha_{TIPS} - \gamma \cdot quality_{TIPS}\right)}{d^2 \cdot \left(s^2 - 2\right) - 2s^2 + 8}. \quad 10.19$$

$$price^*_{IS} = \frac{\left(\gamma \cdot d \cdot quality_{TP} + \gamma \cdot d^2 \cdot quality_{IS} - d \cdot \alpha_{TP}\right) \cdot \left(s^2 - 2\right) + \gamma \cdot d \cdot s \cdot quality_{TIPS} + \left(4 - s^2\right) \cdot \gamma \cdot quality_{IS} - d \cdot s \cdot \alpha_{TIPS}}{d^2 \cdot \left(s^2 - 2\right) - 2s^2 + 8}. \quad 10.20$$

$$price^*_{TIPS} = \frac{\gamma \cdot \left[4 - d^2 + \left(d^2 - 2\right) \cdot s^2\right] \cdot quality_{TIPS} - 2\gamma \cdot s \cdot quality_{TP} - \gamma \cdot d \cdot s \cdot quality_{IS} + 2s \cdot \alpha_{TP} + \left(4 - d^2\right) \cdot \alpha_{TIPS}}{d^2 \cdot \left(s^2 - 2\right) - 2s^2 + 8}. \quad 10.21$$

By following the similar steps in Scenario 1, we can get the optimal quality levels under the concave situations using Eqs. 10.22–10.24 (Lee et al., 2016):

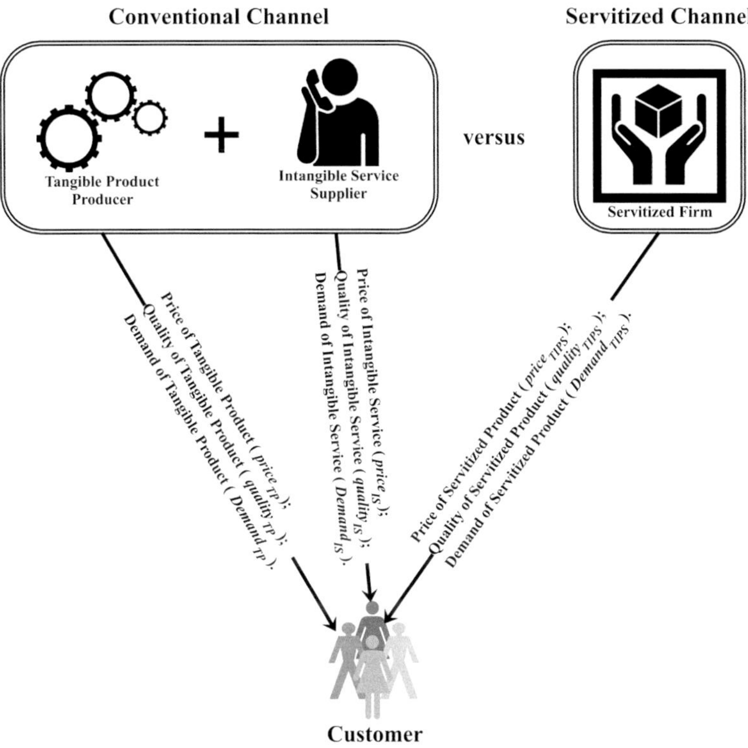

Figure 10.5: Channel competition: Scenario 2.

$$quality^*_{TP} = \frac{\begin{pmatrix} \gamma \cdot D_4 \cdot S_2 \cdot \alpha_{TIPS} \cdot \left(2\gamma^2 \cdot (D_2 \cdot S_1 + 2) - (D_2 \cdot S_2 + 4) \cdot \eta_{TIPS}\right) \cdot \Gamma_{2IS} \\ + \alpha_{TP} \left(\left(2\gamma^2 (D_2 \cdot S_1 + 2)\right) \cdot \left(\Gamma_{2IS} - d^2 \cdot S_1 \cdot \Gamma_{IS}\right) - (D_2 \cdot S_2 + 4) \cdot \eta_{TIPS} \cdot \left(2\Gamma_{2IS} - d^2 \cdot S_2 \cdot \Gamma_{IS}\right)\right) \end{pmatrix}}{\begin{pmatrix} \gamma \cdot D_4 \cdot S_2 \cdot \left(2\gamma^2 \cdot S_1 \cdot (D_2 \cdot S_1 + 2) - S_2 \cdot (D_2 \cdot S_2 + 4) \cdot \eta_{TIPS}\right) \cdot \Gamma_{2IS} + \\ \eta_{TP} \cdot \begin{pmatrix} 2\gamma^4 \cdot \left(d^2 \cdot S_1 - 2S_2\right) \cdot \left(d^2 \cdot (3S_2 + 2) + 2S_2 \cdot S_4\right) \\ + \left(2S_4 - d^2 \cdot S_2\right) \cdot \left(2\gamma^2 \cdot \left(S_4^2 + 2d^2 \cdot S_2\right) \cdot \eta_{TIPS} + \left(2\eta^2 \cdot (D_2 \cdot S_1 + 2)^2 - (D_2 \cdot S_2 + 4)^2 \cdot \eta_{TIPS}\right) \cdot \eta_{IS}\right) \end{pmatrix} \end{pmatrix}}. \quad 10.22$$

$$quality^*_{IS} = \frac{- \left(\gamma \cdot d \cdot \begin{pmatrix} \left(D_4 \cdot s \cdot S_2 \cdot \alpha_{TIPS} \cdot \eta_{TP}\right) \cdot \left(2\gamma^2 \cdot (D_2 \cdot S_1 + 2) - (D_2 \cdot S_2 + 4) \cdot \eta_{TIPS}\right) \\ + \alpha_{TP} \begin{pmatrix} \left(2\gamma^4 \cdot D_4 \cdot S_2 \cdot S_1 \cdot \left(d^2 \cdot S_1 - 2S_2\right) + 2\gamma^2 \cdot \left(d^2 \cdot S_1 - 2S_2\right) \cdot \begin{pmatrix} 2S_2 \cdot (S_4 - 2) + d^4 \cdot S_2 \cdot S_1 \\ -2d^2 \cdot \left(7 - 8s^2 + 2s^4\right) \end{pmatrix} \cdot \eta_{TP} \right) \\ + \left(2S_4 - d^2 \cdot S_2\right) \cdot \left(\gamma^2 \cdot D_4 \cdot S_2^2 + \begin{pmatrix} d^4 \cdot S_2^2 + d^2 \cdot \left(22s^2 - 4s^4 - 28\right) \\ +2\left(24 - 12s^2 + s^4\right) \end{pmatrix} \cdot \eta_{TP} \right) \cdot \eta_{TIPS} \end{pmatrix} \end{pmatrix} \right)}{\begin{pmatrix} \gamma \cdot D_4 \cdot S_2 \cdot \left(2\gamma^2 \cdot S_1 \cdot (D_2 \cdot S_1 + 2) - S_2 \cdot (D_2 \cdot S_2 + 4) \cdot \eta_{TIPS}\right) \cdot \Gamma_{2IS} + \\ \eta_{TP} \cdot \begin{pmatrix} 2\gamma^4 \cdot \left(d^2 \cdot S_1 - 2S_2\right) \cdot \left(d^2 \cdot (3S_2 + 2) + 2S_2 \cdot S_4\right) \\ + \left(2S_4 - d^2 \cdot S_2\right) \cdot \left(2\gamma^2 \cdot \left(S_4^2 + 2d^2 \cdot S_2\right) \cdot \eta_{TIPS} + \left(2\eta^2 \cdot (D_2 \cdot S_1 + 2)^2 - (D_2 \cdot S_2 + 4)^2 \cdot \eta_{TIPS}\right) \cdot \eta_{IS}\right) \end{pmatrix} \end{pmatrix}}. \quad 10.23$$

$$quality^*_{TIPS} = \frac{\left(2\gamma\cdot\left(d^2\cdot S_1-2S_2\right)\cdot\left(\begin{array}{l} s\cdot\alpha_{TP}\left(\gamma^2\cdot D_4\cdot S_2\cdot\Gamma_{2IS}+\eta_{TP}\cdot\left(\gamma^2\left(2S_4+2d^2\cdot S_3-d^4\cdot S_2\right)+2(D_2S_2+4)\cdot\eta_{IS}\right)\right)\\ +D_4\cdot\alpha_{TIPS}\cdot\left(\gamma^2\cdot S_2\cdot\Gamma_{2IS}-\eta_{TP}\left(\gamma^2\cdot S_4+(D_2\cdot S_2+4)\eta_{IS}\right)\right)\end{array}\right)\right)}{\left(\begin{array}{l}\gamma\cdot D_4\cdot S_2\cdot\left(2\gamma^2\cdot S_1(D_2\cdot S_1+2)-S_2\cdot(D_2\cdot S_2+4)\cdot\eta_{TIPS}\right)\cdot\Gamma_{2IS}+\\ \eta_{TP}\cdot\left(\begin{array}{l}2\gamma^4\left(d^2\cdot S_1-2S_2\right)\cdot\left(d^2\cdot(3S_2+2)+2S_2\cdot S_4\right)+\\ \left(2S_4-d^2\cdot S_2\right)\cdot\left(2\gamma^2\cdot\left(S_4^2+2d^2\cdot S_2\right)\cdot\eta_{TIPS}+\left(2\gamma^2\cdot(D_2\cdot S_1+2)^2-(D_s\cdot S_2+4)^2\cdot\eta_{TIPS}\right)\cdot\eta_{IS}\right)\end{array}\right)\end{array}\right)}. \quad 10.24$$

Where $\Gamma_{2TIPS} = \gamma^2 - 2\eta_{TIPS}$, $\Gamma_{TIPS} = \gamma^2 - \eta_{TIPS}$, $\Gamma_{2IS} = \gamma^2 - 2\eta_{IS}$, $\Gamma_{IS} = \gamma^2 - \eta_{IS}$, $\Gamma_{2TP} = \gamma^2 - 2\eta_{TP}$, $\Gamma_{TP} = \gamma^2 - \eta_{TP}$, $S_1 = s^2 - 1$, $S_2 = s^2 - 2$, $S_4 = s^2 - 4$, $D_4 = d^2 - 4$, and $D_2 = d^2 - 2$.

Accordingly, the optimal price levels can also be obtained using Eqs. 10.25–10.27 (Lee et al., 2016):

$$price^*_{TP} = \frac{-\left(\left(d^2\cdot S_2-2S_4\right)\cdot\eta_{TP}\cdot\left(\begin{array}{l}s\cdot\alpha_{TIPS}\cdot\left(2\gamma^2\cdot(D_2\cdot S_1+2)-(D_2\cdot S_2+4)\cdot\eta_{TIPS}\right)\cdot\Gamma_{2IS}+\\ \alpha_{TP}\cdot\left(\begin{array}{l}\left(2\gamma^2\cdot(D_2\cdot S_1+2)\right)\cdot\left(\Gamma_{2IS}-d^2\cdot S_1\cdot\Gamma_{IS}\right)-\\ (D_2\cdot S_2+4)\cdot\eta_{TIPS}\cdot\left(2\Gamma_{2IS}-d^2\cdot S_2\cdot\Gamma_{IS}\right)\end{array}\right)\end{array}\right)\right)}{\left(\begin{array}{l}\gamma\cdot D_4\cdot S_2\cdot\left(2\gamma^2\cdot S_1(D_2\cdot S_1+2)-S_2\cdot(D_2\cdot S_2+4)\cdot\eta_{TIPS}\right)\cdot\Gamma_{2IS}+\\ \eta_{TP}\cdot\left(\begin{array}{l}2\gamma^4\left(d^2\cdot S_1-2S_2\right)\cdot\left(d^2\cdot(3S_2+2)+2S_2\cdot S_4\right)+\\ \left(2S_4-d^2\cdot S_2\right)\cdot\left(2\gamma^2\cdot\left(S_4^2+2d^2\cdot S_2\right)\cdot\eta_{TIPS}+\left(2\gamma^2\cdot(D_2\cdot S_1+2)^2-(D_s\cdot S_2+4)^2\cdot\eta_{TIPS}\right)\cdot\eta_{IS}\right)\end{array}\right)\end{array}\right)}. \quad 10.25$$

$$price^*_{IS} = \frac{-\left(d\cdot\left(s\cdot\alpha_{TIPS}\cdot\eta_{TP}\cdot\left(\begin{array}{l}\left(2\gamma^2\cdot(D_s\cdot S_1+2)-(D_2\cdot S_2+4)\cdot\eta_{TIPS}\right)\cdot\left(\gamma^2\cdot(D_3\cdot S_2+2)-(D_2\cdot S_2+4)\cdot\eta_{TIPS}\right)+\\ \alpha_{TP}\cdot\left(\begin{array}{l}\left(\begin{array}{l}2\gamma^4\cdot D_4\cdot S_2\cdot S_1\cdot\left(d^2\cdot S_1-2D_2\right)+\\ 2\gamma^2\cdot\left(d^2\cdot S_1-2S_2\right)\cdot\left(\begin{array}{l}2S_2\cdot(S_4-2)+d^4\cdot S_2\cdot S_1\\ -2d^2\cdot\left(2s^4-8s^2+7\right)\end{array}\right)\cdot\eta_{TP}\end{array}\right)+\\ \left(2S_4-d^2\cdot S_2\right)\cdot\left(\gamma^2\cdot D_4\cdot S_2^2+\left(\begin{array}{l}d^4\cdot S_2^2+d^2\left(22s^2-4s^4-28\right)+\\ 2\left(s^4-12s^2+24\right)\end{array}\right)\eta_{TP}\right)\cdot\eta_{TIPS}\end{array}\right)\end{array}\right)\right)\right)}{\left(\begin{array}{l}\gamma\cdot D_4\cdot S_2\cdot\left(2\gamma^2\cdot S_1(D_2\cdot S_1+2)-S_2\cdot(D_2\cdot S_2+4)\cdot\eta_{TIPS}\right)\cdot\Gamma_{2IS}+\\ \eta_{TP}\cdot\left(\begin{array}{l}2\gamma^4\left(d^2\cdot S_1-2S_2\right)\cdot\left(d^2\cdot(3S_2+2)+2S_2\cdot S_4\right)+\\ \left(2S_4-d^2\cdot S_2\right)\cdot\left(2\gamma^2\cdot\left(S_4^2+2d^2\cdot S_2\right)\cdot\eta_{TIPS}+\left(2\gamma^2\cdot(D_2\cdot S_1+2)^2-(D_s\cdot S_2+4)^2\cdot\eta_{TIPS}\right)\cdot\eta_{IS}\right)\end{array}\right)\end{array}\right)}. \quad 10.26$$

$$price^*_{TIPS} = \frac{\left(\eta_{TIPS}\cdot\left(d^2\cdot S_2-2S_4\right)\cdot\left(\begin{array}{l}s\cdot\alpha_{TP}\cdot\left(\gamma^2\cdot D_4\cdot S_2\cdot\Gamma_{2IS}+\eta_{TP}\left(\gamma^2\left(2S_4+2d^2\cdot S_3-d^4\cdot S_2\right)+2(D_2\cdot S_2+4)\cdot\eta_{IS}\right)\right)+\\ D_4\cdot\alpha_{TIPS}\cdot\left(\gamma^2\cdot S_2\cdot\Gamma_{2IS}-\eta_{TP}\cdot\left(\gamma^2\cdot S_4+(D_2\cdot S_2+4)\cdot\eta_{IS}\right)\right)\end{array}\right)\right)}{\left(\begin{array}{l}\gamma\cdot D_4\cdot S_2\cdot\left(2\gamma^2\cdot S_1(D_2\cdot S_1+2)-S_2\cdot(D_2\cdot S_2+4)\cdot\eta_{TIPS}\right)\cdot\Gamma_{2IS}+\\ \eta_{TP}\cdot\left(\begin{array}{l}2\gamma^4\left(d^2\cdot S_1-2S_2\right)\cdot\left(d^2\cdot(3S_2+2)+2S_2\cdot S_4\right)+\\ \left(2S_4-d^2\cdot S_2\right)\cdot\left(2\gamma^2\cdot\left(S_4^2+2d^2\cdot S_2\right)\cdot\eta_{TIPS}+\left(2\gamma^2\cdot(D_2\cdot S_1+2)^2-(D_s\cdot S_2+4)^2\cdot\eta_{TIPS}\right)\cdot\eta_{IS}\right)\end{array}\right)\end{array}\right)}. \quad 10.27$$

Based on the equilibrium prices obtainable from these set of calculations, we can observe the following under the setting of Scenario 2:

- **Base Market Size–Servitized Channel is Smaller than Conventional Channel (i.e., $\alpha_{TIPS} < \alpha_{TP}$):** Integrated solution is often better off in comparison to the servitized one.
- **Base Market Size–Servitized Channel is Larger than Conventional Channel (i.e., $\alpha_{TIPS} > \alpha_{TP}$):** Integrated solution is typically worse off in comparison to the servitized one.

These observations confirm that tangible product producers should implement domain dependent service policies, as suggested in (Cusumano et al., 2015).

10.3.2 Summary

The work done by (Lee et al., 2016) demonstrated that the strategy of servitization is a feasible option for a producer under two conditions: when the service dependency of tangible products is high, and the channel substitutability between channels is also high.

10.4 Quantum Games for Question 10.2

For the purpose of addressing Question 10.2, a pioneer work was conducted in (Martínez-Martínez, 2014) by combining different inspiring resources, e.g., theoretical studies in the area of quantum game theory (Eisert et al., 1999), quantum decision theory (Yukalov and Sornette, 2008; Yukalov and Sornette, 2009) and quantum cognition model (Pothos and Busemeyer, 2009). This was done in order to propose a concept of Hamiltonian (Hamill, 2014) strategic interaction, which was demonstrated to be able to account for the entanglement involved in a decision-maker's strategic state.

10.4.1 Quantum Decision Theory

Quantum decision theory (QDT) (Yukalov and Sornette, 2008) is a newly developed decision-making theory which is built on the Hilbert spaces mathematics, a renowned framework in the world of physics due to its remarkable application in addressing quantum mechanics. Under the framework of Hilbert spaces, the uncertainty notion and various notable effects found in the processes of cognition can be formalized for a rigorous examination of decision-making. In essence, a decision maker's choice can be described by QDT as an event that happens stochastically with a probability equal to the summation of two factors, namely, objective utility and subjective attraction. Accordingly, a quantitative prediction (i.e., so-called quarter law) regarding a person's decisions, which is influenced by the average effect of his/her subjectivity, is computable (Favre et al., 2016).

10.4.1.1 Classical Utility Formulation

The classical methodologies for addressing decision-making problems are built on the utility theory (Von Neumann and Morgenstern, 2004; Savage, 1954). When the states of nature are mixed with uncertainty, statistical decision theory is typically employed in order to formalize the process of decision-making (Lindgren, 1971; White, 1976; Longford, 2013; Raiffa and Schlaifer, 2000; Bather, 2000; Berger, 1985; Marshall and Oliver, 1995; Hastings and Mello, 1978; Rivett, 1980; Faber, 2012).

Under this framework, maximizing utility is a crucial step where we can define a probability measure, denoted by $d\mu(state)$, over a set of states of nature (denoted by $state \in States$). Apart from this definition, we can further assume that every single action (denoted by $action \in Actions$) is able to be characterized by its utility measure, represented by $U(action)$. Accordingly, the preferred action (indicated by $action^*$) should meet the conditions expressed using Eq. 10.28 (Martínez-Martínez, 2014):

$$action^* \in \arg\max_{action \in Actions} U(action). \qquad 10.28$$

In practice, uncertainty associated with the decision problem is often introduced due to the knowledge shortage of different states of nature's realization. Therefore, we can define an action-taking's utility as a state of nature's function, denoted by $U(action, state)$. Consequently, the classical decision theory can be turned into a statistical one where a joint probability measure (either discrete or continuous), denoted by $d\mu(state,action)$, is used to calculate its value.

Suppose we have a probability space covering all the actions, $\{Actions, States, d\mu(state,action)\}$, an action's utility level can be formulated using Eq. 10.29 (Martínez-Martínez, 2014):

$$U(action) = \int_{States} U(action, state)\, d\mu(state,action). \qquad 10.29$$

where the joint measure $d\mu(state,action)$ is representable in terms of $d\mu(state)$ using defining Eq. 10.30 (Martínez-Martínez, 2014):

$$d\mu(state,action) = \Pr(action|state)\, d\mu(state). \qquad 10.30$$

where the condition probability of an action under a state of nature is denoted by $\Pr(action|state)$.

Though two competing actions are comparable under classical decision theory, we might encounter some difficulties (e.g., irrational notions' quantitative measure) and various irregularities (e.g., emotions, subjective biases, etc.) (Marwala, 2015) or an obstacle of mathematically defining the random states of nature (Yukalov and Sornette, 2008). Bearing this in mind, one can replace scale-dependent utility functions, represented by $U(action)$, with a normalized and scale-independent utility $\Pr(action)$, as expressed in Eq. 10.31 (Martínez-Martínez, 2014):

$$\Pr(action) \in [0,1];\ \forall action \in Actions, \text{ and } \sum_{action \in Actions} \Pr(action) = 1. \qquad 10.31$$

Equation 10.31 stands for a decision-maker's subjective preferences and a probability measure itself, while Eqs. 10.29 and 10.30 represent the criterion of selection under classical decision theory.

10.4.1.2 States Space

However, during the course of interpreting the classical utility theory and applying it to real processes involved in human decision, various paradoxes are pinpointed and discussed in the literature, e.g., (Machina, 2008; Zeckhauser, 2006). Under this circumstance, quantum decision theory extends the above-mentioned classical framework and allocates the largest probability of being chosen to the most preferred action. Under this concept, the normalized utility is re-formulated using Eq. 10.32 (Yukalov and Sornette, 2008):

$$\Pr(action) \equiv \frac{U(action)}{\sum_{action \in Actions} U(action)}. \qquad 10.32$$

However, no matter how we perform monotonic transformation of a utility function, the same preference order is always obtainable. Therefore, an implication drawn from interpreting Eq. 10.32 reveals that, when trying to model a concrete problem, a range of ambiguity (often non-desirable) can be introduced.

When a decision-maker encounters various intentions, each intention-representation stands for a concrete realization of the intention. For simplicity, an "intention" can be regarded as nothing but a decision that has to be made (e.g., crowdfunding to introduce), while the "intention-representation" denotes all available choices (say, different models of crowdfunding). Let the set of the elementary representation states (denoted by $|n_{intentions_i}\rangle$) form a representation basis, represented by $\{|n_{intentions_i}\rangle\}$, which is orthonormal and exhausts the whole intentions set. The corresponding intention space is then expressed using Eq. 10.33 (Martínez-Martínez, 2014):

$$H_{intentions_i} \equiv \bar{L}\ \{|n_{intentions_i}\rangle\}. \qquad 10.33$$

where $|\cdot\rangle$ denotes any element of vector of the states space (called ket-vector). Likewise, the dual space of the bra-vectors is represented by $\langle\cdot|$. Therefore, Eq. 10.33 depicts the representation basis in the form of a closed linear envelope. In other words, a Hilbert space is defined through Eq. 10.33. Consequently, the intention state can be expanded with respect to the basis, as expressed in Eq. 10.34 (Martínez-Martínez, 2014):

$$|\psi\rangle = \sum_{n_{intentions_i}} c_{n_{intentions_i}} \left|n_{intentions_i}\right\rangle. \qquad 10.34$$

By normalizing intention states, we can get Eq. 10.35 (Martínez-Martínez, 2014):

$$\|\psi\| = 1, \text{ and } \sum_{n_{\text{intentions}_i}} \left| c_{n_{\text{intentions}_i}} \right|^2 = 1 . \tag{10.35}$$

A set of several intentions can be called a "prospect" and the associated prospect space represents the set of all workable combinations in a decision-maker's mind (Yukalov and Sornette, 2008). Therefore, a prospect-representation which denotes a solid implementation of a prospect is the representation states' tensor product, as shown in Eq. 10.36 (Martínez-Martínez, 2014):

$$\left| n_{intentions_i} \right\rangle = \bigotimes_{intentions_i} \left| n_{intentions_i} \right\rangle = \left| n_{intentions_1} n_{intentions_2} n_{intentions_3} \cdots \right\rangle . \tag{10.36}$$

Meanwhile, the basis of prospect-representation, as indicated by $\{|n\rangle\}$, is also orthonormal, as formulated in Eq. 10.37 (Martínez-Martínez, 2014):

$$\langle m|n \rangle = \prod_{intentions_i} \left\langle m_{intentions_i} \middle| n_{intentions_i} \right\rangle = \delta_{mn} = \prod_{intentions_i} \delta_{m_{intentions_i} n_{intentions_i}} . \tag{10.37}$$

Furthermore, the intention spaces' tensor product determines the prospect space using Eq. 10.38 (Martínez-Martínez, 2014):

$$H \equiv \bar{L}\,\{|n\rangle\} = \bigotimes_{intentions_i} H_{intentions_i} . \tag{10.38}$$

Here, we can simply treat the prospect space as a "mind" and its dimensionality is defined using Eq. 10.39 (Martínez-Martínez, 2014):

$$d_H = \prod_{intentions_i} R_{intentions_i} . \tag{10.39}$$

where the ith intention's amount of representations is denoted by $R_{intentions_i}$.

10.4.1.3 Mind and Entanglement

In quantum systems, entanglement represents a property of exhibiting at least two degree-of-freedom (DoF) which are correlated in such a manner that it is impossible to describe the condition of some of them independently of the remaining parts of the system (Streltsov, 2015; Xing and Marwala, 2018a). A correlation of non-classical nature is shown by these quantum systems. Accordingly, we can define the following (Martínez-Martínez, 2014):

- **Disentangled Mind:** A prospect state is regarded as disentangled if it can be expressed as the tensor product of intentions states using Eq. 10.40 (Martínez-Martínez, 2014):

$$|\psi\rangle = \bigotimes_{intentions_i} \psi_{intentions_i} . \tag{10.40}$$

In other words, a collection of all legitimate disentangled prospect states represents the disentangled mind, denoted by $D = \{|\psi\rangle : |\psi_{intentions}\rangle \in H_{intentions}\}$.

- **Entangled Mind:** Similarly, any other prospect state ($|\psi\rangle \in H$), defined using Eq. 10.41 (Martínez-Martínez, 2014), is treated as being entangled.

$$|\psi\rangle = \sum_n c_n |n\rangle \text{ and } \sum_n |c_n|^2 = 1. \qquad 10.41$$

A notable feature of Eq. 10.41 lies in the fact that it cannot be simplified as a form of tensor product, as shown in Eq. 10.40. Therefore, an entangled mind refers to the disentangled mind's complement regarding to the product of intention spaces. Therefore, an entangled mind refers to the disentangled mind's complement with respect to the product of intention spaces (denoted by $H \backslash D$).

10.4.1.4 Process of Decision-Making and the Strategic State

Based on the above understanding, the decision-making process can essentially be divided into the following steps (Martínez-Martínez, 2014):

- **Step 1:** Admissible choices' evaluation among varied actions; under an agent's state-of-mind, the probability of effectuating a prospect-representation (denoted by n) with the state of prospect-representation (denoted by $|n\rangle$) is called the prospect-probability and is formulated in Eq. 10.42 (Martínez-Martínez, 2014):

$$\begin{gathered} \Pr_n = |\langle n|\psi\rangle|^2 = |c_n|^2 \\ \text{where } \sum_n |c_n|^2 = 1 \ (\text{normalization condition}) \end{gathered}. \qquad 10.42$$

- **Step 2:** Optimal action selection; the aimed action associating with the maximum probability.
- **Step 3:** New prospect-representation's selection (based on the obtained probabilities) at the point of decision-making, the aimed action associating with the maximum probability. Two factors (i.e., decision-making and decision-realization) force us to perform the state renormalization by treating the realized action as a certain fact.

By now, one can compute the probabilities (defined using Eq. 10.31) through Eq. 10.42. Suppose an individual decision-maker's mind is featured by a particular fixed state (denoted by $|state\rangle \in H$) which in turn paints a unique image for each decision-maker based on his/her beliefs, habits, etc. In (Martínez-Martínez, 2014), this reference state is termed as the "strategic state". Accordingly, each mind contains a unique strategic state, while different decision-makers' minds hold distinct strategic states.

Built on this construction, a utility factor, denoted by $p_0(\psi)$, can be defined for each prospect-representation (represented by ψ). The utility factor is computable using Eq. 10.43 (Martínez-Martínez, 2014):

$$p_0(\psi) = \sum_n \langle state|P_n P_\psi P_n|state\rangle = \sum_n |c_n|^2 \langle state|P_n|state\rangle . \qquad 10.43$$

where the role of utilities is played by $\langle state|P_n P_\psi P_n|state\rangle$ which is weighted by coefficients $|c_n|^2$. These coefficients can be calculated by projecting the strategic state on the mind basis. The projects $|n\rangle\langle n|$ and $|\psi\rangle\langle\psi|$ are represented by parameters P_n, and P_ψ, respectively. The utility factor can then be normalized using Eq. 10.44 (Martínez-Martínez, 2014):

$$\sum_{\psi\in\Psi} p_0(\psi) = 1 . \qquad 10.44$$

where all prospect states associated with all reasonable prospects compose a set, represented by $\{|\psi\rangle\}$, and its complete form is denoted by Ψ. In other words, Ψ represents a sub-set of the mind space, i.e., the prospect-state set (indicated by $\Psi \in H$).

By now, the decision-maker's observable prospect-probability is defined using Eq. 10.45 (Martínez-Martínez, 2014):

$$\Pr(\psi) = \langle state|P_\psi|state\rangle. \qquad 10.45$$

According to (Yukalov and Sornette, 2009), by introducing the probabilistic decision concept, when $\Pr(\psi)$, defined in Eq. 10.45, reaches its maximum, the corresponding prospect shows the best representation in terms of a decision-maker's strategic mind state.

10.4.2 Quantum Games and Quantum Strategies

Some paradoxical results found in two representative psychological tests, i.e., the two-stage gambling game and the Prisoner's Dilemma game, have not been explained under classical decision theory for many years. A quantum probabilistic explanation offers us a chance at explaining the contraventions of rational decision theory simply and elegantly (Pothos and Busemeyer, 2009).

10.4.2.1 Classical EWL Model

The symmetric Prisoner's Dilemma game (Marwala and Hurwitz, 2017; Umbhauer, 2016; Mazalov, 2014) is a game with two players (denoted by $Player_A$ and $Player_B$) who have two action options: either collaborate ($action_C$) or defect ($action_D$). The normal form of this game can be defined using its canonical payoff matrix, as expressed in Eq. 10.46 (Martínez-Martínez, 2014):

$$\begin{array}{lll} Player_B / Player_A & action_C & action_D \\ action_C & (payoff_{reward}, payoff_{reward}) & (payoff_{loser}, payoff_{temptation}) \\ action_C & (payoff_{temptation}, payoff_{loser}) & (payoff_{punishment}, payoff_{punishment}) \end{array} \quad .10.46$$

where $payoff_{temptation} > payoff_{reward} > payoff_{punishment} > payoff_{loser}$ determines that mutual collaboration represents the Pareto optimal situation. Since both players' dominant strategy is defection, the Nash equilibrium of this game is mutual defection.

By following the EWL model, i.e., quantum game introduced in (Eisert et al., 1999), the quantum formulation can be written using the assignment of classical strategies ($action_C$ and $action_D$) to two possible outcomes which are represented by two basis vectors ($|action_C\rangle$ and $|action_D\rangle$) in a two-state system's Hilbert space, i.e., a qubit, as shown in Eq. 10.47 (Martínez-Martínez, 2014):

$$|action_C\rangle = \begin{pmatrix} 1 \\ 0 \end{pmatrix}, \ |action_D\rangle = \begin{pmatrix} 0 \\ 1 \end{pmatrix}. \quad 10.47$$

Under this formulation, the mind defined using Eq. 10.38 can be regarded as the Hilbert space formed by two spaces' tensor product (e.g., space that corresponds to $Player_A$ selected strategy and space that corresponds to $Player_A$'s evaluation of $Player_B$'s strategy). Accordingly, $H = H_{PlayerA} \otimes H_{PlayerB}$ can be explained by the basis vector set of $\{|action_C action_C\rangle$, $|action_C action_D\rangle$, $|action_D action_C\rangle$, $|action_D action_D\rangle\}$, as defined using Eq. 10.48 (Martínez-Martínez, 2014):

$$\begin{array}{ll} |action_C action_C\rangle = \begin{pmatrix} 1 \\ 0 \\ 0 \\ 0 \end{pmatrix}, & |action_C action_D\rangle = \begin{pmatrix} 0 \\ 1 \\ 0 \\ 0 \end{pmatrix}, \\ |action_D action_C\rangle = \begin{pmatrix} 0 \\ 0 \\ 1 \\ 0 \end{pmatrix}, & \text{and } |action_D action_D\rangle = \begin{pmatrix} 0 \\ 0 \\ 0 \\ 1 \end{pmatrix}. \end{array} \quad 10.48$$

Suppose we have the initial state of $|\psi_0\rangle = |action_C action_C\rangle$, when it comes to the decision of each player regarding whether or not to collaborate, the associated qubit state should change to portray the finalized decision. Here we can employ unitary operators in order to represent quantum operations, which also hint at the evolution of a system's state.

Given the strategic moves of two players (i.e., $Player_A$ and $Player_B$) which are represented by $\hat{S}_{Player_A}$ and $\hat{S}_{Player_B}$, respectively, as suggested in (Eisert et al., 1999), the quantum strategies' set is given using Eq. 10.49 (Martínez-Martínez, 2014):

$$\hat{U}(\xi,\phi)=\begin{pmatrix} e^{i\phi}\cos\frac{\xi}{2} & \sin\frac{\xi}{2} \\ -\sin\frac{\xi}{2} & e^{-i\phi}\cos\frac{\xi}{2} \end{pmatrix}. \quad 10.49$$

Here, Eq. 10.49 denotes 2 × 2 unitary matrices with 2-parameter set, where $\xi \in [0,\pi]$ and $\phi \in \left[0,\frac{\pi}{2}\right]$. In (Martínez-Martínez, 2014), the author focused mainly on the classical strategies' sub-set, i.e., $S_0 \equiv \{\hat{S}(\xi) = \hat{U}(\xi, 0) | \xi \in [0,\pi]\}$. Accordingly, two pure strategies (i.e., collaboration and defect) are given via Eq. 10.50 (Martínez-Martínez, 2014):

$$\hat{action_C} \equiv \hat{S}(0)=\begin{pmatrix} 1 & 0 \\ 0 & 1 \end{pmatrix} \text{ and } \hat{action_D} \equiv \hat{S}(\pi)=\begin{pmatrix} 0 & 1 \\ -1 & 0 \end{pmatrix}. \quad 10.50$$

By now, the quantum formulated game's state can be defined through Eq. 10.51 (Martínez-Martínez, 2014):

$$|\psi_1\rangle = \left(\hat{S}_{Player_A} \otimes \hat{S}_{Player_B}\right)|\psi_0\rangle. \quad 10.51$$

Here, Eq. 10.51 stands for a disentangled prospect state defined by Eq. 10.40. We can see that instead of performing strategic reasoning, the restraining of $Player_A$ from defection lies on whether $\hat{S}_{Player_A} = \hat{action_c}$ is selected or not, indicating nothing more than $Player_A$'s intention for collaboration. Although this setup resembles the classical Prisoner's Dilemma game, it depicts players' choices via qubit pair coupled with several quantum gates acting on them (see Fig. 10.6 for a simplified illustration).

Suppose a decision process can occur in two distinct layers: one corresponds to rational considerations (denoted by $\hat{S}_{Player_A} \otimes \hat{S}_{Player_B}$), while the other one representing a spontaneous process during the course of decision-making (indicated by $\hat{S}'$). Hence, a feasible solution can be given through Eq. 10.52 (Martínez-Martínez, 2014; Eisert et al., 1999):

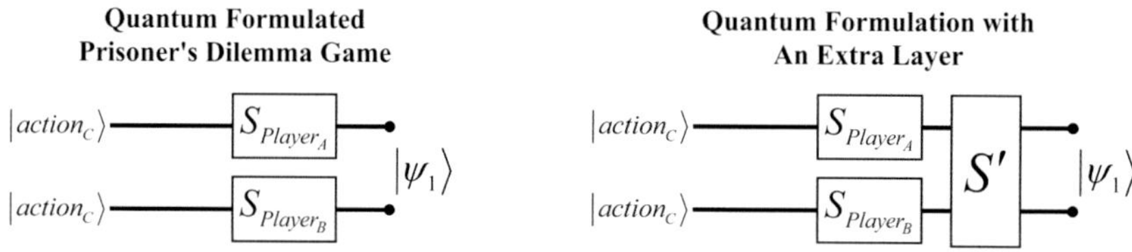

Figure 10.6: Illustration of quantum strategies.

$$\hat{S}' = \exp\left\{i\frac{\gamma}{2}\left(\hat{action}_D \otimes \hat{action}_D\right)\right\}. \quad 10.52$$

where γ stands for a real parameter that falls within the range of $\left[0, \frac{\pi}{2}\right]$ and tunes the mind's entanglement. Explicitly, we can have Eq. 10.53 (Martínez-Martínez, 2014):

$$\hat{S}' = \begin{pmatrix} \cos\frac{\gamma}{2} & 0 & 0 & i\sin\frac{\gamma}{2} \\ 0 & \cos\frac{\gamma}{2} & -i\sin\frac{\gamma}{2} & 0 \\ 0 & -i\sin\frac{\gamma}{2} & \cos\frac{\gamma}{2} & 0 \\ i\sin\frac{\gamma}{2} & 0 & 0 & \cos\frac{\gamma}{2} \end{pmatrix}. \quad 10.53$$

Then, a general case of an entangled prospect state (defined by Eq. 10.41) is given by Eq. 10.54 (Martínez-Martínez, 2014):

$$|\psi_2\rangle = \hat{S}'\left(\hat{S}_{Player_A} \otimes \hat{S}_{Player_B}\right)|\psi_0\rangle. \quad 10.54$$

where Eq. 10.54 can be expressed using Eq. 10.40 when the following conditions hold: $\gamma = 0$ or π.

For $Player_A$ who intends to defect, one can calculate the probability (defined via Eq. 10.42) of effectuation (given S' is introduced) via Eq. 10.55 (Martínez-Martínez, 2014):

$$\Pr_{Player_A}\left(action_D\right) = \cos^2\frac{\gamma}{2} \leq 1. \quad 10.55$$

where deviations can be naturally observed for an entangled mind (i.e., $\gamma \neq 0$).

In (Martínez-Martínez, 2014), the author further introduced a more generic example where $Player_A$ needs to determine collaboration or defection after being informed about which action $Player_B$ will choose. This extended game is also found in (Pothos and Busemeyer, 2009), where another player's information is not considered as the result of a game's theoretical reasoning but rather as an informational input for a problem encountered by an individual decision-maker. This setting is related to the violations of the Sure Thing Principle for uncertainty-involved individual choice, as discussed in (Shafir and Tversky, 1992):

(1) For $\hat{S}_{Player_B} = \hat{action}_D$, it is assured that $Player_B$ will defect,

(2) For $\hat{S}_{Player_B} = \hat{action}_c$, it is assured that $Player_B$ will cooperate,

(3) For $\hat{S}_{Player_B} = \hat{S}(\theta)$, $Player_B$'s intention is unknown and, thus, $Player_A$'s perception of the situation represents the most generic scenario where no additional assumption is made against the combination of strategies.

Since the rational decision made by $Player_A$ should keep utility maximized (i.e., $\hat{S}_{Player_A} = \hat{action}_D$), the computation can be performed using Eq. 10.56 (Martínez-Martínez, 2014):

$$|\psi_2'\rangle = S'\left(\hat{action}_D \otimes \hat{S}(\theta)\right)|action_C action_C\rangle = \begin{pmatrix} i\sin\frac{\gamma}{2}\sin\frac{\theta}{2} \\ -i\sin\frac{\gamma}{2}\cos\frac{\theta}{2} \\ \cos\frac{\gamma}{2}\cos\frac{\theta}{2} \\ \cos\frac{\gamma}{2}\cos\frac{\theta}{2} \end{pmatrix}. \quad 10.56$$

together with the density matrix defined through Eq. 10.57 (Martínez-Martínez, 2014):

$$\rho = |\psi_2'\rangle\langle\psi'|$$
$$= \begin{pmatrix} \sin^2\frac{\gamma}{2}\sin^2\frac{\theta}{2} & -\frac{1}{2}\sin^2\frac{\gamma}{2}\sin\theta & \frac{1}{4}i\sin\gamma\sin\theta & \frac{1}{2}i\sin\gamma\sin^2\frac{\theta}{2} \\ -\frac{1}{2}i\sin\frac{\gamma}{2}\sin\theta & \sin^2\frac{\gamma}{2}\cos^2\frac{\theta}{2} & -\frac{1}{2}i\sin\gamma\cos^2\frac{\theta}{2} & -\frac{1}{4}i\sin\gamma\sin\theta \\ -\frac{1}{4}i\sin\gamma\sin\theta & \frac{1}{2}i\sin\gamma\cos^2\frac{\theta}{2} & \cos^2\frac{\gamma}{2}\cos^2\frac{\theta}{2} & \frac{1}{2}\cos^2\frac{\gamma}{2}\sin\theta \\ -\frac{1}{2}i\sin\gamma\sin^2\frac{\theta}{2} & \frac{1}{4}i\sin\gamma\sin\theta & \frac{1}{2}\cos^2\frac{\gamma}{2}\sin\theta & \cos^2\frac{\gamma}{2}\sin^2\frac{\theta}{2} \end{pmatrix}. \quad 10.57$$

Accordingly, the state of $Player_A$ is obtainable through the operation of partial trace over $Player_B$'s subspace, as given by Eq. 10.58 (Martínez-Martínez, 2014):

$$\rho^{Player_A} \equiv Trace_{Player_B}\,\rho. \quad 10.58$$

In (Martínez-Martínez, 2014), the density matrix was further reduced to a 2 × 2 matrix with the elements defined using Eq. 10.59 (Martínez-Martínez, 2014):

$$\left(\rho^{Player_A}\right)_{ij} = \sum_{k\in\{action_C, action_D\}} \langle ik|\rho|jk\rangle. \quad 10.59$$

where $i, j \in \{action_C, action_D\} \Rightarrow \rho^{Player_A} = \begin{pmatrix} \sin^2\frac{\gamma}{2} & 0 \\ 0 & \cos^2\frac{\gamma}{2} \end{pmatrix}$.

After introducing the projector (denoted by $Projector_{action_D} = |action_D\rangle\langle action_D|$), $Player_A$'s probability of defecting is computable using Eq. 10.60 (Martínez-Martínez, 2014):

$$\Pr_{Player_A}(action_D) = Trace\left(\rho^{Player_A} Projector_{action_D}\right) = \cos^2\frac{\gamma}{2} \le 1. \quad 10.60$$

where the result of Eq. 10.60 is in agreement with what we get from Eq. 10.55.

From Eq. 10.59, the entropy of entanglement between two subspaces (associated with $Player_A$ and $Player_B$, respectively, and located within the mind space) can be defined using Eq. 10.61 (Martínez-Martínez, 2014):

$$
\begin{aligned}
s_{Players_{AB}} &= Trace\left(-\rho^{Player_A}\log\rho^{Player_A}\right) \\
&= -\cos^2\frac{\gamma}{2}\log\left(\cos^2\frac{\gamma}{2}\right)-\sin^2\frac{\gamma}{2}\log\left(\sin^2\frac{\gamma}{2}\right)
\end{aligned}. \qquad 10.61
$$

In a model where quantum gates are involved, $Player_A$'s final decision is influenced by the released information if such information has impact on the entanglement in the state of mind of $Player_A$.

10.4.2.2 Generalized N Strategies

The parameter $\hat{S}'$, which is the layer that represents a spontaneous process during the course of decision-making, can be expressed as in Eq. 10.62 (Martínez-Martínez, 2014):

$$
\begin{aligned}
\hat{S}' &= \exp\left\{i\frac{\gamma}{2}\left(\hat{action}_D \otimes \hat{action}_D\right)\right\} \\
&= \exp\left\{i\frac{\gamma}{2}\begin{pmatrix}0 & 1\\ -1 & 0\end{pmatrix}\otimes\begin{pmatrix}0 & 1\\ -1 & 0\end{pmatrix}\right\} \\
&= \exp\left\{i\frac{\gamma}{2}(2i)\begin{pmatrix}0 & -\frac{i}{2}\\ \frac{i}{2} & 0\end{pmatrix}\otimes(2i)\begin{pmatrix}0 & -\frac{i}{2}\\ \frac{i}{2} & 0\end{pmatrix}\right\} \\
&= \exp\left\{-i\theta\left(J_{2,y}\otimes J_{2,y}\right)\right\}
\end{aligned}. \qquad 10.62
$$

where $\theta \in [0,\pi]$, and $J_{2,y}$ stands for the matrix that represents the angular momentum operator (denoted by J_y) corresponding to a two-state system observed in quantum mechanics. Given an N-component system whose total angular momentum is associated with a states space, the canonical basis, defined as Eq. 10.63 (Martínez-Martínez, 2014), can be selected as the operator J_z's eigenstates.

$$
e_1 \equiv \begin{pmatrix}1\\0\\ \vdots\\0\end{pmatrix},\ e_2 \equiv \begin{pmatrix}0\\1\\ \vdots\\0\end{pmatrix},\ldots,\ e_N \equiv \begin{pmatrix}0\\ \vdots\\0\\1\end{pmatrix} \in \mathbb{C}^N. \qquad 10.63
$$

Accordingly, J_y is characterized by its anti-symmetrical and pure imaginary properties and its matrix elements can be obtained using Eq. 10.64 (Martínez-Martínez, 2014):

$$\left(J_{N,y}\right)_{m,n}=\frac{1}{2i}\left[\begin{array}{l}\delta_{m+1,n}\sqrt{\frac{N-1}{2}\left(\frac{N-1}{2}+1\right)-\left(m-\frac{N+1}{2}\right)\left(m-\frac{N+1}{2}+1\right)}-\\ \delta_{m-1,n}\sqrt{\frac{N-1}{2}\left(\frac{N-1}{2}+1\right)-\left(m-\frac{N+1}{2}\right)\left(m-\frac{N+1}{2}-1\right)}\end{array}\right] \quad .10.64$$

where m and n belong to $\{1,\ldots,N\}$. One can find that $J_{2,y}$ matrices (given by Eq. 10.62) are also obtainable using Eq. 10.64. The schematic representation of this situation is shown in Fig. 10.7.

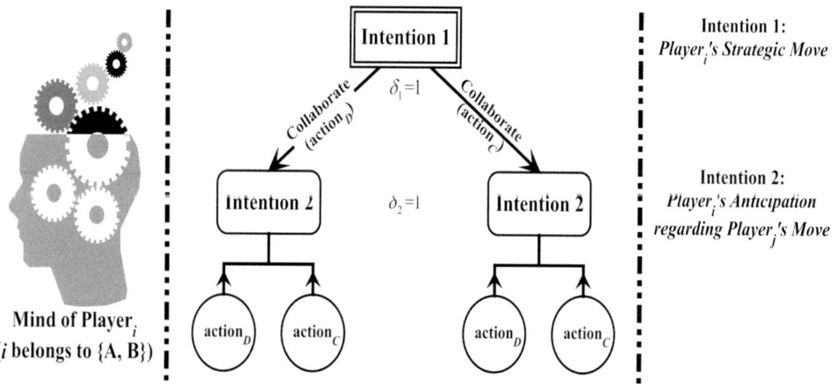

Figure 10.7: Schematic decision-making process representation of two-player involved Prisoner's Dilemma game.

10.4.2.3 Hamiltonian Strategic Interaction

The unitary evolutionary operator (denoted by $\hat{S}'$), indicating a system's state at time t and evolving from the original $t_0 = 0$ time step, can be expressed with respect to a time-independent Hamiltonian (represented by $\hat{H}$), as given by Eq. 10.65 (Martínez-Martínez, 2014):

$$\hat{U}(t,0)=\exp\left\{-i\hat{H}t\right\}. \qquad 10.65$$

According to Eq. 10.62, and introducing the dimensionless parameter $\tilde{\theta} \equiv \frac{\theta}{t}$ into the temporal coordinate, we have Eq. 10.66 (Martínez-Martínez, 2014):

$$\hat{S}'=\exp\left\{-i\hat{H}_{SI}\right\}. \qquad 10.66$$

where the Hamiltonian of strategic interaction (HSI) is denoted by $\hat{H}_{SI}$, which is equal to $\tilde{\theta}\ (J_{2,y} \otimes J_{2,y})$. The HSI corresponds to the mind depicting the Prisoner's Dilemma game.

A generic description of modelling these strategic situations can be summarized as follows (Martínez-Martínez, 2014):

- **Stage 1:** Suppose there is a decision-making problem faced by a person who has to make a selection from K number of intentions. Each of them is tagged by k and contains a number of N_k probable realizations. Here, an assumption was made in (Martínez-Martínez, 2014) that the targeted decision-maker has the capability of ranking the order of all possible realizations.
- **Stage 2:** Each intention is associated with a Hilbert space (denoted by H_k) which has N_k dimensionality and is spanned by a canonical basis (defined by Eq. 10.63). For instance, the best and the second-best ranked realizations for the kth intention are denoted by $|e_1^{(k)}\rangle$ and $|e_2^{(k)}\rangle$, respectively.
- **Stage 3:** For the strategic state leading to the best realization for all intentions, we have Eq. 10.67 (Martínez-Martínez, 2014):

$$\left|state_{best}\right\rangle = \bigotimes_k \left|e_1^{(k)}\right\rangle. \tag{10.67}$$

Based on the definition of the space basis, the following relationship expressed in Eq. 10.68 (Martínez-Martínez, 2014) holds:

$$\left|state_{best}\right\rangle \equiv \begin{pmatrix} 1 \\ 0 \\ \vdots \\ 0 \end{pmatrix} \in \mathbb{C}^{d_H}. \tag{10.68}$$

where d_H equals to $\prod_k N_k$.

- **Stage 4:** Recognizing game information set in order to learn what decision options are available at each time step of the game. Take the Prisoner's Dilemma game, two players making decisions simultaneously is often the case.
- **Stage 5:** HSI calculation uses two components, i.e., the angular momentum and identity matrices, which are defined for encountered intentions and the associated different spaces. In other words, to-be-made decisions are typically influenced by $J_{N_{k'},y}$, while for those uninfluenced ones, identity matrix (denoted by I_{N_k}) can be introduced as the operator. Accordingly, the generic form of HSI can be defined using Eq. 10.69 (Martínez-Martínez, 2014):

$$\hat{H}_{SI} = \tilde{\theta} \bigotimes_{k=1}^{K} \left[\delta_k J_{N_{k,y}} + (1-\delta_k) I_{N_K} \right]. \tag{10.69}$$

where an indicator function is denoted by δ_k, which normally has two situations: one or zero, meaning at any time step of the game that we are contemplating, the decision regarding the *k*th intention is about to be made or discarded.

- **Stage 6:** Finally, the decision-maker's strategic state can be given by Eq. 10.70 (Martínez-Martínez, 2014):

$$\left|state_{best}\right\rangle = S'\left|state_{best}\right\rangle . \qquad 10.70$$

where S' is given using Eqs. 10.66 and 10.69, and the prospect probabilities are computable through Eq. 10.42.

10.4.2.4 Ultimatum Game Illustration

In (Martínez-Martínez, 2014), the Ultimatum game (Umbhauer, 2016) was employed in order to further illustrate the HSI formulation. In this game, a certain amount of money will be shared between two players, where the first player (say, $Player_A$) has the option of making an offer regarding how the money is going to be divided, while the second player (i.e., $Player_B$) can determine to consent ($action_C$) or reject ($action_R$). If consent is the case, the money is divided accordingly, otherwise both of them get nothing. Suppose we have a number of $N = 10$ coins, the dominant strategy for $Player_B$ is to accept any offer with greater than zero number of coins, while $Player_A$'s dominant strategy is to give an offer with near-zero number (i.e., as close to zero as possible) of coins, that is $n = 1$ in this case.

- **Strategic States Computation:** Take $Player_A$'s decision process, there are two intentions that $Player_A$ has to deal with: (1) How many coins does $Player_A$ intend to offer? and (2) How does $Player_A$ hope $Player_B$ to respond to the proposed offer? For the first intention, we can have ten distinct representations regarding the number of coin offerings, ranging from $n = 1$ to $n = 10$. A natural monotonic action ranking can be given by Eq. 10.71 (Martínez-Martínez, 2014):

$$n = 1 \succ n = 2 \succ \cdots \succ n = 10 . \qquad 10.71$$

Here, Eq. 10.71 enables us to link $\tilde{\theta}$ (a dimensionless parameter) to the deviation from standard ideal rationality in a positive manner. Hence, the ranking of elements of the basis found in the first intention's Hilbert space is also obtained as $H_1 = \mathbb{C}^{10}$. In terms of the second intention, the relationship of $action_C \succ action_R$ is preferred. Likewise, the ranking of elements of the basis found in the second intention's Hilbert space is $H_1 = \mathbb{C}^2$. Therefore, the strategic state (its general form can be found in Eq. 10.68) of this game is given by Eq. 10.72 (Martínez-Martínez, 2014):

$$|state_{best}\rangle \equiv \begin{pmatrix} 1 \\ 0 \\ \vdots \\ 0 \end{pmatrix} \in C^{20}. \quad 10.72$$

The schematic representation of two intentions is shown in Fig. 10.8.

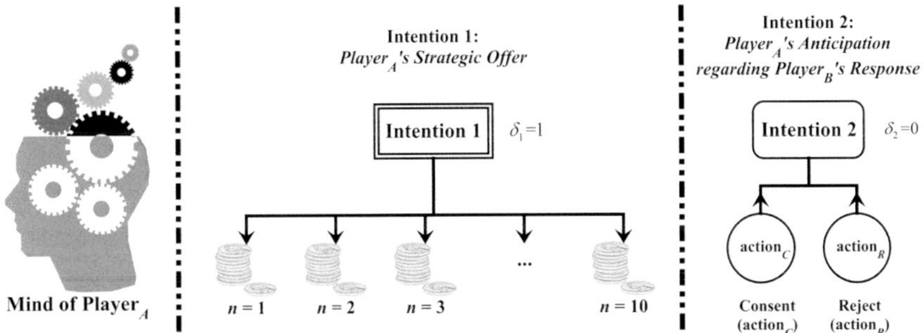

Figure 10.8: Schematic decision-making process representation of $Player_A$ involved in the Ultimatum game.

Since this game is characterised by its sequentiality, its HSI (its general form was defined using Eq. 10.69) is given by Eq. 10.73 (Martínez-Martínez, 2014):

$$\hat{H}_{SI} = \tilde{\theta}\left(J_{10,y} \otimes I_2\right). \quad 10.73$$

where the following formulations, as given by Eqs. 10.74 and 10.75, hold (Martínez-Martínez, 2014):

$$J_{10,y} = \begin{pmatrix} 0 & -\frac{3i}{2} & 0 & 0 & 0 & 0 & 0 & 0 & 0 & 0 \\ \frac{3i}{2} & 0 & -2i & 0 & 0 & 0 & 0 & 0 & 0 & 0 \\ 0 & 2i & 0 & -\frac{i\sqrt{21}}{2} & 0 & 0 & 0 & 0 & 0 & 0 \\ 0 & 0 & \frac{i\sqrt{21}}{2} & 0 & -i\sqrt{6} & 0 & 0 & 0 & 0 & 0 \\ 0 & 0 & 0 & i\sqrt{6} & 0 & -\frac{5i}{2} & 0 & 0 & 0 & 0 \\ 0 & 0 & 0 & 0 & \frac{5i}{2} & 0 & -i\sqrt{6} & 0 & 0 & 0 \\ 0 & 0 & 0 & 0 & 0 & i\sqrt{6} & 0 & -\frac{i\sqrt{21}}{2} & 0 & 0 \\ 0 & 0 & 0 & 0 & 0 & 0 & \frac{i\sqrt{21}}{2} & 0 & -2i & 0 \\ 0 & 0 & 0 & 0 & 0 & 0 & 0 & 2i & 0 & -\frac{3i}{2} \\ 0 & 0 & 0 & 0 & 0 & 0 & 0 & 0 & \frac{3i}{2} & 0 \end{pmatrix}. \quad 10.74$$

$$I_2 = \begin{pmatrix} 1 & 0 \\ 0 & 1 \end{pmatrix}. \tag{10.75}$$

Finally, $Player_A$'s strategic state with respect to $\tilde{\theta}$ is given using Eq. 10.76 (Martínez-Martínez, 2014):

$$|s\rangle = S'|s_b\rangle = \begin{pmatrix} \cos\left(\frac{\tilde{\theta}}{2}\right)^9 \\ 0 \\ 3\cos\left(\frac{\tilde{\theta}}{2}\right)^8 \sin\left(\frac{\tilde{\theta}}{2}\right) \\ 0 \\ \frac{3}{64}\operatorname{cosec}\left(\frac{\tilde{\theta}}{2}\right)^5 \sin\tilde{\theta}^7 \\ 0 \\ \frac{\sqrt{21}}{32}\operatorname{cosec}\left(\frac{\tilde{\theta}}{2}\right)^3 \sin\tilde{\theta}^6 \\ 0 \\ \frac{3}{16}\sqrt{\frac{7}{2}}\operatorname{cosec}\left(\frac{\tilde{\theta}}{2}\right)\sin\tilde{\theta}^5 \\ 0 \\ \frac{3}{8}\sqrt{\frac{7}{2}}\sin\left(\frac{\tilde{\theta}}{2}\right)\sin\tilde{\theta}^4 \\ 0 \\ \frac{\sqrt{21}}{4}\sin\left(\frac{\tilde{\theta}}{2}\right)^3 \sin\tilde{\theta}^3 \\ 0 \\ \frac{3}{2}\sin\left(\frac{\tilde{\theta}}{2}\right)^5 \sin\tilde{\theta}^2 \\ 0 \\ \frac{3}{2}\sin\left(\frac{\tilde{\theta}}{2}\right)^7 \sin\tilde{\theta} \\ 0 \\ \sin\left(\frac{\tilde{\theta}}{2}\right)^9 \\ 0 \end{pmatrix}. \tag{10.76}$$

Then, we can employ a superposition of states, which is weighted in a manner defined using Eq. 10.77 (Martínez-Martínez, 2014) to represent such strategic state:

$$\cos\left(\tfrac{\tilde{\theta}}{2}\right)^9 |1action_C\rangle + 3\cos\left(\tfrac{\tilde{\theta}}{2}\right)^8 \sin\left(\tfrac{\tilde{\theta}}{2}\right)|2action_C\rangle + \cdots + \sin\left(\tfrac{\tilde{\theta}}{2}\right)^9 |10action_C\rangle. \tag{10.77}$$

Take the case of a number $n = 1$ coin, $Player_A$ will propose that offer anticipating that $Player_B$ will consent with the probability of $\left[\cos\left(\frac{\tilde{\theta}}{2}\right)^9\right]^2$. Accordingly, we can observe that the probabilities for distinct offers can be regarded as a function of parameter $\tilde{\theta}$. Considering $Player_A$, if we let $\tilde{\theta} = \theta_{Player_A}$, the probability of proposing an offer with $n > 1$ is non-disappearing, unless θ_{Player_A} equals to zero. Once an offer decision is made, the game's state is reduced to the selected offer's subspace, and then $Player_B$ has to decide between consent or rejection. In this case, the HSI of $Player_B$ is given using Eq. 10.78 (Martínez-Martínez, 2014):

$$\hat{H}_{SI} = \tilde{\theta}\left(\mathrm{I}_{10} \otimes J_{2,y}\right). \tag{10.78}$$

where an identity matrix with the size of 10×10 is denoted by I_{10}, and J_{2y} is equal to $\begin{pmatrix} 0 & -\frac{i}{2} \\ \frac{i}{2} & 0 \end{pmatrix}$. The final $Player_B$'s strategic state is, thus, defined using Eq. 10.79 (Martínez-Martínez, 2014):

$$\cos\left(\tfrac{\tilde{\theta}}{2}\right)\left|naction_C\right\rangle + \sin\left(\tfrac{\tilde{\theta}}{2}\right)\left|naction_R\right\rangle, \text{ for various realizations of the offer } n\,. \tag{10.79}$$

Since $Player_B$'s strategic state is another mind's belongings, i.e., $\tilde{\theta} \equiv \theta_{Player_B}$, the probability for $Player_B$ to consent is $\cos^2\left(\frac{\theta_{PlayerB}}{2}\right) < 1$, unless θ_{Player_B} equals to zero.

- **Observable Outcomes Computation:** Although $Player_A$'s strategic state can be formulated as a function in terms of the value of $\tilde{\theta}_{Player_A}$, it is always impractical to assume that this value is known *a priori*. In other words, a player's strategic state is still a probabilistic description regarding an agent's behaviour, which limits its observability. Therefore, Martínez-Martínez (2014) took one more step to address this issue. Basically, a less restrictive assumption was introduced in order to consider $\tilde{\theta}$ as a random variable whose distribution function is denoted by $f_{\tilde{\theta}}$. In the focal Ultimatum game, the appearance frequency corresponding to each individual offer (ranging from $n = 1$ to $n = 10$) is often observable, denoted by $\Pr(n)$. Accordingly, the probability of $Player_A$ proposing an offer that contains a number of n coins can be defined using Eq. 10.80 (Martínez-Martínez, 2014):

$$\Pr_{Player_A}(n) = \int P_{Player_A}\left(n \middle| \theta_{Player_A}\right) f_{\theta_{Player_A}}\, d\theta_{Player_A}\,. \tag{10.80}$$

Likewise, the probability of $Player_B$ deciding to take an offer that has an amount of n coins is given by Eq. 10.81 (Martínez-Martínez, 2014):

$$\Pr^{(n)}_{Player_A}\left(action_C\right) = \int P_{Player_B}\left(action_C \middle| \theta_{Player_B}\right) f_{\theta_{Player_B}} d\theta_{Player_B} . \qquad 10.81$$

In order to provide a detailed illustration, a Beta distribution was further employed in (Martínez-Martínez, 2014) but is outside the scope of this chapter. Interested readers please refer to (Martínez-Martínez, 2014) for parameter settings and graphical comparison.

10.4.3 Summary

The motivation of this modelling framework stems from the quantization solution for the Prisoner's Dilemma game proposed by (Eisert et al., 1999). Through a set of simulations, Martínez-Martínez (2014) demonstrated a way to calculate deviations from the classical Nash Equilibrium found in games.

10.5 Conclusions

Making a decision about launching an individual campaign using more than one crowdfunding channel or running several projects across different channels is not always an easy endeavour. In the literature, a vast amount of smart computing techniques, e.g., (Xing and Gao, 2014b; Xing and Marwala, 2018l; Xing and Gao, 2014c; Xing and Gao, 2014q; Xing and Gao, 2014m; Xing and Gao, 2014x; Xing et al., 2013; Xing and Gao, 2014ff; Xing and Gao, 2014aa; Xing and Gao, 2014bb; Xing and Gao, 2014cc; Xing and Gao, 2014d; Xing and Gao, 2014dd; Xing and Gao, 2014e; Xing and Gao, 2014f; Xing and Gao, 2014g; Xing and Gao, 2014h; Xing and Gao, 2014i; Xing and Gao, 2014j; Xing and Gao, 2014k; Xing and Gao, 2014l; Xing and Gao, 2014n; Xing and Gao, 2014o; Xing and Gao, 2014p; Xing and Gao, 2014r; Xing and Gao, 2014s; Xing and Gao, 2014t; Xing and Gao, 2014u; Xing and Gao, 2014v; Xing and Gao, 2014w; Xing and Gao, 2014y; Xing and Gao, 2014z), have helped people to conquer various tough, decision-making tasks. These include reconfigurable manufacturing (Xing et al., 2006a; Xing et al., 2009; Xing et al., 2006b), smart maintenance (Xing and Marwala, 2018m) for human–robot interaction (Xing and Marwala, 2018k; Xing and Marwala, 2018n) and the associated various perspectives (Xing and Marwala, 2018e; Xing and Marwala, 2018g; Xing and Marwala, 2018f; Xing and Marwala, 2018d; Xing and Marwala, 2018b; Xing and Marwala, 2018c; Xing and Marwala, 2018j; Xing and Marwala, 2018h; Xing and Marwala, 2018i; Xing, 2016c; Xing, 2016b; Xing and Marwala, 2018a), rational decision-making (Marwala, 2014f), group decision-making (Marwala, 2014e), vehicle routing (Xing et al., 2010c), layout design (Xing et al., 2010e), militarized interstate dispute (Marwala, 2014b; Marwala, 2014d; Marwala and Lagazio, 2011; Marwala, 2014h), part-machine clustering (Xing et al., 2010d), epileptic activity (Marwala, 2014c), antenna optimization (Xing, 2015b; Marwala, 2009; Marwala, 2014g), in-pipe inspection robot conceptual design (Xing, 2015a), supply chain management (Xing, 2014), network neutrality debate (Xing, 2016a) and power system economic load dispatch (Xing, 2015c).

In this chapter, we introduced game theory and quantum formulations to address crowdfunding channel competition. This is by no means a sole rigor solution, but rather an exploratory journey in search for alternative possible choices that satisfy different needs. The implications drawn from this chapter can help crowdfunding platform operators optimize their channel offerings, which in turn attracts more customers and generates more gains (both monetary and intellectual).

References

Agrawal A, Catalini C and Goldfarb A. (2014) Some simple economics of crowdfunding. *Innovation Policy and the Economy* 14: 63–97.

Alois J. (2016) *The end of pebble watch, retrieved from https://www.crowdfundinsider.com/2016/12/93205-end-pebble-watch/, accessed on 09 October 2017.*

Arkrot W, Unger A and Åhlström E. (2017) Crowdfunding from a marketing perspective. *International Business School.* Jönköping University.

Bajarin T. (2014) *Pebble and Glyph: How crowdfunding is creating disruptive new products, retrieved from http://time.com/2564/pebbleandglyphhowcrowdfundingiscreatingdisruptivenewproducts/, accessed on 09 October 2017.*

Banker RD, Khosla I and Sinha KK. (1998) Quality and competition. *Management Science* 44: 1179–1192.

Bather J. (2000) *Decision theory: An introduction to dynamic programming and sequential decisions,* Wiley, ISBN 978-0-471-97649-3.

Bayus B. (2013) Crowdsourcing new product ideas over time: An analysis of the Dell IdeaStorm community. *Management Science* 59: 226–244.

Berger JO. (1985) *Statistical decision theory and Bayesian analysis,* New York: Springer-Verlag New York, Inc. 978-1-4419-3074-3.

Bergl S. (2013) Go your own way. *Fast Company, November 2013, Issue 180*, p. 66.

Bernstein F and Federgruen A. (2004) A general equilibrium model for industries with price and service competition. *Operations Research* 52: 868–886.

Boyd S and Vandenberghe L. (2004) *Convex optimization,* The Edinburgh Building, Cambridge, CB2 8RU, UK: Cambridge University Press, ISBN 978-0-521-83378-3.

Brooke E. (2013) *Health scanner Scanadu Scout breaks Indiegogo's crowdfunding record, retrieved from https://techcrunch.com/2013/07/12/health-scanner-scanadu-scout-breaks-indiegogos-crowdfunding-record/, accessed on 11 october 2017.*

Brown TE, Boon E and Pitt LF. (2017) Seeking funding in order to sell: Crowdfunding as a marketing tool. *Business Horizons* 60: 189–195.

Butenko S and Pardalos PM. (2014) *Numerical methods and optimization: An introduction,* 6000 Broken Sound Parkway NW, Suite 300, Boca Raton, FL 33487-2742: CRC Press, Taylor & Francis Group, LLC, ISBN 978-1-4665-7778-7.

Camerer CF, Loewenstein G and Rabin M. (2004) *Advances in behavioral economics,* 41 William Street, Princeton, New Jersey 08540: Princeton University Press, ISBN 0-691-11681-4.

Cartwright E. (2011) *Behavioral economics,* 2 Park Square, Milton Park, Abingdon, Oxon OX14 4RN: Routledge, ISBN 978-0-415-57309-2.

Chafkin M. (2015) *Why Facebook's $2 billion bet on Oculus Rift might one day connect everyone on earth, retrieved from https://www.vanityfair.com/news/2015/09/oculus-rift-mark-zuckerberg-cover-story-palmer-luckey, accessed on 11 October 2017.*

Chang A. (2015) *Pebble watch crowdfunding hits record on Kickstarter, retrieved from https://www.cnbc.com/2015/03/30/pebble-watch-funding-hits-record--.html, accessed on 09 October 2017.*

Cusumano MA, Kahl SJ and Suarez FF. (2015) Services, industry evolution, and the competitive strategies of product firms. *Strategic Management Journal* 36: 559–575.

Dell. (2017) *IdeaStorm can help take your idea and turn it into reality, retrieved from http://www.ideastorm.com, accessed on 09 December 2017.*

Dolan DN. (2016) *Whirlpool corporation crowdfunded a beer dispenser and nailed it, retrieved from https://go.indiegogo.com/blog/2016/07/whirlpool-corporation-crowdfunded-beer-dispenser-and-nailed-it.html, accessed on 09 October 2017.*

Eisert J, Wilkens M and Lewenstein M. (1999) Quantum games and quantum strategies. *Physical Review Letters* 83: 3077–3080.

Faber MH. (2012) *Statistics and probability theory: In pursuit of engineering decision support,* Springer Dordrecht Heidelberg London New York: Springer Science+Business Media B.V., ISBN 978-94-007-4055-6.

Fang E, Palmatier RW and Steenkamp J-BEM. (2008) Effect of service transition strategies on firm value. *Journal of Marketing* 72: 1–14.

Favre M, Wittwer A, Heinimann HR, et al. (2016) Quantum decision theory in simple risky choices. *PLoS ONE* 11. e0168045.

Gerber E and Hui J. (2013) Crowdfunding: Motivations and deterrents for participation. *ACM Transactions on Computer-Human Interaction* 20: 1–32.

Goodman M. (2013) Kick the Kickstarter habit: A DIY crowdfunding app cuts out the middleman. *Entrepreneur, October 2013, Vol. 41 Issue 10,* p. 102.

Hamill P. (2014) *A student's guide to Lagrangians and Hamiltonians,* University Printing House, Cambridge CB2 8BS, United Kingdom: Cambridge University Press, ISBN 978-1-107-04288-9.

Han S, Kuruzovich J and Ravichandran T. (2013) Service expansion of product firms in the information technology industry: an empirical study. *Journal of Management Information Systems* 29: 127–158.

Hastings NAJ and Mello JMC. (1978) *Decision networks,* 111 River Street, Hoboken, NJ 07030-5774: John Wiley & Sons Ltd, ISBN 978-0-471-99531-9.

Hernando JR. (2017) Crowdfunding: the collaborative economy for channelling institutional and household savings. *The Spanish Review of Financial Economics* 15: 12–20.

Huang J, Leng M and Parlar M. (2013) Demand functions in decision modeling: A comprehensive survey and research directions. *Decision Sciences* 44: 557–609.

Indiegogo. (2013) *Scanadu scout, retrieved from https://www.indiegogo.com/projects/scanadu-scout#/, accessed on 11 october 2017.*

Kastrenakes J. (2016) *Indiegogo wants huge companies to crowdfund their next big products, retrieved from https://www.theverge.com/2016/1/6/10691100/indiegogo-enterprise-crowdfunding-announced-ces-2016, accessed on 10 September 2017.*

Kickstarter. (2012) *OUYA, a new kind of video game console, retrieved from https://www.kickstarter.com/projects/ouya/ouya-a-new-kind-of-video-game-console, accessed on 11 october 2017.*

Kickstarter. (2013) *Neurodome: A dome-format film that explores the brain, retrieved from https://www.kickstarter.com/projects/1100424922/neurodome-a-dome-format-film-that-explores-the-bra, accessed on 11 october 2017.*

Kickstarter. (2014a) *Marina Abramovic Institute, retrieved from https://www.kickstarter.com/projects/maihudson/marina-abramovic-institute-the-founders, accessed on 11 october 2017.*

Kickstarter. (2014b) *Oculus Rift: Step into the game, retrieved from https://www.kickstarter.com/projects/1523379957/oculus-rift-step-into-the-game, accessed on 11 october 2017.*

Lee S, Yoo S and Kim D. (2016) When is servitization a profitable competitive strategy. *International Journal of Production Economics* 173: 43–53.

Levine DK. (2012) *Is behavioral economics doomed? The ordinary versus the extraordinary,* 40 Devonshire Road, Cambridge, CB1 2BL, United Kingdom: Open Book Publishers CIC Ltd., ISBN 978-1-906924-93-5.

Lindgren BW. (1971) *Elements of decision theory,* London, England: Collier Macmillan Ltd, 978-0-023-70880-0.

Longford NT. (2013) *Statistical decision theory,* Heidelberg New York Dordrecht London: Springer Science+Business Media, ISBN 978-3-642-40432-0.

Luckerson V. (2014) *This new kind of Kickstarter could change everything, retrieved from http://business.time.com/2014/01/20/ibm-internal-enterprise-crowdfunding-mimics-kickstarter/print/, accessed on 09 October 2017.*

Machina MJ. (2008) Non-expected utility theory. In: Durlauf SN and Blume LE (eds) *The New Palgrave Dictionary of Economics.* 2nd ed. New York, USA: Palgrave Macmillan, ISBN 978-0-230-22620-3.

Marchegiani L. (in press) From mecenatism to crowdfunding: Engagement and identification in cultural-creative projects. *Journal of Heritage Tourism* https://doi.org/10.1080/1743873X.2017.1337777.

Marshall KT and Oliver RM. (1995) *Decision making and forecasting*: McGraw-Hill Companies, ISBN 978-0-070-48027-8.

Martínez-Martínez I. (2014) A connection between quantum decision theory and quantum games: The Hamiltonian of strategic interaction. *Journal of Mathematical Psychology* 58: 33–44.

Marwala T. (2009) *Computational intelligence for missing data imputation, estimation and management: Knowledge optimization techniques,* New York, USA: IGI Global, ISBN 978-1-60566-336-4.

Marwala T and Lagazio M. (2011) *Militarized conflict modeling using computational intelligence,* London, UK: Springer-Verlag, ISBN 978-0-85729-789-1.

Marwala T and Xing B. (2011) The role of remanufacturing in building a developmental state. *The Thinker: For the Thought Leaders (www.thethinker.co.za).* South Africa: Vusizwe Media, 18–20.

Marwala T. (2013) Multi-agent approaches to economic modeling: Game theory, ensembles, evolution and the stock market. In: Marwala T (ed) *Economic Modeling Using Artificial Intelligence Methods.* Springer London Heidelberg New York Dordrecht: Springer-Verlag London, ISBN 978-1-4471-5009-1, Chapter 11, pp. 195–213.

Marwala T. (2014a) *Artificial intelligence techniques for rational decision making,* Springer Cham Heidelberg New York Dordrecht London: Springer International Publishing Switzerland, ISBN 978-3-319-11423-1.

Marwala T. (2014b) Causal function for rational decision making: Application to militarized interstate dispute. In: Marwala T (ed) *Artificial Intelligence Techniques for Rational Decision Making.* Springer Cham Heidelberg New York Dordrecht London: Springer International Publishing Switzerland, ISBN 978-3-319-11423-1, Chapter 2, pp. 19–37.

Marwala T. (2014c) Correlation function for rational decision making: Application to epileptic activity. In: Marwala T (ed) *Artificial Intelligence Techniques for Rational Decision Making.* Springer Cham Heidelberg New York Dordrecht London: Springer International Publishing Switzerland, ISBN 978-3-319-11423-1, Chapter 3, pp. 39–54.

Marwala T. (2014d) Flexibly-bounded rationality in interstate conflict. In: Marwala T (ed) *Artificial Intelligence Techniques for Rational Decision Making.* Springer Cham Heidelberg New York Dordrecht London: Springer International Publishing Switzerland, ISBN 978-3-319-11423-1, Chapter 6, pp. 91–109.

Marwala T. (2014e) Group decision making. In: Marwala T (ed) *Artificial Intelligence Techniques for Rational Decision Making.* Springer Cham Heidelberg New York Dordrecht London: Springer International Publishing Switzerland, ISBN 978-3-319-11423-1, Chapter 8, pp. 131–150.

Marwala T. (2014f) Introduction to rational decision making. In: Marwala T (ed) *Artificial Intelligence Techniques for Rational Decision Making.* Springer Cham Heidelberg New

York Dordrecht London: Springer International Publishing Switzerland, ISBN 978-3-319-11423-1, Chapter 1, pp. 1–17.

Marwala T. (2014g) Missing data approaches for rational decision making: Application to antenatal data. In: Marwala T (ed) *Artificial Intelligence Techniques for Rational Decision Making*. Springer Cham Heidelberg New York Dordrecht London: Springer International Publishing Switzerland, ISBN 978-3-319-11423-1, Chapter 4, pp. 55–71.

Marwala T. (2014h) Rational counterfactuals and decision making: Application to interstate conflict. In: Marwala T (ed) *Artificial Intelligence Techniques for Rational Decision Making*. Springer Cham Heidelberg New York Dordrecht London: Springer International Publishing Switzerland, ISBN 978-3-319-11423-1, Chapter 5, pp. 73–89.

Marwala T. (2015) *Causality, correlation and artificial intelligence for rational decision making*, 5 Toh Tuck Link, Singapore 596224: World Scientific Publishing Co. Pte. Ltd, ISBN 978-9-81463-086-3.

Marwala T and Hurwitz E. (2017) *Artificial intelligence and economic theory: Skynet in the market*, Gewerbestrasse 11, 6330 Cham, Switzerland: Springer International Publishing AG, ISBN 978-3-319-66103-2.

Mathieu P, Beaufils B and Brandouy O. (2006) *Artificial economics: Agent-based methods in finance, game theory and their applications*, Germany: Springer-Verlag Berlin Heidelberg, ISBN 978-3-540-28578-6.

Mazalov V. (2014) *Mathematical game theory and applications*, The Atrium, Southern Gate, Chichester, West Sussex, PO19 8SQ, United Kingdom: John Wiley & Sons, Ltd, ISBN 978-1-118-89962-5.

Mixon FG, Asarta CJ and Caudill SB. (2017) Patreonomics: Public goods pedagogy for economics principles. *International Review of Economics Education* 25: 1–7.

Mustafa SE and Adnan HM. (2017) Crowdsourcing: A platform for crowd engagement in the publishing industry. *Publishing Research Quarterly* 33: 283–296.

Ordanini A, Miceli L, Pizzetti M, et al. (2011) Crowd-funding: Transforming customers into investors through innovative service platforms. *Journal of Service Management* 22: 443–470.

Paykacheva V. (2014) Crowdfunding as a customer engagement channel. *School of Business and Administration*. Kajaanin Ammattikorkeakoulu University of Applied Sciences.

Pothos EM and Busemeyer JR. (2009) A quantum probability explanation for violations of 'rational' decision theory. *Proceedings of the Royal Society B* 276: 2171–2178.

Procter & Gamble. (2017) *connect + develop^SM^, retrieved from http://www.pgconnectdevelop.com/, accessed on 09 December 2017*.

Raiffa H and Schlaifer R. (2000) *Applied statistical decision theory*: Wiley ISBN 978-0-471-38349-9.

Rivett P. (1980) *Model building for decision analysis*, 111 River Street, Hoboken, NJ 07030-5774: John Wiley & Sons Ltd, ISBN 978-0-471-27654-8.

Roma P, Petruzzelli AM and Perrone G. (2017) From the crowd to the market: The role of reward-based crowdfunding performance in attracting professional investors. *Research Policy* 46: 1606–1628.

Savage LJ. (1954) *The foundations of statistics*: John Wiley & Sons, Inc., ISBN 978-0-486-62349-8.

Shafir E and Tversky A. (1992) Thinking through uncertainty: Nonconsequential reasoning and choice. *Cognitive Psychology* 24: 449–474.

Sheldon RC and Kupp M. (2017) A market testing method based on crowd funding. *Strategy & Leadership* 45: 19–23.

Simon HA. (1955) A behavioral model of rational choice. *The Quarterly Journal of Economics* 69: 99–118.

Streltsov A. (2015) *Quantum correlations beyond entanglement and their role in quantum information theory*, Cham Heidelberg New York Dordrecht London: Springer, ISBN 978-3-319-09655-1.

Suarez FF, Cusumano M and Kahl S. (2013) Services and the business models of product firms: an empirical analysis of the software industry. *Management Science* 59: 420–435.

Sustainable Brands. (2016) *RB, Indiegogo launch 'healthier tomorrow challenge' for health, well-being startups, retrieved from http://www.sustainablebrands.com/news_and_views/startups/sustainable_brands/rb_indiegogo_launch_healthier_tomorrow_challenge_health, accessed on 09 October 2017.*

Umbhauer G. (2016) *Game theory and exercise,* 2 Park Square, Milton Park, Abingdon, Oxon OX14 4RN: Routledge, ISBN 978-0-415-60421-5.

Vanderbilt T. (2015) How an industrial giant thinks small. *Wired, December.* UK, 146–153.

Von Neumann J and Morgenstern O. (2004) *Theory of games and economic behavior,* 41 William Street, Princeton, New Jersey 08540: Princeton University Press, ISBN 978-0-691-11993-9.

White DJ. (1976) *Fundamentals of decision theory*: North-Holland, ISBN 978-0-720-48605-6.

Xing B, Bright G, Tlale NS, et al. (2006a) Reconfigurable manufacturing systems for agile mass customization manufacturing. *Proceedings of the 22nd ISPE International Conference on CAD/CAM, Robotics and Factories of the Future (CARs&FOF 2006), Vellore, India, July 2006.* pp. 473–482.

Xing B, Eganza J, Bright G, et al. (2006b) Reconfigurable manufacturing system for agile manufacturing. *Proceedings of the 12th IFAC Symposium on Information Control Problems in Manufacturing, May 2006, Saint-Etienne, France, pp. on CD.*

Xing B, Nelwamondo FV, Battle K, et al. (2009) Application of artificial intelligence (AI) methods for designing and analysis of reconfigurable cellular manufacturing system (RCMS). *Proceedings of the 2nd International Conference on Adaptive Science & Technology (ICAST), 14–16 December, Accra, Ghana,* pp. 402–409. IEEE.

Xing B, Gao W-J, Nelwamondo FV, et al. (2010a) Ant colony optimization for automated storage and retrieval system. *Proceedings of The Annual IEEE Congress on Evolutionary Computation (IEEE CEC), 18–23 July, CCIB, Barcelona, Spain,* pp. 1133–1139. IEEE.

Xing B, Gao W-J, Nelwamondo FV, et al. (2010b) Artificial intelligence in reverse supply chain management: the state of the art. *Proceedings of the Twenty-First Annual Symposium of the Pattern Recognition Association of South Africa (PRASA), 22–23 November, Stellenbosch, South Africa*, pp. 305–310.

Xing B, Gao W-J, Nelwamondo FV, et al. (2010c) Can ant algorithms make automated guided vehicle system more intelligent? A viewpoint from manufacturing environment. *Proceedings of IEEE International Conference on Systems, Man, and Cybernetics (IEEE SMC), 10–13 October, Istanbul, Turkey*, pp. 3226–3234. IEEE.

Xing B, Gao W-J, Nelwamondo FV, et al. (2010d) Part-machine clustering: The comparison between adaptive resonance theory neural network and ant colony system. In: Zeng Z and Wang J (eds) *Advances in Neural Network Research & Applications, LNEE 67,* pp. 747–755. Berlin Heidelberg: Springer-Verlag.

Xing B, Gao W-J, Nelwamondo FV, et al. (2010e) Two-stage inter-cell layout design for cellular manufacturing by using ant colony optimization algorithms. In: Tan Y, Shi Y and Tan KC (eds) *Advances in Swarm Intelligence, Part I, LNCS 6145,* pp. 281–289. Berlin Heidelberg: Springer-Verlag.

Xing B, Gao W-J and Marwala T. (2013) An overview of cuckoo-inspired intelligent algorithms and their applications. *IEEE Symposium Series on Computational Intelligence (IEEE SSCI), 15–19 April, Singapore,* pp. 85–89. IEEE.

Xing B. (2014) Computational intelligence in cross docking. *International Journal of Software Innovation* 4: 78–124.

Xing B and Gao W-J. (2014a) *Computational intelligence in remanufacturing,* 701 E. Chocolate Avenue, Suite 200, Hershey PA 17033: IGI Global, ISBN 978-1-4666-4908-8.

Xing B and Gao W-J. (2014aa) Chemical-reaction optimization algorithm. In: Xing B and Gao W-J (eds) *Innovative Computational Intelligence: A Rough Guide to 134 Clever Algorithms.* Cham Heidelberg New York Dordrecht London: Springer International Publishing Switzerland, ISBN: 978-3-319-03403-4, Chapter 25, pp. 417–428.

Xing B and Gao W-J. (2014b) *Innovative computational intelligence: A rough guide to 134 clever algorithms,* Cham Heidelberg New York Dordrecht London: Springer International Publishing Switzerland, ISBN: 978-3-319-03403-4.

Xing B and Gao W-J. (2014bb) Emerging chemistry-based CI algorithms. In: Xing B and Gao W-J (eds) *Innovative Computational Intelligence: A Rough Guide to 134 Clever Algorithms.* Cham Heidelberg New York Dordrecht London: Springer International Publishing Switzerland, ISBN: 978-3-319-03403-4, Chapter 26, pp. 429–437.

Xing B and Gao W-J. (2014c) Introduction to computational intelligence. In: Xing B and Gao W-J (eds) *Innovative Computational Intelligence: A Rough Guide to 134 Clever Algorithms.* Cham Heidelberg New York Dordrecht London: Springer International Publishing Switzerland, ISBN: 978-3-319-03403-4, Chapter 1, pp. 3–17.

Xing B and Gao W-J. (2014cc) Base optimization algorithm. In: Xing B and Gao W-J (eds) *Innovative Computational Intelligence: A Rough Guide to 134 Clever Algorithms.* Cham Heidelberg New York Dordrecht London: Springer International Publishing Switzerland, ISBN: 978-3-319-03403-4, Chapter 27, pp. 441–444.

Xing B and Gao W-J. (2014d) Bacteria inspired algorithms. In: Xing B and Gao W-J (eds) *Innovative Computational Intelligence: A Rough Guide to 134 Clever Algorithms.* Cham Heidelberg New York Dordrecht London: Springer International Publishing Switzerland, ISBN: 978-3-319-03403-4, Chapter 2, pp. 21–38.

Xing B and Gao W-J. (2014dd) Emerging mathematics-based CI algorithms. In: Xing B and Gao W-J (eds) *Innovative Computational Intelligence: A Rough Guide to 134 Clever Algorithms.* Cham Heidelberg New York Dordrecht London: Springer International Publishing Switzerland, ISBN: 978-3-319-03403-4, Chapter 28, pp. 445–448.

Xing B and Gao W-J. (2014e) Bee inspired algorithms. In: Xing B and Gao W-J (eds) *Innovative Computational Intelligence: A Rough Guide to 134 Clever Algorithms.* Cham Heidelberg New York Dordrecht London: Springer International Publishing Switzerland, ISBN: 978-3-319-03403-4, Chapter 4, pp. 45–80.

Xing B and Gao W-J. (2014ee) Introduction to remanufacturing and reverse logistics. In: Xing B and Gao W-J (eds) *Computational Intelligence in Remanufacturing.* 701 E. Chocolate Avenue, Suite 200, Hershey PA 17033: IGI Global, ISBN 978-1-4666-4908-8, Chapter 1, pp. 1–17.

Xing B and Gao W-J. (2014f) Bat inspired algorithms. In: Xing B and Gao W-J (eds) *Innovative Computational Intelligence: A Rough Guide to 134 Clever Algorithms.* Cham Heidelberg New York Dordrecht London: Springer International Publishing Switzerland, ISBN: 978-3-319-03403-4, Chapter 3, pp. 39–44.

Xing B and Gao W-J. (2014ff) Overview of computational intelligence. In: Xing B and Gao W-J (eds) *Computational Intelligence in Remanufacturing.* 701 E. Chocolate Avenue, Suite 200, Hershey PA 17033: IGI Global, ISBN 978-1-4666-4908-8, Chapter 2, pp. 18–36.

Xing B and Gao W-J. (2014g) Biogeography-based optimization algorithm. In: Xing B and Gao W-J (eds) *Innovative Computational Intelligence: A Rough Guide to 134 Clever Algorithms.* Cham Heidelberg New York Dordrecht London: Springer International Publishing Switzerland, ISBN: 978-3-319-03403-4, Chapter 5, pp. 81–91.

Xing B and Gao W-J. (2014gg) Used products return pattern analysis using agent-based modelling and simulation. In: Xing B and Gao W-J (eds) *Computational Intelligence in Remanufacturing.* 701 E. Chocolate Avenue, Suite 200, Hershey PA 17033: IGI Global, ISBN 978-1-4666-4908-8, Chapter 3, pp. 38–58.

Xing B and Gao W-J. (2014h) Cat swarm optimization algorithm. In: Xing B and Gao W-J (eds) *Innovative Computational Intelligence: A Rough Guide to 134 Clever Algorithms.* Cham Heidelberg New York Dordrecht London: Springer International Publishing Switzerland, ISBN: 978-3-319-03403-4, Chapter 6, pp. 93–104.

Xing B and Gao W-J. (2014hh) Used product collection optimization using genetic algorithms. In: Xing B and Gao W-J (eds) *Computational Intelligence in Remanufacturing.* 701 E. Chocolate Avenue, Suite 200, Hershey PA 17033: IGI Global, ISBN 978-1-4666-4908-8, Chapter 4, pp. 59–74.

Xing B and Gao W-J. (2014i) Cuckoo inspired algorithms. In: Xing B and Gao W-J (eds) *Innovative Computational Intelligence: A Rough Guide to 134 Clever Algorithms.* Cham Heidelberg New York Dordrecht London: Springer International Publishing Switzerland, ISBN: 978-3-319-03403-4, Chapter 7, pp. 105–121.

Xing B and Gao W-J. (2014ii) Used product remanufacturability evaluation using fuzzy logic. In: Xing B and Gao W-J (eds) *Computational Intelligence in Remanufacturing*. 701 E. Chocolate Avenue, Suite 200, Hershey PA 17033: IGI Global, ISBN 978-1-4666-4908-8, Chapter 5, pp. 75–94.

Xing B and Gao W-J. (2014j) Luminous insect inspired algorithms. In: Xing B and Gao W-J (eds) *Innovative Computational Intelligence: A Rough Guide to 134 Clever Algorithms*. Cham Heidelberg New York Dordrecht London: Springer International Publishing Switzerland, ISBN: 978-3-319-03403-4, Chapter 8, pp. 123–137.

Xing B and Gao W-J. (2014jj) Used product pre–sorting system optimization using teaching–learning-based optimization. In: Xing B and Gao W-J (eds) *Computational Intelligence in Remanufacturing*. 701 E. Chocolate Avenue, Suite 200, Hershey PA 17033: IGI Global, ISBN 978-1-4666-4908-8, Chapter 6, pp. 95–112.

Xing B and Gao W-J. (2014k) Fish inspired algorithms. In: Xing B and Gao W-J (eds) *Innovative Computational Intelligence: A Rough Guide to 134 Clever Algorithms*. Cham Heidelberg New York Dordrecht London: Springer International Publishing Switzerland, ISBN: 978-3-319-03403-4, Chapter 9, pp. 139–155.

Xing B and Gao W-J. (2014kk) Used product delivery optimization using agent-based modelling and simulation. In: Xing B and Gao W-J (eds) *Computational Intelligence in Remanufacturing*. 701 E. Chocolate Avenue, Suite 200, Hershey PA 17033: IGI Global, ISBN 978-1-4666-4908-8, Chapter 7, pp. 113–133.

Xing B and Gao W-J. (2014l) Frog inspired algorithms. In: Xing B and Gao W-J (eds) *Innovative Computational Intelligence: A Rough Guide to 134 Clever Algorithms*. Cham Heidelberg New York Dordrecht London: Springer International Publishing Switzerland, ISBN: 978-3-319-03403-4, Chapter 10, pp. 157–165.

Xing B and Gao W-J. (2014m) Fruit fly optimization algorithm. In: Xing B and Gao W-J (eds) *Innovative Computational Intelligence: A Rough Guide to 134 Clever Algorithms*. Cham Heidelberg New York Dordrecht London: Springer International Publishing Switzerland, ISBN: 978-3-319-03403-4, Chapter 11, pp. 167–170.

Xing B and Gao W-J. (2014n) Group search optimization algorithm. In: Xing B and Gao W-J (eds) *Innovative Computational Intelligence: A Rough Guide to 134 Clever Algorithms*. Cham Heidelberg New York Dordrecht London: Springer International Publishing Switzerland, ISBN: 978-3-319-03403-4, Chapter 12, pp. 171–176.

Xing B and Gao W-J. (2014o) Invasive weed optimization algorithm. In: Xing B and Gao W-J (eds) *Innovative Computational Intelligence: A Rough Guide to 134 Clever Algorithms*. Cham Heidelberg New York Dordrecht London: Springer International Publishing Switzerland, ISBN: 978-3-319-03403-4, Chapter 13, pp. 177–181.

Xing B and Gao W-J. (2014p) Music inspired algorithms. In: Xing B and Gao W-J (eds) *Innovative Computational Intelligence: A Rough Guide to 134 Clever Algorithms*. Cham Heidelberg New York Dordrecht London: Springer International Publishing Switzerland, ISBN: 978-3-319-03403-4, Chapter 14, pp. 183–201.

Xing B and Gao W-J. (2014q) Imperialist competitive algorithm. In: Xing B and Gao W-J (eds) *Innovative Computational Intelligence: A Rough Guide to 134 Clever Algorithms*. Cham Heidelberg New York Dordrecht London: Springer International Publishing Switzerland, ISBN: 978-3-319-03403-4, Chapter 15, pp. 203–209.

Xing B and Gao W-J. (2014r) Teaching–learning-based optimization algorithm. In: Xing B and Gao W-J (eds) *Innovative Computational Intelligence: A Rough Guide to 134 Clever Algorithms*. Cham Heidelberg New York Dordrecht London: Springer International Publishing Switzerland, ISBN: 978-3-319-03403-4, Chapter 16, pp. 211–216.

Xing B and Gao W-J. (2014s) Emerging biology-based CI algorithms. In: Xing B and Gao W-J (eds) *Innovative Computational Intelligence: A Rough Guide to 134 Clever Algorithms*. Cham Heidelberg New York Dordrecht London: Springer International Publishing Switzerland, ISBN: 978-3-319-03403-4, Chapter 17, pp. 217–317.

Xing B and Gao W-J. (2014t) Big bang–big crunch algorithm. In: Xing B and Gao W-J (eds) *Innovative Computational Intelligence: A Rough Guide to 134 Clever Algorithms*.

Cham Heidelberg New York Dordrecht London: Springer International Publishing Switzerland, ISBN: 978-3-319-03403-4, Chapter 18, pp. 321–331.

Xing B and Gao W-J. (2014u) Central force optimization algorithm. In: Xing B and Gao W-J (eds) *Innovative Computational Intelligence: A Rough Guide to 134 Clever Algorithms.* Cham Heidelberg New York Dordrecht London: Springer International Publishing Switzerland, ISBN: 978-3-319-03403-4, Chapter 19, pp. 333–337.

Xing B and Gao W-J. (2014v) Charged system search algorithm. In: Xing B and Gao W-J (eds) *Innovative Computational Intelligence: A Rough Guide to 134 Clever Algorithms.* Cham Heidelberg New York Dordrecht London: Springer International Publishing Switzerland, ISBN: 978-3-319-03403-4, Chapter 20, pp. 339–346.

Xing B and Gao W-J. (2014w) Electromagnetism-like mechanism algorithm. In: Xing B and Gao W-J (eds) *Innovative Computational Intelligence: A Rough Guide to 134 Clever Algorithms.* Cham Heidelberg New York Dordrecht London: Springer International Publishing Switzerland, ISBN: 978-3-319-03403-4, Chapter 21, pp. 347–354.

Xing B and Gao W-J. (2014x) Gravitational search algorithm. In: Xing B and Gao W-J (eds) *Innovative Computational Intelligence: A Rough Guide to 134 Clever Algorithms.* Cham Heidelberg New York Dordrecht London: Springer International Publishing Switzerland, ISBN: 978-3-319-03403-4, Chapter 22, pp. 355–364.

Xing B and Gao W-J. (2014y) Intelligent water drops algorithm. In: Xing B and Gao W-J (eds) *Innovative Computational Intelligence: A Rough Guide to 134 Clever Algorithms.* Cham Heidelberg New York Dordrecht London: Springer International Publishing Switzerland, ISBN: 978-3-319-03403-4, Chapter 23, pp. 365–373.

Xing B and Gao W-J. (2014z) Emerging physics-based CI algorithms. In: Xing B and Gao W-J (eds) *Innovative Computational Intelligence: A Rough Guide to 134 Clever Algorithms.* Cham Heidelberg New York Dordrecht London: Springer International Publishing Switzerland, ISBN: 978-3-319-03403-4, Chapter 24, pp. 375–414.

Xing B. (2015a) Knowledge management: Intelligent in-pipe inspection robot conceptual design for pipeline infrastructure management. In: Kahraman C and Onar SÇ (eds) *Intelligent Techniques in Engineering Management: Theory and Applications.* Cham Heidelberg New York Dordrecht London: Springer International Publishing Switzerland, ISBN 978-3-319-17905-6, Chapter 6, 129–146.

Xing B. (2015b) Novel nature-derived intelligent algorithms and their applications in antenna optimization. In: Matin MA (ed) *Wideband, Multiband, and Smart Reconfigurable Antennas for Modern Wireless Communications.* 701 E. Chocolate Avenue, Hershey PA 17033: IGI Global, ISBN 978-1-4666-8645-8, Chapter 10, pp. 296–339.

Xing B. (2015c) Optimization in production management: Economic load dispatch of cyber physical power system using artificial bee colony. In: Kahraman C and Onar SÇ (eds) *Intelligent Techniques in Engineering Management: Theory and Applications.* Cham Heidelberg New York Dordrecht London: Springer International Publishing Switzerland, ISBN 978-3-319-17905-6, Chapter 12, 275–293.

Xing B and Gao W-J. (2015a) The applications of swarm intelligence in remanufacturing: A focus on retrieval. In: Khosrow-Pour M (ed) *Encyclopedia of Information Science and Technology.* 3rd ed. New York, USA: Information Science Ref. – IGI Global, ISBN 978-1-4666-5888-2, Chapter 7, pp. 66–74.

Xing B and Gao W-J. (2015b) Offshore remanufacturing. In: Khosrow-Pour M (ed) *Encyclopedia of Information Science and Technology.* 3rd ed. New York, USA: Information Science Ref. – IGI Global, ISBN 978-1-4666-5888-2, Chapter 374, pp. 3795–3804.

Xing B and Gao W-J. (2015c) A SWOT analysis of intelligent product enabled complex adaptive logistics systems. In: Khosrow-Pour M (ed) *Encyclopedia of Information Science and Technology.* 3rd ed. New York, USA: Information Science Ref. – IGI Global, ISBN 978-1-4666-5888-2, Chapter 490, pp. 4970–4979.

Xing B. (2016a) Network neutrality debate in the internet of things era: A fuzzy cognitive map extend technology roadmap perspective. In: Mahmood Z (ed) *Connectivity Frameworks for Smart Devices: The IoT Distributed Computing Perspective.* Cham Heidelberg New York

Dordrecht London: Springer International Publishing Switzerland, ISBN 978-3-319-33122-5, Chapter 10, pp. 235–257.

Xing B. (2016b) Smart robot control via novel computational intelligence methods for ambient assisted living. In: Ravulakollu KK, Khan MA and Abraham A (eds) *Trends in Ambient Intelligent Systems.* Switzerland: Springer International Publishing Switzerland, ISBN 978-3-319-30184-6, Chapter 2, pp. 29–55.

Xing B. (2016c) The spread of innovatory nature originated metaheuristics in robot swarm control for smart living environments. In: Espinosa HEP (ed) *Nature-Inspired Computing for Control Systems.* Cham Heidelberg New York Dordrecht London: Springer International Publishing Switzerland, ISBN 978-3-319-26228-4, Chapter 3, pp. 39–70.

Xing B and Marwala T. (2018a) Conclusion. In: Xing B and Marwala T (eds) *Smart Maintenance for Human–Robot Interaction: An Intelligent Search Algorithmic Perspective.* Gewerbestrasse 11, 6330 Cham, Switzerland: Springer International Publishing AG, ISBN 978-3-319-67479-7, Chapter 13, pp. 299–305.

Xing B and Marwala T. (2018b) Cyberware capacity–applications layer perspective. In: Xing B and Marwala T (eds) *Smart Maintenance for Human–Robot Interaction: An Intelligent Search Algorithmic Perspective.* Gewerbestrasse 11, 6330 Cham, Switzerland: Springer International Publishing AG, ISBN 978-3-319-67479-7, Chapter 8, pp. 173–191.

Xing B and Marwala T. (2018c) Cyberware capacity–energy autonomy perspective. In: Xing B and Marwala T (eds) *Smart Maintenance for Human–Robot Interaction: An Intelligent Search Algorithmic Perspective.* Gewerbestrasse 11, 6330 Cham, Switzerland: Springer International Publishing AG, ISBN 978-3-319-67479-7, Chapter 9, pp. 193–216.

Xing B and Marwala T. (2018d) Cyberware capacity–platform and middleware layers perspective. In: Xing B and Marwala T (eds) *Smart Maintenance for Human–Robot Interaction: An Intelligent Search Algorithmic Perspective.* Gewerbestrasse 11, 6330 Cham, Switzerland: Springer International Publishing AG, ISBN 978-3-319-67479-7, Chapter 7, pp. 143–171.

Xing B and Marwala T. (2018e) Hardware capacity–beginning of life perspective. In: Xing B and Marwala T (eds) *Smart Maintenance for Human–Robot Interaction: An Intelligent Search Algorithmic Perspective.* Gewerbestrasse 11, 6330 Cham, Switzerland: Springer International Publishing AG, ISBN 978-3-319-67479-7, Chapter 4, pp. 67–91.

Xing B and Marwala T. (2018f) Hardware capacity–end of life perspective. In: Xing B and Marwala T (eds) *Smart Maintenance for Human–Robot Interaction: An Intelligent Search Algorithmic Perspective.* Gewerbestrasse 11, 6330 Cham, Switzerland: Springer International Publishing AG, ISBN 978-3-319-67479-7, Chapter 6, pp. 111–139.

Xing B and Marwala T. (2018g) Hardware capacity–middle of life perspective. In: Xing B and Marwala T (eds) *Smart Maintenance for Human–Robot Interaction: An Intelligent Search Algorithmic Perspective.* Gewerbestrasse 11, 6330 Cham, Switzerland: Springer International Publishing AG, ISBN 978-3-319-67479-7, Chapter 5, pp. 93–110.

Xing B and Marwala T. (2018h) Human capacity–biopsychosocial perspective. In: Xing B and Marwala T (eds) *Smart Maintenance for Human–Robot Interaction: An Intelligent Search Algorithmic Perspective.* Gewerbestrasse 11, 6330 Cham, Switzerland: Springer International Publishing AG, ISBN 978-3-319-67479-7, Chapter 11, pp. 249–270.

Xing B and Marwala T. (2018i) Human capacity–exposome perspective. In: Xing B and Marwala T (eds) *Smart Maintenance for Human–Robot Interaction: An Intelligent Search Algorithmic Perspective.* Gewerbestrasse 11, 6330 Cham, Switzerland: Springer International Publishing AG, ISBN 978-3-319-67479-7, Chapter 12, pp. 271–295.

Xing B and Marwala T. (2018j) Human capacity–physiology perspective. In: Xing B and Marwala T (eds) *Smart Maintenance for Human–Robot Interaction: An Intelligent Search Algorithmic Perspective.* Gewerbestrasse 11, 6330 Cham, Switzerland: Springer International Publishing AG, ISBN 978-3-319-67479-7, Chapter 10, pp. 219–247.

Xing B and Marwala T. (2018k) Introduction to human robot interaction. In: Xing B and Marwala T (eds) *Smart Maintenance for Human–Robot Interaction: An Intelligent Search Algorithmic Perspective.* Gewerbestrasse 11, 6330 Cham, Switzerland: Springer International Publishing AG, ISBN 978-3-319-67479-7, Chapter 1, pp. 3–19.

Xing B and Marwala T. (2018l) Introduction to intelligent search algorithms. In: Xing B and Marwala T (eds) *Smart Maintenance for Human–Robot Interaction: An Intelligent Search Algorithmic Perspective.* Gewerbestrasse 11, 6330 Cham, Switzerland: Springer International Publishing AG, ISBN 978-3-319-67479-7, Chapter 3, pp. 33–64.

Xing B and Marwala T. (2018m) Introduction to smart maintenance. In: Xing B and Marwala T (eds) *Smart Maintenance for Human–Robot Interaction: An Intelligent Search Algorithmic Perspective.* Gewerbestrasse 11, 6330 Cham, Switzerland: Springer International Publishing AG, ISBN 978-3-319-67479-7, Chapter 2, pp. 21–31.

Xing B and Marwala T. (2018n) *Smart maintenance for human–robot interaction: An intelligent search algorithmic perspective,* Gewerbestrasse 11, 6330 Cham, Switzerland: Springer International Publishing AG, ISBN 978-3-319-67479-7.

Yeh A. (2016) *Announcing the citis project by Heineken and Indiegogo, retrieved from https://go.indiegogo.com/blog/2016/05/heineken-cities-project.html, accessed on 09 October 2017.*

Yukalov VI and Sornette D. (2008) Quantum decision theory as quantum theory of measurement. *Physics Letters A* 372: 6867–6871.

Yukalov VI and Sornette D. (2009) Processing information in quantum decision theory. *Entropy* 11: 1073–1120.

Yukalov VI and Sornette D. (2011) Decision theory with prospect interference and entanglement. *Theory and Decision* 70: 283–328.

Zeckhauser R. (2006) Investing in the unknown and unknowable. *Capitalism and Society* 1: Article 5.

Part VII
Epilogue

CHAPTER 11

Outlook of Crowdfunding

11.1 A Metaphor from the 'Remembrance of Earth's Past' Trilogy

In 2015, the Hugo Award (International) for Best Novel went to Mr. Cixin Liu. His science fiction novel includes the following three volumes:

- Volume 1–The Three-Body Problem (Liu, 2014),
- Volume 2–The Dark Forest (Liu, 2015a), and
- Volume 3–Death's End (Liu, 2015b).

The trilogy has now become an international cultural phenomenon and has been studied by many business and political leaders (including former US president Barack Obama, Facebook's founder Mark Zuckerberg and Xiaomi's CEO Jun Lei).

The second book of Mr. Liu's trilogy, *The Dark Forest*, focuses on explaining the main problem associated with the attempt at communicating with a possible alien civilization. The simplified logic goes like this: Suppose there is an advanced civilization on Planet A. When it detects the existence of another civilization on Planet B, what should Planet A's civilization do? Since it knows very little about Planet B's civilization (i.e., goals, capabilities, future development degree, intentions, etc.), decision-makers on Planet A might conclude that the most reasonable and safe course of action is to remove the threat of Planet B completely while the window of opportunity is still open. Unfortunately, the same thinking logic is exactly followed by Planet B's decision-makers.

11.2 Is Future Crowdfunding A 'Dark Forest'?

Now, let's turn our attention to SingularityNET (SingularityNET, 2017a), a concept of creating a global, decentralized artificial intelligence (AI) marketplace. The marketplace is dedicated to increasing the public's accessibility to all sorts of AI techniques, while offering new means for the developers (behind these techniques) to monetize their work. According to a whitepaper (SingularityNET, 2017b) published by the SingularityNET team, anyone is able to add an AI or machine learning service to its blockchain-backed network and the remuneration for such contribution will be in AGI (standing for artificial general intelligence) tokens. The ICO of SingularityNET started on 8 December 2017 and was planned to stop once 500 million tokens (equivalent to $36 m) are sold. This originally settled hard cap was quickly dwarfed by the actual $150 m pledges.

One of the key team members of SingularityNET is its Chief Humanoid Officer, the so-called 'Sophia Robot'. The aim of this project is great, enabling AI-as-a-Service, but will this be a preliminary version of Skynet envisioned by Marwala and Hurwitz (2017a)? Through the lens of dark forest logic, every civilization sees its survival as the top priority. Since the dark forest closely resembles the mindset of realists—that one party inevitably competes with the others in order to maximize its own power because of the fear associated with the other parties' rising power—can the potential conflict between mankind and AI-powered creatures be resolved peacefully? Is it possible to reach a win-win state between human and robot (Marwala and Hurwitz, 2017b; Xing and Marwala, 2018l; Xing and Marwala, 2018k; Xing and Marwala, 2018e; Xing and Marwala, 2018g; Xing and Marwala, 2018f; Xing and Marwala, 2018d; Xing and Marwala, 2018b; Xing and Marwala, 2018c; Xing and Marwala, 2018j; Xing and Marwala, 2018h; Xing and Marwala, 2018i; Xing and Marwala, 2018a)?

11.3 The Beginning of the End?

The above concerns force us to think abstractly about the prospect of the crowdfunding ecosystem. The following explanation, partially inspired by the first book in Mr. Liu's trilogy, might provide one of many possible avenues.

11.3.1 N-Body Problem

Isaac Newton made a remarkable contribution to modern science by formulating the three laws of motion, the law of gravity, and his resolution to some two-body problems (e.g., sun and one planet). Nevertheless, the involvement of the moon (the third body) gave him so many headaches, as he remarked to the astronomer John Machine (Meyer and Offin, 2017), that "his head never ached but with his studies on the moon."

11.3.1.1 Newton's Laws and Two-Body Problem

Newton's second law, the law of dynamics, can be rewritten as Eq. 11.1 under a Cartesian coordinate system (Musielak and Quarles, 2017):

$$\mathbf{F} = M\frac{d^2\mathbf{R}}{dt^2}. \tag{11.1}$$

where **R** denotes the distance from a mass (represented by M) to the origin of Cartesian coordinate system, and **F** indicates a force.

Meanwhile, the Law of gravity allows us to find a gravitational interaction-induced force (indicated by $\mathbf{F}_g$) by using Eq. 11.2 under a Cartesian coordinate system (Musielak and Quarles, 2017):

$$\mathbf{F}_g = G\frac{M_1 M_2}{R_{12}^3}\mathbf{R}_{12}. \tag{11.2}$$

where a universal gravitational constant is represented by G.

The relative simplicity of the two-body problem lies in that it consists of only two objects (with random masses) interacting via gravitational force. Based on Eq. 11.1, we can formulate the two-body's motions using Eq. 11.3 (Musielak and Quarles, 2017):

$$\frac{d^2\mathbf{R}_{12}}{dt^2} + G\frac{M_1 M_2}{R_{12}^3}\mathbf{R}_{12} = 0. \tag{11.3}$$

By defining $\mu = G(M_1 + M_2)$ and $\mathbf{r} \equiv \mathbf{R}_{12}$, Eq. 11.3 can be reformulated as Eq. 11.4 (Musielak and Quarles, 2017):

$$\ddot{\mathbf{r}} + \frac{\mu}{r^3}\mathbf{r} = 0. \tag{11.4}$$

where the second derivative with respect to time is denoted by $\ddot{\mathbf{r}}$. After performing the cross-product operation on Eq. 11.4 with **r**, we arrive at Eq. 11.5 (Musielak and Quarles, 2017):

$$\mathbf{r}\times\ddot{\mathbf{r}} + \mathbf{r}\times\frac{\mu}{r^3}\mathbf{r} = 0. \tag{11.5}$$

The integration of Eq. 11.5 gives us Eq. 11.6 (Musielak and Quarles, 2017):

$$\mathbf{r}\times\dot{\mathbf{r}} = \mathbf{k}. \tag{11.6}$$

where **k** denotes a constant.

Furthermore, we can also perform the cross-product operation on Eq. 11.4 with **k** and get Eq. 11.7 (Musielak and Quarles, 2017):

$$\ddot{\mathbf{r}} \times \mathbf{k} = \mu \frac{d\left(\frac{\mathbf{r}}{r}\right)}{dt}. \quad 11.7$$

The integration of Eq. 11.7 gives us Eq. 11.8 (Musielak and Quarles, 2017):

$$\dot{\mathbf{r}} \times \mathbf{k} = \mu\left(\frac{\mathbf{r}}{r} + \mathbf{e}\right). \quad 11.8$$

where $\mathbf{e}$ denotes an integration constant.

The solution of Eq. 11.8 is given by Eq. 11.9 (Musielak and Quarles, 2017):

$$r = \frac{\left(\frac{k^2}{\mu}\right)}{1 + e\cos f}. \quad 11.9$$

where f represents the angle between $\mathbf{r}$ and $\mathbf{e}$.

11.3.1.2 General Three-Body Problem

When we move to three-dimensional space, where three random masses are considered, the equation of motions can be rewritten as Eq. 11.10 (Musielak and Quarles, 2017):

$$M_i \frac{d^2\mathbf{R}_i}{dt^2} = G\sum_{j=1}^{3} \frac{M_i M_j}{r_{ij}^3}\mathbf{r}_{ij}. \quad 11.10$$

In order to find the motion's integrals, we can sum Eq. 11.10 over i and get Eq. 11.11 (Musielak and Quarles, 2017):

$$\sum_{j=1}^{3} M_i \frac{d^2\mathbf{R}_i}{dt^2} = 0. \quad 11.11$$

The integration of Eq. 11.11 gives us Eq. 11.12 (Musielak and Quarles, 2017):

$$\sum_{i=1}^{3} M_i \frac{d\mathbf{R}_i}{dt} = \mathbf{C}_1. \quad 11.12$$

where $\mathbf{C}_1$ is a constant. By performing another integration on Eq. 11.12, we can get Eq. 11.13 (Musielak and Quarles, 2017):

$$\sum_{i=1}^{3} M_i \mathbf{R}_i = \mathbf{C}_1 t + \mathbf{C}_2. \quad 11.13$$

where $\mathbf{C}_2$ is also a constant.

Furthermore, by taking a vector product of $\mathbf{R}_i$ with Eq. 11.10, we can have Eq. 11.14 (Musielak and Quarles, 2017):

$$\sum_{i=1}^{3} M_i \mathbf{R}_i \times \frac{d^2\mathbf{R}_i}{dt^2} = 0. \qquad 11.14$$

The integration of Eq. 11.14 gives us Eq. 11.15 (Musielak and Quarles, 2017):

$$\sum_{i=1}^{3} M_i \mathbf{R}_i \times \frac{d\mathbf{R}_i}{dt} = \mathbf{C}_3. \qquad 11.15$$

where $\mathbf{C}_3$ also denotes a constant.

The renowned general three-body problem leaves us 10 integrals of motions, 18 (though the number can be reduced to 8) first-order ordinary differential equations, and two extra integrals of motion. The implication of all these is that no solutions can be found for the general three-body problem, at least by standard mathematical quadrature (Musielak and Quarles, 2017). We can learn that the motion is happening either on a line or on a conic section, but we cannot provide accurate time information. Imagine the following scenario: You ask the train station master "Where is our train?" and he/she replies "On this track." Are you ok with this answer? It would be great if we could give the location of our interested object as a simple function of t, but this is truly easier said than done (Meyer and Offin, 2017).

11.4 Conclusions

The details of the *N*-body problem's mathematical formulation are certainly outside the scope of this chapter. But finding an inertial frame reference in space becomes problematic because it has to be non-accelerated or non-rotated. Ideally, an inertial frame (built on three mutually perpendicular vectors found in space) should be based on the so-called 'fixed stars', but identifying this is an impossible mission since the stars are all believed to be moving with respect to one another. Though Newton's laws fail in certain extreme scenarios (where Einstein's general theory of relativity proves itself), as suggested by Meyer and Offin (2017), if someone wants to follow a spy satellite or send human beings to Mars, it is still better to rely on Newton and leave Einstein alone.

Will the crowdfunding ecosystem (with so many players already participating and the unknown number of technological breakthroughs to be involved) behave like our universe, where no pivot can be found? Who knows. In this book, we used smart computing as a lens with two different filters: Service innovation and bricolage. The structure of this book is summarized and depicted in Fig. 11.1. Interested parties may use this book as a referential roadmap in order to find their own inertial frame in crowdfunding space.

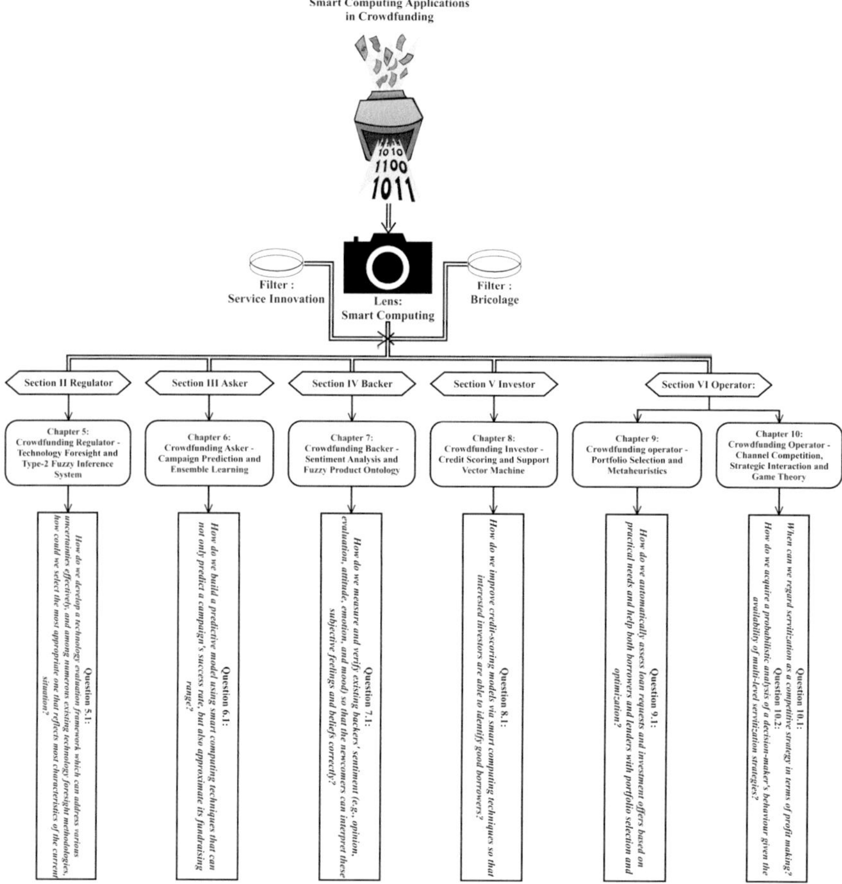

Figure 11.1: Structure of smart computing applications in crowdfunding.

References

Liu C. (2014) *The three-body problem,* 175 Fifth Avenue, New York, NY 10010: Tom Doherty Associates, LLC, ISBN 978-0-7653-7706-7.

Liu C. (2015a) *The dark forest,* 175 Fifth Avenue, New York, NY 10010: Tom Doherty Associates, LLC, ISBN 978-1-4668-5343-0.

Liu C. (2015b) *Death's end*: Macmillan, ISBN 978-1-4668-5345-4.

Marwala T and Hurwitz E. (2017a) *Artificial intelligence and economic theory: Skynet in the market,* Gewerbestrasse 11, 6330 Cham, Switzerland: Springer International Publishing AG, ISBN 978-3-319-66103-2.

Marwala T and Hurwitz E. (2017b) Introduction to man and machines. In: Marwala T and Hurwitz E (eds) *Artificial intelligence and economic theory: Skynet in the market.* Gewerbestrasse 11, 6330 Cham, Switzerland: Springer International Publishing AG, ISBN 978-3-319-66103-2, Chapter 1, pp. 1–14.

Meyer KR and Offin DC. (2017) *Introduction to hamiltonian dynamical systems and the N-body problem,* Gewerbestrasse 11, 6330 Cham, Switzerland: Springer International Publishing AG, ISBN 978-3-319-53690-3.

Musielak Z and Quarles B. (2017) *Three body dynamics and its applications to exoplanets,* Gewerbestrasse 11, 6330 Cham, Switzerland: Springer International Publishing AG, ISBN 978-3-319-58225-2.

SingularityNET. (2017a) *SingularityNET, retrieved from https://singularitynet.io, accessed on 25 December 2017.*

SingularityNET. (2017b) SingularityNET: A decentralized, open market and inter-network for AIs. SingularityNET, retrieved from http://public.singularitynet.io/whitepaper.pdf, accessed on 25 December 2017.

Xing B and Marwala T. (2018a) Conclusion. In: Xing B and Marwala T (eds) *Smart Maintenance for Human–Robot Interaction: An Intelligent Search Algorithmic Perspective.* Gewerbestrasse 11, 6330 Cham, Switzerland: Springer International Publishing AG, ISBN 978-3-319-67479-7, Chapter 13, pp. 299–305.

Xing B and Marwala T. (2018b) Cyberware capacity–applications layer perspective. In: Xing B and Marwala T (eds) *Smart Maintenance for Human–Robot Interaction: An Intelligent Search Algorithmic Perspective.* Gewerbestrasse 11, 6330 Cham, Switzerland: Springer International Publishing AG, ISBN 978-3-319-67479-7, Chapter 8, pp. 173–191.

Xing B and Marwala T. (2018c) Cyberware capacity–energy autonomy perspective. In: Xing B and Marwala T (eds) *Smart Maintenance for Human–Robot Interaction: An Intelligent Search Algorithmic Perspective.* Gewerbestrasse 11, 6330 Cham, Switzerland: Springer International Publishing AG, ISBN 978-3-319-67479-7, Chapter 9, pp. 193–216.

Xing B and Marwala T. (2018d) Cyberware capacity–platform and middleware layers perspective. In: Xing B and Marwala T (eds) *Smart Maintenance for Human–Robot Interaction: An Intelligent Search Algorithmic Perspective.* Gewerbestrasse 11, 6330 Cham, Switzerland: Springer International Publishing AG, ISBN 978-3-319-67479-7, Chapter 7, pp. 143–171.

Xing B and Marwala T. (2018e) Hardware capacity–beginning of life perspective. In: Xing B and Marwala T (eds) *Smart Maintenance for Human–Robot Interaction: An Intelligent Search Algorithmic Perspective.* Gewerbestrasse 11, 6330 Cham, Switzerland: Springer International Publishing AG, ISBN 978-3-319-67479-7, Chapter 4, pp. 67–91.

Xing B and Marwala T. (2018f) Hardware capacity–end of life perspective. In: Xing B and Marwala T (eds) *Smart Maintenance for Human–Robot Interaction: An Intelligent Search Algorithmic Perspective.* Gewerbestrasse 11, 6330 Cham, Switzerland: Springer International Publishing AG, ISBN 978-3-319-67479-7, Chapter 6, pp. 111–139.

Xing B and Marwala T. (2018g) Hardware capacity–middle of life perspective. In: Xing B and Marwala T (eds) *Smart Maintenance for Human–Robot Interaction: An Intelligent Search Algorithmic Perspective.* Gewerbestrasse 11, 6330 Cham, Switzerland: Springer International Publishing AG, ISBN 978-3-319-67479-7, Chapter 5, pp. 93–110.

Xing B and Marwala T. (2018h) Human capacity–biopsychosocial perspective. In: Xing B and Marwala T (eds) *Smart Maintenance for Human–Robot Interaction: An Intelligent Search Algorithmic Perspective.* Gewerbestrasse 11, 6330 Cham, Switzerland: Springer International Publishing AG, ISBN 978-3-319-67479-7, Chapter 11, pp. 249–270.

Xing B and Marwala T. (2018i) Human capacity–exposome perspective. In: Xing B and Marwala T (eds) *Smart Maintenance for Human–Robot Interaction: An Intelligent Search Algorithmic Perspective.* Gewerbestrasse 11, 6330 Cham, Switzerland: Springer International Publishing AG, ISBN 978-3-319-67479-7, Chapter 12, pp. 271–295.

Xing B and Marwala T. (2018j) Human capacity–physiology perspective. In: Xing B and Marwala T (eds) *Smart Maintenance for Human–Robot Interaction: An Intelligent*

Search Algorithmic Perspective. Gewerbestrasse 11, 6330 Cham, Switzerland: Springer International Publishing AG, ISBN 978-3-319-67479-7, Chapter 10, pp. 219–247.

Xing B and Marwala T. (2018k) Introduction to human robot interaction. In: Xing B and Marwala T (eds) *Smart Maintenance for Human–Robot Interaction: An Intelligent Search Algorithmic Perspective.* Gewerbestrasse 11, 6330 Cham, Switzerland: Springer International Publishing AG, ISBN 978-3-319-67479-7, Chapter 1, pp. 3–19.

Xing B and Marwala T. (2018l) *Smart maintenance for human–robot interaction: An intelligent search algorithmic perspective,* Gewerbestrasse 11, 6330 Cham, Switzerland: Springer International Publishing AG, ISBN 978-3-319-67479-7.

Appendix A

Crowdfunding Further Readings

Crowdfunding has experienced an exponential development during the past few years. In the literature, there are several surveys/reviews available (Bouncken et al., 2015; Moritz and Block, 2016; Pichler and Tezza, 2016; McKenny et al., 2017; Kuppuswamy and Bayus, 2015; Short et al., 2017; Hossain, 2015). In this appendix, we outline a non-exhaustive list of crowdfunding further readings in the following forms.

A.1 Book

1) *Access to bank credit and SME financing: (Rossi, 2017)*
2) *Archetypes of crowdfunding platforms: a multidimensional comparison: (Danmayr, 2014)*
3) *Banker's guide to new small business finance: venture deals, crowdfunding, private equity, and technology: (Green, 2014)*
4) *Crowdfunding in Europe: state of the art in theory and practice: (Brüntje and Gajda, 2016)*
5) *Crowdfunding: a guide to raising capital on the Internet: (Dresner, 2014)*
6) *Crowdfunding: an introduction: (Wicks, 2013)*
7) *Equity crowdfunding for investors: a guide to risks, returns, regulations, funding portals, due diligence, and deal terms: (Freedman and Nutting, 2015)*
8) *Financing the underfinanced: online lending in China: (Wang et al., 2015)*
9) *FinTech in Germany: (Dorfleitner et al., 2017)*
10) *Hardware startup: (DiResta et al., 2015)*
11) *Kickstarter for Dummies: (Cebulski, 2013)*
12) *Marketplace lending, financial analysis, and the future of credit: (Akkizidis and Stagars, 2016)*
13) *The crowd funding services handbook: raising the money you need to fund your business, project, or invention: (Rich, 2014)*
14) *The crowdfunding bible: how to raise money for any startup, video game, or project: (Steinberg and DeMaria, 2015)*

15) *The crowdfunding revolution: how to raise venture capital using social media:* (Marom and Lawton, 2012)
16) *The everything guide to crowdfunding: learn how to use social media for small-business funding:* (Young, 2013)
17) *The future of FinTech: integrating finance and technology in financial services:* (Nicoletti, 2017)
18) *The JOBS Act: crowdfunding for small businesses and startups:* (Cunningham, 2012)
19) *The relevance of crowdfunding: the impact on the innovation process of small entrepreneurial firms:* (Scholz, 2015)

A.2 Thesis

- **Doctoral Thesis:**
  1) *Access to credit by the poor: the role of P2P lending via not-for-profit organizations:* (Ahmed, 2011)
  2) *An economic analysis of microcredit lending:* (Wu, 2010)
  3) *An empirical examination of factors influencing participant behavior in crowdfunded markets:* (Burtch, 2013)
  4) *Collective action and the financing of innovation: evidence from crowdfunding:* (Carr, 2013)
  5) *Discrimination, trust and social capital: three essays in applied public economics:* (Theseira, 2009)
  6) *Essays in financial economics:* (Shue, 2011)
  7) *Essays on determinants of financial behavior of individuals:* (Barasinska, 2011)
  8) *Essays on search and herding:* (Berkovich, 2009)
  9) *Essays on strategic social interactions: evidence from microfinance and laboratory experiments in the field:* (Breza, 2012)
  10) *Information asymmetry in relationship versus transactional debt markets: evidence from peer-to-peer lending:* (Everett, 2011)
  11) *Institutional lending models, mission drift, and microfinance institutions:* (Paris, 2013)
  12) *Networking on the margins: the regulation of payday lending in Canada:* (Kobzar, 2012)
  13) *Rebranding gay: new configurations of digital media and commercial culture:* (Ng, 2013)
  14) *Software innovations: the influence of quality, diversity and structure of network ties:* (Singh, 2010)
  15) *The impact of commercial peer-to-peer lending Websites on the finance of small business ventures:* (Kgoroeadira, 2014)
  16) *The impact of information and communication technology on intermediation, outreach, and decision rights in the microfinance industry:* (Weber, 2012)
  17) *Two-sided markets in the online world: an empirical analysis:* (Hildebrand, 2011)

- **Master Thesis:**

1) *A content analysis of Kickstarter: the influence of framing and reward motivations on campaign success: (Sauro, 2014)*
2) *Active consumerism: measuring the importance of crowdfunding factors on backers' decisions to financially support Kickstarter campaigns: (Duvall, 2014)*
3) *An empirical study into the field of crowdfunding: (Zhang, 2012)*
4) *An exploration in funding independent film: (Strader, 2014)*
5) *Branding post-capitalism: (Bailey, 2013)*
6) *Business and legal issues of small business oriented crowdfunding platforms: (Arko, 2014)*
7) *Characteristics of content and social spread strategy on the IndieGoGo crowdfunding platform: (Stern, 2013)*
8) *Civic crowdfunding: a financial kickstart to urban area development: (Veelen, 2015)*
9) *Civic crowdfunding: participatory communities, entrepreneurs and the political economy of place: (Davies, 2014)*
10) *Combining loan requests and investment offers: (Christensen, 2007; Martinho, 2009)*
11) *Connected strangers: manipulating social perceptions to study trust: (Pao, 2010)*
12) *Crowd funding & renewable energy projects: contribution of crowd funding to renewable energy projects in The Netherlands: (Meeuwsen, 2013)*
13) *Crowdfunding – a multifaceted phenomenon: (Dehling, 2013)*
14) *Crowdfunding - Startups' alternative funding source beyond banks, business angels and venture capitalists: (Metelka, 2014)*
15) *Crowdfunding - the development of an online community: (Kobberø, 2014)*
16) *Crowdfunding and its utilization for startup finance in Finland: factors of a successful campaign: (Härkönen, 2014)*
17) *Crowdfunding and the regulatory hurdles in the United States and Europe: (Ebisch, 2012)*
18) *Crowdfunding for Chinese social and environmental small and growing businesses: an investigation of the feasibility of using U.S.-based crowdfunding platforms: (Yip, 2014)*
19) *Crowdfunding for financing new ventures: consequences of the financial model on operational decisions: (Voorbraak, 2011)*
20) *Crowdfunding for innovations: a qualitative research on resources, capabilities and stakes: (Bakker-Rakowska, 2014)*
21) *Crowdfunding in Finland – a new alternative disruptive funding instrument for businesses: (Lasrado, 2013)*
22) *Crowdfunding in the banking industry: adjusting to a digital era: (Setälä, 2017)*
23) *Crowdfunding sustainability: how do entrepreneurs of sustainability projects utilize the potential crowdfunding for fundraising: (Kleppe and Nilsen, 2017)*
24) *Crowdfunding: a healthy practice: (Reid, 2016)*
25) *Crowdfunding: critical factors to finance a project successfully: (Leite, 2012)*
26) *Crowdfunding: drivers and barriers in an international perspective: (Nagymihaly, 2013)*

27) *Crowdfunding: material incentives and perform: (Matos, 2012)*
28) *Crowdfunding: why museums should cultivate the millennial online donor: (Nagymihaly, 2013)*
29) *Crowdinvestor investment decision-making: a study of motivation, investment process and criteria: (Scheder and Arbøll, 2014)*
30) *Crowdsourcing and participatory mechanisms in crowdfunded design projects: (Pape, 2014)*
31) *Crowdsourcing and the law: (Wolfson, 2012)*
32) *Crowdsourcing in community participatory planning in China: case studies in four communities in Shenzhen: (Zhang, 2014b)*
33) *Democratizing commercial real estate investing: the impact of the JOBS Act and crowdfunding on the commercial real estate market: (Burgett, 2013)*
34) *Design of online peer-to-peer investment platform offering incentive-compatible revenue-sharing innovation: (Vorapatchaiyanont, 2013)*
35) *Development 2.0? The case of Kiva.org and online social lending for development: (Carlman, 2010)*
36) *Drivers and barriers for cross-border crowdfunding: evidence from Belgian stakeholders: (Aurelie, 2013)*
37) *Effect of social media on crowdfunding project results: (Moisseyev, 2013)*
38) *Exploratory study on technology related successfully crowdfunding projects' post online market presence: (Samanci and Kiss, 2014)*
39) *Exploring the Seattle museum community's perceptions toward crowdfunding: (Alejandre, 2013)*
40) *Failure to prosper: risk premia in peer to peer lending: (Christensen, 2007)*
41) *Financing success through equity crowdfunding: the case of start-ups and SMEs funded on European crowdfunding platform: (Medziausyte and Neugebauer, 2017)*
42) *Geography of pledging and application of funds of crowdfunding platforms and the impact on their online notoriety: (Jesus, 2013)*
43) *Independent film marketing in the digital age: (Kvernmoen, 2014)*
44) *Innovation space: (Schneider-Sikorsky, 2014)*
45) *Is your peer a lemon? Relative assessment of risk remuneration on the P2P lending: (Golubnicijs, 2012)*
46) *Know your crowd: the drivers of success in reward-based crowdfunding: (Wechsler, 2013)*
47) *Long-term study of crowdfunding platform: predicting project success and fundraising amount: (Chung, 2014)*
48) *Main drivers of crowdfunding success: a conceptual framework and empirical analysis: (Evers, 2012)*
49) *Microfinance and the use of peer to peer social lending to increase sustainability: case study Kiva - a peer to peer social lending organization: (Marchitti, 2013)*
50) *Online marketing and fundraising strategies for non-profit organisations in the Cape Town health sector: (Booth, 2013)*
51) *Online peer to peer lending: clustering borrowers using self-organizing maps: (Kangas, 2014)*

52) *Opportunities and limitations of (social) entrepreneurial approaches: a case study of the recycling sector in Cape Town, South Africa: (Linnay, 2013)*
53) *Peer-to-peer lending: the effects of institutional involvement in social lending: (Claesson and Tengvall, 2015)*
54) *Social innovation: an analysis of its drivers and of the crowdfunding phenomena: (Reynolds, 2013)*
55) *Sports inspire: a crowd-funding community for sports philanthropy: (Battcher, 2013)*
56) *Support vector machines for credit scoring: (Haltuf, 2014)*
57) *The antecedents of interest in social crowdfunding: exploring the 8 mechanisms that drive charitable giving: (Srkoc and Zarim, 2013)*
58) *The best place for cartoons: (Crowell, 2014)*
59) *The consumer potential of collaborative consumption: identifying (the) motives of Dutch collaborative consumers & measuring the consumer potential of collaborative consumption within the municipality of Amsterdam: (Glind, 2013)*
60) *The crowdfunding investor (CFI) - a new species in an existing ecosystem: (Stokar, 2014)*
61) *The impact of group lending on participants' entrepreneurial resources: (Teinemaa, 2013)*
62) *The influence of perceived consumer values on future intention to contribute to crowdfunding: (Avakyan, 2013)*
63) *The motivational foundations of lenders: social lending through crowdfunding platforms: (Estrela, 2013)*
64) *The use of crowdfunding as an alternative way to finance small businesses in France: (Couffinhal, 2014)*
65) *Transitioning international development projects to sustainable businesses: the challenges in commercializing a research project and the engineer-entrepreneurs behind the product: (Cho, 2012)*
66) *Understanding crowd funding: cost of capital and factors for success: (Teo, 2013)*
67) *What characteristics of crowdfunding platforms influence the success rate? An empirical study into relevant determinants in order to explain differences in success rates between crowdfunding platforms worldwide: (Willems, 2013)*
68) *What drives motivation to participate financially in a crowdfunding community: (Harms, 2007)*

- **Others:**
    1) *A comparison and assessment of acquiring funds of corporate entities through crowdfunding and through a bank credit: (Bammatter, 2012)*
    2) *Analysis of crowdfunding descriptions for technology projects: (Fernandes, 2013)*
    3) *Crowdfunding as a customer engagement channel: (Paykacheva, 2014)*
    4) *Crowdfunding as a way of financing start-ups in Poland: (Niźnik-Klocek, 2012)*
    5) *Crowdfunding as an alternative way of raising funds for NGOs: (Makýšová, 2017)*
    6) *Crowdfunding from a Marketing Perspective: (Arkrot et al., 2017)*
    7) *Crowdfunding in museums: (Bump, 2014)*

8) *Crowdfunding: an empirical and theoretical model of non-profit support: (Read, 2013)*
9) *Exploiting the crowd: the New Zealand response to equity crowd funding: (Hillind, 2014)*
10) *Key determinants of successful crowdfunding campaigns in the gaming industry: (Vettenranta, 2017)*
11) *Less risk and more reward: applying the crowdfunding model to local and regional theater organizations: (Laughlin, 2013)*
12) *The rise of peer-to-peer platforms and the change in the borrower's rational choice: (Lim, 2013)*

A.3 Journal Article

1) *3D robotics: disrupting the drone market: (Stuart and Anderson, 2015)*
2) *A class exercise to explore crowdfunding: (Smith and Green, 2015)*
3) *A conceptualized investment model of crowdfunding: (Tomczak and Brem, 2013)*
4) *A co-utility approach to the mesh economy: the crowd-based business model: (Turi et al., 2017)*
5) *A cross-cultural comparative analysis of crowdfunding projects in the United States and South Korea: (Cho and Gawon Kim, 2017)*
6) *A decision support system for borrower's loan in P2P lending: (Wu and Xu, 2011)*
7) *A decision tree model for herd behavior and empirical evidence from the online P2P lending market: (Luo and Lin, 2013)*
8) *A market testing method based on crowd funding: (Sheldon and Kupp, 2017)*
9) *A research agenda: crowdfunding at the service of responsible social entrepreneurship and innovations: (Ingham and Assadi, 2016)*
10) *A rewarding experience: exploring how crowdfunding is affecting music industry business models: (Gamble et al., 2017)*
11) *A taxonomy of crowdsourcing based on task complexity: (Nakatsu et al., 2014)*
12) *A very quiet revolution: a primer on securities crowdfunding and Title III of the JOBS Act: (Knight et al., 2012)*
13) *Academic research on crowdfunders: what's been done and what's to come?: (Macht and Weatherston, 2015)*
14) *Advantages of crowdfunding as an alternative source of financing of small and medium-sized enterprises: (Golić, 2013)*
15) *An empirical investigation of signaling in reward-based crowdfunding: (Kunz et al., 2017)*
16) *An exploratory analysis of Title II equity crowdfunding success: (Mamonov et al., 2017)*
17) *An overview study on P2P lending: (Gao and Feng, 2014)*
18) *Analysis and outlook of applications of blockchain technology to equity crowdfunding in China: (Zhu and Zhou, 2016)*
19) *Analysis of financing sources for start-up companies: (Čalopa, 2014)*
20) *Appealing to the crowd: ethical justifications in Canadian medical crowdfunding campaigns: (Snyder et al., in press)*

21) *Are promises meaningless in an uncertain crowdfunding environment?: (Hauge and Chimahusky, 2016)*
22) *Attributes of angel and crowdfunded investment as determinants of VC screening decisions: (Drover et al., 2017)*
23) *Auctions for social lending: a theoretical analysis: (Chen et al., 2014)*
24) *Book review: the crowdfunding revolution: how to raise venture capital using social media: (Ma, 2013)*
25) *Book reviews: crowdsouring: (Aitamurto, 2014)*
26) *Borrower decision aid for people-to-people lending: (Puro et al., 2010)*
27) *Budgeting for crowdfunding rewards: (Buff and Alhadeff, 2013)*
28) *Can microfinance crowdfunding reduce financial exclusion? Regulatory issues: (Marakkath and Attuel-mendes, 2015)*
29) *Cash from the crowd: (Colombo et al., 2015a)*
30) *Choose wisely: crowdfunding through the stages of the startup life cycle: (Paschen, 2017)*
31) *Cinetics: fueling entrepreneurial innovations through crowdfunding: (Li et al., 2016)*
32) *Citizen participation, open innovation, and crowdsourcing: challenges and opportunities for planning: (Seltzer and Mahmoudi, 2012)*
33) *Civic crowdfunding research: challenges, opportunities, and future agenda: (Stiver et al., 2015)*
34) *Clinical research: should patients pay to play?: (Emanuel et al., 2015)*
35) *Comparing crowdsourcing initiatives: toward a typology development: (Ali-Hassan and Allam, 2016)*
36) *Competition against common sense: insights on peer-to-peer lending as a tool to allay financial exclusion: (Loureiro and Gonzalez, 2015)*
37) *Complementing the local and global: promoting sustainability action through linked local-level and formal sustainability funding mechanisms: (Merritt and Stubbs, 2012)*
38) *Configurational paths to sponsor satisfaction in crowdfunding: (Xu et al., 2016)*
39) *Convincing the crowd: entrepreneurial storytelling in crowdfunding campaigns: (Manning and Bejarano, 2017)*
40) *Corporate disclosure as a transaction cost: the case of SMEs: (Ferrarini and Ottolia, 2013)*
41) *Cost estimation of building individual cooperative housing with crowdfunding model: case of Beijing, China: (Liu et al., 2017)*
42) *Could peer-to-peer loans substitute for payday loans?: (Livingston, 2012)*
43) *Crowd funding: a case study at the intersection of social media and business ethics: (Padgett and Rolston, 2014)*
44) *Crowded identity: managing crowdsourcing initiatives to maximize value for participants through identity creation: (Fedorenko et al., 2017)*
45) *Crowdfrauding: avoiding Ponzi entrepreneurs when investing in new ventures: (Baucus and Mitteness, 2016)*

46) *Crowd-funded micro-grants for genomics and "big data": an actionable idea connecting small (artisan) science, infrastructure science, and citizen philanthropy: (Özdemir et al., 2013)*
47) *Crowdfunding 2.0: the next-generation philanthropy: (Özdemir et al., 2015a)*
48) *Crowdfunding affordable homes with tax credit investment partnerships: (Pickering, 2013)*
49) *Crowdfunding and an innovator's access to capital: (Kitch, 2014)*
50) *Crowdfunding and diaspora philanthropy: an integration of the literature and major concepts: (Flanigan, 2017)*
51) *Crowdfunding and non-profit media: the emergence of new models for public interest journalism: (Carvajal et al., 2012)*
52) *Crowdfunding and social networks in the music industry: implications for entrepreneurship: (Martínez-Cañas et al., 2012)*
53) *Crowdfunding and sport: how soon until the fans own the franchise? (Fallone, 2014)*
54) *Crowdfunding and the expansion of access to startup capital: (Stanberry and Aven, 2014)*
55) *Crowdfunding and the Federal Securities Laws: (Bradford, 2012)*
56) *Crowdfunding and the revitalisation of the early stage risk capital market: catalyst or chimera?: (Harrison, 2013)*
57) *Crowdfunding and Value Creation: (Meyskens and Bird, 2015)*
58) *Crowdfunding and wine business: some insights from Fundovino experience: (Mariani et al., 2017)*
59) *Crowdfunding as a fast-expanding market for the creation of capital and shared value: (Baumgardner et al., 2015)*
60) *Crowdfunding as an emerging fundraising tool: with special reference to the Malaysian regulatory framework: (Abdullah, 2006)*
61) *Crowdfunding cleantech: (Cumming et al., 2017)*
62) *Crowdfunding drug development: the state of play in oncology and rare diseases: (Dragojlovic and Lynd, 2014)*
63) *Crowdfunding effort identifies the causative mutation in a patient with Nystagmus, Microcephaly, Dystonia and Hypomyelination: (Isakov et al., 2015)*
64) *Crowdfunding for biotechs: how the SEC's proposed rule may undermine capital formation for startups: (Farnkoff, 2013)*
65) *Crowdfunding for cardiovascular research: (Krittanawong et al., 2018)*
66) *Crowdfunding for environmental ventures: an empirical analysis of the influence of environmental orientation on the success of crowdfunding initiatives: (Hörisch, 2015)*
67) *Crowdfunding for renewable and sustainable energy projects: an exploratory case study approach: (Lam and Law, 2016)*
68) *Crowdfunding for video games: factors that influence the success of and capital pledged for campaigns: (Cha, in press)*
69) *Crowdfunding genomics and bioinformatics: (Cameron et al., 2013)*
70) *Crowdfunding human capital contracts: (Vogel, 2015)*

71) *Crowdfunding in a prosocial microlending environment: examining the role of intrinsic versus extrinsic cues: (Allison et al., 2014)*
72) *Crowdfunding in France: a new revolution?: (Sannajust et al., 2014)*
73) *Crowdfunding in libraries, archives and museums: (Riley-Huff et al., 2016)*
74) *Crowdfunding in the U.S. and abroad: what to expect when you're expecting: (Weinstein, 2013)*
75) *Crowdfunding in the United Kingdom: a cultural economy: (Langley, 2016)*
76) *Crowdfunding independent and freelance journalism: negotiating journalistic norms of autonomy and objectivity: (Hunter, 2015)*
77) *Crowdfunding innovative ideas: how incremental and radical innovativeness influence funding outcomes: (Chan and Parhankangas, 2017)*
78) *Crowdfunding microstartups: it's time for the Securities and Exchange Commission to approve a small offering exemption: (Pope, 2011)*
79) *Crowdfunding nostalgia: kickstarter and the revival of classic PC game genres: (Gilbert, 2017)*
80) *Crowdfunding our health: economic risks and benefits: (Renwick and Mossialos, 2017)*
81) *Crowdfunding social ventures: a model and research agenda: (Lehner, 2013)*
82) *Crowdfunding the Azolla fern genome project: a grassroots approach: (Li and Pryer, 2014)*
83) *Crowdfunding the movies: a business analysis of crowdfinanced moviemaking in small geographical markets: (Braet et al., 2013)*
84) *Crowdfunding to generate crowdsourced R&D: the alternative paradigm of societal problem solving offered by second generation innovation and R&D: (Callaghan, 2014)*
85) *Crowdfunding, cascades and informed investors: (Parker, 2014)*
86) *Crowdfunding: (Goins and Little, 2013)*
87) *Crowdfunding: (Kuti and Madarász, 2014)*
88) *Crowdfunding: a New Media & Society special issue: (Bennett et al., 2015)*
89) *Crowdfunding: a new option for funding health projects: (Otero, 2015)*
90) *Crowdfunding: a new paradigm in start-up financing: (Manchanda and Muralidharan, 2014)*
91) *Crowdfunding: a new source of money for the healthcare sector: (Botzenhart-Eggstein, 2015)*
92) *Crowdfunding: a Spimatic application of digital fandom: (Booth, 2015)*
93) *Crowdfunding: geography, social networks, and the timing of investment decisions: (Agrawal et al., 2015)*
94) *Crowdfunding: online charity or a modern tool for innovative projects implementation?: (Profatilov et al., 2015)*
95) *Crowdfunding: tapping the right crowd: (Belleflamme et al., 2014)*
96) *Crowdfunding: the collaborative economy for channeling institutional and household savings: (Hernando, 2017)*
97) *Crowdfunding: the collaborative economy for channeling institutional and household savings: (Hernando, 2016)*

98) *Crowdfunding: towards a redefinition of the artist's role – the case of MegaTotal: (Galuszka and Brzozowska, 2015)*
99) *Crowd-funding: transforming customers into investors through innovative service platforms: (Ordanini et al., 2011)*
100) *Crowdfunding's impact on start-up IP strategy: (O'Connor, 2014)*
101) *Crowdsourcing and organizational forms: emerging trends and research implications: (Palacios et al., 2016)*
102) *Crowdsourcing and outsourcing: the impact of online funding and distribution on the documentary film industry in the UK: (Sørensen, 2012)*
103) *Crowdsourcing as a new instrument in the government's arsenal: explorations and considerations: (Dutil, 2015)*
104) *Crowdsourcing, innovation and firm performance: (Xu et al., 2015)*
105) *Crowdsourcing: a new conceptual view for food safety and quality: (Soon and Saguy, 2017)*
106) *Crowdsourcing: a platform for crowd engagement in the publishing industry: (Mustafa and Adnan, 2017)*
107) *Cultural differences and geography as determinants of online prosocial lending: (Burtch et al., 2014)*
108) *Cyberfinance: liberating the financial markets: (Wales, 2015)*
109) *Demand-driven securities regulation: evidence from crowdfunding: (Cumming and Johan, 2013)*
110) *Democratizing innovation and capital access: the role of crowdfunding: (Mollick and Robb, 2016)*
111) *Detecting Fraudulent Behavior on crowdfunding platforms: the role of linguistic and content-based cues in static and dynamic contexts: (Siering et al., 2016)*
112) *Determinants of backers' funding intention in crowdfunding: social exchange theory and regulatory focus: (Zhao et al., 2017)*
113) *Digital social innovation: the crowdfunding model: (Dulaurans, 2014)*
114) *Do small businesses still prefer community banks?: (Berger et al., 2014)*
115) *Does heart or head rule donor behaviors in charitable crowdfunding markets: (Gleasure and Feller, 2016a)*
116) *Does my contribution to your crowdfunding project matter: (Kuppuswamy and Bayus, 2017)*
117) *Does the crowdfunding platform matter: risks of negative attitudes in two-sided markets: (Lacan and Desmet, in press)*
118) *Does the possibility to make equity investments in crowdfunding projects crowd out reward-based investments: (Cholakova and Clarysse, 2015)*
119) *Effects of the price of charitable giving: evidence from an online crowdfunding platform: (Meer, 2014)*
120) *Emerging technologies and the democratisation of financial services: a metatriangulation of crowdfunding research: (Gleasure and Feller, 2016b)*
121) *Entrepreneurial finance and innovation: an introduction and agenda for future research: (Chemmanur and Fulghieri, 2014)*
122) *Equity crowdfunding and peer-to-peer lending in New Zealand: the first year: (Murray, 2015)*

123) *Equity crowdfunding: a new phenomena: (Vulkan et al., 2016)*
124) *Equity-based crowdfunding: potential implications for small business capital: (Taylor, 2015)*
125) *Ex ante crowdfunding and the recording industry: a model for the US: (Kappel, 2009)*
126) *Exploring agency dynamics of crowdfunding in start-up capital financing: (Ley and Weaven, 2011)*
127) *Exploring the determinants of crowdfunding: the influence of the banking system: (Paulet and Relano, 2017)*
128) *External supports in reward-based crowdfunding campaigns: a comparative study focused on cultural and creative projects: (Bao and Huang, 2017)*
129) *Far from the maddening crowd: does the JOBS Act provide meaningful redress to small investors for securities fraud in connection with crowdfunding offerings?: (James, 2013)*
130) *Fast forward on crowdfunding: (Sheik, 2013)*
131) *Financing creativity: crowdfunding as a new approach for theatre projects: (Boeuf et al., 2014)*
132) *Financing social enterprises: (Kickul and Lyons, 2015)*
133) *Freedom technologists and the new protest movements: a theory of protest formulas: (Postill, 2014)*
134) *From big data analysis to personalized medicine for all: challenges and opportunities: (Alyass et al., 2015)*
135) *From the crowd to the market: the role of reward-based crowdfunding performance in attracting professional investors: (Roma et al., 2017)*
136) *From the wisdom of crowds to my own judgment in microfinance through online peer-to-peer lending platforms: (Yum et al., 2012)*
137) *From value chains to technological platforms: the effects of crowdfunding in the digital game industry: (Nucciarelli et al., 2017)*
138) *Fund my treatment: a call for ethics-focused social science research into the use of crowdfunding for medical care: (Snyder et al., 2016)*
139) *Funders' positive affective reactions to entrepreneurs' crowdfunding pitches: the influence of perceived product creativity and entrepreneurial passion: (Davis et al., 2017)*
140) *Funding biotech start-ups in a post-VC world: (Bains et al., 2014)*
141) *Funding for research? Look to the crowd: (Cadogan, 2014)*
142) *Funding from the crowd: an Internet-based crowdfunding platform to support business set-ups from universities: (Wieck et al., 2013)*
143) *Funding news freedom: (Nevill, 2014)*
144) *Gender differences in the contribution patterns of equity-crowdfunding investors: (Mohammadi and Shafi, in press)*
145) *Going offline broadening crowdfunding research beyond the online context: (Gras et al., 2017)*
146) *Government-incentivized crowdfunding for one-belt, one-road enterprises: design and research issues: (Lee et al., 2016)*

147) *Grassroots capitalism or: how I learned to stop worrying about financial risk in the exempt market and love equity crowdfunding: (Figliomeni, 2014c)*
148) *Guidelines for successful crowdfunding: (Forbes and Schaefer, 2017)*
149) *Harnessing the crowd to accelerate molecular medicine research: (Smith and Merchant, 2015)*
150) *Hedonic value and crowdfunding project performance: a propensity score matching-based analysis: (Zhao and Vinig, 2017)*
151) *Herding behavior in online P2P lending: an empirical investigation: (Lee and Lee, 2012)*
152) *Historical forms of user production: (Hamilton, 2014)*
153) *How accounts shape lending decisions through fostering perceived trustworthiness: (Sonenshein et al., 2010)*
154) *How do firms make money selling digital goods online: (Lambrecht et al., 2014)*
155) *How friendship networks work in online P2P lending markets: (Li et al., 2015)*
156) *How low can you go? — Overcoming the inability of lenders to set proper interest rates on unsecured peer-to-peer lending markets: (Mild et al., 2015)*
157) *How should crowdfunding research evolve? A survey of the entrepreneurship theory and practice: (McKenny et al., 2017)*
158) *How to overcome the barriers between economy and sociology with open innovation, open evaluation and crowdfunding?: (Freund, 2010)*
159) *Human equity? Regulating the new income share agreement: (Oei and Ring, 2015)*
160) *Identify innovative business models: (Pisano et al., 2015)*
161) *Implications of online funding regulations for small businesses: (Yeoh, 2014)*
162) *Individual crowdfunding practices: (Belleflamme et al., 2013)*
163) *Information cascades among investors in equity crowdfunding: (Vismara, 2016)*
164) *Instance-based credit risk assessment for investment decisions in P2P lending: (Guo et al., 2016)*
165) *Internal social capital and the attraction of early contributions in crowdfunding: (Colombo et al., 2015b)*
166) *Introduction to the special issue on crowdfunding and FinTech: (Ma and Liu, 2017)*
167) *Investor and market protection in the crowdfunding era: disclosing to and for the "crowd": (Heminway, 2014)*
168) *Investor communication in equity-based crowdfunding: a qualitative-empirical study: (Moritz et al., 2015)*
169) *Is crowdfunding a viable source of clinical trial research funding?: (Sharma et al., 2015)*
170) *Is crowdfunding different? Evidence on the relation between gender and funding success from a German peer-to-peer lending platform: (Barasinska and Schäfer, 2014)*
171) *Is crowding out due entirely to fundraising? evidence from a panel of charities: (Andreoni and Payne, 2011)*
172) *Is the crowd sensitive to distance?—how investment decisions differ by investor type: (Guenther et al., in press)*
173) *JOBS Act eases securities-law regulation of smaller companies: (Parrino and Romeo, 2012)*

174) *Journal of Internet Banking and Commerce: (Chen and Han, 2012)*
175) *Kicking off social entrepreneurship: how a sustainability orientation influences crowdfunding success: (Calic and Mosakowski, 2016)*
176) *Kickstarter my heart: extraordinary popular delusions and the madness of crowdfunding constraints and Bitcoin bubbles: (Groshoff, 2014)*
177) *Kickstarting trans*: the crowdfunding of gender/sexual reassignment surgeries: (Farnel, 2015)*
178) *Knowledge and assessment of crowdfunding in communication: the view of journalists and future journalists: (Sanchez-González and Palomo-Torres, 2014)*
179) *Knowledge is power: how implementing affirmative disclosures under the JOBS act could promote and protect benefit corporations and their investors: (Farley, 2015)*
180) *Less risk and more reward: applying the crowdfunding model to local and regional theater organizations: (Laughlin, 2013)*
181) *Lessons from the first patent infringement lawsuit filed against a crowdfunded project on a crowdfunding portal: (Postolski and Nowotarski, 2013)*
182) *Like oil and water: equity crowdfunding and securities regulation: (Hogan, 2014)*
183) *Linguistic style and crowdfunding success among social and commercial entrepreneurs: (Parhankangas and Renko, 2017)*
184) *Living labs for user-driven innovation: (Guzmán et al., 2013)*
185) *Lonely rebel or pioneer of the future? Towards an understanding of moral stakeholder framing of activist brands: (Stoeckl, 2014)*
186) *Mad money: rethinking private placements: (Cable, 2014)*
187) *Market mechanisms and funding dynamics in equity crowdfunding: (Hornuf and Schwienbacher, in press)*
188) *Microfinance, crowdfunding, and sustainability: a case study of telecenters in a South Asian developing country: (Royal and Windsor, 2014)*
189) *Micro-lending institutions: using social networks to create productive capabilities: (Anthony, 1997)*
190) *Mind the gap: crowdfunding and the role of seed money: (Deutsch et al., 2017)*
191) *More than three's a crowd . . . in the best interest of companies! Crowdfunding as zeitgeist or ideology?: (Bouaiss et al., 2015)*
192) *New financial alternatives in seeding entrepreneurship: microfinance, crowdfunding, and peer-to-peer innovations: (Bruton et al., 2014)*
193) *Non-bank financing in Ireland: a comparative perspective: (O'Toole et al., 2015)*
194) *Non-profit differentials in crowd-based financing: evidence from 50,000 campaigns: (Pitschner and Pitschner-Finn, 2014)*
195) *Not just an ego-trip: exploring backers' motivation for funding in incentive-based crowdfunding: (Bretschneider and Leimeister, in press)*
196) *Opinion leaders as intermediaries in audience building for independent films in the Internet age: (Meißner, 2014)*
197) *Crowdsourcing in public policy: technologies, subjects and its socio-political role: (Morozova, 2015)*
198) *Peer-to-peer fundraising and crowdfunding in health care philanthropy: (Granville, 2016)*

199) *Persuasion in crowdfunding: an elaboration likelihood model of crowdfunding performance: (Allison et al., 2017)*
200) *Philanthropic innovation and creative capitalism: a historical and comparative perspective on social entrepreneurship and corporate social responsibility: (Rana, 2013)*
201) *Power to the people: how the SEC can empower the crowd: (Saunders, 2014)*
202) *Pricing shares in equity crowdfunding: (Hornuf and Neuenkirch, in press)*
203) *Producing a worthy illness: personal crowdfunding amidst financial crisis: (Berliner and Kenworthy, 2017)*
204) *Project description and crowdfunding success: an exploratory study: (Zhou et al., in press)*
205) *Protecting the crowd through escrow: three ways that the SEC can protect crowdfunding investors: (Carni, 2014)*
206) *Pure and hybrid crowds in crowdfunding markets: (Chen et al., 2016)*
207) *Raising money for scientific research through crowdfunding: (Wheat et al., 2013)*
208) *Regulatory focus and information cues in a crowdfunding context: (Ciuchta et al., 2016)*
209) *Remarkable advocates: an investigation of geographic distance and social capital for crowdfunding: (Kang et al., 2017)*
210) *Research on angel investments: the intersection of equity investments and entrepreneurship: (Zachary and Mishra, 2013)*
211) *Research on crowdfunding: reviewing the (very recent) past and celebrating the present: (Short et al., 2017)*
212) *Resistance to crowdfunding among entrepreneurs: an impression management perspective: (Gleasure, 2015)*
213) *Resolving information asymmetry: signaling, endorsement, and crowdfunding success: (Courtney et al., 2017)*
214) *Revolution of securities law in the Internet age: a review on equity crowd-funding: (Huang and Zhao, in press)*
215) *Reward-based crowdfunding of entrepreneurial projects: the effect of local altruism and localized social capital on proponents' success: (Giudici et al., in press)*
216) *Risk assessment in social lending via random forests: (Malekipirbazari and Aksakalli, 2015)*
217) *Russian political crowdfunding: (Sokolov, 2015)*
218) *Science by the masses: is crowdfunding the future for biotech start-ups?: (Brenan, 2014)*
219) *SE(c)(3): a catalyst for social enterprise crowdfunding: (Reiser and Dean, 2015)*
220) *Search and herding effects in peer-to-peer lending: evidence from prosper.com: (Berkovich, 2011)*
221) *SEC adopts "crowdfunding" rules for start-up businesses: an easy way to bet on the next Google?: (Gelfond and Eren, 2016)*
222) *Seeking funding in order to sell: crowdfunding as a marketing tool: (Brown et al., 2017)*
223) *Selecting early-stage ideas for radical innovation: tools and structures: (Nicholas et al., 2015)*

224) *Serial crowdfunding, social capital, and project success: (Buttice et al., 2017)*
225) *Should crowdfunding investors rely on lists of the best private companies: (Jensen et al., 2015)*
226) *Should securities regulation promote equity crowdfunding: (Hornuf and Schwienbacher, 2017)*
227) *Should securities regulation promote equity crowdfunding?: (Hornuf and Schwienbacher, 2017)*
228) *Signaling in equity crowdfunding: (Ahlers et al., 2015)*
229) *Social business in online financing: crowdfunding narratives of independent documentary producers in Turkey: (Koçer, 2015)*
230) *Social responsibility and crowdfunding businesses: a measurement development study: (Marom, 2017)*
231) *Some simple economics of crowdfunding: (Agrawal et al., 2014)*
232) *Squalls in the safe harbor: investment advice & regulatory gaps in regulation crowdfunding: (Olson, 2015)*
233) *Strategic Herding Behavior in Peer-to-Peer Loan Auctions: (Herzenstein et al., 2011a)*
234) *Strategies for reward-based crowdfunding campaigns: (Kraus et al., 2016)*
235) *Study on crowdfunding's promoting effect on the expansion of electric vehicle charging piles based on game theory analysis: (Zhu et al., 2017)*
236) *Success drivers of online equity crowdfunding campaigns: (Lukkarinen et al., 2016)*
237) *Success in the management of crowdfunding projects in the creative industries: (Hobbs et al., 2016)*
238) *Success of crowd-based online technology in fundraising: an institutional perspective: (Kshetri, 2015)*
239) *Supply and demand on crowdlending platforms: connecting SME-sized enterprise borrowers and consumer investors: (Maier, 2016)*
240) *Team rivalry and lending on crowdfunding platforms: an empirical analysis: (Ge and Luo, 2016)*
241) *Tell me a good story and I may lend you money: the role of narratives in peer-to-peer lending decisions: (Herzenstein et al., 2011b)*
242) *The anti-crowd pleaser: fixing the crowdfund act's hidden risks and inadequate remedies: (Mashburn, 2013)*
243) *The backer–developer connection: exploring crowdfunding's influence on video game production: (Smith, 2015)*
244) *The barriers facing artists' use of crowdfunding platforms: personality, emotional labor, and going to the well one too many times: (Davidson and Poor, 2015)*
245) *The benefits of online crowdfunding for fund-seeking business ventures: (Macht and Weatherston, 2014)*
246) *The cartographic ambiguities of HarassMap: crowdmapping security and sexual violence in Egypt: (Grove, 2015)*
247) *The case of crowdfunding in financial inclusion: a survey: (Kim and Moor, 2017)*
248) *The case of crowdfunding in financial inclusion: a survey: (Kim and Moor, 2017)*
249) *The Crowdfund Act's strange bedfellows: democracy and start-up company investing: (Wroldsen, 2014)*

250) *The determinants of crowdfunding success: a semantic text analytics approach: (Yuan et al., 2016)*
251) *The different ways of collaboration between a retail bank and crowdfunding: (Attuel-Mendes, 2017)*
252) *The dynamics of crowdfunding: an exploratory study: (Mollick, 2014)*
253) *The economics of crowdfunding platforms: (Belleflamme et al., 2015)*
254) *The effect of virtuous and entrepreneurial orientations on microfinance lending and repayment: a signaling theory perspective: (Moss et al., 2015)*
255) *The emergence and effects of fake social information: evidence from crowdfunding: (Wessel et al., 2016)*
256) *The evolving entrepreneurial finance landscape: (Wright et al., 2016)*
257) *The financing options for new small and medium enterprises in South Africa: (Fatoki, 2014)*
258) *The Fintech Trilemma: (Brummer and Yadav, 2017)*
259) *The funding gap and the role of financial return crowdfunding: some evidence from European platforms: (Borello, 2015)*
260) *The global institutionalization of microcredit: (Auguard, 2011)*
261) *The global significance of crowdfunding: solving the SME funding problem and democratizing access to capital: (Pekmezovic and Walker, 2016)*
262) *The impact of sentiment orientations on successful crowdfunding campaigns through text analytics: (Wang et al., 2017)*
263) *The impact of the JOBS Act on independent film finance: (Chaudry, 2014)*
264) *The impacts of fundraising periods and geographic distance on financing music production via crowdfunding in Brazil: (Mendes-Da-Silva et al., 2016)*
265) *The influence of internal social capital on serial creator's success in crowdfunding: (Skirnevskiy et al., 2017)*
266) *The influence of online information on investing decisions of reward-based crowdfunding: (Bi et al., 2017)*
267) *The JOBS Act and crowdfunding: harnessing the power–and money–of the masses: (Stemler, 2013)*
268) *The JOBS act and crowdfunding: how narrowing the secondary market handicaps fraud plaintiffs: (Morsy, 2014)*
269) *The lemons problem in crowdfunding: (Tomboc, 2013)*
270) *The moral economy of crowdfunding and the transformative capacity of fan-ancing: (Scott, 2015)*
271) *The new ways to raise capital: an exploratory study of crowdfunding: (Rossi, 2014)*
272) *The pastoral crowd: exploring self-hosted crowdfunding using activity theory and social capital: (Gleasure and Morgan, in press)*
273) *The people's NIH? Ethical and legal concerns in crowdfunded biomedical research: (Perry, 2015)*
274) *The place of crowdfunding in the discovery of scientific and social value of medical research: (Savio, 2017)*
275) *The political economy of state capitalism and shadow banking in China: (Tsai, 2015)*
276) *The rise of crowdfunding: social media, big data, cloud technologies: (Colgren, 2014)*

277) *The risk of money laundering through crowdfunding: a funding portal's guide to compliance and crime fighting: (Robock, 2014)*
278) *The role of charismatic rhetoric in crowdfunding: an examination with computer-aided text analysis: (Anglin et al., 2014)*
279) *The role of community in crowdfunding success: evidence on cultural attributes in funding campaigns to "save the local theater": (Josefy et al., 2017)*
280) *The role of crowdfunding in entrepreneurial finance: (Mitra, 2012)*
281) *The role of multidimensional social capital in crowdfunding: a comparative study in China and US: (Zheng et al., 2014)*
282) *The role of trust management in reward-based crowdfunding: (Zheng et al., 2016)*
283) *The siren call of equity crowdfunding: (Dorff, 2014)*
284) *The social network and the crowdfund act: Zuckerberg, Saverin, and venture capitalists' Dilution of the crowd: (Wroldsen, 2013)*
285) *The strategic challenges of a social innovation: the case of rang de in crowdfunding: (Ashta et al., 2015)*
286) *The strategic challenges of a social innovation: the case of Rang De in crowdfunding: (Ashta et al., 2015)*
287) *The United States' new crowdfunding rules: a Pandora's box?: (Neslund, 2014)*
288) *The wisdom of the crowd in funding: information heterogeneity and social networks of crowdfunders: (Polzin et al., in press)*
289) *Three is a crowd? Exploring the potential of crowdfunding for renewable energy in the Netherlands: (Vasileiadou et al., 2016)*
290) *To crowdfund research, scientists must build an audience for their work: (Byrnes et al., 2014)*
291) *Toward a better understanding of crowdfunding, openness and the consequences for innovation: (Stanko and Henard, 2017)*
292) *Towards an integrated crowdsourcing definition: (Estellés-Arolas and González-Ladroón-de-Guevara, 2012)*
293) *Trends in the crowdfunding of educational technology startups: (Antonenko et al., 2014)*
294) *Trust Building in Online Peer-to-Peer Lending: (Zhang et al., 2014a)*
295) *Unlocked and loaded: government censorship of 3D-printed firearms and a proposal for more reasonable regulation of 3D-printed goods: (Bryans, 2015)*
296) *Upcrowding energy co-operatives - evaluating the potential of crowdfunding for business model innovation of energy co-operatives: (Dilger et al., 2017)*
297) *Valuation of crowdfunding: benefits and drawbacks: (Valančienė and Jegelevičiūtė, 2013)*
298) *Value-in-context in crowdfunding ecosystems: how context frames value co-creation: (Quero et al., 2017)*
299) *Veronica Mars, fandom, and the 'effective economics' of crowdfunding poachers: (Hills, 2015)*
300) *What can crowdsourcing do for decision support? (Chiu et al., 2014)*
301) *What do we know about entrepreneurial finance and its relationship with growth: (Fraser et al., 2015)*

302) *What is crowdfunding? Bringing the power of kickstarter to your entrepreneurship research and teaching activities: (Voelker and McGlashan, 2013)*
303) *What will the crowd fund: preferences of prospective donors for drug development fundraising campaigns: (Dragojlovic and Lynd, 2016)*
304) *When do investors forgive entrepreneurs for lying?: (Pollack and Bosse, 2014)*
305) *When social enterprises engage in finance: agents of change in lending relationships, a Belgian typology: (Périlleux, 2015)*
306) *Which updates during an equity crowdfunding campaign increase crowd participation?: (Block et al., in press)*
307) *Why do borrowers pledge collateral? New empirical evidence on the role of asymmetric information: (Berger et al., 2011)*
308) *Why supporters contribute to reward-based crowdfunding: (Steigenberger, 2017)*
309) *Wisdom of the intermediary crowd: what the proposed rules mean for ambitious crowdfunding intermediaries: (Deschler, 2014)*

A.4 Report

1) *A framework for European crowdfunding: (De Buysere et al., 2012)*
2) *A review of the regulatory regime for crowdfunding and the promotion of non-readily realisable securities by other media: (Financial Conduct Authority, 2015)*
3) *Alternative finance crowdfunding in Canada: unlocking real value through FinTech and crowd innovation: (The National Crowdfunding Association (NCFA) of Canada, 2016)*
4) *Alternative lending: a regulatory approach to peer-to-peer lending: (Grant Thornton UK LLP, 2014)*
5) *An introduction to crowdfunding: (Nesta, 2012)*
6) *Best practice for crowdfunding: (Deloitte Legal, 2016)*
7) *Creating a crowdfunding ecosystem in Chile: (Multiateral Investment Fund, 2015)*
8) *Crowdfunding 2.0: the next-generation philanthropy: (Özdemir et al., 2015b)*
9) *Crowdfunding and charities: information for charities, donors, and fundraisers about the use of crowdfunding: (Australian Charities and Not-for-Profits Commission, 2017)*
10) *Crowdfunding and P2P lending: which opportunities for microfinance?: (Savarese, 2015)*
11) *Crowdfunding and peer-to-peer lending: legal framework and risks: (Nowak and Korn, 2013)*
12) *Crowdfunding from an investor perspective: (Oxera, 2015)*
13) *Crowdfunding guide: for nonprofits, charities and social impact projects: (Ania and Charlesworth, 2015)*
14) *Crowdfunding in 360°: alternative financing for the digital era: (Cuesta et al., 2015)*
15) *Crowdfunding in China: the financial inclusion dimension: (CGAP, 2017)*
16) *Crowdfunding industry report 2015/2016: (Massolution, 2015)*
17) *Crowdfunding monitoring Switzerland 2017: (Dietrich and Amrein, 2017)*
18) *Crowdfunding: reshaping the crowd's engagement in culture: (Voldere and Zeqo, 2017)*

19) *Crowdfunding's potential for the developing world: (infoDev and The World Bank, 2013)*
20) *Financial inclusion in the People's Republic of China: an analysis of existing research and public data: (Sparreboom and Duflos, 2012)*
21) *How to...crowdfunding for local authorities: (Glover, 2017)*
22) *Innovations in access to finance for SMEs: (ACCA, 2014)*
23) *Moving mainstream: the European alternative finance benchmarking report: (Wardrop et al., 2015)*
24) *Opinion of the European banking authority on lending-based crowdfunding: (European Banking Authority, 2015)*
25) *P2P lending – a threat to traditional players: (Danmayr, 2014; Infosys Technologies Limited, 2009)*
26) *P2P lending: is financial democracy a click away: (Powers et al., 2008)*
27) *Peer pressure: how P2P lending platforms are transforming the consumer lending industry: (PwC, 2015)*
28) *Peer to peer lending ("P2P") platforms: (BG Consulting Group Ltd., 2014)*
29) *Peer-to-peer crowdfunding: information and the potential for disruption in consumer lending: (Morse and NBER, 2015)*
30) *Peer-to-peer lending: industry overview & understanding the marketplace: (Assetz Capital, 2014)*
31) *Real estate crowdfunding: gimmick or game changer?: (IPF, 2016)*
32) *Review of crowdfunding for development initiatives: (Gajda and Walton, 2013)*
33) *State of SME finance in the United States: (Firoozmand et al., 2015)*
34) *Survey on the access to finance of enterprises in the Euro area: October 2014 to March 2015: (European Central Bank, 2015)*
35) *The financial advisor's guide to P2P investing: (Albright et al., 2015)*
36) *The misregulation of person-to-person lending: (Verstein, 2011)*
37) *The sharing economy: accessibility based business models for peer-to-peer markets: (Cebulski, 2013; Dervojeda et al., 2013)*
38) *The voice of the people: insights from the latest peer-to-peer lending research: (Zhang, 2014a)*
39) *Trends in lending: (Bank of England, 2014)*
40) *UK peer-to-peer (P2P) finance: (Equity Development Limited, 2015)*
41) *Understanding alternative finance: the UK alternative finance industry report 2014: (Baeck et al., 2014)*
42) *Understanding peer to peer lending: (Renton, 2015)*
43) *Where are they now? A report into the status of companies that have raised finance using equity crowdfunding in the UK: (Nabarro, 2015)*
44) *Working the crowd: a short guide to crowdfunding and how it can work for you: (Baeck and Collins, 2013)*

A.5 Other Publications

1) *Banking by another name: (Foroohar, 2014)*
2) *Bruce seeks crowdfunding: (Anonymous, 2015a)*

3) *Buying in to the next big thing: crowdfunding gives a boost to developers with cutting-edge designs:* (Bertolucci, 2015)
4) *Can crowdfunding be more than a fad?:* (Eisen, 2014)
5) *China crowdfunding report:* (Zhang et al., 2014b)
6) *Community chest: a zero-interest lending platform appeals to a local crowd:* (Goodman, 2015)
7) *Co-op launches crowdfunding campaign to save honeybees: agriculture industry is suffering from unexpected loss of hives:* (Dewey, 2015)
8) *Crowd funding — should Australia embrace the growing crowd?:* (Colla and Wong, 2013)
9) *Crowd funding sites may put power of the people behind business ideas:* (Allen, 2012)
10) *Crowd funding turns great ideas into great jobs:* (Polis, 2014)
11) *Crowd funding: what all nonprofit organizations should know:* (Hartnett and Matan, 2015)
12) *Crowd mentality: can crowdfunding draw new money into commercial real estate?:* (Mattson-Teig, 2014a)
13) *Crowdfunding 101: the life cycle of a crowdfunding campaign:* (Arsenault, 2015)
14) *Crowdfunding 2.0: want funding from the crowd? Get ready to bare your soul:* (Dahl, 2014)
15) *Crowdfunding a vaccine to rid the world of AIDS:* (Anonymous, 2014a)
16) *Crowdfunding enters realm of placemaking:* (Dewey, 2014)
17) *Crowdfunding for charities and social enterprises:* (Roche-Saunders and Hunt, 2017)
18) *Crowdfunding for municipal projects:* (Feinberg, 2014)
19) *Crowdfunding for real estate development: is it even legal?:* (Lem, 2014)
20) *Crowdfunding platforms for microfinance: a new way to eradicate poverty through the creation of a global hub?:* (Attuel-Mendes, 2014)
21) *Crowdfunding provisions under the new rule 506(c): new opportunities for real estate capital formation:* (Levine and Feigin, 2014)
22) *Crowdfunding site sees deal surge: Localstake awaits rules that will expand investor base:* (Council, 2014)
23) *Crowdfunding smarts: getting the public to invest in your business can breed loyal customers—if you don't flub the pitch:* (Richardson, 2014a)
24) *Crowdfunding success: the short story - analyzing the mix of crowdfunded ventures:* (Lichtig, 2015)
25) *Crowdfunding watch: Beacon Reader:* (Anonymous, 2014b)
26) *Crowdfunding watch: CounterCrop:* (Anonymous, 2015b)
27) *Crowdfunding watch: Funds2Orgs:* (Anonymous, 2015c)
28) *Crowdfunding watch: powEARTHful:* (Anonymous, 2013a)
29) *Crowdfunding watch: ZooShare Biogas:* (Anonymous, 2014c)
30) *Crowdfunding, decoded:* (Grossman, 2015)
31) *Crowdfunding: a new gateway for indie artists:* (Silverberg, 2014)

32) *Crowdfunding: a threat or opportunity for university research funding?: (Baskerville and Cordery, 2015)*
33) *Crowdfunding: Kickstarter is funding the commercialization of new technologies: (Greenwald, 2012)*
34) *Crowdfunding: opportunities and potential roadblocks for startups: (Royse and Davis, 2014)*
35) *Crowdfunding: reaching a new class of angel investors: (Sieck, 2012)*
36) *Crowdfunding: the new way to finance franchises? Franchisors need to give considerable thought to how crowdfunding may impact their franchise systems: (Schneiderman and Chiu, 2014)*
37) *Crowdfunding: the stars are the limit: (The Economist, 2015a)*
38) *Crowdfunding: what is crowdfunding and what does it mean for the future of financing?: (Belshe, 2014)*
39) *Defying the traditional model: crowdfunding in science fiction and fantasy: (Jones, 2014)*
40) *Digital economy offers exciting prospects: from China's experiences, less-developed countries can see how technology changes lives and transforms societies: (Ben and Chen, 2018)*
41) *Driving access to the global property market: (Williams, 2015)*
42) *Equity capital raising: winemaker plans IPO using crowdfunding: (Platt, 2015)*
43) *Fintech and the financing of entrepreneurs: from crowdfunding to marketplace lending: (Fenwick et al., 2017)*
44) *Firms aim to corner auto lending market via crowdfunding: (Samaad, 2014)*
45) *Get the money to launch your dream: (Anonymous, 2013b)*
46) *Griffin Technology tries its hand at crowdfunding: (Johnston, 2015)*
47) *House Caucus urges SEC to act on 'crowdfunding': (Anonymous, 2014d)*
48) *How crowd wisdom closes the gender gap: (Robb, 2015)*
49) *Is crowdfunding a viable startup option? HVAC companies consider alternative fundraising through Indiegogo, Kickstarter, and others: (Turpin, 2015)*
50) *Is crowdfunding the savior of games development?: (Husin, 2015)*
51) *Join the crowd: seniors housing is the latest real estate sector to try crowdfunding: (Carr, 2014)*
52) *Join the right crowd: everybody's doing it. But what kind of crowdfunding is the right kind for your business?: (Alsever, 2015)*
53) *Lessons taken on board in Scottish crowdfunding project: (Anonymous, 2015d)*
54) *Libraries find success in crowdfunding: (Cottrell, 2014)*
55) *Making big FinTech footprints: (Ben, 2017)*
56) *Making money from "free": your music has value, even if you give it away: (Chertkow and Feehan, 2015)*
57) *Michigan banks can benefit from the MILE Act: (Coke, 2014)*
58) *Million-donor movies: more and more independent regional filmmakers are looking at crowdfunding to make movies: (Mishra, 2014)*
59) *More cash for start-ups: new bills could expand crowdfunding options: (Dahl, 2012)*

60) *Muller taps ideas of the rank and file via crowdfunding: (DiDio, 2014)*
61) *Mutually beneficial: venture capital and crowdfunding learn to play nice: (Hogg, 2014)*
62) *New Ontario rules for equity crowdfunding could boost social finance: (Anonymous, 2014f)*
63) *Panel to discuss crowdfunding phenomenon: (York, 2014)*
64) *Peer pressure: crowdfunding firms are successfully raising minimum buy-in amounts: (Mattson-Teig, 2014b)*
65) *Peer-to-peer lending the new alternative: (Foroohar, 2014)*
66) *Peer-to-peer lending: (The Economist, 2015b)*
67) *Please crowd me: how crowdfunding will meld into community banking: (Nichols, 2013)*
68) *Pointers to the future: (The Economist, 2014)*
69) *Pre-selling the future: (Anonymous, 2013c)*
70) *Rating the platforms: a crowdfunding review site helps 'treps choose wisely: (Goodman, 2014)*
71) *Single-family, crowdfunded: entrepreneurs hope to build Simon-esque real estate empire with novel approach: (Council, 2015)*
72) *Small offer delivers big future: relaxed securities rules are making it easier to fund growing business, but it's not just crowdfunding that's cause for excitement: (Summerfield and Parke, 2014)*
73) *SMEs have alternative to bank loans for funding: (Sparreboom and Duflos, 2012; Cohen, 2015)*
74) *Successful failures: established businesses use crowdfunding to test new products: (Richardson, 2014b)*
75) *Swimming upstream: crowdfunding as a new type of pay TV: (Giuffre, 2014)*
76) *Tapping the crowds for research funding: (Weigmann, 2013)*
77) *The art of money: a new bank and app address the specialized financial needs of creatives: (Newman, 2015)*
78) *The bank of Bob: (Gregory, 2015)*
79) *The entrepreneurial artist: how crowdfunding can jump-start your project Part 1: (Figliomeni, 2014a)*
80) *The entrepreneurial artist: what can equity crowdfunding do for me? Part 2: (Figliomeni, 2014b)*
81) *The ins and outs of kickstarter: how crowdfunding brings textiles and home furnishings projects to life: (Anonymous, 2014e)*
82) *The regulation of crowdfunding in Ireland: (Houlihan and Curneen, 2017)*
83) *The rise and rise of crowdfunding: (Gloria, 2013)*
84) *Turning private assets into a P2P business model: an analysis of the evolving (and stagnating) discussion on leadership: (Boukouray, 2015)*
85) *Voices from the crowdfunding crowd: (Stewart, 2015)*
86) *What does the crowdfunding revolution mean for the UK business rescue market: (Landers and Hood, 2015)*

References

Aagaard P. (2011) The global institutionalization of microcredit. *Regulation & Governance* 5: 465–479.

Abdullah A. (2006) Crowdfunding as an emerging fundraising tool: with special reference to the Malaysian regulatory framework. *Islam and Civilisational Renewal*: 98–119.

ACCA. (2014) Innovations in access to finance for SMEs. The Association of Chartered Certified Accountants, pp. 1–12.

Agrawal A, Catalini C and Goldfarb A. (2014) Some simple economics of crowdfunding. *Innovation Policy and the Economy* 14: 63–97.

Agrawal A, Catalini C and Goldfarb A. (2015) Crowdfunding: geography, social networks, and the timing of investment decisions. *Journal of Economics & Management Strategy* 24: 253–274.

Ahlers GKC, Cumming D, Günther C, et al. (2015) Signaling in equity crowdfunding. *Entrepreneurship: Theory and Practice* 39: 955–980.

Ahmed R. (2011) Access to credit by the poor: the role of P2P lending via not-for-profit organizations. The University of Texas at Dallas.

Aitamurto T. (2014) Book reviews: crowdsouring. *New Media & Society* 16: 692–699.

Akkizidis I and Stagars M. (2016) *Marketplace lending, financial analysis, and the future of credit: integration, profitability, and risk management*, The Atrium, Southern Gate, Chichester, West Sussex, PO19 8SQ, United Kingdom: John Wiley & Sons Ltd, ISBN 978-1-119-09916-1.

Albright D, Jones JA and Staples C. (2015) The financial advisor's guide to P2P investing. DaraAlbright.com, Wealth360Advisors.com, and IRAeXchange.net, pp. 1–21.

Alejandre CC. (2013) Exploring the Seattle museum community's perceptions toward crowdfunding. University of Washington.

Ali-Hassan H and Allam H. (2016) Comparing crowdsourcing initiatives: toward a typology development. *Canadian Journal of Administrative Sciences* 33: 318–331.

Allen M. (2012) Crowd funding sites may put power of the people behind business ideas. *San Diego Business Journal, 16 July 2012, Vol. 33 Issue 29*, pp. 4–62.

Allison TH, Davis BC, Short JC, et al. (2014) Crowdfunding in a prosocial microlending environment: examining the role of intrinsic versus extrinsic cues. *Entrepreneurship: Theory and Practice* 39: 53–73.

Allison TH, Davis BC, Webb JW, et al. (2017) Persuasion in crowdfunding: an elaboration likelihood model of crowdfunding performance. *Journal of Business Venturing* 32: 707–725.

Alsever J. (2015) Join the right crowd: everybody's doing it. But what kind of crowdfunding is the right kind for your business? *Inc., July/August 2015*, pp. 59–62.

Alyass A, Turcotte M and Meyre D. (2015) From big data analysis to personalized medicine for all: challenges and opportunities. *BMC Medical Genomics* 8: 1–22.

Andreoni J and Payne AA. (2011) Is crowding out due entirely to fundraising? evidence from a panel of charities. *Journal of Public Economics* 95: 334–343.

Anglin AH, Allison TH, McKenny AF, et al. (2014) The role of charismatic rhetoric in crowdfunding: an examination with computer-aided text analysis. *Social Entrepreneurship and Research Methods* 9: 19–48.

Ania A and Charlesworth C. (2015) Crowdfunding guide: for nonprofits, charities and social impact projects. 215 Spadina Ave, Suite 400, Toronto, ON, M5T 2C7: HiveWire & Centre for Social Innovation, Document Version: 1701 - D.

Anonymous. (2013a) Crowdfunding watch: powEARTHful. *Corporate Knights Magazine, Nov 2013, Vol. 12 Issue 4*, pp. 14.

Anonymous. (2013b) Get the money to launch your dream. *Consumer Reports Money Adviser, September 2013, Vol. 10 Issue 9*, p. 13.

Anonymous. (2013c) Pre-selling the future. *Trends Magazine, July 2013, Issue 123*, pp. 8–13.

Anonymous. (2014a) Crowdfunding a vaccine to rid the world of AIDS. *Inc., May 2014*, p. 60.

Anonymous. (2014b) Crowdfunding watch: Beacon Reader. *Corporate Knights Magazine, Spring 2014, Vol. 13 Issue 2*, pp. 11–12.

Anonymous. (2014c) Crowdfunding watch: ZooShare Biogas. *Corporate Knights Magazine, Winter 2014, Vol. 13 Issue 1*, pp. 13–14.

Anonymous. (2014d) House Caucus urges SEC to act on 'crowdfunding'. *Telecommunications Reports, 1 September 2014, Vol. 80 Issue 17*, p. 26.

Anonymous. (2014e) The ins and outs of kickstarter: how crowdfunding brings textiles and home furnishings projects to life. *Home Textiles Today, February 2014, Vol. 35 Issue 3*, pp. 22–23.

Anonymous. (2014f) New Ontario rules for equity crowdfunding could boost social finance. *Corporate Knights Magazine, Summer 2014, Vol. 13 Issue 3*, p. 13.

Anonymous. (2015a) Bruce seeks crowdfunding. *Publican's Morning Advertiser, 26 March 2015, Issue 186*, p. 10.

Anonymous. (2015b) Crowdfunding watch: CounterCrop. *Corporate Knights Magazine, Winter 2015, Vol. 14 Issue 1*, p. 14.

Anonymous. (2015c) Crowdfunding watch: Funds2Orgs. *Corporate Knights Magazine, Spring 2015, Vol. 14 Issue 2*, p. 13.

Anonymous. (2015d) Lessons taken on board in Scottish crowdfunding project. *Horticulture Week, 21 August 2015*, p. 6.

Anthony DL. (1997) Micro-lending institutions: using social networks to create productive capabilities. *The International Journal of Sociology and Social Policy* 17: 156–178.

Antonenko PD, Lee BR and Kleinheksel AJ. (2014) Trends in the crowdfunding of educational technology startups. *TechTrends* 58: 36–41.

Arko A. (2014) Business and legal issues of small business oriented crowdfunding platforms. Bucerius Law School.

Arkrot W, Unger A and Åhlström E. (2017) Crowdfunding from a marketing perspective. *International Business School*. Jönköping University.

Arsenault A. (2015) Crowdfunding 101: the life cycle of a crowdfunding campaign. Culture Days, 1–43.

Ashta A, Assadi D and Marakkath N. (2015) The strategic challenges of a social innovation: the case of Rang De in crowdfunding. *Strategic Change* 24: 1–14.

Assetz Capital. (2014) Peer-to-peer lending: industry overview & understanding the marketplace. Assetz House, Newby Road, Stockport, Cheshire, SK7 5DA Assetz Capital, pp. 1–24.

Attuel-Mendes L. (2014) Crowdfunding platforms for microfinance: a new way to eradicate poverty through the creation of a global hub? *Cost Management, March/April 2014, Vol. 28 Issue 2*, pp. 38–47.

Attuel-Mendes L. (2017) The different ways of collaboration between a retail bank and crowdfunding. *Strategic Change* 26: 213–225.

Aurelie B. (2013) Drivers and barriers for cross-border crowdfunding: evidence from Belgian stakeholders. *Louvain School of Management*. Universite Catholique de Louvain.

Australian Charities and Not-for-Profits Commission. (2017) Crowdfunding and charities: information for charities, donors, and fundraisers about the use of crowdfunding. Australian Government, pp. 1–9.

Avakyan K. (2013) The influence of perceived consumer values on future intention to contribute to crowdfunding. *Faculty of Economics and Business*. University of Amsterdam.

Baeck P and Collins L. (2013) Working the crowd: a short guide to crowdfunding and how it can work for you. 1 Plough Place, London EC4A 1DE, UK: Nesta, retrieved from http://www.nesta.org.uk/sites/default/files/working_the_crowd.pdf, accessed on 23 June 2014.

Baeck P, Collins L and Zhang B. (2014) Understanding alternative finance: the UK alternative finance industry report 2014. 1 Plough Place, London, EC4A 1DE Nesta, pp. 1–95.

Bailey P. (2013) Branding post-capitalism. University of London.

Bains W, Wooder S and Guzman DRM. (2014) Funding biotech start-ups in a post-VC world. *Journal of Commercial Biotechnology* 20: 10–27.
Bakker-Rakowska J. (2014) Crowdfunding for innovations: a qualitative research on resources, capabilities and stakes. Universiteit Twente.
Bammatter M. (2012) A comparison and assessment of acquiring funds of corporate entities through crowdfunding and through a bank credit. *School of Business.* University of Applied Sciences Northwestern Switzerland.
Bank of England. (2014) Trends in lending. Bank of England, pp. 1–17.
Bao Z and Huang T. (2017) External supports in reward-based crowdfunding campaigns: a comparative study focused on cultural and creative projects. *Online Information Review* 41: 626–642.
Barasinska N. (2011) Essays on determinants of financial behavior of individuals. *Fachbereich Wirtschaftswissenschaft.* Freie Universität Berlin.
Barasinska N and Schäfer D. (2014) Is crowdfunding different? Evidence on the relation between gender and funding success from a German peer-to-peer lending platform. *German Economic Review* 15: 436–452.
Baskerville RF and Cordery CJ. (2015) Crowdfunding: a threat or opportunity for university research funding? *Alternative Investment Analyst Review* Spring: 41–52.
Battcher PA. (2013) Sports inspire: a crowd-funding community for sports philanthropy. *Faculty of the USC Graduate School.* University of Southern California.
Baucus MS and Mitteness CR. (2016) Crowdfrauding: avoiding ponzi entrepreneurs when investing in new ventures. *Business Horizons* 59: 37–50.
Baumgardner T, Neufeld C, Huang PC-T, et al. (2015) Crowdfunding as a fast-expanding market for the creation of capital and shared value. *Thunderbird International Business Review* 59: 115–126.
Belleflamme P, Lambert T and Schwienbacher A. (2013) Individual crowdfunding practices. *Venture Capital* 15: 313–333.
Belleflamme P, Lambert T and Schwienbacher A. (2014) Crowdfunding: tapping the right crowd. *Journal of Business Venturing* 29: 585–609.
Belleflamme P, Omrani N and Peitz M. (2015) The economics of crowdfunding platforms. *Information Economics and Policy* 33: 11–28.
Belshe B. (2014) Crowdfunding: what is crowdfunding and what does it mean for the future of financing? *Smart Business Northern California, June 2014, Vol. 7 Issue 7,* p. 22.
Ben S. (2017) Making big FinTech footprints. *China Daily: Africa Weekly.* 30.
Ben S and Chen X. (2018) Digital economy offers exciting prospects: from China's experiences, less-developed countries can see how technology changes lives and transforms societies. *China Daily: Africa Weekly.* 11.
Bennett L, Chin B and Jones B. (2015) Crowdfunding: a New Media & Society special issue. *New Media & Society* 17: 141–148.
Berger AN, Espinosa-Vega MA, Frame WS, et al. (2011) Why do borrowers pledge collateral? New empirical evidence on the role of asymmetric information. *Journal of Financial Intermediation* 20: 55–70.
Berger AN, Goulding W and Rice T. (2014) Do small businesses still prefer community banks? *Journal of Banking & Finance* 44: 264–278.
Berkovich E. (2009) Essays on search and herding. *Economics.* University of Pennsylvania.
Berkovich E. (2011) Search and herding effects in peer-to-peer lending: evidence from prosper.com. *Annals of Finance* 7: 389–405.
Berliner LS and Kenworthy NJ. (2017) Producing a worthy illness: personal crowdfunding amidst financial crisis. *Social Science & Medicine* 187: 233–242.
Bertolucci J. (2015) Buying in to the next big thing: crowdfunding gives a boost to developers with cutting-edge designs. *Kiplingers Personal Finance, June 2015,* p. 70.
BG Consulting Group Ltd. (2014) Peer to peer lending ("P2P") platforms. BG Consulting Group Ltd., pp. 1–6.

Bi S, Liu Z and Usman K. (2017) The influence of online information on investing decisions of reward-based crowdfunding. *Journal of Business Research* 71: 10–18.

Block J, Hornuf L and Moritz A. (in press) Which updates during an equity crowdfunding campaign increase crowd participation? *Small Business Economics* DOI 10.1007/s11187-017-9876-4.

Boeuf B, Darveau J and Legoux R. (2014) Financing creativity: crowdfunding as a new approach for theatre projects. *International Journal of Arts Management* 16: 33–48.

Booth EA. (2013) Online marketing and fundraising strategies for non-profit organisations in the Cape Town health sector. *Faculty of Informatics and Design.* Cape Town, South Africa: Cape Peninsula University of Technology.

Booth P. (2015) Crowdfunding: a Spimatic application of digital fandom. *New Media & Society* 17: 149–166.

Borello G. (2015) The funding gap and the role of financial return crowdfunding: some evidence from European platforms. *Journal of Internet Banking and Commerce* 20: 1–20.

Botzenhart-Eggstein E. (2015) Crowdfunding: a new source of money for the healthcare sector? *BIOPRO Baden-Württemberg GmbH, retrieved on 16 September 2015 from https://www.bio-pro.de/en/fachartikel/crowdfunding-anew-source-of-money-for-the-healthcare-sector/.*

Bouaiss K, Maque I and Meric J. (2015) More than three's a crowd . . . in the best interest of companies! Crowdfunding as zeitgeist or ideology? *Society and Business Review* 10: 23–39.

Boukouray J. (2015) Turning private assets into a P2P business model: an analysis of the evolving (and stagnating) discussion on leadership. *Entrepreneur Middle East, July 2015,* pp. 44–45.

Bouncken RB, Komorek M and Kraus S. (2015) Crowdfunding: the current state of research. *International Business & Economics Research Journal* 14: 407–415.

Brüntje D and Gajda O. (2016) *Crowdfunding in Europe: state of the art in theory and practice,* Cham Heidelberg New York Dordrecht London: Springer International Publishing Switzerland, ISBN 978-3-319-18016-8.

Bradford SC. (2012) Crowdfunding and the federal securities laws. *Columbia Business Law Review* 1: 1–150.

Braet O, Spek S and Pauwels C. (2013) Crowdfunding the movies: a business analysis of crowdfinanced moviemaking in small geographical markets. *Journal of Media Business Studies* 10: 1–23.

Brenan J. (2014) Science by the masses: is crowdfunding the future for biotech start-ups? *IEEE Pulse* January/February: 59–62.

Bretschneider U and Leimeister JM. (in press) Not just an ego-trip: exploring backers' motivation for funding in incentive-based crowdfunding. *Journal of Strategic Information Systems* http://dx.doi.org/10.1016/j.jsis.2017.02.002.

Breza EL. (2012) Essays on strategic social interactions: evidence from microfinance and laboratory experiments in the field. *Department of Economics.* Massachusetts Institute of Technology.

Brown TE, Boon E and Pitt LF. (2017) Seeking funding in order to sell: crowdfunding as a marketing tool. *Business Horizons* 60: 189–195.

Brummer C and Yadav Y. (2017) The Fintech Trilemma. *SSRN Electronic Journal* https://ssrn.com/abstract=3054770.

Bruton G, Khavul S, Siegel D, et al. (2014) New financial alternatives in seeding entrepreneurship: microfinance, crowdfunding, and peer-to-peer innovations. *Entrepreneurship: Theory and Practice* 39: 9–26.

Bryans D. (2015) Unlocked and loaded: government censorship of 3D-printed firearms and a proposal for more reasonable regulation of 3D-printed goods. *Indiana Law Journal* 90: 901–934.

Buff LA and Alhadeff P. (2013) Budgeting for crowdfunding rewards. *MEIEA Journal* 13: 27–44.

Bump MR. (2014) Crowdfunding in museums. *College of Arts and Sciences.* Seton Hall University.

Burgett BL. (2013) Democratizing commercial real estate investing: the impact of the JOBS Act and crowdfunding on the commercial real estate market. Massachusetts Institute of Technology.

Burtch G. (2013) An empirical examination of factors influencing participant behavior in crowdfunded markets. Temple University.

Burtch G, Ghose A and Wattal S. (2014) Cultural differences and geography as determinants of online prosocial lending. *MIS Quarterly* 38: 773–794.

Buttice V, Colombo MG and Wright M. (2017) Serial crowdfunding, social capital, and project success. *Entrepreneurship Theory and Practice* March: 183–207.

Byrnes JEK, Ranganathan J, Walker BLE, et al. (2014) To crowdfund research, scientists must build an audience for their work. *PLoS ONE* 9: 1–29.

Cable AJB. (2014) Mad money: rethinking private placements. *Washington & Lee Law Review* 71: 22–53.

Cadogan D. (2014) Funding for research? look to the crowd. *College & Research Libraries News, May 2014, Vol. 75 Issue 5*, pp. 268–271.

Calic G and Mosakowski E. (2016) Kicking off social entrepreneurship: how a sustainability orientation influences crowdfunding success. *Journal of Management Studies* 53: 738–767.

Callaghan CW. (2014) Crowdfunding to generate crowdsourced R&D: the alternative paradigm of societal problem solving offered by second generation innovation and R&D. *International Business & Economics Research Journal* 13: 1499–1514.

Čalopa MK. (2014) Analysis of financing sources for start-up companies. *Management: Journal of Contemporary Management Issues* 19: 19–43.

Cameron P, Corne DW, Mason CE et al. (2013) Crowdfunding genomics and bioinformatics. *Genome Biology* 14: 1–5.

Carlman A. (2010) Development 2.0? The case of Kiva.org and online social lending for development. *Department of Sociology and Social Anthropology, Faculty of Arts and Social Sciences.* Stellenbosch University.

Carni US. (2014) Protecting the crowd through escrow: three ways that the SEC can protect crowdfunding investors. *Journal of Corporate & Financial Law* 19: 681–706.

Carr R. (2014) Join the crowd: seniors housing is the latest real estate sector to try crowdfunding. *National Real Estate Investor, September/October 2014, Vol. 56 Issue 5*, pp. 51–52.

Carr SD. (2013) Collective action and the financing of innovation: evidence from crowdfunding. University of Virginia.

Carvajal M, García-Avilés JA and González JL. (2012) Crowdfunding and non-profit media: the emergence of new models for public interest journalism. *Journalism Practice* 6: 638–647.

Cebulski A. (2013) *Kickstarter for Dummies,* 111 River Street, Hoboken, NJ 07030-5774: John Wiley & Sons, Inc., ISBN 978-1-118-50543-4.

CGAP. (2017) Crowdfunding in China: the financial inclusion dimension. CGAP, pp. 1–4.

Cha J. (in press) Crowdfunding for video games: factors that influence the success of and capital pledged for campaigns. *International Journal of Media Management* http://dx.doi.org/10.1080/14241277.2017.1331236.

Chan CSR and Parhankangas A. (2017) Crowdfunding innovative ideas: how incremental and radical innovativeness influence funding outcomes. *Entrepreneurship Theory and Practice* March: 237–263.

Chaudry S. (2014) The impact of the JOBS Act on independent film finance. *DePaul Business & Commercial Law Journal* 12: 215–234.

Chemmanur TJ and Fulghieri P. (2014) Entrepreneurial finance and innovation: an introduction and agenda for future research. *Review of Financial Studies* 27: 1–19.

Chen D and Han C. (2012) A comparative study of online P2P lending in the USA and China. *Journal of Internet Banking and Commerce* 17: 1–15.

Chen L, Huang Z and Liu D. (2016) Pure and hybrid crowds in crowdfunding markets. *Financial Innovation* 2: 1–18.
Chen N, Ghosh A and Lambert NS. (2014) Auctions for social lending: a theoretical analysis. *Games and Economic Behavior* 86: 367–391.
Chertkow R and Feehan J. (2015) Making money from "free": your music has value, even if you give it away. *Electronic Musician, September 2015, Vol. 31 Issue 9*, pp. 60–64.
Chiu C-M, Liang T-P and Turban E. (2014) What can crowdsourcing do for decision support? *Decision Support Systems* 65: 40–49.
Cho HS. (2012) Transitioning international development projects to sustainable businesses: the challenges in commercializing a research project and the engineer-entrepreneurs behind the product. *Department of Mechanical Engineering*. Massachuetts Institute of Technology.
Cho M and Gawon Kim MA. (2017) A cross-cultural comparative analysis of crowdfunding projects in the United States and South Korea. *Computers in Human Behavior* 72: 312–320.
Cholakova M and Clarysse B. (2015) Does the possibility to make equity investments in crowdfunding projects crowd out reward-based investments? *Entrepreneurship Theory and Practice* 39: 145–172.
Christensen CR. (2007) Failure to prosper: risk premia in peer to peer lending. Laramie, Wyoming: University of Wyoming.
Chung J. (2014) Long-term study of crowdfunding platform: predicting project success and fundraising amount. Logan, Utah: Utah State University.
Ciuchta MP, Letwin C, Stevenson RM, et al. (2016) Regulatory focus and information cues in a crowdfunding context. *Applied Psychology: An International Review* 65: 490–514.
Claesson G and Tengvall M. (2015) Peer-to-peer lending: the effects of institutional involvement in social lending. *Jönköping International Business School*. Jönköping: Jönköping University.
Cohen S. (2015) SMEs have alternatives to bank loans for funding. *SMEs Influence, March 2015*, pp. 47.
Coke T. (2014) Michigan banks can benefit from the MILE Act. *Michigan Banker, May 2014, Vol. 26 Issue 5*, p. 14.
Colgren D. (2014) The rise of crowdfunding: social media, big data, cloud technologies. *Strategic Finance, April 2014, Vol. 96 Issue 4*, pp. 56–57.
Colla A and Wong T. (2013) Crowd funding—should Australia embrace the growing crowd? *Keeping Good Companies, April 2013, Vol. 65, No. 3*, pp. 154–158.
Colombo MG, Franzoni C and Rossi-Lamastra C. (2015a) Cash from the crowd. *Science* 348: 1201–1202.
Colombo MG, Franzoni C and Rossi-Lamastra C. (2015b) Internal social capital and the attraction of early contributions in crowdfunding. *Entrepreneurship Theory and Practice* 39: 75–100.
Cottrell M. (2014) Libraries find success in crowdfunding. *American Libraries, May 2014, Vol. 45 Issue 5*, pp. 12–13.
Couffinhal B. (2014) The use of crowdfunding as an alternative way to finance small businesses in France. Dublin Business School and Liverpool John Moore's University.
Council J. (2014) Crowdfunding site sees deal surge: Localstake awaits rules that will expand investor base. *Indianapolis Business Journal, 15–21 September 2014, Vol. 35 Issue 29*, pp. 1–30.
Council J. (2015) Single-family, crowdfunded: entrepreneurs hope to build Simon-esque real estate empire with novel approach. *Indianapolis Business Journal, 13 April 2015, Vol. 36 Issue 7*, pp. 3–26.
Courtney C, Dutta S and Li Y. (2017) Resolving information asymmetry: signaling, endorsement, and crowdfunding success. *Entrepreneurship Theory and Practice* March: 265–290.
Crowell M. (2014) The best place for cartoons. *College of Mass Communication and Media Ar.* Southern Illinois University Carbondale.

Cuesta C, Lis SFd, Roibas I, et al. (2015) Crowdfunding in 360°: alternative financing for the digital era. *Digital Economy Watch.* BBVA Research, 1–25.

Cumming D and Johan S. (2013) Demand-driven securities regulation: evidence from crowdfunding. *Venture Capital* 15: 361–379.

Cumming DJ, Leboeuf G and Schwienbacher A. (2017) Crowdfunding cleantech. *Energy Economics* 65: 292–303.

Cunningham WM. (2012) *The JOBS Act: crowdfunding for small businesses and startups,* 233 Spring Street, 6th Floor, New York, NY 10013: Apress, Springer-Verlag New York, Inc., ISBN 978-1-4302-4755-5.

Dahl D. (2012) More cash for start-ups: new bills could expand crowdfunding options. *Inc., December 2011 / January 2012, Vol. 33 Issue 10,* p. 28.

Dahl D. (2014) Crowdfunding 2.0: want funding from the crowd? Get ready to bare your soul. *Inc. Magazine, February 2014,* pp. 72.

Danmayr F. (2014) *Archetypes of crowdfunding platforms: a multidimensional comparison*: Springer Fachmedien Wiesbaden, ISBN 978-3-658-04558-6.

Davidson R and Poor N. (2015) The barriers facing artists' use of crowdfunding platforms: personality, emotional labor, and going to the well one too many times. *New Media & Society* 17: 289–307.

Davies R. (2014) Civic crowdfunding: participatory communities, entrepreneurs and the political economy of place. *Department of Comparative Media Studies.* Massachusetts Institute of Technology.

Davis B, Hmieleski KM, Webb JW, et al. (2017) Funders' positive affective reactions to entrepreneurs' crowdfunding pitches: the influence of perceived product creativity and entrepreneurial passion. *Journal of Business Venturing* 32: 90–106.

De Buysere K, Gajda O, Kleverlaan R, et al. (2012) *A framework for European crowdfunding*: European Crowdfunding Network, ISBN 978-3-00-040193-0.

Dehling S. (2013) Crowdfunding—a multifaceted phenomenon. *School of Management and Governance.* University of Twente.

Deloitte Legal. (2016) Best practice for crowdfunding. Law Firm Deloitte Legal Oü, Roosikrantsi 2, 10119 Tallinn, Estonia: FinanceEstonia & Law Firm Deloitte Legal Estonia, pp. 1–4.

Dervojeda K, Verzijl D, Nagtegaal F, et al. (2013) The sharing economy: accessibility based business models for peer-to-peer markets. Business Innovation Observatory, pp. 1–20.

Deschler GD. (2014) Wisdom of the intermediary crowd: what the proposed rules mean for ambitious crowdfunding intermediaries. *Louis University Law Journal* 58: 1145–1187.

Deutsch J, Epstein GS and Nir A. (2017) Mind the gap: crowdfunding and the role of seed money. *Managerial and Decision Economics* 38: 53–75.

Dewey C. (2014) Crowdfunding enters realm of placemaking. *Grand Rapids Business Journal, 7 July 2014, Vol. 32 Issue 27,* pp. 1–2.

Dewey C. (2015) Co-op launches crowdfunding campaign to save honeybees: agriculture industry is suffering from unexpected loss of hives. *Grand Rapids Business Journal, 28 June 2015, Vol. 33 Issue 26,* pp. 5–6.

DiDio L. (2014) Muller taps ideas of the rank and file via crowdfunding. *Communications of the ACM* 57: 25.

Dietrich A and Amrein S. (2017) Crowdfunding monitoring Switzerland 2017. Grafenauweg 10, Postfach 7344, 6302 Zug: Institut für Finanzdienstleistungen Zug (IFZ), ISBN 978-3-906877-11-2.

Dilger MG, Jovanović T and Voigt K-I. (2017) Upcrowding energy co-operatives—evaluating the potential of crowdfunding for business model innovation of energy co-operatives. *Journal of Environmental Management* 198: 50–62.

DiResta R, Forrest B and Vinyard R. (2015) *Hardware startup: building your product, business & brand,* 1005 Gravenstein Highway North, Sebastopol, CA 95472: O'Reilly Media, Inc., ISBN 978-1-449-37103-6.

Dorff MB. (2014) The siren call of equity crowdfunding. *Journal of Corporation Law* 39: 493–524.

Dorfleitner G, Hornuf L, Schmitt M, et al. (2017) *FinTech in Germany*: Springer International Publishing AG, ISBN 978-3-319-54665-0.

Dragojlovic N and Lynd LD. (2014) Crowdfunding drug development: the state of play in oncology and rare diseases. *Drug Discovery Today* 19: 1775–1780.

Dragojlovic N and Lynd LD. (2016) What will the crowd fund? preferences of prospective donors for drug development fundraising campaigns. *Drug Discovery Today* 21: 1863–1868.

Dresner S. (2014) *Crowdfunding: a guide to raising capital on the Internet,* 111 River Street, Hoboken, NJ 07030: John Wiley & Sons, Inc., ISBN 978-1-118-49297-0.

Drover W, Wood M and Zacharakis A. (2017) Attributes of angel and crowdfunded ventures as determinants of VC screening decisions. *Entrepreneurship Theory and Practice* May: 323–347.

Dulaurans M. (2014) Digital social innovation: the crowdfunding model. *Revista de Comunicación Vivat Academia* XVII: 72–82.

Dutil P. (2015) Crowdsourcing as a new instrument in the government's arsena: explorations and considerations. *Canadian Public Administration* 58: 363–383.

Duvall KM. (2014) Active consumerism: measuring the importance of crowdfunding factors on backers' decisions to financially support Kickstarter campaigns. *Department of Journalism.* West Virginia University.

Ebisch J. (2012) Crowdfunding and the regulatory hurdles in the United States and Europe. Tilburg. Tilburg University.

Eisen D. (2014) Can crowdfunding be more than a fad? *Hotel Management, 16 June 2014, Vol. 229 Issue 8*, pp. 8–56.

Emanuel EJ, Joffe S, Grady C, et al. (2015) Clinical research: should patients pay to play? *Perspective* 7: 1–4.

Equity Development Limited. (2015) UK peer-to-peer (P2P) finance. Equity Development, 15 Eldon Street, London, EC2M 7LD Equity Development Limited, pp. 1–11.

Estellés-Arolas E and González-Ladroón-de-Guevara F. (2012) Towards an integrated crowdsourcing definition. *Journal of Information Science* 38: 189–200.

Estrela F. (2013) The motivational foundations of lenders: social lending through crowdfunding platforms. Universidade Católica Portuguesa.

European Banking Authority. (2015) Opinion of the European banking authority on lending-based crowdfunding. European Banking Authority (EBA)/Op/2015/03, pp. 1–40.

European Central Bank. (2015) Survey on the access to finance of enterprises in the Euro area: October 2014 to March 2015. 60640 Frankfurt am Main, Germany European Central Bank, ISSN 1831-9998, pp. 1–51.

Everett CR. (2011) Information asymmetry in relationship versus transactional debt markets: evidence from peer-to-peer lending. West Lafayette, Indiana: Purdue University.

Evers M. (2012) Main drivers of crowdfunding success: a conceptual framework and empirical analysis. *Rotterdam School of Management.* Erasmus University.

Fallone EA. (2014) Crowdfunding and sport: how soon until the fans own the franchise? *Marquette Sports Law Review* 25: 7–37.

Farley LA. (2015) Knowledge is power: how implementing affirmative disclosures under the JOBS Act could promote and protect benefit corporations and their investors. *Minnesota Law Review* 99: 1507–1570.

Farnel M. (2015) Kickstarting trans*: the crowdfunding of gender/sexual reassignment surgeries. *New Media & Society* 17: 215–230.

Farnkoff BJ. (2013) Crowdfunding for biotechs: how the SEC's proposed rule may undermine capital formation for startups. *The Journal of Contemporary Health Law and Policy* XXX: 131–183.

Fatoki O. (2014) The financing options for new small and medium enterprises in South Africa. *Mediterranean Journal of Social Sciences* 5: 748–755.

Fedorenko I, Berthon P and Rabinovich T. (2017) Crowded identity: managing crowdsourcing initiatives to maximize value for participants through identity creation. *Business Horizons* 60: 155–165.

Feinberg M. (2014) Crowdfunding for municipal projects. *Government Procurement, December/January 2014, Vol. 21 Issue 6*, p. 14.

Fenwick M, McCahery JA and Vermeulen EPM. (2017) Fintech and the financing of entrepreneurs: from crowdfunding to marketplace lending. *SSRN Electronic Journal* http://ssrn.com/abstract_id=2967891.

Fernandes R. (2013) Analysis of crowdfunding descriptions for technology projects. Massachusetts Institute of Technology.

Ferrarini G and Ottolia A. (2013) Corporate disclosure as a transaction cost: the case of SMEs. *European Review of Contract Law* 9: 363–386.

Figliomeni M. (2014a) The entrepreneurial artist: how crowdfunding can jump-start your project Part 1. *Canadian Musician, July/August 2014, Vol. 36 Issue 4*, p. 62.

Figliomeni M. (2014b) The entrepreneurial artist: what can equity crowdfunding do for me? Part 2. *Canadian Musician, September/October 2014, Vol. 36 Issue 5*, p. 62.

Figliomeni M. (2014c) Grassroots capitalism or: how I learned to stop worrying about financial risk in the exempt market and love equity crowdfunding. *Dalhousie Journal of Legal Studies* 23: 105–129.

Financial Conduct Authority. (2015) A review of the regulatory regime for crowdfunding and the promotion of non-readily realisable securities by other media. Financial Conduct Authority.

Firoozmand S, Haxel P, Jung E, et al. (2015) State of SME finance in the United States. TradeUp Capital Fund and Nextrade Group, LLC, pp. 1–39.

Flanigan ST. (2017) Crowdfunding and diaspora philanthropy: an integration of the literature and major concepts. *Voluntas* 28: 492–509.

Forbes H and Schaefer D. (2017) Guidelines for successful crowdfunding. *Procedia CIRP* 60: 398–403.

Foroohar R. (2014) Banking by another name. *Time, 20 October 2014*, p. 26.

Fraser S, Bhaumik SK and Wright M. (2015) What do we know about entrepreneurial finance and its relationship with growth. *International Small Business Journal* 33: 70–88.

Freedman DM and Nutting MR. (2015) *Equity crowdfunding for investors: a guide to risks, returns, regulations, funding portals, due diligence, and deal terms*, 111 River Street, Hoboken, NJ 07030: John Wiley & Sons, Inc., ISBN 978-1-1188-5356-6.

Freund R. (2010) How to overcome the barriers between economy and sociology with open innovation, open evaluation and crowdfunding? *International Journal of Industrial Engineering and Management* 1: 105–109.

Gajda O and Walton J. (2013) Review of crowdfunding for development initiatives. IMC Worldwide for Evidence and Demand, DOI: http://dx.doi.org/10.12774/eod_hd061.jul2013.gadja;walton, pp. 1–31.

Galuszka P and Brzozowska B. (2015) Crowdfunding: towards a redefinition of the artist's role—the case of MegaTotal. *International Journal of Cultural Studies*: 1–15.

Gamble JR, Brennan M and McAdam R. (2017) A rewarding experience? exploring how crowdfunding is affecting music industry business models. *Journal of Business Research* 70: 25–36.

Gao R and Feng J. (2014) An overview study on P2P lending. *International Business and Management* 8: 14–18.

Ge L and Luo X. (2016) Team rivalry and lending on crowdfunding platforms: an empirical analysis. *Financial Innovation* 2: 1–8.

Gelfond S and Eren B. (2016) SEC adopts "crowdfunding" rules for start-up businesses: an easy way to bet on the next Google? *Journal of Investment Compliance* 17: 117–121.

Gilbert AS. (2017) Crowdfunding nostalgia: kickstarter and the revival of classic PC game genres. *The Computer Games Journal* 6: 17–32.

Giudici G, Guerini M and Rossi-Lamastra C. (in press) Reward-based crowdfunding of entrepreneurial projects: the effect of local altruism and localized social capital on proponents' success. *Small Business Economics* DOI 10.1007/s11187-016-9830-x.

Giuffre L. (2014) Swimming upstream: crowdfunding as a new type of pay TV. *Metro Magazine, Spring 2014, Issue 182*, pp. 125–126.

Gleasure R. (2015) Resistance to crowdfunding among entrepreneurs: an impression management perspective. *Journal of Strategic Information Systems* 24: 219–233.

Gleasure R and Feller J. (2016a) Does heart or head rule donor behaviors in charitable crowdfunding markets? *International Journal of Electronic Commerce* 20: 499–524.

Gleasure R and Feller J. (2016b) Emerging technologies and the democratisation of financial services: a metatriangulation of crowdfunding research. *Information and Organization* 26: 101–115.

Gleasure R and Morgan L. (in press) The pastoral crowd: exploring self-hosted crowdfunding using activity theory and social capital. *Information Systems* DOI: 10.1111/isj.12143: 1–27.

Glind Pvd. (2013) The consumer potential of collaborative consumption: identifying (the) motives of Dutch collaborative consumers & measuring the consumer potential of collaborative consumption within the municipality of Amsterdam. *Faculty of Geosciences*. Utrecht University.

Gloria T. (2013) The rise and rise of crowdfunding. *B&T Magazine, 5/24/2013, Vol. 63 Issue 2792*, pp. 31.

Glover J. (2017) How to...crowdfunding for local authorities. Third Floor, 251 Pentonville Road, London N1 9NG: LGiu: the Local Democracy Think Tank & Spacehive, pp. 1–16.

Goins T and Little E. (2013) Crowdfunding. *The Investment Lawyer* 21: 3–7.

Golić Z. (2013) Advantages of crowdfunding as an alternative source of financing of small and medium-sized enterprises. *Proceedings of the Faculty of Economics in East Sarajevo* 8: 39–48.

Golubnicijs D. (2012) Is your peer a lemon? Relative assessment of risk remuneration on the P2P lending market. Stockholm School of Economics.

Goodman M. (2014) Rating the platforms: a crowdfunding review site helps 'treps choose wisely. *Entrepreneur, January 2014, Vol. 42 Issue 1*, p. 68.

Goodman M. (2015) Community chest: a zero-interest lending platform appeals to a local crowd. *Entrepreneur* 43: 53.

Grant Thornton UK LLP. (2014) Alternative lending: a regulatory approach to peer-to-peer lending. Grant Thornton UK LLP, pp. 1–12.

Granville VJ. (2016) Peer-to-peer fundraising and crowdfunding in health care philanthropy. *Healthcare Philanthropy Journal* Spring: 30–35.

Gras D, Nason RS, Lerman M, et al. (2017) Going offline: broadening crowdfunding research beyond the online context. *Venture Capital* 19: 217–237.

Green CH. (2014) *Banker's guide to new small business finance: venture deals, crowdfunding, private equity, and technology*, Hoboken, New Jersey: John Wiley & Sons, Inc., ISBN 978-1-118-83787-0.

Greenwald T. (2012) Crowdfunding: Kickstarter is funding the commercialization of new technologies. *Technology Review, May/June 2012, Vol. 115 Issue 3*, p. 46.

Gregory S. (2015) The bank of Bob. *Time*. pp. 50–52.

Groshoff D. (2014) Kickstarter my heart: extraordinary popular delusions and the madness of crowdfunding constraints and Bitcoin bubbles. *William & Mary Business Law Review* 5: 489–557.

Grossman P. (2015) Crowdfunding, decoded. *Real Simple, May 2015, Vol. 16 Issue 5*, p. 144.

Grove NS. (2015) The cartographic ambiguities of HarassMap: crowdmapping security and sexual violence in Egypt. *Security Dialogue* 46: 345–364.

Guenther C, Johan S and Schweizer D. (in press) Is the crowd sensitive to distance? how investment decisions differ by investor type. *Small Business Economics* DOI 10.1007/s11187-016-9834-6.

Guo Y, Zhou W, Luo C, et al. (2016) Instance-based credit risk assessment for investment decisions in P2P lending. *European Journal of Operational Research* 249: 417–426.

Guzmán JG, Carpio AFd, Colomo-Palacios R, et al. (2013) Living labs for user-driven innovation: a process reference model. *Research Technology Management, May/June 2013, Vol. 56 Issue 3*, pp. 29–39.

Hörisch J. (2015) Crowdfunding for environmental ventures: an empirical analysis of the influence of environmental orientation on the success of crowdfunding initiatives. *Journal of Cleaner Production* 107: 636–645.

Härkönen J. (2014) Crowdfunding and its utilization for startup finance in Finland: factors of a successful campaign. *School of Business.* Lappeenranta: Lappeenranta University of Technology.

Haltuf M. (2014) Support vector machines for credit scoring. *Department of Banking and Insurance, Faculty of Finance.* Prague: University of Economics in Prague.

Hamilton JF. (2014) Historical forms of user production. *Media, Culture & Society* 36: 491–507.

Harms M. (2007) What drives motivation to participate financially in a crowdfunding community? Amsterdam: Vrije Universitaet Amsterdam.

Harrison R. (2013) Crowdfunding and the revitalisation of the early stage risk capital market: catalyst or chimera? *Venture Capital* 15: 283–287.

Hartnett B and Matan R. (2015) Crowd funding: what all nonprofit organizations should know. Sobel & Co. LLC, pp. 1–17.

Hauge JA and Chimahusky S. (2016) Are promises meaningless in an uncertain crowdfunding environment? *Economic Inquiry* 54: 1621–1630.

Heminway JM. (2014) Investor and market protection in the crowdfunding era: disclosing to and for the "crowd". *Vermont Law Review* 38: 827–848.

Hernando JR. (2016) Crowdfunding: the collaborative economy for channelling institutional and household savings. *Research in International Business and Finance* 38: 326–337.

Hernando JR. (2017) Crowdfunding: the collaborative economy for channelling institutional and household savings. *The Spanish Review of Financial Economics* 15: 12–20.

Herzenstein M, Dholakia UM and Andrews RL. (2011a) Strategic herding behavior in peer-to-peer loan auctions. *Journal of Interactive Marketing* 25: 27–36.

Herzenstein M, Sonenshein S and Dholakia UM. (2011b) Tell me a good story and I may lend you money: the role of narratives in peer-to-peer lending decisions. *Journal of Marketing Research* 48: S138–S149.

Hildebrand T. (2011) Two-sided markets in the online world: an empirical analysis. *Wirtschaftswissenschaftlichen Fakultät.* Humboldt-Universität zu Berlin.

Hillind HW. (2014) Exploiting the crowd: the New Zealand response to equity crowd funding. *Faculty of Law.* Victoria University of Wellington.

Hills M. (2015) Veronica Mars, fandom, and the 'affective economics' of crowdfunding poachers. *New Media & Society* 17: 183–197.

Hobbs J, Grigore G and Molesworth M. (2016) Success in the management of crowdfunding projects in the creative industries. *Internet Research* 26: 146–166.

Hogan J. (2014) Like oil and water: equity crowdfunding and securities regulation. *Lewis & Clark Law Review* 18: 1091–1116.

Hogg S. (2014) Mutually beneficial: venture capital and crowdfunding learn to play nice. *Entrepreneur, December 2014, Vol. 42 Issue 12,* p. 86.

Hornuf L and Neuenkirch M. (in press) Pricing shares in equity crowdfunding. *Small Business Economics* DOI 10.1007/s11187-016-9807-9.

Hornuf L and Schwienbacher A. (2017) Should securities regulation promote equity crowdfunding? *Small Business Economics* 49: 579–593.

Hornuf L and Schwienbacher A. (in press) Market mechanisms and funding dynamics in equity crowdfunding. *Journal of Corporate Finance* http://dx.doi.org/10.1016/j.jcorpfin.2017.08.009.

Hossain M. (2015) Crowdsourcing in business and management disciplines: an integrative literature review. *Journal of Global Entrepreneurship Research* 5: 1–19.

Houlihan C and Curneen D. (2017) The regulation of crowdfunding in Ireland. Dillon Eustace, 1–4.

Huang T and Zhao Y. (in press) Revolution of securities law in the Internet age: a review on equity crowd-funding. *Computer Law & Security Review.*

Hunter A. (2015) Crowdfunding independent and freelance journalism: negotiating journalistic norms of autonomy and objectivity. *New Media & Society* 17: 272–288.

Husin S. (2015) Is crowdfunding the savior of games development? *HWM, August 2015,* pp. 35.

infoDev and The World Bank. (2013) Crowdfunding's potential for the developing world. 1818 H Street NW, Washington DC 20433: Information for Development Program (infoDev)/The World Bank, pp. 1–104.

Infosys Technologies Limited. (2009) P2P lending—a threat to traditional players. Plot No. 44, Electronics City, Hosur Road, Bangalore—560100, India Infosys Technologies Limited, pp. 1–6.

Ingham M and Assadi D. (2016) A research agenda: crowdfunding at the service of responsible social entrepreneurship and innovations. *Cahiers du CEREN* 49: 27–51.

IPF. (2016) Real estate crowdfunding: gimmick or game changer? IPR Research, pp. 1–35.

Isakov O, Lev D, Blumkin L, et al. (2015) Crowdfunding effort identifies the causative mutation in a patient with Nystagmus, Microcephaly, Dystonia and Hypomyelination. *Journal of Genetics and Genomics* 42: 79–81.

James TG. (2013) Far from the maddening crowd: does the JOBS Act provide meaningful redress to small investors for securities fraud in connection with crowdfunding offerings? *Boston College Law Review* 54: 1767–1801.

Jensen MRH, Marshall BB and Jahera JS. (2015) Should crowdfunding investors rely on lists of the best private companies? *Corporate Finance Review, May/June 2015*, pp. 4–15.

Jesus DOFd. (2013) Geography of pledging and application of funds of crowdfunding platforms and the impact on their online notoriety. *Católica Lisbon School of Business and Economics.* Universidade Católica Portuguesa.

Johnston L. (2015) Griffin technology tries its hand at crowdfunding. *Twice: this week in consumer electronics, 20 April 2015, Vol. 30 Issue 8*, p. 28.

Jones MM. (2014) Defying the traditional model: crowdfunding in science fiction and fantasy. *Publishers Weekly, 24 November 2014, Vol. 261 Issue 48*, pp. 32–34.

Josefy M, Dean TJ, Albert LS, et al. (2017) The role of community in crowdfunding success: evidence on cultural attributes in funding campaigns to "save the local theater". *Entrepreneurship Theory and Practice* March: 161–182.

Kang L, Jiang Q and Tan C-H. (2017) Remarkable advocates: an investigation of geographic distance and social capital for crowdfunding. *Information & Management* 54: 336–348.

Kangas R. (2014) Online peer to peer lending: clustering borrowers using self-organizing map. *School of Business.* Lappeenranta University of Technology.

Kappel T. (2009) Ex ante crowdfunding and the recording industry: a model for the US. *Loyola of Los Angeles Entertainment Law Review* 29: 375–385.

Kgoroeadira R. (2014) The impact of commercial peer-to-peer lending Websites on the finance of small business ventures. *Cranfield School of Management.* Cranfield University.

Kickul J and Lyons TS. (2015) Financing social enterprises. *Entrepreneur Research Journal* 5: 83–85.

Kim H and Moor LD. (2017) The case of crowdfunding in financial inclusion: a survey. *Strategic Change* 26: 193–212.

Kitch EW. (2014) Crowdfunding and an innovator's access to capital. *George Mason Law Review* 21: 887–894.

Kleppe IN and Nilsen E. (2017) Crowdfunding sustainability: how do entrepreneurs of sustainability projects utilise the potential of crowdfunding for fundraising? Norwegian School of Economics, NHH.

Knight TB, Leo H and Ohmer AA. (2012) A very quiet revolution: a primer on securities crowdfunding and Title III of the JOBS Act. *Michigan Journal of Private Equity & Venture Capital Law* 2: 135–153.

Koçer S. (2015) Social business in online financing: crowdfunding narratives of independent documentary producers in Turkey. *New Media & Society* 17: 231–248.

Kobberø JC. (2014) Crowdfunding—the development of an online community. Copenhagen Business School.
Kobzar O. (2012) Networking on the margins: the regulation of payday lending in Canada. *Centre for Criminology and Sociolegal Studies*. University of Toronto.
Kraus S, Richter C, Brem A, et al. (2016) Strategies for reward-based crowdfunding campaigns. *Journal of Innovation & Knowledge* 1: 13–23.
Krittanawong C, Zhang HJ, Aydar M, et al. (2018) Crowdfunding for cardiovascular research. *International Journal of Cardiology* 250: 268–269.
Kshetri N. (2015) Success of crowd-based online technology in fundraising: an institutional perspective. *Journal of International Management* 21: 100–116.
Kunz MM, Bretschneider U, Erler M, et al. (2017) An empirical investigation of signaling in reward-based crowdfunding. *Electronic Commerce Research* 17: 425–461.
Kuppuswamy V and Bayus BL. (2015) A review of crowdfunding research and findings. *SSRN Electronic Journal* http://ssrn.com/abstract=2685739.
Kuppuswamy V and Bayus BL. (2017) Does my contribution to your crowdfunding project matter? *Journal of Business Venturing* 32: 72–89.
Kuti M and Madarász G. (2014) Crowdfunding. *Public Finance Quarterly* 59: 355–366.
Kvernmoen S. (2014) Independent film marketing in the digital age. *Department of Media Arts*. Island University, Brooklyn Campus.
Lacan C and Desmet P. (in press) Does the crowdfunding platform matter? risks of negative attitudes in two-sided markets. *Journal of Consumer Marketing* https://doi.org/10.1108/JCM-03-2017-2126.
Lam PTI and Law AOK. (2016) Crowdfunding for renewable and sustainable energy projects: an exploratory case study approach. *Renewable and Sustainable Energy Reviews* 60: 11–20.
Lambrecht A, Goldfarb A, Bonatti A, et al. (2014) How do firms make money selling digital goods online? *Marketing Letters* 25: 331–341.
Landers J and Hood N. (2015) What does the crowdfunding revolution mean for the UK business rescue market. *Corporate Rescue and Insolvency, April 2015*, pp. 76–77.
Langley P. (2016) Crowdfunding in the United Kingdom: a cultural economy. *Economic Geography* 92: 301–321.
Lasrado LA. (2013) Crowdfunding in Finland—a new alternative disruptive funding instrument for businesses. Tampere: Tampere University of Technology.
Laughlin L. (2013) Less risk and more reward: applying the crowdfunding model to local and regional theater organizations. *School of Public and Environmental Affairs*.
Lee CH, Zhao JL and Hassna G. (2016) Government-incentivized crowdfunding for one-belt, one-road enterprises: design and research issues. *Financial Innovation* 2: 1–14.
Lee E and Lee B. (2012) Herding behavior in online P2P lending: an empirical investigation. *Electronic Commerce Research and Applications* 11: 495–503.
Lehner OM. (2013) Crowdfunding social ventures: a model and research agenda. *Venture Capital* 15: 289–311.
Leite PdM. (2012) Crowdfunding: critical factors to finance a project successfully. *Faculdade de Economia e Gestão*. Universidade do Porto.
Lem MJ. (2014) Crowdfunding for real estate development: is it even legal? *Building, February/March 2014, Vol. 64 Issue 1*, pp. 12–13.
Levine ML and Feigin PA. (2014) Crowdfunding provisions under the new rule 506(c): new opportunities for real estate capital formation. *The CPA Journal, June 2014, Vol. 84 Issue 6*, pp. 46–51.
Ley A and Weaven S. (2011) Exporing agency dynamics of crowdfunding in start-up capital financing. *Academy of Entrepreneurship Journal* 17: 85–110.
Li F-W and Pryer KM. (2014) Crowdfunding the Azolla fern genome project: a grassroots approach. *GigaScience* 3: 1–4.
Li S, Lin Z, Qiu J, et al. (2015) How friendship networks work in online P2P lending markets. *Nankai Business Review International* 6: 42–67.

Li Z, Jarvenpaa SL and Pattan N. (2016) Cinetics: fueling entrepreneurial innovations through crowdfunding. *Journal of Information Technology Teaching Cases* 6: 75–83.

Lichtig B. (2015) Crowdfunding success: the short story - analyzing the mix of crowdfunded ventures. *Wharton Research Scholars Journal* Paper 121, http://repository.upenn.edu/wharton_research_scholars/121.

Lim T. (2013) The rise of peer-to-peer platforms and the change in the borrower's rational choice. University of Puget Sound.

Linnay J. (2013) Opportunities and limitations of (social) entrepreneurial approaches: a case study of the recycling sector in Cape Town, South Africa. Kingston, Ontario, Canada: Queen's University.

Liu J, Li X, Wu D, et al. (2017) Cost estimation of building individual cooperative housing with crowdfunding model: case of Beijing, China. *Journal of Intelligent Manufacturing* 28: 749–757.

Livingston LS. (2012) Could peer-to-peer loans substitute for payday loans? *Accounting & Taxation* 4: 77–94.

Loureiro YK and Gonzalez L. (2015) Competition against common sense: insights on peer-to-peer lending as a tool to allay financial exclusion. *International Journal of Bank Marketing* 33: 605–623.

Lukkarinen A, Teich JE, Wallenius H, et al. (2016) Success drivers of online equity crowdfunding campaigns. *Decision Support Systems* 87: 26–38.

Luo B and Lin Z. (2013) A decision tree model for herd behavior and empirical evidence from the online P2P lending market. *Information Systems and e-Business Management* 11: 141–160.

Ma SS. (2013) Book review: the crowdfunding revolution: how to raise venture capital using social media. *Journal of Commercial Biotechnology* 19: 76–77.

Ma Y and Liu D. (2017) Introduction to the special issue on crowdfunding and FinTech. *Financial Innovation* 3: 1–4.

Macht SA and Weatherston J. (2014) The benefits of online crowdfunding for fund-seeking business ventures. *Strategic Change* 23: 1–14.

Macht SA and Weatherston J. (2015) Academic research on crowdfunders: what's been done and what's to come? *Strategic Change* 24: 191–205.

Maier E. (2016) Supply and demand on crowdlending platforms: connecting small and medium-sized enterprise borrowers and consumer investors. *Journal of Retailing and Consumer Services* 33: 143–153.

Makýšová L. (2017) Crowdfunding as an alternative way of raising funds for NGOs. *Faculty of Economics and Administration.* Masaryk University.

Malekipirbazari M and Aksakalli V. (2015) Risk assessment in social lending via random forests. *Expert Systems with Applications* 42: 4621–4631.

Mamonov S, Malaga R and Rosenblum J. (2017) An exploratory analysis of Title II equity crowdfunding success. *Venture Capital* 19: 239–256.

Manchanda K and Muralidharan P. (2014) Crowdfunding: a new paradigm in start-up financing. *Global Conference on Business and Finance Proceedings* 9: 369–374.

Manning S and Bejarano TA. (2017) Convincing the crowd: entrepreneurial storytelling in crowdfunding campaigns. *Strategic Organization* 15: 194–219.

Marakkath N and Attuel-mendes L. (2015) Can microfinance crowdfunding reduce financial exclusion? Regulatory issues. *International Journal of Bank Marketing* 33: 624–636.

Marchitti T. (2013) Microfinance and the use of peer to peer social lending to increase sustainability: case study Kiva—a peer to peer social lending organization. *Department of History, International and Social Studies.* Aalborg University.

Mariani A, Annunziata A, Aprile MC, et al. (2017) Crowdfunding and wine business: some insights from Fundovino experience. *Wine Economics and Policy* 6: 60–70.

Marom D and Lawton K. (2012) *The crowdfunding revolution: how to raise venture capital using social media*: McGraw-Hill Education, ISBN 978-0-0717-9045-1.

Marom S. (2017) Social responsibility and crowdfunding businesses: a measurement development study. *Social Responsibility Journal* 13: 235–249.

Martínez-Cañas R, Ruiz-Palomino P and Pozo-Rubio Rd. (2012) Crowdfunding and social networks in the music industry: implications for entrepreneurship. *International Business & Economics Research Journal* 11: 1471–1476.

Martinho LPdCB. (2009) Combining loan requests and investment offers. *Faculdade de Engenharia.* Universidade do Porto.

Mashburn D. (2013) The anti-crowd pleaser: fixing the crowdfund act's hidden risks and inadequate remedies. *Emory Law Journal* 63: 127–174.

Massolution. (2015) Crowdfunding industry report 2015/2016. Massolution, retrieved from http://crowdexpert.com/crowdfunding-industry-statistics/, accessed on 01 September 2017.

Matos HNFdC. (2012) Crowdfunding: material incentives and performance. Universidade Católica Portuguesa.

Mattson-Teig B. (2014a) Crowd mentality: can crowdfunding draw new money into commercial real estate? *National Real Estate Investor, January/February 2014, Vol. 56 Issue 1*, pp. 30–31.

Mattson-Teig B. (2014b) Peer pressure: crowdfunding firms are successfully raising minimum buy-in amounts. *National Real Estate Investor, July/August 2014, Vol. 56 Issue 4*, p. 12.

McKenny AF, Allison TH, Ketchen DJ, et al. (2017) How should crowdfunding research evolve? A survey of the entrepreneurship theory and practice editorial board. *Entrepreneurship Theory and Practice* March: 291–304.

Medziausyte J and Neugebauer P. (2017) Financing success through equity crowdfunding: the case of start-ups and SMEs funded on European crowdfunding platform. *International Business School.* Jönköping University.

Meer J. (2014) Effects of the price of charitable giving: evidence from an online crowdfunding platform. *Journal of Economic Behavior & Organ* 103: 113–124.

Meeuwsen RJMM. (2013) Crowd funding & renewable energy projects: contribution of crowd funding to renewable energy projects in The Netherlands. Delft University of Tech.

Meißner N. (2014) Opinion leaders as intermediaries in audience building for independent films in the Internet age. *Convergence: The International Journal of Research into New Media Technologies*: 1–24.

Mendes-Da-Silva W, Rossoni L, Conte BS, et al. (2016) The impacts of fundraising periods and geographic distance on financing music production via crowdfunding in Brazil. *Journal of Cultural Economics* 40: 75–99.

Merritt A and Stubbs T. (2012) Complementing the local and global: promoting sustainability action through linked local-level and formal sustainability funding mechanisms. *Public Administration and Development* 32: 278–291.

Metelka A. (2014) Crowdfunding—Startups' alternative funding source beyond banks, business angels and venture capitalists. *School of Management.* Blekinge Institute of Technology.

Meyskens M and Bird L. (2015) Crowdfunding and value creation. *Entrepreneur Research Journal* 5: 155–166.

Mild A, Waitz M and Wöckl J. (2015) How low can you go? Overcoming the inability of lenders to set proper interest rates on unsecured peer-to-peer lending markets. *Journal of Business Research* 68: 1291–1305.

Mishra A. (2014) Million-donor movies: more and more independent regional filmmakers are looking at crowdfunding to make movies. *Business Today, 2 February 2014, Vol. 23 Issue 2*, pp. 98–101.

Mitra D. (2012) The role of crowdfunding in entrepreneurial finance. *Delhi Business Review* 13: 67–72.

Mohammadi A and Shafi K. (in press) Gender differences in the contribution patterns of equity-crowdfunding investors. *Small Business Economics* DOI 10.1007/s11187-016-9825-7.

Moisseyev A. (2013) Effect of social media on crowdfunding project results. Lincoln, Nebraska: University of Nebraska.

Mollick E. (2014) The dynamics of crowdfunding: an exploratory study. *Journal of Business Venturing* 29: 1–16.

Mollick E and Robb A. (2016) Democratizing innovation and capital access: the role of crowdfunding. *California Management Review* 58: 72–87.

Moritz A, Block J and Lutz E. (2015) Investor communication in equity-based crowdfunding: a qualitative-empirical study. *Qualitative Research in Financial Markets* 7: 309–342.

Moritz A and Block JH. (2016) Crowdfunding: a literature review and research directions. In: Brüntje D and Gajda O (eds) *Crowdfunding in Europe: state of the art in theory and practice*. Springer Cham Heidelberg New York Dordrecht London: Springer International Publishing Switzerland, ISBN 978-3-319-18016-8, Part I, Chapter 2, pp. 25–53.

Morozova EV. (2015) Crowdsourcing in public policy: technologies, subjects and its socio-political role. *Asian Social Science* 11: 111–121.

Morse A and NBER. (2015) Peer-to-peer crowdfunding: information and the potential for disruption in consumer lending. University of California, Berkeley, pp. 1–30.

Morsy S. (2014) The JOBS act and crowdfunding: how narrowing the secondary market handicaps fraud plaintiffs. *Brooklyn Law Review* 79: 1373–1405.

Moss TW, Neubaum DO and Meyskens M. (2015) The effect of virtuous and entrepreneurial orientations on microfinance lending and repayment: a signaling theory perspective. *Entrepreneurship: Theory and Practice* 39: 27–52.

Multiateral Investment Fund. (2015) Creating a crowdfunding ecosystem in Chile. 1300 New York Avenue NW, Washington, DC 20005: Inter-American Development Bank, pp. 1–84.

Murray J. (2015) Equity crowdfunding and peer-to-peer lending in New Zealand: the first year. *JASSA The Finsia Journal of Applied Finance* 2: 5–10.

Mustafa SE and Adnan HM. (2017) Crowdsourcing: a platform for crowd engagement in the publishing industry. *Publishing Research Quarterly* 33: 283–296.

Nabarro. (2015) Where are they now? A report into the status of companies that have raised finance using equity crowdfunding in the UK. AltFiData, pp. 1–32.

Nagymihaly G. (2013) Crowdfunding: drivers and barriers in an international perspective. *Aarhus School of Business*. University of Aarhus.

Nakatsu RT, Grossman EB and Iacovou CL. (2014) A taxonomy of crowdsourcing based on task complexity. *Journal of Information Science* 40: 823–834.

Neslund K. (2014) The United States' new crowdfunding rules: a Pandora's box? *Singidunum Journal of Applied Sciences* 2014 Supplement: 31–37.

Nesta. (2012) An introduction to crowdfunding. 1 Plough Place, London EC4A 1DE Nesta, pp. 1–4.

Nevill G. (2014) Funding news freedom. *Index on Censorship* 43: 63–66.

Newman M. (2015) The art of money: a new bank and app address the specialized financial needs of creatives. *Entrepreneur* 43: 76.

Ng EC. (2013) Rebranding gay: new configurations of digital media and commercial culture. *Department of Communication*. University of Massachusetts Amherst.

Nicholas J, Ledwith A and Bessant J. (2015) Selecting early-stage ideas for radical innovation: tools and structures. *Research Technology Management* 58: 36–43.

Nichols C. (2013) Please crowd me: how crowdfunding will meld into community banking. *ABA Banking Journal, October 2014, Vol. 106 Issue 10*, pp. 10–11.

Nicoletti B. (2017) *The future of FinTech: integrating finance and technology in financial services*: Palgrave Macmillan, ISBN 978-3-319-51414-7.

Niźnik-Klocek M. (2012) Crowdfunding as a way of financing start-ups in Poland. *School of Business and Services Management*. Jamk University of Applied Sciences.

Nowak GJ and Korn B. (2013) Crowdfunding and peer-to-peer lending: legal framework and risks. The New York Times Building, 620 Eighth Avenue, New York, New York 10018 Pepper Hamilton LLP, pp. 1–26.

Nucciarelli A, Li F, Fernandes KJ, et al. (2017) From value chains to technological platforms: the effects of crowdfunding in the digital game industry. *Journal of Business Research* 78: 341–352.

O'Toole CM, Lawless M and Lambert D. (2015) Non-bank financing in Ireland: a comparative perspective. *The Economic and Social Review* 46: 133–161.

O'Connor SM. (2014) Crowdfunding's impact on start-up IP strategy. *George Mason Law Review* 21: 895–918.

Oei S-Y and Ring D. (2015) Human equity? Regulating the new income share agreement. *Vanderbilt Law Review* 63: 681–760.

Olson E. (2015) Squalls in the safe harbor: investment advice & regulatory gaps in regulation crowdfunding. *Journal of Corporation Law* 40: 539–564.

Ordanini A, Miceli L, Pizzetti M, et al. (2011) Crowd-funding: transforming customers into investors through innovative service platforms. *Journal of Service Management* 22: 443–470.

Otero P. (2015) Crowdfunding: a new option for funding health projects. *Arch Argent Pediatr* 113: 154–157.

Oxera. (2015) Crowdfunding from an investor perspective. Park Central, 40/41 Park End Street, Oxford, OX1 1JD, UK.: Oxera Consulting LLP, ISBN: 978-92-79-46659-5.

Özdemir V, Badr KF, Dove ES, et al. (2013) Crowd-funded micro-grants for genomics and "big data": an actionable idea connecting small (artisan) science, infrastructure science, and citizen philanthropy. *OMICS A Journal of Integrative Biology* 17: 161–172.

Özdemir V, Faris J and Srivastava S. (2015a) Crowdfunding 2.0: the next-generation philanthropy. *EMBO Reports* 16: 267–271.

Özdemir V, Faris J and Srivastava S. (2015b) Crowdfunding 2.0: the next-generation philanthropy. *European Molecular Biology Organization (EMBO) Reports* 16: 267–271.

Périlleux A. (2015) When social enterprises engage in finance: agents of change in lending relationships, a Belgian typology. *Strategic Change* 24: 285–300.

Padgett BL and Rolston C. (2014) Crowd funding: a case study at the intersection of social media and business ethics. *Journal of the International Academy for Case Studies* 20: 61–66.

Palacios M, Martinez-Corral A, Nisar A, et al. (2016) Crowdsourcing and organizational forms: emerging trends and research implications. *Journal of Business Research* 69: 1834–1839.

Pao S-Y. (2010) Connected strangers: manipulating social perceptions to study trust. *School of Architecture and Planning*. Massachusetts Institute of Technology.

Pape M. (2014) Crowdsourcing and participatory mechanisms in crowdfunded design projects. *Faculty of Graduate and Postdoctoral Affairs*. Ottawa, Ontario: Carleton University.

Parhankangas A and Renko M. (2017) Linguistic style and crowdfunding success among social and commercial entrepreneurs. *Journal of Business Venturing* 32: 215–236.

Paris BL. (2013) Institutional lending models, mission drift, and microfinance institutions. *Martin School of Public Policy and Administration*. Lexington, Kentucky: University of Kentucky.

Parker SC. (2014) Crowdfunding, cascades and informed investors. *Economics Letters* 125: 432–435.

Parrino RJ and Romeo PJ. (2012) JOBS act eases securities-law regulation of smaller companies. *Journal of Investment Compliance* 13: 27–35.

Paschen J. (2017) Choose wisely: crowdfunding through the stages of the startup life cycle. *Business Horizons* 60: 179–188.

Paulet E and Relano F. (2017) Exploring the determinants of crowdfunding: the influence of the banking system. *Strategic Change* 26: 175–191.

Paykacheva V. (2014) Crowdfunding as a customer engagement channel. *School of Business and Administration*. Kajaanin Ammattikorkeakoulu University of Applied Sciences.

Pekmezovic A and Walker G. (2016) The global significance of crowdfunding: solving the SME funding problem and democratizing access to capital. *William & Mary Business Law Review* 7: 347–458.

Perry JE. (2015) The people's NIH? Ethical and legal concerns in crowdfunded biomedical research. *Notre Dame Journal of Law, Ethics & Public Policy* 29: 453–470.

Pichler F and Tezza I. (2016) Crowdfunding as a new phenomenon: orignins, features and literature review. In: Bottiglia R and Pichler F (eds) *Crowdfunding for SMEs: A European Perspective.* Palgrave Macmillan, ISBN 978-1-137-56020-9, Chapter 2, pp. 5–43.

Pickering S. (2013) Crowdfunding affordable homes with tax credit investment partnerships. *Review of Banking & Financial Law* 33: 937–1005.

Pisano P, Pironti M and Rieple A. (2015) Identify innovative business models: can innovative business models enable players to react to ongoing or unpredictable trends. *Entrepreneurship Research Journal* 5: 181–199.

Pitschner S and Pitschner-Finn S. (2014) Non-profit differentials in crowd-based financing: evidence from 50,000 campaigns. *Economics Letters* 123: 391–394.

Platt G. (2015) Equity capital raising: winemaker plans IPO using crowdfunding. *Global Finance, March 2015, Vol. 29 Issue 3*, p. 66.

Polis J. (2014) Crowd funding turns great ideas into great jobs. *Boulder County Business Report, 27 April–10 May 2012, Vol. 31 Issue 10*, p. 11A.

Pollack JM and Bosse DA. (2014) When do investors forgive entrepreneurs for lying? *Journal of Business Venturing* 29: 741–754.

Polzin F, Toxopeus H and Stam E. (in press) The wisdom of the crowd in funding: information heterogeneity and social networks of crowdfunders. *Small Business Economics* DOI 10.1007/s11187-016-9829-3.

Pope N. (2011) Crowdfunding microstartups: it's time for the Securities and Exchange Commission to approve a small offering exemption. *University of Pennsylvania Journal of Business Law* 13: 101–129.

Postill J. (2014) Freedom technologists and the new protest movements: a theory of protest formulas. *Convergence: The International Journal of Research into New Media Technologies* 20: 402–418.

Postolski D and Nowotarski M. (2013) Lessons from the first patent infringement lawsuit filed against a crowdfunded project on a crowdfunding portal. *The Licensing Journal* 33: 18–25.

Powers J, Magnoni B and Knapp S. (2008) Person-to-person lending: is financial democracy a click away? Ea Consultants and Abt Associates, pp. 1–48.

Profatilov DA, Bykova ON and Olkhovskaya MO. (2015) Crowdfunding: online charity or a modern tool for innovative projects implementation? *Asian Social Science* 11: 146–151.

Puro L, Teich JE, Wallenius H, et al. (2010) Borrower decision aid for people-to-people lending. *Decision Support Systems* 49: 52–60.

PwC. (2015) Peer pressure: how P2P lending platforms are transforming the consumer lending industry. PricewaterhouseCoopers LLP, pp. 1–18.

Quero MJ, Ventura R and Kelleher C. (2017) Value-in-context in crowdfunding ecosystems: how context frames value co-creation. *Service Business* 11: 405–425.

Rana S. (2013) Philanthropic innovation and creative capitalism: a historical and comparative perspective on social entrepreneurship and corporate social responsibility. *Shruti Alabama Law Review* 64: 1121–1174.

Read A. (2013) Crowdfunding: an empirical and theoretical model of non-profit support. University of Puget Sound.

Reid S. (2016) Crowdfunding: a healthy practice? : Mount Saint Vincent University.

Reiser DB and Dean SA. (2015) SE(c)(3): a catalyst for social enterprise crowdfunding. *Indiana Law Journal* 90: 1091–1129.

Renton P. (2015) Understanding peer to peer lending. LendAcademy.com, pp. 1–13.

Renwick MJ and Mossialos E. (2017) Crowdfunding our health: economic risks and benefits. *Social Science & Medicine* 191: 48–56.

Reynolds CC. (2013) Social innovation: an analysis of its drivers and of the crowdfunding phenomena. Otto Beischeim School of Management.

Rich JR. (2014) *The crowd funding services handbook: raising the money you need to fund your business, project, or invention*, Hoboken, New Jersey: John Wiley & Sons, Inc., ISBN 978-1-118-85300-9.
Richardson V. (2014a) Crowdfunding smarts: getting the public to invest in your business can breed loyal customers—if you don't flub the pitch. *Restaurant Business, January 2014, Vol. 113 Issue 1*, pp. 25–26.
Richardson V. (2014b) Successful failures: established businesses use crowdfunding to test new products. *Entrepreneur's Startups* 29: 16.
Riley-Huff DA, Herrera K, Ivey S, et al. (2016) Crowdfunding in libraries, archives and museums. *The Bottom Line* 29: 67–85.
Robb A. (2015) How crowd wisdom closes the gender gap. *Inc., December 2014/January 2015, Vol. 36 Issue 10*, pp. 112.
Robock Z. (2014) The risk of money laundering through crowdfunding: a funding portal's guide to compliance and crime fighting. *Michigan Business & Entrepreneurial Law Review* 4: 113–129.
Roche-Saunders G and Hunt O. (2017) Crowdfunding for charities and social enterprises. *Charity and Social Enterprise Update* Summer: 7–9.
Roma P, Petruzzelli AM and Perrone G. (2017) From the crowd to the market: the role of reward-based crowdfunding performance in attracting professional investors. *Research Policy* 46: 1606–1628.
Rossi M. (2014) The new ways to raise capital: an exploratory study of crowdfunding. *International Journal of Financial Research* 5: 8–18.
Rossi SPS. (2017) *Access to bank credit and SME financing*: Palgrave Macmillan, ISBN 978-3-319-41362-4.
Royal C and Windsor GSS. (2014) Microfinance, crowdfunding, and sustainability: a case study of telecenters in a South Asian developing country. *Strategic Change* 23: 425–438.
Royse R and Davis C. (2014) Crowdfunding: opportunities and potential roadblocks for startups. *California CPA, November 2014, Vol. 83 Issue 5*, pp. 18–19.
Sørensen IE. (2012) Crowdsourcing and outsourcing: the impact of online funding and distribution on the documentary film industry in the UK. *Media, Culture & Society* 34: 726–743.
Samaad MA. (2014) Firms aim to corner auto lending market via crowdfunding. *Credit Union Times. 14 May 2014, Vol. 25 Issue 19*, p. 8.
Samanci M and Kiss G. (2014) Exploratory study on technology related successfully crowdfunding projects' post online market presence. *School of Economics and Management*. Lund University.
Sanchez-González M and Palomo-Torres M-B. (2014) Knowledge and assessment of crowdfunding in communication: the view of journalists and future journalists. *Media Education Reseacrh Journal* 43: 101–110.
Sannajust A, Roux F and Chaibi A. (2014) Crowdfunding in France: a new revolution? *The Journal of Applied Business Research* 30: 1919–1928.
Saunders RK. (2014) Power to the people: how the SEC can empower the crowd. *Vanderbilt Journal of Entertainment and Technology Law* 16: 945–975.
Sauro JJ. (2014) A content analysis of Kickstarter: the influence of framing and reward motivations on campaign success. San Diego State University.
Savarese C. (2015) Crowdfunding and P2P lending: which opportunities for microfinance? Rue de I'Industie 10 - 1000 Brussels, Belgium: European Microfinance Network (EMN) aisbl, pp. 1–34.
Savio LD. (2017) The place of crowdfunding in the discovery of scientific and social value of medical research. *Bioethics* 31: 384–392.
Scheder B and Arbøll CK. (2014) Crowdinvestor investment decision-making: a study of motivation, investment process and criteria. Copenhagen Business School.
Schneider-Sikorsky PA. (2014) Innovation spaces. *School of Management*. Massachusetts Institute of Technology.

Schneiderman J and Chiu L. (2014) Crowdfunding: the new way to finance franchises? Franchisors need to give considerable thought to how crowdfunding may impact their franchise systems. *Franchising World, October 2014, Vol. 46 Issue 10*, pp. 37–38.

Scholz N. (2015) *The relevance of crowdfunding: the impact on the innovation process of small entrepreneurial firms,* Wiesbaden, Germany: Springer Gabler, ISBN 978-3-658-09836-0.

Scott S. (2015) The moral economy of crowdfunding and the transformative capacity of fan-ancing. *New Media & Society* 17: 167–182.

Seltzer E and Mahmoudi D. (2012) Citizen participation, open innovation, and crowdsourcing: challenges and opportunities for planning. *Journal of Planning Literature* 28: 3–18.

Setälä K. (2017) Crowdfunding in the banking industry: adjusting to a digital era. *School of Business and Economics.* Jyväskylä University.

Sharma A, Khan J and Devereaux PJ. (2015) Is crowdfunding a viable source of clinical trial research funding? *The Lancet* 386: 338.

Sheik S. (2013) Fast forward on crowdfunding. *The Computer & Internet Lawyer* 30: 17–21.

Sheldon RC and Kupp M. (2017) A market testing method based on crowd funding. *Strategy & Leadership* 45: 19–23.

Short JC, Ketchen DJ, McKenny AF, et al. (2017) Research on crowdfunding: reviewing the (very recent) past and celebrating the present. *Entrepreneurship Theory and Practice* March: 149–160.

Shue K. (2011) Essays in financial economics. *Graduate School of Arts and Sciences.* Cambridge, Massachusetts: Harvard University.

Sieck DR. (2012) Crowdfunding: reaching a new class of angel investors. *Business NH Magazine, Nov 2012, Vol. 29 Issue 11*, pp. 34–35.

Siering M, Koch J-A and Deokar AV. (2016) Detecting fraudulent behavior on crowdfunding platforms: the role of linguistic and content-based cues in static and dynamic contexts. *Journal of Management Information Systems* 33: 421–455.

Silverberg D. (2014) Crowdfunding: a new gateway for indie artists. *Broken Pencil, Summer 2014*, pp. 16–19.

Singh H. (2010) Software innovations: the influence of quality, diversity and structure of network ties. University of Connecticut.

Skirnevskiy V, Bendig D and Brettel M. (2017) The influence of internal social capital on serial creators' success in crowdfunding. *Entrepreneurship Theory and Practice* March: 209–236.

Smith AN. (2015) The backer–developer connection: exploring crowdfunding's influence on video game production. *New Media & Society* 17: 198–214.

Smith MW and Green KM. (2015) A class exercise to explore crowdfunding. *Business Education Innovation Journal* 7: 33–43.

Smith RJ and Merchant RM. (2015) Harnessing the crowd to accelerate molecular medicine research. *Trends in Molecular Medicine* 21: 403–405.

Snyder J, Crooks VA, Mathers A, et al. (in press) Appealing to the crowd: ethical justifications in Canadian medical crowdfunding campaigns. *Journal of Medical Ethics* doi:10.1136/medethics-2016-103933.

Snyder J, Mathers A and Crooks VA. (2016) Fund my treatment!: a call for ethics-focused social science research into the use of crowdfunding for medical care. *Social Science & Medicine* 169: 27–30.

Sokolov AV. (2015) Russian political crowdfunding. *Demokratizatsiya* 23: 117–149.

Sonenshein S, Herzenstein M and Dholakia UM. (2010) How accounts shape lending decisions through fostering perceived trustworthiness. *Organizational Behavior and Human Decision Processes* 115: 69–84.

Soon JM and Saguy IS. (2017) Crowdsourcing: a new conceptual view for food safety and quality. *Trends in Food Science & Technology* 66: 63–72.

Sparreboom P and Duflos E. (2012) Financial inclusion in the People's Republic of China: an analysis of existing research and public data. CGAP & Working Group on Inclusive Finance in China, pp. 1–53.

Srkoc M and Zarim RA. (2013) The antecedents of interest in social crowdfunding: exploring the 8 mechanisms that drive charitable giving. Copenhagen: Copenhagen Business School.

Stanberry K and Aven F. (2014) Crowdfunding and the expansion of access to startup capital. *International Research Journal of Applied Finance* V: 1382–1391.

Stanko MA and Henard DH. (2017) Toward a better understanding of crowdfunding, openness and the consequences for innovation. *Research Policy* 46: 784–798.

Steigenberger N. (2017) Why supporters contribute to reward-based crowdfunding. *International Journal of Entrepreneurial Behavior & Research* 23: 336–353.

Steinberg S and DeMaria R. (2015) *The crowdfunding bible: how to raise money for any startup, video game, or project,* www.asmallbusinessexpert.com: Overload Entertainment, LLC, ISBN 978-1-105-72628-6.

Stemler AR. (2013) The JOBS Act and crowdfunding: harnessing the power–and money–of the masses. *Business Horizons* 56: 271–275.

Stern JS. (2013) Characteristics of content and social spread strategy on the IndieGoGo crowdfunding platform. The University of Texas at Austin.

Stewart H. (2015) Voices from the crowdfunding crowd. *Opera News, March 2015, Vol. 79 Issue 9,* p. 64.

Stiver A, Barroca L, Minocha S, et al. (2015) Civic crowdfunding research: challenges, opportunities, and future agenda. *New Media & Society* 17: 249–271.

Stoeckl VE. (2014) Lonely rebel or pioneer of the future? Towards an understanding of moral stakeholder framing of activist brands. *Advances in Consumer Research* 42: 371–376.

Stokar D. (2014) The crowdfunding investor (CFI) - a new species in an existing ecosystem. *Chair of Entrepreneurial Risks.* Scheuchzerstrasse 7, SEC F 7, CH - 8092 Zurich, Switzerland: ETH Zürich.

Strader L. (2014) An exploration in funding independent film. University of Akron.

Stuart T and Anderson C. (2015) 3D robotics: disrupting the drone market. *California Management Review* 57: 91–112.

Summerfield N and Parke C. (2014) Small offer delivers big future: relaxed securities rules are making it easier to fund growing business, but it's not just crowdfunding that's cause for excitement. *NZ Business, July 2014, Vol. 28 Issue 6,* pp. 48–49.

Taylor R. (2015) Equity-based crowdfunding: potential implications for small business capital. *Issue Brief, 14 April 2015, Number 5,* pp. 1–8.

Teinemaa T. (2013) The impact of group lending on participants' entrepreneurial resources—a case study of microcredit in rural Estonia. Copenhagen Business School.

Teo LT. (2013) Understanding crowd funding: cost of capital and factors for success. *System Design and Management Program.* Massachusetts Institute of Technology.

The Economist. (2014) Pointers to the future. *The Economist* 413, No. 8909: 72.

The Economist. (2015a) Crowdfunding: the stars are the limit. *The Economist* 414, No. 8925: 58.

The Economist. (2015b) Peer-to-peer lending. *The Economist Special Report International Banking: Slings and arrows* 415, No. 8937: 6–9.

The National Crowdfunding Association (NCFA) of Canada. (2016) Alternative finance crowdfunding in Canada: unlocking real value through FinTech and crowd innovation. 1240 Bay Street, Suite 501, Toronto, ON M5R 2A7: The National Crowdfunding Association (NCFA) of Canada, pp. 1–20.

Theseira WE. (2009) Discrimination, trust and social capital: three essays in applied public economics. *Business and Public Policy.* University of Pennsylvania.

Tomboc GFB. (2013) The lemons problem in crowdfunding. *John Marshall Journal of Information Technology & Privacy Law* XXX: 253–279.

Tomczak A and Brem A. (2013) A conceptualized investment model of crowdfunding. *Venture Capital* 15: 335–359.

Tsai KS. (2015) The political economy of state capitalism and shadow banking in China. *Issues & Studies* 51: 55–97.

Turi AN, Domingo-Ferrer J, Sánchez D, et al. (2017) A co-utility approach to the mesh economy: the crowd-based business model. *Review of Managerial Science* 11: 411–442.

Turpin JR. (2015) Is crowdfunding a viable startup option? HVAC companies consider alternative fundraising through Indiegogo, Kickstarter, and others. *Air Conditioning Heating & Refrigeration News, Vol. 254 Issue 3*, pp. 1–6.

Valančienė L and Jegelevičiūtė S. (2013) Valuation of crowdfunding: benefit and drawbacks. *Economics and Management* 18: 39–48.

Vasileiadou E, Huijben JCCM and Raven RPJM. (2016) Three is a crowd? Exploring the potential of crowdfunding for renewable energy in the Netherlands. *Journal of Cleaner Production* 128: 142–155.

Veelen Tv. (2015) Civic crowdfunding: a financial kickstart to urban area development? An analysis of the institutional structure in which civic crowdfunding in urban area development is embedded. *Systems Engineering, Policy Analysis and Management.* Delft University of Technology.

Verstein A. (2011) The misregulation of person-to-person lending. University of California, Davis, pp. 445–503.

Vettenranta T. (2017) Key determinants of successful crowdfunding campaigns in the gaming industry. *School of Business.* Aalto University.

Vismara S. (2016) Information cascades among investors in equity crowdfunding. *Entrepreneurship Theory and Practice* November: 1–31.

Voelker TA and McGlashan R. (2013) What is crowdfunding? Bringing the power of kickstarter to your entrepreneurship research and teaching activities. *Small Business Institute Journal* 9: 11–22.

Vogel M. (2015) Crowdfunding human capital contracts. *Cardozo Law Review* 36: 1577–1609.

Voldere ID and Zeqo K. (2017) Crowdfunding: reshaping the crowd's engagement in culture. European Union, ISBN 978-92-79-67975-9, 1–203.

Voorbraak KJPM. (2011) Crowdfunding for financing new ventures: consequences of the financial model on operational decisions. *School of Industrial Engineering.* Eindhoven University of Technology.

Vorapatchaiyanont R. (2013) Design of online peer-to-peer investment platform offering incentive-compatible revenue-sharing innovation. *Department of Economics.* Stanford University.

Vulkan N, Åstebro T and Sierra MF. (2016) Equity crowdfunding: a new phenomena. *Journal of Business Venturing Insights* 5: 37–49.

Wales K. (2015) Cyberfinance: liberating the financial markets. *The Capco Institute Journal of Financial Transformation* 41: 11–22.

Wang JG, Xu H and Ma J. (2015) *Financing the underfinanced: online lending in China,* Heidelberg New York Dordrecht London: Springer-Verlag Berlin Heidelberg, ISBN 978-3-662-46524-0.

Wang W, Zhu K, Wang H, et al. (2017) The Impact of sentiment orientations on successful crowdfunding campaigns through text analytics. *The Institution of Engineering and Technology (IET) Software* 11: 229–238.

Wardrop R, Zhang B, Rau R, et al. (2015) Moving mainstream: the European alternative finance benchmarking report. Wardour, Drury House 34–43 Russell Street, London WC2B 5HA: University of Cambridge and EY, pp. 1–44.

Weber DM. (2012) The impact of information and communication technology on intermediation, outreach, and decision rights in the microfinance industry. Arizona State University.

Wechsler J. (2013) Know your crowd: the drivers of success in reward-based crowdfunding. *Department of Mass Media and Communication Research.* University of Fribourg.

Weigmann K. (2013) Tapping the crowds for research funding. *European Molecular Biology Organization (EMBO) Reports* 14: 1043–1046.

Weinstein RS. (2013) Crowdfunding in the U.S. and abroad: what to expect when you're expecting. *Cornell International Law Journal* 46: 427–453.

Wessel M, Thies F and Benlian A. (2016) The emergence and effects of fake social information: evidence from crowdfunding. *Decision Support Systems* 90: 75–85.

Wheat RE, Wang Y, Byrnes JE, et al. (2013) Raising money for scientific research through crowdfunding. *Trends in Ecology & Evolution* 28: 71–72.

Wicks M. (2013) *Crowdfunding: an introduction,* 3552 Promenade Crescent, Victoria, B.C. V9C 4L2: Blue Beetle Books Inc.

Wieck E, Bretschneider U and Leimeister JM. (2013) Funding from the crowd: an Internet-based crowdfunding platform to support business set-ups from universities. *International Journal of Cooperative Information Systems* 22: 1–12.

Willems W. (2013) What characteristics of crowdfunding platforms influence the success rate? An empirical study into relevant determinants in order to explain differences in success rates between crowdfunding platforms world wide. Erasmus Universiteit Rotterdam.

Williams G. (2015) Driving access to the global property market. *Finweek, 9 July 2015,* pp. 29–31.

Wolfson SM. (2012) Crowdsourcing and the law. University of Texas at Austin.

Wright M, Lumpkin T, Zott C, et al. (2016) The evolving entrepreneurial finance landscape. *Strategic Entrepreneurship Journal* 10: 229–234.

Wroldsen JS. (2014) The Crowdfund Act's strange bedfellows: democracy and start-up company investing. *Kansas Law Review* 62: 257–401.

Wroldsen JSJ. (2013) The social network and the crowdfund act: Zuckerberg, Saverin, and venture capitalists' Dilution of the crowd. *Vanderbilt Journal of Entertainment & Technology Law* 15: 583–635.

Wu H. (2010) An economic analysis of microcredit lending. *Department of Bioresource Policy, Business & Economics.* Saskatoon, Saskatchewan: University of Saskatchewan.

Wu J and Xu Y. (2011) A decision support system for borrower's loan in P2P lending. *Journal of Computers* 6: 1183–1190.

Xu B, Zheng H, Xu Y, et al. (2016) Configurational paths to sponsor satisfaction in crowdfunding. *Journal of Business Research* 69: 915–927.

Xu Y, Ribeiro-Soriano DE and Gonzalez-Garcia J. (2015) Crowdsourcing, innovation and firm performance. *Management Decision* 53: 1158–1169.

Yeoh P. (2014) Implications of online funding regulations for small businesses. *Journal of Financial Regulation and Compliance* 22: 349–364.

Yip C. (2014) Crowdfunding for Chinese social and environmental small and growing businesses: an investigation of the feasibility of using U.S.-based crowdfunding platforms. Duke University.

York T. (2014) Panel to discuss crowdfunding phenomenon. *San Diego Business Journal, 10 February 2014, Vol. 35 Issue 6,* p. 3.

Young TE. (2013) *The everything guide to crowdfunding: learn how to use social media for small-business funding,* 57 Littlefield Street, Avon, MA 02322 U.S.A.: Adams Media, ISBN 978-1-4405-5033-1.

Yuan H, Lau RYK and Xu W. (2016) The determinants of crowdfunding success: a semantic text analytics approach. *Decision Support Systems* 91: 67–76.

Yum H, Lee B and Chae M. (2012) From the wisdom of crowds to my own judgment in microfinance through online peer-to-peer lending platforms. *Electronic Commerce Research and Applications* 11: 469–483.

Zachary R and Mishra CS. (2013) Research on angel investments: the intersection of equity investments and entrepreneurship. *Entrepreneurship Research Journal* 3: 160–170.

Zhang B. (2014a) The voice of the people: insights from the latest peer-to-peer lending research. LendIt Europe Conference University of Cambridge, pp. 1–41.

Zhang Q. (2014b) Crowdsourcing in community participatory planning in China: case studies in four communities in Shenzhen. *Department of Urban Studies and Planning.* Massachusetts Institute of Technology.

Zhang T, Tang M, Lu Y, et al. (2014a) Trust building in online peer-to-peer lending. *Journal of Global Information Technology Management* 17: 250–266.

Zhang T, Yip C, Wang G, et al. (2014b) China crowdfunding report. China Impact fund (CIF) of Dao Ventures, pp. 1–64.

Zhang Y. (2012) An empirical study into the field of crowdfunding. *School of Economics and Management*. Lund: Lund University.

Zhao L and Vinig T. (2017) Hedonic value and crowdfunding project performance: a propensity score matching-based analysis. *Review of Behavioral Finance* 9: 169–186.

Zhao Q, Chen C-D, Wang J-L, et al. (2017) Determinants of backers' funding intention in crowdfunding: social exchange theory and regulatory focus. *Telematics and Informatics* 34: 370–384.

Zheng H, Li D, Wu J, et al. (2014) The role of multidimensional social capital in crowdfunding: a comparative study in China and US. *Information & Management* 51: 488–496.

Zheng H, Hung J-L, Qi Z, et al. (2016) The role of trust management in reward-based crowdfunding. *Online Information Review* 40: 97–118.

Zhou M, Lu B, Fan W, et al. (in press) Project description and crowdfunding success: an exploratory study. *Information Systems Frontiers* DOI 10.1007/s10796-016-9723-1.

Zhu H and Zhou ZZ. (2016) Analysis and outlook of applications of blockchain technology to equity crowdfunding in China. *Financial Innovation* 2: 1–11.

Zhu L, Zhang Q, Lu H, et al. (2017) Study on crowdfunding's promoting effect on the expansion of electric vehicle charging piles based on game theory analysis. *Applied Energy* 196: 238–248.

Index

Prof. Ben's Bio-Sketch

Professor of Banking & Finance

Dean, Academy of Internet Finance, Zhejiang University

Before joining Zhejiang University as a full-time professor in May 2014, Dr. Ben was a veteran banker with ABN AMRO, HSBC and lastly JP Morgan Chase as CEO, JP Morgan Chase Bank (China) Co Ltd.

Some of Dr. Ben's various other positions with external organizations include: National Executive Committee member of All China Federation of Industry & Commerce; Joint Chairman of Zhejiang Association of Internet Finance; Executive Director, International Monetary Institute, Renmin University of China, and a Board Member of Sino-German Center of Finance & Economics in Gothe University Frankfurt; Counselor to Zhejiang Provincial Government and a consultant to the central government in China as well as non-executive director of leading companies including China International Capital Corporation, Tsingtao Brewery and Bank of Ningbo.

As a well-sought-after speaker on forums both home and abroad, Dr. Ben has published many articles and books in fintech and international finance, including In Pursuit of Presence & Prominence - Chinese Banks Going Global (Zhejiang University Press and Springer 2017).

Dr. Ben has a bachelor's degree in engineering from Tsinghua University, a Master's degree in management from Renmin University of China and Ph.D. degree in Economics from Purdue University, USA. His research areas include: International Finance, New Finance (Internet Finance, FinTech, Entrepreneurial Finance), Global Business.

About the Authors

Bo Xing, D.Ing., is an Associate Professor at the Institute for Intelligent Systems, University of Johannesburg, South Africa. Prior to this, he was an Associate Professor at the Department of Computer Science, School of Mathematical and Computer Science, University of Limpopo, South Africa. He served as a senior lecturer under the division of Center for Asset Integrity Management (C-AIM) at the Department of Mechanical and Aeronautic Engineering, Faculty of Engineering, Built Environment and Information Technology, University of Pretoria, South Africa. Dr. Xing earned his DIng degree (Doctorate in Engineering with a focus on soft computing and remanufacturing) in early 2013 from the University of Johannesburg, South Africa. He also obtained his BSc and MSc degree both in Mechanical Engineering from the Tianjin University of Science and Technology, P.R. China, and the University of KwaZulu-Natal, South Africa, respectively. He was a scientific researcher at the Council for Scientific and Industrial Research (CSIR), South Africa. He has published 3 books, over 50 research papers in the form of international journals, and international conference proceedings. His current research interests lie in applying various nature-inspired computational intelligence methodologies towards big data analysis, miniature robot design and analysis, advanced mechatronics system, and e-maintenance.

Tshilidzi Marwala, Ph.D., is the Vice-Chancellor and Principal of the University of Johannesburg. He was previously the Deputy Vice Chancellor for Research and Internationalisation as well as the Dean of Engineering at the University of Johannesburg. He was a full Professor of Electrical Engineering, the Carl and Emily Fuchs Chair of Systems and Control Engineering as well as the SARChI Chair of Systems Engineering at the University of the Witwatersrand. He was also an executive assistant to the technical director at the South African Breweries. He holds a Bachelor of Science in Mechanical Engineering (magna cum laude) from Case Western Reserve University (USA), a Master of Mechanical Engineering from the University of Pretoria, a Ph.D. in Engineering from Cambridge University, was a post-doctoral research associate at the Imperial College

(London) and completed a Program for Leadership Development at Harvard Business School. He is a registered professional engineer, a Fellow of TWAS, the World Academy of Sciences, the Academy of Science of South Africa, the African Academy of Sciences and the South African Academy of Engineering. He is a Senior Member of the IEEE (Institute of Electrical and Electronics Engineering) and a distinguished member of the ACM (Association for Computing Machinery). His research interests are multi-disciplinary, and they include the theory and application of computational intelligence to engineering, computer science, finance, social science and medicine. He has an extensive track record in human capacity development, having supervised 47 Masters and 21 Ph.D. students to completion. Some of these students have proceeded with their doctoral and post-doctoral studies at leading universities such as Harvard, Oxford, Cambridge, British Columbia, Rutgers, Purdue and Keio. He has published 12 books (one has been translated into Chinese), over 280 papers in journals, proceedings, book chapters and magazines and holds three international patents. He is an associate editor of the International Journal of Systems Science (Taylor and Francis Publishers) and has been a reviewer for more than 40 ISI journals. He has been a visiting scholar at Harvard University, University of California at Berkeley, Wolfson College of University of Cambridge and Nanjing Tech University as well as member of the programming council of the Faculty of Electrical Engineering at the Silesian University of Technology. He has received more than 45 awards including the Order of Mapungubwe. His writings and opinions have appeared in the New Scientist, The Economist and Time Magazine.